Plant Genetics, Genomics

Volume 1

AF181843

Plant Genetics and Genomics: Crops and Models

Series Editor: Richard A. Jorgensen

Forthcoming and planned volumes

Vol. 1 Genomics of Tropical Crop Plants (eds: Paul Moore/Ray Ming)
Vol. 2 Genetics and Genomics of Soybean (ed: Gary Stacey)
Vol. 3 Genetics and Genomics of Cotton (ed: Andy Paterson)
Vol. 4 Plant Cytogenetics: Genome Structure and Chromosome Function (eds: Hank Bass/Jim Birchler)
Vol. 5 Plant Cytogenetics: Methods and Instruction (eds: Hank Bass/Jim Birchler)
Vol. 6 Genetics and Genomics of the Rosaceae (eds: Kevin Folta/Sue Gardiner)
Vol. 7 Genetics and Genomics of the Triticeae (ed: Catherine Feuillet/Gary Muehlbauer)
Vol. 8 Genomics of Poplar (ed: Stefan Janssen et al.)

Paul H. Moore · Ray Ming
Editors

Genomics of Tropical Crop Plants

Foreword by Deborah Delmer

Springer

Editors

Paul H. Moore
USDA-ARS, PBARC
Hawaii Agriculture Research Center
99-193 Aiea Heights Drive
Aiea 96701
phmoore@hawaii.edu
p.moore@harc-hspa.com

Ray Ming
Department of Plant Biology
University of Illinois at Urbana-Champaign
Champaign, IL, USA
1201 W. Gregory Drive
Urbana 61801
288 ERML, MC-051
rming@life.uiuc.edu

ISBN: 978-1-4419-2433-9 e-ISBN: 978-0-387-71219-2

© 2010 Springer Science+Business Media, LLC
All rights reserved. This work may not be translated or copied in whole or in part without the written permission of the publisher (Springer Science+Business Media, LLC, 233 Spring Street, New York, NY 10013, USA), except for brief excerpts in connection with reviews or scholarly analysis. Use in connection with any form of information storage and retrieval, electronic adaptation, computer software, or by similar or dissimilar methodology now known or hereafter developed is forbidden.
The use in this publication of trade names, trademarks, service marks, and similar terms, even if they are not identified as such, is not to be taken as an expression of opinion as to whether or not they are subject to proprietary rights.

Cover illustration: Small holders in tropics: Photograph by Gisela Orjeda, Bioversity International. Most of bananas grown in the tropics are produced by smallholders for home consumption or selling in local markets.

Printed on acid-free paper.

9 8 7 6 5 4 3 2 1

springer.com

Foreword

Having spent most of my life in Academia teaching and carrying out fundamental research on plant form and function, I found this collection of essays to be of considerable interest as they expanded my knowledge of genomics to plants beyond the well-studied model systems of Arabidopsis, rice, and temperate maize. It was even more valuable to me in terms of my more recent interest in international agriculture—in particular, my personal interest in promoting the integration of findings from the advanced plant sciences into current approaches to crop improvement for the benefit of poor, small-scale farmers in the developing world. In my experience, one of the greatest challenges to such integration is the relatively weak interaction among scientists working at the forefront of genomics research and those involved in the improvement of crops important to such farmers. Since most of the crops important to the poor are of tropical origin, one hopes that this very readable collection of essays will help bridge that gap as they should be of interest to both types of scientists. My own experience is interesting in this regard—I had thought that it might be useful to read just a few of these essays and ended up wanting to read them all!

Reading this collection forced me to consider several important issues. The first is how clearly we all now are indebted to the foresight of scientists like Chris Somerville who championed the little weed Arabidopsis as the first best model system. Since dicots and monocots (in particular the cereals) are different in so many ways, the wisdom of The Rockefeller Foundation and its Asian partners in promoting rice as a second best model system is also worth recognizing. In addition to having a relatively small genome, rice had the additional advantage of being a crop important to millions of poor people around the world. In virtually every essay in this book, one can see how those trying to develop genomics for tropical crops are relying upon the vast resources developed for these two plants.

The second issue that comes to mind is the very different emphasis the private sector has put upon certain crops compared to the public sector. This is only natural and is not meant as a criticism. The large private sector companies have their largest markets in the temperate climes of the world, and crops adapted to those climes, in particular, temperate hybrid maize, has been a major focus for them. In fact, the two words "temperate" and "hybrid" are key to distinguishing their efforts from those of the major public sector efforts that have focused much more on non-hybrid tropical species that have less commercial appeal but are critically important to the developing world.

The third issue that comes to mind from my own experience is the difficulty of communicating the value of structural and functional genomics to public sector breeders in many parts of the developing world. There are exceptions to this. Because rice has been a model for genomics and so much information is available, it has been much easier for breeders to get a head start in learning how to use this information wisely—one can see very sophisticated use of rice genomics now in places like China and the Philippines. Wheat may be another example, driven largely by the fact that there are many breeders in the developed world who also have a keen interest in this crop. The situation for maize is very instructive—vast genomic resources have been accumulated both by the private and public sectors—yet, as pointed out in one of the chapters in this book, these are dominated by information that comes from temperate maize—a very different beast from tropical maize. One passion that I personally have is a belief that we are now at a stage where genomics can be used to define all the critical loci that define the key differences between a temperate and a tropical version of a crop—not only for maize but for other crops like sorghum and sugar beet. This could vastly speed the use of temperate information for tropical crop improvement, and now as we face serious climate change, it could also have great benefit for the enhancement of tolerance of temperate crops to global warming.

Finally, what also strikes me are the many challenges presented to breeders by these many tropical crops. Many, like the tropical trees, have very long life cycles; many like cassava and oil palm are terribly heterozygous and breeders lack inbred lines; and many lack robust protocols for transformation. And, critical for the focus of this book, some lack even the most basic DNA sequence information needed to develop markers and maps for use in characterization of genetic diversity and for breeding. This is a bit of a "chicken and egg" situation—without such information, breeders have little opportunity to discover the value of genomics for their improvement efforts—and with no demonstrated value yet of genomics for most of these crops, there is little high level pressure to develop these critical tools. To those who support the development of such crops, it seems critical to break this cycle by first promoting the sequencing of hundreds of thousands of ESTs in a number of genetically-diverse lines of each of these crops, to use these for development of SNPs and other markers, and to promote efforts to make very clear how these may be useful to the relevant breeders. Of course there are many other critical needs, but this seems a good start at least.

Thus, it is my hope that the essays in this book will indeed be read by a wide range of scientists and donors—public and private sector, developed and developing world—and can serve as a good starting point to bring all together to promote the more rapid improvement of these vast genetic resources for the benefit of all.

Professor Emeritus, UC Davis Deborah Delmer
Program Officer of The Rockefeller Foundation (retired)
Consultant on the role of advanced sciences for developing world agriculture

Preface

Tropical crop plants are one of the world's most valuable assets. They provide food, feeds, clothing, and shelter for a large portion of the world's population. They are also the source for energy, industrial biomaterials, and pharmaceutical products. Tropical crop plants are widely diverse in form and environmental requirements, thus possessing characteristics, genes, and genetic elements of potential value for numerous temperate and tropical crops. However, the genetic resources of tropical plants are not only underutilized, they are also in danger of being lost due to the destruction of natural habitats at their sites of origin, the high costs of conservation programs, and our lack of appreciation of the worth of these precious resources.

Considering the tropical origin of much of the biological diversity that is responsible for genes and phenotypes of temperate crops, there is a critical need for assessing the genomics of tropical plant species. Remarkable progress has been made in several tropical crop plants, notably sorghum and papaya that are in the final stages of whole genome sequencing. International consortia or networks have been established for a number of other tropical crops to mobilize and coordinate resources and efforts towards generating genomic tools and eventual sequencing of the genome for basic biological research and crop improvement. These crops include sugarcane, banana, coffee, citrus, millet, cacao, and peanut. The genomic information generated by these international consortia will enhance the capacity for identification, characterization, and cloning of agronomically important genes of tropical crop plants.

At the same time, the rapid advance in genomics has provided new tools for tropical crops to solve some problems that have long been obstacles to traditional breeding. Marker-assisted selection has been practiced effectively to increase yield and stress tolerance. Transformation has successfully generated viral and fungal disease resistant varieties. Ever-improving DNA sequencing technology has provided and will continue to provide a vast amount of sequence data, and increase the capacity to generate new markers for genome and QTL mapping and genome analysis and genome-wide expression analysis. Genomic analysis of tropical crop plants promises to add new dimensions to growing information available for temperate crops in learning more about morphology, physiology, and parallel evolution in diverse plant lineages.

This book summarizes, for 20 tropical crop plants, recent progress on genomic research, including the development of molecular markers, genomic and cDNA libraries, expressed sequence tags (ESTs), genetic and physical maps, gene expression profiles, and whole genome sequences. The first section of the book provides background information about the evolutionary origin and environments of tropical crop species, international programs that are addressing the needs of tropical agriculture, and the potential for new technologies to increase the productivity and value of tropical crops. This introductory section is followed by chapters detailing progress and potential for the 20 specific tropical crop species.

In assembling a book of this scope, the question naturally arises: how were crops selected for inclusion? Given enough time and room to cover a wider range of tropical crops, it would have been useful to include even more species than we were able to do in the present volume. Ultimately, the decision rested on two primary criteria: did we consider the crop tropical in origin and production? and if so, was the genomic information on the crop sufficiently new and unreported elsewhere to warrant inclusion. Given this last criterion, we obviously omitted some major crops, notably the important staple crops rice and cassava.

We wish to thank the 106 tropical plant scientists involved in this effort as authors, and the nearly equal number of anonymous reviewers who served as experts on the fields outside our own areas of expertise. We also wish to thank Jinnie Kim of Springer and Richard Jorgenson, University of Arizona, who envisioned and encouraged us in this undertaking.

In July 2007, Norman Borlaug was awarded the United States Congressional Gold Medal, the country's highest civilian honor, for contributions he made to the "Green Revolution" to solve an earlier threat to food security. The citation for the award described the great challenges that await agriculture in the 21st century as "persistent poverty and environmental degradation in developing countries (editor's note: read this as primarily tropical countries), changing global climate patterns, and the use of food crops to produce biofuels." We share this view of our challenges going forward. As you read the chapters of this book, it will become evident that their authors share it as well. Furthermore, we all believe that the science of genomics will enable what Borlaug described as "the advent of a Gene Revolution that stands to equal, if not exceed, the Green Revolution of the 20th Century."

Paul H. Moore
Honolulu, Hawaii

Ray Ming
Urbana, Illinois

September 2007

Contents

1 **Tropical Environments, Biodiversity, and the Origin of Crops** 1
Paul Gepts

2 **International Programs and the Use of Modern Biotechnologies for Crop Improvement** 21
Jean-Marcel Ribaut, Philippe Monneveux, Jean-Cristophe Glaszman, Hei Leung, Theo Van Hintum, and Carmen de Vicente

3 **Transgenics for New Plant Products, Applications to Tropical Crops** 63
Samuel S.M. Sun

4 **Genomics of Banana and Plantain (*Musa* spp.), Major Staple Crops in the Tropics** 83
Nicolas Roux, Franc-Christophe Baurens, Jaroslav Doležel, Eva Hřibová, Pat Heslop-Harrison, Chris Town, Takuji Sasaki, Takashi Matsumoto, Rita Aert, Serge Remy, Manoel Souza, and Pierre Lagoda

5 **Genomics of *Phaseolus* Beans, a Major Source of Dietary Protein and Micronutrients in the Tropics** 113
Paul Gepts, Francisco J.L. Aragão, Everaldo de Barros, Matthew W. Blair, Rosana Brondani, William Broughton, Incoronata Galasso, Gina Hernández, James Kami, Patricia Lariguet, Phillip McClean, Maeli Melotto, Phillip Miklas, Peter Pauls, Andrea Pedrosa-Harand, Timothy Porch, Federico Sánchez, Francesca Sparvoli, and Kangfu Yu

6 **Genomics of *Theobroma cacao*, "the Food of the Gods"** 145
Mark J. Guiltinan, Joseph Verica, Dapeng Zhang, and Antonio Figueira

7 **Chickpea, a Common Source of Protein and Starch in the Semi-Arid Tropics** 171
Fred J. Muehlbauer and P.N. Rajesh

8 Genomics of Citrus, a Major Fruit Crop of Tropical and Subtropical Regions 187
Mikeal L. Roose and Timothy J. Close

9 Genomics of Coffee, One of the World's Largest Traded Commodities 203
Philippe Lashermes, Alan Carvalho Andrade, and Hervé Etienne

10 Cowpea, a Multifunctional Legume 227
Michael P. Timko and B.B. Singh

11 Genomics of *Eucalyptus*, a Global Tree for Energy, Paper, and Wood 259
Dario Grattapaglia

12 Ginger and Turmeric, Ancient Spices and Modern Medicines 299
David R. Gang and Xiao-Qiang Ma

13 Genomics of Macadamia, a Recently Domesticated Tree Nut Crop 313
Cameron Peace, Ray Ming, Adele Schmidt, John Manners, and Vasanthe Vithanage

14 Genomics of Tropical Maize, a Staple Food and Feed across the World 333
Yunbi Xu and Jonathan H. Crouch

15 Molecular Research in Oil Palm, the Key Oil Crop for the Future 371
Sean Mayes, Farah Hafeez, Zuzana Price, Don MacDonald, Norbert Billotte, and Jeremy Roberts

16 Genomics of Papaya, a Common Source of Vitamins in the Tropics 405
Ray Ming, Qingyi Yu, Andrea Blas, Cuixia Chen, Jong-Kuk Na, Paul H. Moore

17 Genomics of Peanut, a Major Source of Oil and Protein 421
Mark David Burow, Michael Gomez Selvaraj, Hari Upadhyaya, Peggy Ozias-Akins, Baozhu Guo, David John Bertioli, Soraya Cristina de Macedo Leal-Bertioli, Marcio de Carvalho Moretzsohn, and Patricia Messenberg Guimarães

18 Genomics of Pineapple, Crowning The King of Tropical Fruits 441
Jose Ramon Botella and Mike Smith

19 Genomics of Tropical Solanaceous Species: Established and Emerging Crops 453
Richard C. Pratt, David M. Francis, and Luz S. Barrero Meneses

20 Genomics of Sorghum, a Semi-Arid Cereal and Emerging Model for Tropical Grass Genomics 469
Andrew H. Paterson, John E. Bowers, and F. Alex Feltus

21 Sugarcane: A Major Source of Sweetness, Alcohol, and Bio-energy 483
Angélique D'Hont, Glaucia Mendes Souza, Marcelo Menossi, Michel Vincentz, Marie-Anne Van-Sluys, Jean Christophe Glaszmann, and Eugênio Ulian

22 Genomics of Wheat, the Basis of Our Daily Bread 515
Manilal William, Peter Langridge, Richard Trethowan, Susanne Dreisigacker, and Jonathan Crouch

23 Genomics of Yams, a Common Source of Food and Medicine in the Tropics 549
Hodeba D. Mignouna, Mathew M. Abang, and Robert Asiedu

Subject Index 571

Contributors

Mathew M. Abang
International Center for Agricultural Research in the Dry Areas (ICARDA), Aleppo, Syria

Rita Aert
Katholieke Universiteit Leuven, Kasteelpark Arenberg 13, B-3001, Leuven, Belgium

Alan Carvalho Andrade
Embrapa - Recursos Genéticos e Biotecnologia, Parque Estação Biológica, CP 02372, 70770-900 Brasilia DF Brazil

Francisco J.L. Aragão
EMBRAPA Recursos Geneticos & Biotecnologia, Laboratorio Introdução & Expressão Genes, PqEB W5 Norte, BR-70770900 Brasilia, DF Brazil

Robert Asiedu
International Institute of Tropical Agriculture (IITA), Ibadan, Nigeria

Everaldo de Barros
Universidade Federal de Viçosa, BIOAGRO/DBG, BR-36571000 Viçosa, Brazil

Franc-Christophe Baurens
Centre de Coopération Internationale pour la Recherche en Agriculture et le Développement UMR DAP, TA-A 96/03 Avenue Agropolis, 34098 Montpellier Cedex 5, France

David John Bertioli
Universidade Católica de Brasília, Pró-Reitoria de Pós-Graduação e Pesquisa, Campus II, SGAN 916, Brasilia DF, Brasil

Norbert Billotte
CIRAD-CP, Avenue Agropolis, 34398 Montpellier Cedex, France

Matthew W. Blair
Centro Internacional de Agricultura Tropical, Apartado Aéreo 6713, Cali, Colombia

Andrea Blas
Hawaii Agriculture Research Center, 99-193 Aiea Heights Drive, Aiea, HI 96744, USA
Department of Molecular Biosciences and Bioengineering, University of Hawaii at Manoa, 1955 East-West Road, Honolulu, HI 96822, USA

Jose Ramon Botella
Plant Genetic Engineering Laboratory, School of Integrative Biology, University of Queensland, Brisbane 4072, Australia.

John E. Bowers
Plant Genome Mapping Laboratory, University of Georgia, 111 Riverbend Road Rm 228, Athens GA 30602

Rosana Brondani
EMBRAPA Rice and Beans, Laboratory of Biotechnology, BR-74001970 Goiânia, Brazil

William Broughton
Université de Genève, Laboratoire de Biologie Moléculaire des Plantes Supérieures, CH-1292 Genève, Switzerland

Mark David Burow
Texas A&M University, Texas Agricultural Experiment Station; 1102 East FM 1294, Lubbock, TX 79403 USA; and Texas Tech University, Department of Plant and Soil Science, 15th and Detroit, Lubbock, TX 79409 USA

Cuixia Chen
Department of Plant Biology, University of Illinois at Urbana-Champaign, 1201 W. Gregory Drive, Urbana, IL 61801, USA

Timothy J. Close
Department of Botany & Plant Sciences, University of California, Riverside, CA 92521 USA

Jonathan H. Crouch
Generation Challenge Programme (GCP), c/o International Maize and Wheat Improvement Center (CIMMYT), Carretera Méx-Veracruz, El Batán, Texcoco, México CP56130

Angélique D'Hont
CIRAD, (Centre de Coopération Internationale en Recherche Agronomique pour le Développement), UMR1096, Avenue Agropolis, TA40/03, F-34398 Montpellier, France

Jaroslav Doležel
Institute of Experimental Botany, Laboratory of Molecular Cytogenetics and Cytometry, Sokolovská 6, CZ-7720 Olomouc, Czech Republic

Susanne Dreisigacker
Genetic Resources and Enhancement Unit, International Maize and Wheat Improvement Center (CIMMYT), Carretera Méx-Veracruz, El Batán, Texcoco, México CP56130

Hervé Etienne
Centre de Coopération Internationale en Recherche Agronomique pour le Développement, BIOS, UMR RPB, TA A-98/IRD, 34394 Montpellier Cedex 5, France

F. Alex Feltus
Plant Genome Mapping Laboratory, University of Georgia, 111 Riverbend Road Rm 228, Athens GA 30602

Antonio Figueira
Centro de Energia Nuclear na Agricultura Av. Centenário 303 Caixa Postal 96, 13400-970 Piracicaba, SP, Brazil

David M. Francis
The Ohio State University, Department of Horticulture and Crop Science, Ohio Agricultural Research and Development Center, 1680 Madison Ave., Wooster, Ohio 44691-4096

Incoronata Galasso
Consiglio Nazionale delle Ricerche, Istituto di Biologia e Biotecnologia Agraria, I-20133 Milan, Italy

David R. Gang
University of Arizona, Department of Plant Sciences and BIO5 Institute, 1657 E. Helen Street, Tucson, AZ 85719

Paul Gepts
University of California, Department of Plant Sciences / MS1, Section of Crop and Ecosystem Sciences, Davis, CA 95616-8780, USA

Jean Christophe Glaszmann
CIRAD, (Centre de Coopération Internationale en Recherche Agronomique pour le Développement), UMR1096, Avenue Agropolis, TA40/03, F-34398 Montpellier, France. Generation Challenge Programme

Dario Grattapaglia
Plant Genetics Laboratory, Embrapa - Recursos Genéticos e Biotecnologia, Parque Estação Biológica, Brasília 70770-970 DF, and Graduate Program in Genomic Sciences and Biotechnology, Universidade Catolica de Brasília – SGAN 916 modulo B, Brasília 70790-160 DF, Brazil

Mark J. Guiltinan
The Pennsylvania State University, Department of Horticulture, Life Sciences Building, University Park, PA 16802-5807, USA

Patricia Messenberg Guimarães
Empresa Brasileira de Pesquisa Agropecuária (EMBRAPA), Recursos Genéticos e Biotecnologia, Parque Estação Biológica, CP 02372, Brasilia DF, Brasil

Baozhu Guo
USDA-ARS, Crop Protection and Management Research Unit, 2747 Davis Road, Tifton, GA 31793 USA

Farah Hafeez
Department of Genetics, University of Cambridge, The Downing Site, Cambridge, Cambridgeshire, CB2 3EF, UK

Gina Hernández
Universidad Nacional Autónoma de México, Centro de Ciencias Genómicas, Ap. Postal 565-A, Cuernavaca 62191, Morelos, Mexico

Pat Heslop-Harrison
University Leicester, Department of Biology, Leicester LE1 7RH, UK

Eva Hřibová
Institute of Experimental Botany, Laboratory of Molecular Cytogenetics and Cytometry, Sokolovská 6, CZ-7720 Olomouc, Czech Republic

James Kami
University of California, Department of Plant Sciences / MS1, Section of Crop and Ecosystem Sciences, 1 Shields Avenue, Davis, CA 95616-8780, USA

Pierre Lagoda
Joint FAO/IAEA Division International Atomic Energy Agency, Plant Breeding and Genetic Section, Wagramer Strasse 5, PO Box 100 A-1400 Vienna, Austria

Peter Langridge
Australian Centre for Plant Functional Genomics, School of Agriculture, Food & Wine, The University of Adelaide, Waite Campus, Australia (ACPFG)

Patricia Lariguet
Université de Genève, Laboratoire de Biologie Moléculaire des Plantes Supérieures, CH-1292 Genève, Switzerland

Philippe Lashermes
Institut de Recherche pour le Développement, UMR RPB, BP 64501, 34394 MontpellierCedex 5, France

Soraya Cristina de Macedo Leal-Bertioli
Empresa Brasileira de Pesquisa Agropecuária (EMBRAPA), Recursos Genéticos e Biotecnologia, Parque Estação Biológica, CP 02372, Brasilia DF, Brasil

Hei Leung
GCP- IRRI, Manila 1099, P.O. Box 933, The Philippines. Generation Challenge Programme

Phillip McClean
North Dakota State University, Department of Plant Sciences, Fargo, ND 58105, USA

Xiao-Qiang Ma
University of Arizona, Department of Plant Sciences and BIO5 Institute, 1657 E. Helen Street, Tucson, AZ 85719

Don MacDonald
Department of Genetics, University of Cambridge, The Downing Site, Cambridge, Cambridgeshire, CB2 3EF, UK

John Manners
CSIRO Plant Industry, Queensland Bioscience Precinct, St. Lucia, Brisbane, QLD 4067, Australia

Takashi Matsumoto
National Institute of Agrobiological Sciences, 2-1-2 Kannondai, Tsukuba, Ibaraki, Japan 305-8602

Sean Mayes
School of Biosciences, Sutton Bonington Campus, Nottingham University, Loughborough, Leicestershire, LE12 5RD, UK

Maeli Melotto
University of Texas at Arlington, Department of Biology, Arlington, TX 76109, USA

Luz S. Barrero Meneses
Corporación Colombiana de Investigación Agropecuaria (CORPOICA), Bogota, Colombia

Marcelo Menossi
Departmento de Genetica e Evolução IB- Unicamp, Centro de Biologia Moleculare Engenharia Genética, Universidade Estadual de Campinas, C.P. 6010, CEP 13083-970, Campinas, SP, Brazil

Hodeba D. Mignouna
African Agricultural Technology Foundation (AATF), Nairobi, Kenya

Phillip Miklas
USDA-ARS, Vegetable and Forage Crop Research Unit, Prosser, WA 99350, USA

Ray Ming
Department of Plant Biology, University of Illinois at Urbana-Champaign, 1201 W. Gregory Drive, Urbana, IL 61801, USA

Philippe Monneveux
Generation Challenge Programme – Genetic Resources and Enhancement Unit, International Maize and Wheat Improvement Center (CIMMYT), Carretera Méx-Veracruz, El Batán, Texcoco, México CP56130

Paul H. Moore
USDA-ARS, PBARC, Hawaii Agriculture Research Center, 99-193 Aiea Heights Drive, Aiea, HI 96744, USA

Marcio de Carvalho Moretzsohn
Empresa Brasileira de Pesquisa Agropecuária (EMBRAPA), Recursos Genéticos e Biotecnologia, Parque Estação Biológica, CP 02372, Brasilia DF, Brasil

Fred J. Muehlbauer
USDA-ARS, 303 Johnson Hall, Washington State University, Pullman, WA 99164, USA

Jong-Kuk Na
Department of Plant Biology, University of Illinois at Urbana-Champaign, 1201 W. Gregory Drive, Urbana, IL 61801, USA

Peggy Ozias-Akins
University of Georgia, Department of Horticulture, and National Environmentally Sound Production Agriculture Laboratory (NESPAL), 2356 Rainwater Road, Tifton, GA 31794, USA

Andrew H. Paterson
Plant Genome Mapping Laboratory, University of Georgia, 111 Riverbend Road Rm 228, Athens GA 30602

Peter Pauls
University of Guelph, Department of Plant Agriculture, Guelph, ON N1G 2W1, Canada

Cameron Peace
Department of Horticulture and Landscape Architecture, Washington State University, Pullman, WA 99164, US

Andrea Pedrosa-Harand
Universidade Federal de Pernambuco, Department of Botany, Laboratory of Plant Cytogenetics, BR-50670420 Recife, PE, Brazil

Timothy Porch
USDA-ARS, Tropical Agriculture Research Station, 2200 PA Campos Ave, Suite 201, Mayagüez, PR 00680 USA

Richard C. Pratt
The Ohio State University, Department of Horticulture and Crop Science, Ohio Agricultural Research and Development Center, 1680 Madison Ave., Wooster, Ohio 44691-4096

Zuzana Price
Department of Genetics, University of Cambridge, The Downing Site, Cambridge, Cambridgeshire, CB2 3EF, UK

P.N. Rajesh
Department of Crop and Soil Sciences, 301 Johnson Hall, Washington State University, Pullman, WA 99164, USA

Serge Remy
Katholieke Universiteit Leuven, Kasteelpark Arenberg 13, B-3001, Leuven, Belgium

Jean-Marcel Ribaut
Generation Challenge Programme – Genetic Resources and Enhancement Unit, International Maize and Wheat Improvement Center (CIMMYT), Carretera Méx-Veracruz, El Batán, Texcoco, México CP56130

Jeremy Roberts
School of Biosciences, Sutton Bonington Campus, Nottingham University, Loughborough, Leicestershire, LE12 5RD, UK

Mikeal L. Roose
Department of Botany & Plant Sciences, University of California, Riverside, CA 92521 US

Nicolas Roux
Bioversity International, Parc Scientifique, Agropolis II, 34397 Montpellier Cedex 5, France

Federico Sánchez
Universidad Nacional Autónoma de México, Instituto de Biotecnología, Departamento Biología Molecular de Plantas, Apartado Postal 510-3, Cuernavaca 62271, Morelos, Mexico

Takuji Sasaki
National Institute of Agrobiological Sciences, 2-1-2 Kannondai, Tsukuba, Ibaraki, Japan 305-8602

Adele Schmidt
CSIRO Plant Industry, Queensland Bioscience Precinct, St. Lucia, Brisbane, QLD 4067, Australia

Michael Gomez Selvaraj
Texas Tech University, Department of Plant and Soil Science, 15th and Detroit, Lubbock, TX 79409 USA

B.B. Singh
Department of Genetics and Plant Breeding, G.B. Pant University of Agriculture and Technology, Pantnagar 263145, Uttaranchal State, India

Mike Smith
Department of Primary Industries & Fisheries, Maroochy Research Station, Nambour 4560, Queensland, Australia

Glaucia Mendes Souza
Departamento de Bioquímica, Instituto de Química, Universidade de São Paulo, Av. Prof. Lineu Prestes 748, 05508-900, São Paulo, SP, Brazil

Manoel Souza
Empresa Brasileira de Pesquisa Agropecuária (EMBRAPA) LABEX Europa, Plant Research International (PRI), Wageningen University & Research Centre (WUR), Wageningen, The Netherlands

Francesca Sparvoli
Consiglio Nazionale delle Ricerche, Istituto di Biologia e Biotecnologia Agraria, I-20133 Milan, Italy

Samuel S.M. Sun
Department of Biology, the Chinese University of Hong Kong, Hong Kong, China

Michael P. Timko
Department of Biology, University of Virginia, Charlottesville, VA 22904 USA

Chris Town
The J. Craig Venter Institute, 9712 Medical Center Drive, Rockville, MD 20850, USA

Richard Trethowan
Plant Breeding Institute, University of Sydney, PMB11, Camden, NSW 2570, Australia

Eugênio Ulian
Monsanto do Brasil Ltda, Av. Nações Unidas 12901, 04578-000 São Paulo SP, Brazil

Hari Upadhyaya
International Crops Research Institute for the Semi-arid Tropics (ICRISAT), Genetic Resources, Patancheru 502 324, Andhra Pradesh, India

Theo Van Hintum
Wageningen University, 6700 AA Wageningen, P.O. Box 16, The Netherlands. Generation Challenge Programme

Marie-Anne Van-Sluys
Departamento de Botânica, Instituto de Biociências, Universidade de São Paulo, Rua do Matão 277, 05508-090, São Paulo, SP, Brazil

Joseph Verica
The Pennsylvania State University, Department of Horticulture, Life Sciences Building, University Park, PA 16802-5807, USA

Carmen de Vicente
Generation Challenge Programme – Genetic Resources and Enhancement Unit, International Maize and Wheat Improvement Center (CIMMYT), Carretera Méx-Veracruz, El Batán, Texcoco, México CP56130

Michel Vincentz
Departmento de Genetica e Evolução IB- Unicamp, Centro de Biologia Moleculare Engenharia Genética, Universidade Estadual de Campinas, C.P. 6010, CEP 13083-970, Campinas, SP, Brazil

Vasanthe Vithanage
CSIRO Plant Industry, Queensland Bioscience Precinct, St. Lucia, Brisbane, QLD 4067, Australia

Manilal William
Genetic Resources and Enhancement Unit, International Maize and Wheat Improvement Center (CIMMYT), Carretera Méx-Veracruz, El Batán, Texcoco, México CP56130

Yunbi Xu
Genetic Resources and Enhancement Unit, International Maize and Wheat Improvement Center (CIMMYT), Carretera Méx-Veracruz, El Batán, Texcoco, México CP56130

Kangfu Yu
Agriculture and Agri-Food Canada, Greenhouse and Processing Crops Research Centre, Harrow, ON N0R 1G0, Canada

Qingyi Yu
Hawaii Agriculture Research Center, 99-193 Aiea Heights Drive, Aiea, HI 96744, USA

Dapeng Zhang
USDA/ARS/PSI, Sustainable Perennial Crops Laboratory, 10300 Baltimore Av., Bldg 001 Rm 223, BARC-West, Beltsville MD 20705, USA

Introduction

The disciplines of plant genetics and plant genomics are revolutionizing plant biology research by fundamentally changing the way plant biologists perform research and the way they view and understand plants. Not only are tools and data sets expanding more rapidly than they can be analyzed and understood, but also plant biologists must assimilate, assess and interpret data from diverse sources and in many plant species. No longer can plant biologists focus on a single discipline, such as physiology, development, or biochemistry – all now depend on each other and all are deeply rooted in genetics and genomics. As a result, it is often said, "we are all *biologists* now!", meaning that it is no longer meaningful to think of ourselves as physiologists, biochemists or geneticists, but rather we must think of as multi-disciplinary biologists, working across and breaking down traditional disciplinary boundaries. This creates a great need for basic information that is accessible to all plant biologists, which is purpose of this series – to bring novel, fundamental information about genetics and genomics of plants to the entire plant biology community.

The book series *Plant Genetics and Genomics: Crops and Models* is designed to provide current overviews and summaries of the state of the art in genetics and genomics for all of the major crops or groups of crops, as well as for each major genetic model system for which a significant need exists. Most volumes will focus on a single crop, species, group of close relatives, or group of plants with similar biology (such as Tropical Crops). Other volumes will have a specific disciplinary or technological focus, such as cytogenetics, comparative genomics, translational genomics, and epigenomics, encompassing the plant kingdom. In this way, we hope that all the most important areas of interests to both basic and applied plant scientists will be covered.

Richard A. Jorgensen
Series Editor, Plant Genetics and Genomics

Chapter 1
Tropical Environments, Biodiversity, and the Origin of Crops

Paul Gepts

So long as freedom from hunger is only half achieved, so long as two thirds of the nations have food deficits, no citizen, no nation can afford to be satisfied. We have the ability, as members of the human race, we have the means, we have the capacity to eliminate hunger from the face of the earth in our lifetime. We only need the will.

President J. F. Kennedy, 1963

We believe that it is indeed possible to end world hunger by the year 2000. More than ever before, humanity possesses the resources, capital, technology and knowledge to promote development and to feed all people, both now and in the foreseeable future. . . . Only a modest expenditure is needed each year - a tiny fraction of total expenditure which amounts to $650 billion a year. What is required is the political will to put first things first and to give absolute priority to freedom from hunger.

FAO World Food Colloquium, 1992

Food supplies have also been made more vulnerable by our reliance on a very small number of species: just 30 crop species dominate food production and 90 per cent of our animal food supply comes from just 14 mammal and bird species – species which themselves rely on biodiversity for their productivity and survival. There has been a substantial reduction in crop genetic diversity in the field and many livestock breeds are threatened with extinction . . . .I urge individuals and institutions alike to give greater attention to biodiversity as a key theme in our efforts to fight the twin scourges of hunger and poverty and achieve the Millennium Development Goals.

U.N. Secretary-General K. Annan, 2004

P. Gepts
University of California, Department of Plant Sciences / MS1, Section of Crop and Ecosystem Sciences, Davis, CA 95616-8780, USA.
e-mail: plgepts@ucdavis.edu

P.H. Moore, R. Ming (eds.), *Genomics of Tropical Crop Plants*

© Springer 2008

Abstract Plants play an important, but often insufficiently recognized. role in human societies, chiefly as providers of food, feed, and fiber, but for other uses as well such as drugs and building materials. In those cases where demand for a particular plant product exceed natural supply, humans initiated cultivation of those plants 10,000 years ago, resulting in the domestication of a limited number of species in several areas of the world, generally located in tropical and subtropical areas. Tropical and subtropical areas consist of several ecozones defined by climate, soil, and vegetation and fauna. One of the major factors distinguishing tropical ecozones is the distribution of rainfall and particularly the length of the dry season, if any. From a biological standpoint, most of biodiversity is concentrated in tropical areas. This may explain in part why the majority of crops discussed in this volume originated in either tropical or subtropical ecozones with summer rains or in the tropical ecozone with year-round rains. The fundamental contribution of genomics to plant breeding is to provide information on the genotypic basis of phenotypic variation. Based on this information, marker-assisted selection systems can be developed that can increase the efficiency at identifying agronomically useful diversity and transferring it into improved cultivars. In turn, marker-assisted selection is expected to increase the efficiency of plant breeding. To achieve this goal, however, genomic resources have to be developed, not only in model species, but especially in target species as illustrated in this volume. Having access to a diverse set of improved crops is a critical element of strategies to alleviate food insecurity and poverty, which affect disproportionately rural populations. Recently, the goal of increasing crop productivity has taken added urgency because of the combined impact of the focus on biofuels and global warming. Genomics is a crucial tool in raising crop productivity for the foreseeable future.

1.1 Introduction

Plants have played and still play an important role in human societies, from traditional, indigenous societies to modern societies with advanced technology. Although the role of plants may not be so readily apparent in our increasingly urbanized societies, examples abound of the lasting impact of plants in human activities (Balick and Cox 1996; Lewington 2003; Marinelli 2005). For example, a survey by Estrada-Lugo (1989) of the citation frequencies of various plant uses in the Códice Florentino, an account of pre-Columbian life by Bernardo de Sahagún in the early 16th century, shows a myriad of uses. These include, but are not limited to, uses as dyes, drugs, fibers, food, forage, fuel, medicines, ornamentals, resins, sleeping aids, spices, and venom. Plants fulfill various roles in ceremonies, construction, rituals, and taxation. In our contemporary societies, the use of plants has not diminished. Either alone, or in conjunction with animal or industrial products, they still play an important role in our daily lives, from building materials for our houses, tires and fuel for transportation, personal care products, an ever-broadening range of vegetables and fruits, paper products, and medical drugs.

Many plant products are derived from wild-growing plants, often at the risk of eventual extinction of the native populations of these plants; some products, however, are harvested from cultivated plants, when the demand has surpassed the production capacity of the original, wild populations. This can be seen as one of the stimuli leading to the origins of agriculture, namely as a response to an imbalance between the local supply and demand of a plant (or animal) product.

The switch from hunting-gathering to agriculture, which took place first some 10,000 years ago more or less simultaneously in several areas of the tropics and semi-tropics (Smith 1995), is one of the most momentous accomplishments of human evolution. Without agriculture, humans would not be able to feed and clothe themselves. The transition to agriculture involved several important changes such as the adoption of a sedentary lifestyle, the development of villages, the acquisition of pottery-making ability, and, not least, the domestication of plants and animals. This transition has, therefore, rightfully been called the Neolithic Revolution because of the all-encompassing changes it induced in human societies.

The specifics of where, when, and why agriculture originated is still the subject of passionate debates, but is beyond the scope of this volume on genomics of tropical crops (Smith 1995; Gepts 2004a; Bellwood 2005; Zeder 2006; Barker 2006). What is of interest, however, are the geographic aspects of crop domestication and cultivation, and particularly their relationship to the tropical environment. I define the tropical environment to include not only those areas that straddle the equator, but also those straddling the Tropics of Cancer (23.5°N) and Capricorn (23.5°S). The Tropic of Cancer runs through Mexico, northern Africa, the Peninsula of Arabia, India, and Indochina. The Tropic of Capricorn runs through southern Brazil, Namibia and South Africa, and Australia. In this introduction, I will present the physical features of the tropical environments, the biogeographical features of the tropical environments, and the geographic distribution of centers of agricultural origins and domestication. I will then proceed discussing – in a general way - the contributions of genomics to plant breeding and the contributions of crop improvement to hunger and poverty alleviation, a context relevant to the tropical focus of this book.

1.2 Tropical Environments

The terrestrial landmasses of Earth are divided into a limited number of ecozones, which are large areas in which physical factors such as climate and soil interact to establish a characteristic environment where an assemblage of plants grow and provide a habitat for animals. Schultz (2005) provides an excellent overview of the nine ecozones that have been recognized. His treatment will be followed in this and the next section. These ecozones form broad, sometimes fragmented, east-west bands between the poles and the equator in the northern and southern hemispheres. In addition to the Polar, Boreal, Temperate Midlatitude, and Dry Midlatitude ecozones, which cover the polar to temperate environments, five additional ecozones are located in subtropical and tropical zones (Table 1.1).

Table 1.1 Semi-tropical and tropical ecozones (modified from Schultz 2005)

Ecozone	Plant formation (climax formations)	Examples of geographic distribution	Proportion of terrestrial distribution (%)
Subtropics with winter rains (Mediterranean)	Sclerophyllous forest and shrub formation	Mediterranean, California, S.W. Australia, S.W. South Africa, Chile	2
Subtropics with year-round rains	Subtropical rain forest	S.E. US, Uruguay and S. Brazil, S. China, E. Australia, E. South Africa	4
Dry subtropics and tropics	Desert and semi-deserts Winter-wet grass & shrub steppes (subtropics) Summer-wet thorn savanna (tropics) and thorn steppe	N. Mexico, N. Argentina, Bolivia, Paraguay, Sahara, Southern Africa, Central Australia	21
Tropics with summer rains	Dry to moist savanna (includes tropical deciduous forest)	S. Mexico and Central America, S. America (N. and S. of Amazon), W. Africa, India, S.E. Asia, N. Australia	16
Tropic with year-round rain	Tropical rainforest	Amazon, C. America, Central Africa, Indonesia, Papua New Guinea	8

Climate is the fundamental factor determining the geographic extent and boundaries of these ecozones and their geomorphological processes, soil formation, plant growth, and land use potential. The most important climatic variables are solar radiation (especially the photosynthetically active radiation between 400 and 700 nm), the mean air temperature, and the length of the growing season, which are the determinants of the annual primary production by vegetation. In addition, the extreme values of these climatic variables and their frequency is also important in determining the distribution of the ecozones. The growing season is defined as the annual total of months with a mean temperature above 5 °C and a precipitation (p, expressed in mm) of double the monthly mean value of temperature (expressed in °C): $p\ (mm) \geq 2\ t_{mon}\ (°C)$. In tropical regions, plant growth can be interrupted by drought; in contrast, at middle to high latitudes, plant growth is interrupted by cold.

Climate and vegetation play an important, but not exclusive, role in determining soil type. The chemical fertility of soils refers to the quantity of basic nutrients and how they are bound to soil particles (soluble and exchangeable fractions), which depends on the soil mineral (clay type) and organic matter (humus) composition. The amount of rainfall influences the extent to which certain mineral elements stay within the soil horizons exploited by plant roots or are leached away. The organic matter content in soils depends on how much and what type of litter is decomposed and under which conditions. The soil structure determines the amount of water available in the root area, which is derived from the difference in water content between the field moisture capacity (maximum of water held in the soil after excess moisture has flowed away) and the permanent wilting point (an equilibrium point between root suction potential and soil water potential).

There is a close correlation between ecozones and major plant formations because of convergent evolution of different plant species as they have become adapted to local conditions. This adaptation process has led to the development of a limited number of life forms (e.g., evergreen broadleaf, dry season deciduous, succulents, and grasses, as proposed by Raunkiaer 1934). In turn, plant formations consist of one or more life forms in characteristic proportions (Table 1.1). The different plant formations show great variation in primary productivity, which is due primarily to the size and structure of the above-ground biomass and climatic and edaphic factors. The larger the biomass, the larger the productivity, with the exception of grasslands in steppes and grassland savannas, which show a higher production capacity from a smaller amount of biomass. The structure of biomass also plays a role in that it determines the size and distribution of leaf surfaces, which play an important role in carbon assimilation.

The leaf area index (total leaf area/total ground surface underneath the leaf cover) is a measure of light interception and varies from 5-6 in deciduous forests to 7-8 in subtropical rainforests and 9-10 in tropical rainforests. Generally, biomass production increases with the leaf area index.

Environmental factors involved in primary productivity include the length of the growing season, solar radiation, temperature, water availability, and soil nutrient availability. In general, biomass productivity is a function of climate latitudinal changes. The highest productivity measures are observed in tropical rainforests because they have year-round rain, high precipitation totals, high temperatures, and intense solar radiation.

1.3 Major Subtropical and Tropical Ecozones (based on Schultz 2005)

1.3.1 Subtropics with Winter Rains

The Subtropic with Winter Rains ecozone (also known as the Mediterranean ecozone) is the smallest (2% of landmass) and most fragmented ecozone. It is present on the five continents, on their western side between 30-40° latitudes between the dry tropics and subtropics (towards the equator) and temperate midlatitudes (on the polar side). Precipitation ranges from 300 to 800 mm, mostly in winter and with a few summer months without rainfall at all. The mean monthly temperature is above 18 °C for at least four months. Plant growth is interrupted mostly in the summer because of drought rather in than in the winter because of cold.

This ecozone is rich in plant diversity with many endemic species. It is dominated by evergreen sclerophyllous forests (now often replaced by sclerophyllous shrubs, because of excessive logging over time). Sclerophylly manifests itself as hardened leaves, with a thickened epidermis, a shiny, waxy surface, and other adaptations to summer drought. In addition to trees and shrubs, the vegetation is also rich in hemicryptophytes (e.g., perennial rosette plants), geophytes (bulbs, rhizomes), and therophytes (seed-propagated plants such as annual legumes or grasses). The primary productivity of this ecozone ranges from 3.5 t biomass dry weight ha^{-1} $year^{-1}$ (in shrub vegetation like the garrigue and chaparral) to 6.5 t ha^{-1} $year^{-1}$ (in oak forest) (Schultz 2005). This ecozone has become a very important player in the global trade of vegetables and fruits.

1.3.2 Subtropics with Year-Round Rain

This ecozone represents 4% of the landmass. It is located between 25° and 35° latitude on the southeastern side of the Americas, Africa, Asia, and Australia. Towards the equator, it borders with the Tropical ecozone with year-round rains and towards the poles with the Temperate Midlatitude ecozone. Extended dry periods are infrequent but growth can be limited by occasional frost spells. Summers are hot because of high insolation and temperatures similar to those in the tropics.

The vegetation in this ecozone is a dense rainforest, which evolves towards the west into a semi-evergreen moist forest and eventually a deciduous forest as precipitation gradually decreases. Productivity ranges from 14 to 23 t biomass dry weight ha^{-1} $year^{-1}$. The natural vegetation has largely been destroyed and replaced by urban areas, industries, agriculture, and forestry.

1.3.3 Dry Subtropics and Tropics

Located in the subtropical high pressure belts in the northern and southern hemispheres, this ecozone consists of three subdivisions, which in total represent 21% of the terrestrial landmass. Deserts and semi-deserts are characterized by grass cover over less than 50% of their surface. In contrast, thorn savannas and steppes (summer rains) and grass and shrub steppes (winter rains) have a grass cover on more than 50% of their surface. Precipitation is low, between 100 and 500 mm per year. There is a high level of solar radiation, but a large proportion is also reflected (albedo of 25–30%).

Vegetation in this ecozone shows adaptation to drought stress. There is an increase in the frequency of woody plants as drought increases. In grass steppes and thorn savannas, hemicryptophytes such as perennial herbaceous plants (e.g., grasses) predominate. The primary productivity is low, ranging from below 0.2 t biomass dry weight ha^{-1} $year^{-1}$ to $2.5 - 3.0$ t ha^{-1} $year^{-1}$.

1.3.4 Tropics with Summer Rains

This ecozone (16% of landmass) is located between the Tropical ecozone with Year-round Rains towards the equator and the Dry Subtropics and Tropics ecozone towards the poles. All months have mean temperatures above 18 °C. The mean temperature is lowest in the winter dry season, which lasts 2.5 to 7.5 months. The annual precipitation ranges from 500 to 1500 mm.

The vegetation of this ecozone is divided into tree, shrub, and grass savannas. Tree density ranges from a near-absence to an almost continuous tree cover, depending on the amount of moisture. The moist savanna shows denser stands of taller trees and taller and denser grasses as well. Soils in the drier savanna have a higher mineral exchange capacity and base saturation. They are also richer in humus. They can support permanent cultivation. Soils in the moister savanna are developed on deeply weathered bedrock. Organic matter decomposes quickly and there is a high level of leaching. Soils are therefore poor in humus and nutrients. Agriculture in moister savannas consists therefore of shifting cultivation with a fallow period of several years. Because of population increases, the length of the fallow period has gradually been reduced with an ensuing reduction in soil fertility.

This ecozone is the most densely settled and agriculturally most intensely used area in the tropics. It has several advantages compared to the Tropics with Year-round Rains. Its soil fertility is higher, the winter dry period allows clearing by fire, extensive grasslands can be used for cattle grazing, and the end of the growing season shows a high intensity of solar radiation. In addition, the duration of the rainy season is sufficient for cultivation of a wide range of crops. The primary productivity of this ecozone ranges from 10 to 21 t biomass dry weight ha^{-1} $year^{-1}$.

1.3.5 Tropics with Year-Round Rain

About 8% of the terrestrial landmass belongs to this ecozone, which extends from the Equator to 10° northern and southern latitudes. The boundary with the Subtropics with Year-round Rains is the 18 °C isotherm for the coldest month and with the Tropics with Summer Rains the 1500 mm precipitation line. It is characterized by year-round rainfall (up to 2000–4000 mm), a strong solar radiation year-round, and a nearly constant temperature (25–27 °C). As a consequence, plant growth continues throughout the year. This ecozone shows many similarities with the Moist Savannas for soil characteristics, vegetation, and land use. The primary productivity is highest, with 20-30 t biomass dry weight ha^{-1} $year^{-1}$, but may require shifting cultivation because of pool soil fertility. Declines in yield are due to loss of nutrients, fixation of phosphates, and increase in aluminum toxicity. Fallows of 15-30 years are required but are increasingly cut short because of population increases. Slash and burn agriculture can be replaced by continuous agriculture with fertilizers and liming.

1.4 Geographic Distribution of Biodiversity

Biodiversity can be defined as the sum total of all living organisms on Earth, including plants and animals, but also fungi, protozoans, bacteria, mycoplasma, and viruses, as individuals, populations, species, communities of organisms, and ecosystems. One way of quantifying biodiversity, but by no means the only one, is to count the number of species. Surprisingly, there is little definitive knowledge about this important data point. Up to 30 million species may exist although only 1.8 million have been described (http://www.earthscape.org/t1/ wie01/new_species.html). In plants, some 250,000 species have been described but by some estimates 300,000 to over 400,000 species may exist (Bramwell 2002; Govaerts 2003).

Biodiversity is unevenly distributed. Generally, many more species are distributed in tropical environments. This gradient has existed since before the time of the dinosaurs and is best documented for animals, including mammals, birds, frogs, and butterflies. For plants, the same general trend holds as well, although other factors may play a role such as aridity. Currie and Paquin (1987) showed that on a global scale plant species richness showed the strongest correlation with climate (evapotranspiration) and net primary productivity (NPP), the two factors being correlated. In turn, areas of high NPP also have a complex vegetation structure. For example, tropical forests shows several layers of vegetation. The tropical rainforest shows four layers at three, six, 30, and 50 m. In contrast, temperate forests only have two layers. Thus, climate (especially temperature and precipitation) is a major determinant of biodiversity (Kleidon and Mooney 2000).

On a more local scale, however, infertile soils can lead to species richness and endemism (e.g., serpentine soils: Brady et al. 2005). In contrast, plant species richness may decrease with increasing productivity presumably because of competition (Mittelbach et al. 2001). Other local attributes attempting to explain the distribution of biodiversity or more detailed analyses of certain variables or processes invoke differences between tropical and temperate areas (Partel et al. 2007), differences in scale of analysis (Sarr et al. 2005), energy flows through different habitats (Clarke and Gaston 2006), or energy–water interactions (Hawkins et al. 2003), differential rates of speciation and extinction in the tropics vs. other areas (Mittelbach et al. 2007), heterogeneity in topography and soils (Nichols et al. 1998), and landscape age and history (Sarr et al. 2005).

The distribution pattern of families of flowering plants reveals an interesting pattern as well. Some 30% of families are widespread, 20% are mainly temperate, and 50% are tropical. This distribution led Crane and Lidgard (1989) to suggest that flowering plants arose first in the tropics. The Ice Age affected species distribution mainly at higher latitudes, but also affected tropical rainforests, which became fragmented. In turn, this fragmentation may have favored speciation as well. Thus, there are many potential factors accounting for the higher diversity in tropical environments.

One consequence of the uneven distribution of biodiversity is the existence of biodiversity "hotspots," areas that are especially rich in species (Myers 1990). Some 25 such hotspots have been identified (Myers et al. 2000). These hotspots contain

44% of all plant species worldwide on 1.4% of the terrestrial surface of Earth. Fifteen hotspots are located in the tropical ecozone with year-round rains, five hotspots in the subtropical ecozone with winter rains (Mediterranean), and nine hotspots are mainly or completely made up of islands. Sixteen hotspots are in the tropics. One of the major threats to these hotspots is human population growth.

A major reason for presenting the different ecozones and the distribution of biodiversity is to examine a possible relationship with centers of agricultural origins and crop domestication.

1.5 Centers of Agricultural Origins and Crop Domestication

It has been known since the 19th century that agriculture originated in specific areas of the world. The current consensus about these areas relies mainly on the three centers (smaller, well-circumscribed areas) and three non-centers (larger, wide-ranging areas) identified by Harlan (1971, 1992), to which a few additional centers are added to account for additional, often more recent results. The three centers are Mesoamerica (southern half of Mexico and northern half of Central America), the Fertile Crescent (an arc of mountainous areas roughly surrounding Mesopotamia and including, from west to east, the Levant [Israel, Palestine, Lebanon, and western Syria], southwestern Turkey, and western Iran), and the north Chinese center (centered around the Huanghe or Yellow river).

The three non-centers include the Andes and areas to the east of this mountain chain, included in the Tropics with Summer Rain and Dry Tropics ecozones (eastern Bolivia, western and central Brazil, Paraguay, Uruguay, and NW Argentina). They also include a broad east-west swath comprising the Sahelian and Sudanian savannas and the Ethiopian highlands. The Asian non-center includes the eastern part of India, the Indochinese Peninsula, the Indonesian and Philippines archipelagos, and New Guinea. These six centers and non-centers are the most important areas of agricultural origins and domestication. There are, however, some areas outside these six centers that have played a role as well. For example, agriculture was initiated independently in the eastern half of the U.S. and gave rise to sunflower, before Mesoamerican crops such as beans, maize, and squash spread to North America. Central Asia witnessed the domestication of the apple and pomegranate.

Crop domestication centers are located disproportionately near or in biodiversity hotspots as defined by Myers et al. (2000) (Table 1.2). About 28% of the surface area of domestication centers lies within these hotspots, whereas the hotspots themselves constitute at most 12% of the subtropical and tropical ecozones (as defined by Schultz 2005), where the major domestication centers are located, or 6% of the total landmass. This non-random location of areas of domestication within the subtropical or tropical ecozones probably reflects the reliance of hunter-gatherers and early farmers on biodiversity for their daily subsistence, which would have been facilitated by an abundance of different species with different life cycles, adaptations, and useful products. Plants that were eventually domesticated were already

Table 1.2 Relative abundance of domestication centers in biodiversity hotspots

	Biodiversity hotspots[1]		Biodiversity non-hotspots	
	Location	Area ($\times 10^3$ km^2)	Location	Area ($\times 10^3$ km^2)
Domestication centers	Tropical Andes	1,258,000	Lowland South America[2]	3,276,145
	Mesoamerica	1,155,000	North America[3]	3,942,627
	Chocó/Darién/ W. Ecuador	260,600		
	Mediterranean Basin (including Levant and S.E. Turkey)	2,362,000	Sahel[4]	3,000,000
			Ethiopia	1,104,300
			China (except South-Central China)	8,840,821
	Caucasus	500,000		
	Indo-Burma	2,060,000	New Guinea	872,840
	South-Central China	800,000		
Total area domestication centers	In biodiversity hotspots	8,395,600	In biodiversity non-hotspots	21,036,733
Non-domestication centers	As listed in Myers et al. 2000	9,048,700	Subtropical and tropical ecozones[5]	76,500,000
	Total biodiversity hotspots	17,444,300	Total landmass	148,939,063

[1] From Myers et al. 2000
[2] Lowland South America: E. Bolivia, Paraguay, western Brazil (Acre, Rondônia, Matto Grosso, Goiás, Tocantins), Uruguay, NW Argentina (Jujuy, Salta, Tucumán)
[3] North America: Eastern half of the 48 contiguous states
[4] Between 100 and 600 mm isohyets; 400–600 km in width over a length of 6000 km
[5] From Schultz (2005)

harvested by hunter-gatherers. Hence, the abundance of species may have allowed the first farmers to choose those species that were most amenable to cultivation.

Harlan (1992) observed that the different ecozones and biomes they harbor had contributed to different extents to our array of crops. In his survey, the two major biomes in this respect were the Subtropics with Winter Rains (Mediterranean) and Tropics with Summer Rains (Savanna) ecozones. Ecozones showing an interruption of vegetation growth, for example due to drought whether in the winter or the summer, are thought to have stimulated transition to a farming economy because seasonal scarcities created a need to establish reserves for storable food such as grains. The majority of crops discussed here originated either in the Tropics with Summer Rains (Savanna to Dry Forest vegetation) or Tropics with Year-round Rains (Table 1.3), reflecting the focus and scope of this volume. It should be kept in mind, however, that while some crops are still cultivated in their original ecozone: e.g., cacao and coffee, other crops have shown an ecological expansion into more temperate ecozones, such as *Phaseolus* beans, maize, and sorghum.

Table 1.3 Taxonomic classification and geographic and ecological origin of crops discussed in this volume

Crop	Order[1]	Family	Main organ harvested	Geographic center(s) of origin	Ecozone origin
Banana/plantain, *Musa* spp.	M	Musaceae	Fruit	Southeast Asia	Tropics with summer rains
Cacao, *Theobroma cacao*	D	Sterculiaceae	Grain	Mesoamerica	Tropics with year-round rains
Chickpea, *Cicer arietinum*	D	Fabaceae	Grain	Southwest Asia	Subtropics with winter rains
Cowpea, *Vigna unguiculata*	D	Fabaceae	Grain	Africa	Tropics with summer rains
Citrus, *Citrus* spp.	D	Rutaceae	Fruit	China	Subtropics with year-round rains
Coffee, *Coffea* spp.	D	Rubiaceae	Grain	Ethiopia	Tropics with summer rains
Eucalyptus, *Eucalyptus* spp.	D	Myrtaceae	Wood	Australia	Widely distributed
Ginger, *Zingiber officinale*, and Turmeric, *Curcuma longa*	M	Zingiberaceae	Rhizome	China and S. Asia, respectively	Tropics with summer rains
Macadamia, *Macadamia* spp.	D	Proteaceae	Nut	Australia	Subtropics with year-round rains
Maize (tropical), *Zea mays*	M	Poaceae	Grain	Mesoamerica	Tropics with summer rains
Oil palm, *Elaeis guineensis*	M	Arecaceae	Fruit	Africa	Tropics with year-round rains
Papaya, *Carica papaya*	D	Caricaceae	Fruit	Mesoamerica	Tropics with year-round rains
Peanut, *Arachis hypogea*	D	Fabaceae	Grain	S.W. Brazil	Tropics with summer rains
Phaseolus beans, *Phaseolus* spp.	D	Fabaceae	Grain	Mesoamerica, Andes	Tropics with summer rains
Pineapple, *Ananas comosus*	M	Bromeliaceae	Fruit	S.W. Brazil	Tropics with summer rains
Sorghum, *Sorghum bicolor*	M	Poaceae	Grain	Africa	Tropics with summer rains
Sugarcane, *Saccharum officinarum*	M	Poaceae	Stem	S.E Asia	Tropics with year-round rains
Yam, *Dioscorea* spp.	M	Dioscoraceae	Root	Africa, S.E. Asia	Tropics with summer rains

[1] D: Dicotyledonae; M: Monocotyledonae

1.6 Genomics of Tropical Crops and Food Security

1.6.1 A Brief Overview of Genomics

The genome of a living organism is the sum total of the information contained in its genetic material, including its biochemical and structural organization, and its expression at the RNA, protein, and metabolite levels. Most genetic material is located in the nucleus but some is also located in cytoplasmic organelles (chloroplast and mitochondria) where they specify functions that are essential for the survival of living organisms. At its most basic, this genetic material is constituted by DNA. The nucleotide sequence of DNA provides the primary level of information responsible for coding enzymatic or structural proteins and ribosomal RNA ("DNA code"). In the nucleus, DNA is packaged with proteins (particularly histones) into chromatin, which, in turn, is the basic constituent of chromosomes. Various reversible chemical modifications, such as acetylations and phosphorylations, affect gene expression and specify chromosomal functional domains. Thus, there is a "histone or epigenetic code," whose effect on gene expression is superimposed onto that of the genetic code specified by the primary DNA sequence (van Driel et al. 2003; Lam et al. 2005). Furthermore, gene expression takes place in several steps, including transcription, splicing, and translation, each of which can have a major effect on trait expression (e.g., Yamaguchi and Mayfield 2005). Additional levels of complexity in gene expression are attributable to epistatic interactions, reflecting the fact that many traits have a complex inheritance. The involvement of more than one gene in a specific trait is the rule rather than the exception. In addition, environmental effects, which are often unpredictable, also play an important role in trait expression. It should, therefore, be clear that the expression of any trait is the outcome of a complex chain of events. Approaches focusing on one gene at a time have limited power to generate a complete picture of biochemical, developmental, and other pathways leading to the expression of economically important traits.

Genomics then is an ensemble of high-throughput analytical methods developed to study the genome of living organisms. Genomics can be further subdivided into areas depending on the target of the inquiry. Structural genomics investigates the DNA sequence of an organism, the distribution of coding sequences within the genome, the features such as centromeres and telomeres responsible for the function of chromosomes, micro- and macro-rearrangements in sequences. The best known examples of investigations in this area are the complete genome sequences of several organisms including plants such as Arabidopsis (Arabidopsis Genome Initiative 2000), rice (Goff et al. 2002; Yu et al. 2002), poplar (Tuskan et al. 2006), and *Medicago truncatula* (http://www.medicago.org/genome/ downloads/Mt1/). Progress has been made, and continues to be made, in the efficiency with which genomes can be sequenced (Hall 2007), suggesting that substantial sequencing will continue and that many more organisms will be sequenced to answer basic scientific questions and understand the molecular basis of economically important traits.

Functional genomics describes the products of genes, including RNA (transcriptomics), proteins (proteomics), and metabolites (metabolomics). Transcriptomics relies on isolation of mRNA and reverse transcription of these messages into a DNA form to generate either partial- (expressed sequence tags [ESTs]) or full-length sequences of genes expressed in different tissues or in response to different biotic or abiotic external stimuli. In turn, gene indices have been created that list the different genes identified and their redundancies (assemblies) (e.g., plant gene indices at TIGR: http://www.tigr.org/tdb/tgi/plant.shtml). Large collections of ESTs have been established for some species such as Arabidopsis, eucalyptus, rice, soybean, and wheat, and the number for other species is increasing as well (http://plantta.tigr.org/cgi-bin/plantta_release.pl).

The high-throughput nature of genomics generates a large amount of data. To keep up with the data flow and allow its analysis, special software tools have been created, which fall under the label bioinformatics. These tools can, for example, identify similarities in sequence motifs between a query sequence and a database (e.g., BLAST: Altschul and Gish 1996), identify repetitive or microsatellite sequences (e.g., MicrosatDesign: Singan and Colbourne 2005), and align large-scale sequences such as bacterial artificial chromosome (BAC) sequences (ACT or Artemis Comparison Tool: http://www.sanger.ac.uk/Software/ACT).

1.6.2 Contributions of Genomics to the Improvement of Tropical Crops

How can genomics contribute to the genetic improvement of crops, in general, and tropical crops, in particular? This question can be considered both from a short-term perspective (how can the development of improved cultivars benefit from the tools and information provided by genomics?) and a broader view (how does crop improvement fit in the development process of lesser developed countries, particularly in the alleviation of food insecurity and poverty?).

The most fundamental contribution of genomics to plant breeding is to provide information on the genotypic or molecular basis of phenotypic variation for crop biodiversity conservation and genetic improvement (Gepts 2006). Plant breeding has been successful in recombining and deploying new genetic variation based on phenotypic evaluations of suites of genes responsible for the expression of agronomic traits, many of which are under quantitative control. Plant breeders evaluate the expression of suites of genes in the aggregate but, with a few exceptions, cannot select individual loci and allelic variation at these loci. The molecular information provided by genomic approaches consists of the number of loci, the magnitude of the effect of different alleles at these loci, the interactions among loci, the linkage relationships with other genes (coding for the same or other traits) and the environmental effects on gene expression, and, ultimately, the ever-important gene x environment interactions.

A large part of the natural variation of crop plants and their wild relatives has not been used so far in plant breeding (Tanksley and McCouch 1997; Gepts 2000; Gur and Zamir 2004). The high-throughput nature of genomics makes possible an extensive molecular evaluation of genetic diversity in exotic germplasm, i.e., unadapted landraces (farmer-improved domesticated lines) and wild relatives. To be applicable to plant breeding, however, these evaluations of molecular variation have to be accompanied with phenotypic evaluation in the field or other locales, a.k.a. phenotyping. These phenotypic evaluations become the rate-limiting factor because multi-year, multi-location trials are required to obtain an accurate estimate of the phenotypic value in the face of a variable climatic and edaphic environment. Further dissection of a trait into component sub-traits can narrow down the genetic control and increase the heritability of the trait (Varshney et al. 2005). Unless plant breeding education is strengthened (Gepts and Hancock 2006), phenotypic evaluations are becoming a lost science as well.

The lack of utilization of exotic germplasm can be attributed to several causes. First, the lack of adaptation of this germplasm prevents its thorough evaluation in local conditions and therefore the discovery of useful variation (e.g., photoperiod sensitivity). Second, certain traits may also prevent evaluations (e.g., seed shattering and viny growth habit of wild legumes make yield evaluations difficult). Third, approaches to characterizing the genetic basis of agronomic traits has emphasized analysis of the progeny of pedigreed crosses between two parents. This reduces the number of germplasm accessions that can be characterized at any one time. Recently, more emphasis has been placed on association mapping (Mackay and Powell 2007), which relies on the analysis of linkage disequilibrium (Flint-Garcia et al. 2003) in existing populations such as germplasm collections. Association mapping by its very nature can lead to a broader analysis of existing genetic diversity. Fourth, germplasm banks may have funds for germplasm maintenance but not evaluation because of underfunding. Fifth, the exchange of germplasm is increasingly subject to ownership and sovereignty issues arising from international treaties such as the Convention on Biological Diversity and the Trade-Related Intellectual Property Rights (TRIPS) agreement of the World Trade Organization (Gepts 2004b 2006). It is important to note, however, that none of these difficulties is insurmountable and that genomics allows a more efficient extraction of useful genes from exotic germplasm.

Genomics can assist plant breeding in two major ways. In the short term, it provides a DNA sequence resource that can be used to develop a large number of polymorphic markers, such as microsatellites and single-nucleotide polymorphisms (SNPs). This sequence resource can originate from various sources, including ESTs, BAC-end sequences, BAC sequences, and whole-genome sequences in some cases. Initially, most markers were random markers representing anonymous sequences. Increasingly, however, markers are now also derived from candidate gene sequences. The large number of markers allows for a saturation of the genome and detailed linkage mapping of genes of interest, including quantitative trait loci (QTLs), either by analysis of pedigreed populations (resulting from a cross between known genotypes) or natural populations (association mapping). Once the location of genes and QTLs is determined, further research is generally necessary to identify additional markers near these genes or QTLs to allow application of marker-assisted selection (MAS) (Collard et al. 2005). This contribution of genomics relies on the development of genomic sequence resources within the target species because polymerase chain reaction (PCR)-based markers such as microsatellites and SNPs depend on precise DNA sequences to develop primers for efficient and unequivocal amplification in the PCR reaction.

The second way in which genomics can contribute to breeding is to identify the specific genes responsible for specific traits. This approach requires a considerably larger investment in genomic resources than the first. It involves the development and utilization of functional genomics tools, such as array technology, map-based cloning, transcript profiling, targeting induced local lesions in genomes (TILLING), and transformation (Sreenivasulu et al. 2007). This identification then allows for the screening of germplasm to discover allelic variants and the development of allele-specific markers for MAS. In contrast with the first approach, model systems such as Arabidopsis, *Medicago truncatula*, tomato, and rice can provide information that helps identify the molecular basis of shared traits with the target species (Morgante and Salamini 2003).

Marker-assisted selection is most suitable in the following selection situations (Xu et al. 2005): to bypass a testcross or progeny test or a laborious field or lab test, to conduct a selection independent of a normal test environment, to test a progeny at an earlier breeding stage or for multiple genes or traits, or to conduct a whole-genome selection. Thus, the main benefit of MAS is to facilitate and accelerate breeding operations in specific situations. It is an additional tool for plant breeders but does not replace regular plant breeding operations, especially field evaluations. It does not replace all types of selection, especially in the case of genetically complex traits, conditioned by a large number of genes with small effects and for which it is difficult to establish a good correlation between markers and phenotypes. Further discussion on the application of MAS is provided by Knapp (1998), Francia et al. (2005), Davies et al. (2006), Ribaut and Ragot (2007), and chapters in this volume. An additional challenge is the existence of gene interactions (epistasis) and epigenetic phenomena (Morgante and Salamini 2003; Varshney et al. 2005; Valliyodan and Nguyen 2006), which, although they have generally been difficult to deal with in genetic analyses (Carlborg and Haley 2004), need to be taken into account to obtain a realistic representation of the genetic control of a trait (e.g., Johnson and Gepts 2002)

In summary, addition of information on variation at individual gene loci and their interactions with other genes and the environment through the use of genomics promises to boost the success of plant breeding to new heights.

1.6.3 Crop Improvement and Food Insecurity and Poverty

Food security has been defined by the Food and Agricultural Organization (FAO) as: "*Food security is a situation that exists when all people, at all times, have physical, social, and economic access to sufficient, safe, and nutritious food that meets their dietary needs and food preferences for an active and healthy life.*" Some 800 million people remain currently under-nourished. According to FAO statistics, up to two billion people lack food security intermittently due to varying degrees of poverty (FAO: http://www.fao.org/es/ess/faostat/foodsecurity/index_en.htm). The majority of people undernourished or lacking food security live in southern Asia and sub-Saharan Africa. It has now become clear that food security does not depend only on sufficient production of food, particularly of those providing a dietary energy supply such as carbohydrate crops (e.g., cereal crops, cassava). Actually, the concept of food insecurity is a multi-dimensional problem, as it involves production of a diversified food supply that will supply a wide range of macro- and micro-nutrients based on biodiverse ecosystems, including the harvesting of wild and underutilized species, growing locally adapted varieties, and eating from local ecosystems (Toledo and Burlingame 2006). Most of the crops discussed in this volume are part of a foundation for a nutritious diet because they provide one or more macro-nutrients (Table 1.4) in addition to micronutrients and can complement each other nutritionally (e.g., complementation for essential amino acids between legumes and cereals). In addition, a biodiverse agroecosystem also contributes to productivity and sustainability of agriculture by exploiting different environmental niches and diversifying income sources (Hawtin 2000).

Table 1.4 Macronutrient contributions of crops discussed in this volume (per sample of 100 g[1])

Crop	Water (g)	Energy (Kcal)	Protein (g)	Lipids (g)	Carbohydrates (g)
Banana: Raw	75	89	1	~ 0	23
Plantain: Raw	65	122	1	~ 0	32
Cacao: Dry powder	3	229	20	14	54
Chickpea: Mature seeds, raw	12	364	19	6	61
Cowpea: Mature seeds, raw	12	336	24	1	60
Citrus: Orange juice, raw	88	188	1	~ 0	10
Coffee: Brewed	100	1	~ 0	~ 0	0
Eucalyptus	NA	NA	NA	NA	NA
Ginger: Ground spice	9	347	9	6	71
Turmeric: Ground spice	11	354	8	10	65
Macadamia: Raw nut	1	718	8	76	14
Maize: Whole-grain flour	11	361	2	2	77
Oil palm: Oil	0	884	0	100	0
Papaya: Raw	89	39	1	~ 0	10
Peanut: All types, raw	7	567	26	49	16
Vegetable oil	0	884	0	100	0
Phaseolus beans:					
Snap (green)	90	31	2	~ 0	7
Dry (kidney)	12	333	24	1	60
Pineapple: Raw	87	48	1	~ 0	13
Sorghum	9	339	11	3	75
Sugarcane: Granulated sugar	~ 0	387	0	0	~ 100
Yam: Raw	70	118	2	~ 0	28

[1] USDA Nutrient Data Laboratory: http://www.nal.usda.gov/fnic/foodcomp/search/ ; all numbers rounded to nearest integer

Furthermore, there is a strong correlation between agricultural productivity, hunger, and poverty (von Braun et al. 2003). Seventy-five percent of the world's poor live in rural areas and make their living from agriculture. Hunger and child malnutrition are greater in rural than urban areas. The higher the proportion of rural population that obtains its income solely from subsistence farming, the higher the frequency of malnutrition. Malnutrition erodes children's ability to learn and reduces the ability of adults to work and give birth to healthy children. Many of the consequences of childhood malnutrition are seen only much later in adulthood. Thus, malnutrition is part of a self-perpetuating vicious circle that needs to be broken if hunger is to be eliminated.

Rosegrant and Cline (2003) have argued that achieving global food security will require policy and investment reforms on multiple fronts, including human resources and education (for better farming or facilitating careers outside of agriculture), rural infrastructure (e.g., roads, safe drinking water, sewage, health care), water resources, and agricultural research. The latter usually provides a high return on investment. The public sector, charitable organizations, and the civil sector all play important roles in agricultural research, especially in tropical agriculture because of the limited market potential and the capability to develop publicly accessible (i.e., non-proprietary) technologies. Goals of agricultural research include increasing biomass, harvest index, and tolerance to external stresses, most recently drought stress.

One of the main challenges of agricultural research is to address increasing water scarcity. This scarcity is due in part to increasing demand for water from a burgeoning world population but also from agriculture, if increased yields are to become reality. Rockström et al. (2007) have estimated that an additional 1,850 km3/year will be required to produce the food needed to eradicate hunger. Part of the water management solution is to develop new varieties with increased water use efficiency (WUE) and tolerance to drought. Genomics can help obtain such new varieties by identifying drought tolerance or WUE genes that can become targets for selection (e.g., Ribaut and Ragot 2007).

An additional dimension to the contribution of genomics to plant breeding is the changing climatic and economic environment in which agriculture operates (Cassman 2007). Agriculture has until recently been able to keep up with human population growth on Earth as a source of food, feed, and fiber. Currently, however, the balance between food supply and demand is rapidly shifting from surplus to deficit. Due to the rapid rise in prices of petroleum, there is now a global expansion of biofuel production based on maize, oil crops (e.g., oil palm, soybean, Jatropha), and sugar crops (mainly sugarcane). Farmers in some countries will enjoy higher prices for these commodities, but the urban and rural poor will pay much higher prices for basic food staples (as is happening already in Mexico: M. Roig-Franzia: http://www.washingtonpost.com/wp-dyn/content/article/2007/01/26/ AR2007012601896_pf.html). Further uncertainty is caused by global climate change. Increases in production due to higher CO_2 concentration in the atmosphere are more than offset by reductions in yield by increased temperature (Lobell and Field 2007). Alleviating hunger will, therefore, no longer be a matter of poverty alleviation and more equitable food distribution, but according to Cassman (2007), will depend on accelerating gain in crop yields and overall food production capacity. Plant breeding, aided by genomics, will play an important role in this endeavor.

Acknowledgments Research on bean genomics in my group has been funded by the USDA CSREES NRI Plant Genome program.

References

Altschul SF, Gish W (1996) Local alignment statistics. Computer Methods for Macromolecular Sequence Analysis, pp. 460–480

Arabidopsis Genome Initiative (2000) Analysis of the genome sequence of the flowering plant *Arabidopsis thaliana*. Nature 408:796–815

Balick MJ, Cox PA (1996) Plants, people, and culture: the science of ethnobotany. Freeman, New York

Barker G (2006) The agricultural revolution in prehistory. Oxford University Press, Oxford

Bellwood P (2005) First farmers: the origins of agricultural societies. Blackwell, Malden, MA

Brady KU, Kruckeberg AR, Bradshaw HD (2005) Evolutionary ecology of plant adaptation to serpentine soils. Annual Review of Ecology Evolution and Systematics 36:243–266

Bramwell D (2002) How many plant species are there? Plant Talk http://www.plant-talk.org/stories/28bramw.html (Verified June 8, 2007)

Carlborg Ö, Haley CS (2004) Epistasis: too often neglected in complex trait studies? Nature Rev Genetics 5:618–U614

Cassman KG (2007) Climate change, biofuels, and global food security. Environ Res Lett 2:011002

Clarke A, Gaston KJ (2006) Climate, energy and diversity. Proc Royal Soc B-Biol Sci 273:2257–2266

Collard BCY, Jahufer MZZ, Brouwer JB, Pang ECK (2005) An introduction to markers, quantitative trait loci (QTL) mapping and marker-assisted selection for crop improvement: the basic concepts. Euphytica 142:169–196

Crane PR, Lidgard S (1989) Angiosperm diversification and paleolatitudinal gradients in Cretaceous floristic diversity. Science 246:675–678

Currie DJ, Paquin V (1987) Large-scale biogeographical patterns of species richness of trees. Nature 329:326–327

Davies J, Berzonsky WA, Leach GD (2006) A comparison of marker-assisted and phenotypic selection for high grain protein content in spring wheat. Euphytica 152:117–134

Estrada Lugo EIJ (1989) El Códice Florentino: su informaci6n etnobotánica. Colegio de Postgraduados, Chapingo, México

Flint-Garcia SA, Thornsberry JM, Buckler IV ES (2003) Structure of linkage disequilibrium in plants. Ann Rev Plant Biol 54:357–374

Francia E, Tacconi G, Crosatti C, Barabaschi D, Bulgarelli D, et al. (2005) Marker assisted selection in crop plants. Plant Cell Tiss Organ Cult 82:317–342

Gepts P (2000) A phylogenetic and genomic analysis of crop germplasm: a necessary condition for its rational conservation and utilization. In: Gustafson J (ed) Proc Stadler Symp. Plenum, New York, pp. 163–181

Gepts P (2004a) Domestication as a long-term selection experiment. Plant Breed Rev 24 (Part 2): 1–44

Gepts P (2004b) Who owns biodiversity and how should the owners be compensated? Plant Physiol 134:1295–1307

Gepts P (2006) Plant genetic resources conservation and utilization: The accomplishments and future of a societal insurance policy. Crop Sci 46:2278–2292

Gepts P, Hancock J (2006) The future of plant breeding Crop Sci 46:1630–1634
Goff SA, Ricke D, Lan T-H, Presting G, Wang R, et al. (2002) A draft sequence of the rice genome (*Oryza sativa* L. ssp. *japonica*). Science 296:92–100
Govaerts R (2003) How many species of seed plants are there? - a response. Taxon 52:583–584
Gur A, Zamir D (2004) Unused natural variation can lift yield barriers in plant breeding. PLoS Biology 2:1610–1615
Hall N (2007) Advanced sequencing technologies and their wider impact in microbiology. J Exp Biol 210:1518–1525
Harlan JR (1971) Agricultural origins: centers and non-centers. Science 174:468–474
Harlan JR (1992) Crops and man, 2nd edn. American Society of Agronomy, Madison, WI
Hawkins BA, Field R, Cornell HV, Currie DJ, Guegan JF, et al. (2003) Energy, water, and broad-scale geographic patterns of species richness. Ecology 84:3105–3117
Hawtin GC (2000) Genetic diversity and food security. UNESCO The Courier http://www.unesco.org/courier/2000_Johnson WC, Gepts P05/uk/doss23.htm (Verified July 13, 2007)
Johnson WC, Gepts P (2002) The role of epistasis in controlling seed yield and other agronomic traits in an Andean x Mesoamerican cross of common bean (*Phaseolus vulgaris* L.). Euphytica 125:69–79
Kleidon A, Mooney HA (2000) A global distribution of biodiversity inferred from climatic constraints: results from a process-based modelling study. Global Change Biol 6:507–523
Knapp SJ (1998) Marker-assisted selection as a strategy for increasing the probability of selecting superior genotypes. Crop Sci 38:1164–1174
Lam AL, Pazin DE, Sullivan BA (2005) Control of gene expression and assembly of chromosomal subdomains by chromatin regulators with antagonistic functions. Chromosoma (Berlin) 114:242–251
Lewington A (2003) Plants for people. Transworld, London
Lobell DB, Field CB (2007) Global scale climate - crop yield relationships and the impacts of recent warming. Environmental Res Lett 2:014002
Mackay I, Powell W (2007) Methods for linkage disequilibrium mapping in crops. Trends Plant Sci 12:57–63
Marinelli J (ed) (2005) Plant. Dorling Kindersley, New York
Mittelbach GG, Steiner CF, Scheiner SM, Gross KL, Reynolds HL, et al. (2001) What is the observed relationship between species richness and productivity? Ecology 82:2381–2396
Mittelbach GG, Schemske DW, Cornell HV, Allen AP, Brown JM, et al. (2007) Evolution and the latitudinal diversity gradient: speciation, extinction and biogeography. Ecol Lett 10:315–331
Morgante M, Salamini F (2003) From plant genomics to breeding practice. Curr Opinion Biotechnol 14:214–219
Myers N (1990) The biodiversity challenge: Expanded hot-spots analysis. The Environmentalist 10:243–256
Myers N, Mittermeier RA, Mittermeier CG, da Fonseca GAB, Kent J (2000) Biodiversity hotspots for conservation priorities. Nature 403:853–858
Nichols WF, Killingbeck KT, August PV (1998) The influence of geomorphological heterogeneity on biodiversity II. A landscape perspective. Conservation Biol 12:371–379
Partel M, Laanisto L, Zobel M (2007) Contrasting plant productivity-diversity relationships across latitude: The role of evolutionary history. Ecology 88:1091–1097
Raunkiaer C (1934) The life forms of plants and statistical plant geography. Oxford University Press, Oxford
Ribaut JM, Ragot M (2007) Marker-assisted selection to improve drought adaptation in maize: the backcross approach, perspectives, limitations, and alternatives. J Exp Bot 58:351–360
Rockström J, Lannerstad M, Falkenmark M (2007) Assessing the water challenge of a new green revolution in developing countries. Proc Natl Acad Sci USA 104:6253–6260
Rosegrant MW, Cline SA (2003) Global food security: Challenges and policies. Science 302:1917–1919
Sarr DA, Hibbs DE, Huston MA (2005) A hierarchical perspective of plant diversity. Quart Rev Biol 80:187–212

Schultz J (2005) The ecozones of the world, 2nd edn. Springer, Berlin
Singan V, Colbourne JK (2005) MicrosatDesign is a pipeline for transforming sequencer trace files into DNA markers. CGB Technical Report 2005–01. The Center for Genomics and Bioinformatics, Indiana University, Bloomington http://cgb.indiana.edu/files/articles/CGB-TR-200501.pdf (Verified July 13, 2007)
Smith B (1995) The emergence of agriculture. Scientific American Library, New York
Sreenivasulu N, Sopory S.K, Kishor PBK (2007) Deciphering the regulatory mechanisms of abiotic stress tolerance in plants by genomic approaches. Gene 388:1–13
Tanksley S, McCouch S (1997) Seed banks and molecular maps: unlocking genetic potential from the wild. Science 277:1063–1066
Toledo A, Burlingame B (2006) Biodiversity and nutrition: A common path toward global food security and sustainable development. J Food Composition Anal 19:477–483
Tuskan GA, DiFazio S, Jansson S, Bohlmann J, Grigoriev I, et al. (2006) The genome of black cottonwood, *Populus trichocarpa* (Torr. & Gray). Science 313:1596–1604
Valliyodan B, Nguyen HT (2006) Understanding regulatory networks and engineering for enhanced drought tolerance in plants. Curr Opin Plant Biol 9:189–195
van Driel R, Fransz PF, Verschure PJ (2003) The eukaryotic genome: a system regulated at different hierarchical levels. J Cell Sci 116:4067–4075
Varshney RK, Graner A, Sorrells ME (2005) Genomics-assisted breeding for crop improvement. Trends Plant Sci 10:621–630
von Braun J, Swaminathan MS, Rosegrant MW (2003) 2003–2004 IFPRI annual report essay: Agriculture, food security, nutrition and the Millennium Development Goals. IFPRI http://www.ifpri.org/pubs/books/ar2003/ar2003_essay.htm (Verified July 13, 2007)
Xu YB, McCouch SR, Zhang QF (2005) How can we use genomics to improve cereals with rice as a reference genome? Plant Mol Biol 59:7–26
Yamaguchi K, Mayfield SP (2005) Transcriptional and translational regulation of photosystem II gene expression. Advances in Photosynthesis and Respiration:The light-driven water: Plastoquinone oxireductase, pp. 649–668
Yu J, Hu S, Wang J, Wong GK-S, Li S, et al. (2002) A draft sequence of the rice genome (*Oryza sativa* L. ssp. *indica*). Science 296:79–92
Zeder MA (2006) Central questions in the domestication of plants and animals. Evolutionary Anthropology 15:105–117

Chapter 2
International Programs and the Use of Modern Biotechnologies for Crop Improvement

Jean-Marcel Ribaut, Philippe Monneveux, Jean-Cristophe Glaszman, Hei Leung, Theo Van Hintum, and Carmen de Vicente

Abstract This chapter describes some of the key steps taken during the development of the international agriculture research effort over the last century and presents an overview of the various players involved. The role and niche of foundations, the private sector, and national programs are discussed in the context of accessing and using genomics resources. A description of the Generation Challenge Programme (GCP) is presented as a detailed case study of an international initiative aiming to develop and utilize genomics resources to enable plant breeders in the developing world to produce better crop varieties for resource-poor farmers. The challenges faced by the international programs, and the GCP in particular, to positively impact crop productivity in marginal environments are discussed.

2.1 Introduction

Combating hunger by raising agricultural productivity through the use of modern technologies to improve crop varieties for marginal, or less favorable, environments is one of the major challenges faced by international programs focusing on agriculture and development (Toenniessen et al. 2003). As we come to understand more about the complex issues inherent to the transfer and application of new plant biotechnologies to developing countries, we recognize that many solutions can be found only through innovative partnerships and collaboration. International programs have been critical to the support and coordination of activities related to tropical crop improvement, addressing various challenges faced by international agricultural research.

During the 1960s, many countries, particularly the newly independent African nations, experienced difficult agricultural and food situations. Based on lessons learned from the Green Revolution, several international initiatives were developed

J.-M. Ribaut
Generation Challenge Programme – Genetic Resources and Enhancement Unit, International Maize and Wheat Improvement Center (CIMMYT), Carretera Méx-Veracruz, El Batán, Texcoco, México CP56130
e-mail: j.ribaut@cgiar.org

© Springer 2008

in an intensive effort to support research aimed at assuring food supplies and improving tropical agriculture (Pardey et al. 2003). The mission and activities of these international programs have evolved over the last four decades, starting with the development and dissemination of improved germplasm and moving towards the development of more holistic and integrative strategies, emphasizing self-reliance, sustainability, and access to new technologies (Conway and Toenniessen 2003). Thanks to the recent technological revolution, genomics approaches have begun to dramatically improve our capacity to explore the genetic constitution of many of the agriculturally important species. Powerful tools are available that provide breeders of major crops with the means to understand the genetic basis underlying the quantitative inheritance of many traits of agronomic interest. As a consequence, many scientists, particularly those involved in public sector-supported plant research have a strong desire to see the fruits of their research translated into concrete benefits. However, opportunities to interact with scientists involved in international agriculture for development are scarce and funding is limited compared to the investments made in fundamental research. The issue of connecting innovation in plant research to downstream applications (Delmer 2005) is one of the biggest challenges faced by international programs. Overcoming these challenges is essential to ensure that advances in science and the large amounts of data generated by the 'omics' technologies over the last years can have a positive impact on plant breeding in developing countries.

2.2 Some History

2.2.1 The Post-World War II Period

When the end of World War II was still far from certain, 44 governments committed themselves in 1943 to create an international organization to work for food and agriculture security (http://www.fao.org/UNFAO/histo-e.htm). The report of the interim commission and the draft constitution of the Marshall Plan served as the main inputs for the foundation, in October 1945, of the Food and Agriculture Organization (FAO) as a specialized agency of the United Nations (UN). From its birth, FAO was considered as a multidisciplinary institution "concerned with that large sector represented by the world's farms, forests, and fisheries, and by the needs of human beings for their products." The Marshall Plan, which provided some US$13 billion to rebuild the shattered infrastructure and cities of Europe, created a precedent for the delivery of large-scale international aid, and this was used as a model for aid delivery to Asia, Latin America, and Africa (FAO 2005). Ever since, the FAO has played a fundamental role in making inventories of the land, water, fisheries, and forestry resources of many countries. It has conducted agricultural censuses that have been used to inform the planning of development policies and strategies. FAO has also been conspicuous for its practical development work and for its timely intervention during emergencies.

The year 1943 is also considered as the dawn of the Green Revolution, with the establishment of the Office of Special Studies (OSS). This was a collaborative research program involving the Rockefeller Foundation (RF), established in 1913, and the presidential administration of Mexico and researchers from the United States and Mexico (Perkins 1997). The major initiative of the OSS was the development of high-yielding maize and wheat varieties for Mexico. The adoption of these new seed varieties soon became widespread, allowing Mexico to become not just self-sufficient in wheat production by 1951 but also to evolve into a wheat exporter shortly thereafter (Conway 1997).

2.2.2 The Food Situation and International Research in the 1960s

As the process of global decolonization gained pace during the 1960s, an increasing number of independent nations emerged as members of the UN and its agencies. Institution-building support by FAO has been fundamental in establishing national government structures for the agriculture, forestry, and fisheries sectors in many developing countries, from the moment of their birth as independent nations. It was felt that solutions to the problems of mass poverty and malnutrition were best to be found through the application of development policies directed primarily at the rapid growth of the domestic economy. The core of this strategy was to concentrate investment, generally financed from external sources, into the relatively modernized sectors of the economy, centered on primary export products (agriculture and mines). It was argued that this should generate export income, which would, in turn, favor financial development, notably through the purchase of machinery, and thereby kick-start the process of industrialization.

In Africa and in some areas of Asia and Latin America, the production of export crops was encouraged, often to the detriment of food crops (Berry 1984). Between 1960 and 1970, annual growth rate in the production of basic foodstuffs (cereals, root crops and legumes) was about 2.5%, whereas that of non-food products mainly intended for export (coffee, tobacco, cotton, rubber, etc.) was about 4% (Gakou 1987). Imported inputs were targeted mainly to export crops where, as for example in Kenya, 80% of fertilizer was used for coffee and tea production in commercial plantations, in Uganda, 84% for tea and sugar production, and in Senegal, 52% for groundnut production (Gakou 1987).

The unfavorable climatic conditions of the late 1960s and early 1970s only worsened the situation, with drought, in particular, severely affecting the production of the basal food crops. In the Sahel region, annual millet and sorghum (the basis of the local diet for a very large proportion of the population) production over the period 1969–70 to 1973–74 fell by 33%. At the same time, the production level of cash crops was maintained or even increased, thanks to governmental support (the best lands in well-watered areas, hydro-agricultural installations, access to inputs, etc.). The resulting food crisis led to a dislocation of many national socio-economic systems. Simultaneously, this period was associated with a major increase in population growth rate, due to a significant decrease in mortality. As a result, urbanization intensified, as millions of farmers were forced to move to the cities to either seek work or to access food aid. Unemployment grew sharply in the large cities, while the gradual depopulation of the countryside only aggravated the decline in agricultural production which, in Africa, was based primarily on a plentiful supply of labor. As a consequence, famines were common in many African countries.

The international agricultural research community responded in various ways to this crisis. Starting in the 1960s, the World Bank and the regional development banks progressively strengthened their portfolios in agriculture and rural development, and countries began to establish specialized ministries for development and cooperation. With the experience of agricultural development in Mexico judged as a success, RF sought to spread the Green Revolution to other nations.

2.2.3 The Green Revolution

The OSS became an informal international research institution in 1959, and in 1963 it formally became known as The International Maize and Wheat Improvement Center (CIMMYT), an international center dedicated to the furthering of genetic progress in wheat and maize (Conway 1997). A close collaboration between the Ford Foundation (FF) and RF led to what has been termed the "Green Revolution." This program cost around $600 million, and was able to bring new farming technologies and increased crop productivity to Latin America and Asia from the 1940s to the 1960s. The FF and the Indian government collaborated to import a huge quantity of wheat seed from CIMMYT. In 1961, RF and the FF proceeded to jointly establish The International Rice Research Institute (IRRI) in the Philippines. Wheat and rice high yielding varieties (HYVs) spread rapidly throughout Latin America, Asia, and North Africa. Most carried semi-dwarfing genes, which were of strategic importance in the development of the Green Revolution. Global yields of rice, maize, and wheat increased steadily during the period 1961–1985, doubling in developing countries. In Asia, these gains in rice productivity have been attributed roughly equally to improved irrigation, higher levels of fertilizer, and genetically superior varieties.

The major achievement of the Green Revolution has been to reduce the occurrence of famine, especially given the increase in the world population of ca. 4 billion since its inception. Unprecedented harvests from the new varieties of rice, wheat, and maize, developed by international research, were recorded in many parts of Asia and Latin America. The impact was particularly spectacular in India, where cereal yields increased by 146% between 1961 and 2000 (Spitz 1987). Between 1973 and 1994, the average real income of small farmers in southern India rose by 90%, and over the same period, the income of landless laborers rose by 125% (Dev 1998).

The Green Revolution also had its down sides. In certain regions, crop land traditionally used for subsistence agriculture was redirected to the production of grain for export and/or for animal feed, leading to a loss in food security (Hazell et al. 1991). In India, Green Revolution wheat occupied much of the land previously used for pulse production, even though wheat did not constitute a large proportion of the local diet. The Green Revolution has also been criticized for causing environmental damage (Conway 1997). Excessive and inappropriate use of fertilizers and pesticides has polluted waterways, poisoned agricultural workers, and killed beneficial insects and other wildlife. Irrigation practices have led to soil salinization and in some cases to abandonment of farming land. The heavy dependence on a few major cereal varieties led to some loss of biodiversity. In reality, some of these outcomes were due to inadequate extension and training, an absence of effective management of water, and subsidy policies that encouraged excessive use of modern technologies in the absence of the corresponding capacity to manage them. The Green Revolution also had major sociological impacts on rural communities. The transition from traditional agriculture, in which inputs were generated on-farm, to Green Revolution agriculture, which required the purchase of inputs, meant that a number of smaller farmers fell into debt, resulting in the loss of their land. In addition, the increased level of mechanization on the larger farms made possible by the Green Revolution curtailed an important source of employment in the rural economy.

2.3 The Various Partners in International Agriculture

Recent decades have seen a redefinition of the role and composition of the partners involved in international agriculture. In most countries, the state has moved away from many areas o f past activity, such as the marketing of agricultural produce or farm inputs and the management of agro-industries, to concentrate its effort on the provision of essential services and infrastructure and on the construction of legal, institutional, and policy frameworks to encourage the participation of non-state actors. On May 19 1971, the World Bank led the effort, along with FAO and the UN Development program (UNDP), to create the Consultative Group on International Agricultural Research (CGIAR). In all, 18 governments and organizations attended as members, and 10 as observers, although none of these were from developing countries. The founding meeting defined the objectives, composition, and organizational structure of CGIAR. Equally significant have been the growth and diversification of private foundations to support agricultural development and move people out of poverty. The most recent such event has been the commitment of the Bill & Melinda Gates Foundation to devote $100 million to African agriculture over the next five years, in combination with a decision by RF to contribute an additional $50 million. The ambition of this initiative is set to spark a new Green Revolution on the continent that benefited least from the previous one.

The last decades have also been marked by a multiplication in the number of interested civic institutions, especially non-governmental organizations (NGOs), both national and international. Civic society's increasing access to information and awareness, accompanied by a growing public scientific debate in the past year, has provided a unique opportunity to promote food security and food safety through equitable and sustainable agriculture. Over this same period, the private sector has grown to be an increasingly important player in national economies, often becoming the major supplier of technologies, inputs, services, and markets for producers, a situation which demands a creative redefinition of the roles of the public and private sectors in development (Spielman and von Grebmer 2004). Many NGOs were created to fill the gap between the state and the private sector. As their resources have grown, their role has expanded to providing development assistance (with several having a larger presence in developing countries than does the FAO).

2.3.1 The CGIAR System

A series of high-level consultations is currently exploring how the international community can best protect and strengthen the international agricultural research centers that have contributed so much to past development. In particular, the emphasis is on measures to consolidate and spread the benefits of international agricultural research beyond Asia and to participate in an intensive international effort to support research specializing in food supply and tropical agriculture.

2.3.1.1 The First Decade (1971–1980)

The founding CGIAR objective was to "increase the pile of rice" in tropical countries that faced serious scarcity. The highest priority was given to research on the major cereals, but the portfolio was broadened to include cassava, chickpea, sorghum, potato, and the millets, so that it now encompasses 27 crops. The founding resolution of CGIAR aimed to take into account not just technical, but also ecological, economic and social factors. CGIAR has consequently developed several new areas of activity, including livestock research, farming systems, conservation of genetic resources, plant nutrition, water management, policy research, and services to the national agricultural research centers in developing countries. As the scope of research has widened, the number of international centers in the CGIAR family has grown from four to 13.

2.3.1.2 The Second Decade (1981–1990)

The objective of the research during this period was to improve the nutritional level and general economic well-being of the poor by increasing sustainable food production. This approach implied a shift in focus towards poverty alleviation and gave a high level of priority to the protection of biodiversity, land, and water. Four major programs were identified: enhancing sustainability through resource conservation and management, increasing the productivity of commodity production systems, improving the policy environment, and strengthening national research capabilities.

2.3.1.3 The Third Decade (1991–2000)

CGIAR extended its research focus in these later years to include agro-forestry, forestry, fisheries, water management, and banana/plantain. The number of centers fell to the current level of 16. The CGIAR mission statement was re-formulated to read "through international research and related activities, and in partnership with national research systems, to contribute to sustainable improvements in the productivity of agriculture, forestry and fisheries in developing countries in ways that enhance nutrition and well-being, especially of low-income people." Productivity and natural resource management became the twin pillars of research focusing on aquatic resources, the conservation of genetic resources (biodiversity), food crops, forestry/agro-forestry, livestock, soil and water nutrients, water management, and policy research. Poverty reduction remained the touchstone of CGIAR-supported research, and the protection of biodiversity was a major concern, given that CGIAR is the custodian of one of the world's largest ex situ collections of plant genetic resources. (This collection includes over 600,000 accessions representing more than 3,000 crop, forage, and pasture species.) An important additional objective was to mobilize biotechnology through research alliances to maximize its contribution to a more sustainable rate of agricultural growth in developing countries.

2.3.1.4 The Fourth Decade (2001–2010)

Based on the consensus that emerged from various working groups, CGIAR's mid-term meeting (held at Durban in May 2001) decided to develop decision-making, with full use of information technology, to both share information and reach decisions, to improve the level of its scientific advice with the creation of a science council, to establish a system office and an integrated communication strategy, and to improve the coherence, efficiency, and cost-effectiveness of the services provided to the CGIAR system. The "Agroecosystem Analysis and Farming System Research" approach and other similar methods have been adopted to promote a more holistic view of agriculture. The "Rapid Rural Appraisal" and the "Participatory Rural Appraisal" methods are employed to help scientists understand the problems faced by farmers and even to give farmers a role in the development process. Finally, CGIAR initiated the formulation and implementation of the high-impact, focused, and time-bound Challenge Programs (CPs).

2.3.2 Private Foundations

Private foundations have developed as major players in supporting international agriculture and developing local networks. Although most support research and development to enhance international agriculture and to promote the use of new technologies (in a broad sense, from agronomic practices to biotechnology) as an aid to crop breeding, their strategies and foci are diverse, depending inter alia, on the level of resources available. An extensive presentation of all the foundations involved in international agriculture is beyond the scope of this chapter, but in the following section, a few representative examples are described. The particular case of RF, considered by many as a model institution in the field of support of international agriculture, is highlighted.

The W.K. Kellogg Foundation, established in 1930, ranks today among the world's largest private foundations and emphasizes the development of food systems and food quality. Grants are awarded in the United States, Latin America, and the Caribbean, and seven southern African countries – Botswana, Lesotho, Malawi, Mozambique, South Africa, Swaziland, and Zimbabwe. The food systems brief focuses on "catalyzing efforts that lead to a safe, wholesome food supply for this and future generations, while ensuring that food production and food-related business systems are economically viable, environmentally sensitive, sustainable long-term, and socially responsible."

The McKnight Foundation, founded in 1953, seeks to increase food security for resource-poor people in developing countries through its Collaborative Crop Research Program (CCRP). The CCRP, which has committed $53.5 million since 1993, combines elements of research and development, seeking innovative solutions to real problems surrounding the availability, access, and utilization of nutritious food by the poorest rural people. The CCRP acts as a competitive grants program which focuses strictly on neglected or under-researched crops (e.g., roots and tubers, legumes, and less-studied cereals, such as finger millet). Along with its support for the development of biotechnology, it promotes the development and use of molecular markers to complement phenotypic selection in plant breeding. A typical project is seeking to improve the ability of finger millet to resist drought and blast, using a comparative genetics approach to identify target genomic regions of interest. Both conventional breeding and marker-assisted selection will help produce hardier varieties that will suit local farmers' preferences.

The Bill & Melinda Gates Foundation, founded in 2000, is currently the largest foundation in the world, with an endowment of some $31.9 billion. It seeks to reduce inequities and improve lives around the world. In developing countries, its primary briefs are to enhance health care and reduce extreme poverty. Although concentrating primarily on medical research, a decision was taken in 2006 to extend support to agricultural projects. It has recently partnered with RF to enhance agricultural science and small-farm productivity in Africa, aiming to move tens of millions of people out of extreme poverty and to significantly reduce the incidence of hunger. The joint Program for Africa's Seed Systems (PASS) aims to support the development of improved varieties of African crops and the training of a new generation of African crop scientists and to ensure that genetically improved seeds reach the small-holder farmers via a network of African agro-dealers.

The Syngenta Foundation for Sustainable Agriculture, founded in 2001, has an annual budget of about $3 million. It supports research leading to sustainable food security in the poorest regions of the world, as well as promoting public discussion in the area of nutrition. Since its inception, it has been a consistent supporter of the use of biotechnology in plant breeding, and in particular, encourages the sharing of technologies, products, and expertise developed by the parent Syngenta Company with public institutions. An example of its work is the support given since 1999 to the Insect Resistant Maize for Africa joint research project, carried out by the Kenya Agricultural Research Institute and CIMMYT. This project aims to produce maize resistant to key insect pests, adapted to various Kenyan agroecological zones.

The Kirkhouse Trust represents an example of how an organization with limited resources ($3–4 million per year) can make an impact in a specific research area. The trust was founded by the leading geneticist Ed Southern, who is a strong promoter of the use of modern technology to complement plant breeding. The trust was formed in 2000 with the aim of bringing biotechnology, especially molecular markers, to breeding programs in India and sub-Saharan Africa. Because it is led by scientists who understand well the potential benefits, but also the limitations, of biotechnology, and because of its limited funding, the trust has focused on a small number of under-studied crops and on the development of human resources and capacities to promote the development of molecular breeding. This has included the gene-space sequencing of the cowpea genome to accelerate the development of molecular markers in this crop. The trust is providing equipment for marker assisted selection to several groups in Africa and is supporting a molecular marker network.

Few trusts and foundations focusing on agriculture and development are based in the South, but the Barwale Foundation in India is an example of these. It was established in 1986 by Dr. B.R. Barwale, its present chairman, who was a winner of the World Food Prize in 1998, and who established the Maharashtra Hybrid Seeds Company Limited in 1964. The mission of the foundation is "to promote research, technology and knowledge in the areas of agriculture, health care and education for human welfare," clearly indicating its support for modern technology. Under its food security theme, the foundation supports multidisciplinary research through the application of biotechnology tools towards the improvement of the major crops. Currently, the focus is on marker-assisted selection in rice to breed for traits which confer enhanced yield potential, hybrid seed production efficiency, and biotic/abiotic stress resistance.

2.3.3 The Rockefeller Foundation for Rice Improvement

RF has a long, complex, and rich history in promoting agricultural development throughout the developing world. It began its major field-based program in Mexico in the 1940s, and these led to the series of technologies, insights, and processes collectively known as the Green Revolution. During the 1950s, its success in Mexico led RF to establish similar programs in Colombia, Chile, and India. The 1960s saw RF establish, jointly with FF, four international agricultural research centers. Finally, in 1971, it helped establish the CGIAR as a consortium of donors to support the international centers. An external review conducted in 1982 strongly recommended that RF explore the potential of molecular biology for improving plant breeding. Over the next two years, RF officers consulted experts and assessed the relative status and merits of a program focusing either on a few, or on a single crop species. In late 1984, and in the context of a large program which targeted several crops and continents, a decision was taken to implement a comprehensive rice program, ranging from fundamental research through to the application of new molecular-based techniques in breeding (Toenniessen and Herdt 1988).

The RF Board of Trustees approved its rice program in December 1984 and was aware from the outset of the high risk involved and the probable 10–15-year time frame that would be needed to accomplish its objectives. A long-term program of rice biotechnology research was designed, based on a two-year survey and an analysis of the prospects for the genetic improvement of the world's major food crops. At that time, the rice genome was still totally uncharted and even its size was uncertain; no DNA molecular markers/maps were available; and since no cereal species had yet been regenerated from protoplast culture, there was no evidential support for the idea that transformation could ultimately become a tool for rice genetic improvement. During the first five to seven years, the RF program laid the scientific basis for "rice biotechnology" in the shape of the International Program on Rice Biotechnology (IPRB). Its early successes were the first DNA-based rice genetic map, protocols for rice regeneration and transformation, and the use of genomic information from rice pathogens to understand host-plant resistance. These and other advances changed the way in which rice geneticists dealt with breeding objectives such as insect resistance, abiotic stress tolerance, and hybrid rice. Later, it became clear that rice held a pivotal position in the evolution of all the cereal species. Over the ensuing seven to eight years, the program shifted its focus to technology transfer to institutions in the major rice-producing and consuming countries, a task that required a strengthening of both the physical and human resources within the national and international rice research systems in Asia, Africa, and Latin America. The RF program management then sought to support further technology development and application, while promoting international collaborative research-cum-training, its greatest asset. This linking of fledgling national rice biotechnology programs with those at advanced research institutes (ARIs) in the United States, Europe, Japan, and Australia became the hallmark of RF's management strategy.

Over its lifetime, the IPRB dispersed about $105 million, at an average of about $6 million per year (Table 2.1). The proportion of funding allocated to research and training was approximately 70% and 30%, respectively. However, since the training program was fully integrated into the research priorities, much of the training support contributed directly to research outcomes. In the same way, the allocation of 30% of the funding to high-income country and 47% to low-income country institutions is misleading. The level of integration in the program allowed the remaining 23% of funds to be used to create "bridging" elements, via the support of meetings and workshops, which have contributed significantly to the creation of lasting and close linkages among the high-income countries, low-income countries, and the international centers. During the program's 17-year life, over 400 (primarily Asian) rice scientists have been trained in this manner. The successful fusion of research and training produced many of the long-term collaborative relationships that have outgrown their dependence on RF support and continue today (for example, the IRRI-managed Asian Rice Biotechnology Network). The rapid research progress in rice plant molecular biology and genomics (Wang et al. 2005) has also managed to attract significant amounts of other financial support for rice-centered research, resulting in the situation today where rice has become the genomic model for the

Table 2.1 The Rockefeller Foundation's International Program on Rice Biotechnology expenditures x $1,000 (O'Toole et al. 2000)

Year	Basic research (HIC)a	Applied research (LIC)b	International Centers	Social science	Meetings/ Administration	Fellowships training	Total
2000	157	1,400	500	0	64	810	2,931
1999	468	1,689	1,006	55	466	2,266	5,950
1998	561	2,480	729	50	288	2,305	6,413
1997	566	2,068	936	0	523	2,418	6,511
1996	2,073	1,462	1,161	289	346	2,173	7,504
1995	1,974	1,845	1,289	280	240	2,071	7,699
1994	1,263	2,139	1,622	100	614	1,878	7,616
1993	2,400	1,537	1,857	307	176	2,525	8,802
1992	2,474	1,591	1,088	405	499	2,305	8,362
1991	2,081	1,309	800	69	385	2,160	6,804
1990	3,100	1,847	1,092	196	284	2,050	8,569
1989	3,049	2,811	773	181	372	1,038	8,224
1988	1,689	718	655	467	100	635	4,264
1987	4,753	170	621	1,217	100	368	7,229
1986	1,530	125	746	0	155	364	2,920
1985	859	131	15	0	34	427	1,466
1984	2,780	131	50	0	15	488	3,464
Total	31,777	23,453	14,940	3,616	4,661	26,281	104,728

HIC = high-income countries of the industrialized world. LIC = low-income countries of the developing world.

monocot species, complementing the longer established dicot model *Arabidopsis thaliana*.

RF finally closed this truly international program in 2000. The rationale for closure was explained by John O'Toole, the IPRD leader, by the statement: "We are confident that the rice research community, particularly in Asia, now has the capacity to keep rice at the forefront of biotechnology research and to produce the new rice varieties the world urgently needs, as long as the community continues to share and work collaboratively" (O'Toole et al. 2000).

2.3.4 The Evolution of National Agricultural Research Systems

During the Green Revolution, national agricultural research systems (NARS) found themselves very much at the downstream end of the delivery chain of international agriculture. Most, if not all of the NARS were passive recipients of products generated by international programs which consisted mainly of improved germplasm and efforts to train scientists and improve the infrastructure of their breeding programs. In more recent times, however, the capacity of the NARS, in terms of their financial resources, infrastructure, and expertise, has evolved in a somewhat country-specific manner, reflecting the health of their domestic economies. Thus, in some countries, capacity has degraded, while in others there have been major improvements, as evidenced by a change from requiring training and support from large international programs to becoming mutual partners in agricultural research. Examples of this latter situation are in Brazil, China, Mexico, South Africa, and Thailand, where the capacity to use modern biotechnology has burgeoned over the last decades, to the extent that their NARS or research centers are now recognized at the international level as leaders in plant genomics (Huang et al. 2002). Teams from China and Thailand were heavily involved in the international rice sequencing project that started 10 years ago. A major factor in this progress has been the provision of significant levels of financial support from respective governments. Thus, for instance, the annual Empresa Brasileira de Pesquisa Agropecuária (EMBRAPA, the Brazilian Agricultural Research Corporation) budget of about US$400 million is larger than that of the entire CGIAR system (about US$350 million). These "large" NARS are beginning to communicate with one another, as illustrated by the 2006 agreement between Brazil, China, and India to collaborate in the area of agriculture, including the exchange of genetic resources and joint efforts in plant biology and breeding.

However, the situation remains uneven across the developing world. For the majority of NARS, the capacity to conduct research in plant biology and support plant breeding remains rather limited, and in some cases it has even decreased over time. Although there has been a strong focus on agricultural development in Africa in recent years, many of the African NARS, especially those in Sub-Saharan Africa, remain dependent on international support. As a result, the funding strategy of international programs has to be carefully differentiated, focusing on those NARS requiring the highest level of support demand and engaging the large NARS as active partners in research and development. These more needy NARS tend to be in countries whose population has a high proportion of resource-poor people; thus building the capacities of breeding programs and seed systems in these countries is vital to achieving any improvement in the ability of poor farmers to grow improved germplasm.

2.3.5 The Role of the Private Sector

There is universal agreement that partnerships between public research institutions, private companies, and civic organizations represent the best means to exploit the strengths of the various players to channel knowledge and resources into those areas where they can most effectively increase food production and therefore significantly reduce poverty. The exploration of public-private partnerships (PPP) is critical, given that private sector investment in agricultural research is growing so rapidly and is providing new technologies and market opportunities for developing countries. These trends imply that it is unlikely that the UN's Millennium Development Goals can be achieved without a concerted effort to enlist private sector support and collaboration in agricultural research and development.

The level of private-sector investment in agricultural research is estimated to account for some 35% of global investment in agricultural research and development

(Pardey and Beintema 2001). Approximately 13% of this sum is directed into advanced research in agricultural biotechnology by the leading multinationals and other smaller biotechnology companies (Byerlee and Fischer 2001). Little of this investment is directly or intentionally pro-poor, given that private enterprises have to rely on revenue from the sale of their product. As a result, the vast majority of private-sector investment in agricultural research is aimed at those crops, traits, and technologies that are relevant to farming in advanced, industrialized countries, as this is the route to achieve the level of profitability sufficient to guarantee an adequate return on investment. The few resources devoted to developing countries (either directly or in the form of research spill-over) tend to be concentrated in large countries with a highly commercialized agricultural sector, and so the benefit of this research to small-scale, resource-poor farmers is at best only marginal.

Nevertheless, some private seed companies have, over the last decade, been signaling clearly that they would welcome the opportunity to collaborate with the public sector to apply some of their know-how to benefit poor farmers. Such a joint effort could extend beyond technology and information transfer, as the public sector can also benefit from the private sector culture, where success is measured in terms of usable product. As a result, private sector scientists tend to be more adept at working at the interface between basic and applied biology. Deborah Delmer has recently proposed an interesting PPP model (Delmer 2005), based on the observation that beneficial traits such as disease and pest resistance or drought tolerance in major commercial crops could perhaps be advantageously addressed by the private sector, while the public sector maintains a range of locally adapted germplasm relevant to the poor farmer. In such a PPP, the public sector would support an effort to transfer valuable private-sector traits/genes into a range of locally adapted varieties suitable for low-input agriculture, while the private sector concentrated on varieties that could be sold to the large-scale farmers in developed economies.

To explore such options and to lay out the landscape for the development of PPP, the International Food Policy Research Institute organized a workshop in September 2005 entitled: "An Internal Dialogue on Pro-Poor Public-Private Partnerships for Food and Agriculture". The proceedings of this meeting are available at (http://www.ifpri.org/ events/conferences/2005/PPP/pppproc.asp). The objectives of the dialog were to take stock, identify new opportunities, address key policy issues, and develop practical and sustainable solutions that could foster more effective pro-poor partnerships. Over 90 key players attended the event, including representatives of the leading multinational and developing country agribusiness firms, developing-country governments, scientific research institutes, and NGOs. The fruitful discussions identified relevant points to promote PPP and made some recommendations in the form of a bold agenda to promote a greater presence of pro-poor partnerships in food and agriculture.

Although there is clear desire on both the public and private sides to work together, the current level of PPP in international agriculture remains far from its full potential. The experience of the 2005 dialog highlighted the limitations of moving from general agreement and good intentions to concrete action and commitments. To progress, specific common areas of interest, i.e., win-win situations must be identified, and each negotiation will need to be conducted on a case-by-case basis. Although in most cases, especially when dealing with research and technology transfer, the PPP will have to be negotiated with a single private partner to avoid conflicts of interest, there have already been a few examples of successful collaborations between public institutions and multiple private sector players. An example is the effort to transfer apomixis, the asexual reproduction of plants through seed, into maize from its wild relative *Tripsacum*. Since 1990, a joint project involving CIMMYT and the French Institut de Recherche pour le Développement (IRD) (formerly ORSTOM) has focused on understanding the genetic basis of apomixis and the formulation of strategies to transfer the trait (Savidan et al. 2000). To accelerate progress in such a potentially revolutionary area, CIMMYT and IRD entered into formal research collaboration with three seed companies (Pioneer Hi-Bred International, Groupe Limagrain, and Novartis Seeds) during the latter part of 1999. For the seed-producing partners, the control of apomixis could create new options for the multiplication of high quality seed, while for CIMMYT and IRD, the transfer of apomixis to maize would provide a means to deliver superior hybrid crop traits, such as improved disease resistance and higher yield, to resource-poor farmers. Although this research has yet to make any direct impact on breeding, the collaboration has generated a body of relevant knowledge surrounding the genetics of apomixis (e.g., Grimanelli et al., 2005).

The basis of any PPP collaboration is dependent on the nature of the private partner(s). It is reasonable to expect that the multinationals will increase their contribution to the international effort by providing technologies and traits/genes, while medium and small local enterprises will play a key role in crop improvement, seed production, and seed distribution for local markets. A leading example of the different levels of interaction possible with the private sector is provided by Winrock International (http://www.winrock.org/initiative.asp?topic=Public-Private %20Partnerships&topicid=75), an international non-profit organization dedicated to promoting PPP. Winrock International's agriculture initiative is focused on improvements in farm productivity and the sustainable use and conservation of natural resources while introducing new approaches that connect the farmers to their markers and production to consumer demand. Their goal is to build PPPs for program implementation and to promote similar inter-sector collaborations to foster long-term sustainability.

Private institutions are in principle willing to contribute to facilitate research processes and based on their experience, to share information about how to do things and, perhaps even more importantly, about "what not to do." Within the Generation Challenge Program (GCP: see below, section 2.4), scientists from both Pioneer Hi-Bred International and Syngenta have agreed to participate in some planning meetings and have actively helped identify optimal approaches to address specific issues related to research and/or data management. Over the last two years, a panel of scientists from Pioneer Hi-Bred International has provided continuous feedback on proposals for GCP commissioned work, and one Syngenta scientist is currently a member of the GCP management team review and advisory panel. Scientists from Limagrain have also offered to facilitate information access and sharing between the GCP and GENOPLANTE (http://www.genoplante.com/) databases. Note that the latter program aims to encourage joint research projects between both public (European Union-funded projects and others) and private partners in the arena of plant genomic research and provides a further example of a successful PPP.

2.4 The Generation Challenge Program as a Study Case

2.4.1 The Four Challenge Programs

The Challenge Programs (CPs) were designed to be time-bound, independently governed projects addressing themes of overwhelming global importance and need: nutrition, water use efficiency and conservation, crop genetic diversity, and the Sub-Saharan Africa region. The hallmarks of the CPs are their collaborative approach, thematic orientation, and their time limit of 10 years. The first CPs were Water and Food (aimed at addressing water productivity in nine river basins), HarvestPlus (the improvement via breeding in the micronutrient content of six staple crops), and Generation (the exploitation of crop genetic diversity via the application of comparative genetics; the discussion from section 2.4.2 to the end of section 2.4 focuses on this particular program, the GCP); they were launched on a pilot basis in 2003, following CGIAR-approved processes and guidelines for their development and implementation. The first two years experience in implementing the three CPs provided initial indications of whether the research objectives would be met, and helped to improve program design and development. In 2004, Sub-Saharan Africa was approved by the CGIAR as a fourth CP. The CPs respond directly to the major issues in the global development agenda. The CPs are intended to creatively mobilize resources (human, financial, knowledge, and technology) to address major global or regional problems. They encourage broad-based partnerships to harness front-line science within and outside the CGIAR system to benefit the poor, protect the environment, and strengthen the social network. International interest in and support for the CPs, and thereby CGIAR, has grown within both the private and public sectors.

2.4.2 The GCP Mission and Research Strategy

The GCP is a research and capacity building network that uses plant genetic diversity, advanced genomic science, and comparative biology to develop tools and technologies to help plant breeders in the developing world produce better crop varieties for resource-poor farmers. One of the strong rationales for the creation of the GCP was to develop a strong interface between fundamental and applied research in support of global agriculture. A cornerstone of the GCP is to link laboratories in developed economies with user communities in developing countries to accelerate the use of elite genetic stocks and new marker technologies for crop breeding.

The GCP mission involves both a broadening of the knowledge base underpinning the breeding of the world's staple food crops, along with the development of specific products that are useful for plant breeders to improve the livelihoods of resource-poor farmers in marginal, drought-prone environments. Knowledge generation requires a freedom to experiment with new ideas across disciplines and crops, while product development demands a clear roadmap for translating knowledge into tangible products. The GCP research approach is based on these two pillars to ensure that suitable knowledge is generated and that potential products are tested and validated in target environments, within the larger global context of producing useful products for resource-poor farmers. This scheme is illustrated in the form of a set of vertically aligned activities, starting with discovery and moving up through validation, application, and use (Fig. 2.1).

The discovery phase expands the knowledge base with respect to cross-cutting biological questions at various levels of plant architecture and across a broad set of crops, to promote the development and refinement of methodologies involving genetic and genomic resources. Its aim is to explore genetic diversity to define and identify useful alleles, involving the large-scale screening with molecular markers of germplasm in gene banks, the selection of reference sets, and their subsequent phenotyping. Novel diversity can also be created through recombination, wide crossing, and mutagenesis. Candidate genes/genomic regions involved in the plant response to stressful environments are to be identified by the application of molecular genetics, genomics, and comparative biology, in conjunction with reliable plant phenotyping. This phase also includes the development of new methodologies, protocols, and databases.

The validation phase seeks to translate information arising from the discovery phase into an understanding of gene function in priority crops and target

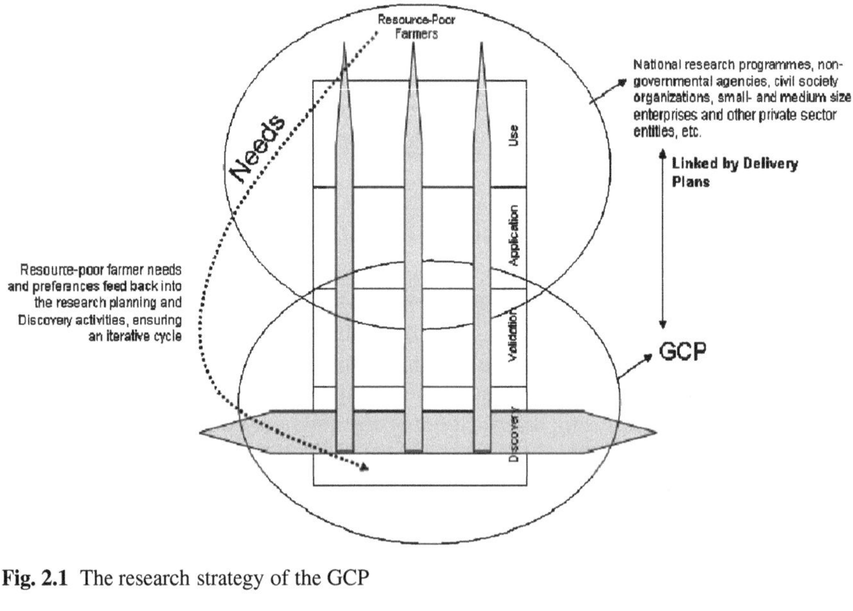

Fig. 2.1 The research strategy of the GCP

environments. It is, by necessity, both crop- and environment-specific, and typically focuses on multiple traits. The process involves the creation and/or selection of germplasm contrasting for key traits, and using these contrasts to confirm the predicted effects of genes or genetic regions on the plant phenotype under specific experimental conditions. Activities in the validation phase represent test cases for integrating and applying new knowledge and tools and for identifying gaps and bottlenecks in the application of new discoveries to plant breeding in target environments.

The application phase is focused on deploying validated products (e.g., markers, screening methodologies, information tools, etc.) to improve existing or to develop new germplasm adapted to local conditions within the target environments via large-scale breeding. This phase includes the incorporation of traits to promote drought adaptation or to meet local demand.

The use phase aims at bringing improved varieties to the farmers' fields. This involves seed multiplication and distribution, as well as socio-economic tasks related to the acceptability, potential benefits, and enabling conditions for the take-up of new varieties (issues that ideally should be considered when breeding priorities are first determined, and then revisited when farmers use, or fail to use, new varieties).

The GCP limits its activity to the discovery and validation phases, with the application and use phases expected to be carried out by inter alia NARS, NGOs, civic organizations, and small- and medium-size private enterprises. The GCP links itself to the application and use arena by requiring all research projects to develop a delivery plan to identify the downstream players that are most likely to use GCP products in their part of the delivery chain. In addition to identifying these user groups, GCP research projects are also required to incorporate them to the maximum extent possible in project planning and execution.

2.4.3 The Overall Scientific Approach

The application of genomics is dramatically improving our capacity to explore the genetic constitution of many agriculturally important species. Together with the growing number of partial and complete genomic sequences, high-throughput and massively parallel genotyping platforms are increasingly coming on stream, allowing the analysis of transcripts, proteins, and pathways and, more importantly, helping to extract useful variability from the wealth of resources stored in germplasm banks.

The GCP represents a novel approach to research-for-development by bringing genomic science to bear on the agricultural constraints of farmers in the world's poorest countries. More than ever, varietal improvement relies on a profound understanding of the genetic basis of functional diversity that will serve to broaden crop adaptation and improve productivity by the stacking of new or existing favorable alleles. The GCP scientific approach is based on the concept that genetic resources (which provide the raw materials) should interact with both advanced biological exploration (which provides an understanding of the genetic basis of traits) and breeding programs (which realize value by applying conventional and advanced methods to produce new and improved varieties). Because of its exceptional network of partners, consisting of nine CGIAR centers, more than 30 ARIs situated both in the North and in the South, and about 35 NARS, the GCP is uniquely positioned to support activities throughout the pipeline presented in Fig. 2.2. Germplasm curators operate at the level of large numbers of accessions, markers, and data points and organize screens and funnels to optimize access to pre-existing diversity; molecular physiologists and geneticists couple functional analyses at various scales between the cell, the plant, and the crop, and analyze populations to identify those genes responsible for useful variability; plant breeders intercross specific germplasm accessions, recombine and tag useful alleles, and, together with agronomists, apply the best phenotyping methods to enable the most effective selection under an appropriate range of field conditions.

To achieve its research agenda, generate added value to its products, and ensure their delivery, the GCP is organized into five subprograms (SPs), which span the spectrum of research and development from germplasm, genomics, and bioinformatics to molecular breeding for agricultural development.

SP1: *Genetic diversity of global genetic resources*. This subprogram is charged with exploring the genetic diversity of the various germplasm collections of the CGIAR mandate crops. The information collected as a result of its activity forms the raw material for all the other GCP research and research products.

SP2: *Comparative genomics for gene discovery*. This subprogram focuses on developing genomic tools, technologies, and approaches to better understand the genetic basis of key traits in crop species important to developing countries. The

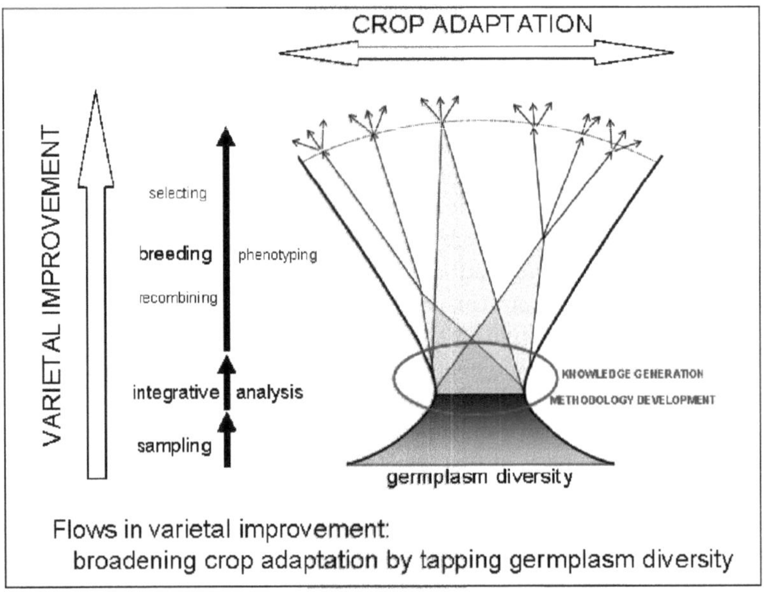

Fig. 2.2 The global scientific approach of the GCP

chief role of the subprogram is to discover and validate the function of key genes involved in stress adaptation, notably drought tolerance.

SP3: *Trait capture for crop improvement.* This subprogram focuses on the use of newly developed technologies in conjunction with well proven methods, to increase the efficiency, speed, and scope of plant breeding. Its particular goal is to ensure that the GCP generates products that are actually used in breeding programs.

SP4: *Genetic resources, genomic and crop information systems, and bioinformatics.* This subprogram aims to develop information systems, analytical tools, protocols, and other products, as well as to ensure their integration into the GCP network working from a coherent and easily accessible information gateway.

SP5: *Capacity building and enabling delivery.* This subprogram expands the capacity of researchers to accomplish the cutting-edge research agenda and make use of GCP products.

The challenges and current achievements of the subprograms are presented in detail in the following sections.

2.4.4 Diversity of Global Genetic Resources

SP1 is focused on genetic diversity and is tasked with the identification of novel, diverse, and superior variants (alleles) of genes involved in the determination of target traits. It interacts strongly with SP2, which identifies the candidate genes, and delivers them to SP3 as broadly-based germplasm and marker-assisted selection assays. In accordance with the overarching priority of the GCP, particular emphasis is given to drought tolerance. The bulk of activity surrounds the evaluation of germplasm collections, the purpose of which is to greatly extend available descriptions of crop genetic diversity, particularly via the widespread application of molecular markers. Particular attention is devoted to developing and validating simple markers, so that the technology can be readily adopted by the NARS programs for use in local germplasm. This activity has established a global genotyping facility which emphasizes efficiency, throughput, flexibility, and accessibility. At the same time, the focus has remained on establishing links between genotype and drought tolerance, a particularly complex and challenging trait. Germplasm evaluation for such a phenotype must be carried out in a way that best uncovers the key genetic factors (genes/alleles/haplotypes) determining variation. Currently this implies a joint description of molecular polymorphisms and phenotypic variation followed by the identification of phenotype/genotype associations. This component thus rests on complementary crop-specific and methodological modules, which have to be coordinated in the interests of efficiency and managed in a totally open fashion to attract the interest of national programs.

2.4.4.1 Understanding the Diversity of the Major World Food Crops

A basic prerequisite for all breeding programs is an adequate description of the diversity available in relevant germplasm. Knowledge of the structure of diversity present in a germplasm collection guides strategies for both breeding and characterization, and is certainly necessary for selecting a representative sample in which an in-depth survey is to be undertaken. Molecular markers are key to the acquisition of this knowledge and have been the focus from the beginning of the GCP program across all the mandate crops.

The general strategy can be broken down into three steps (Fig. 2.3). The first involves the assembly of a representative sample of accessions on the basis of passport information, exploiting, where available, outcomes of any pre-existing attempts to identify such collections (see Upadhyaya et al. 2006 for an example); this resembles the notion of the core collection as defined by Brown (1989), although the materials have a composite origin from diverse collections. The second features an extensive genotyping effort, based mostly on microsatellites. This type of marker has broad value for germplasm characterization; some loci are informative of structure at the species level, because they have evolved at a very slow pace; others reflect recent variation and so provide resolution even within narrowly based germplasm. Given the magnitude of this task, the responsibility for each crop has generally been shared between GCP consortium members, coordinated by the CGIAR center which holds the mandate for each particular crop. The genotypic data, where possible in conjunction with other available information, is then used to define the reference sample chosen for in-depth evaluation. These genotyping projects have varied in

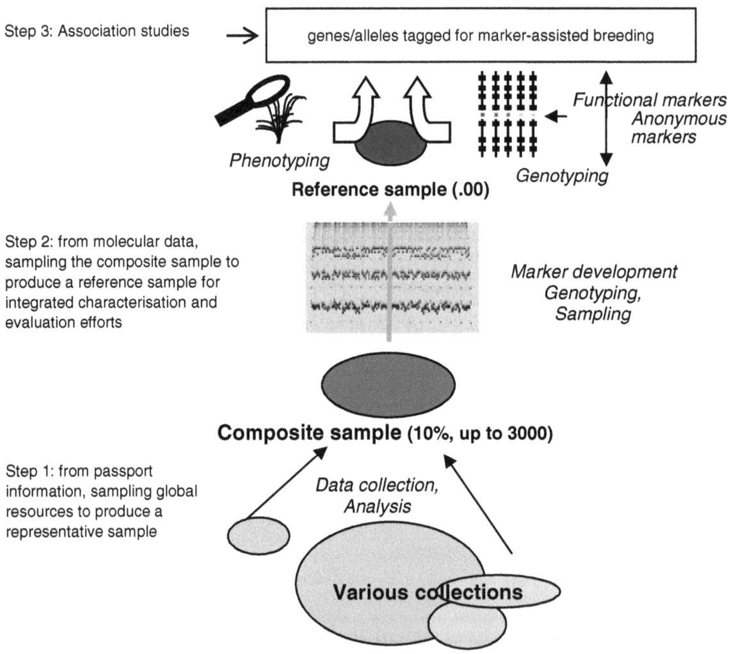

Fig. 2.3 The three steps involved in the definition of a germplasm reference sample, used for gene and allele mining and the identification of informative markers for molecular breeding

size, depending on the size of the complete collection and the numbers of markers available in each species. The range has been from 3,000 accessions genotyped at 50 microsatellite loci, to 350 accessions at 20 loci (Table 2.2). Half of the projects were completed in 2006, with the remaining half set to be completed in 2007. The third step is to use the phenotypic and in-depth genotypic (using both anonymous markers and/or specific candidate genes to ensure whole-genome coverage) description of the reference sample to identify associations resulting from either linkage disequilibrium (LD) or direct gene involvement.

A major output of this exercise has been the identification and release of reference samples for each crop. These are handled as genetic stocks in the respective CGIAR centers to facilitate their rapid and reliable distribution to breeders and germplasm specialists.

2.4.4.2 High Throughput Genotyping Techniques in Reference Laboratories

The GCP and its partners need access to facilities where diverse and preferably high throughput genotyping can be conducted on their materials. The first round of genotyping has mostly involved microsatellites. This will change, as future needs are likely to fall into two classes: (1) genome-wide surveys with anonymous markers,

Table 2.2 Genotyping activities conducted within the GCP between 2004 and 2007

Crop	Genotyping target		Project coordinator
	N° acc	N° loci	
Cereals			
Rice	3000	50	k.mcnally@cgiar.org
Wheat	3000	50	m.warburton@cgiar.org
Maize	1775	50	m.warburton@cgiar.org
Sorghum	3000	50	t.hash@cgiar.org
Barley	3000	50	m.baum@cgiar.org
Pearlmillet	1000	20	h.upadhyaya@cgiar.org
Fingermillet	500	20	h.upadhyaya@cgiar.org
Foxtail millet	500	20	h.upadhyaya@cgiar.org
Legumes			
Common bean	3000	50	m.blair@cgiar.org
Cowpea	2000	50	m.ferguson@cgiar.org
Chickpea	3000	50	h.upadhyaya@cgiar.org
Pigeon pea	1000	20	h.upadhyaya@cgiar.org
Lentil	1000	30	b.furman@cgiar.org
Groundnut	1000	20	h.upadhyaya@cgiar.org
Faba bean	1000	20	b.furman@cgiar.org
Roots, tubers, others			
Cassava	3000	36	m.fregene@cgiar.org
Potato	1000	50	m.ghislain@cgiar.org
Yam	350	20	r.asiedu@cgiar.org
Sweet potato	500	50	w.gruneberg@cgiar.org
Musa	200	50	n.roux@cgiar.org
Coconut	1000	22	patricia.lebrun-turquay@cirad.fr

and (2) the monitoring of allelic variation at candidate genes. The former aims to locate key genes on the basis of LD analysis, either in simple segregating populations or in collections of germplasm. Single nucleotide polymorphisms (SNPs) are heavily exploited in rice at the French National Genotyping Centre (CNG) in a project which should provide a case study for a species where sufficient sequence information is available. The Diversity Array Technology (DArT) platform (Wenzl et al. 2004) provides an alternative technology for species having a lesser amount of sequence information and is currently being applied to sorghum, cassava, banana, and coconut (in collaboration with DArT Pty. Ltd., Canberra, Australia). The rationale for monitoring allelic variation at candidate genes is to validate the involvement of these genes in the determination of the target trait, and then to identify the best-performing alleles. EcoTILLING (Comai et al. 2004) - a procedure which identifies allelic diversity at the sequence level in natural populations – based on agarose gel electrophoresis is currently being validated as a SNP discovery platform at IRRI (Raghavan et al. 2007). International Center for Agricultural Research in the Dry Areas leads a barley project that aims to reveal SNPs which affect the expression of the alleles rather than changes in encoded proteins. The method relies on contrasts in the relative abundance of allelic transcripts in heterozygotes. This may point to heritable variation in gene expression, now considered to be a fundamental mechanism responsible for determining the genetic control of complex traits. Where high throughput sequencing is possible, resequencing of candidate genes is currently the preferred approach to allele discovery, since it generates full length sequence data. In a joint (with SP2) activity of this type, three laboratories work on the identification of orthologous genes (in partnership with CNG) and five provide crop-specific information and materials. A database detailing the allelic variation among seven crop species will be generated and this will provide the genotypic basis for establishing associations with phenotype.

2.4.4.3 The search for Superior Alleles (or Haplotypes) via Association Studies

Association studies require robust descriptions of both genotypic and phenotypic diversity. In a CIMMYT-Cornell University maize project, the variation in a set of candidate genes involved in the response to water limitation (related to the biosynthesis of particular carbohydrates, ABA, and polyamines) was aligned with the quantification of specific metabolites in the leaves and silks under drought conditions to identify haplotypes correlated with desired plant phenotype. For slow generation crops, it is of particular interest to exploit the materials and evaluation data that are routinely produced in mainstream breeding activities. A requirement for this approach is the existence of a significant level of LD (otherwise, it is not possible to correlate phenotype and genotype). To assess feasibility and propose case studies, therefore, the population structure and the level of LD are currently being assessed in breeding populations of a range of crop species (potato, cassava, yam, banana, and coconut). By extending this strategy to wider genetic bases, the targeted creation of introgression materials should generate many opportunities for the fine-scale genetic analysis of traits, since it widens the range of alleles present. The International

Center for Tropical Agriculture (CIAT) is leading this type of activity in rice, in a program to produce chromosome segment substitution lines from four wild species.

SP1 will evolve in the coming years in two main directions. First, along with SP4, it will help germplasm managers develop and apply methodologies to supplement routine characterization and evaluation using the most appropriate and informative genotyping. The choice of genotyping targets will be driven, in part at least, by the outcomes of the SP2 program. The product of this activity will be genetic information that can be directly applied to breeding (by SP3). Second, as an integrative activity across all the SPs, it must aim to stimulate genetic base-broadening activities with hybridization approaches that will result in the production of populations giving a high level of genetic resolution, which are also of direct value in breeding.

2.4.5 Comparative Genomics for Gene Discovery

SP2 research is directed at bridging the genotype-phenotype gap via the use of genomic tools and comparative biology (Fig. 2.4). It seeks to apply modern analytical tools, brings together genetic resources, and explores the relationship between gene expression and phenotypic variation. The aim is to relate these both to variation at the genomic level emerging from SP1 research and to phenotypic performance (with SP3). The consortium style of the GCP unites the CGIAR and NARS programs, a blend of research organizations with the capacity to produce and record massive amounts of phenotypic data across crop species and environments and to investigate

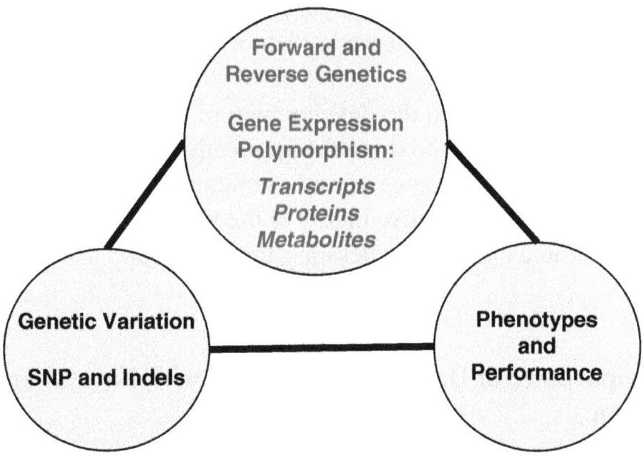

Fig. 2.4 GCP research activities aimed at the identification of genes of large and small effects, exploiting genetic and genomic analyses to connect DNA variation with phenotype and performance. Forward and reverse genetics determine the causal relationships between individual genes and phenotype, whereas gene expression polymorphisms on a genome-wide scale offer a view of genome regulation not possible from the examination of individual genes

variation at the molecular level using genetic mapping or association analysis approaches. For most of the CGIAR mandate crops, relatively well-populated genetic maps are already available, while the most advanced comparative maps have been assembled within the cereals (rice, sorghum, maize, barley, millet, and wheat). Beyond mapping, an expanding amount of DNA sequence information, coupled with improved bioinformatics tools, is enabling the identification of orthologous genes, which can be exploited to predict gene function across plant species.

The SP2 research portfolio focuses on developing genomic tools, technologies, and approaches to achieve an understanding of the inheritance of stress tolerance, notably drought, in a range of major crop species important for developing countries. The primary role is to discover and validate the function of key genes involved in the stress response by examining patterns of gene expression across phenotypically informative crops (or genotypes within a crop). A critical follow-on task involves the validation of causal relationships between candidate gene expression and phenotype using appropriate genetic stocks (targeted mutations, over- and under-expression systems). Structural and functional conservation in regulatory circuits is particularly relevant for identifying genes with a large effect on phenotype. The manipulation of regulatory elements can bring about dramatic changes in phenotype, which are often thought to be controlled by many genes, each responsible for only a minor effect. Recognizing that final crop varieties need to be tolerant to multiple stresses, we are interested in exploring the interactive effects, both synergistic and antagonistic, between stress responses.

SP2 is positioned to exploit innovations in medical and biological practice and future conceptual advances in crop genomics. A major effort is devoted to improving our genetic understanding of stress tolerance (both drought and other stresses) through a combination of bioinformatic analysis and experimental studies based on the advanced genetic stocks available to the GCP. We are also committed to developing high-quality phenotyping of selected genetic materials for the validation of gene function. This will include the systematic phenotyping of mutants and the evaluation of near isogenic lines. Finally, as active participants in the research consortium, we will continue to contribute to the development of genomic tools and "designer" genetic stocks in selected crops to drive the application of successful approaches to improve crops across a diverse range of environments. Since the implementation of the GCP research agenda, the relative merits of the various experimental approaches are beginning to become clear. Three lessons are particularly salient for guiding the future agenda.

2.4.5.1 The Importance of Designer Germplasm with Natural and Induced Variation

Well-designed genetic stocks greatly facilitate the process of gene discovery and validation. Segregating populations provide the materials needed to identify key chromosomal regions. Mutants, in which a genetic lesion and an altered phenotype co-segregate, allow gene function to be precisely assigned to a genetic locus. In the past, mutagenesis in crop plants (with a few exceptions) has been only rarely used to identify or validate gene function, but it is currently enjoying a renaissance with the development of genomic tools able to efficiently identify sequence variation. Mutagenized populations represent a valuable resource for the validation of candidate genes, and induced variation can complement natural variation in both forward genetics and breeding. The GCP recognizes the importance of well-designed genetic stocks for gene identification and expects to play a catalytic role in supporting the production and use of such genetic resources by the research community. Examples of these activities are shown in Table 2.3.

Improved and low-cost methodologies to make use of mutant collections are also being sought. The TILLING (Targeting Induced Local Lesions IN Genomes, McCallum et al. 2000) procedure has been streamlined by using agarose gel electrophoresis to detect heteroduplexes (Raghavan et al. 2007). In removing the need for labeled primers and a costly infrastructure, TILLING can therefore be carried out in NARS institutions. This "TILLING-made-simple" approach has made the analysis of mutagenized populations both feasible and appealing. With genotyping technology becoming more affordable and robust, the future bottleneck in the dissection of plant traits may well be the availability of well-designed genetic stocks. For GCP investment to be effective, it is important to assure institutional commitments for stock development, maintenance, and distribution. Ideally, genetic stock development for each crop should be coordinated through breeding networks, or be linked with research consortia where the GCP can play a facilitating role.

Table 2.3 Examples of GCP-supported activities in the development and use of specialized genetic stocks in various mandate crops

Crop	Project Activity	Status	Contact
Wheat	Assembly of wheat genetic stocks	Approximately 600 wheat genetic stocks assembled at CIMMYT. Seed is being multiplied for systematic phenotyping and distribution.	Suzanne Dreisigacker, CIMMYT s.dreisigacker@cgiar.org
Rice	Phenotyping network to identify mutations in stress-associated genes using available international rice mutant collections	OryGenes DB (http://orygenesdb.cirad.fr) reports about 140,000 insertion tags; available insertion mutants give a 50% success rate of finding a tagged mutation.	Andy Pereira, Virginia Polytechnic University, pereiraa@vbi.vt.edu
Bean	Production of mutagenised bean populations	M_2/M_3 progeny of common bean (about 3000) produced and propagated for TILLing	Mathew Blair, CIAT m.blair@cgiar.org
Potato	Production of true-seed mutants of potato relatives	EMS-induced mutants of *S. verrucosum* (2n = 24).	Marc Ghislain, CIP m.ghislain@cgiar.org

2.4.5.2 Genome Regulation is Responsible for Large and Small Quality Trait Loci

Comparative mapping based on syntenic relationships was one of the key approaches in the original concept for the SP2 research program. Map comparisons have advanced rapidly because of the expanding amount of genome sequence information and the increasing use of expressed sequences as genetic markers (the latter are much more informative for comparative mapping purposes than anonymous markers). However, physical maps alone are not necessarily fully informative as to how a plant responds to environmental stresses or perturbations. Variation in the level of gene expression between genotypes provides an additional layer of complexity to the relationship between structural and functional variation (Kliebenstein et al. 2006). Common patterns of gene expression observed across species under similar experimental treatments may also reveal the functional significance of particular genes.

Thanks to the complete genome map of rice, transcriptome analysis is able to generate a global view of gene expression in a way that has not been possible in the past. Already two interesting phenomena have emerged from such analyses. The first relates to clusters of genes that appear to be differentially expressed under salinity stress at the panicle initiation stage (Walia et al. 2007). Some of these clusters, which span 1-2 Mbp, overlap with known quality trait loci (QTL) for salinity tolerance. The second has arisen from a comparative gene expression analysis involving a drought-sensitive and a tolerant variety, in which 14 regions across the genome (each containing about 20 genes) showed patterns of correlated expression. Some of these regions co-localized with reported QTL for drought tolerance (M Raveendran and R. Mauleon, IRRI; K. Satoh and S. Kikuchi, National Institute of Agrobiological Science, unpublished data).

The results obtained to date emphasize the value of combining outputs from the various SP2 gene expression projects. The bioinformatics pipeline and computing infrastructure provided by SP4 has enabled the identification of gene sets, based not only on the extent of differential expression, but also on the identification of correlated patterns of expression in a chromosomal context. By aligning differentially expressed genes and regions of correlated gene expression with QTL maps, a relatively small set of genes can now be identified, resulting in a short-cut in the identification of candidate stress tolerance genes.

2.4.5.3 Using Genotypes with Field-Proven Phenotypes for Gene Discovery

Investigating genotypes associated with favorable field performance ensures that the biological processes under study are agronomically relevant. While this may seem a trivial consideration, it is far from trivial in practice, as it requires the assembly of the right genetic resources, agronomic conditions for evaluation (relevant phenotypic assays), and experimental expertise into a single project. By facilitating collaboration with multiple partners, the GCP facilitates access to field-proven materials and the expertise for phenotyping under greenhouse or field conditions.

A number of SP2 projects has demonstrated the usefulness of convergent approaches to reduce the number of candidate genes. We anticipate that this approach will be particularly relevant for identifying the genes underlying complex traits. Given the increased robustness of transcriptome analysis, along with the declining cost in conducting such experiments, it is expected that genome-wide analyses will be more widely used in the coming years. Comparative analysis across crops requires cross "mapping" at several levels - synteny, orthology and expression patterns - so that there is a need to further strengthen the integrated analysis of mapping and expression data. Several SP2 projects are close to delivering promising practical products, including advanced germplasm and cloned genes (hence perfect markers). A focused investment to build capacity in the NARS to use these products is needed to ensure uptake. This can be most effectively achieved by enhancing existing research and breeding networks, a task that is undertaken by SP5.

2.4.6 Trait capture for Crop Improvement

Activities grouped under SP3 aim to increase the efficiency, speed, and scope of plant breeding, both by adopting new technological advances and by linking upstream research outputs with practical product development. SP3 has implemented a product management process designed to ensure the optimal flow of upstream research outputs to the more applied research activity within the GCP. This is to be accomplished by integrating the research outputs from SP1 and SP2 and by facilitating the diffusion and use of the data sets organized and analysed by SP4. A particular focus of SP3 is the validation and refinement of molecular breeding systems and the resulting enhanced germplasm by the fine-tuning of the technology needed to deliver more efficient approaches and tools (e.g., validated markers) to breeders and enhanced germplasm to farmers. The GCP research related to aluminum toxicity tolerance is representative of this approach. In acid soils, toxic levels of aluminum are released into the soil solution, where they impair root growth and function. Collaboration between Cornell University (USA) and EMPRAPA (Brazil) has isolated Alt_{SB}, a major gene governing aluminum tolerance in sorghum. A survey of Alt_{SB} alleles across germplasm will now be possible as the prelude to an association mapping exercise aiming to identify the most effective allele(s) for deployment into sorghum breeding programs. The plan is to develop low-cost, easy-to-use SNP assays for marker-assisted selection. The phenotyping facilities developed by EMBRAPA will be used to evaluate the impact of the various Alt_{SB} alleles on sorghum yield in acid soils. Using a combination of genomics and statistical approaches, this program is already bridging the gap between upstream research programs and applied breeding programs. Its ultimate goal is to generate genotypes expressing enhanced aluminum toxicity tolerance to be distributed to farmers affected by acid soils in Africa and other developing countries.

SP3 is committed (in conjunction with SP5) to provide appropriate technical assistance to breeding, programs to enable them to take advantage of molecular breeding in tropical staple crops. It aims to develop communities of practice, supported by regional centers of excellence and state-of-the-art technologies, which will allow for a system of centralized validation and the refinement of new technologies delivering protocols for routine application in the NARS institutes. SP3 plays a vital role in creating community linkages with plant breeders committed to the evaluation, validation, and refinement of molecular breeding technologies generated by the GCP. For example, a multi-country rice project is currently being supported to exploit the expertise and facilities developed by Biotec, the Thai National Center for Genetic Engineering and Biotechnology. The project aims to develop a backcross marker-assisted selection program to introduce specific traits in the most popular rice varieties of the region. Each country has chosen its particular trait(s) of interest. These are: seed quality, transferred into a drought resistant background (Cambodia), salt tolerance into a high quality background (Myanmar), seed quality into a widely adapted background (Laos), brown plant-hopper resistance into elite irrigated rice, blast resistance into glutinous rice, and seed quality into widely adapted and drought tolerance backgrounds (Thailand). After generating the introgression lines, trait validation will be carried out in Thailand, Cambodia, Myanmar, and Laos. The rationale of the program is that the development of materials will be accelerated, and this will make a significant contribution to rice improvement, to the welfare of the farmers through increased rice production and greater cash income, and more generally, to the economic development of the Mekong region.

The selection of appropriate background genotypes for molecular breeding programs is critically important to assure the widespread impact of new genes and traits. SP3 collates, collects, and generates the most appropriate breeding lines for this purpose. New and innovative approaches are to be developed to increase genetic diversity, giving opportunities to access new sources of drought tolerance and disease resistance. In collaboration with scientists working in SP1, SP2, and SP4, SP3 is also generating protocols for low-cost trait diagnosis and high throughput array-based genotyping. Significant progress has already been made in low-cost assay technologies for gene-based, marker-assisted selection of grain quality in maize and pest and disease resistance in rice and cassava. Low technology methods, such as dot-blot and microtiter plate-based assays, and high throughput techniques for hub laboratories, such as micro-assay based genotyping and fluorescence-based assays, have been refined, tested and are ready to be disseminated to breeding programs.

The development of effective selection systems for the improvement of complex traits such as drought tolerance has been elusive to date. Recent developments in genomics, computation, and biometrics now offer a genuine opportunity to dissect drought tolerance into its component traits, which should simplify the manipulation of this critically important trait. Phenotyping is recognized as the major bottleneck in many genomic studies, and the definition of robust protocols is an absolute necessity. Drought tolerance phenotyping remains difficult because of limited capacity, inadequate protocols, extensive phenotypic diversity, and problems in the design of controlled stress experiments. Any accurate phenotyping protocol must include a precise characterization of the test environment (which should ideally be as representative as possible of the target environment), a definition of the secondary traits resulting from the imposed environmental constraints, and a full understanding of the adaptive behavior and the critical periods for the crop (Fig. 2.5). The timing, intensity, and homogeneity of the application of stress are particularly important issues to be addressed. An important aim of SP3 is consequently to reinforce phenotyping capacity. Thus the GCP supports the efforts of EMPRAPA to develop a drought phenotyping platform that can be used as a case-study for the development of a standardized phenotyping methodology. It is also funding a project involving Agropolis (France) and the Commonwealth Scientific and Industrial Research Organization (Australia), in which the objectives are to: (1) provide model-based methodologies to analyze multi-annual/multi-site climatic data sets and to define drought patterns within crop production regions and breeding sites, (2) use plant model parameters instead of directly measured growth traits, and (3) improve models connecting genetic information with model parameters. Documents, workshops, and courses are used to disseminate functional methods and protocols, which can better define the targeted and testing environment and which provide a better choice and more accurate measurement of relevant traits.

The activities in this SP require a coordinated input from scientists of different disciplines, eco-regions, and types of institution. SP3 aims to stimulate the emergence of holistic teams involving NARS, ARIs and CGIAR centers, and favors strong collaborations across disciplines, crops, and institutions as well as linkages between genomics, genetics, physiology, and biometrics. The product management implementation conducted by SP3, based on a capture of interdisciplinary synergies and end-user feedback on priorities and outputs, is expected to play an essential role in the validation and diffusion of products and their further delivery to breeders in the NARS.

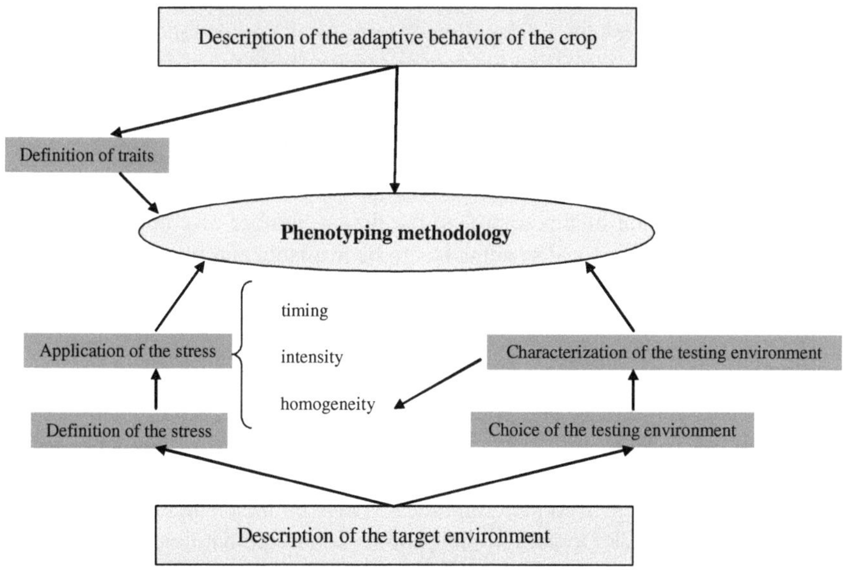

Fig. 2.5 The definition of phenotyping methodologies - a complex process

2.4.7 Genetic Resources, Genomic, and Crop Information Systems

One of the major strengths of the GCP is the involvement of a wide range of actors spread all over the world, allowing it to tap knowledge and use resources wherever they are available. Clearly, this can cause difficulties in relation to coordination, communication, and data flow. The latter is tackled in SP4, the task of which is to link and integrate information components and analytical tools relevant to the GCP into a coherent information gateway. Substantial amounts of information are being generated from numerous labs and fields as output of the first three SPs. The value of this information is significantly dependent on the way in which it is managed, analyzed, and made accessible. This in turn depends on the way analytical tools and other information resources are made available. SP4 aims to maximize the value of the information by allowing optimal management, access, and analysis of data relevant to the GCP.

The data exchange within the GCP is based primarily on web-service technology, which also allows the rest of the world to access GCP data, analytical tools, and work flows. Web services, as defined by the World Wide Web Consortium, are software systems designed to support inter-operable machine-to-machine interaction over a network. In the GCP context this means that the computers hosting databases with data generated by the GCP or relevant to the GCP, and computers with tools created by or relevant to the GCP, are connected in a way that allows them to be accessed as if they were part of a single integrated system. Thus, a scientist who needs access to data or analytical tools need not be concerned where the data and tools are maintained, but can use them online. The technology that allows this is elegantly simple: the component systems are "wrapped up" in a software package that displays whatever is inside in a uniform way, allowing the communication software to access the information or tools of all components via a standard protocol. The software extracting the data or services from the component systems then only needs to understand the standard protocol. At the same time, this allows each component system (i.e., partner institute) to make its own choice of hardware configuration, operating system, database management software, database structure, computer language, etc. (Fig. 2.6).

The implementation of this approach requires a number of components: technology for wrapping the local systems has to be available and implemented on the component systems, a common language for communication between the middleware and the components has to be defined, and the middleware layer and application layer have to be developed. All these components are covered by SP4. The technology used in the GCP is existing technology: BioMoby and BioCASE (see www.biomoby.open-bio.org and www.biocase.org). GCP has contributed to these technologies by developing a system which defines a simple architecture and its Java implementation for creating BioMoby webservices accessing BioCASE-aware databases. The language used to communicate between the elements of the system is derived from the GCP Domain Models and its connected ontologies. These were (and continue to be) developed in collaboration with a wide range of players from inside and outside the GCP using the principle that "if it exists, it is used or adapted;

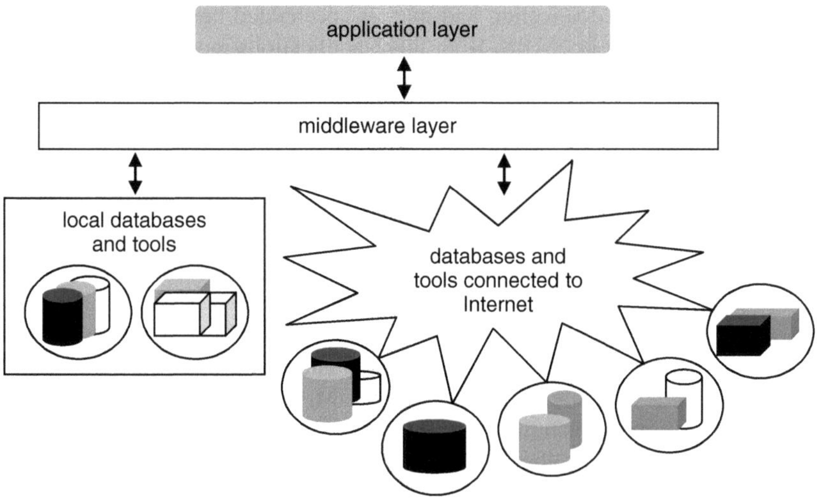

Fig. 2.6 The GCP Information Platform: databases and tools, available either locally or on the internet, are wrapped and connected to a middleware layer which allows software to deal with the components as if it were a single virtual database and toolkit

while if it does not, it is created." Technology for the development of the domain models was selected and editorial teams formed to work on specific information domains. The resulting models are used not only in the web services, but also in the development of the software for middleware and the applications, the third component of the platform. The GCP Information Platform will be available via the internet and, with a more extensive functionality, via a workbench which needs to be installed locally.

Developing such a platform takes time, and since the data generated in GCP projects need to be available as soon as possible, a much simpler approach was followed in parallel to the development of the web services-based platform. Based on the domain models, templates were developed which allow GCP scientists to store their data in Excel spreadsheets in a standardized manner, assuring ease of interpretation by any other scientist. These dataset files are available from a central website, the GCP Data Repository, and can be searched and downloaded by any interested scientist (see gcpcr.grinfo.net). It is expected that in future, all data will be available both via web services and as downloadable files.

Apart from these activities to ensure proper access to data and tools, SP4 is also involved in a wide range of other activities. It supports the local capture and management of information, monitors and improves data quality, and develops new methodology and software for bioinformatics and quantitative genetics analysis. The latter serves the needs of GCP scientists as identified by the leaders of the first three SPs.

An interesting example of software development, which also illustrates the principle of "use it when it exists," is the development of iMAS, a decision support tool for marker assisted selection and marker-accelerated breeding for use by local breeding programs in NARS. Both these applications require the definition of simply inherited markers closely linked to genetic factors determining components of yield and polygenic traits of value. This is generally achieved by either genetic mapping or an association mapping approach, the latter typically being based on a sample of unrelated individuals (Fig. 2.7). For both approaches, many steps separate the initial phenotyping and genotyping of individuals from the identification and application of the markers in molecular breeding. The choice of appropriate experimental design and data analysis is critical, but this choice is clouded by a plethora of relevant software packages. The GCP therefore created iMAS, which aims to provide a set of simple-to-follow guidelines to allow the user to make the most appropriate choice of experimental design and data analysis, and provide the optimal associated software. The advantage of iMAS is that it frees the user from having to make time-consuming data format conversions to fit the output of one program to the input of another. Based on a technology similar to the web services described above, all information is transformed so that the user can easily switch between the component programs.

The development of iMAS involved the following steps: first an inventory of potentially useful free software was made. The quality of this software was tested and compared, and the best product for each particular function was selected. Permission to use the software for iMAS was obtained from the authors, and the software linking the components and interfacing with the program was written, tested, and refined. Currently iMAS contains elements of IRRISTAT, GMendel, PlabQTL, Win QTL-Cartographer, PopMin, GGT, and TASSEL. Online decision guides were

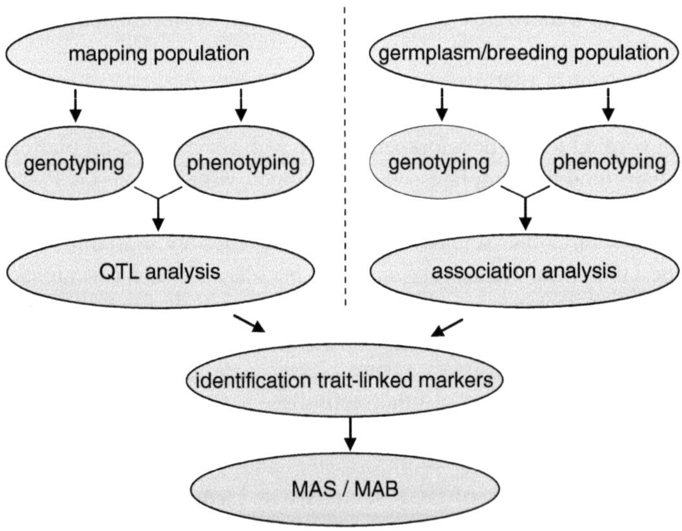

Fig. 2.7 Workflow of iMAS: an integrated decision support system to facilitate marker-assisted plant breeding

developed and incorporated into the system, and a user manual was written. Finally, potential users were exposed to the system, and their feedback is currently being used to improve and expand the system.

Overall, SP4 has generated a solid foundation in terms of human capacity, methodology, architecture, software, etc., and this has been used to build a wide spectrum of user-oriented products. We are presently in the exciting phase where the first products are being released to the end users.

2.4.8 Capacity Building and Enabling Delivery

Capacity building and enabling the delivery of research outputs is the focus and the major cross-sectional theme of the whole GCP. A wide gap in technology development, access, and adoption divides the high-income from the low-income countries. This disparity affects the extent to which use is made of many of the GCP research activities, such as genetic resources, developments in genomics, access to current research information, funding opportunities, and the knowledge required to merge new approaches with traditional crop improvement practices. In addition, clear policies covering issues of intellectual property and access and benefit sharing are required both in the developing and the developed world.

The success of the GCP is very dependent on the adoption, adaptation, and application of its research outputs for the ultimate benefit of the resource-poor farmers. Its impact in their fields relies on NARS personnel being familiar with the science emerging from the GCP, and then exploiting it in their breeding programs to address the needs of the farmers and consumers for whom they work. Thus the effective delivery of GCP research and products implies that strong links must exist with the NARS, and that substantial effort be dedicated to enhancing their capacity.

The SP5 effort is spread over a diverse portfolio of activities. A platform of training resources, which includes training courses and learning materials in phenotyping, marker assisted breeding, genomics, bioinformatics, and policies, has been created. Phenotyping has been a rather neglected activity in the past, but is now re-emerging in the genomics era as a crucial component. Because of the intimate relationship between phenotype and environment, the role of the NARS is particularly important. Courses and learning materials in the form of a technical guide for basic concepts and phenotyping procedures, plus the exploration of techniques in different crops to ensure consistent implementation are important assets for the advancement of GCP research.

SP5 support is also provided to promising NARS scientists, who, it is hoped, will go on to become future research and breeding leaders in their home institutions. The intention is to further the education of these scientists by exposing them to a high level research environment, where they will learn the value of the scientific method, experience a culture of collaboration and research success and failure, and explore unfamiliar avenues of research. Several fellowships have already been awarded: for example, a Senegalese student was trained at CIRAD (France) in ecoTILLING. The technique was subsequently used to produce new genetic markers in sorghum. A second fellow, from Nigeria, was assigned to CIAT (Colombia), which curates the largest world collection of cassava, to learn gene tagging and marker-assisted selection. Nigeria's root and tuber expansion program is now benefiting from a molecular breeding laboratory established through a GCP project. Awards are also given to researchers to spend short study periods at another research organization to acquire hands-on experience in a new skill or methodology. In an example of this activity, a researcher from the International Potato Center (Peru) visited International Crops Research Institute for the Semi-Arid Tropics (ICRISAT, India) to learn how to measure certain physiological responses of plant roots.

SP5 also seeks to facilitate the flow of products from the scientists in the laboratory to the breeder in the field. An example is the promotion of microsatellite kits that were developed from the crop genotyping activity in SP1. The kits provide a set of reference markers designed to assess germplasm diversity and to facilitate direct comparisons across institutions and germplasm collections. The Genotyping Support Service (GSS) was launched to simplify access to genotyping technology by NARS breeding programs, bridging the gap between the science and the field. The GSS sources cost-efficient genotyping worldwide, provides access to data, and assists with its interpretation. The cassava breeding program at the Mickocheni Agricultural Research Institute (Tanzania) has developed populations segregating for resistance to cassava mosaic disease (CMD) from crosses between improved germplasm from CIAT and susceptible Tanzanian-grown varieties. Marker-assisted selection, using four markers linked to a major CMD resistance gene, has now been performed with the support of the GSS. The GSS has also contributed to the analysis of genetic diversity present in the EMBRAPA banana collection, which houses around 400 accessions. This analysis will help to determine the representation of EMBRAPA banana germplasm in worldwide collections.

As a contribution to access and benefit sharing, SP5 recently developed the IP "Helpdesk," an online resource for the GCP community and its partners. The Helpdesk resource include articles and presentations, and is dedicated to providing support to researchers in matters related to germplasm access, IP rights over research materials, and legal agreements within and outside the GCP. The Helpdesk relies on the expertise of a panel that can respond to questions posed by GCP members, partners, and stakeholders.

A major achievement of SP5 is the definition of the GCP delivery strategy, predicated on the principle that all GCP products are created to respond to need identified by the users. This means that all projects are required to identify clear objectives, the attainment of which can be measured through quantifiable outputs. Capacity building has become an integral component of the delivery strategy by ensuring that partners engaged in GCP research projects include appropriate user groups able to exploit research outputs. The delivery philosophy of the GCP shapes the main objectives of capacity building efforts, i.e., helping meet intra-project capacity needs to ensure product delivery. The delivery plan kit is an important part of the implementation of the strategy. It consists of a series of templates, designed to facilitate a straightforward compilation of expected outputs and products, expected applications and users, constraints for project success, capacity needs to be fulfilled, forms of distribution of the product, and a time-line for monitoring progress.

The GCP operates under the philosophy that the effectiveness of every link in the value chain affects the overall success of its research. Thus a major lesson learned by SP5 is that its activities have to be tailored to the real needs of the GCP end users, and that it must attract and retain the commitment of the NARS institutions. This implies the nurturing of institutional as well as personal relationships. What has emerged is a novel concept - capacity building à la carte - in which capacity building efforts are tailored to the personalized training and research support of researchers from a chosen group of NARS institutions. The rationale of this choice is to identify, within ongoing GCP projects, those researchers who show the greatest level of commitment and ability to exploit research outputs and to transfer them into the field. In the longer term, the vision is that these institutions will act as future hubs for GCP product delivery.

The GCP is well aware of how difficult it is to change the modus operandi of research institutions, research teams, and researchers themselves. In the conventional model, upstream research in the public (both national and international) sector is planned and conducted independently of downstream users. This mode, which is considered to be compatible with the production of high quality science, has failed to achieve a significant level of adoption of its outputs, at least by those who may have benefited had they been engaged in the planning and implementation phases of the research. The GCP, via its delivery strategy, is attempting to change the mindset of upstream research communities, in forcing research proposals to take into account users and applications, to anticipate the constraints to success, and to devise potential solutions to overcome these.

Finally, but most importantly in the context of delivery, is the notion of targeting impact, in terms of geographical area, crop, and trait(s). SP5, on behalf of the GCP, invests in a number of socio-economic studies chosen to inform the decision-makers responsible for ensuring that GCP efforts are directed where they are needed most and where they have the highest chance of success.

2.5 Challenges and Perspectives

2.5.1 The Challenge and Perspective of Food Security

Despite three decades of rapidly expanding global food supplies, almost 800 million people (about 1 in 8 of the world population) are still not free from hunger (Sanchez and Swaminathan 2005). The total food available per person in the world has risen by 11% over this period, while the estimated number of hungry people has fallen by 16%, from 942 to 786 million. However, if China is excluded from the analysis, the number of hungry people has actually increased by over 11% over the same period (from 536 to 597 million). Globally, food supplies have more than doubled in the last 40 years, faster than the population has increased, resulting in a 15% increase in the per capita supply of calories; however, these global averages hide significant regional variations. For example Sub-Saharan Africa, with 16 of the 18 most undernourished countries in the world, remains the only region where per capita food production continues to worsen year by year (Rockefeller Foundation 2006).

The world's population is expected to increase by an additional 50% between now and 2050. Per capita food consumption in countries suffering from shortages will probably have increased, and diets will have become diversified to eliminate well understood nutritional deficiencies. These changes will have a heavy impact on food production systems, natural resources, and the environment. In addition, the response to a higher food demand has already caused substantial damage, some of it irreversible, to the world's natural resources; and the specter of imminent climate change complicates predictions of trends in food supply. The main challenge is whether improvements in food production and the available natural resources will be enough to cope in a sustainable manner until 2050, after which the world population is projected to stabilize.

To address problems of world hunger, the Millennium Project was commissioned by the United Nations secretary-general to recommend the best strategies for meeting the millennium development goals (MDGs). The Hunger Task Force was established in October 2002 to determine how to meet the hunger MDG of reducing the proportion of hungry people by half from 1990 to 2015. Task force members came from diverse backgrounds in science, policy, the private sector, civil society, UN agencies, and government, with broad representation from developed and developing countries. After analysis, stakeholder consultations, and observation, the task force recently produced its report, which is summarized in Sanchez and Swaminathan (2005).

2.5.2 The Challenges for International Programs

The very essence of an international program means that efficiency, not to mention survival, is a major issue. There are few hard and fast rules for international agricultural centers and networks, and the logistical aspects of this work are constantly in flux.

Coordination of effort. Because international agriculture is globally so important, a significant effort is needed to minimize (and if possible eliminate) the duplication of work among different programs. Constant communication and vigilance enhances efficiency, but a major imperative is to coordinate the work of relevant international organizations. By working together, programs can combine collective knowledge, and at the same time avoid project replication. Proper coordination should deliver optimal resource allocation throughout the delivery chain, from basic research to the development of seed systems, and will help to deal with the inevitable bottlenecks and breakdowns which compromise the entire system. Although several examples of coordination among programs are forthcoming (the alliance between The Bill & Melinda Gates Foundation and RF for Sub-Saharan Africa being a good example), flexibility to adjust the activities of independent organizations is limited. Because each is governed by their own particular agenda, they have independent governance, and they understandably want to maintain their specific identity.

Long-term versus short-term objectives. Diverse donors, including governmental, private, and philanthropic organizations, support international agriculture. In the 1980s, most donors chose to support programs or institutions with few or no strings attached, but today the norm for donors is to implement their individual agendas, limitations, politics and rules, which can result in making the completion of scientific goals or a program's focus more challenging than it used to be. Donors are increasingly interested in making a short-term impact, and so emphasize rapid progress. The establishment of a clear framework and milestones is essential for all research projects, but applying pressure to ensure short-term impact can easily force the research agenda into a "low hanging fruits" mode, where only the readily achievable, less complex goals are set. Maintaining an appropriate balance between long-term and short-term objectives is challenging, especially given the increasing pressure on the scientific community to demonstrate that modern biotechnology can make a significant impact. This is particularly problematical for plant breeding, since progress there is so heavily governed by plant generation time. For most crops the development of a new variety requires a period of about 10 years.

Partnership. On top of the challenges presented by coordination among programs and time-frame of the research agenda, a further essential element is to assemble the right combination of partners to maximize complementarity of expertise necessary to ensure delivery. Unfortunately, there is no magic formula for the optimal proportion of each ingredient (public and private institutions, North and South, NARS, and civic societies) to ensure success, so the appropriate team composition becomes heavily dependent on crop and region. International organizations need to play a catalytic role in this domain, bringing the different players together and fostering research in such a way that product flows are optimized within the scientific community. Only in this way can products be adopted, adapted, and applied for the ultimate benefit of resource-poor farmers.

Human resources and facilities. International agriculture requires time and in-place systems to ensure success. A lack of local infrastructure and expertise is a major bottleneck for the use of modern plant breeding tools in the South. Thus a challenge lies in retaining trained scientists in the South, since low salaries and poor infrastructure are the main reasons for the unwillingness of so many NARS researchers studying at ARIs in the North to return to their home countries. To address this issue, international programs have supported the development of high quality education hubs in the South. A good example of this activity is the African Centre for Crop Improvement at the University of KwaZulu-Natal (South Africa), which trains African students in modern breeding to Ph. D. level by supporting their thesis work at their home institution. RF (and more recently PASS) which supports this center, is fully committed to funding returnees in their research. By building capacity within Africa, programs such as this will not only help African scientists to learn, but can also put in place the infrastructure to make a significant impact. Most international programs seek strong partnerships with institutions embedded in the target region. Where human resource is a clear limitation for international agriculture, the number of potential partners is of course reduced. So a further challenge is how to cope with the common situation where a small number of scientists become over-committed across a broad set of diverse activities and end up with ever increasing administrative responsibilities.

2.5.3 The Particular Challenge of Policy and Intellectual Property Issues

In bringing food security and self-sufficiency to impoverished nations, the Green Revolution had to overcome formidable technical, scientific, and cultural challenges. But at least there were few legal obstacles to this important work. Germplasm passed from one country to another with little more than a handshake, and it was recognized that the greatest value of most food crops grown in developing countries lay not in their commercial potential but in their capacity to alleviate hunger and malnutrition. The world has changed since then.

The enormous growth of biotechnology, in terms of knowledge and products, over recent years has coincided with an expansion of legal and policy doctrines. These have made technology transfer to developing countries far more complicated than it used to be. The provisions of the Convention on Biological Diversity require the advance informed consent of national governments prior to the transfer of genetic resources, and this has clearly impeded a significant body of humanitarian purpose-oriented research. The full implementation of the recent International Treaty on Plant Genetic Resources for Food and Agriculture should mitigate some of these difficulties, but since the treaty does not apply to all agriculturally significant crops, it remains uncertain as to quite how effective it will be in eliminating legal bottlenecks.

In addition, intellectual property issues now reduce the ease with which biological technologies can be transferred to developing countries. As originally conceived, patents and plant variety rights were intended to stimulate innovation by allowing, for limited periods of time, inventors and innovators the right to exclude others from using their work. Patent law was based on familiar types of inventions in the mechanical, chemical, and other traditional industries. However, with growing intensity since companies began to perceive the commercial value of genetic technologies, many countries have extended patenting to plant-related technologies. In the 1990s, the substance of patent and plant variety rights laws shifted from being mostly of national concern to become an issue of international policy. Multilateral agreements (such as the North American Free Trade Agreement and the World Trade Organization's Agreement on Trade and Intellectual Property Related Aspects of Intellectual Property Rights) have set international minimum standards for the protection of plant-related intellectual property, requiring that signatory states adopt legislation which at least provides an effective form of protection for novel plant varieties. More recently, a host of bilateral trade agreements has committed many developing countries to making patents available for plants themselves, or at least to make the effort to pass legislation enabling the patenting of plants. As a result, public sector agricultural researchers must increasingly analyze the patent status of technologies they wish to use, even in developing countries.

A further result of the proliferation of legal and policy regimes is that many holders of genetic resources (or useful data relating to genetic resources) are less willing than they were in the past to share them freely for humanitarian purposes. Even where biological materials are not protected by a patent, many who possess the materials require downstream users to agree to highly restrictive terms and conditions, which often defeat the purpose for which the materials were requested. Accessing genomics databases often requires entry into a contract which limits the utility of the tools for humanitarian research. Negotiating over such terms, monitoring agreements, and ensuring compliance with them costs money, requires special expertise, and demands concentrated attention in a manner that many public institutions find unfamiliar and difficult.

Fortunately, public agricultural research institutions have begun to develop the capacity to address these legal and policy challenges in a constructive manner. They are learning that it is not inevitable that these issues will always prevent the conduct of important research. By devoting adequate resources, planning and ongoing attention to these matters, research organizations can, in most cases, still achieve their goals.

2.6 Conclusions

From a human welfare standpoint, the greatest benefits of plant biotechnology will surely result from the adoption of improved crop varieties in the countries where so many people still depend on agriculture for their livelihood. Despite the increasing level of industrialization achieved by many countries in the South, and the consequent rise in living standards enjoyed by these populations, the need to increase the supply of affordable food to many millions of people remains urgent. Environmental degradation and the specter of global climate change have only reinforced the need to accelerate and improve crop breeding (Foley et al. 2005), since these forces will, quite apart from the familiar problem of population increase, demand a much faster change in technology than has ever been experienced in the past. The global nature of these challenges and the increasing realization that the world is truly a "global village" require not a national or even a regional but a genuinely international response. Despite all the challenges, the international agricultural programs represent a powerful and key driver of technological change in the developing world. Several international programs, including the GCP, have shown how it is possible to overcome many of the barriers that have historically frustrated innovation relevant to agriculture in the South. It is clear that the establishment of structures which reward a collaborative culture between all partners and allow the research outputs from these programs to be widely available can make substantial strides towards the alleviation of hunger and poverty for the world's neediest people.

References

Berry SS (1984) The food crisis and agrarian change in Africa: a review essay, African Studies Review 27:59–112

Brown AHD (1989) Core collections: a practical approach to genetic resources management. Genome 31:818–824

Byerlee D, Fischer K (2001) Accessing modern science: Policy and institutional options in developing countries. IP Strategy Today 1. <http://www.biodevelopments.org/ ip/ipst1n.pdf>. Accessed December 19, 2003.

Comai L, Young K, Till BJ, Reynolds SH, Greene EA, et al. (2004) Efficient discovery of DNA polymorphisms in natural populations by ecotilling. Plant J 37:778–786

Conway G (1997) The doubly Green Revolution; Ithaca: Cornell University Press. 335 pp.

Conway G, Toenniessen G (2003) Science for African food security. Science 299:1187–1188

Delmer DP (2005) Agriculture in the developing world: Connecting innovations in plant research to downstream applications. Proc Natl Acad Sci USA 102:15739–15746

Dev M (1998) Regional disparities in agricultural labour productivity and rural poverty. Indian Econ Rev 23:167–205

FAO (2005) FAO Reform. A vision for the twenty-first century. FAO, Roma, Italy

Foley JA, Defries R, Asner GP, Barford C, Bonan G, et al. (2005) Global consequences of land use. Science 309:570–574

Gakou ML (1987) The crisis in African agriculture. Studies in African political economy. The United Nations University Zed Books Ltd. London and New Jersey. xii + 100 pp.

Grimanelli D, Perotti E, Ramirez J, Leblanc O (2005) Timing of the maternal-to-zygotic transition during early seed development in maize. Plant Cell 17:1–12

Hazell PBR, Ramasamy C, Rajagopalan V, Aiyasamy PK, Bliven N (1991) Economic changes among village households. In: Hazell PBR, Ramasamy C (eds) The Green Revolution Reconsidered: The Impact of High-Yielding Rice Varieties in South India, Baltimore: Johns Hopkins University Press pp. 29–56

Huang J, Rozelle S, Pray C, Wang Q (2002) Plant biotechnology in China. Science 295:674–677

Kliebenstein DJ, West MAL, van Leeuwen H, Kim K, Doerge RW, et al. (2006) Genomic survey of gene expression diversity in *Arabidopsis thaliana*. Genetics 172:1179–1189

McCallum CM, Comai L, Greene EA, Henikoff S (2000) Targeting induced local lesions IN genomes (TILLING) for plant functional genomics. Plant Physiol 123(2):439–442

O'Toole JC, Toenniessen GH, Murashige T, Harris RR, Herdt RW (2001) The Rockefeller Foundation's International Program on Rice Biotechnology. In: Khush GS, Brar DS, Hardy B (eds) Rice genetics IV. Proc 4th Intl Rice Genetics Symp. International Rice Research Institute. pp. 39–59

Pardey PG, Beintema NM (2001) Slow magic: Agricultural R&D a century after Mendel (technical report 36). Washington, DC: Agricultural Science and Technology Indicators/International Food Policy Research Institute

Pardey P, Roseboom J, Beintema N (2003) Agricultural research in Africa: three decades of development. ISNAR Briefing Paper

Perkins JH (1997) Geopolitics and the green revolution: wheat, genes and the cold war. Oxford University Press US. 352 pp.

Raghavan C, Naredo E, Wang H, Atienza G, Liu B, et al. (2007) Rapid method for detecting SNPs on agarose gels and applications for candidate gene mapping. Mol Breed 19:87–101

Rockefeller Foundation (2006) Africa's turn: a new green revolution for the 21st Century. The Rockefeller Foundation publication 12 pp.

Sanchez PA, Swaminathan MS (2005) Cutting world hunger in half. Science 307:357–359

Savidan Y, Carman JG, Dresselhaus T (2000) The flowering of apomixis: from mechanisms to genetic engineering. CIMMYT/IRD/EC. Xii, 243 pp.

Spielman DJ, von Grebmer K (2004) Public-private partnerships in agricultural research: An analysis of challenges facing industry and the consultative group on international agricultural research. EPTD Discussion Paper No. 113. International Food Policy Research Institute, 63 pp.

Spitz P (1987) The Green Revolution re-examined in India. In: Glaeser (ed). The Green Revolution Revisited: Critique and Alternatives. In: Allen & Unwin (ed), London, pp. 56–75

Toenniessen GH, Herdt RW (1988) The Rockefeller Foundation's International Program on rice biotechnology. Presented at USAID-sponsored conference on Strengthening Collaboration in Biotechnology: International Agriculture and the Private Sector, 17–21 April 1988, Washington, D.C. P 291–317

Toenniessen GH, O'Toole JC, DeVries J (2003) Advances in plant biotechnology and its adoption in developing countries. Curr Opin Plant Biol 6:191–198

Upadhyaya HD, Furman BJ, Dwivedi SL, Udupa SM, Gowda CLL, et al. (2006) Development of a composite collection for mining germplasm possessing allelic variation for beneficial traits in chickpea. Plant Genetic Resources: Characterisation and Utilisation 4:13–19

Walia H, Wilson C, Zeng L, Ismail AM, Condamine P, et al. (2007) Genome-wide transcriptional analysis of salinity stressed japonica and indica rice genotypes during panicle initiation stage. Plant Mol Biol 69:609–623

Wang Y, Xue Y, Li J (2005) Towards molecular breeding and improvement of rice in China. Trends in Plant Sc. 10:610–614

Wenzl P, Carling J, Kudrna D, Jaccoud D, Huttner E, et al. (2004) Diversity Array Technology (DArT) for whole-genome profiling of barley. Proc Natl Acad Sci USA 101:9915–9920

Chapter 3
Transgenics for New Plant Products, Applications to Tropical Crops

Samuel S.M. Sun

Abstract Advancements in plant science and agricultural technology now allow the direct transfer of gene(s) from diverse origins into target crops for improvement, with the advantages of breaking cross-species barriers and saving time in comparison to conventional breeding and selection. Transgenic technology has been used and commercialized since 1994 to produce new crop products with herbicide tolerance, insect resistance, virus resistance, and improved post-harvest quality. These input traits are characteristic of first generation transgenic crops that continue to be widely adopted by farmers globally. Numerous transgenic crop new products, with increased emphasis on output traits such as improved and novel product quality (which are more appealing and directly beneficial to the consumers), are under development and field testing. Activities in developing crops with new and better agronomic properties and using plants as bioreactors to produce high value products are also on the rise. While tropical plant germplasm, with its rich biodiversity increasingly revealed through gene discovery through genomics and associated technologies, can offer novel genes and regulatory mechanisms for crop improvement, transgenic technology provides a complementary approach with new possibilities for improving tropical crops to assure food security and nutritional well-being of the people in the tropics.

3.1 Introduction

Plants are the primary source of food for humans and feed for their livestock. Through domestication and activities of breeding and selection, plants have been developed into crops that serve as the major source of dietary carbohydrates, proteins, lipids, vitamins, and minerals for humans and livestock. To enhance agricultural productivity, plant characteristics contributing to crop yield, quality, and production economics are identified and objectives established to increase total biomass, harvestable yield, nutritional quality, resistance to biotic and abiotic stresses, and to improve plant architecture, response to photoperiod and inputs, and processing characteristics. With continuous progress in plant sciences, plant genes were cloned

S.S.M. Sun
Department of Biology, the Chinese University of Hong Kong, Hong Kong, China

© Springer 2008

and the first transgenic plant expressing a foreign gene was realized in 1983. Since then, a powerful new agricultural biotechnology, in addition to the conventional breeding and selection methods, has been available for application to crop improvement. This new technology has the advantages of breaking the cross-species barrier for gene introduction and is less time consuming for generating new traits/products when compared to conventional crop improvement methods. Through recombinant DNA and transgenic techniques, the new method offers unlimited source of genes and methods for their regulation that are more specific, precise, and time saving than are the traditional methods for crop improvement. With this promise and capability, the first ever transgenic tomato with improved post-harvest quality was approved for marketing in 1994; the first transgenic crop product with insect resistance was commercialized in 1996. By the year 2006, 11 years after the first commercialization of any biotech crop, 10.3 million farmers in 22 countries had planted 102 million hectares of transgenic crops, a 60-fold increase over the global biotech crop planting area in 1996. Transgenic soybean ranked top among crop species with 58.6 million hectares (m ha) (57%), followed by maize (25.2 m ha, 25%), cotton (13.4 m ha, 13%) and canola (4.5 m ha, 5%). In terms of transgenic traits, herbicide tolerance dominates (69.9 m ha, 68%), followed by Bt insect resistance (19.0 m ha, 19%) and stacked traits (herbicide tolerance plus insect resistance), 13.1 m ha (13%). From 1996 to 2005, global accumulated impact of transgenic crops in terms of net economic benefits to biotech farmers was 27 billion, and the accumulated reduction in pesticides was 224,300 million tons of active ingredients, equivalent to a 15% reduction in the associated environmental impact of pesticide use on these crops (James 2007).

The transgenic crop products currently on market, including the major traits of herbicide tolerance, insect resistance, and virus resistance, have all been generated by single gene transfer and manipulation to influence the performance of crop in the field and its production economics. These first generation biotech crops are more beneficial to the growers and developers with fewer inputs so that the traits involved are referred to as input traits (Castle et al. 2006), even though the environmental benefits of the traits benefit society at large. In the subsequent stage of crop biotechnology development, the transfer and manipulation of genes are targeted for increasing the quality of food, feed, and other products catering to the needs of the end users. Traits such as improved food nutritive content benefit the consumers more directly and are referred to as output traits, a characteristic of second generation biotech crops (Willmitzer 1999). Transgenic manipulation of output traits often involves more than a single gene and is thus a process of metabolic engineering. During the development of input traits, the concept of using plants as bioreactors to produce useful products arose. It has since been an area of active research and development in plant biotechnology. Recently, rapid advancements in genomics and related high throughput technologies have produced unprecedented opportunities for the discovery of genes and their regulatory mechanisms, which will accelerate the use of transgenic technology for crop improvement.

Tropical and subtropical regions of the world house over 50% of its biodiversity (Kochhar 1981), offering the greatest wealth of forms and genes from wild and cultivated plants. For crop biotechnology, tropical biodiversity represents a rich and invaluable source of useful genes and associated regulatory elements/mechanisms for crop improvement. Tropical and subtropical regions are also the location of many developing countries having dense populations and large food needs. Many of the staple foods in these regions are starchy roots and tubers that provide calories, but are poor in nutritional quality due to their low content of protein and micronutrients. Thus, while the tropics can offer the world a wealth of novel genetic elements for crop improvement, and recent advances in tropical crop genomics coupled with transgenic technology have the potential to contribute to improving the adequacy and nutritional state of food for the tropics, there remains much to be done for this to be accomplished.

This chapter will cover transgenic new products from crops (emphasizing those already commercialized or under development) and will analyze the future prospects of plant transgenics for improving tropical crop plants.

3.2 Transgenic Plant Products

3.2.1 Commercialized Products

Transgenic crops such as soybean, cotton, corn, canola, squash, and papaya with input traits of herbicide tolerance, insect resistance, and virus resistance are the major transgenic products commercialized since 1996. Plants with these traits meet the farmers' desire of high yields with fewer inputs such as reduced herbicide and pesticide use while having reduced environmental impact. Studies have shown that when applications of agricultural chemicals were reduced as a result of growing insect resistant cotton, incidences of farmers' health problems were also reduced (Huang et al. 2002). Further, the first generation transgenic products were generated by relatively simple transgenic manipulation of single genes. All these factors contribute to the success of first generation transgenic crops.

3.2.1.1 Herbicide Tolerance

Weeds decrease crop yields, accounting for a 13% loss of total world crop production. Annual herbicide production and sale is the largest component in agrochemical business. In the US, the annual cost of herbicides is approximately US $5 billion. Developing crops having the herbicide tolerance input trait was thus one of the earliest targets of research and development on agricultural biotechnology, and the transgenic product became one of the first and most widely adopted commercialized biotech accomplishments. Most of the herbicide-tolerant crops now on market were engineered through two approaches: 1) the herbicide target molecules (either enzymes or other cell components) were engineered for over production, or to become insensitive to the herbicide, and 2) the crops were engineered with gene(s) or a pathway to degrade or detoxify the herbicide.

a. Glyphosate-tolerant crops: Glyphosate is a broad-spectrum, leaf applicable, non-selective, non-toxic to animals, organic phosphate herbicide that is easily degraded in soil. Glyphosate inhibits the enzyme 3-enolpyruvateshikimate- 5-phosphate synthase (EPSPS) of the aromatic amino acid synthesis pathway. A modified *Agrobacterium* gene encoding EPSPS (named CP4 EPSPS) was developed, and the gene CP4 *epsps* was introduced into soybean, generating transgenic soybeans tolerant to herbicide glyphosate (Padgette et al. 1966). Because glyphosate herbicides are relatively inexpensive and broadly toxic to nearly all broadleaf and grass weeds, and the use of herbicide-tolerant crops allow reduced- and no-till practices, transgenic glyphosate-tolerant soybeans are welcome by farmers and widely adopted. Currently over 85% of US soybeans and 56% of soybeans globally are glyphosate tolerant. Similar approaches have been applied to cotton, canola, and corn and these transgenic new products are increasingly adopted by farmers (Castle et al. 2006).
b. Crops tolerant to other herbicides: Phosphinothricin or bialaphos-based herbicides (glufosinate) are broad-spectrum and non-selective organic phosphate herbicides that break down rapidly in the soil. These herbicides strongly inhibit glutamine synthase activity in plants, resulting in the accumulation of toxic ammonium in the cells that kills the plants. To engineer glufosinate-tolerant crops, two approaches can be used, either over-express the glutamine synthase gene, or introduce a gene to deactivate the herbicide. The enzyme phosphinothricin acetyltransferase (PAT or BAR) modifies phosphinothricin into an inactive form through acetylation. The *pat* gene was isolated from *Streptomyces viridichromogenes* while the *bar* gene from S. *hygroscopicus.* Using either of these two genes, phosthinothricin-tolerant transgenic cotton, corn and canola were developed (De Block et al. 1987). Bromoxynil herbicides inhibit electron transport in photosynthesis. Introduction of nitrilase can detoxify bromoxynil. A gene encoding BXN nitrilase from *Klebsiella pneumoniae* (Stalker et al. 1988) was introduced into cotton and canola to generate resistance. However, transgenic crops with tolerance to these two classes of herbicides, especially the bromoxynil herbicides, are not as popular as those that are glyphosate tolerant since glyphosate costs less and controls more weed species (Castle et al. 2006). Using the same molecular approaches, tolerance has been generated in plants against the sulphonylurea and imidazolinone herbicides, which inhibit the branched-chain amino acid biosynthesis pathway, by introducing a mutant acetolactate synthase (*ALS*) gene that is resistant to the herbicides. For the herbicide atrazine, which inhibits photosystem II, resistance can be engineered by introducing a mutant gene for Q8 protein, or by introducing the gene encoding glutathione-S-transferase to detoxify the atrazine.

3.2.1.2 Insect Resistance

Insect pests cause approximately a 13% loss of world crop production even though the annual worldwide expenditure on insecticides amounts to US $8 billion. Crops are affected and damaged by diverse species of insects, often with specificity to crops at a certain stage of development or specific organs of the crops. To engineer new crop products resistant to insects, genes conferring this input traits must first be identified. Most of the resistance genes discovered thus far target the digestive system of insects, either as an anti-feedant or as a toxin. Examples of such genes are those encoding proteinase inhibitors that block insect digestive enzymes, or induce hypersecretion of digestive enzymes leading to depletion of essential amino acids; amylase inhibitors that inhibit carbohydrate digestion, which in turn, affect insect larval development; lectins that can bind to the midgut epithelial cell; and chitinase that may affect the formation of chitin, a structural component of insects. Many of these genes have been introduced into a variety of crops where they have demonstrated different degrees of plant protection. However, it is the gene encoding a crystal protein (Cry) or delta-endotoxin in *Bacillus thuringiensis* (*Bt*), a gram-positive, spore forming soil bacterium, that has received the most attention and reached the highest level of usage in producing insect-tolerant transgenic crops. During sporulation, *Bt* produces crystal proteins that are highly toxic to a broad range of insects, but are not harmful to mammals. The Cry protein has a molecular weight about 130 kDa, although some truncated forms also occur. In the insect midgut, the Cry protein is processed into an active N-terminal 65–70 kDa truncated form that causes the death of insect. Cry proteins consist of a family of homologous forms exhibiting diversity in insecticidal specificity. Numerous Cry proteins and their genes have been identified and cloned, and used to generate crops for resistance to specific insect pests (Schuler et al. 1998, De Maagd et al. 1999, De Maagd et al. 2003, Whalon and Wingerd 2003, Federici 2005). Insect resistant cotton and corn transformed with the Bt genes *cry1Ac* and *cry1Ab*, respectively, were commercialized in 1996 and have since been widely adopted by farmers. It is worth noting that in addition to these private sector efforts, the public institution, the Chinese Academy of Agricultural Sciences, China, has also developed insect resistant cotton, by combining Bt gene *cry1Ac* with the trypsin inhibitor gene *CpTI* from cowpea, that is widely adopted by farmers in China (Wu and Guo, 2005).

3.2.1.3 Disease Resistance

Like weeds and insect pests, plant diseases also cause about a 13% loss of the total world crop production. Plant virus infections lead to a range of diseases causing significant economic damage to most of the world's major crops (Agrios 1997). Since there are no effective chemical viricides available, effort has been made from the inception of plant biotechnology to apply this new technology to develop virus resistance crops. Several approaches have been demonstrated to confer virus resistance to target crops, including the genes encoding viral coat proteins (CPs), replicases, movement proteins, proteinases, defective interfering RNA, and satellite RNAs. Recent studies suggest that plant protection against the viruses is, in most cases, by an RNA-based post-transcriptional gene silencing mechanism (O'Brien and Forster 1994, Cooper et al. 1995, Lomonossoff 1995, Dawson 1996, Baulcombe 1996, Fuchs and Gonsalves 1997, Beachy 1997, Malpica et al. 1998, Waterhouse et al. 2001). The use of genes or gene sequences derived from viral genomes to confer virus resistance in transgenic plants is known as pathogen-derived resistance (PDR, Sanford and Johnston 1985). The first transgenic virus resistant crop commercialized was squash resistant to watermelon mosaic virus (WMV) and zucchini yellow mosaic virus (ZYMV) through the introduction and expression of the coat protein genes of WMV2 and ZYMV (Fuchs and Gonsalves 1995). Other crops with resistance to a variety of viruses have also been developed. However, the most successful and widely known virus-resistant transgenic product, jointly developed in 1997 by the public institutions Cornell University, University of Hawaii, and the United States Department of Agriculture, along with an industrial entity, Upjohn Company, is papaya, which is resistant to papaya ringspot virus (PRSV) through expression of the PRSV coat protein gene. Farmer and consumer acceptance of the PRSV resistant transgenic papaya varieties (Sunrise and Rainbow) has contributed greatly towards reviving the papaya industry in Hawaii that had been decimated by this virus (Ferreira et al. 2002).

3.2.1.4 Improved Post-harvest Quality

Ripening is a normal maturation process of many fruits and vegetables. Delayed ripening by transgenic technology will allow farmers more flexibility in marketing their products and offer consumers produce near maximum freshness. Delayed ripening benefits both farmers and consumers and thus can be classified as both an input and output trait. Transgenic approaches developed to control the ripening process include aspects of ethylene metabolism including the suppression of 1-aminocyclopropane-1-carboxylic acid (ACC) synthase, which converts S-adenosylmethionine (SAM) to ACC during ethylene synthesis, or suppression of ACC oxidase, which catalyzes the oxidation of ACC to ethylene, by anti-sense technology, or by introduction of a truncated copy of the synthase gene (Theologis et al. 1993, Ayub et al. 1996), and the reduction of the amount of ACC available for ethylene synthesis by an ACC deaminase transgene, which converts ACC to alpha-ketobutylic acid (Klee et al. 1991). Control of fruit softening during maturation was also demonstrated by the suppression of polygalacturonase (PG) enzyme through antisense technology (Sheehy et al. 1988), or by introduction of a truncated copy of the PG gene to delay the cell wall pectin from degradation during fruit maturation. Tomato engineered with improved post-harvest quality by the PG technology received approval for marketing in 1994, under the trade name Flavr-Savr^{TM}, marking the first ever transgenic crop approved for marketing.

3.2.2 Products Under Development

Numerous proof-of-concept transgenic plant products have been generated and reported by academic and industrial laboratories throughout the world. While certain of them may in the future turn into new products for commercialization, it is not the aim of this chapter to account for these events. Instead, prototype products that have been applied for field tests in the US, indicating that they are further down the path towards application, are summarized here. As the United States is most active in plant transgenics research, development, and application, this information will provide insight relevant to the status and development of plant transgenic activities.

The database of APHIS/USDA's Biotechnology Regulatory Services (www.isb.vt.edu/cfdocs/ISBlists1.cfm) reveals, as of August 10, 2007, an accumulation of 16,814 field test permits have been approved since 1987. Among these plant transgenic activities, a total of 866 phenotypes are involved that can be grouped into 10 categories: agronomic properties (AP), bacterial resistance (BR), fungal resistance (FR), herbicide tolerance (HT), insect resistance (IR), marker gene (MG), nematode resistance (NR), other (OO), product quality (PQ), and virus resistance (VR). The category of HT ranks top, with 4,330 permits approved for field testing, representing 26% of all approved permits, followed by IR (3,698, 22%) and PQ (3,079, 18%), while NR comes in last (42, near 0%) (Figure 3.1). Thus transgenic products with input traits, mainly HT and IR, have dominated the scene the last 20 years. However, when the number of permits for each phenotype category approved for field testing is compared over 10 year intervals (between 1988 and 1997 and 2006), a declining trend is noted for all major input traits including HT, IR, and VR, while the output trait PQ and category of AP are on significant rise during this 20 year period (Figure 3.2).

While these trends may reflect the more mature, steady, and maintenance state of those major input trait products, it is clear that increasing interest is turning toward the output traits that are more appealing to consumers. The data on major individual phenotypes approved for field testing (45 permits or more), though generally in agreement with the category distributions, provide further details on the individual phenotypes under different categories (Table 3.1). For example, both glyphosate and phosphinothricin tolerance are dominating phenotypes, however, the former is far more active in development, attesting to certain extent of its popularity in the market. Alteration of carbohydrate, oil, and protein quality and

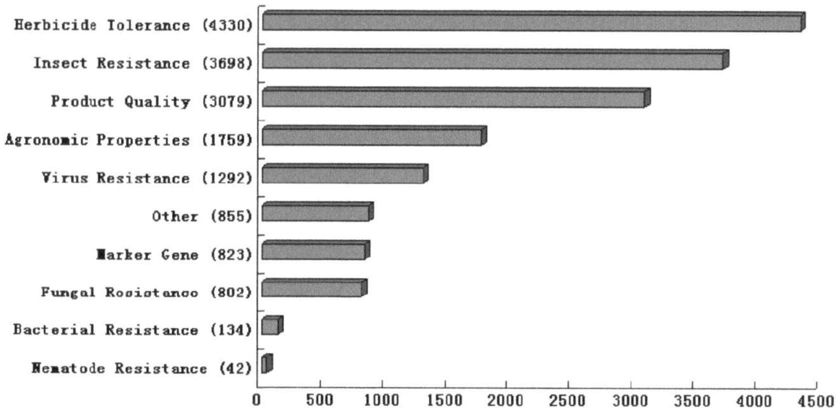

Fig. 3.1 Number of Permits for Each Phenotype Category Approved for Field Testing

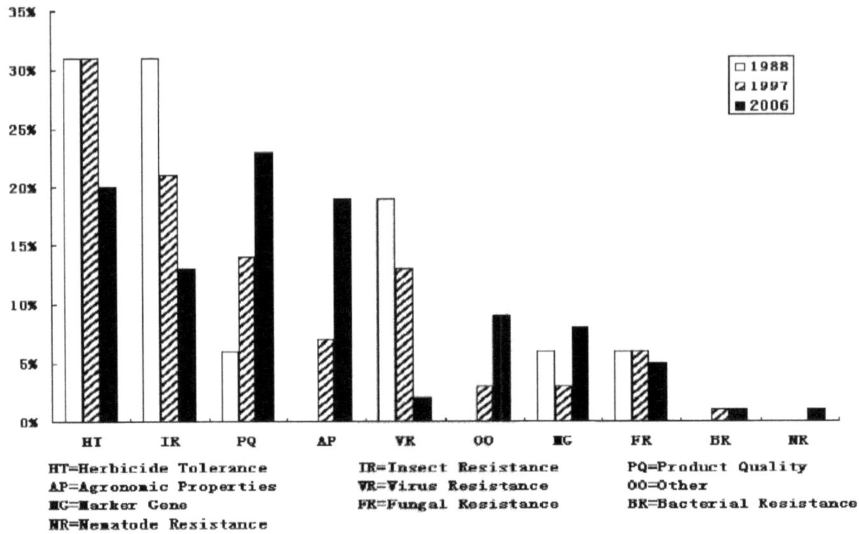

Fig. 3.2 Number of Permits (%) for Each Phenotype Category Approved for Field Testing

Table 3.1 Number of Permits for Each Phenotype Approved for Field Testing (Listed are Numbers ≧ 45)

Phenotype	Number	Phenotype	Number	Phenotype	Number
Glyphosate Tolerant	2506	European Corn Borer Resistant	139	Phytophthora Resistant	70
Lepidopteran Resistant	1736	Lysine Level Increased	138	Storage Protein Altered	70
Coleopteran Resistant	1356	Fertility Altered	134	Salt Tolerance Increased	68
Phosphinothricin Tolerant	1114	Altered Amino Acid Composition	124	Tryptophan Level Increased	67
Visual Marker	586	Growth Rate Altered	123	Solids Increased	64
Yield Increased	432	Zymv Resistant	120	Isoxazole Tolerant	63
CBI	403	Wmv2 Resistant	118	Lepidopteran Resistant/ Coleopteran Resistant	62
Carbohydrate Metabolism Altered	256	Fruit Ripening Altered	113	Bromoxynil Tolerant	58
Oil Profile Altered	252	Novel Protein Produced	108	Prsv Resistant	55

Table 3.1 (continued)

Phenotype	Number	Phenotype	Number	Phenotype	Number
Pvy Resistant	232	Imidazolinone Tolerant	103	Nitrogen Metabolism Altered	53
Drought Tolerant	229	Fusarium Resistant	92	Ear Mold Resistant	53
Seed Composition Altered	210	Pharmaceutical Proteins Produced	89	Glyphosate Tolerance	50
Colorado Potato Beetle Resistant	189	Fruit Ripening Delayed	89	Bruising Reduced	49
Cmv Resistant	179	Sclerotinia Resistant	86	Sulfonylurea Tolerant	48
Protein Quality Altered	158	Protein Altered	84	Lysine Level Alterered	47
Male Sterile	147	Gene Expression Altered	76	Lepidopteran Resistance	45
Plrv Resistant	140	Oil Quality Altered	74	Drought Tolerance	45

properties are major activities in the category of product quality, while developing plants as bioreactors to produce novel proteins is also notable. A total of 1,395 genes and 301 institutes, of both public and private sectors, are involved in the research and development of these prototype transgenic products. These statistical data document the very active efforts and innovation directed toward the development of new crop products through transgenic technology and reveal the transgenic new products that may appear in the future (www.isb.vt.edu/cfdocs/ISBlists1.cfm; www.agbios.com/dbase.php?action=synopsis).

3.2.2.1 Products and Approaches Relevant to Tropical Crops

While the input and output traits and the transgenic technology involved (reviewed above) are relevant to both temperate and tropical crop improvement, some products and transgenic approaches are of special interest to tropical crops.

Post-harvest Quality

Banana, papaya, pineapple, and mango are important climacteric fruits of the tropics. The proven transgenic technology to delay fruit and vegetable ripening, as reviewed in Section 3.2.1.4, can be applied to benefit these tropical fruits that are globally favorites of many, by enhancing their shelf life and retaining their freshness, appearance, flavor, and nutrition for long distance export.

Nutritional Quality

Crops provide the proteins, vitamins, and minerals needed for human nutrition. However, in certain staple crops, especially roots and tubers that are major staple foods in the tropics, plant proteins are often low in content and deficient in essential amino acids. When these staple crops are used as sole or major source of dietary proteins, such deficiencies will cause adverse effects on human nutrition and health (Young and Pellett 1994). In general, cereal proteins are low in lysine (1.5–4.5% vs. 5.5% of WHO recommendation) while legume, root, tuber, and most vegetable proteins are deficient in the sulfur amino acids (methionine and cysteine, 1.0–2.0% vs. 3.5% of the WHO reference protein) (Sun 1999). Several transgenic approaches have been attempted to enhance the essential amino acid contents in plants, including enhancing the protein bound fraction by (i) modifying the protein sequence (De Clercq et al. 1990, Dickinson et al. 1990, Zuo 1993, Takaiwa et al. 1995, Marcellino et al. 1996, Tu et al. 1998, Katsube et al. 1999); (ii) producing a synthetic protein (Jaynes et al. 1986, Yang et al. 1989, Kim et al. 1992, Keeler et al., 1997, Zhang et al. 2003); (iii) expressing a heterologous protein (Sun and Larkins 1993, Sun et al. 2000, Sun and Liu 2004); and (iv) manipulating homologous protein expression (Coleman et al. 1997, Singh et al. 2000, Maruta et al. 2001, Lai and Messing 2002), and (v) increasing the pool of a specific free essential amino acid, such as lysine, through metabolic engineering (Mazur et al. 1999, Galili et al. 2002, Galili et al. 2002, for review). Significant enhancements, especially for methionine and lysine contents, have been demonstrated in the target transgenic plants through these approaches, for a review see (Sun and Liu 2004).

Vitamins and minerals are essential food components for human health. Deficiency in dietary micronutrients such as vitamin A, iron, iodine, or zinc, will result in micronutrient malnutrition and various deficiency diseases. An adequate and diverse diet of fruits, vegetables and animal products is the best solution in obtaining sufficient micronutrients. However, for people, especially children, in many poor developing countries who rely solely or mostly on a single staple food crop, such as rice, sorghum, cassava, or banana, will suffer from nutrient deficiency diseases, a major source of morbidity and mortality worldwide (Toenniessen 2002). Biofortification through transgenic technology to increase the micronutrient contents in staple crops is a promising approach. Transgenic and associated molecular technologies have been demonstrated, or are under investigation and development, to enhance the synthesis and bioavailability of vitamins and minerals in plants, for example iron and zinc (Lucca et al. 2001, Zimmermann and Hurrell 2002, Ghandilyan et al. 2006, Lucca et al. 2006), carotenoids including provitamin A (Botella-Pavia and Rodriguez-Concepcion 2006, Lucca et al., 2006), vitamin C (Agius et al. 2003, Ishikawa et al. 2006), vitamin E (Shintani and DellaPenna 1998, Van Eenennaam et al. 2003, Dellapenna and Last 2006), folates (Rebeille et al. 2006), and pantothenate (vitamin B_5) (Chakauya et al. 2006). A good example is the fortification of pro-vitamin A (β-carotene) in rice through transgenic technology. Ye et al. (2000) genetically engineered the first generation of pro-vitamin enriched rice, named *Golden Rice 1* (*GR1*), by transferring and expressing the *psy* gene, encoding phytoene synthase from daffodil (*Narcissus pseudonarcissus*), and the *crtI* gene, encoding phytoene desaturase from the bacterium *Erwinia uredovora,* in rice endosperm to achieve a yield of 1.6 μg pro-vitamin A/g in the endosperm. Efforts were made to further increase the content of pro-vitamin A in *Golden Rice*. In 2005, Paine et al. developed the second generation *Golden Rice*, *GR2*, by replacing the phytoene synthase coding sequence of daffodil in *GR1* with that of maize to achieve a yield of 31 μg pro-vitamin A/g in the endosperm, a 20-fold enhancement. With this improvement, the daily vitamin A allowance of a 1–3 year-old child could be provided by 72 g of *GR2*, which is within the range of 100–200 g of rice consumed per child per day in the target countries (www.goldenrice.org).

Four international consortia, with the support from the Bill & Melinda Gates Foundation, under the Grand Challenges in Global Health (GCGH) Initiative (www.gcgh.org), are undertaking research projects to engineer nutrient-rich staple crops, namely banana, cassava, rice, and sorghum, through a combination of transgenic, genomic, molecular marker-assisted (and conventional) technologies. The common goal is to develop new varieties of staple crops with high β-carotene, vitamin E, protein, and enhanced bioavailability of iron and zinc, for populations in the developing countries who rely on these crops as major staple foods, especially those in the tropical Africa and Southern Asia regions (www.gcgh.org/projects/improveNutrition/NutrientRichPlants/default.htm); www.goldenrice.org/Content5-GCGH/GCGH1.html).

The *Golden Rice* and GCGH projects serve as examples that transgenic and associated molecular technologies do offer new possibilities, in addition to the existing breeding and selection methods, for improving tropical crop plants. Since no variety of rice has ever been found to contain β-carotene in its endosperm, the β-carotene biosynthesis pathway can be introduced for synthesizing pro-vitamin A in the new rice product, only through transgenic technology.

Plants as Bioreactors

Transgenic plants have emerged as an attractive bioreactor platform for large-scale production of industrial enzymes, pharmaceutical proteins, and other biomolecules (Goddijin and Pen 1995, Daniell et al. 2001, Sparrow et al. 2007). When compared to bioreactors based on other systems (such as bacteria, yeast, transfected animal cell lines, or transgenic animals) the bioreactors based on plants hold several advantages, including: low capital and operating costs, easy to scale up, eukaryote post-translational modifications, low risk of human and animal pathogen contaminations, and a relatively high protein yield (Fischer et al. 1999, Fischer and Emans 2000, Fischer et al. 2004). Proof-of-concept production of diverse biomolecules in plants, including carbohydrates, lipids, and proteins (such as high-value pharmaceutical polypeptides and industrial enzymes) has been demonstrated (Goddijn and Pen 1995, Hood and Jilka 1999, Mercernier et al. 2001). The first commercialization of plant-produced recombinant proteins was egg white avidin from maize (Hood et al. 1997) that is marketed by Sigma Chemical Company. Another example commercialized plant-derived recombinant protein is hirudin, an anticoagulant used to treat thrombosis, produced in transgenic oilseed rape (Boothe et al. 1997). More recently, bovine trypsin was produced at commercial levels in transgenic maize, with functional equivalence to native bovine pancreatic trypsin (Woodard et al. 2003) and a recombinant antibody against hepatitis B was commercially produced in tobacco plants in Cuba (Pujol et al. 2005). Other products including oligopeptides, sugar oligomers, starch, fatty acids, oils, secondary compounds, and degradable polymers have been demonstrated as feasible to manufacture in transgenic plants, while the composition and property of oil, starch, and protein can be modified in the production (Willmitzer and Topfer 1992, Galun and Breiman 1997, Broun et al. 1999, Herbers and Sonnewald 1999, Slattery et al. 2000, Napier et al. 2006). As reviewed above, literature indicates that plant transgenic technology can certainly be applied to develop tropical plants as bioreactors for production of naturally occurring, or logically designed high value bio-products.

3.2.2.2 Tropical Germplasm as Source of Transgenes

Tropical soil, water, and climate sustain a vast assemblage of plants, wild and cultivated, with wide range of genetic variability. These plants provide us foods and materials for construction, clothing, medicine, and industry uses. Tropical plants also offer invaluable biological systems for studying plant biology and evolution under unique tropical environments, and they are a rich source of genes and their regulatory elements/ mechanisms for crop improvement. A few examples are given to illustrate the exploitation of tropical plants for genes with agronomic importance. Sugarcane, a tropical crop known for its highest productivity in the world, was the plant system that contributed to our understanding of C4 photosynthesis. Through transgenic approaches, the phosphoenolpyruvate carboxylase (PEPC) genes of C4 photosynthesis was cloned from C4 maize and transferred into C3 rice. Results revealed that transgenic rice plants with high level of expression of the maize PEPC enzyme exhibited reduced sensitivity of photosynthesis to O_2inhibition (Ku et al. 1999). The gene encoding the methionine-rich 2S albumin protein in the seeds of tropical Brazil nut tree (*Bertholletia excelsa* H.K.B.) was cloned (Altenback et al. 1987), transferred, and expressed in tobacco (Altenback et al., 1989). Results demonstrated that it is feasible to enhance the essential amino acid methionine (by 30%) in the tobacco seeds, representing the first successful transgenic approach to significantly increase the essential amino acid methionine content of seeds (Willmitzer and Topfer 1992). In our more recent search for genes encoding high lysine protein for nutritional improvement, a gene encoding a 18-kDa protein with 10.8 mol % lysine was identified in winged bean, an edible tropical legume of the tropics, and the lysine-rich protein gene was cloned (Sun et al. 1993) and expressed in rice, resulting in 20% increase in lysine content (Liu 2002). A seed albumin, AmA1, with nutritionally balanced amino acid composition was identified and cloned from a tropical grain amaranths (*Amaranthus hypochondriacus*) and expressed in potato, resulting in a significant increase in most of the essential amino acids and total tuber protein (Chakraborty et al. 2000).

3.3 Future Prospects

3.3.1 Transgenics for Crop Improvement

Transgenic approaches and technologies have produced a generation of crops with improved input and output traits that are beneficial to farmers, consumers, and the environment. These modified crops are already on global markets and have been widely adopted globally for more than 10 years. Many other crops are under development with new phenotypes with improved biotic and abiotic stress tolerance (especially drought resistance in view of concerns over future water supplies), agronomic properties, product quality, and therapeutic and industrial bioreactor products. These advances strongly demonstrate that transgenic technology offers innovations and new possibilities that conventional breeding and selection methods can not achieve, so they compliment conventional methods for crop improvement. Recent rapid progress in genomics including complete sequence information, genetic maps, arrays of molecular markers, ESTs, and bacterial artificial chromosome libraries, first available for model plants such as *Arabidopsis* and rice, are becoming available for many crops, including those of the tropics (see crop examples of this book). With these biological data, resources, and genetic materials, and the application of knowledge from synteny, functional genomics, and bioinformatics, tens of thousands of plant genes/alleles and their regulatory elements/mechanisms will be discovered. Through transgenic technology and approaches, these plant genes and their regulator elements, either in native or modified form, can be used to produce improved and novel products, from single or stacked traits (a current and future trend in plant transgenics), for improved crop function. The potential and future prospects of producing improved and new crop products through transgenic technology are very promising indeed.

3.3.2 Untapped Food Sources and Novel Genes from Tropical Germplasm

With the ever growing world population and shrinking agricultural lands, future food security is a growing human concern. Plants of the tropics, with their great genetic diversity, offer potential underexploited, and unidentified food resources. Especially in need are plant foods rich in protein and micronutrients. With advances in plant transgenic technology and genomics of tropical crops (this book), we are now more ready than ever to identify and clone genes by tapping into tropical germplasm, especially that of unexploited and unidentified species, for improved and novel traits.

Then through transgenic and technology to generate new crop products for the world as well as the tropics.

3.3.3 Transgenics to Improve Tropical Crops

The new genomics and transgenic technologies applied to agriculture could bring great benefits to developing countries to fight hunger (Delmer 2005, Borlaug 2007). Many people in the tropics rely on starchy root and tuber crops as staple food supplies. These people frequently suffer from hunger and malnutrition since yields of these crops are vulnerable to unfavorable biotic and abiotic stresses that may be prevalent and unique to tropical environments, and whose food value are notably deficient in proteins, essential amino acids, vitamins, and minerals. Through molecular approach, genes target for overcoming these environmental stresses and nutritional deficiencies can be identified and transgenic technology can be applied for local crop improvement.

Plant bioreactors are a transgenic application with potential to produce high value pharmaceutical and industry products in large quantities at low costs. Tropical plants could be developed into highly efficient bioreactors. For example, tropical starchy roots and tubers, efficient in producing starch, would be good candidate bioreactors to produce starch with novel properties and functions. In addition, roots and tubers often grow robustly, even in poor soils, to produce high biomass and they are generally propagated asexually, thus can avoid issues of pollen transfer and seed spillage. These traits are all advantageous for their use as bioreactors, i.e. high production efficiency and fewer biosafety concerns (Sparrow et al. 2007). Likewise other tropical plants could be developed as bioreactors to produce high value products, for example, tropical oil plants (e.g. palms) for production of specialty oils; tropical legumes for production of therapeutic proteins and industrial enzymes; and tropical medicinal plants for production of bioactive compounds. It would be highly preferable that tropical plants selected for use as bioreactors be non-food crops, having high biomass, and are asexually reproduced or, if they are seed crop, they be self-pollinated.

An ability to transform the selected plant species is a requirement of transgenic technology. Many tropical crops are known as being recalcitrant in regeneration and transformation. Thus, it is important to develop and establish efficient transformation systems for the target crops.

During the 20 years since the first commercialization of biotech crops, only a few transgenic tropical crop products, among the 16,814 total permits approved, have been approved for field testing in the U.S. (Table 3.2). With the rapidly increasing interest and progress in tropical plant biology and genomics, there has been a growing effort to improve local tropical crops for food security and nutritional well-being of the people in the tropics. Applications of transgenic technology to further improve tropical crop productivity and the emergence of new crop products of tropical origin are expected to rise in the future.

Table 3.2 Number of Permits for Regulated Organism of Tropical Plant Approved for Field Testing

Tropical Plant	Number	Tropical Plant	Number
Banana	3	Ginger	0
Cavendish banana	1	Macadamia	0
Cacao	0	Papaya	28
Chickpea	1	Peanut	45
Citrus sinensis X Poncirus trifoliata	2	Pineapple	6
Coconut Palm	0	Rubber tree	0
Coffee	3	Sorghum	6
Eucalyptus grandis	32	Sugarcane	55
Eucalyptus hybrid	8	Yam	0
Eucalyptus camaldulensis	5	Tropical Maize	0
Eucalyptus urophylla	1		

References

Agius F, Gonzalex-Lamothe R, Caballero JL, Munoz-Blanco J, Botella MA, et al. (2003) Engineering increased vitamin C levels in plants by overexpression of a D-galacturonic acid reductase. Nat Biotech 21:177–181

Agrios GN (1997) Plant Pathology, 4th edition. Academic Press, Inc., San Diego, p. 655

Altenbach SB, Pearson KW, Leung FW, Sun SSM (1987) Cloning and sequence analysis of a cDNA encoding a Brazil nut protein exceptionally rice in methionine. Plant Mol Biol 8:239–250

Altenbach SB, Pearson KW, Meeker G, Staraci LC, Sun SSM (1989) Enhancement of the methionine content of seed proteins by the expression of a chimeric gene encoding a methionine-rich protein in transgenic plants. Plant Mol Biol 13:513–522

Ayub R, Guis M, Ben Amor M, Gillot L, Roustan JP, et al. (1996) Expression of ACC oxidase antisense gene inhibits ripening of cantaloupe melon fruits. Nat Biotech 14:862–866

Baulcombe DC (1996) Mechanisms of pathogen-derived resistance to viruses in transgenic plants. Plant Cell 8:1833–1844

Beachy RN (1997) Mechanisms and application of pathogen-derived resistance in transgenic plants. Curr Opin Plant Biotech 8:215–220

Boothe JG, Parmenter DL, Saponja JA (1997) Molecular farming in plants: oilseeds as vehicles for the production of pharmaceutical proteins. Drug Develop Res 42:172–181

Borlaug, N (2007) Sixty-two years of fighting hunger: personal recollections. Euphytica online (http://www.springerlink.com/content/d023617827754266)

Botella-Pavia P, Rodriguez-Concepcion M (2006) Carotenoid biotechnology in plants for nutritionally improved foods. Physiol Plant 126:369–381

Broun P, Gettner S, Somerville C (1999) Genetic engineering of plant lipids. Annu Rev Nutr 19:197–216

Castle LA, Wu G and McElroy D (2006) Agricultural input traits: past, present and future. Curr Opin Biotech 17:105–112

Chakauya E, Coxon KM, Whitney HM, Ashurst JL, Abell C, et al. (2006) Pantothenate biosynthesis in higher plants: advance and challenges. Physiol Plant 126:319–329

Chakraborty S, Chakraborty N, Datta A (2000) Increased nutritive value of transgenic potato by expressing a nonallergenic seed albumin gene from *Amaranthus hypochondriacus*. Proc Natl Acad Sci USA 97:3724–3729

Coleman CE, Clore AM, Ranch JP, Higgins R, Lopes MA, et al. (1997) Expression of a mutant alpha-zein creates the floury2 phenotype in transgenic maize. Proc Natl Acad Sci USA 94:7094–7097

Cooper B, Lapidot M, Heick JA, Dodds JA, Beachy RN (1995) A defective movement protein of TMV in transgenic plants confers resistance to multiple viruses whereas the functional analog increases susceptibility. Virology 206:307–313

Daniell H, Streatfield SJ, Wycoff K (2001) Medical molecular farming: production of antibodies, biopharmaceuticals and edible vaccines in plants. Trends Plant Sci 6:219–226

Dawson WD (1996) Gene silencing and virus resistance: A common mechanism. Trends Plants Sci 1:107–108

De Block M, Botterman J, Vandewiele M, Dockx J, Thoen C, et al. (1987) Engineering herbicide resistance in plants by expression of a detoxifying enzyme. EMBO J 6:2513–2518

De Clercq A, Vandewlele M, Van Damme J, Guerche P, Van Montagu M, et al. (1990) Stable accumulation of modified 2S albumin seed storage protein with higher methionine contents in transgenic plants. Plant Physiol 94:970–979

De Maagd RA, Bosch D, Stiekema W (1999) *Bacillus thuringiensis* toxin-mediated insect resistance in plants. Trends Plant Sci 4:9–13

De Maagd RA, Bravo A, Berry C, Crickmore N, Schnepf HE (2003) Structure, diversity, and evolution of protein toxins from spore-forming entomorpathogenic bacteria. Annu Rev Genet 37:409–433

DellaPenna D, Last RL (2006) Progress in the dissection and manipulation of plant vitamin E biosynthesis. Physiol Plant 126:356–368

Delmer DP (2005) Agriculture in the developing world: Connecting innovations in plant research to downstream application. Proc Natl Acad Sci USA 102:15739–15746

Dickinson CD, Scott MO, Hussein EHA, Argos P, Nielsen NC (1990) Effect of structural modification on the assembly of a glycinin subunit. Plant Cell 2:403–413

Federici BA (2005) Insecticidal bacteria: an overwhelming success for invertebrate pathology. J Invertebr Pathol 89:30–38

Ferreira SA, Pitz KY, Manshardt R, Zee F, Fitch M, et al. (2002) Virus coat protein transgenic papaya provides practical control of papaya ringspot virus in Hawaii. Plant Dis 86: 101–105

Fischer R, Emans N (2000) Molecular farming of pharmaceutical proteins. Transgenic Res 9:279–299

Fischer R, Drossarf J, Commandeur U, Schillberg S, Emans N (1999) Towards molecular farming in the future: moving from diagnostic protein and antibody production in microbes to plants. Biotech Appl Biochem 30:101–108

Fischer R, Stoger E, Schillberg S, Christou P, Twyman RM (2004) Plant-based production of biopharmaceuticals. Curr Opin Plant Biol 7:152–158

Fuchs M, Gonsalves D (1995) Resistance of transgenic hybrid squash ZW-20 expressing the coat protein genes of zhucchini yellow mosaic virus and watermelon mosaic virus 2 to mixed infections by both potyviruses. Bio/Tech 13:1466–1473

Fuchs M, Gonsalves D (1997) Genetic engineering. In: Environmentally Safe Approaches to Crop Disease Control. CRC Lewis, Boca Raton, pp. 333–363

Galili G, Galili S, Lewinsohn E, Tadmor Y (2002) Genetic, molecular, and genomic approaches to improve the value of plant foods and feeds. Crit Rev Plant Sci 21:167–204

Galili G, Hoefgen R (2002) Metabolic engineering of amino acids and storage protein in plants. Metab Engng 4:3–11

Galun E, Breiman A (1997) Transgenic plants. Imperial College Press, London, pp. 234–248

Ghandilyan A, Vreugdenhil D, Aarts MGM (2006) Progress in the genetic understanding of plant iron and zinc nutrition. Physiol Plant 126:407–417

Goddijn OJM, Pen J (1995) Plants as bioreactors. Trends Biotech 13:379–387

Herbers K, Sonnewald U (1999) Production of new/modified proteins in transgenic plants. Curr Opin Biotech 10:163–168

Hood EE, Jilka JM (1999) Plant-based production of xenogenic protein. Curr Opin Biotech 10:382–386

Hood EE, Witcher D, Maddock S, Meyer T, Baszynski C, et al. (1997) Commercial production of avidin from transgenic maize: characterization of transformant, production, processing, extraction and purification. Mol Breeding 3:291–306
Huang J, Rozelle S, Pray C, Wang Q (2002) Plant biotechnology in China. Science 295:674–677
Ishikawa T, Dowdle J, Smirnoff N (2006) Progress in manipulating ascorbic acid biosynthesis and accumulation in plants. Physiol Plant 126:343–355
James C (2007) Global status of commercialized biotech/GM crops: 2006. ISAAA Brief No. 35–2006
Jaynes JM, Yang MS, Espinoza NO, Dodds JH (1986) Plant protein improvement by genetic engineering: use of synthetic genes. Trends Biotech 4:314–320
Katsube T, Kurisaka N, Ogawa M, Maruyama N, Ohtsuka R, et al. (1999) Accumulation of soybean glycinin and its assembly with the glutelins in rice. Plant Physiol 120:1063–1073
Keeler SJ, Maloney CL, Webber PY, Patterson C, Hirata LY, et al. (1997) Expression of de novo high-lysine alpha-helical coiled-coil proteins may significantly increase the accumulated levels of lysine in mature seeds of transgenic tobacco plants. Plant Mol Biol 34:15–29
Kim JH, Cetiner S, Jaynes JM (1992) Enhancing the nutritional quality of crop plants: design, construction and expression of an artificial plant storage protein gene. In: Bhatnagar D, Cleveland TE (eds) Molecular Approaches to Improving Food Quality and Safety, AVI Book, New York, pp. 1–36
Klee HJ, Hayford MB, Kretzmer KA, Barry GF, Kishore GM (1991) Control of ethylene synthesis by expression of a bacterial enzyme in transgenic tomato plants. Plant Cell 3:1187–1193
Kochhar, SL (1981) Tropical Crops: A Textbook of Economic Botany. Macmillan Publishers, 3rd Edition. pp. 1–17
Ku MSB, Agarie S, Nomura M, Fukayama H, Tsuchida H, et al. (1999) High-level expression of maize phosphoenolpyruvate carboxylase in transgenic rice plants. Nat Biotech 17:76–80
Lai JS, Messing J (2002) Increasing maize seed methionine by mRNA stability. Plant J 30:295–402
Liu QQ (2002) Genetic engineering rice for increased lysine. PhD thesis, Yangzhou University and the Chinese University of Hong Kong, Hong Kong
Lomonossoff GL (1995) Pathogen-derived resistance to plant viruses. Annu Rev Phytopathol 33:323–343
Lucca P, Hurrell R, Potrykus I (2001) Genetic engineering approaches to improve the bioavailabilty and the level of iron in rice grains. Theor Appl Genet 102:392–397
Lucca P, Poletti S, Sautter C (2006) Genetic engineering approaches to enrich rice with iron and vitamin A. Physiol Plant 126:291–303
Malpica CA, Cervera MT, Simoens C, van Mobtagu M (1998) Engineering resistance against viral diseases in plants. Subcell Biochem 29:287–320
Marcellino LH, Neshich G, Grossi de Sa MF, Krebbers E, Gander ES (1996) Modified 2S albumins with improved tryptophan content are correctly expressed in transgenic tobacco plants. FEBS Lett 385:154–158
Maruta Y, Ueki J, Saito H, Nitta N, Imaseki H (2001) Transgenic rice with reduced glutelin content by transformation with glutelin antisense gene. Mol Breed 8:273–284
Mazur B, Krebbers E, Tingey S (1999) Gene discovery and production development for grain quality traits. Science 285:372–375
Mercenier A, Wiedermann U, Breiteneder H (2001) Edible genetically modified microorganisms and plants for improved health. Curr Opin Biotech 12:510–515
Napier JA, Haslam R, Caleron MV, Michaelson LV, Beaudoin F, et al. (2006) Progress towards the production of very long-chain polyunsaturated fatty acid in transgenic plants: plant metabolic engineering comes of age. Physiol Plant 126:398–406
O'Brien IEW, Forster RLS (1994) Disruption of virus movement confers broad-spectrum resistance against systemic infection by plant viruses with a triple gene block. Proc Natl Acad Sci USA 91:10310–10314.
Padgette SR, Re DB, Barry GF, Eichholtz DE, Delannay X, et al. (1996) New weed control opportunities: development of soybeans with a Roundup ReadyR gene. In: Duke SO (ed) Herbicide-

resistant Crops: Agricultural, Economic, Environmental, Regulatory and Technological Aspects. Lewis Publishers, pp. 53–84
Paine JA, Shipton CA, Chaggar S, Howells RM, Kennedy MJ, et al. (2005) A new version of Golden Rice with increased pro-vitamin A content. Nat Biotech 23:482–487
Pujol M, Ramirez NI, Ayala M, Gavilondo JV, Rodriguez M, et al. (2005) An integral approach towards a practical application for a plant-made monoclonal antibody in vaccine purification. Vaccine 23:1833–1837
Rebeille F, Ravanel S, Jabrin S, Douce R, Storozhenko S, et al. (2006) Folates in plants: biosynthesis, distribution, and enhancement. Physiol Plant 126:330–342
Sanford JC, Johnston SA (1985) The concept of parasite-derived resistance derived resistance genes from the parasite's own genome. J Theor Biol 113:395–405
Schuler TH, Poppy GM, Kerry BR, Denholm I (1998) Insect-resistant transgenic plants. TIBECH 16:168–175
Sheehy R, Kramer M, Hiatt W (1988) Reduction of polygalacturonase activity in tomato fruit by antisense RNA.. Proc Natl Acad Sci USA 85:8805–8809
Shintani D, DellaPenna D (1998) Elevating the vitamin E content of plants through metabolic engineering. Science 282:2098–2010
Singh J, Sharp PJ, Skerritt JH (2000) A new candidate protein for high lysine content in wheat grain. J Sci Food Agric 81:216–226
Slattery CJ, Kavakli IH, Okita TW (2000) Engineering starch for increased quantity and quality. Trends Plant Sci 5:291–298
Sparrow PAC, Irwin JA, Dale PJ, Twyman RM, Ma JKC (2007) Pharma-planta: Road testing the developing regulatory guidelines for plant-made pharmaceuticals. Transgenic Res 16:147–161
Stalker DM, McBride KE, Malyj LD (1988) Herbicide resistance in transgenic plants expressing a bacterial detoxification gene. Science 242:419–423
Sun SSM (1999) Methionine enhancement in plants. In: Singh BK (ed) Plant Amino Acids: biochemistry and biotechnology, Marcel Dekker, pp. 509–522
Sun SSM, Larkins BA (1993) Transgenic plants for improving seed storage protein. In: Kung SD, Wu R (eds) Transgenic Plants (Vol 1) Engineering and Utilization. Academic Press, New York, pp 339–371
Sun SSM, Wang ML, Tu HM, Zuo WN, Xiong LW, et al. (2000) Transgenic approach to improve crop quality. In: Lin ZP (ed) Green Genes for the 21st Century. Sci Pub., Beijing, pp. 207–219
Sun SSM, Liu QQ (2004) Transgenic approaches to improve the nutritional quality of plant proteins. In Vitro Cell Dev Biol –Plants 40:55–162
Sun SSM, Xiong LW, Jing YX, Liu BL (1993) Lysine rich protein from winged bean. US patent #5,270,200
Takaiwa F, Katsube T, Kitagawa S, Higasa T, Kito M, et al. (1995) High level accumulation of soybean glycinin in vacuole-derived protein bodies in the endosperm tissue of transgenic tobacco seed. Plant Sci 111:39–49
Theologis A, Oeller PW, Wong LM, Rpttmann WH, Gantz DM (1993) Use of a tomato mutant constructed with reverse genetics to study fruit ripening, a complex developmental process. Develop Genet 14:282–295
Toenniessen GH (2002) Crop genetic improvement for enhanced human nutrition. J Nutr 132:2943S–2946S
Tu HM, Godfrey LW, Sun SSM (1998) Expression of the Brazil nut methionine-rich protein and mutants with increased methionine in transgenic potato. Plant Mol Biol 37:829–838
Van Eenennaam AL, Lincoln K, Durrett TP, Valentin HE, Shewmaker CK, et al. (2003) Engineering vitamin E content:from Arabidopsis mutant to soy oil. Plant Cell 15:3007–3019
Waterhouse PM, Wang MB, Lough T (2001) Gene silencing as an adaptive defense against viruses. Nature 411:834–842
Whalon ME and Wingerd BA (2003) BT: mode of action and use. Arch Insect Biochem Physiol 54:200–211
Willmitzer L (1999) Plant biotechnology: output traits – the second generation of plant biotechnology products is gaining momentum. Curr Opin Biotech 10:161–162

Willmitzer L, Topfer R (1992) Manipulation of oil, starch and protein composition. Curr Opin Biotech 3:176–180

Woodard SL, Mayor JM, Bailey MR, Bailey MR, Barker DK, et al. (2003) Maize (*Zea mays*)-derived bovine trypsin: characterization of the first large-scale, commercial protein product from transgenic plants. Biotech Appl Biochem 38:123–130

Wu KM and Guo YY (2005) The evolution of cotton pest management practices in China. Annu Rev Entomol 50:31–52

Yang MS, Espinoza NO, Nagplala PG, Doods JH, White FF, et al. (1989) Expression of a synthetic gene for improved protein quality in transformed potato plants. Plant Sci 64:99–111

Ye X, Al-Babili S, Kloti A, Zhang J, Lucca P, et al. (2000) Engineering the provitamin A (beta-carotene) biosynthesis pathway into (carotene-free) rice endosperm. Science 287:303–305

Young VR, Pellett PL (1994) Plant proteins in relation to human protein and amino acid nutrition. Am J Clin Nutr 59:1203S–1212S

Zhang P, Jaynes JM, Potrykus I, Gruissem W, Puonti-Kaerlas J. (2003) Transfer and expression of an artificial storage protein (*ASP1*) gene in cassava (*Manihot esculata* Cranz). Transgenic Res 12:243–250

Zimmermann MB, Hurrell RF (2002) Improving iron, zinc and vitamin A nutrition through plant biotechnology. Curr Opin Biotech 13:142–145

Zuo WN (1993) Sulfur-rich 2S proteins in Lechydidaceae and their methionine-enriched forms of transgenic plants. PhD thesis, University of Hawaii

Chapter 4
Genomics of Banana and Plantain (*Musa* spp.), Major Staple Crops in the Tropics

Nicolas Roux, Franc-Christophe Baurens, Jaroslav Doležel, Eva Hřibová, Pat Heslop-Harrison, Chris Town, Takuji Sasaki, Takashi Matsumoto, Rita Aert, Serge Remy, Manoel Souza, and Pierre Lagoda

Abstract This chapter on *Musa* (banana and plantain) genomics covers the latest information on activities and resources developed by the Global Musa Genomics Consortium. Section 4.1 describes the morphology of the plant, its socio-economical importance and usefulness as an experimental organism. Section 4.2 describes the complexity of *Musa* taxonomy and the importance of genetic diversity. Section 4.3 details the genetic maps which have recently been developed and those that are currently being developed. Section 4.4 presents the five BAC libraries which are now publicly available from the Musa Genome Resource Centre and can be distributed in various forms under a material transfer agreement. Section 4.5 gives an overview of cytogenetics and genome organization, showing that the genus *Musa* has a quite high proportion of repetitive DNA; the discovery of the first pararetrovirus integrated in the genome makes it unique. Section 4.6 explains the first attempts to sequence the genome by BAC end sequencing, whole BAC sequencing, and reduced representation sequencing. Section 4.7 addresses functional genomics with the description of cDNA libraries, gene validation using gene trapping, mutation induction and tilling techniques, as well as genetic transformation. Section 4.8 draws overall conclusions. This chapter demonstrates that by organizing the Global Musa Genomics Consortium (currently comprising 33 member institutions from 23 countries), duplication of effort can be minimized and the results of *Musa* genomics research are rapidly made accessible to taxonomists, breeders and the biotechnology community.

4.1 Introduction

Banana and plantain (*Musa* spp.), which we will refer to as "bananas" or *Musa* in this chapter, are giant herbaceous plants, perennial but monocarpic, from two to 15 meters high. The underground stem (corm, rhizome) has short internodes. The

---

N. Roux
Bioversity International, Parc Scientifique, Agropolis II, 34397 Montpellier Cedex 5, France.
e-mail: n.roux@cgiar.org

P.H. Moore, R. Ming (eds.), *Genomics of Tropical Crop Plants*
© Springer 2008

corm's terminal growing point produces leaves in a spiral succession. Normal leaves consist of a sheath, a petiole, and a blade. The sheaths are nearly circular and tightly packed into a non-woody pseudostem. The petiole is 30–90 cm long and U-shaped in cross-section. The blade emerges from the middle of the stem as a rolled cylinder ("cigar") that then unfurls. In a healthy plant, older leaves are pushed aside until they hang down and their blades shrivel. A shoot flowers only once and dies back to ground level after it has borne fruit. Yet the plant is perennial, as the corm's life is perpetuated by suckers; a whole clump may thus develop over many years or even decades.

The inflorescence develops at a certain stage of plant development, usually after about 25–50 leaves have been produced. About a month later it emerges in the leaf crown and hangs down. It is a complex spike consisting of a stout stalk with flower clusters in a spiral. Each cluster has 12–20 flowers in two rows, covered by a large reddish bract. In the first five to 15 clusters the flowers are functionally female and give rise to fruit, while subsequent flowers are male. One by one the bracts rise to expose the flowers; they usually fall after one or two days, but in some cultivars are retained and dry on the stalk. At the lower end they form a bulbous "male bud."

The banana fruit is a berry. It contains many ovules but, in cultivated varieties, no seeds; the fruit develops by means of parthenocarpy, i.e. without fertilization. Whereas the bunch grows downwards, the fruits curve up. A fruit cluster is generally called a "hand" and a single fruit a "finger." Fingers differ from cultivar to cultivar in characteristics such as shape, size, color of skin, and flavor. A good bunch of a commercial dessert banana consists of eight hands of 15 fingers, each with an average weight of 150 g, so that the fruit weighs 18 kg and the entire bunch 20 kg. However, a bunch weight of 30 kg or more is not at all rare (Samson 1986). Traditional banana cultivars and plantains tend to have much smaller bunches.

4.1.1 Economic, Agronomic, and Societal Importance of Bananas

Banana is one of the most important, but undervalued, food crops in the world. World production of bananas reaches approximately 106 million tonnes per year (FAO 2005). Around 120 countries produce bananas throughout the tropical and sub-tropical regions with approximately one-third being produced in each of the African, Asian-Pacific, and Latin American-Caribbean regions (INIBAP 1999). Around 87% of all the bananas grown worldwide are produced by small-scale farmers for home consumption or for sale in local and regional markets, while the remaining 13%, mainly dessert bananas, are traded internationally. Bananas provide a staple food for millions of people, particularly in Africa, an area where the green revolution has had little influence. Bananas are an important food security crop, providing a cheap and easily produced source of energy. These perennial plants often survive where conflict or natural disasters have adversely affected the production of annual, arable crops. In addition, bananas are rich in certain minerals and in vitamins A, C, and B6. It has been estimated that the highest consumption rates are on the island of New Guinea and in the Great Lakes region of East Africa, where bananas form a large proportion of the diet and consumption amounts to 200–250 kg person^{-1} year^{-1} — whereas in Europe and North America consumption is approximately 15–16 kg person^{-1} year^{-1} (INIBAP 1992). Although today bananas and plantains are best known as a food crop, almost every part of the plant can be used in one way or another. This may explain why in India the banana is popularly known as "kalpatharu," meaning "herb with all imaginable uses" (Anon. 2005). Bananas provide an important source of fiber (for instance, abaca/manila hemp, derived from *Musa textilis*, is particularly important in the Philippines), and among other uses, can be fermented to produce alcohol. Bananas have also been proposed as a useful means to deliver edible vaccines: the fruit can be eaten uncooked, is sterile until peeled, and is often the first solid food eaten by babies.

4.1.2 Banana as an Experimental Organism

The *Musa* genome is relatively small, with a haploid genome of 500 to 600 Mbp (only 25% larger than rice) divided among 11 chromosomes (in most genotypes). Like rice and the other cereals, *Musa* is a monocotyledon. However, the Poales (to which the cereals belong) and Zingiberales (including *Musa*, ginger, and various ornamentals) diverged well over 100 million years ago and only a limited number of documented cases of microsynteny between the *Musa* and rice genomes have so far been reported (Aert et al. 2004). Thus the oft-recited proposition that "rice can serve as a model for other monocot genomes" should be corrected to state that "rice can serve as a model for other genomes in the monocot order Poales."

Musa offers an interesting model for genetic studies, as it is one of the few plant species with bi-parental cytoplasmic inheritance: paternal inheritance of mitochondria and maternal inheritance of chloroplasts (Carreel et al. 2002). In the center of origin and diversity of bananas in Southeast Asia there are many sterile clones that have been genomically static for thousands of years of vegetative propagation in the same environment. There are also partially fertile and highly fertile wild diploid equivalents that have been actively evolving for the same period in the same environment. Moreover, they have thus co-evolved with most of the *Musa* pathogens. This co-evolution reinforces the position of *Musa* as an ideal model to study plant and pathogen evolution at a genomic level.

The phenomena of parthenocarpy and sterility that have combined in the derivation of the typical edible banana are of interest in a wide range of other fruit crops, though banana is of special interest in that there are relatively few parthenocarpic monocotyledons. Interestingly, *Musa* was the first species where a pararetrovirus was shown to be integrated in the plant genome with the capacity to give rise to episomal banana streak badnavirus (Harper et al. 1999). Understanding the mechanism behind this phenomenon may lead to important applications, such as gene targeting (Global Musa Genomics Consortium 2002).

4.2 Genetic Diversity

Bananas belong to the family Musaceae, which consists of two genera: *Musa*, that includes some 25 species, and *Ensete*, with some seven species; some taxonomists recognize a third genus, *Musella*, with just a single species. *Musa* species have been divided into four sections: Australimusa, Callimusa, Rhodochlamys, and Eumusa (Simmonds and Shepherd 1955). We believe that this division is still the most appropriate (Taxonomic Advisory Group 2006), despite propositions for minor modification or fusions (Wong et al. 2002). The Eumusa section is the most widely geographically represented and contains the two major species *M. acuminata* and *M. balbisiana,* which are at the origin of the great majority of the edible bananas. *M. acuminata* has been divided into eight subspecies (banksii, burmannica, burmannicoides, malaccensis, microcarpa, truncata, siamea, zebrina), whereas *M. balbisiana* is less morphologically diversified. Reproduction within these wild species occurs both by seed and vegetatively.

A second group of diploid, semi-fertile pre-cultivars is considered as the ancestors of present-day edible bananas (e.g., soft seeded banana). In this group vegetative propagation is predominant. The remainder of known banana diversity resides in some 1000 distinct cultivars that have arisen by hybridization and mutation during the course of several thousand years of domestication. Cultivars may be diploid or polyploid and most are a combination of the genomes of *M. acuminata* (denoted A) and *M. balbisiana* (denoted B) with at least one A genome (i.e., AA, AB, AAA, AAB, ABB). Another distinct group of cultivated bananas is the Fehi (or Fe'i) cultivars, from Pacific islands, which have characteristic erect bunches and orange pulp. Their genomic composition remains to be fully elucidated, but they are believed to originate from the Australimusa section (including *M. textilis* and denoted by the T genome) and appear to be allied to *M. maclayi* and/or *M. lododensis. Musa schizocarpa* (S genome) also appears to be a parent of some traditional cultivars. More recently, tetraploid hybrids with different genomic constitutions have also been produced through breeding programs (e.g., AAAA: Goldfinger [FHIA-01]; AAAB: CRBP 39, and ABBT: Yawa 2 fiber banana). Reproduction within this group is strictly vegetative.

The genetic diversity of wild diploid bananas has been investigated using molecular markers since the 1980s. Different genera of the Musaceae and the representatives of the different sections within *Musa* species are well differentiated (Jarret and Litz 1986; Horry 1988, 1989; Lanaud et al. 1992; Carreel 1994; Carreel 1994; Baurens et al. 1997a, b; Pillay et al. 2000; D'Hont et al. 2000; Creste et al. 2003). Intra-species genetic diversity has been extensively studied in both *M. acuminata* and *M. balbisiana.* A strong genetic differentiation between them has been confirmed many times using isozymes, RFLPs, AFLPs, and microsatellites and recently using DArTs (unpublished results).

M. balbisiana appears to have a relatively low genetic diversity but to date this species is under-represented in collections and in the samples which have been studied. On the other hand, geographical structuring in *M. balbisiana* wild populations from south China has been reported using highly polymorphic SSR markers (Ge et al. 2005).

M. acuminata is more diverse and a global organization of the genetic diversity into four groups has been proposed (Carreel 1994). These four groups or poles of diversity include representatives of *banksii/errans, malaccensis, zebrina/microcarpa* and*burmannica/burmannicoides/siamea,* which can be represented as the four apices of a tetrahedral pyramid, other *M. acuminata* subspecies being intermediate. This global representation of genetic diversity within wild species of *M. acuminata* has been confirmed using STMS markers (Grapin et al. 1998). Today, the number of wild diploid species available in the international banana collections is far from representing the whole existing genetic diversity; meanwhile, the exchange of banana DNA material has been slowed by the perception of genetic diversity as a national asset. However, recent advances in high-throughput techniques have facilitated the development of international projects for extensive study of the global genetic diversity of the banana germplasm using microsatellites and DArTs (Wenzl et al. 2004).

Diploid and polyploid cultivars of banana have also been studied using molecular markers either to evaluate the genetic diversity within a group of cultivars or between the different groups of cultivars of defined geographical origins (Horry 1988, 1989; Jarret et al. 1992; Lanaud et al. 1992; Lebot et al. 1993; Carreel 1994; Carreel 1994; Baurens et al. 1997a, b; Pillay et al. 2000; D'Hont et al. 2000; Carreel et al. 2002; Ude et al. 2002; Creste et al. 2003; Nair et al. 2005) but also to better understand the relationship between wild species, pre-cultivars, and cultivars (Carreel 1994; Baurens et al. 1996; Grapin et al. 1998; Raboin et al. 2005). In this last study, the diploid ancestors of the most popular commercial dessert cultivars, Cavendish and Gros Michel, have been investigated using molecular markers. Based on the hypothesis of a cross between a diploid pre-cultivar giving a non-reduced 2n gamete and a haploid donor, these two very popular triploid clones could have been produced 2000 years ago (Baurens 1997) from 'Akondro Mainty' a cultivar from Madagascar, and 'Khai Naï On' from Thailand (Raboin et al. 2005).

Another facet of the genetic diversity of banana cultivars is highlighted by the study of a group of cooking bananas called plantains, which are triploid cultivars of AAB genomic composition. Genetic diversity within this group comprising more than a hundred different clones is very low (Horry 1988; Carreel 1994; Ude et al. 2002; Noyer et al. 2005). Despite some contradictory data (Ude et al. 2003; De Langhe et al. 2005), all the plantain cultivars are believed to originate from an extremely narrow genetic basis. However, this limited genetic diversity is very important in terms of agronomic characteristics (e.g. plant height, bunch size, number of hands) but also in terms of sugar and vitamin contents.

Though some interesting hypotheses involving DNA methylation are being studied to explain this apparent paradox (Noyer et al. 2005), the relationship between genotype and environment in determining phenotype is poorly understood in banana. Understanding this interaction will constitute one challenge of banana genomics, along with helping to create new, stress-resistant, and productive varieties.

4.3 Genetic Mapping in *Musa*

The idea of genetic mapping in *Musa* originated in 1990 for the banana breeding programs. The first mapping populations, studied by Fauré et al. (1993b), involved a cross between the wild diploid *M. acuminata* ssp. *banksii* and a diploid AA cultivated type, SF265. This first attempt to map the banana genome highlighted that translocation events are of major concern in *Musa* genetics studies. A second genetic map has been developed using a selfed progeny of M53, a diploid AA cooking type (Noyer et al. 1997). This was the first map of banana showing 11 linkage groups putatively corresponding to the 11 pairs of chromosomes of the banana. In this mapping population, also, the presence of translocations was suspected to be responsible for a high level of segregation distortion. Vilarinhos (2004) compared a map from a new segregating F2 population, AFCAM, with the two previous genetic maps to better understand translocation events in banana. These genetic maps are available on the Tropgene database developed by CIRAD (http://tropgenedb.cirad.fr/). This comparison confirms that, in the different crosses studied, linkage groups are grossly conserved.

These genetic mapping attempts brought to light the complexity of gamete segregation in *Musa* even at the diploid level. Some studies involving polyploid donors have been reported (Crouch et al. 1998, 1999) and show highly skewed segregation ratios.

Most of the teams that have been involved in mapping programs found it difficult to produce and maintain large populations, mainly because of the presence of translocations. The most recent attempt to build a genetic map of banana is still ongoing and involves a cross between *M. acuminata* ssp. *microcarpa* 'Borneo', and *M. acuminata* ssp. *malaccensis* 'Pisang lilin', which are thought to differ by a single translocation event. This mapping project will serve as a foundation for a core set of microsatellite markers for future use in mapping efforts in banana.

4.4 BAC Cloning and Utilization

Gene cloning and genome sequencing require a nuclear genome to be available in the form of DNA fragments that can be maintained for long periods of time and that together ideally represent the entire genome. This requirement is fulfilled by recombinant DNA libraries in which the fragments are contained within self-replicating vectors. Large DNA fragments can be propagated in yeast artificial chromosome (YAC) and bacterial artificial chromosome (BAC) vectors. The latter are usually preferred due to their relatively large insert size (typically 100 – 150 kb), stability, and ease of manipulation. The BAC vector is based on the Factor F of *Escherichia coli* and allows for strict copy number control of DNA clones so that they are stably maintained at 1–2 copies per cell (Shizuya et al. 1992).

Construction of a BAC library requires microgram amounts of high molecular weight (HMW) DNA (i.e., megabase-sized), which is accessible to restriction enzymes. Preparation of such DNA in banana is hampered by the presence of high levels of polyphenols and carbohydrates in tissue homogenates. Despite the difficulties, Vilarinhos et al. (2003) succeeded in constructing a genomic BAC library (MA4) from *M. acuminata* 'Calcutta 4' (Table 4.1). Safár et al. (2004) reported on the construction of BAC library from *M. balbisiana* (MBP), the second most frequent progenitor of cultivated banana (Table 4.1). This species exhibits an even higher content of phenolics and polysacharides than *M. acuminata*, and the authors used two strategies for preparation of HMW DNA. The first involved an improved protocol of Vilarinhos et al. (2003), while the second was based on the method of Simková et al. (2003), in which intact nuclei are purified using flow cytometric sorting. The latter strategy resulted in high quality DNA and reduced library contamination with cytoplasmic DNA.

The third published library was developed from *M. acuminata* 'Tuu Gia', a black Sigatoka-resistant clone (Ortiz-Vázquez et al. 2005). The library was cloned in a transformation-competent binary bacterial artificial chromosome (BIBAC) vector (Table 4.1), which makes the inserts suitable for *Agrobacterium*-mediated transformation. There are two other *Musa* BAC libraries, which are publicly available but until now have not been published in a peer-reviewed journal. The C4BAM library (Table 4.1) was created in 1999 from *M. acuminata* 'Calcutta 4' and complements the MA4 library by using the *Bam*HI cloning site (James et al. unpublished). The second unpublished library, referred to as MAC, was developed in CIRAD from *M. acuminata* 'Grande Naine' (Piffanelli et al. unpublished). It is the first genomic BAC library constructed from a triploid commercial cultivar of dessert banana (Table 4.1).

A large-scale sequencing project on banana is currently being planned but so far the BAC libraries have not been used to develop clone-based physical maps, nor are we aware of any ongoing positional gene cloning project. Despite that, the availability of BAC libraries facilitated the analysis of the *Musa* genome at a level not attainable before. The breakthrough involved a complete sequencing of two BAC clones (82 kb and 73 kb long) randomly selected from the C4BAM library (Aert et al. 2004). This work provided the first insights into the genome organization and revealed the average G+C content of 38–39% and gene density of one per 6.9 and 10.5 kb. Further insights into the *Musa* genome structure were obtained after sequencing a higher number of BAC clones and ends of BAC clones selected from the MA4 library (sections 4.6.1 and 4.6.2).

Another attractive use of BAC libraries includes their screening for the presence of clones carrying conserved domains of resistance gene analogues (James et al. 2006; Miller et al. 2006). A large-scale project utilizing the conserved gene sequences and synteny between *Musa* and rice to characterize genome structure and to identify stress-related genomic regions and exploitable genes is under way within the Generation Challenge Programme (http://www.generationcp.org). Finally, selected BAC clones were located to mitotic chromosomes by fluorescence *in situ* hybridization (FISH) with the aim of linking genetic and physical maps and identifying chromosome translocations (Vilarinhos et al. 2006).

The average insert size and the number of clones in *Musa* BAC libraries indicate that the technical difficulties with their construction can be overcome and high

Table 4.1 Publicly available BAC libraries

Musa accession		BAC library*							Reference
Species	Clone	Name	Vector	Cloning site	Average insert size(kb)	Genome coverage	Number of clones	Clones with cytoplasmic DNA	
M. acuminata	Calcutta 4	MA4	pIndigoBAC-5	*Hind*III	100	9x	55,152	1.5 %	Vilarinhos et al. (2003)
M. balbisiana	Pisang Klutuk Wulung	MBP	pIndigoBAC-5	HindIII	135	9x	36,864	3.3%	Safár et al. (2004)
M. acuminata	Tuu Gia	TGBIBAC	pCLD04541	*Hind*III	100	5.1x	30,700	1.9%	Ortiz-Vázquez et al. (2005)
M. acuminata	Calcutta 4	C4BAM	pECBAC-1	*Bam*HI	110	3x	17,280	?	James et al. (unpublished)
M. acuminata	Grande Naine	MAC	pIndigoBAC-5	*Hind*III	145	4.5x	55,296	?	Piffanelli et al. (unpublished)

*) The quality of BAC libraries is typically defined by the average insert size, number of clones, and presence of cytoplasmic DNA. Average insert size and the number of clones then determines the genome coverage and probability of recovering any DNA sequence in the library (Clarke and Carbon 1976).

quality BAC libraries can be produced. In fact, it is highly probable that more BAC libraries will be needed in the near future. All currently available libraries were made using only one cloning site (Table 4.1) and thus some genome regions may be under-represented. Until recently, *M. acuminata* 'Calcutta 4' was considered a model for genomics of *Musa*. However, this clone is heterozygous, a feature that would complicate the assembly of genomic sequences during a sequencing project. Therefore different genotypes are being considered and the best solution seems to be to sequence a dihaploid individual of *M. acuminata* 'Pahang'. Consequently, an ordered genomic BAC library will have to be constructed from this genotype.

4.5 Cytogenetics and Genome Organization

4.5.1 Molecular Cytogenetics

A karyotype, which describes a chromosome complement in terms of number, size, and form of chromosomes, is one of the basic characteristics of any species. In addition to plant phenotype, chromosome number was a criterion used by Cheesman (1947) to suggest the division of genus *Musa* into four sections: Eumusa ($x = 11$), Rhodochlamys ($x = 11$), Callimusa ($x = 10$) and Australimusa ($x = 10$). Although generally followed, this division has been questioned based on recent molecular data and basic chromosome numbers in some other species, such as *M. beccarii* with $2n = 2x = 18$ (Bartoš et al. 2005).

Chromosome number and size can be established reliably only by observing chromosomes in mitosis or meiosis. A need for dividing cells at a specific cell cycle phase limits the tissues suitable for the analysis. As a response to this, flow cytometric estimation of nuclear DNA content was introduced (Dolezel et al. 1994). Flow cytometric ploidy screening was found useful in classifying *Musa* germplasm (Horry et al. 1998, Dolezelová et al. 2005), to follow ploidy levels during breeding programs, during culture*in vitro* (Roux et al. 2001), and in experiments aiming at creating novel ploidy levels (van Duren et al. 1996). Flow cytometry was sensitive enough to detect aneuploidy (Roux et al. 2003) and was also used to determine genome size in various *Musa* species, which ranged from approximately 530 Mbp/1C to 798 Mbp/1C (Lysák et al. 1999, Bartoš et al. 2005).

Since the early days of *Musa* cytogenetics, researchers and breeders have struggled to identify individual chromosomes. Due to similarity in size and form of *Musa* chromosomes, these attempts achieved only limited success. The only option was to analyze chromosome behavior during meiosis. This analysis provided useful data on the presence of translocations and inversions and the level of structural heterozygosity in individual clones and hybrids (Wilson 1946a,b,c; Fauré et al. 1993a). Shepherd (1999) provides examples of such analyses.

The ability of molecular cytogenetic techniques to localize DNA sequences on chromosomes opened new avenues for analyzing genome structure in *Musa*. Early reports of Dolezelová et al. ((1998)) and Osuji et al. (1998) employed FISH to study the genomic arrangement of ribosomal rDNA loci. Both studies revealed only one pair of loci in a diploid chromosome set both in *M. acuminata* and *M. balbisiana*. The number of 5S rDNA loci varied from 4 to 8 in diploids. These results imply that, in both species, a probe for 45S rDNA can be used to identify a particular chromosome, while a probe for 5S rDNA identifies only a group of chromosomes. The analysis of rDNA loci was extended by Bartoš et al. (2005) to the representatives of other sections of the *Musa* genus and provided a more complex picture.

FISH with probes for 45S rDNA highlights the nucleolus organizing region (NOR) and allows a chromosome to be linked with its satellite, which is often spatially separated on metaphase spreads (Dolezel et al. unpublished). Traditional chromosome staining protocols do not visualize DNA of the extended NOR and the separated satellite can be mistaken for an extra ('mini') chromosome. In fact, there have been several reports describing the occurrence of mini chromosomes in *Musa* (Shepherd 1996a, b). A conclusion that at least some reported mini chromosomes were actually spatially separated satellites was confirmed by D'Hont et al. (2000).

In addition to rDNA, a few other DNA sequences were mapped to *Musa* chromosomes using FISH including telomeric repeats (Osuji et al. 1998), a gypsy-like retrotransposon *monkey* (Balint-Kurti et al. 2000), and a set of repetitive DNA sequences called *Radka* (Valárik et al. 2002). The ability of FISH to localize DNA sequences was utilized by Harper et al. (1999) to demonstrate that banana streak badnavirus (BSV) was integrated in the *Musa* nuclear genome.

Although these studies provided first insights into the molecular organization of *Musa* chromosomes, they did not provide the additional markers for FISH needed to identify specific chromosomes. Inspired by the successful use of BAC clones in other crops (Harper and Cande 2000), BACs were tested as probes for FISH in *Musa*. Unfortunately, only a very few BAC clones could be localized to specific loci due to the presence of dispersed repeats or the inability to detect sites of BAC DNA hybridization. Nevertheless, Vilarinhos et al. (2006) reported successful FISH mapping of four marker-tagged BAC clones selected from a genetic linkage group II.

Genomic *in situ* hybridization (GISH) is a variant of FISH in which genomic DNA is used as a probe. GISH allows identification of parental chromosomes and their recombination products in a hybrid. Osuji et al. (1997) used GISH to assess genomic constitution of various hybrids between *M. acuminata* (A genome) and *M. balbisiana* (B genome). D'Hont et al. (2000) were able to identify not only A and B genomes in hybrid *Musa* clones using GISH, but also the chromosomes of the S genome of *M. schizocarpa* and the T genome of the section Australimusa. This work revealed the presence of the S and T genomes in some banana hybrids.

4.5.2 Repetitive Part of the Genome

Although the nuclear genome of *Musa* is relatively small, a significant part consists of repetitive DNA (Table 4.2). The methods used to reveal the molecular organization of this genome component included screening of DNA libraries with genomic

Table 4.2 Sequences type in the *Musa* genome

Sequence Type	Fraction of Genome
Non-transposon repeats	23%
Transposons	13%
Coding regions	11%
Other regions (introns and non-repetitive intergenic)	53%

DNA to identify clones with abundant sequences (Valárik et al. 2002), and with probes homologous to different parts of retrotransposons (Baurens et al. 1998; Teo et al. 2002). Cot-analysis (see section 4.6.3) has been used to separate highly and moderately repeated parts of the genome (Hřibová et al. 2006). With the falling sequencing costs, *in silico* analysis of sequences from complete BAC clones or their ends (sections 4.6.1 and 4.6.2) becomes an important tool to study the repetitive landscape of the genome.

Probably the first repetitive DNA sequence in *Musa* was characterized by Baurens et al. (1997a), who isolated a fragment of a *Copia*-like repetitive element from *M. acuminata* ssp. *banksii*. They showed that the element was homologous to the *pol* gene of other *Copia*-like retrotransposons. The same authors characterized a repetitive DNA family called Brep 1, which was not homologous to known DNA sequences. The family was found to be dispersed in the genomes of the Musaceae with various copy numbers. Balint-Kurti et al. (2000) identified a Ty3/gypsy-like retrotransposon, *monkey,* in a genomic DNA library of a triploid clone Grande Naine (AAA). *Monkey* was found distributed throughout the genome with preferential clustering in the NOR region. The authors estimated that *monkey* represented about 1.2–3Mbp or 0.2 – 0.5% of the nuclear genome (1Cx). Valárik et al. (2002) screened partial genomic DNA libraries from *M. acuminata* and *M. balbisiana* with the aim of isolating and characterizing the most abundant repetitive DNA sequences in *Musa*. From a set of 22 repetitive clones collectively termed Radka, they selected 12 repetitive clones for detailed analysis that included determination of genomic distribution by FISH. In general the study indicated similar genomic organization of repetitive DNA in *M. acuminata* and *M. balbisiana*. Higher copy numbers observed in *M. acuminata* were in line with the larger genome size of this species.

To provide a more complete picture of the repetitive part of the *Musa* genome, Hřibová et al. (2006) used Cot analysis to isolate a highly repetitive part of the genome of *M. acuminata* 'Calcutta 4'. DNA fragments sequences thus obtained were used to construct a DNA library from which 614 clones with insert sizes ranging from 300 to 900bp were sequenced. Of these, 48% of the clones represented novel and until now undescribed sequences. The remaining clones carried sequences homologous with DNA sequences deposited in GenBank. Despite the attempts to remove all clones hybridizing with probes for Radka sequences prior to sequencing, 9.5% of sequenced clones showed homology to rDNA and 3.7% of clones were homologous to other Radka repetitive sequences. A majority (24%) of sequenced clones showed homology to various types of retrotransposons, the Ty3/gypsy type *monkey* retrotransposon (Balint-Kurti et al. 2000) being the most frequent among them (16%). Dot-plot analysis revealed that 87 (14%) of the sequenced clones contained various (semi)-tandem and palindrome repeated sequences. Nevertheless, only a few perfect tandem repeats were isolated, indicating a low abundance of this class of repeats in the banana genome.

4.5.3 Retro-elements and BSV

DNA sequences without a clear function for the host represent the majority of genomic DNA in most eukaryotic species including banana: less than 10% of the nuclear DNA is made up of gene-coding sequences. DNA belonging to the class of retroelements makes up a major proportion of this non-coding DNA fraction. These elements, typically 5 to 10 kb long, amplify in the genome through an RNA intermediate that is reverse transcribed into DNA and reinserted into the nuclear genome. Retroelements encode reverse transcriptase and other enzymes related to their excision and reinsertion, and the RNA transcripts act as templates for translation to proteins and for reverse transcription; however, they have no clear function for their host genome. The replicative mode of amplification and reinsertion means that retroelements can increase to very high copy numbers in plant genomes, and reinsertion means that retroelements are widely dispersed through the genome. In *Musa*, as in other species, a major group of retroelements is characterized by long terminal repeats (LTRs), which flank the core coding domains, and have been found from both BAC sequencing and analysis of the repetitive DNA in the genome. The LTR retrotransposons are divided into two classes, the Ty1-copia-like Pararetroviridae family which was reported by Baurens et al. (1997a), and the Ty3-gypsy-like elements in the Metaviridae family (Balint-Kurti et al. 2000). More recently, additional variants of both families have been analyzed by Valárik et al. (2002) and Teo et al. (2002). In most elements, stop codons have been found indicating that the elements sequenced are not translated, but it is presumed that other elements are intact and may reverse-transcribe degenerated elements. Many of the sequenced BACs from *Musa* (www.*Musa*genomics.org and the Genbank/EMBL databases) include fragments of retrotransposons, often annotated by the names of their protein components including the polyprotein (pol) and reverse transcriptase (RT) domains. One of the first *Musa* BACs analyzed in detail, MuG9, was shown to include multiple fragments of retroelements, including characteristic flanking regions (Aert et al. 2004), and a third of recognizable sequence motifs were clearly related to retroelements.

An important group of viruses, the badnaviruses including the Banana Streak Virus, BSV, are related by sequence to the retroelements. Harper et al. (1999) showed that the BSV-related sequences are integrated within the nuclear genome, and since then it has become clear that the elements can be expressed and give rise to a viral infection, although integration is not an essential part of the viral life cycle

(Hull 2002). Hull et al. (2000) and others have speculated that the presence of integrated copies may be related to virus resistance through induction of transcriptional or post-transcriptional gene silencing of homologous sequences.

The abundance and wide genomic distribution of retrotransposons, and their fast evolution rate, make them valuable DNA markers for biodiversity and phylogenetic studies. When insertional polymorphisms are found between varieties, the presence or absence of an element at a particular genomic site can be used for variety identification or for grouping varieties with a common ancestor which includes that particular insertion. More universally, polymorphisms in retroelements can be detected by PCR using primers facing outwards from their distal sequences (often a characteristic LTR), the interretroelement amplified polymorphism or IRAP method. Because of the high copy number of the conserved retroelement sequences, and their presence within the distance that can be amplified by PCR, a single PCR amplification will generate 5 to 20 fragments between different elements from a particular *Musa* genomic DNA. The presence and absence of these fragments can then be scored and used as markers to evaluate the diversity and relationships between different accessions (Teo et al. 2002; Nair et al. 2005).

4.6 Genome Sequencing

4.6.1 BAC End Sequencing

Sequencing the ends of bacterial artificial chromosomes (BAC end sequencing) can provide significant clues about the species' genomics. Although the sequences sampled are not truly random, due to the requirement for a specific restriction enzyme site for library construction, the sequences nevertheless provide a first-pass estimate of the composition of the genome. In addition, BAC end sequences can be mined for simple sequence repeats (SSRs) and other genetic markers that can be used to anchor both the BACs, and the finger-print contigs (FPC) in which they are embedded (if available) to the genetic map. Furthermore, end-sequenced BACs play an important role in scaffolding sequence assemblies in a whole genome shotgun (WGS) sequencing project.

Sequencing of 3,456 clones from the 'Calcutta 4' *Hin*dIII library MA4 generated 6,376 reads (GenBank Accessions DX451990-DX45835050, Cheung and Town 2007). After cleaning and filtering, 6,252 high quality reads representing 4,420,944 bp, 2,979 clone pairs with an average length of 707 bp, were available for further analysis. Searching these reads against several databases showed significant homology to mitochondria and chloroplast (10%), transposons (13%), repetitive sequences (23%), proteins (11%), and approximately 2,000 *Musa* ESTs (0.02%). Based on these results, the genome is projected to be composed of regions and elements as shown in Table 4.2.

A total of 352 potential SSR markers have been uncovered. The most abundant simple sequence repeats in four size categories were AT-rich. The BAC end sequences can also be used to investigate possible microsynteny between *Musa* and other genomes such as rice (*Oryza sativa*). Amongst the BAC-end-sequences, 2,646 had a significant BLAST match to *O. sativa* genome sequences; of these, 593 had both members of the pair matching and after filtering the mitochondria and chloroplast matches, 55 matched the same chromosome. Of these, a total of 10 were shown to span the *O. sativa* genome with a separation of between 9 and 500 kb and might thus represent examples of microsynteny, whereas the 41 that exceeded this separation may represent translocations.

4.6.2 Whole BAC Sequencing

As sequencing technology becomes increasingly innovative, acquisition of massive nucleotide sequences from any organism can be accomplished in a much more efficient and accelerated way. The genome sequence information is not only useful for gene isolation and subsequent characterization but it is also indispensable for developing novel crop breeding strategies. So far, genome-wide DNA sequence information is not yet available for *Musa*. However, BAC libraries of several *Musa* cultivars have been constructed for genomic characterization of specific regions of the genome (refer to section 4). The BAC sequences could be used to characterize the complete structure of the target gene and its *cis* elements such as the sequences responsible for transcription regulation. Additional information such as the existence of other related genes, simple sequence repeats and transposable elements could also be clarified. Furthermore, comparative bioinformatic analysis of the *Musa* BAC sequences and corresponding regions of other crops with available sequence, such as rice, may reveal the genome organization of the target region in detail that will provide insights on the evolution of monocotyledons.

Within a project of the Generation Challenge Programme entitled: "Musa genome frame-map construction and connection with the rice sequence," *Musa* BAC clones with target genes or marker probes were subjected to shotgun-sequencing. For each BAC clone, a total of 4000 sequences from both ends of subclones were analyzed. The short sequences comprising 500–600 bases were assembled with a Phred-Phrap assembler (Ewing and Green 1998) to generate the original BAC sequence as sequence contigs. Usually, the total amount of nucleotide sequences corresponds to 10–15 times the analyzed BAC clone. The sequence gaps were resolved by full-sequencing of subclones to bridge the gaps between two contigs.

As an initial attempt to sequence the *Musa* genome, seven BAC clones were sequenced generating 626,864 bps nucleotide sequences. These sequences were then analyzed using the Rice Genome Automated Annotation System (RiceGAAS, http://ricegaas.dna.affrc.go.jp/), a fully automated annotation system that was originally designed for rice genome sequences. RiceGAAS integrates the results of several gene prediction programs such as GENSCAN and FGENESH with blast analysis against rice ESTs and proteins to reveal the most plausible protein coding regions on the sequence. The functions of the predicted gene models are then assigned using the GFSelector program in RiceGAAS. In the near future, RiceGAAS will be tuned-up for *Musa* genome annotation by incorporating *Musa* ESTs and protein information to facilitate a more accurate analysis. From the 626,864 bps nucleotide sequence accumulated so far, 157 gene models have been predicted. These include many hypothetical or unknown proteins, but also include genes with important biological functions such as MAPK/ERK-kinase, Myb transcription factor, and ATP-binding cassette. The average gene density of these seven BACs is about 4 kb per gene. The previously reported density was 8.7 kb per gene (Aert et al. 2004). A similar study has been conducted recently by Lescot et al. 2007.

4.6.3 Reduced Representation Sequencing

Gene enrichment strategies for plant genomes are invaluable for complementing whole-genome sequencing, as the latter is technically difficult and costly. Currently, three major reduced representation techniques are being utilized in plant studies. The first one is expressed sequence tag (EST) sequencing (Venter 1993) where large numbers of end sequences of cDNA clones are generated. ESTs can rapidly identify coding regions, but are limited by expression bias.

A second strategy is methylation filtration (Rabinowicz et al. 1999) where genomic libraries enriched in hypo-methylated, presumably gene-coding, sequences are created. However, as methylation can change drastically during major developmental transitions or in response to abiotic stresses, genes involved in development and stress responses may be lost with the methylation filtration strategy.

A third way to selectively isolate and study important or interesting sequences is Cot-based cloning and sequencing (CBCS) (Peterson et al. 2002). This technique is rooted in the principles of DNA renaturation kinetics and allows separation of DNA fractions into low-copy or high-copy sequences. The most repetitive DNA renatures first and the double-stranded DNA can be separated from the lower copy number, unrenatured DNA. Cot fractionation is completely independent of gene expression and methylation patterns. The efficacy of this technique has been demonstrated in sorghum (Peterson et al. 2002), maize (Yuan et al. 2003; Whitelaw et al. 2003) and wheat (Lamoureux et al. 2005).

We studied the ability of Cot-filtration to enrich genes and filter out repetitive DNA in *M. acuminata* 'Calcutta 4'. After determining the Cot cloning value, a Cot100 library was constructed. Simultaneously, an unfiltered library was created. Data analysis showed that Cot-filtration at a Cot100 value resulted in a 1.9-fold gene-enrichment, a 1.75-fold enrichment in unknown low-copy sequences, and a 2.5-fold reduction in repetitive DNA. Comparison of these data with results obtained in three other plant genomes showed that the efficiency of Cot-filtration in banana is comparable with that observed in other species (Table 4.3). These results suggest that Cot-filtration is a highly useful tool to enable sequencing of the fraction of banana DNA that contains genes.

Table 4.3 Categories and frequencies (%) of Cot filtration sequences in four different plant genomes

	Banana (Cot100)	Maize (Cot466)	Wheat (Cot1600)	Sorghum (Cot10000)
Repetitive DNA	22	31.8	31.3	22
Genes	23	23	34.3	25.6
Non-hits	52.9	44.3	34.4	67.9

4.7 Functional Genomics and Gene Validation

4.7.1 cDNA Libraries and ESTs

The transcriptome, defined as the complete set of RNA transcripts produced by the genome at any one time, is dynamic and changes under different temporal and/or spatial circumstances due to different patterns of gene expression. Transcriptomics is the name given to the study of the transcriptome of an organism.

The production of a database of ESTs is the initial step in the transcriptomics of a given organism. An EST is a unique DNA sequence derived from a cDNA library (therefore from a sequence which has been transcribed in some tissue or at some stage of development), and can be derived from a transcribed protein-coding or non-protein-coding nucleotide sequence. It was originally intended as a way to identify gene transcripts, but has since been instrumental in gene discovery and sequence determination. Some authors use the term "EST" to describe genes for which no further information exists besides the tag.

The identification of ESTs has proceeded rapidly, with approximately 40 million ESTs now available in public databases (e.g. GenBank 10/2006). An EST is produced by one-shot sequencing of a cloned mRNA (i.e. sequencing several hundred base pairs from one or both ends of cDNA clones taken from a cDNA library). The resulting sequence is a relatively low quality fragment whose length is limited by current technology to approximately 500 to 800 nucleotides. Because these clones consist of DNA that is complementary to mRNA, the ESTs represent expressed portions of genes. Often, they are expressed only in certain tissues at certain points in time; however, some of them can be constitutively expressed.

Once identified, an EST can be mapped, by a combination of genetic mapping procedures, to a unique locus in the genome and serves to identify and characterize that gene locus. ESTs are also useful for designing probes for DNA microarrays that can be used to study the pattern of gene expression in one specific scenario of interest, such as during biotic or abiotic stress. By identifying the genes up- or down-regulated during that specific moment, and further characterizing them, one can then better understand that specific phenomenon, and consequently improve the chances of modifying it.

The production of a database of banana ESTs is one of the strategies used by the GMGC for the characterization of the genome of this socially and economically important fruit crop. The first known study on banana transcriptome was done before the creation of the GMGC, by Syngenta, which produced and characterized two cDNA libraries of the Cavendish variety Grand Naine (AAA) (Table 4.4). These two cDNA libraries are available to members of the GMGC through signing a material transfer agreement.

Another transcriptomics study of banana was a collaborative one done by EMBRAPA, the Catholic University of Brasilia (UCB), and CIRAD, with the support of the National Council for Scientific and Technological Development (CNPq) in Brazil. Information on this work can be found at http://genoma.embrapa.br/*Musa*/en/index.html, as well as in Santos et al. (2005) and Souza Jr et al. (2005). This common effort permitted the production and characterization of seven cDNA libraries (Table 4.4), all from *M. acuminata* (genome A). EMBRAPA also produced and characterized, with the support of the Generation Challenge Programme, two cDNA libraries of the Pisang Klutuk Wulung (PKW) variety, an *M. balbisiana*

Table 4.4 Banana cDNA libraries which EST sequences are available through the Global *Musa* Genomics Consortium. Availability for some of these ESTs is restricted to members of the GMGC and requires signature of MTA

Genus	Species	Vernacular name	Library ID	Description	Constructor
Musa	*acuminata*	Grande Naine	ESTSYN-F	mRNA obtained from a mixed sample of fruit tissues: (a) Early fruit emergence, development and cell expansion (peel and pulp), 0 weeks after shooting and 2 weeks after shooting; (b) Starch accumulation (peel and pulp), 5 weeks after shooting; (c) Fully mature pre-climacteric green fruit (pulp); (d) Early climacteric mature fruit 24 hours post ethylene exposure (pulp); (e) Climacteric mature fruit 65 hours post ethylene exposure (pulp).	Syngenta
Musa	*acuminata*	Grande Naine	ESTSYN-L	mRNA obtained from a mixed sample of leaves from a plant grown under greenhouse conditions: (a) Immature 'cigar' leaf; (b) First fully expanded leaf; (c) Mature fully expanded leaf.	Syngenta
Musa	*acuminata*	Grande Naine	MACVLINFLS	mRNA obtained from a mixed samples of leaf tissues collected between 25 and 33 days after *in vitro* infection with *Mycosphaerella fijiensis*.	Embrapa & CIRAD
Musa	*acuminata*	Calcutta 4	MA4LINFES	mRNA obtained from a mixed samples of leaf tissues collected between 3 and 10 days after *in vitro* infection with *Mycosphaerella fijiensis*.	Embrapa & CIRAD

Table 4.4 continued

Genus	Species	Vernacular name	Library ID	Description	Constructor
Musa	*acuminata*	Calcutta 4	MUC4LHot	mRNA obtained from a mixed sample of leaves collected at nine different moments at three different temperatures (25 °C, 35 °C, and 45 °C) – See Santos et al. (2005).	Embrapa
Musa	*acuminata*	Calcutta 4	MUC4LCold	mRNA obtained from a mixed sample of leaves collected at nine different moments at three different temperatures (25 °C, 15 °C, and 5 °C) – See Santos et al. (2005).	Embrapa
Musa	*acuminata*	Calcutta 4	MUC4Root	mRNA obtained from roots collected from *in vitro* plants of 10 cm high.	Embrapa
Musa	*acuminata*	Calcutta 4	MUC4Flower	mRNA obtained from male flowers collected from plants grown under field conditions.	Embrapa
Musa	*acuminata*	Calcutta 4	MUC4Peel	mRNA obtained from peel removed from green fruits collected from plants grown under field conditions.	Embrapa
Musa	*balbisiana*	PKW	MbPKW_Leaf	mRNA obtained from leaves collected from hydroponics' plants of 30 cm high.	Embrapa
Musa	*balbisiana*	PKW	MbPKW_Root	mRNA obtained from roots collected from hydroponics' plants of 30 cm high.	Embrapa

diploid banana (genome B) (Table 4.4). So far, considering these nine cDNA libraries developed and characterized in this Brazilian-French initiative, and with sequencing support from the National Institute of Agrobiological Sciences (NIAS) in Japan, about 45,000 reads have been generated. These reads are being transferred to GMGC and will be made available for analysis by the members of the consortium in the near future.

4.7.2 Gene Trapping

In contrast to model plant species like Arabidopsis and rice that have entered the post-sequencing era, genomics in banana is still in its infancy, even though significant progress has been made in the last few years. While bioinformatic tools integrated with global gene expression analyses allow modeling of gene function, experimental confirmation by mutation and phenotypic analysis are still required for most genes. Classical knock-out mutagenesis is not suitable for functionally redundant, essential, or pleiotropic genes. In addition, low abundant transcripts or genes that are expressed in a relatively small number of cells are unlikely to be identified in gene expression screens. To overcome these limitations, several 'gene trapping' systems have been developed (reviewed by Springer 2000).

To discover and functionally characterize novel genes and promoters, several promoter trap vectors have been constructed and introduced into banana via a high throughput *Agrobacterium* transformation system (see section 4.7.4). In this type of gene trap, a promoterless reporter gene is linked to a T-DNA border and, after integration is activated, by flanking promoters. Despite the use of the sensitive luciferase (luc) reporter gene system and the screening of tens of thousands of transgenic cell colonies, the activation frequency (the number of colonies showing luciferase activity expressed as a percentage of the total number of colonies screened) initially did not exceed 0.1%. By improving the tagging vector with the codon-optimized luciferase (luc+) gene reporter gene placed near the right T-DNA border, luciferase expression significantly increased and the activation frequency grew to 2.5% (Remy et al. 2005). This technique has also recently been applied to tag developmentally regulated promoters (Santos et al. unpublished results). In contrast to these knock-out mutations that frequently cause non-visible phenotype activation, tagging creates semi-dominant gain-of-function mutations by enhancer-driven transcriptional activation of nearby genes that can even be functionally redundant or essential for survival. Moreover, by its design, this trapping system allows simple selection for the activated gene(s), which makes it an attractive forward genomic tool for the discovery of trait-specific genes even in non-model plants like *Catharanthus roseus* (van der Frits et al. 2001) and *Petunia hybrida* (Zubko et al. 2002). Based on these results and the fact that the cauliflower mosaic virus 35S enhancer is effective in banana (Sági et al. 1995) several activation tagging constructs were introduced into banana (Remy et al. unpublished results).

Application of gene trapping for large-scale reverse genetics in banana might be premature due to the lack of the genome sequence. Considering an average genome size of 600 Mb (Lysák et al. 1999) and assuming an average gene length of 3 kb, approximately 300,000 T-DNA inserts would be required to reach a 95% probability of finding an insertion in any given banana gene. This number of T-DNA tagged lines required to saturate every banana gene is likely an overestimation because the average number of T-DNA inserts is greater than one and T-DNA insertions occur preferentially in gene-rich regions (Barakat et al. 2000). As tens of thousands of tagged cell colonies can be produced within two to three months (Remy et al. 2005), reverse genetic studies might be feasible in banana provided that tagged lines can be maintained at an *in vitro* stage in this vegetatively propagated crop.

4.7.3 Mutagenesis, TILLING

Mutation induction has played a major role in the development of superior crop varieties. This translates into a tremendous economic impact on agriculture and food production that is currently valued in billions of dollars and millions of cultivated hectares (Ahloowalia et al. 2004). Worldwide, more than 2500 new crop varieties were officially released to farmers (http://www-mvd.iaea.org/). The prime strategy in mutation-based breeding is to upgrade well-adapted, local plant varieties and landraces by altering one or two major traits that currently limit their productivity or reduce their quality value. Radiation-based mutation induction has been the method most frequently used (89%). Interestingly, more than 60% of mutant varieties were released after 1985in the era of transgenics in plant breeding. Thus physical mutation induction presents itself as a cheap, safe, and clean alternative. In fact, one might state that there is no difference between artificially produced induced mutants and spontaneous mutants found in nature. Compared to cross-breeding, special care is taken in selecting homozygous, non-chimerical mutant lines. To achieve this, induced mutants are passed through several generations of selfing (in order to achieve homozygosity); or clonal propagation, usually through *in vitro* techniques (in order to dissociate chimeras). These steps mimic what happens in nature (through evolution) and leads to fixation of the mutation events. Thus, all what plant breeders do when using mutation induction is to emulate nature, to broaden genetic variation in the breeding germplasm in a reduced timeframe, compatible with the breeder's life expectancy.

In vegetatively propagated crops, where genetic variation may be limited due to lack of sexual recombination, and especially in edible *Musa* (banana and plantain), due to sterility and polyploidy, mutation induction is a tool of choice to be promoted. This statement does not diminish the value of selection of spontaneous mutations, which have played an essential role in the speciation and domestication of banana (Buddenhagen 1987). Neither does it constitute a negative appraisal of ongoing breeding programs based on mass selection, residual fertility of polyploid cultivars, or reconstruction of polyploids from diploid cultivars (Persley and DeLanghe 1987). Mutation induction allows breeders to escape the deadlock of sterility and parthenocarpy by creating useful variants.

Three mutant banana varieties are worth mentioning here: 'Novaria' in Malaysia (early flowering) (Mak et al. 1996), 'Klue Hom Thong' KU1 in Thailand (bunch size and cylindrical shape) (Anon. 1990), and 'Albeely' in Sudan (higher yield) (Anon. 2006). More mutants are in the pipeline for varietal release in Costa Rica, Cuba, Malaysia, The Philippines, and Sri Lanka. The targeted traits are improved agronomical characters such as reduced height, earliness, and larger fruit size. Although disease resistance seems to be more difficult to achieve through mutation induction techniques, tolerance to the juglone toxin from *Mycosphaerella fijiensis* and tolerance to *Fusarium oxysporum* f.sp. *cubense* were obtained, and mutants are about to be field screened (Roux 2004). One of the limiting factors of mutation induction for banana is the random nature of the mutations, which implies extensive screening to identify the useful mutants. However, field screening of thousands of bananas is laborious, expensive, and given the cycle of the banana plant and the size of plantations, site specific. Thus a phased strategy of *in vitro* pre-screening, greenhouse confirmation screening and field testing ought to be devised to enhance efficiency (Roux et al. 2004). Another obstacle lies in the high degree of chimerism produced through the traditional shoot tip mutation induction techniques. Recent advances in improving cell suspension procedures and strategies, reducing the formation of chimeric plantlets, allow us to enhance the formation of more stable and useful mutant variants (Roux et al. 2004).

In recent years there has been increased interest in understanding the genome of all organisms. This goes in parallel with the explosion of fundamental and strategic research to understand gene structure and function, especially in crop and model plants. Banana is a good candidate to become a model for vegetatively propagated plants not only due to its small genome size (Global Musa Genomics Consortium 2002). On the other hand, mutation induction seems to be one of the most efficient and cost effective tools for functional genomics projects dealing with both direct and reverse genetics strategies: to obtain the full range of phenotypes we need to increase both the breadth and depth of the mutant resources, i.e. we need to mine new loci and we need to mine new alleles in known loci to close the so called phenotype gap, "the gulf between the available mutant resource and the full range of phenotypes that is essential to exploit fully investigated species" (Brown and Peters 1996). The time seems ripe now, because in recent years targeting induced local lesions in genomes (TILLING) has established itself as a powerful tool for functional genomics and reverse genetics (Henikoff et al. 2004), to move on and to exploit the genomics resources banana has to offer as a tropical crop. This is a high throughput methodology, the conjunction of a high throughput molecular reverse genetics technique and mutation induction, allowing for the mining of new alleles in a known locus, and can be performed at a very early stage. Negative results on the same phenotype would hint at new genes to characterize. These mutants could then be subjected to expression screens in order to identify these unknown genes. Consequently, at the IAEA Biotechnology Laboratories Plant Breeding Unit at Seibersdorf (Austria), a mutation grid and TILLING platform for banana is being planned, based on *M. acuminata* 'Calcutta 4'. Given the recent advances at identifying potential diploid ancestors of the triploid Cavendish and Gros Michel cultivars (Raboin et al. 2005), TILLING, might open up new ways of breeding bananas. This is all the more promising because the potential of TILLING, already established as a successful functional genomics discovery platform in model organisms, has recently been verified as a proof of concept of a technology for crop improvement in allohexaploid wheat (Slade et al. 2005).

4.7.4 Genetic Transformation

Before genomics became part of the banana research agenda, a paper on protoplast electroporation (Sági et al. 1994) marked the start of the effort to generate a range of efficient-transformation protocols in banana. The first of these methods avoids tedious protoplast manipulations by particle-bombardment of embryogenic cell suspensions (ECS) directly and is combined with efficient antibiotic selection and regeneration of transformed cells (Sági et al. 1995; Becker et al. 2000). An average of 10 to 20 independent transgenic plants can usually be obtained from 50 mg of cells, and efficient delivery of multiple genes has been accomplished (Remy et al. 1998a). However, the method requires the time-consuming task of establishing highly regenerable ECS (Côte et al. 1996; Strosse et al. 2006). This bottleneck was circumvented in a second method developed by May et al. (1995) with particle bombardment of meristems to induce a wound response and via infection with *Agrobacterium tumefaciens*. Whereas transgenic plants can be generated in a few months, this method suffers from a relatively low stable transformation frequency (STF) of typically five putative transformants from 100 bisected meristems and from the potential generation of chimeric plants. Following the observations that *A. tumefaciens* is attracted and can bind to various banana tissues (Pérez Hernández et al. 1999), recent reports describe transgenic banana production via *A. tumefaciens*-mediated transformation of meristematic tissue (Acereto-Escoffié et al. 2005; Pei et al. 2005; Tripathi et al. 2005), though no or limited proof for stable integration of transgenes was provided, and the data presented did not allow an assessment of STF or chimerism. On the other hand, developing a relatively genotype-independent banana transformation protocol remains worth pursuing in view of the many different cultivars grown by smallholders. With an average STF of approximately 50 independent transgenic plants per 50 mg of cells, the third method of *A. tumefaciens* transformation of ECS is so far the most efficient (Khanna et al. 2004; Pérez Hernández et al. 2006a).

The most obvious application of banana transformation is to enhance fungal disease resistance for which multiple strategies are available (reviewed by Sági 2000). For example, expression of antimicrobial peptides of plant (Remy et al. 1998b) and non-plant (Chakrabarti et al. 2003) origin for fungal disease control have been reported. In addition, genetic transformation has allowed the characterization of novel, banana virus-derived promoters (Hermann et al. 2001; Schenk et al. 2001) and the improvement of transgene expression. Finally, vaccine delivery has been investigated by the expression of hepatitis B surface antigen in transgenic fruit (Sunil Kumar et al. 2005).

Recent advances in analysis of the banana genome (Aert et al. 2004; Coemans et al. 2005; Remy et al. 2005; Santos et al. 2005) and the sequence data that are becoming available (www.*Musa*genomics.org) pose a new challenge. For reliable functional analysis of the multiple transgenic lines required for each gene or promoter, an efficient transformation pipeline is a prerequisite. To this end, *A. tumefaciens*-mediated transformation of ECS is the preferred method because of its efficiency and the relatively low number of transgene inserts that range from 1 to 4 (Khanna et al. 2004; Pérez Hernández et al. 2006b). The available fast PCR-based techniques for detailed characterization of T-DNA insertions and the corresponding flanking regions (Remy et al. 2005; Pérez Hernández et al. 2006b) will also help correct interpretation of the expression data.

4.8 Perspective

The Global Musa Genomics Consortium aims to apply genomics to the sustainable improvement of *Musa*, a crop of global importance. The Consortium believes that newly available genomics technologies, which cover the analysis and sequencing of all the DNA, its genes, their expression, recombination, and diversity, can now be applied directly to the sustainable improvement of this major crop. The Consortium aims to develop freely accessible resources for *Musa* genomics and use the new knowledge and tools to enable both targeted conventional breeding and transgenic strategies. The genomics strategy also encourages better utilization and maintenance of *Musa* biodiversity. Futhermore, knowledge of *Musa* genomics, a monocotyledon cultivated as a polyploid, will provide a model resource for the exploitation of the genomes of other important species, increasing the utility of all genomic information. The Consortium will benefit from the successes of current genome and genomics programs and enabling technologies in delivering new genetic knowledge and tools, towards the development of new varieties. Information will be leveraged from the complete genomic sequences of Arabidopsis and rice, as well as the extensive sequence tags of other species, and these will be applied to *Musa* improvement. High throughput technologies, developed primarily for human genome analysis but available widely, will be accessed to put in place rapidly the resources and tools needed.

The overall aims of the Global *Musa* Genomics Consortium lie within the context of increasing the world's prosperity through fighting poverty and food insecurity in developing countries. The Consortium will achieve this aim by using the tools and expertise at its disposal to increase the productivity of the developing world's fourth most important crop, a staple food and key cash crop for nearly a billion people. As a result of genomics research, banana and plantain productivity can be increased in ways which will remain sustainable, particularly in the face of changing economic, social, and environmental conditions. The Consortium believes that such increases in productivity gained through fundamental knowledge and application of genomics will help to ensure future food and income security for millions of men, women, and children in the developing world.

Acknowledgments We are grateful to Dr. Richard Markham, director of the commodities for livelihood program, Bioversity International, for providing helpful comments on this chapter. Much of the work described in this chapter was undertaken under the auspices of the Global Musa Genomics Consortium and was supported by the Generation Challenge Programme, the United States Agency for International Development, the International Atomic Energy Agency, the national governments of France, Belgium, Brazil and Czech Republic, as well as the unrestricted funding of Bioversity International (formerly the International Plant Genetic Resources Institute and The International Network for the Improvement of Banana and Plantain).

References

Acereto-Escoffié POM, Chi-Manzanero BH, Echeverría-Echeverría S, Grijalva R, James Kay A, et al. (2005) *Agrobacterium*-mediated transformation of *Musa acuminata* cv. "Grand Nain" scalps by vacuum infiltration. Sci Hortic 105:359–371

Aert R, Ság L, Volckaert G (2004) Gene content and density in banana (*Musa acuminata*) as revealed by genomic sequencing of BAC clones. Theor Appl Genet 109:129–139

Ahloowalia BS, Maluszynski M, Nichterlein K (2004) Global impact of mutation-derived varieties. Euphytica 135:187–204

Anonymous (1990) List of new mutant cultivars; *Musa* sp. (banana). Mutation Breed Newsl 35:32–41

Anonymous (2005) All is good with the banana tree. Spore (FRA) 118:3

Anonymous (2006) Technical Cooperation Projects Highlights. Plant Breed Genet Newsl 117:13

Balint-Kurti PJ, Clendennen SK, Dolezelová M, Valárik M, Dolezel J, et al. (2000) Identification and chromosomal localization of the monkey retrotransposon in *Musa* sp. Mol Gen Genet 263:908–915

Barakat A, Gallois P, Raynal M, Mestre-Orteg D, Sallaud C, et al. (2000) The distribution of T-DNA in the genomes of transgenic Arabidopsis and rice. FEBS Lett 471:161–164

Bartoš J, Alkhimova O, Dolezelová M, De Langhe E, Dolezel J (2005) Nuclear genome size and genomic distribution of ribosomal DNA in *Musa* and Ensete (*Musaceae*): taxonomic implications. Cytogenet Genome Res 109:50–57

Baurens FC (1997) Identification par PCR des espèces impliquées dans la composition génomique des cultivars de bananier à l'aide de séquences répétées. PhD, Université Paul Sabatier, Toulouse, France

Baurens FC, Noyer JL, Lanaud C Lagoda PJL (1996) Use of competitive PCR to assay copy number of repetitive elements in banana. Mol Gen Genet 253:57–64

Baurens FC, Noyer JL, Lanaud C, Lagoda PJL (1997a) A repetitive sequence family of banana (*Musa* sp.) shows homology to Copia-like elements. J Genet Breed 51:135–142

Baurens FC, Noyer JL, Lanaud C, Lagoda PJL (1997b) Assessment of a species-specific element (Brep 1) in banana. Theor Appl Genet 95:922–931

Baurens FC, Noyer JL, Lanaud C, Lagoda PJL (1998) Inter-Alu PCR like genomic profiling in banana. Euphytica 99:137–142

Becker DK, Dugdale B, Smith MK, Harding RM, Dale JL (2000) Genetic transformation of Cavendish banana (*Musa* spp. AAA group) cv. 'Grand Nain' via particle bombardment. Plant Cell Rep 19:229–234

Brown SD, Peters J (1996) Combining mutagenesis and genomics in the mouse-closing the phenotype gap. Trends Genet 12:433–435

Buddenhagen IW (1987) Disease Susceptibility and Genetics in Relation to Breeding of Bananas and Plantains. In: Persley GJ DeLanghe EA (eds) Banana and Plantain Breeding Strategies. ACIAR, Canberra 21, pp 95–109

Carreel F (1994) Etude de la diversité génétique des bananiers (genre *Musa*) à l'aide de marqueurs RFLP. PhD, Institut National Agronomique, Paris-Grignon, France

Carreel F, Faure S, Gonzalez de Leon D, Lagoda PJL, Perrier X, et al. (1994) Evaluation de la diversité génétique chez les bananiers diploides (*Musa* sp.). Genet Sel Evol 26:125–136

Carreel F, Gonzalez de Leon D, Lagoda PJL, Lanaud C, Jenny C, et al. (2002) Ascertaining maternal and paternal lineage within *Musa* by chloroplast and mitochondrial DNA RFLP analyses. Genome 45:679–692

Chakrabarti A, Ganapathi TR, Mukherjee PK, Bapat VA (2003) MSI-99, a magainin analogue, imparts enhanced disease resistance in transgenic tobacco and banana. Planta 216:587–596

Cheesman EE (1947) Classification of the bananas II. The Genus *Musa* L. Kew Bull 2:106–117

Clarke L, Carbon J (1976) A colony bank containing synthetic Col El hybrid plasmids representative of the entire *E. coli* genome. Cell 9: 91–99.

Coemans B, Matsumura H, Terauchi R, Remy S, Swennen R, et al. (2005) SuperSAGE combined with PCR walking allows global gene expression profiling of banana (*Musa acuminata*), a non-model organism. Theor Appl Genet 111:1118–1126

Côte FX, Domergue R, Mommarson S, Schwendiman J, Teisson C, et al. (1996) Embryogenic cell suspensions from the male flower of *Musa* AAA. Physiol Plant 97:285–290

Creste S, Tulmann Neto A, De Oliveira Silva S, Figueira A (2003) Genetic characterization of banana cultivars (*Musa* spp.) from Brazil using microsatellite markers. Euphytica 132:259–268

Crouch HK, Crouch JH, Jarret RL, Cregan PB, Ortiz R (1998) Segregation at Microsatellite loci in Haploid and Diploid gametes of *Musa*. Crop Sci 38:211–217

Crouch JH, Crouch HK, Constandt H, Van Gysel A, Breyne P, et al. (1999) Comparison of PCR-based marker analyses of *Musa* breeding populations. Mol Breed 5:233–244

D'Hont A, Paget-Goy A, Escoute J, Carreel F (2000) The interspecific genome structure of cultivated banana, *Musa* spp. revealed by genomic DNA in situ hybridization. Theor Appl Genet 100:177–183

De Langhe E, Pillay M, Tenkouano A, Swennen R (2005) Integrating morphological and molecular taxonomy in *Musa*: the african plantains (*Musa* spp. AAB group). Plant Syst Evol 255: 225–236

Dolezel J, Dolezelová M, Novák FJ (1994) Flow cytometric estimation of nuclear DNA amount in diploid bananas (*Musa acuminata* and *Musa balbisiana*). Biol Plant 36:351–357

Dolezelová M, Dolezel J, Van den Houwe I, Roux N, Swennen R (2005) Focus on the *Musa* collection: Ploidy levels revealed. Info*Musa* 14:34–36

Dolezelová M, Valárik M, Swennen R, Horry JP, Dolezel J (1998) Physical mapping of the 18S–25S and 5S ribosomal RNA genes in diploid bananas. Biol Plant 41:497–505

Ewing B, Green P (1998) Base calling of automated sequencer traces using phred. II. Error probabilities. Genome Res 8:186–194

FAO, (2005) http://faostat.fao.org/

Fauré S, Bakry F, De León DG (1993a) Cytogenetic studies of diploid bananas In: Ganry J (ed) Proc Int Symp Breeding Banana and Plantain for Resistance to Diseases and Pests. CIRAD, Montpellier, pp 77–92

Fauré S, Noyer JL, Carreel F, Horry JP, Bakry F, et al. (1993b) A molecular marker-based linkage map of diploid bananas (*Musa acuminata*). Theor Appl Genet 87:517–526

Cheung F, Town CD (2007) A BAC end view of the *Musa acuminata* genome. BMC Plant Biology 7:29

Ge XJ, Liu MH, Wang K, Schaal BA, Chiang TY (2005) Population structure of wild bananas, *Musa balbisiana*, in China determined by SSR fingerprinting and cpDNA PCR-RFLP. Mol Ecol 14:933–944

Global *Musa* Genomics Consortium (2002) Beyond Arabidopsis and Rice, Strategy for the Global *Musa* Genomics Consortium. Report of a meeting held in Arlington, USA, 17–20 July 2001, INIBAP, Pro*MUSA*, pp 25–30

Grapin A, Noyer JL, Dambier D, Carreel F, Lanaud C, et al. (1998) Diploid *Musa acuminata* genetic diversity with Sequence Tagged Microsatellite Sites. Electrophoresis 19:1374–1380

Harper G, Osuji JO, Heslop-Harrison JSP, Hull R (1999) Integration of banana streak badnavirus into the *Musa* genome: molecular and cytogenetic evidence. Virology 255:207–213

Harper LC, Cande WZ (2000) Mapping a new frontier development of integrated cytogenetic maps in plants. Rev Funct Integr Genom 1:89–98

Henikoff S, Till BJ, Comai L (2004) TILLING. Traditional mutagenesis meets functional genomics. Plant Physiol 135:630–636

Hermann SR, Becker DK, Harding RM, Dale JL (2001) Promoters derived from banana bunchy top virus-associated components S1 and S2 drive transgene expression in both tobacco and banana. Plant Cell Rep 20:642–646

Horry JP (1988) Distribution of anthocyanins in wild and cultivated banana varieties. Phytochemistry 27:2667–2672

Horry JP (1989) Chimio taxonomie et organisation génétique dans le genre *Musa*. I, II, III. Fruits 44:455–475, 509–520, 573–578

Horry JP, Dolezel J, Dolezelová M, Lysák MA (1998) Do natural AxB tetraploid bananas exist? Info*Musa* 7:5–6

Hřibová E, Macas J, Dolezelová M, Dolezel J (2006) Characterization of the highly repeated part of the banana genome. In: Abstracts of the 5th Plant Genomics European Meetings. The Italian Plant Genomics Network, Venice, p 205

Hull R (2002) Matthews' Plant Virology, 4th edn. San Diego, Academic Press

Hull R, Harper G, Lockhart B (2000) Viral sequences integrated into plant genomes. Trends Plant Sci 5:362–365

IAEA, http://www-mvd.iaea.org/

INIBAP (1992) Banana and Plantain – food for thought. In: Annual report 1992, INIBAP, Montpellier, France, pp 7–11

INIBAP (1999) Musa production around the world – trends, varieties and regionnal importance. In: Annual Report 1998, INIBAP, Montpellier, pp 42–47

James AC, Jimenez-Martinez R, Canto-Canche B, Peraza-Echeverria S (2006) Discovery and characterization of disease resistance gene homologues in a plant transformation competent BIBAC library of the banana *Musa acuminata* cv Tuu Gia (AA). Plant and Animal Genome XIV Abst W6, p 8

Jarret RL, Litz RE (1986) Enzyme polymorphism in *Musa acuminata* Colla. J Hered 77:183–186

Jarret RL, Gawel N, Whittemore A, Sharrock S (1992) RFLP based phylogeny of *Musa* species in Papua New Guinea. Theor Appl Genet 84:579–584

Khanna H, Becker D, Kleidon J, Dale J (2004) Centrifugation assisted *Agrobacterium tumefaciens*-mediated transformation (CAAT) of embryogenic cell suspensions of banana (*Musa* spp Cavendish AAA and Lady finger AAB). Mol Breed 14:239–252

Lamoureux D, Peterson DG, Li W, Fellers JP, Gill BS (2005) The efficacy of Cot-based gene enrichment in wheat (*Triticum aestivum* L.). Genome 48:1120–1126

Lanaud C, Tezenas du Montcel H, Jolivot MP, Glaszmann JC, González de León D (1992) Variation of ribosomal gene spacer length among wild and cultivated banana. Heredity 68:147–156

Lebot V, Aradhya KM, Manchardt R, Meilleur B (1993) Genetic relationships among cultivated bananas and plantains from Asia and the Pacific. Euphytica 67:163–175

Lescot M, Piffanellil P, Ciampi AY, Ruiz M, Blanc G, et al. (2007) Molecular insights into the Musa genome: syntenic relationships to rice and between Musa species. BMC Genomics (under review)

Lysák MA, Dolezelová M, Horry JP, Swennen R, Dolezel J (1999) Flow cytometric analysis of nuclear DNA content in *Musa*. Theor Appl Genet 98:1344–1350

Mak C, Ho YW, Tan YP, Ibrahim R (1996) Novaria - A new banana mutant induced by gamma irradiation. Infor*Musa* 5:35–36

May GD, Afza R, Mason HS, Wiecko A, Novak FJ, et al. (1995) Generation of transgenic banana (*Musa acuminata*) plants via *Agrobacterium*-mediated transformation. Bio/Technol 13:486–492

Miller RNG, Pappas GJ, Souza MT, Bertioli DJ (2006) Analysis of resistance gene analogs in *Musa acuminata* subps *burmanicoides* var Calcutta 4. Plant and Animal Genome XIV, Abst W7, p 8

Nair AS, Teo CH, Schwarzacher T, Heslop Harrison P (2005) Genome classification of banana cultivars from South India using IRAP markers. Euphytica 144:285–290

Noyer JL, Causse S, Tomekpe K, Bouet A, Baurens FC (2005) A new image of plantain diversity assessed by SSR, AFLP and MSAP markers. Genetika 124:61–69

Noyer JL, Dambier D, Lanaud C, Lagoda PJL (1997) The saturated map of diploid banana *Musa acuminata*. Plant and Animal Genome V, Abst P335, p 138

Ortiz-Vázquez E, Kaemmer D, Zhang HB, Muth J, Rodríguez-Mendiola M, et al. (2005) Construction and characterization of a plant transformation-competent BIBAC library of the black Sigatoka-resistant banana *Musa acuminata* cv. Tuu Gia (AA). Theor Appl Genet 110:706–713

Osuji JO, Crouch J, Harrison G, Heslop-Harrison JS (1998) Molecular cytogenetics of *Musa* species, cultivars and hybrids: location of 18S-5.8S-25S and 5S rDNA and telomere-like sequences. Ann Bot 82:243–248

Osuji JO, Harrison G, Crouch J, Heslop-Harrison JS (1997) Identification of the genomic constitution of *Musa* L. lines (bananas, plantains and hybrids) using molecular cytogenetics. Ann Bot 80:787–793

Pei XW, Chen SK, Wen RM, Ye S, Huang JQ, et al. (2005) Creation of transgenic bananas expressing human lysozyme gene for Panama Wilt resistance. J Integr Plant Biol 47:971–977

Pérez Hernández JB, Remy S, Galán Saúco V, Swennen R, Sági L (1999) Chemotactic movement and attachment of *Agrobacterium tumefaciens* to banana cells and tissues. J Plant Physiol 155:245–250

Pérez Hernández JB, Remy S, Swennen R, Sági L (2006a) Banana (*Musa* sp.). In: Wang K (ed) Methods in Molecular Biology, vol 344: Agrobacterium Protocols, vol 2. Humana Press Inc, Totowa, NJ, pp 167–176

Pérez Hernández JB, Swennen R, Sági L (2006b) Number and accuracy of T-DNA insertions in transgenic banana (*Musa* spp.) plants characterized by an improved anchored PCR technique. Transgenic Res 15:139–150

Persley GJ, DeLanghe EA (1987) Banana and Plantain Breeding Strategies ACIAR, Canberra 21

Peterson DG, Schulze SR, Sciara EB, Lee SA, Bowers JE, et al. (2002) Integration of Cot analysis, DNA cloning, and high-throughput sequencing facilitates genome characterization and gene discovery. Genome Res 12:795–807

Pillay M, Nwakanma DC, Tenkouano A (2000) Identification of RAPD markers linked to A and B genome sequence in *Musa* L. Genome 43:763–767

Rabinowicz PD, Schutz K, Dedhia N, Yordan C, Parnell LD, et al. (1999) Differential methylation of genes and retrotransposons facilitates shotgun sequencing of the maize genome. Nature Genet 23:305–308

Raboin LM, Carreel F, Noyer J-L, Baurens F-C, Horry J-P, et al. (2005) Diploid Ancestors of Triploid Export Banana Cultivars: Molecular Identification of 2n Restitution Gamete Donors and n Gamete Donors. Molecular Breed 16:333–341

Remy S, Buyens A, Cammue BPA, Swennen R, Sági L (1998a) Production of transgenic banana plants expressing antifungal proteins. Acta Hort 490:433–436

Remy S, François I, Cammue BPA, Swennen R, Sági L (1998b) Co-transformation as a potential tool to create multiple and durable resistance in banana (*Musa* spp.). Acta Hort 461:361–365

Remy S, Thiry E, Coemans B, Windelinckx S, Swennen R, Sági L (2005) Improved T-DNA vector for tagging plant promoters via high-throughput luciferase screening. BioTechniques 38:763–770

Roux N, Dolezel J, Swennen R, Zapata-Arias FJ (2001) Effectiveness of three micropropagation techniques to dissociate cytochimeras in *Musa* spp. Plant Cell Tissue Organ Cult 66:189–197

Roux N, Toloza A, Radecki Z, Zapata-Arias FJ, Dolezel J (2003) Rapid detection of aneuploidy in *Musa* using flow cytometry. Plant Cell Rep 21:483–490

Roux NS (2004) Mutation Induction in *Musa* – Review. In: Jain SM, Swennen R (eds) Banana Improvement: Cellular, Molecular Biology and Induced Mutations. Science Publishers, Inc, Enfield, pp 23–32

Roux NS, Toloza A, Dolezel J, Panis B (2004) Usefulness of Embryonic Cell Suspension Cultures for the Induction and Selection of Mutations in *Musa* spp. In: Jain SM, Swennen R (eds) Banana Improvement: Cellular, Molecular Biology and Induced Mutations. Science Publishers, Inc, Enfield, pp 33–44

Safár J, Noa-Carrazana JC, Vrána J, Bartoš J, Alkhimova O, et al. (2004) Creation of a BAC resource to study the structure and evolution of the banana (*Musa balbisiana*) genome. Genome 47:1182–1191

Sági L (2000) Engineering resistance to diseases caused by fungi. In: Jones D (ed) Diseases of Banana, Abacá and Enset. CABI, Wallingford, UK, pp 482–491

Sági L, Panis B, Remy S, Schoofs H, De Smet K, et al. (1995) Genetic transformation of banana and plantain (*Musa* spp.) via particle bombardment. Bio/Technology 13:481–485

Sági L, Remy S, Panis B, Swennen R, Volckaert G (1994) Transient gene expression in electroporated banana (*Musa* spp., cv. 'Bluggoe', ABB group) protoplasts isolated from regenerable embryogenic cell suspensions. Plant Cell Rep 13:262–266

Samson JA (1986) Tropical fruits, 2nd edn. Longman Scientific and Technical, Harlow, UK, 335 pp

Santos CMR, Martins NF, Hörberg HM, de Almeida ERP, Coelho MCF, et al. (2005) Analysis of expressed sequence tags from *Musa acuminata* spp. *burmannicoides*, var. Calcutta 4 leaves submitted to temperature stresses. Theor Appl Genet 110: 1517–1522

Schenk PM, Remans T, Sági L, Elliott AR, Dietzgen RG, et al. (2001) Promoters for pregenomic RNA of banana streak badnavirus are active for transgene expression in monocot and dicot plants. Plant Mol Biol 47:399–412

Shepherd K (1999) Cytogenetics of the genus *Musa*. International Network for the Improvement of Banana and Plantain, Montpellier, France

Shepherd K, da Silva KM (1996a) Mitotic instability in banana varieties. Aberrations in conventional triploid plants. – Fruits 51:99–103

Shepherd K, da Silva KM (1996b) Mitotic instability in banana varieties. I. Plants from callus and shoot tip cultures. – Fruits 51:5–11

Shizuya H, Birren B, Kim UJ, Mancino V, Slepak T, et al. (1992) Cloning and stable maintenance of 300-kilobase-pair fragments of human DNA in Escherichia coli using an F-factor-based vector. Proc Natl Acad Sci (USA) 89:8794–8797

Šimková H, Cíhalíková J, Vrána J, Lysák MA, Dolezel J (2003) Preparation of HMW DNA from plant nuclei and chromosomes isolated from root tips. Biol Plant 46:369–373

Simmonds NW, Shepherd K (1955) The taxonomy and origins of the cultivated bananas. J Linn Soc London (Bot) 5:302–312

Slade AJ, Fuerstenberg SI, Loeffler D, Steine MN, Facciotti D (2005) A reverse genetic, nontransgenic approach to wheat crop improvement by TILLING. Nature Biotechnol 23:75–81

Souza Jr MT, Santos CM, Martins NF, da Silva FR, Togawa RC, et al. (2005) Transcriptoma de *Musa acuminata* no DATA*Musa* Boletim de Pesquisa e Desenvolvimento (109) da Embrapa Reccursos Genéticos e Biotecnologia. 21 pp (http://www.cenargen.embrapa.br/publica/download.html#bp2005)

Springer PS (2000) Gene traps: tools for plant development and genomics. Plant Cell 12:1007–1020

Strosse H, Schoofs H, Panis B, Andre E, Reyniers K, et al. (2006) Development of embryogenic cell suspensions from meristematic tissue in bananas and plantains (*Musa* spp.). Plant Sci 170:104–112

Sunil Kumar GB, Ganapathi TR, Revathi CJ, Srinivas L, Bapat VA (2005) Expression of hepatitis B surface antigen in transgenic banana plants. Planta 222:484–493

Taxonomic Advisory Group (TAG) (2006) Launching the Taxonomic Advisory Group: Developing a strategic approach to the conversation and use of *Musa* diversity. Report of a meeting held in Limbe, Cameroon, 29 May–3 June 2006, INIBAP

Teo CH, Tan SH, Othman YR, Schwarzacher T (2002) The cloning of Ty 1-copia-like retrotransposons from 10 varieties of banana (*Musa* sp). J Biochem Mol Biol Biophys 6:193–201

Tripathi L, Tripathi JN, Hughes Jd'A (2005) *Agrobacterium*-mediated transformation of plantain (*Musa* spp.) cultivar Agbagba. Afr J Biotechnol 4:1378–1383

Ude G, Pillay M, Ogundiwin E, Tenkouano A (2002) Analysis of genetic diversity and sectional relationships in *Musa* using AFLP markers. Theor Appl Genet 104:1239–1245

Ude G, Pillay M, Ogundiwin E, Tenkouano A (2003) Genetic diversity in African plantain core collection using AFLP and RAPD markers. Theor Appl Genet 107:248–255

Valárik M, Šimková H, Hribová E, Šafár J, Dolezelová M, et al. (2002) Isolation, characterization and chromosome localization of repetitive DNA sequences in banana (*Musa* spp.). Chromosome Res 10:89–100

van der Frits L, Hilliou F, Memelink J (2001) T-DNA activation tagging as a tool to isolate regulators of a metabolic pathway from a genetically non-tractable plant species. Transgenic Res 10:513–521

Van Duren M, Morpurgo R, Dolezel J, Afza R (1996): Induction and verification of autotetraploids in diploid banana (*Musa acuminata*) by *in vitro* techniques. Euphytica 88:25–34

Venter JC (1993) Identification of new human receptor and transporter genes by high throughput cDNA (EST) sequencing. J Pharm Pharmacol 45 (Suppl 1):355–360

Vilarinhos A, Carreel F, Rodier M, Hippolyte I, Benabdelmouna A, et al. (2006) Characterization of translocations in banana by FISH of BAC clones anchored to a genetic map. Plant and Animal Genome XIV, Abst W4, p 8

Vilarinhos AD (2004) Cartographie génétique et cytogénétique chez le bananier : caractérisation des translocations. PhD thesis, Ecole Nationale Agronomique de Montpellier, Ecole doctorale biologie integrative, Montpellier, France

Vilarinhos AD, Piffanelli P, Lagoda P, Thibivilliers S, Sabau X, et al. (2003) Construction and characterization of a bacterial artificial chromosome library of banana (*Musa acuminata* Colla). Theor Appl Genet 106:1102–1106

Wenzl P, Carling J, Kudrna D, Jaccoud D, Huttner E, et al. (2004) Diversity arrays technology (DArT) for whole-genome profiling of barley. Proc Natl Acad Sci (USA) 101:9915–9920

Whitelaw CA, Barbazuk WB, Pertea G, Chan AP, Cheung F, et al. (2003) Enrichment of gene-coding sequences in maize by genome filtration. Science 302:2118–2120
Wilson GB (1946a) Cytological studies in the *Musae*. I. Meiosis in some triploid clones. Genetics 31:241–258
Wilson GB (1946b) Cytological studies in the *Musae*. II. Meiosis in some diploid clones. Genetics 31:475–482
Wilson GB (1946c) Cytological studies in the *Musae*. III. Meiosis in some seedling clones. Genetics 31:483–493
Wong C, Kiew R, Argent G, Set O, Lee SK, et al. (2002) Assessment of the validity of the section in *Musa* (*Musaceae*) using AFLP. Ann Bot 90:231–238
Yuan Y, SanMiguel PJ, Bennetzen JL (2003) High-Cot sequence analysis of the maize genome. Plant J 34:249–255
Zubko E, Adams CJ, Machácková I, Malbeck J, Scollan C, et al. (2002) Activation tagging identifies a gene from *Petunia hybrida* responsible for the production of active cytokinins in plants. Plant J 29:797–808

Chapter 5
Genomics of *Phaseolus* Beans, a Major Source of Dietary Protein and Micronutrients in the Tropics

Paul Gepts, Francisco J.L. Aragão, Everaldo de Barros, Matthew W. Blair, Rosana Brondani, William Broughton, Incoronata Galasso, Gina Hernández, James Kami, Patricia Lariguet, Phillip McClean, Maeli Melotto, Phillip Miklas, Peter Pauls, Andrea Pedrosa-Harand, Timothy Porch, Federico Sánchez, Francesca Sparvoli, and Kangfu Yu

Abstract Common bean is grown and consumed principally in developing countries in Latin America, Africa, and Asia. It is largely a subsistence crop eaten by its producers and, hence, is underestimated in production and commerce statistics. Common bean is a major source of dietary protein, which complements carbohydrate-rich sources such as rice, maize, and cassava. It is also a rich source of minerals, such as iron and zinc, and certain vitamins. Several large germplasm collections have been established, which contain large amounts of genetic diversity, including the five domesticated *Phaseolus* species and wild species, as well as an incipient stock collection. The genealogy and genetic diversity of *P. vulgaris* are among the best known in crop species through the systematic use of molecular markers, from seed proteins and isozymes to simple sequence repeats, and DNA sequences. Common bean exhibits a high level of genetic diversity, compared with other selfing species. A hierarchical organization into gene pools and ecogeographic races has been established. There are over 15 mapping populations that have been established to study the inheritance of agronomic traits in different locations. Most linkage maps have been correlated with the core map established in the BAT93 x Jalo EEP558 cross, which includes several hundreds of markers, including Restriction Fragment Length Polymorphisms, Random Amplified Polymorphic DNA, Amplified Fragment Length Polymorphisms, Short Sequence Repeats, Sequence Tagged Sites, and Target Region Amplification Polymorphisms. Over 30 individual genes for disease resistance and some 30 Quantitative Trait Loci for a broad range of agronomic traits have been tagged. Eleven BAC libraries have been developed in

P. Gepts
University of California, Department of Plant Sciences / MS1, Section of Crop and Ecosystem Sciences, 1 Shields Avenue, Davis, CA 95616-8780, USA.
e-mail: plgepts@ucdavis.edu

P.H. Moore, R. Ming (eds.), *Genomics of Tropical Crop Plants*
© Springer 2008

genotypes that represent key steps in the evolution before and after domestication of common bean, a unique resource among crops. Fluorescence in situ hybridization provides the first links between chromosomal and genetic maps. A gene index based on some *P. vulgaris* 21,000 expressed sequence tags (ESTs) has been developed. ESTs were developed from different genotypes, organs, and physiological conditions. They resolve currently in some 6,500–6,800 singletons and 2,900 contigs. An additional 20,000 embryonic *P. coccineus* ESTs provides an additional resource. Some 1,500 M_2 Targeting Local Lesions In Genomes populations exist currently. Finally, transformation methods by biolistics and Agrobacterium have been developed, which can be applied for genetic engineering. Root transformation via *A. rhizogenes* is also possible. Thus, the Phaseomics community has laid a solid foundation towards its ultimate goal, namely the sequencing of the *Phaseolus* genome. These genomic resources are a much-needed source of additional markers of known map location for marker-assisted selection and the accelerated improvement of common bean cultivars.

5.1 Introduction

5.1.1 Economic, Agronomic, and Societal Importance of Phaseolus Beans

FAO statistics provide insight into the economic status of common bean (*Phaseolus vulgaris*) throughout the world (http://faostat.fao.org/site/408/default.aspx; verified December 5, 2006). Dry beans were grown on 27.7 million ha in 148 countries in 2004 and total production was 18.7 million metric tons (MT). *Phaseolus* beans are the most important grain legume for direct human consumption. Bean production is more than twice that of chickpea, which is the second most important grain legume. Importantly, eight of the top ten producers are considered to be developing countries. When developmental status is considered further, it is seen that developing countries produce 86% of worldwide production of beans. The main production areas of common bean are Latin America (with Brazil and Mexico as the most important producers) and eastern Africa (where per capita consumption is the highest in the world in countries such as Burundi, Rwanda, and Uganda) (Broughton et al. 2003). Bean production is increasing in some Asian countries such as China and Myanmar, primarily for export purposes (see below). Green bean production amounted to 6.4 million metric tons on 890,000 ha in 2004.

In many of the countries in the world, especially developing countries, common bean is consumed as an important part of the diet and not exported. An extreme example is Brazil, the third leading producer, where less than 0.1% of the 2 million MT produced is exported. The importance of bean to diets in the developing world is reflected in the fact that for developing countries only 13% of production is exported. This contrasts with developed countries, which export 31% of their production. From an economic perspective, dry bean exports generate US$812 million for developing countries, whereas developed countries receive US$460 million for their exports. The two major exporters are China and Myanmar. For Myanmar, beans (whether *Phaseolus* or other dry bean species) are its most important agricultural export generating US$253 million in 2004. For this very poor country, the bean export income is significant when balanced against their total of only US$2.9 billion in exports in 2004–2005.

Common bean is an important dietary component especially in some developing countries. For some, such as Brazil and Mexico, it is a major source of protein. In some of the least developed countries, such as Burundi, Rwanda, and Uganda, bean provides 40%, 31%, and 15% of the daily intake of total protein, respectively. For other developing countries, Nicaragua, Cuba, and North Korea, the percent total protein derived from bean is 19%, 13%, and 11%, respectively. For a major producer, like Brazil, beans provide 9% of the protein. Qualitatively, bean seed proteins provide sulfur essential amino acids such as lysine, but are poor in methionine, thus complementing cereals in this respect (Bressani 1983; Gepts and Bliss 1985). Not surprisingly, beans are also a major source of calories for residents of these countries. In addition, common bean plays an important role as a source of minerals, especially iron and zinc (Broughton et al. 2003), for which it also complements cereals. Genetic variation for seed content of these minerals has been demonstrated (Islam et al. 2002; Beebe et al. 2000a) and breeding for enhanced mineral content (biofortification) can therefore reasonably be expected (Guzmán-Maldonado et al. 2003).

For those countries where beans are an important crop, yield varies significantly. As expected, on average the yield in developed countries, 1,944 kg/ha, is significantly higher than those in developing countries, 1,035 kg/ha. The situation is more drastic for those least developed countries that depend heavily on beans as a food source. In Burundi, Rwanda, and Uganda the yields are 918 kg/ha, 671 kg/ha, and 638 kg/ha, respectively. Identifying and minimizing yield limiting factors is an ongoing concern for many bean improvement programs. In addition, given the prevalence of bean in these diets, modifying the nutrient content in general to make it a more balanced and nutritious food source is also receiving emphasis.

5.1.2 Phaseolus as an Experimental Organism

The genus *Phaseolus* is a member of the tropical tribe Phaseoleae, which also includes cowpea, pigeon pea, and soybean. The Phaseoleae tribe is part of the Phaseoloid-Millettioid clade, which diverged some 45–50 million years ago from the Hologalegina clade, which contains most temperate crop legumes, such as pea, alfalfa (and *Medicago truncatula*), chickpea, and lentil (Lavin et al. 2005). Synteny between *Phaseolus* and other legumes is negatively correlated with phylogenetic distance. Thus, the highest synteny levels are observed with the genus *Vigna* (cowpea and mung bean), followed by soybean, and distantly, the Hologalegina clade (Boutin et al. 1995; Lee et al. 2001; Yan et al. 2004; Choi et al. 2004; Moffet and Weeden 2006). For example, the region marked by the *Bng122-D0140-Bng*

171-Bng173 markers on linkage group B1 of common bean is syntenic with a region on LG G of soybean. This region harbors a cluster of disease and nematode resistance genes (Freyre et al. 1998; Foster-Hartnett et al. 2002; Kelly et al. 2003).

Phaseolus is a diploid genus with most species having 2n = 2x = 22 chromosomes (some species have 2n = 2x = 20). The genome size of *P. vulgaris* (580 Mbp/haploid genome) is comparable to that of rice (490 Mbp/haploid genome; Bennett and Leitch 2005). In common bean, the levels of duplication and the amount of highly repeated sequences are generally low. Mapping experiments demonstrated that most loci are single copy (Vallejos et al. 1992; Freyre et al. 1998; McClean et al. 2002). Gene families tend to be small, and the traditionally large families such as resistance gene analogs (Rivkin et al. 1999) and protein kinases (Vallad et al. 2001) are of moderate size. Further experiments are needed, however, to confirm these conclusions and compare these results to those of other legumes.

Studies of the evolution of common bean have uncovered several noteworthy features that are of interest to other crop plants. Unique among crop plants, common bean consists of two geographically distinct, evolutionary lineages (Andean and Mesoamerican) that predate domestication and trace back to a common, still extant ancestor located in Ecuador and northern Peru (Debouck et al. 1993; Kami et al. 1995). Patterns of marker diversity and virulence in pathogens and *Rhizobium* parallel those in the bean host, suggesting host-microbe coevolution (Guzmán et al. 1995; Geffroy et al. 1999, 2000; Kelly and Vallejo 2004; Araya et al. 2004; Mkandawire et al. 2004; Aguilar et al. 2004). Geffroy et al. (1999, 2000) have shown that Andean and Mesoamerican resistance specificities appeared in the same, presumably ancestral gene cluster. The inheritance of the domestication syndrome in common bean was the second among all crop plants and the first one in the legumes to be investigated (Koinange et al. 1996). The traits involved, such as growth habit (e.g., determinacy), photoperiod sensitivity and phenology, pod and seed size, and seed color, were not only important in domestication but remain crucial agronomic traits determining farmer and consumer acceptability.

In common bean, sequence variation value based on four genotypes was θ_W (x 10^3) = 4.8 (P. Cregan, personal communication). This compares with θ_W (x 10^3) = 0.97 for *Glycine max* (Zhu et al. 2003), θ_W (x 10^3) = 7.1 for *Arabidopsis thaliana* (Schmid et al. 2005), and θ_W (x 10^3) = 9.6 for *Zea mays* (Tenaillon et al. 2001). The higher polymorphism in common bean relative to soybean is presumably related to its diversification into the two geographic gene pools in the Andes and Mesoamerica, mentioned earlier.

5.2 Bean Genetic Resources

There are a large number of collections of *Phaseolus* germplasm collections, which include the wild and domesticated genotypes of the five domesticated *Phaseolus* species (*P. vulgaris*: common bean; *P. coccineus*: runner bean; *P. dumosus*: year bean; *P. acutifolius*: tepary bean; *P. lunatus*: lima bean) and other wild *Phaseolus* species. The largest and most diverse is the World Phaseolus collection at the Centro Internacional de Agricultura Tropical (CIAT) in Cali, Colombia (http://isa.ciat.cgiar.org/urg/main.do?language=en). Other large collections are those of the USDA in Pullman, WA, USA (http://www.ars.usda.gov/main/ site_main.htm?modecode=53481500), the Institut für Pflanzengenetik und Kulturpflanzenforschung (IPK) in Gatersleben, Germany (http://fox-serv.ipk-gatersleben.de/), and the Centro Nacional de Recursos Genéticos e Biotecnologia (CENARGEN/ EMBRAPA) in Brasilia (http://www.cenargen.embrapa.br/). There are also national collections as well as a taxonomic collection focused on wild members of the Phaseolinae subtribe at the National Botanic Garden of Belgium (http://www.br.fgov.be/research/collections/living/phaseolus/index.html). Recently, the USDA collection at Pullman established a *Phaseolus* stock collection (http://www.ars.usda.gov//Main/site_main.htm?docid=9065). Together, these collections represent a substantial wealth of genetic diversity that is generally freely available for plant genetic and breeding research.

Several types of populations have been used for linkage mapping in common beans. In terms of population structure, the first molecular maps in common bean were based on first backcross and F_2 generation populations (Vallejos et al. 1992; Nodari et al. 1993a). While easy to develop, the backcross and F_2 populations tended to have few total genotypes and seed supply was limiting, problems that were overcome with the development of recombinant inbred (RI) populations from the initial mapping populations (Freyre et al. 1998; Yu et al. 1998). The BAT93 x Jalo EEP558 RI population has become the core mapping population for common bean because markers from other linkage maps have been mapped in this population and linkage groups from different maps can therefore be correlated (Freyre et al. 1998; Vallejos et al. 2001; Blair et al. 2003). In addition, over 15 other recombinant inbred line populations have been created to map individual or multiple traits (reviewed in Broughton et al. 2003; Kelly et al. 2003). The majority of mapping populations have been produced by crosses across gene pool boundaries (i.e., Mesoamerican x Andean) usually with divergent parents showing contrasting phenotypic characteristics (such as morphology and disease resistance) and high genetic polymorphism (Nodari et al. 1992; Broughton et al. 2003). Relatively few populations have been created from within-gene pools crosses (Andean x Andean or Mesoamerican x Mesoamerican: e.g., Frei et al. 2005; Kolkman and Kelly 2003). Compared to other crop species, genetic mapping in *P. vulgaris* has not utilized many inter-specific crosses for the construction of linkage maps except for the introgression of some disease resistance (Bai et al. 1997). Similarly, very few crosses between wild and domesticated common bean have been used for genetic mapping except for one study of a domesticated x wild cross, which was used to determine the inheritance of the domestication syndrome (Koinange et al. 1996) and the recent use of advanced backcrossing to analyze wild beans for positive yield quantitative trait loci (QTL) (Mauro Herrera 2003; Blair et al. 2006b).

5.3 Marker and Sequence Diversity

As with all plant species, diversity levels and organization of genetic diversity ("structure") were initially estimated with a wide array of molecular marker types. The majority of these marker analyses were designed to understand the organization of diversity in the species, which is now one of the best known among crop species. Based on phaseolin seed storage protein variation and partial reproductive isolation, Gepts and colleagues developed the two-gene pool concept for *P. vulgaris* (Gepts and Bliss 1985, 1986; Gepts et al. 1986; Koenig et al. 1990; Gepts 1990). This result was confirmed by an extensive isozyme analysis (wild: Koenig and Gepts 1989b; domesticated: Singh et al. 1991b) and mtDNA restriction fragment length polymorphisms (RFLP) (Khairallah et al. 1990, 1992). Those analyses also found that the within-gene pool variation was less than that found between gene pools. Allozyme diversity was a key component utilized to define three Mesoamerican (Durango, Jalisco, and Mesoamerica) and three Andean (Nueva Granada, Peru, and Chile) domesticated races in common bean (Singh et al. 1991a). More recently, allozyme data provided important ancestry information regarding the origin of genetic materials from southwestern Europe by demonstrating that the patterns of variation in this region are similar to those found in the Americas (e.g., Santalla et al. 2002). Essentially the allozyme data provided an important hypothesis regarding the origin of common bean and the levels of genetic diversity within the species.

Other marker types have provided data for further analysis of the organization of diversity in common bean. Nuclear RFLP comparisons also supported the two-gene pool concept, but unlike isozyme analyses, the levels of diversity within each gene pool were similar (Becerra-Velásquez and Gepts 1994). Using random amplified polymorphic DNA (RAPD) analysis, Freyre et al. (1998) identified a clear separation between Andean and Mesoamerican gene pools but also detected geographic structure within the Mesoamerican gene pool. A RAPD analysis led Beebe et al. (2000b) to observe subsets of landraces within each Mesoamerican race and define a new Guatemalan race within the Mesoamerican gene pool. These results suggest a level of diversity large enough to distinguish races and subraces. This was not the case for an amplified fragment length polymorphism (AFLP) analysis of Andean genotypes where the genotypes formed essentially a single large pool (Beebe et al 2001). AFLP markers, in contrast, proved to be a very powerful fingerprinting tool to distinguish closely related genotypes belonging to the same commercial class, such as the yellow bean class (Pallottini et al. 2004). AFLP and Inter Short Sequence Repeatmarkers have also proven to be very useful to assess the level and direction of gene flow between wild and domesticated bean populations (Papa and Gepts 2003; Papa et al. 2005; Payró de la Cruz et al. 2005; Zizumbo-Villarreal et al. 2005).

A suite of SSR markers, recently developed for common bean (Yu et al. 1999; Gaitán-Solís et al. 2002; Blair et al. 2003), provide a new tool for diversity analyses. Blair et al. (2006a) measured diversity among 44 genotypes with 129 simple sequence repeats (SSRs), and as expected, they observed two gene pools corresponding to the Mesoamerican and Andean genotypes. It was somewhat surprising that the Andean genotypes were more diverse than the Mesoamerican lines. Recently, in a study of 172 landraces that represent a broad cross-section of the phenotypic diversity found in the USDA core collection of common bean, greater SSR diversity among the Andean genotypes was also observed (McClean et al. 2006). Applying a model-based approach to population structure analysis (Pritchard et al. 2000), six Mesoamerican and three Andean subpopulations were observed. In addition, a subpopulation, most similar to other Andean genotypes, consisted of landraces collected from throughout the range of the species. These sequence and marker analyses are forming a foundation on which association mapping can now be applied to common bean. Genome sequencing and extensive genome-wide marker development provide new avenues for crop improvement. Principal among these is association mapping, as an alternative to linkage analysis that uses the natural sequence diversity within a species to define the various loci controlling a complex trait (Jorde 2000; Mackay 2001). Because this approach can uncover potential causative single nucleotide polymorphisms (SNPs; Thornsberry et al. 2001) or markers linked to a gene associated with a trait of interest (Hagenblad et al. 2004), understanding sequence and marker variation is important to applying this approach (Nordborg et al. 2005).

For common bean, 27,000 DNA sequences are deposited in GenBank (http:www.ncbi.nlm.nih.gov; verified July 18, 2006). Among these, the vast majority of these are expressed sequence tag (EST) sequences (see section 5.1). EST sequencing projects typically sample not only different tissues but different genotypes. By sampling genotypic differences, and applying stringent screening criteria, polymorphisms such as SNPs and insertion/deletions (indels) can be discovered that define a level of sequence diversity within a species. Ramírez et al. (2005) analyzed contigs derived from over 21,000 ESTs and compared contigs derived from single genotypes. By comparing sequences from Negro Jamapa (Mesoamerican) and G19833 (Andean), they discovered 421 polymorphisms from 196 contigs between these two genotypes. Most (94%) of the polymorphisms were SNPs.

McConnell et al. (2006) compared DNA sequence differences between BAT93 and Jalo EEP558, the parents of the community-wide mapping population (see section 5.2). For their analysis, they considered 322 genes and compared sequence data from the 3' UTR and exon and intron sequences from the 3' end of the coding region of the genes. Among these genes, 70% were polymorphic; the majority (86%) of the polymorphisms were SNPs. The distributions of the SNPs were similar in exons (44%) and introns (39%). Only 15% of the differences were located in the 3' UTR. In contrast, indels were most often discovered within introns. A SNP occurred every 151 nt, and on average each gene contained 2.7 SNPs. That value is very similar to the 2.8 SNPs per contig reported by Ramírez et al. (2005). This value will certainly increase as the size of the sequenced fragment increases and more genotypes are compared.

An extensive comparison of intron 1 of dihydroflavonol 4-reductase (DFR) measured the sequence variation among 92 genotypes including both landraces and varieties (McClean et al. 2004). Among these genotypes, 20 haplotypes were defined based on 69 polymorphisms. The level of diversity was similar among landraces and varieties. Furthermore, the Middle American gene pool was more diverse than the Andean pool. Recent results with intron 3 of chalcone isomerase show a similar pattern with greater diversity in the Middle American gene pool (McClean and Lee, unpublished results). By contrast, the level of diversity among all genotypes is lower for this region of the genome than for intron 1 of DFR and only 10 haplotypes were observed.

5.4 Genetic Mapping and Tagging in *Phaseolus vulgaris*

5.4.1 Genetic Mapping

Linkage mapping in common beans as in other crops has benefited from a range of molecular technologies that have greatly supplemented the number of genetic markers used in genetic maps for the species. As a result, a large number of markers are in use today for common bean mapping compared to early linkage maps that were based almost entirely on a limited number of morphological markers such as those for flower or seed color or certain pod traits and growth habit characteristics (Bassett 1991). Isozymes and seed proteins, both analyzed based on biochemical assays, were among the first molecular markers to be used in genetic maps but suffered from the limitation of requiring multiple protocols and different source tissues (Gepts 1988; Arndt and Gepts 1989; Koenig and Gepts 1989a; Vallejos and Chase 1991). With the advent of DNA technologies, RFLP markers emerged as the first DNA-based markers in common beans to be used on a large scale (Vallejos et al. 1992; Nodari et al. 1993a; Adam-Blondon et al. 1994). New mapping populations were created at UC Davis and the University of Florida (UF) based on the crosses BAT93 x Jalo EEP558 (F2) and XR-235-1-1 x Calima (BC1) to determine the linkage relationships of large numbers of RFLP markers with totals of 224 Bng and 108 D series clones mapped in these studies. As single copy markers, RFLPs were especially useful for comparative mapping and map integration and led to the creation of the core linkage map of Freyre et al. (1998). In that study, RFLP probes from the initial UC Davis and UF maps plus those of Adam-Blondon et al. (1994) were analyzed to create a core map for the BAT93 x Jalo EEP558 (RIL) population that combined RFLPs with additional marker types including RAPDs, allozymes and known genes. RFLPs were also useful for comparative mapping across legume species as they hybridized well to genomic DNA of soybean and mung bean (Boutin et al. 1995). Sequencing of the Bng probes used to establish the UF map showed them to be rich in gene sequences (Murray et al. 2002).

Polymerase chain reaction (PCR)-based markers added greatly to the efficiency of genetic mapping in common beans by increasing marker throughput, initially with multi-locus marker systems such as those of the RAPD and then AFLP techniques and more recently with single copy marker systems such as STS/SCAR Sequence-Tagged Sites/Sequence-Characterized Amplified Regions and SSR primer pairs. RAPD and AFLP markers were especially useful for saturating RFLP-anchored genetic maps and for creating genetic maps for additional populations

(Freyre et al. 1998; Johnson and Gepts 2002; Tar'an et al. 2002). As genetic maps became more saturated and based on a wide range of molecular marker types, the number of linkage groups coalesced to equal the haploid complement of chromosomes for beans (n = 11). The cumulative genetic distance represented by linkage maps also increased as a greater number of markers was used. Comparative mapping based on RAPD bands was frequently useful for linking genetic maps by correlating the presence of fragments with the evolutionary origin of parental genotypes in either the Mesoamerican or Andean gene pools. The advent of single-copy PCR-based markers such as SSR and STS/SCAR markers has further aided in map comparison and integrating genetic maps. Over 30 STS or SCAR markers developed for many different disease or insect resistance genes and seed color traits have been placed on the core genetic map (McClean et al. 2002; Blair et al. 2006c; Miklas et al. 2006b.

Meanwhile, SSR markers are proving informative due to their multiallelic nature, codominant inheritance, and wide distribution in the genome. So far, SSR markers have been derived from Genbank sequences (Yu et al. 1999, 2000; Métais et al. 2002; Blair et al. 2003; Guerra-Sanz 2004), SSR-enriched genomic libraries (Gaitán-Solís et al. 2002, Yaish and de la Vega 2003; Buso et al. 2006) and bacterial artificial chromosome (BAC) sequences (Caixeta et al. 2005a). The first effort towards the integration of SSR markers into a common bean genetic map was performed by Yu et al. (2000), mapping 15 SSRs onto a framework map based on RAPD and RFLP markers, followed by Blair et al. (2003), integrating 100 SSRs into two linkage maps with AFLP, RAPD, and RFLP markers.

Additional integration of new SSR loci is ongoing at Empresa Brasileira de Pesquisas Agropecuarias Arroz e Feijão (Santo Antônio de Goiás, GO, Brazil) (for BAT93 x Jalo EEP558) and CIAT (for DOR364 x G19833) and will allow the identification of more polymorphic and transferable markers that can be mapped across populations. Furthermore, existing SSRs have been useful in two recent mapping studies by Ochoa et al. (2006) and Blair et al. (2006c) showing their potential for anchoring new genetic maps. Apart from SSR and STS markers, a new set of gene-based markers is being implemented for genetic mapping in common bean that includes TRAP and resistance-gene analogs (RGA)-based markers targeting disease resistance genes (López et al. 2003; Mutlu et al. 2006; Miklas et al. 2006a). Meanwhile, gene-based SNP markers being mapped at CIAT, and mapping of EST sequences at North Dakota State University and at the University of Saskatchewan holds promise for the development of a transcriptional map for beans that could help in the establishment of correlations between candidate genes and specific QTLs.

For the future, use of reference populations, anchor markers, and comparative mapping between populations will continue to assist in the placement of new markers onto genetic maps and the comparison of genetic distances between markers. Comparative mapping has become important for correlating locations of QTL for biotic or abiotic stress resistance, nitrogen fixation, seed characteristics, and root traits (Nodari et al. 1993b; Tar'an et al. 2002; Beebe et al. 2006; Miklas et al. 2006b; Ochoa et al. 2006). In addition, the use of anchor markers, especially those that are highly polymorphic such as SSRs, will allow the genetic mapping of more narrow crosses that are often of interest to plant breeders.

5.4.2 Gene Tagging

More than 30 individual genes for disease resistance and a similar number of genes for QTL underlying major traits with significant impact to common bean agriculture in the tropics have been successfully linked with markers. Such genes tightly linked with markers are referred to as "tagged genes." The primary goal for gene tagging in common bean has been to identify markers tightly linked with disease resistance traits for the purpose of marker-assisted selection. The first linked marker (RAPD A14.1100) was identified for the *Ur-4* rust resistance gene (Miklas et al. 1993) and was used for gene pyramiding and retention of a less effective gene (*Ur-4*) in the presence of an epistatic gene, *Ur-11*, with broad effect against the hypervariable rust pathogen (Stavely et al. 1994). Since then, many resistance genes have been tagged (see recent reviews Miklas et al. 2006b; Kelly et al. 2003; Ragagnin et al. 2005) including *Ur-3*, *Ur-5*, *Ur-6*, *Ur-7*, *Ur-9*, *Ur-11*, and *Ur-13* for resistance to rust; *Co-1*, *Co-1*^{2}, *Co-2*, *Co-4*, *Co-4*^{2}, *Co-5*, *Co-6*, *Co-9*, and *Co-10*, for anthracnose resistance; five *Phg* genes (Caixeta et al. 2005b) for resistance to angular leaf spot; *I*, *bc-1*^{2}, and *bc-3* for bean common mosaic virus; *Bct* for beet curly top virus; *bgm-1* for bean golden yellow mosaic virus (BGYMV); and *Pse-1*, *Pse-2*, and *Pse-3* for resistance to halo bacterial blight. Other yet unnamed R genes for resistance to rust, anthracnose, and angular leaf spot have also been tagged. In addition to R genes, markers tightly linked to QTL with major effect against pathogens causing BGYMV, common bacterial blight, Fusarium root rot, and white mold diseases have been tagged, localized on core linkage maps, and in certain cases used effectively in marker-assisted breeding.

Genetic population structure used to tag genes include: near-isogenic lines (NILs) developed by conventional backcrossing as used to tag *Ur-4* (Miklas et al. 1993) and *Ur-11* (Johnson et al. 1995) genes; NILs developed from heterogeneous inbred lines as used for *Ur-3* (Haley et al. 1994) and *bc-1*^{2} (Miklas et al. 2000); bulked segregant analysis using F_2 or inbred populations including recombinant inbred lines (RILs) as for *bgm-1* (Urrea et al. 1996) and *Bct* (Larsen and Miklas 2004); and a whole map approach as with *Co-2* (Adam-Blondon et al. 1994) and *Ur-9* (Jung et al. 1996) genes. Selective (Miklas et al. 1996, 2003a; Kolkman and Kelly 2003) and whole genome mapping (Jung et al. 1996; Miklas et al. 2001) approaches have been widely used to identify and generate markers for QTL. NILs combined with bulked segregant analysis was used to reduce identification of false positive markers in tagging *Ur-4*, *Ur-11*, and other genes.

Most genes to date have been tagged with RAPD markers which have been converted to SCAR markers (see list by Miklas 2005). Exceptions include the *Ur-13* gene tagged with an AFLP marker (Mienie et al. 2005) and a QTL conferring resistance to white mold linked with the phaseolin seed protein locus (*Phs*) (Miklas et al. 2001). A few codominant RAPD markers have been generated as for *bgm-1* (Urrea et al. 1996), *bc-3* (Johnson et al. 1997), and *Co-4*^{2} (Awale and Kelly 2001), but most markers identified have been dominant and in coupling phase linkage (*cis*) with the R genes and QTL. The predominance of dominant markers reduces the efficiency of marker-assisted selection. Use of dominant markers in repulsion phase (*trans*) linkage alone or in combination with markers in coupling can be used to improve selection efficiency (reviewed by Kelly and Miklas 1998). Codominant markers like SSRs accelerate the identification of homozygous individuals without the necessity for progeny tests, but still too few exist to be effective for gene tagging studies. Ideally, selection of markers flanking the resistance gene would be used to improve marker assisted selection (MAS) efficiency, but such flanking markers are generally not available for the genes and QTL that have been tagged thus far in common bean. For marker-assisted backcrossing, lack of flanking markers can be partially overcome by also selecting for the recurrent parent genomic background outside the target locus using random markers (Hagiwara et al. 2001). Other gene-sequence based marker systems like RGA (López et al. 2003), Resistance Gene Analog Polymorphism (Mutlu et al. 2006), and TRAP (Miklas et al. 2006a) show promise for tagging targeted traits but have not yet lead to MAS. As EST databases expand for common bean, concurrently STS and SNP markers will be generated, leading to additional and more direct marker systems for tagging genes.

The utility of tightly linked markers for MAS has been evaluated by surveying an array of accessions with and without the gene for presence and absence of the marker. These surveys have revealed the utility for many markers to be limited to specific gene pools (Miklas et al. 1993). Once limitations of a marker are understood, MAS can be implemented more effectively as a breeding tool. The utility of QTL for MAS is determined by expression of the QTL in multiple environments and in different populations, and by the effectiveness of MAS itself.

MAS has been implemented in different breeding strategies for common bean improvement. Gene pyramiding, as in the example for marker-assisted detection of *Ur-4* gene in the presence of *Ur-11*, led to release of Beldakmi-RR-7 navy bean (Stavely et al. 1994). Kidney and cranberry bean germplasm releases (Miklas et al. 2002; Miklas and Kelly 2002) benefited from MAS of the hypostatic *I* gene in the presence of *bc-3* to facilitate combining genes for more durable resistance to bean common mosaic virus (BCMV) and bean common mosaic necrosis virus (BCMNV). Marker-assisted backcrossing was used to rapidly deploy resistance genes to combat emerging disease epidemics: *bgm-1* in snap bean RGMR lines (Stavely et al. 2001) to combat BGYMV in Florida and *Co-4*^{2} in USPT-ANT-1 pinto (Miklas et al. 2003b) to combat anthracnose in Minnesota and North Dakota. The former exemplifies the additional benefits of marker-assisted backcrossing of a recessive gene and a gene that would otherwise be difficult to detect by direct pathogen screening. Resistance to BGYMV in the GMR lines (Singh et al. 2000a, 2000b) was selected and characterized by presence of markers for an R gene *(bgm-1)* and a major QTL conditioning resistance (linked with the SW12 marker). MAS for multiple QTL in early generations followed by phenotypic selection with the pathogen in later generations resulted in the release of pinto and kidney bean germplasm lines with improved resistance to common bacterial blight (Miklas et al. 2006c, 2006d). ABCP-8 pinto with enhanced resistance to common bacterial blight benefited from MAS for a single QTL (Mutlu et al. 2005a, 2005b).

The applications above pertain to breeding for resistance to individual pathogens, whereas, an even greater utility for MAS in the future will be simultaneous introgression of R genes conferring resistance to different pathogens in the same genetic background. The concept of multiple disease resistance is relatively straightforward; however, it is not easily accomplished by conventional means due to the inherent difficulty associated with the simultaneous monitoring of several different R genes by direct pathogen screening. Defining reaction symptoms after multiple inoculations with different pathogens is an arduous task as is discerning epistatic interactions between the R genes. The availability of molecular markers tightly linked to the R genes can bypass these difficulties because the markers are not affected by the environment or by epistatic interactions among the genes to which they are linked. Multiple-disease-resistant Mesoamerican bean lines with "carioca-type" grains were developed by MAS for R genes: *Co-4*, *Co-6*, *Co-10*, *Ur-OuroNegro*, and *Phg-1* (Ragagnin et al. 2005). Subsequently, *Co-4* was replaced by the more effective allele, *Co-4*^{2} (Alzate-Marin et al. 2005). Similarly, MAS was used to develop black and red bean cultivars with multiple disease resistance (Costa et al. 2006). Breeding for multiple disease resistance using MAS is a dynamic process, and new resistance genes can be incorporated at any moment as long as the gene of interest is 'tagged' with tightly linked markers and introduced in the appropriate genetic background.

5.5 BAC Cloning and Utilization

5.5.1 Development of BAC Libraries

In the genus *Phaseolus*, eleven BAC libraries have been constructed (Table 5.1). Although most of these libraries have relatively high genome equivalents, it must be noted that this, as with all BAC libraries, represents a statistical assessment based upon the assumption that the restriction sites are completely random throughout the genome. Because it has been demonstrated that this is not the case, (e.g., Nodari et al. 1992), the genome coverage must be viewed only as a probability of a particular region being represented, not as an absolute representation of the number of times a region will appear in a library. For a more complete coverage of a region, several libraries constructed with different enzymes should be constructed and assayed.

The BAC libraries listed here represent a unique phylogenetically ordered set of libraries for use in evolutionary studies as well (Fig. 5.1). In addition to a library in *P. lunatus* cv. Henderson, chosen as an outgroup, all the other libraries were developed in *P. vulgaris*. These *P. vulgaris* libraries represent key steps in the intra-specific evolution and domestication of common bean. DGD1962 is a wild bean from northern Peru, representing the presumed ancestral gene pool of the species (Debouck et al. 1993; Kami et al. 1995). The remainder of the libraries is distributed in the two evolutionary lineages that were domesticated. In the Mesoamerican lineage G02771 and G12946 are wild Mexican beans that contain the three subfamilies of the Arcelin-Phytohemagglutinin-Alpha-amylase inhibitor (APA) seed proteins,

Table 5.1 Some characteristics of BAC libraries developed in *Phaseolus* spp.

Library Species[1]	Genotype	Number of Clones	Restriction Site	Average Clone Size (kB)	Genomes	Empty Clones	Chloroplast (%)	Reference[2]
Pv	Sprite	33,792	*Eco*RI	90	4.79			a
Pv	BAT93	110,592	*Hind*III	125	20.8	< 0.5	0.05	b
Pv	DGD1962	55,296	*Hind*III	105	8.7	< 0.5	0.4	b
Pv	G02771	55,296	*Hind*III	139	12.1	< 0.5	0.08	b
Pv	G19833	55,296	*Hind*III	145	12			c
Pv	G12949	30,720	*Hind*III	135	6.5	-		d
Pv	OAC H45	33,024	*Hind*III	100	5	~ 0		e
Pv	HR67	22,560	*Bam*HI	300	8.1	< 0.5		f
Pv	OAC Rex	31,776	*Bam*HI	150	5.6	< 0.5		f
Pv	G02333	24,960	*Hind*III	125	6			g
Pl	Henderson	55,296	*Hind*III	110	9.5	< 0.5	0.03	b

[1] Pv: *Phaseolus vulgaris*; *Pl*: *Phaseolus lunatus*
[2] a: Vanhouten and Mackenzie 1999; b: Kami et al. 2006; c: M. Blair, pers. comm..; d: Galasso et al. 2005; e: Yu et al. 2006: f: Perry et al. 2006; g: Melotto et al. 2003

including the arcelin subfamily (see section 5.4.2). G02333 is a Mexican landrace highly resistant to anthracnose.

BAT93 and OAC-HR45 and OAC-HR67 are breeding lines and OAC-Rex is a cultivar from the Mesoamerican gene pool. Such an array of BAC libraries can help in describing the structural modifications that have accompanied phenotypic changes occurring during domestication, not only in the genes themselves, but also in adjacent, regulatory regions. Furthermore, they will allow an analysis of

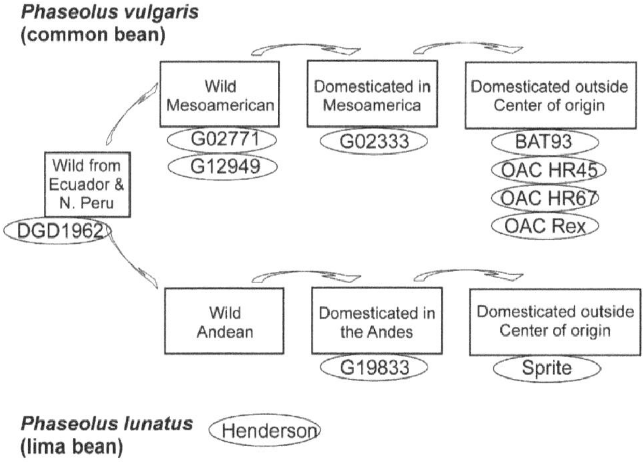

Fig. 5.1 Phylogenetic and genealogical distribution of BAC libraries in *Phaseolus* spp. Boxes represent different segments of the *P. vulgaris* gene pools and the general direction of their evolution. Names surrounded by ellipses are the genotypes in which BAC libraries have been established

the effects, if any, of selection exerted during and after domestication on genetic diversity in chromosome regions harboring domestication genes.

BAC libraries of the common bacterial blight (CBB) resistant variety OAC-Rex (Michaels et al. 2005), derived from OAC95-4 (Tar'an et al. 2001), and the CBB-resistant breeding line HR67 were constructed with a BiBAC2 vector that is designed to be used directly for plant transformation with Agrobacterium (Chang et al. 2003). The OAC-Rex library is being screened with a probe derived from the pvCTT001 microsatellite marker associated with a major CBB resistance locus on the B5 linkage group (Tar'an et al. 2001).

In the Andean gene pool, G19833 is a landrace from Peru, whereas Sprite is a bred variety. Thus, using this array of BAC libraries, it is possible to study overall structural evolution of the genome in *Phaseolus* both prior and after domestication. It is also possible to analyze phenotypic changes resulting from specific structural modification at the genome level. Some of the first information obtained towards this goal is shown in the next section for two disease resistance loci (*Co-4* and *I*) and one pest resistance locus (*APA*).

5.5.2 Whole-BAC Mapping and Sequencing

With the recent availability of BAC libraries in *Phaseolus* sp., a number of BACs have been completely sequenced for analysis. These results demonstrate both the strength and some of the weakness of this technology.

Using BAC clones identified from the Sprite library, Melotto et al. (2004) obtained the complete sequence of a 106,000 bp clone containing part of the *Co-4* locus for resistance to anthracnose. Analysis of the sequence demonstrated five copies of the *COK-4* gene, which is the putative gene responsible for the fungal resistance. Sequence alignment and phylogenetic analysis indicated that this gene has undergone several rounds of duplication and divergence. In addition, 19 other putative genes were identified by a BLAST homology search that had not been isolated from *Phaseolus* prior to this analysis. At least four putative retrotransposon elements were also identified. At the time of publication, this was the largest contiguous DNA sequence available for *P. vulgaris*.

Using BAC-end sequencing and hybridization, Vallejos et al. (2006) were able to construct a 425 kb contig around the *I* gene that conveys resistance to BCMV. This contig contained a large cluster of TIR-NBS-LRR sequences. Susceptible cultivars harboring the recessive *i* allele displayed simpler, more variable haplotypes compared to the resistant cultivars harboring the dominant *I* allele, which showed a single haplotype. Further data suggested that this locus may have expanded during evolution and domestication (Vallejos et al. 2006).

Kami et al. (2006) published the complete sequence for a BAC clone from the G02771 library containing the arcelin, phytohaemagglutinin, alpha-amylase inhibitor (APA) locus that is responsible for resistance to bruchid insects. G02771 is considered to be one of the more recently evolved *Phaseolus vulgaris* genotypes based on the presence of the most recently identified variant of the APA family, the arcelins. As in the previous examples, this gene family has multiple members, two putative arcelins, three phytohaemagglutinins, and one alpha-amylase inhibitor, and has also undergone several duplication and divergence events that correlate with a change in function as well. At least six putative retrotransposons were identified in this clone along with 10 other putative coding regions unrelated to the APA gene family. A separate region was characterized in the clone that has high (>95%) identity to a portion of the chloroplast genome. Although these nuclear encoded chloroplast "splinters" have been documented in the Arabidopsis and rice genomes, this is the first time they have been observed in *Phaseolus* (Shahmuradov et al. 2003). Additional APA-containing clones from different BAC libraries (Table 5.1) will be sequenced in the near future to further understand the evolution of the APA locus. Already, Sparvoli and coworkers have undertaken a similar analysis in the G12949 genotype (Galasso et al. 2005). The BAC library screening resulted in three overlapping clones, containing several arcelin variants, the phytohemaglutinin-L and phytohemagglutinin-E genes and many retrotrotransposons. A fourth BAC clone, containing a second phytohemagglutinin-L gene and the alpha-amylase inhibitor gene, was part of the APA locus, but it was not possible to link it to the other three BACs (Galasso et al. unpublished results). Preliminary comparison with data obtained by Kami et al. (2006) shows a different degree of complexity of the APA locus between the two wild genotypes, demonstrating the utility of different libraries to describe the evolutionary pathway followed by a single complex locus and the need of further BAC sequencing to propose an evolutionary model for the APA gene family.

As these studies demonstrate, in addition to yielding information about the genes of primary interest, whole BAC sequencing can yield a lot of unanticipated information about genomic structure and genetic information. However, much of this information is limited in that there is currently a lack of sequences from other varieties and species from which to derive a useful hypothesis. As the cost of whole BAC sequencing declines, it is hoped more supportive sequence information will become available.

5.6 Molecular Cytogenetics

Common bean is a diploid species with 22 chromosomes (Sarbhoy 1978; Maréchal et al. 1978; and references therein). The chromosomes are small in size and similar in morphology. Therefore, it had not been possible for a long time to recognize all chromosome pairs, even when banding techniques were applied (Zheng et al. 1991). Although common bean, as other species in the genus, develops giant, polytene chromosomes in the suspensor cells of the immature proembryo (Nagl 1969), these chromosomes are not well-suited for detailed cytogenetic studies, because the endoreduplicated sister chromatids do not pair along their entire lengths, giving these chromosomes a loose appearance (Pedrosa 2003).

Major advances in the chromosome analysis in the species came with the application of the fluorescence in situ hybridization technique (FISH). This technique enables the localization of different DNA sequences along the chromosomes, making the identification of particular chromosome pairs possible. The first sequences to be localized on common bean mitotic chromosomes were the 5S and the 45S rDNA gene clusters (Moscone et al. 1999). Combined with chromosome morphology and the distribution of constitutive heterochromatin, these sequences could be used as chromosome markers to identify almost all chromosome pairs of the species. Interestingly, the two cultivars that were analyzed in this work had different number and size of 45S rDNA sites, which was sufficient to change the size of a few chromosome pairs. This analysis was therefore expanded to 37 accessions of common bean, representing the known genetic diversity of the species, and it became clear that this gene family has been amplified in the Andean gene pool, whereas the Mesoamerican gene pool seems to have retained the ancestral number of loci (3) in most of its representatives (Pedrosa-Harand et al. 2006).

By the time Moscone and co-workers published their work, several genetic linkage maps were already available for the species, including a unified core map (Freyre et al. 1998). The correlation between linkage groups and chromosomes was, however, not elucidated. Using pooled RFLP markers from each linkage group of a genetic map, this correlation could be established (Pedrosa et al. 2003). Each pool of markers could be used to identify a chromosome pair and an improved ideogram was constructed. This analysis showed no correspondence between linkage group and chromosome sizes. To understand these differences in recombination among chromosomes, a detailed chromosome map is now being built for common bean, similar to the existing maps of both model legumes (Pedrosa et al. 2002; Kulikova et al. 2001). Genetic markers from each linkage group have been used to select BACs, now available from different libraries. The selected clones are individually mapped to the chromosomes of the species, revealing the localization of the corresponding marker. A BAC clone corresponding to an anthracnose resistance gene locus has been mapped to linkage group B8 according to the common bean core linkage map (Melotto et al. 2004). BACs corresponding to several other genomic regions have been selected and have been used to build up the chromosome map. This map will enable the comparison of genetic and physical distances in several genomic regions and will anchor a future physical map in relation to the major chromosome landmarks: centromere, telomeres, and heterochromatin. The same BACs will be available for comparative studies with closely related species, giving insights into the chromosome evolution mechanisms in this group.

5.7 Functional Genomics

5.7.1 EST Development

Partial sequencing of cDNA inserts or expressed ESTs, obtained from many tissues and organs, has been used as an effective method for gene discovery, molecular marker generation, and transcription pattern characterization. It is an efficient approach for identifying large number of plant genes expressed during different developmental stages and in response to a variety of environmental conditions.

In the last years, the number of *P. vulgaris* EST sequences has increased considerably. As of June 2006, 21,377 *P. vulgaris* EST sequences are available at the GenBank, National Center for Biotechnology Information (NCBI) (http://www.ncbi.nlm.nih.gov/dbEST/). Table 5.2 summarizes the information on publicly available cDNA libraries that have been used to generate EST sequences. The libraries are derived from different *P. vulgaris* genotypes including Mesoamerican and Andean gene pools, different plant organs and biological conditions. Contributions to the development of ESTs come from collaborative projects performed within the frame of Phaseomics among scientists from academic institutions in Brazil, Colombia, Mexico, and the United States.

The complete EST data set, generated by Ramírez et al. (2005) and Melotto et al. (2005), has been assembled into contigs to provide a single *P. vulgaris* gene index containing 20,578 ESTs (Graham et al. 2006). Of these, 6,787 were classified as singletons and the remaining assembled into 2,883 contigs (http://www.ccg.unam.mx/phaseolusest/) (Graham et al. 2006). In addition, in July 2005, The Institute for Genomic Research (TIGR) released version 1.0 of the TIGR Common_bean Gene Index (http://www.tigr.org). The input sequences are 21,290 and the output sequences are classified as 2,906 tentative consensus (TC) sequences and 6,578 singletons ESTs, which are values similar to those reported by Graham et al. (2006).

Table 5.2 Currently available *Phaseolus vulgaris* cDNA libraries and EST sequences

Genotype	Tissue/condition	Total ESTs	Attribution	Reference
Mesoamerican Negro Jamapa	Nodules elicited by *Rhizobium tropici* CIAT899	4,636	UNAM/UM[1]	Ramírez et al. 2005
Mesoamerican Negro Jamapa	Pods	3,667	UNAM/UM[1]	Ramírez et al. 2005
Mesoamerican Negro Jamapa	Roots	4,329	UNAM/UM[1]	Ramírez et al. 2005
Mesoamerican Negro Jamapa	Leaves	3,456	UNAM/UM[1]	Ramírez et al. 2005
Andean G19833	Leaves	4,938	CIAT[2]	Ramírez et al. 2005
Mesoamerican SEL 1308	Leaves	449	USP[3]	Melotto et al. 2005
Mesoamerican SEL 1308	Seedling shoots	2,474	USP[3]	Melotto et al. 2005
Mesoamerican SEL 1308	Seedling shoots inoculated with *Colletotrichum lindemuthianum*	2,332	USP[3]	Melotto et al. 2005

[1] Universidad Nacional Autónoma de México/University of Minnesota
[2] Centro Internacional de Agricultura Tropical, Cali, Colombia
[3] Universidade de São Paulo

The *P. vulgaris* EST sequence resource developed has provided tools for projects oriented to characterize the transcript profile of different bean organs and/or growth conditions. For instance, macroarray technology was used for transcriptome analysis of bean mature nodules elicited by the nitrogen fixing bacteria: *Rhizobium tropici* (Ramírez et al. 2005). This study led to the identification of genes over-expressed in nodules as compared to other plant organs such as roots, leaves, stems, and pods. Nodule transcript profile showed that genes related to nitrogen and carbon metabolism are integrated for ureide production (Ramírez et al. 2005).

In *P. coccineus*, some 20,000 ESTs were isolated from globular stage embryos of developing seed and deposited in GenBank by R. Goldberg (http://www.ncbi.nlm.nih.gov/entrez/query.fcgi?db=Nucleotide&cmd=search&term=phaseolus+and+coccineus+and+embryo+and+est). Because of the close relationship between *P. vulgaris* and *P. coccineus*, these sequences can be used to isolate homologous sequences in common bean (Nanni et al. 2005).

Because genetic and genomic resources of common bean are limited, a comparative genomic approach using bean ESTs and those of other plant species is important to determine the usefulness of sequence information from other plants to study common bean. Bean unigenes share more orthologous sequences with soybean and *Arabidopsis thaliana* unigenes than with *Orzya sativa* and *Zea mays* unigenes (Melotto et al. 2005). However the degree of conservation among genes of bean, soybean, and *A. thaliana* varies between functional categories (Melotto et al. 2005). Single linkage clustering analysis of homologous sequences among five different plant species (*P. vulgaris, Lupinus albus*, soybean, *M. truncatula*, and *A. thaliana*) has been used to identify candidate genes that are relevant for adaptation to phosphorus deficiency in bean and are shared with other species (Graham et al. 2006).

Ongoing projects within the Phaseomics community will contribute to increase the number of *P. vulgaris* ESTs sequences in the near future. Kirstin Bett's group (U. Saskatchewan, Canada) is sequencing 20,000 ESTs from subtraction suppression cDNA libraries prepared from above-ground plant tissue of *P. vulgaris* cv. ICA Pijao and *P. angustissimus* PI535272 subjected to cold treatment (K. Bett, personal communication). Gary Stacey's group (U. of Missouri, USA) is sequencing a total of 20,000 ESTs from normalized subtracted cDNA libraries prepared from leaves of *P. vulgaris* cv. Early Gallatin treated with the fungal pathogen *Uromyces* (G. Stacey, personal communication). Mario Aguilar's group (U. Nacional de La Plata, Argentina) is sequencing 6,000 ESTs from a cDNA library prepared from *P. vulgaris* cv. Negro Jamapa hairy roots harvested after a short time of inoculation with *Rhizobium etli* SC15 (M. Aguilar, personal communication). Comprehensive information on EST sequences will provide relevant genetic and genomic resources that may be used to achieve the ultimate goal of improving bean crop quality and production.

5.7.2 TILLING

Targeted induced local lesions in genomes (TILLING) is a powerful reverse genetic approach that uses gene-specific primers for the identification of mutants of a gene of interest from a large mutagenesis population (McCallum et al. 2000). Theoretically, given a sufficient population size, genome saturation can be achieved and mutants for any gene can be identified. Tagged mutagenesis, such as T-DNA or transposon-based systems, generally requires an effective transformation system. TILLING, however, does not rely on transformation and thus allows for reverse genetic approaches in transformation recalcitrant species, such as common bean (for transformation, however, see section 5.7.3).

Significant advances have been made in the development of a TILLING platform in common bean. The common bean genotype BAT93, from the Mesoamerican gene pool, was selected for TILLING by the common bean research community because it is one of the parents of the core genetic map (Freyre et al. 1998), it has been used in the generation of a large BAC library (Kami et al. 2006; see also section 5.5.1), has broad adaptation, and has desirable characteristics, such as disease resistance (Broughton et al. 2003). Based on genome saturation in other species, such as *Lotus japonicus* (Perry et al. 2003), it was estimated that about 5,000 mutagenesis lines would be required for genome saturation in BAT93. The magnitude of the project required collaboration and a TILLING consortium for tool development was created, including the University of Geneva (Geneva, Switzerland), USDA/ARS/TARS (Mayagüez, Puerto Rico), and CIAT (Cali, Colombia). Studies optimizing mutagen concentration determined that 35–40 mM ethylmethane sulfonate (EMS) resulted in the desired lethality rate of 60–70% in BAT93. Mutagenesis was confirmed using a screen for nodulation mutants and found that 35 mM EMS resulted in 10% putative nodulation-deficient mutants in a population of 348 M2 lines (C. Pankhurst, P. Lariguet, and W. Broughton, unpublished results). Thus, the initial BAT93 mutagenesis population showed a high mutation frequency and proved effective for forward genetic screening.

The TILLING consortium has currently produced about 1,500 M2 families and has advanced about 900 families to the M3 generation. DNA has been extracted from all M2 families and these DNAs will be used to confirm appropriate mutagen concentration and for TILLING analysis. To manage the large number of lines in the mutant population, the template DNA will be multiplexed for TILLING analysis and a database will be created for the organization of phenotypic and genotypic data. A critical component of TILLING is access to a genomic sequence for primer development. Common bean genomic and cDNA sequences of the selected loci will be searched using databases from other species such as *Medicago*, *Lotus*, soybean, pea, and *Arabidopsis* and from the common bean EST collections. Once the sequences of *P. vulgaris* loci of interest are available, gene specific primers will be designed using codons optimized to detect deleterious lesions (CODDLE) , developed for the discovery of deleterious lesions within coding sequences. The consortium expects to identify allelic series, which will allow for the study of gene function, as has been the case in previous TILLING efforts (Henikoff et al. 2004; McCallum et al. 2000).

These initial studies have shown that the common bean EMS mutagenesis protocol is effective at generating mutants and that mutants can be identified using appropriate screening protocols. To expand access to this TILLING platform, the consortium will establish seed banks for screening and distribution of the BAT93 mutagenesis population at its three locations and will provide the mechanisms for reverse genetic screening of the population through TILLING analysis. TILLING in common bean will have wide applications for both basic and applied research, as requests for mutants have already been received.

5.7.3 Transformation

Common bean has been transformed using two main approaches: particle bombardment or biolistics and Agrobacterium transformation. The applicability of the particle bombardment to introduce and transiently express genes into dry bean was demonstrated in the beginning of 1990s (Genga et al. 1992; Aragão et al. 1992, 1993). The bombardment of meristematic cells of embryonic axes revealed that foreign genes could efficiently reach the superficial cell layers, demonstrating the feasibility of producing transgenic bean using this method (Aragão et al. 1993). Russell et al. (1993) reported the production of transgenic navy bean plants using an electrical particle acceleration device. However, the protocol was time-consuming and the frequency of transformation was low (0.03%) and variety-limited. Later, Kim and Minamikawa (1996) reported the recovery of transgenic bean plants (cv. Goldstar) by bombarding embryonic axes. A transformation system was developed for regeneration of transgenic bean plants based on the bombardment of apical meristematic regions associated with an efficient tissue culture protocol for multiple shoot induction, elongation, and rooting (Aragão et al. 1996; Aragão and Rech 1997). The frequency of transformation ranged from 0.2 to 0.9% using linear or circular plasmid vectors, respectively (Aragão et al. 1996; Vianna et al. 2004). Morphology of the explants used during bombardment may influence the generation of transgenic plants because some varieties present the central region of the meristematic region covered by the primordial leaves (Aragão and Rech 1997). As shoot differentiation occurred in the peripheral layers of the meristematic ring, the number of cells that could be reached by the microparticles coated-DNA will be drastically reduced. Consequently, the efficiency of transformation could also be reduced.

One important constraint in the transformation system based on the bombardment of meristematic tissue of embryonic axes is the difficulty of having efficient selection of transformed cells. Using the selective agent imazapyr, an herbicidal molecule of the imidazolinone class capable of systemically translocating and concentrating in the apical meristematic region of the plant, we have been able to increase the recovery of fertile soybean (Aragão et al. 2000) and, more recently, dry bean plants (Bonfirm et al. 2007).

Since the first report on *P. vulgaris* transformation in 1993, a few groups have transformed bean to introduce agronomic traits. Our group has introduced useful genes, such as the *be2s1* gene to improve the methionine content of the seeds. In two of the five transgenic lines, the methionine content was significantly increased by 14 and 23% compared to that in non-transformed plants (Aragão et al. 1999). Russell et al. (1993) introduced into bean the *bar* gene, which encodes the phosphinothricin acetyl transferase (PAT) and confers resistance to both phosphinothricin and the herbicide glufosinate ammonium, and the coat protein gene from the bean golden mosaic geminivirus (BGMV) in an attempt to produce virus-resistant plants. The introduced *bar* gene was shown to confer resistance to the herbicide under glasshouse conditions. However, transgenic bean plants expressing the BGMV coat protein gene did not exhibit virus resistance (D. Maxwell, unpublished results). Recently, we have generated highly BGMV-resistant plants by expression of a mutated *AC1* viral gene (Faria et al. 2006) and using interfering RNA (RNAi) (F.J.L Aragão and J. C. Faria, unpublished results). The transgenic bean lines were introduced into the breeding program to evaluate gene expression in different genetic backgrounds under glasshouse and field conditions. Transgenic bean lines containing the *bar* gene and resistant to the herbicide glufosinate ammonium were tested in field (Aragão et al. 2002). This was the first field release of a transgenic line of *P. vulgaris*.

Agrobacterium tumefaciens-assisted transformation is a standard protocol successfully applied to generate transgenic legume plants of some *Phaseolus* species (Zambre et al. 2005), although the frequency of transformation for *Phaseolus vulgaris* (common bean) has been very low (Zhang et al. 1997). The only report where *A. tumefaciens* was used to achieve expression of a lea gene that conferred abiotic tolerance in *P. vulgaris* is a very recent protocol based on sonication and vacuum assistance (Liu et al. 2005). Although this novel and cheap transformation method looks quite promising, the transformation efficiency must be improved.

Agrobacterium rhizogenes is a soil bacterium responsible for development of hairy root disease on a range of dicotyledonous plants (Tepfer 1990). Infection at the wounding sites by *A. rhizogenes* results in the transfer, integration, and expression of T-DNA from the root-inducing plasmid into the plant cells (Grant et al. 1991). Although *A. rhizogenes*-mediated root transformation has been described for a number of legumes, the induction of composite plants has not been previously reported for common bean (Díaz et al. 2000; Lee et al. 1993; Cheon et al. 1993; Stiller et al. 1997; Boisson-Dernier et al. 2001; and Van de Velde et al. 2003). Provided that root transformation is efficient, such composite plants are a simple, fast and reproducible strategic alternative to stable transformed lines, for functional genomics programs. A protocol originally developed for soybean (Bond and Gresshoff 1993) has been successfully modified for application in bean. After trying several *A. rhizogenes* strains (5), strain K599 (cucumopine-type) was the most efficient to induce hairy roots on several wild accessions (3), landraces (6), and cultivars (2) of *P. vulgaris* and three other species within the genus *Phaseolus* [*P. coccineus* (2), *P. acutifolius* (2), and *P. lunatus* (2)]. High transformation efficiency rates (75–90% frequency) were reproducibly found in three independent experiments with different *P. vulgaris* wild accessions, landraces and cultivars.

Briefly, for hairy root induction, beans seedlings are infected with a fresh culture of *A. rhizogenes* K599 by wounding the cotyledonary nodes with a syringe, and then covered with a plastic lid to increase humidity, because hairy roots are very sensitive to desiccation. Over 5–8 days a globular tumor develops on the stem at the wound site. Visible hairy roots emerged from the tumor as early as one week after infection. At three weeks post-infection, between 85–100% plants had abundant hairy roots.

Stem with primary root was removed from the plant by cutting about 1 cm below the cotyledon nodes. Plants with induced roots were repotted in fresh sterile vermiculite (Estrada-Navarrete et al. 2006). Transgenic roots in the genus *Phaseolus* offers a new strategy for over-expressing or suppressing endogenous genes. This method can be readily scaled up to perform functional genomics focused on root biology and root-microbe interactions.

5.8 Perspective

The ultimate goal of the Phaseomics (Phaseolus Genomics) Initiative is to sequence the common bean genome as a platform for the more efficient improvement of common bean (Broughton et al. 2003). With relatively limited resources, the Phaseolus community has laid the foundation to achieve this goal by developing genomic resources such as an impressive collection of germplasm and genetic stocks, mapping populations, BAC libraries, an incipient EST and TILLING assembly, and the development of tools such as FISH and transformation. Further work is necessary, however, including the development of a full physical map (including BAC-end sequencing) and correlation of this map with the genetic map, further assembly of EST and TILLING collections, establishment of a transcript map, and increasing the efficiency of transformation, among others. Already bean breeders have been very active in translating genomic information into mapping and tagging genes for agronomic interest for marker-assisted selection. Several cultivars and improved germplasm accessions have been released, which resulted from MAS. Currently, the limiting factor for MAS is the number of markers of known map location, primarily because of the lack of DNA sequence information available. A coordinated genomic effort in *Phaseolus* closely matched with breeders is a necessary condition for the full use of MAS and an accelerated development of improved bean cultivars and, hence, a more diverse and nutritious diet.

References

Adam-Blondon A, Sévignac M, Bannerot H, Dron M (1994) SCAR, RAPD and RFLP markers tightly linked to a dominant gene (*Are*) conferring resistance to anthracnose in common bean. Theor Appl Genet 88:865–870

Aguilar OM, Riva O, Peltzer E (2004) Analysis of *Rhizobium etli* and of its symbiosis with wild *Phaseolus vulgaris* supports coevolution in centers of host diversification. Proc Natl Acad Sci USA 101:13548–13553

Alzate-Marin AL, Arruda KM, Souza KA, Barros EG, Moreira MA (2005) Introgression of *Co-4*^{2} and *Co-5* anthracnose resistance genes into "Carioca" common bean cultivars with the aid of MAS. Ann Rep Bean Improv Coop 48:70–71

Aragão FJL, Rech EL (1997) Morphological factors influencing recovery of transgenic bean plants (*Phaseolus vulgaris* L) of a Carioca cultivar. Int J Plant Sci 158:157–163

Aragão FJL, Desa MFG, Almeida ER, Gander ES, Rech EL (1992) Particle bombardment-mediated transient expression of a Brazil nut methionine-rich albumin in bean (*Phaseolus vulgaris* L). Plant Mol Biol 20:357–359

Aragão FJL, Desa MFG, Davey MR, Brasileiro ACM, Faria JC, Rech EL (1993) Factors influencing transient gene-expression in bean (*Phaseolus vulgaris* L) using an electrical particle-acceleration device. Plant Cell Rep 12:483–490

Aragão F, Barros L, Brasileiro A, Ribeiro S, Smith F, Sanford J, Faria J, Rech E (1996) Inheritance of foreign genes in transgenic bean (*Phaseolus vulgaris* L) co-transformed via particle bombardment. Theor Appl Genet 93:142–150

Aragão FJL, Barros LMG, de Sousa MV, Grossi de Sa MF, Almeida ERP, Gander ES, Rech EL (1999) Expression of a methionine-rich storage albumin from the Brazil nut (*Bertholletia excelsa* H.B.K., Lecythidaceae) in transgenic bean plants (*Phaseolus vulgaris* L., Fabaceae). Genet Mol Biol 22:445–449

Aragão FJL, Sarokin L, Vianna GR, Rech EL (2000) Selection of transgenic meristematic cells utilizing a herbicidal molecule results in the recovery of fertile transgenic soybean [*Glycine max* (L.) Merril] plants at a high frequency. Theor Appl Genet 101:1–6

Aragão FJL, Vianna GR, Albino MMC, Rech EL (2002) Transgenic dry bean tolerant to the herbicide glufosinate ammonium. Crop Sci 42:1298–1302

Araya CM, Alleyne AT, Steadman JR, Eskridge KM, Coyne AP (2004) Phenotypic and genotypic characterization of *Uromyces appendiculatus* from *Phaseolus vulgaris* in the Americas. Plant Dis 88:830–836

Arndt GC, Gepts P (1989) Segregation and linkage for morphological and biochemical markers in a wide cross in common bean (*Phaseolus vulgaris*). Ann Rep Bean Improv Coop 32:68–69

Awale HE, Kelly JD (2001) Development of SCAR markers linked to $Co\text{-}4^2$ gene in common bean. Ann Rpt Bean Improv Coop 44:119–120

Bai YH, Michaels TE, Pauls KP (1997) Identification of RAPD markers linked to common bacterial blight resistance genes in *Phaseolus vulgaris* L. Genome 40:544–551

Bassett MJ (1991) A revised linkage map of common bean. HortSci 26:834–836

Becerra-Velásquez VL, Gepts P (1994) RFLP diversity in common bean (*Phaseolus vulgaris* L). Genome 37:256–263

Beebe S, Gonzalez AM, Rengifo J (2000a) Research on trace minerals in the common bean. Food and Nutrition Bull 21:387–391

Beebe S, Skroch PW, Tohme J, Duque MC, Pedraza F, et al. (2000b) Structure of genetic diversity among common bean landraces of Middle American origin based on correspondence analysis of RAPD. Crop Sci 40:264–273

Beebe S, Rengifo J, Gaitan E, Duque MC, Tohme J (2001) Diversity and origin of Andean landraces of common bean. Crop Sci 41:854–862

Beebe SE, Rojas-Pierce M, Yan X, Blair MW, Pedraza F, et al. (2006) Quantitative trait loci for root architecture traits correlated with phosphorus acquisition in common bean. Crop Sci 46:413–423

Bennett M, Leitch I (2005) Angiosperm DNA C-Values database. Release 4.0 http://www.rbgkew.org.uk/cval/database1.html

Blair MW, Pedraza F, Buendia HF, Gaitán-Solís E, Beebe SE, et al. (2003) Development of a genome-wide anchored microsatellite map for common bean (*Phaseolus vulgaris* L). Theor Appl Genet 107, 1362–1374

Blair MW, Giraldo MC, Buendia HF, Tovar E, Duque MC, et al. (2006a) Microsatellite marker diversity in common bean (*Phaseolus vulgaris* L). Theor Appl Genet 113:100–109

Blair MW, Iriarte G, Beebe S (2006b) QTL analysis of yield traits in an advanced backcross population derived from a cultivated Andean x wild common bean (*Phaseolus vulgaris* L) cross. Theor Appl Genet 112:1149–1163

Blair MW, Muñoz C, Garza R, Cardona C (2006c) Molecular mapping of genes for resistance to the bean pod weevil (*Apion godmani* Wagner) in common bean. Theor Appl Genet 112: 913–923

Boisson-Dernier A, Chabaud M, Garcia F, Becard G, Rosenberg C, et al. (2001) Agrobacterium rhizogenes-transformed roots of *Medicago truncatula* for the study of nitrogen-fixing and endomycorrhizal symbiotic associations. Molec Plant-Microbe Inter 14:695–700

Bond JE, Gresshoff PM (1993) Soybean transformation to study molecular physiology. In: Gresshoff PM (ed) Plant responses to the environment. CRC Press, Boca Raton, FL, USA, pp 25–44

Bonfim K, Faria JC, Nogueira EOPL, Mendes EA, Aragão FJL (2007) RNAi-Mediated Resistance to Bean Golden Mosaic Virus in genetically engineered common bean (*Phaseolus vulgaris*) Molec Plant-Microbe Inter 20:717–726

Boutin S, Young N, Olson T, Yu Z, Shoemaker R, et al. (1995) Genome conservation among three legume genera detected with DNA markers. Genetics 38:928–937

Bressani R (1983) Research needs to upgrade the nutritional quality of common beans (*Phaseolus vulgaris*). Qualitas Plantarum Plant Foods Human Nutrition 32:101–110

Broughton WJ, Hernandez G, Blair M, Beebe S, Gepts P, et al. (2003) Beans (*Phaseolus* spp.) – model food legumes. Plant Soil 252:55–128

Buso GSC, Amaral ZPS, Brondani RPV, Ferreira ME (2006) Microsatellite markers for the common bean – *Phaseolus vulgaris*. Mol Ecol Notes 6:252–254

Caixeta ET, Borém A, Kelly JD (2005a) Development of microsatellite markers based on BAC common bean clones. Crop Breed Appl Biotech 5:125–133

Caixeta ET, Borem A, Azate-Marin AL, Fagundes SA, Silva MGM, et al. (2005b) Allelic relationships for genes that confer resistance to angular leaf spot in common bean. Euphytica 145:237–245

Chang YL, Henriquez X, Preuss D, Copenhaver GP, Zhang HB (2003) A plant-transformation-competent BIBAC library from the *Arabidopsis thaliana* Landsberg ecotype for functional and comparative genomics. Theor Appl Genet 106:269–276

Cheon CI, Lee NG, Siddique ABM, Bal AK, Verma DPS (1993) Roles of plant homologs of Rab1p and Rab7p in the biogenesis of the peribacteroid membrane, a subcellular compartment formed de-novo during root-nodule symbiosis. EMBO J 12:4125–4135

Choi H-K, Mun J-H, Kim D-J, Zhu H, Baek J-M, et al. (2004) Estimating genome conservation between crop and model legume species. Proc Natl Acad Sci USA 101:15289–15294

Costa MR, Tanure JPM, Arruda KMA, Carneiro JES, Moreira MA, et al. (2006) Pyramiding of anthracnose, angular leaf spot and rust resistance genes in black and red bean cultivars. Ann Rep Bean Improv Coop 49:187–188

Debouck DG, Toro O, Paredes OM, Johnson WC, Gepts P (1993) Genetic diversity and ecological distribution of *Phaseolus vulgaris* in northwestern South America. Econ Bot 47:408–423

Díaz CL, Spaink HP, Kijne JW (2000) Heterologous rhizobial lipochitin oligosaccharides and chitin oligomers induce cortical cell divisions in red clover roots, transformed with the pea lectin gene. Molec Plant-Microbe Inter 13:268–276

Estrada-Navarrete G, Alvarado-Affantranger X, Olivares J-E, Díaz-Camino C, Santana O, Murillo E, Guillén G, Sánchez-Guevara N, Acosta J, Quinto C, Li D, Gresshoff PM, Sánchez F (2006) *Agrobacterium rhizogenes*-transformation of the *Phaseolus* spp.: A tool for functional genomics. Molec Plant-Micr Inter 19:1385–1393

Faria JC, Albino MMC, Dias BBA, Cancado LJ, da Cunha NB, Silva LD, Vianna GR, Aragão FJL (2006) Partial resistance to bean golden mosaic virus in a transgenic common bean (*Phaseolus vulgaris* L.) line expressing a mutated rep gene. Plant Sci 171:565–571

Foster-Hartnett D, Mudge J, Larsen D, Danesh D, Yan HH, et al. (2002) Comparative genomic analysis of sequences sampled from a small region on soybean (*Glycine max*) molecular linkage group G. Genome 45:634–645

Frei A, Blair MW, Cardona C, Beebe SE, Gu H, et al. (2005) QTL mapping of resistance to *Thrips palmi* Karny in common bean. Crop Sci 45:379–387

Freyre, R, Skroch P, Geffroy V, Adam-Blondon A-F, Shirmohamadali A, et al. (1998) Towards an integrated linkage map of common bean. 4. Development of a core map and alignment of RFLP maps. Theor Appl Genet 97:847–856

Gaitán-Solís E, Duque MC, Edwards KJ, Tohme J (2002) Microsatellite repeats in common bean (*Phaseolus vulgaris*): Isolation, characterization, and cross-species amplification in *Phaseolus* ssp. Crop Sci 42:2128–2136

Galasso I, Lioi L, Lanave C, Campion B, Bollini R, et al. (2005) Identification and sequencing of a BAC clone belonging to the *Phaseolus vulgaris* (L) insecticidal *Arc4* lectin locus. Ann Rep Bean Improv Coop 48:40–40
Geffroy V, Sicard D, de Oliveira J, Sévignac M, Cohen S, et al. (1999) Identification of an ancestral resistance gene cluster involved in the coevolution process between *Phaseolus vulgaris* and its fungal pathogen *Colletotrichum lindemuthianum*. Molec Plant-Microbe Inter 12:774–784
Geffroy V, Sévignac M, De Oliveira J, Fouilloux G, Skroch P, et al. (2000) Inheritance of partial resistance against *Colletotrichum lindemuthianum* in *Phaseolus vulgaris* and co-localization of QTL with genes involved in specific resistance. Molec Plant-Microbe Inter 13:287–296
Genga AM, Allavena A, Ceriotti A, Bollini R (1992) Genetic transformation in *Phaseolus* by high-velocity particle microprojectiles. Acta Hort. 300:309–313
Gepts P (1988) Provisional linkage map of common bean. Annu Rep Bean Improv Coop 31:20–25
Gepts P (1990) Biochemical evidence bearing on the domestication of *Phaseolus* beans. Econ Bot 44(3S):28–38
Gepts P, Bliss FA (1984) Enhanced available methionine concentration associated with higher phaseolin levels in common bean seeds. Theor Appl Genet 69:47–53
Gepts P, Bliss FA (1985) F_1 hybrid weakness in the common bean: differential geographic origin suggests two gene pools in cultivated bean germplasm. J Hered 76:447–450
Gepts P, Bliss FA (1986) Phaseolin variability among wild and cultivated common beans (*Phaseolus vulgaris*) from Colombia. Econ Bot 40:469–478
Gepts P, Osborn TC, Rashka K, Bliss FA (1986) Phaseolin-protein variability in wild forms and landraces of the common bean (*Phaseolus vulgaris*): evidence for multiple centers of domestication. Econ Bot 40:451–468
Graham MA, Ramírez M, Valdés-López O, Lara M, Tesfaye M, et al. (2006) Identification of phosphorus stress induced genes in *Phaseolus vulgaris* L. through clustering analysis across several species. Funct Plant Biol 33:789–797
Grant JE, Dommiss EM, Conner AJ (1991) Gene transfer to plants using Agrobacterium. In: Murray DR (ed) Advanced methods in plant breeding and biotechnology. CAB International, Wallingford, pp 50–73
Guerra-Sanz JM (2004) New SSR markers of *Phaseolus vulgaris* from sequence databases. Plant Breed 123:87–89
Guzmán P, Gilbertson RL, Nodari R, Johnson WC, Temple SR, et al. (1995) Characterization of variability in the fungus *Phaeoisariopsis griseola* suggests coevolution with the common bean (*Phaseolus vulgaris*). Phytopathology 85:600–607
Guzmán-Maldonado SH, Martínez O, Acosta-Gallegos JA, Guevara-Lara F, Paredes-López O (2003) Putative quantitative trait loci for physical and chemical components of common bean. Crop Sci 43:1029–1035
Hagenblad J, Tang CL, Molitor J, Werner J, Zhao K, et al. (2004) Haplotype structure and phenotypic associations in the chromosomal regions surrounding two *Arabidopsis thaliana* flowering time loci. Genetics 168:1627–1638
Hagiwara WE, dos Santos JB, do Carmo LM (2001) Use of RAPD to aid selection in common bean backcross breeding programs. Crop Breed Appl Biotech 1:355–362
Haley S, Afanador L, Miklas P, Stavely J, Kelly J (1994) Heterogeneous inbred populations are useful as sources of near-isogenic lines for RAPD marker localization. Theor Appl Genet 88:337–342
Henikoff S, Till BJ, Comai L (2004) TILLING. Traditional mutagenesis meets functional genomics. Plant Physiol 135:630–636
Islam FMA, Basford KE, Jara C, Redden RJ, Beebe S (2002) Seed compositional and disease resistance differences among gene pools in cultivated common bean. Genetic Res Crop Evol 49:285–293
Johnson E, Miklas P, Stavely J, Martínez-Cruzado J (1995) Coupling- and repulsion-RAPDs for marker-assisted selection of PI 181996 rust resistance in common bean. Theor Appl Genet 90:659–664

Johnson WC, Guzmán P, Mandala D, Mkandawire A, Temple S, et al. (1997) Molecular tagging of the *bc-3* gene for introgression into Andean common bean. Crop Sci 37: 248–254
Johnson WC, Gepts P (2002) The role of epistasis in controlling seed yield and other agronomic traits in an Andean x Mesoamerican cross of common bean (*Phaseolus vulgaris* L.). Euphytica 125:69–79
Jorde LB (2000) Linkage disequilibrium and the search for complex disease genes. Genome Res 10:1435–1444
Jung G, Coyne D, Skroch P, Nienhuis J, Arnaud-Santana E, et al. (1996) Molecular markers associated with plant architecture and resistance to common blight, web blight, and rust in common beans. J Am Soc Hort Sci 121:794–803
Kami J, Becerra Velásquez B, Debouck DG, Gepts P (1995) Identification of presumed ancestral DNA sequences of phaseolin in *Phaseolus vulgaris*. Proc Natl Acad Sci USA 92:1101–1104
Kami J, Poncet V, Geffroy V, Gepts P (2006) Development of four phylogenetically-arrayed BAC libraries and sequence of the APA locus in *Phaseolus vulgaris*. Theor Appl Genet 112:987–998
Kelly J, Miklas PN (1998) The role of RAPD markers in breeding for disease resistance in common bean. Mol Breed 4:1–11
Kelly JD, Vallejo VA (2004) A comprehensive review of the major genes conditioning resistance to anthracnose in common bean. Hortscience 39:1196–1207
Kelly JD, Gepts P, Miklas PN, Coyne DP (2003) Tagging and mapping of genes and QTL and molecular marker-assisted selection for traits of economic importance in bean and cowpea. Field Crop Res 82:135–154
Khairallah MM, Adams MW, Sears BB (1990) Mitochondrial DNA polymorphisms of Malawian bean lines: further evidence for two major gene pools. Theor Appl Genet 80:753–761
Khairallah MM, Sears BB, Adams MW (1992) Mitochondrial restriction fragment polymorphisms in wild *Phaseolus vulgaris* – insights in the domestication of common bean. Theor Appl Genet 84:915–922
Kim JW, Minamikawa T (1996) Transformation and regeneration of French bean plants by the particle bombardment process. Plant Sci 117:131–138
Koenig R, Gepts P (1989a) Segregation and linkage of genes for seed proteins, isozymes, and morphological traits in common bean (*Phaseolus vulgaris*). J Hered 80:455–459
Koenig R, Gepts P (1989b) Allozyme diversity in wild *Phaseolus vulgaris*: further evidence for two major centers of diversity. Theor Appl Genet 78:809–817
Koenig R, Singh SP, Gepts P (1990) Novel phaseolin types in wild and cultivated common bean (*Phaseolus vulgaris*, Fabaceae). Econ Bot 44:50–60
Koinange EMK, Singh SP, Gepts P (1996) Genetic control of the domestication syndrome in common-bean. Crop Sci 36:1037–1045
Kolkman JM, Kelly JD (2003) QTL conferring resistance and avoidance to white mold in common bean. Crop Sci 43:539–548
Kulikova O, Gualtieri G, Geurts R, Kim DJ, Cook D, et al. (2001) Integration of the FISH pachytene and genetic maps of Medicago truncatula. Plant J 27:49–58
Larsen RC, Miklas PN (2004) Generation and molecular mapping of a sequence characterized amplified region marker linked with the *Bct* gene for resistance to *Beet curly top virus* in common bean. Phytopathology 94:320–325
Lavin M, Herendeen PS, Wojciechowski MF (2005) Evolutionary rates analysis of Leguminosae implicates a rapid diversification of the major family lineages immediately following an Early Tertiary emergence. Syst Biol 54:575–594
Lee JM, Grant D, Vallejos CE, Shoemaker RC (2001) Genome organization in dicots. II. Arabidopsis as a 'bridging species' to resolve genome evolution events among legumes. Theor Appl Genet 103:765–773
Lee, NG, Stein B, Suzuki H, Verma DPS (1993) Expression of antisense nodulin-35 RNA in *Vigna aconitifolia* transgenic root nodules retards peroxisome development and affects nitrogen availability to the plant. Plant J 3:599–606
Liu ZC, Park BJ, Kanno A, Kameya T (2005) The novel use of a combination of sonication and vacuum infiltration in Agrobacterium-mediated transformation of kidney bean (*Phaseolus vulgaris* L) with lea gene. Mol Breed 16:189–197

López CE, Acosta IF, Jara C, Pedraza F, Gaitán-Solís E, et al. (2003) Identifying resistance gene analogs associated with resistances to different pathogens in common bean. Phytopathology 93:88–95

Mackay TFC (2001) The genetic architecture of quantitative traits. Ann Rev Genet 35:303–339

Maréchal R, Mascherpa J-M, Stainier F (1978) Etude taxonomique d'un groupe complexe d'espèces des genres *Phaseolus* et *Vigna* (Papilionaceae) sur la base de données morphologiques et polliniques, traitées par l'analyse informatique. Boissiera 28:1–273

Mauro Herrera M (2003) Wild bean populations as source of genes to improve the yield of cultivated *Phaseolus vulgaris* L. Ph.D. thesis, University of California, Davis

McCallum CM, Comai L, Greene EA, Henikoff S (2000) Targeting induced local lesions in genomes (TILLING) for plant functional genomics. Plant Physiol 123:439–442

McClean PE, Lee RK, Otto C, Gepts P, , Bassett MJ (2002) Molecular and phenotypic mapping of genes controlling seed coat pattern and color in common bean (*Phaseolus vulgaris* L.) J Hered 93:148–152

McClean PE, Lee RK, Miklas PN (2004) Sequence diversity analysis of dihydroflavonol 4-reductase intron 1 in common bean. Genome 47:266–280

McClean PE, Lee RD, McConnell MD, Mamidi S, White CP (2006) Sequence and marker-based diversity in common bean. Plant Animal Genome 14:W144, p 40

McConnell M, Mamidi S, Lee R, McClean P (2006) DNA sequence polymorphism among common bean genes. Ann Rept Bean Improv Coop 49:143–144

Melotto M, Fransisco C, Camargo LEA (2003) Towards cloning the *Co-4*^{2} locus using a bean BAC library. Ann Rep Bean Improv Coop 46:51–52

Melotto M, Coelho MF, Pedrosa-Harand A, Kelly JD, Camargo LEA (2004) The anthracnose resistance locus *Co-4* of common bean is located on chromosome 3 and contains putative disease resistance-related genes. Theor Appl Genet 109:690–699

Melotto M, Monteiro-Vitorello CB, Bruschi AG, Camargo LEA (2005) Comparative bioinformatic analysis of genes expressed in common bean (*Phaseolus vulgaris* L.) seedlings. Genome 48:562–570

Métais I, Hamon B, Jalouzot R, Peltier D (2002) Structure and level of genetic diversity in various bean types evidenced with microsatellite markers isolated from a genomic enriched library. Theor Appl Genet 104:1346–1352

Michaels TE, Smith TW, Larsen J, Beattie AD, Pauls KP (2005) OAC Rex common bean. Can J Plant Sci 86:733–736

Mienie C, Liebenberg M, Pretorius Z, Miklas P (2005) SCAR markers linked to the common bean rust resistance gene *Ur-13*. Theor Appl Genet 111:972–979

Miklas PN (2005) List of SCAR marker for disease resistance: Aug 2005 update http://www.ars.usda.gov/SP2UserFiles/Place/53540000/miklas/SCARtable.pdf

Miklas PN, Kelly JD (2002) Registration of two cranberry bean germplasm lines resistant to Bean Common Mosaic and Necrosis Potyviruses: USCR-7 and USCR-9. Crop Sci 42:673–674

Miklas PN, Stavely JR, Kelly JD (1993) Identification and potential use of a molecular marker for rust resistance in common bean. Theor Appl Genet 85:745–749

Miklas PN, Johnson E, Stone V, Beaver JS, Montoya C, et al. (1996) Selective mapping of QTL conditioning disease resistance in common bean. Crop Sci 36:1344–1351

Miklas PN, Larsen RC, Riley R, Kelly JD (2000) Potential marker-assisted selection for *bc-1*^{2} resistance to bean common mosaic potyvirus in common bean. Euphytica 116:211–219

Miklas PN, Johnson WC, Delorme R, Gepts P (2001) QTL conditioning physiological resistance and avoidance to white mold in dry bean. Crop Sci 41:309–315

Miklas PN, Hang AN, Kelly JD, Strausbaugh CA, Forster RL (2002) Registration of three kidney bean germplasm lines resistant to Bean Common Mosaic and Necrosis Potyviruses: USLK-2 Light Red Kidney, USDK-4 Dark Red Kidney, and USWK-6 White Kidney. Crop Sci 42:674–675

Miklas PN, Delorme R, Riley R (2003a) Identification of QTL conditioning resistance to white mold in snap bean. J Am Soc Hort Sci 128:564–570

Miklas PN, Kelly JD, Singh SP (2003b) Registration of anthracnose resistant pinto bean germplasm line USPT ANT 1. Crop Sci 43:1889–1890

Miklas PN, Hu J, Grünwald NJ, Larsen KM (2006a) Potential application of TRAP (Targeted Region Amplified Polymorphism) markers for mapping and tagging disease resistance traits in common bean. Crop Sci 46:910–916

Miklas PN, Kelly JD, Beebe SE, Blair MW (2006b) Common bean breeding for resistance against biotic and abiotic stresses: from classical to MAS breeding. Euphytica 147:106–131

Miklas PN, Smith JR, Singh SP (2006c) Registration of common bacterial blight resistant dark red kidney bean germplasm line USDK-CBB-15 10.2135/cropsci2005.06-0110. Crop Sci 46:1005–1007

Miklas PN, Smith JR, Singh SP (2006d) Release of common bacterial blight resistant pinto bean germplasm lines USPT-CBB-5 and USPT-CBB-6. Ann Rep Bean Improv Coop 49:283–284

Mkandawire ABC, Mabagala RB, Guzman P, Gepts P, Gilbertson RL (2004) Genetic diversity and pathogenic variation of common blight bacteria (*Xanthomonas campestris* pv. *phaseoli* and *X. campestris* pv. *phaseoli* var. *fuscans*) suggests pathogen coevolution with the common bean. Phytopathology 94:593–603

Moffett MD, Weeden NF (2006) Investigation of synteny conservation between *Pisum* and *Phaseolus*. Abstract, Plant & Animal Genome XIV, 2006: http://www.intl-pag.org/14/abstracts/PAG14_P442.html Moscone EA, Klein F, Lambrou M, Fuchs J, Schweizer D (1999) Quantitative karyotyping and dual-color FISH mapping of 5S and 18S-25S rDNA probes in the cultivated *Phaseolus* species (Leguminosae). Genome 42:1224–1233

Murray J, Larsen J, Michaels TE, Schaafsma A, Vallejos CE, et al. (2002) Identification of putative genes in bean (*Phaseolus vulgaris*) genomic (Bng) RFLP clones and their conversion to STSs. Genome 45:1013–1024

Mutlu N, Miklas PN, Coyne DP (2006) Resistance gene analog polymorphism (RGAP) markers co-localize with disease resistance genes and QTL in common bean. Mol Breed 17: 127–135

Mutlu N, Miklas P, Reiser J, Coyne D (2005a) Backcross breeding for improved resistance to common bacterial blight in pinto bean (*Phaseolus vulgaris* L.). Plant Breed 124:282–287

Mutlu N, Miklas PN, Steadman JR, Vidaver AV, Lindgren D, et al. (2005b) Registration of pinto bean germplasm line ABCP-8 with resistance to common bacterial blight. Crop Sci 45:806

Nagl W (1969) Banded polytene chromosomes in the legume *Phaseolus vulgaris*. Nature 221:70–71

Nanni L, Losa A, Bellucci E, Kater M, Gepts P, et al. (2005) Identification and molecular diversity of a genomic sequence similar to SHATTERPROOF (*SHP1*) in *Phaseolus vulgaris* L. Plant Animal Genome XIII:P470, p 188

Nodari RO, Koinange EMK, Kelly JD, Gepts P (1992) Towards an integrated linkage map of common bean. I. Development of genomic DNA probes and levels of restriction fragment length polymorphism. Theor Appl Genet 84:186–192

Nodari RO, Tsai SM, Gilbertson RL, Gepts P (1993a) Towards an integrated linkage map of common bean. II. Development of an RFLP-based linkage map. Theor Appl Genet 85:513–520

Nodari RO, Tsai SM, Guzmán P, Gilbertson RL, Gepts P (1993b) Towards an integrated linkage map of common bean. 3. Mapping genetic factors controlling host-bacteria interactions. Genetics 134:341–350

Nordborg M, Hu TT, Ishino Y, Jhaveri J, Toomajian C, et al. (2005) The pattern of polymorphism in *Arabidopsis thaliana*. PLoS Biology 3:e196

Ochoa IE, Blair MW, Lynch JP (2006) QTL analysis of adventitious root formation in common bean under contrasting phosphorus availability 10.2135/cropsci2005.12-0446. Crop Sci 46:1609–1621

Pallottini L, Garcia E, Kami J, Barcaccia G, Gepts P (2004) The genetic anatomy of a patented yellow bean. Crop Sci 44:968–977

Papa R, Gepts P (2003) Asymmetry of gene flow and differential geographical structure of molecular diversity in wild and domesticated common bean (*Phaseolus vulgaris* L.) from Mesoamerica. Theor Appl Genet 106:239–250

Papa R, Acosta J, Delgado-Salinas A, Gepts P (2005) A genome-wide analysis of differentiation between wild and domesticated *Phaseolus vulgaris* from Mesoamerica. Theor Appl Genet 111:1147–1158

Payró de la Cruz E, Gepts P, Colunga GarciaMarín P, Zizumbo Villareal D (2005) Spatial distribution of genetic diversity in wild populations of *Phaseolus vulgaris* L. from Guanajuato and Michoacán, México. Genet Res Crop Evol 52:589–599

Pedrosa A (2003) Chromosomal organisation and physical mapping in legumes. Ph.D. thesis, University of Vienna, 143 p

Pedrosa A, Sandal N, Stougaard J, Schweizer D, Bachmair A (2002) Chromosomal map of the model legume *Lotus japonicus*. Genetics 161:1661–1672

Pedrosa A, Vallejos C, Bachmair A, Schweizer D (2003) Integration of common bean (*Phaseolus vulgaris* L.) linkage and chromosomal maps. Theor Appl Genet 106:205–212

Pedrosa-Harand A, Almeida CCS, Mosiolek M, Blair MW, Schweizer D, et al. (2006) Extensive ribosomal DNA amplification during Andean common bean (*Phaseolus vulgaris* L.) evolution. Theor Appl Genet 112:924–933

Perry JA, Wang TL, Welham TJ, Gardner S, Pike JM, et al. (2003) A TILLING reverse genetics tool and a web-accessible collection of mutants of the legume *Lotus japonicus*. Plant Physiol 131:866–871

Perry G, Reinprecht Y, Pauls KP (2006) Identification of common bacterial blight resistance genes in *Phaseolus vulgaris*. Abstract P16051, Plant Biology 2006, Joint Annual Meeting of the American Society of Plant Biologists and the Canadian Society of Plant Physiologists, Boston

Pritchard J, Stephens M, Donnelly P (2000) Inference of population structure using multilocus genotype data. Genetics 155:945–959

Ragagnin VA, Alzate-Marin AL, Souza TLPO, Sanglard DA, Moreira MA, et al. (2005) Use of molecular markers to pyramiding multiple genes for resistance to rust, anthracnose and angular leaf spot in the common bean. Ann Rep Bean Improv Coop 48:94–95

Ramírez M, Graham MA, Blanco-López L, Silvente S, Medrano-Soto A, et al. (2005) Sequencing and analysis of common bean ESTs. Building a foundation for functional genomics. Plant Physiol 137:1211–1227

Rivkin M, Vallejos C, McClean P (1999) Disease-related sequences in common bean. Genome 42:41–47

Russell DR, Wallace KM, Bathe JH, Martinell BJ (1993) Stable transformation of *Phaseolus vulgaris* via electric discharge mediated particle acceleration. Plant Cell Rep 12:165–169

Santalla M, Rodino AP, de Ron AM (2002) Allozyme evidence supporting southwestern Europe as a secondary center of genetic diversity for the common bean. Theor Appl Genet 104:934–944

Sarbhoy RK (1978) Cytogenetical studies in genus *Phaseolus* Linn. I and II. Somatic and meiotic studies in fifteen species of *Phaseolus*. Cytologia 43:161–180

Schmid K, Ramos-Onsins S, Ringys-Beckstein H, Weisshaar B, Mitchell-Olds T (2005) A multilocus sequence survey in *Arabidopsis thaliana* reveals a genome-wide departure from a neutral model of DNA sequence polymorphism. Genetics 169:1601–1615

Shahmuradov IA, Akbarova YY, Solovyev VV, Aliyev JA (2003) Abundance of plastid DNA insertions in nuclear genomes of rice and Arabidopsis. Plant Mol Biol 52:923–934

Singh SP, Gepts P, Debouck DG (1991a) Races of common bean (*Phaseolus vulgaris* L., Fabaceae). Econ Bot 45:379–396

Singh SP, Nodari R, Gepts P (1991b) Genetic diversity in cultivated common bean. I. Allozymes. Crop Sci 31:19–23

Singh SP, Morales FJ, Miklas PN, Teran H (2000a) Selection for bean golden mosaic resistance in intra- and interracial bean populations. Crop Sci 40:1565–1572

Singh SP, Morales FJ, Terán H (2000b) Registration of bean golden mosaic resistant dry bean germplasm GMR 1 and GMR 5. Crop Sci 40:1836

Stavely JR, Kelly JD, Grafton KF (1994) BelMiDak-rust-resistant navy dry beans germplasm lines. HortScience 29:709–710

Stavely JR, McMillan RT, Beaver JS, Miklas PN (2001) Release of three McCaslan type, indeterminate, rust and golden mosaic resistant snap bean germplasm lines BelDade RGMR 4, 5 and 6. Ann Rep Bean Improv Coop 44:197–198

Stiller J, Martirani L, Tuppale S, Chian RJ, Chiurazzi M, et al. (1997) High frequency transformation and regeneration of transgenic plants in the model legume *Lotus japonicus*. J Exp Bot 48:1357–1365

Tar'an B, Michaels TE, Pauls KP (2001) Mapping genetic factors affecting the reaction to *Xanthomonas axonopodis* pv. *phaseoli* in *Phaseolus vulgaris* L. under field conditions. Genome 44:1046–1056

Tar'an B, Michaels TE, Pauls KP (2002) Genetic mapping of agronomic traits in common bean. Crop Sci 42:544–556

Tenaillon MI, Sawkins MC, Long AD, Gaut RL, Doebley JF, et al. (2001) Patterns of DNA sequence polymorphism along chromosome 1 of maize (*Zea mays* ssp *mays* L.). Proc Natl Acad Sci USA 98:9161–9166

Tepfer D (1990) Genetic transformation using *Agrobacterium rhizogenes*. Physiol Plant 79:140–146

Thornsberry JM, Goodman MM, Doebley J, Kresovich S, Nielsen D, et al. (2001) *Dwarf8* polymorphisms associate with variation in flowering time. Nat Genet 28:286–289

Urrea C, Miklas P, Beaver J, Riley R (1996) A codominant randomly amplified polymorphic DNA (RAPD) marker useful for indirect selection of bean golden mosaic virus resistance in common bean. J Am Soc Hort Sci 121:1035–1039

Vallad G, Rivkin M, Vallejos C, McClean P (2001) Cloning and homology modelling of a *Pto*-like protein kinase family of common bean (*Phaseolus vulgaris* L.). Theor Appl Genet 103:1046–1058

Vallejos C, Skroch P, Nienhuis J (2001) *Phaseolus vulgaris* – The common bean. Integration of RFLP and RAPD-based linkage maps. In: Phillips R, Vasil I (eds), DNA-based markers in plants. Kluwer, Dordrecht, the Netherlands, pp. 301–317

Vallejos CE, Astua-Monge G, Jones V, Plyler TR, Sakiyama NS, et al. (2006) Genetic and molecular characterization of the *I* Locus of *Phaseolus vulgaris*. Genetics 172:1229–1242 Vallejos CE, Chase CD (1991) Linkage between isozyme markers and a locus affecting seed size in *Phaseolus vulgaris* L. Theor Appl Genet 81:413–419

Vallejos CE, Sakiyama NS, Chase CD (1992) A molecular marker-based linkage map of *Phaseolus vulgaris* L. Genetics 131:733–740

Van de Velde W, Mergeay J, Holsters M, Goormachtig S (2003) *Agrobacterium rhizogenes*-mediated transformation of *Sesbania rostrata*. Plant Sci 165:1281–1288

Vanhouten W, Mackenzie S (1999) Construction and characterization of a common bean bacterial artificial chromosome library. Plant Mol Biol 40:977–983

Vianna GR, Albino MMC, Dias BBA, Silva LdM, Rech EL, Aragão FJL (2004) Fragment DNA as vector for genetic transformation of bean (*Phaseolus vulgaris* L.). Scientia Horticulturae (Amsterdam) 99:371–378

Yaish MWF, de la Vega MP (2003) Isolation of (GA)(n) microsatellite sequences and description of a predicted MADS-box sequence isolated from common bean (*Phaseolus vulgaris* L.). Genet Mol Biol 26:337–342

Yan HH, Mudge J, Kim DJ, Shoemaker RC, Cook DR, et al. (2004) Comparative physical mapping reveals features of microsynteny between *Glycine max*, *Medicago truncatula*, and *Arabidopsis thaliana*. Genome 47:141–155

Yu K, Park S, Poysa V, Gepts P (2000) Integration of simple sequence repeat (SSR) markers into a molecular linkage map of common bean (*Phaseolus vulgaris* L.). J Hered 91:429–434

Yu K, Haffne M, Park SJ (2006) Construction and characterization of a common bean BAC library. Ann Rep Bean Improv Coop 49:61–63

Yu KF, Park SJ, Poysa V (1999) Abundance and variation of microsatellite DNA sequences in beans (*Phaseolus* and *Vigna*). Genome 42:27–34

Yu Z, Stall R, Vallejos C (1998) Detection of genes for resistance to common bacterial blight of beans. Crop Sci 38:1290–1296

Zambre M, Goossens A, Cardona C, Van Montagu M, Terryn N, et al. (2005) A reproducible genetic transformation system for cultivated *Phaseolus acutifolius* (tepary bean) and its use to assess the role of arcelins in resistance to the Mexican bean weevil. Theor Appl Genet 110:914–924

Zhang ZY, Coyne DP, Mitra A (1997) Factors affecting *Agrobacterium*-mediated transformation of common bean. J Am Soc Hort Sci 122:300–305

Zheng J, Nakata M, Uchiyama H, Morikawa H, Tanaka R (1991) Giemsa C-banding patterns in several species of *Phaseolus* L. and *Vigna* Savi, Fabaceae. Cytologia 56:459–466

Zhu YL, Song QJ, Hyten DL, Van Tassell CP, Matukumalli LK, et al. (2003) Single-nucleotide polymorphisms in soybean. Genetics 163:1123–1134

Zizumbo-Villarreal D, Colunga-GarcíaMarín P, Payró de la Cruz E, Delgado-Valerio P, Gepts P (2005) Population structure and evolutionary dynamics of wild–weedy–domesticated complexes of common bean in a Mesoamerican region. Crop Sci 35:1073–1083

Chapter 6
Genomics of *Theobroma cacao*, "the Food of the Gods"

Mark J. Guiltinan, Joseph Verica, Dapeng Zhang, and Antonio Figueira

Abstract *Theobroma cacao*, the chocolate tree, is an important tropical tree-crop that provides sustainable economic and environmental benefits to some of the poorest and most ecologically sensitive areas of the world. Recent progress in the development of genomics tools for cacao is reviewed. These include a reference molecular genetic map, simple sequence repeats and other molecular markers, two germplasm databases with microsatellite DNA fingerprints and other molecular data, many quantitative trait loci mapping projects which have identified disease resistance and yield component loci, several expressed sequence tag resources, a cacao microarray, bacterial artificial chromosome libraries, and a genetic transformation system. The evolutionary relatedness of cacao with other important crops and model plant systems positions cacao genomics to play a significant role in translational plant genomics. The future prospects for the contribution of cacao genomics to improvement of this crop for sustainable cacao production and as a tool for poverty alleviation and environmental stabilization are discussed.

6.1 Introduction

6.1.1 Economic, Agronomic, and Societal Importance of Cacao

Theobroma cacao L. (cacao; Malvaceae *sensu lato*) is a small under-story tree endemic to the lowland rainforests of the Amazon basin (Wood and Lass 1985; Bartley 2005). Today, cacao is grown throughout the humid tropics, often in agroforestry-ecosystems with other fruit and commodity crops. The annual world production of cocoa (the product obtained from dried, fermented cacao seeds) is approximately 3 million tons, with 2/3 being processed into cocoa powder and cocoa butter, and the remaining 1/3 used for cocoa liquor (the flavor and color component of chocolate) (Wood and Lass 1985). Cocoa is the major export commodity

M.J. Guiltinan
The Pennsylvania State University, Department of Horticulture, Life Sciences Building, University Park, PA 16802-5807, USA
e-mail: mjg9@psu.edu

© Springer 2008

of several countries in West Africa (68% of world production), providing major economic resources to Ivory Coast, Cameroon, Nigeria, and Ghana. Other major cocoa exporters include Ecuador, Venezuela, Brazil, Costa Rica, Malaysia, and Indonesia. Worldwide, approximately 5 to 6 million smallholder farmers grow 95% of the world's production. World cocoa export trade is \$5 to \$6 billion/year. In the United States, the use of cocoa and cocoa butter in chocolate manufacturing, cosmetics, and other products drives an approximately \$70 billion dollar market, providing over 60,000 jobs (Morais 2005). US chocolate production also uses large amounts of sugar, nuts, and milk valued at approximately \$3 billion/year in receipts to American farmers.

Cacao-growing regions are largely centered in important biodiversity hotspots, and in proximity to 13 of the world's most biologically diverse regions (Piasentin and Klare-Repnik 2004). Because cacao is a shade-grown perennial tree crop with a cropping cycle of 50+ years, cultivation provides environmental benefits such as enhancement of biodiversity in avian migratory corridors, soil and watershed conservation, and buffer zones near endangered rainforest habitats (Rice and Greenberg 2003; Ruf and Zadi 2003). Development agencies such as USAID, Conservation International, The World Wildlife Federation, and the World Cocoa Foundation are increasingly aware of the role of cacao in stabilizing local economies and environments and have stepped up their involvement with cacao farmers in these regions (Guyton et al. 2003).

Cacao diseases reduce the potential crop by an estimated 810,000 tons annually (30% of world production) and individual farm losses can approach 100% (Keane 1992; Bowers et al. 2001). For example, in Southern Bahia, Brazil, introduction of the witches' broom disease, caused by the fungus *Moniliophthora perniciosa* (Aime and Phillips-Mora 2005), resulted in a decrease in production from 300,000 tons in 1989 to 130,000 tons 10 years later, for an estimated loss of \$220 million each year (Pereira et al. 1990). In addition, economic loss due to cacao disease has caused widespread social disruption among smallholder growers. Widely dispersed cacao pathogens include several species of *Phytophthora* that cause multiple diseases of economic importance including pod rot, trunk canker, and leaf-blight (Appiah et al. 2003; Chowdappa et al. 2003; Appiah et al. 2004). *Phytophthora megakarya*, the most aggressive species, is reported to have entered Ivory Coast, the world's leading cocoa producer (Nyasse et al. 2002; Opoku et al. 2002; Appiah et al. 2003; Risterucci et al. 2003; Efombagn et al. 2004). Other important diseases and pests include frosty pod in Central America caused by *Moniliophthora roreri* (Evans et al. 2003), the cocoa pod borer in Asia (Day 1984; Santoso et al. 2004), and cocoa swollen shoot virus in West Africa (Hanna 1954; Hervé et al. 1991; Muller and Sackey 2005).

6.1.2 Cacao as an Experimental Organism

Theobroma cacao is a simple diploid with ten chromosomes ($2n = 2x = 20$) and a small genome. Published genome size estimates vary from 390 Mb to 415 Mb

(Figueira et al. 1992; Couch et al. 1993); however, recent fluorescent flow-cytometer measurements estimate a genome size of 447 Mb ± 11 Mbp 2C (Arumuganathan, Carlson, and Guiltinan, unpublished). This size represents the average of 10 genotypes measured, with three replicates per genotype. This is a relatively small size for a plant genome and thus makes certain aspects of cacao genomics more feasible, such as the possibility of sequencing the entire genome.

Cacao has several limitations as an experimental organism. For example, its life cycle takes a minimum of 2–3 years from seed to seed, and progeny of crosses must be grown for many years to fully evaluate their productivity and disease resistance characteristics. Many cacao genotypes are self-incompatible, making genetic analysis and breeding strategies labor intensive. Furthermore, the plants require large areas of land and large inputs of labor to maintain and evaluate field tests. In addition, the seeds are recalcitrant, so germplasm must be conserved as living collections in the field or greenhouses. These and other factors combine to make cacao a very difficult and slow experimental system.

6.1.3 Current Status of Cacao Genetics and Breeding

Breeding programs were established in the major cacao growing countries starting in the 1920s (Toxopeus 1969; Bartley 2005). In the 1930s and 40s, valuable germplasm was collected from the Amazon regions of Brazil, Ecuador, and Peru (Pound 1940; Bartley 2005), and clonal descendants of these accessions are maintained in present day germplasm collections (Lockwood and Gyamfi 1979; Engels 1981; Iwaro et al. 2003). Large genetic variation has been identified in wild populations throughout the Amazon, but this diversity has not yet been widely incorporated into cultivated varieties (Bartley 2005). Disease resistance is currently the primary trait targeted by cacao breeders. Other important traits include yield efficiency, flavor characteristics, cocoa butter content (% seed lipid content) and quality (fatty acid saturation), tolerance to abiotic stress, and various horticultural traits such as precocity, rootstock/scion interactions, plant height, and stature.

The fields of cacao genetics, breeding, and biotechnology have been the subject of a number of relatively recent review articles (Bartley 1994; Hughes and Ollennu 1994; Bennett 2003; Silva and Figueira 2005; Guiltinan 2007, Maximova et al. 2007), two books (Dias 2001; Bartley 2005), and one conference proceedings (Eskes 2003).

6.1.4 Cacao Genomic Resources:

Today's cacao genetics research community is well organized, highly collaborative, and poised to make use of new genomics resources (reviewed by Bennett 2003). To formally foster collaboration and communication between cacao breeders and geneticists, the International Group for Genetic Improvement of Cocoa (INGENIC) was formed in 1994 (http://ingenic.cas.psu.edu). It now includes over 300 members, representing 35 developing and developed countries around the world. To coordinate the activities of the INGENIC members interested in molecular approaches, the INGENIC Study Group for Molecular Biology (INGENIC-MOL-BIOL) was formally chartered in October of 2003 (Johnson 2003). An international research symposium is held by INGENIC in a developing country every third year.

Most of the cocoa producing-countries have research facilities funded by international and national organizations that support agricultural research, such as the Cocoa Research Institute of Ghana (CRIG), the Cocoa Research Institute of Nigeria (CRIN), Institute of Agricultural Research for Development (IRAD - Cameroon), the Commissco Executiva do Plano da Lavoura Cacaueira (CEPLAC – Brazil), and the Instituto Nacional Autonomo de Investigaciones Agropecuarias (INIAP – Ecuador). The main strengths of these organizations are the many scientists with strong experiences in cacao agriculture, as well as their extensive field sites and breeding programs. It is essential that researchers in developed and developing countries establish and maintain strong working relationships and collaborative research and training programs to maximize the potential impact of their research on the cacao farmers and the environment. The Cocoa Producers' Alliance (COPAL) also supports basic research in cacao and sponsors the International Cocoa Research Conference (http://www.copal-cpa.org/).

In the United States, the USDA-ARS has established two centers of cacao research in Beltsville, MD, and Miami, FL, which carry out a wide array of research projects with collaborating laboratories worldwide (Pugh et al. 2004). Several Universities have research programs in cacao, including The Pennsylvania State University, which is home to the American Cocoa Institute Endowed Program in the Molecular Biology of Cacao. Several centers of excellence in cacao research in Europe can be found in the Centre de Cooperation Internationale en Recherche Agronomique pour le Developpement (CIRAD), Montpellier, France, The University of Reading, UK, and at other sites. Several centers of excellence in cacao research can also be found in Central and South America; Centro Agronómico Tropical de Investigación y Enseñanza, Costa Rica (CATIE), University of São Paulo, Brazil (USP), Universidade Estadual de Santa Cruz, Brazil (UESC) and others.

6.2 Genomic Mapping and Characterization

6.2.1 Molecular Markers for the Characterization of Cacao Germplasm

In the past two decades, research on the genetics of cacao has benefited enormously from molecular markers. Significant progress has been made in molecular characterization of cacao germplasm. The key objectives of this line of work include: reducing redundancy and mislabeling in cacao gene banks, understanding genetic diversity in *ex situ* collections and in farmer's fields, verifying genealogical information and characterizing germplasm for useful agronomic traits.

Isozymes were the first molecular markers utilized in cacao (Lanaud 1986). Although the available loci and the numbers of polymorphisms typically generated by each isozyme were low, this simple system enabled assessment of genetic diversity and mating system, assisted genotype identification, and contributed to linkage mapping (Ronning and Schnell 1994; Sounigo et al. 2005; Warren et al. 1994; Lachenaud et al. 2004). However, the isozyme markers are outdated because of their low polymorphism and the environmental effect on the "phenotype."

Commonly used DNA markers in cocoa include restriction fragment length polymorphism (RFLP), random amplified polymorphic DNA (RAPD), amplified fragment length polymorphism (AFLP), and simple sequence repeats (SSRs). These markers differ in genomic abundance, level of polymorphism detected, locus specificity, reproducibility, technical requirements, and financial cost. They have been used to answer various research questions in cacao.

RFLP is a non-polymerase chain reaction (PCR) based DNA marker that was first applied in cacao in the early 1990s (Laurent et al. 1994). The polymorphism of RFLP is moderately high in cacao. Its high reproducibility and the co-dominant allelic nature make it a suitable marker for the construction of genetic linkage maps and tagging genes and quantitative trait loci (QTL) linked to characters of agronomic importance (Lanaud et al. 1995, Risterucci et al. 2000) and assessment of genetic diversity (N'Goran et al. 1994; Motamayor et al. 2002). Because RFLP probes are invariably specific to a limited number of loci, RFLP is not a very effective tool for cacao genotype identification. Another major drawback is that RFLP is not amenable to automation and data generation is both laborious and expensive.

RAPD was the first PCR based DNA fingerprinting method to be applied for the genetic characterization of cocoa (Wilde et al. 1992). This system is technically simple to perform but has a low reproducibility between experiments and laboratories. RAPD is not a direct measure of heterozygosity which makes it less useful for actual genotyping as opposed to simply distinguishing between clones. The number of variable markers that can be generated by RAPD is small and not suitable for high-throughput application. Nevertheless, RAPD is a simple-to-use DNA marker for scoring variations between cacao clones. Russell et al. (1993) showed that a minimum of three RAPD primers were able to distinguish 25 cacao accessions according to their geographical origin. Various studies used RAPD for identification of accession mislabeling and duplication (Christopher et al. 1999; Faleiro et al. 2002; Sounigo et al. 2005). RAPD has also been widely used in analyses of genetic diversity (Figueira et al. 1994; N'Goran et al. 1994; Lerceteau et al. 1997; Whitkus et al. 1998; Marita et al. 2001; Yamada et al. 2001; Faleiro et al. 2004; Sounigo et al. 2005).

AFLP is a PCR based fingerprinting protocol that combines the strength of RAPD and RFLP (Vos et al. 1995). AFLP is highly polymorphic with considerable reproducibility within a laboratory. Similar to RAPD, AFLP does not require sequence data for primer construction. However, because of its dominant nature, AFLP is not a direct measure of heterozygosity so it has limited use in genotyping.

Perry et al. (1998) reported identification of parental plants of cacao and their hybrids using AFLP. AFLP was reliable for distinguishing closely related cocoa varieties (Saunders et al. 2001). Queiroz et al. (2003) reported the identification of a major QTL associated with resistance to witches' broom disease, based on AFLP linkage mapping. However, there have been few studies to date using AFLP in cacao germplasm management. The major disadvantages of AFLP include the dominance of alleles and the possible non-homology of co-migrating fragments belonging to different loci, which limit its wide application in cacao.

In the past decade, SSRs, also known as microsatellites, have emerged as the most widely used marker for cacao, enabling great strides to be made in the characterization of cocoa germplasm (Lanaud et al. 1999). SSRs are typically co-dominant and multiallelic, allowing precise discrimination (or matching) of individual clones based on the multi-locus fingerprints. The data analysis and interpretation of results fit the genetic model of cacao. As SSR sequence information can be easily shared between laboratories, data generated in different laboratories can be standardized through a combination of ring testing of reference genotypes, standardized protocol for allele sizing, and the development of marker-specific size ladders containing all the common alleles for a given locus (Sampaio et al. 2003; Cryer et al. 2006a, b). The standardized datasets then can be merged for joint analysis (George et al. 2004; Presson et al. 2006).

Because of these major advantages, SSR has been the system of choice for cacao in the past decade. It has been applied to various aspects of cacao germplasm management, including:

A. *Identification of mislabeled and duplicate accessions in germplasm collections:* Incorrect labeling of accessions has been a significant problem that has hindered the efficient conservation and use of cacao germplasm (Motilal and Butler 2003; Turbull et al. 2004). RAPD and AFLP have sufficient discriminatory power to separate accessions with different genotypes, but these markers are very limited in their ability to determine absolute identity between two individual trees due to artifact polymorphisms. The RAPD and AFLP based conclusion that two individuals are identical was usually only approximate, and no statistical rigor was added to this assertion (Perry et al. 1998; Christopher et al. 1999; Sounigo et al. 2005). This lack of precision limits the application of RAPD and AFLP for cacao germplasm identification.

   SSR markers, used together with high throughput genotyping facilities, enable large-scale assessment of genetic identity in cacao gene banks. In contrast to dominant markers such as AFLP and RAPD, identical cacao genotypes can have an exact match in the multi-locus SSR profiles (Zhang et al. 2006a). Thus, SSR based multi-locus genotyping has been widely applied in various national and international germplasm collections (Boccara and Zhang 2006; Cryer et al. 2006a, b; Zhang et al. 2006a; Johnson et al. 2007). SSR was also used to verify genetic identities in the breeding progenies (Takrama et al. 2005) and to monitor germplasm integrities for accessions maintained in vitro (Rodriguez et al. 2004). An international consortium was formed to identify the cacao genotypes and describe the genetic diversity in the major international and national cocoa collections maintained in the Americas. To date, more than 4,000 cocoa accessions maintained in the two international genebanks (CATIE, Costa Rica, The International Cocoa Genebank, Trinidad (ICG, T) and in several other national collections in the Americas have been genotyped with a set of 15 standard microsatellite (SSR) loci. Based on the multi-locus SSR profiles, duplicated accessions, both within and among different collections, were unambiguously identified. The reference profiles, together with the derived information in genetic diversity, are being submitted to the international databases The International Cocoa Germplasm Database (ICGD) [http://www.icgd.rdg.ac.uk/] and CocoaGenDB).

B. *Investigation of genealogical relationships:* Cacao trees have an open-pollinated breeding system. Their outcrossing nature and insects-mediated pollen dispersal mean that the majority of the progenies are based on the random mating of several parents. Therefore, the co-dominant nature of SSR is highly suitable for the investigation of the alleged genealogical relationship. Parentage and sibling relationship analysis between individuals is based on formal estimates derived from allele frequencies. Statistical rigor can be formally computed for the estimation. Parentage analysis is extremely useful for understanding gene flow in natural populations, monitoring parentage contribution in cacao seed gardens, and breeding programs. Using SSR markers, Schnell et al. (2005) determined the parental population for a group of productive and unproductive seedlings growing in Hawai'i. Seven SSR loci were found to have alleles that were associated with productive or unproductive seedlings. Motamayor et al. (2003) reported SSR allelic evidence that the Trinitario cacao is a result of natural hybridization and subsequent introgressions between Lower Amazon Forastero and ancient Criollo varieties.

   Parentage analysis based on the multi-locus SSR data is now routinely applied to verify the recorded pedigree in cocoa germplasm. For populations known to have family structures but lacking detailed pedigree information (i.e., most of Pound's collections made in the 1930s), re-construction of genealogical relationship is being performed using the multi-locus SSR data.

C. *Assessment of on-farm diversity:* Using SSR markers, Aikpokpodion et al. (2005) assessed the genetic diversity of the cacao cultivated in different agroecologies in Nigeria. A low adoption rate of improved germplasm was revealed in farmers' fields through SSR analysis. They also showed that only a small fraction of genetic diversity in the national germplasm collection has been exploited (Bhattacharjee et al. 2004). Another survey conducted in southern Cameroon found that farm accessions differed by geographical origin and parentage. The progeny of some promising Upper Amazon clones (T-clones) were poorly represented in the cacao farms (Efombagn et al. 2006). An SSR based diversity survey was also carried out in Ivory Coast and Ghana (Opoku et al. 2005).

In a study of trees found near the center of origin of cacao, Zhang et al. (2006b) observed a high allelic diversity in semi-domesticated cocoa trees from the Ucayali valley of Peru. Their results substantiate the hypothesis that the Peruvian

Amazon hosts a high level of cacao genetic diversity and that the diversity has a spatial structure, which highlights the need for additional collection and conservation measures for cacao germplasm in this region.

D. *Investigating phylogeography:* Understanding the spatial patterns of biodiversity, the processes of gene flow, and the historical and climatic effects on the distribution of genetic diversity is essential for sustainable conservation and effective use of cacao germplasm. SSR is an ideal tool for assessing intra-specific phylogeography for cacao. Motamayor et al. (2002) assessed the allelic composition in varieties from Central America (genotypes from Guiana, Amazonia, and Orinoco regions were compared using SSR). Their results support the hypothesis that cacao originated in the upper Amazon and suggest the likely dispersal route from the Amazon to Central America and Mexico. Sereno et al. (2006) compared samples from natural populations from the Brazilian Amazon (Acre, Rondonia, lower Amazon, and upper Amazon). The Brazilian upper Amazon population was found to have the largest genetic diversity and therefore was suggested to be part of the center of diversity for the species. SSR was also used to examine the origin and dispersal of "Nacional" cocoa in Ecuador. The result obtained supports the hypothesis that "National" cacao had an early establishment in the coastal region of Ecuador before other exotic germplasm was introduced into this region (R. Loor and F. Amores, personal communication).

As the sequence data of ESTs are increasingly generated and deposited in public databases, EST derived SSR markers are increasingly becoming available (Borrone et al. 2007). Database availability significantly reduces the cost of identifying and developing useful SSR markers. Moreover, a putative function may be assigned to the locus based upon the sequence homology. Another key achievement that significantly enhances the effectiveness of SSR is the newly available statistical tools that do not require prior assumptions of population boundaries, e.g., maximum likelihood and Bayesian approaches (Pritchard et al. 2000; Piry et al. 2004). These tools, together with powerful computers, greatly increase the potential for using SSR markers in studying population structure, environmental adaptation, genotype and population assignment, and kinship analysis.

RFLP, RAPD, AFLP, and SSR were the markers of choice during the last two decades. Among them, SSR markers are preferred for cacao germplasm characterization and this system will continue to play a major role in cocoa breeding and germplasm research. However, SSR uses electrophoresis based assays and is therefore time consuming and expensive. Moreover, SSR is not an ideal system for conducting association studies in plants because of the occurrence of homoplasy (Rafalski 2002). Emphasis is now shifting towards the development of molecular markers that can be detected through non-gel-based assays. One of the most popular of these is single nucleotide polymorphism (SNP), which detects sites where DNA sequences differ by a single base pair. SNPs are the most abundant class of polymorphisms in plant genomes (Buckler and Thornsberry 2002; Zhang and Hewitt 2003). Assays of SNPs do not require DNA separation by size and, therefore, can be automated in an assay-plate format or on microchips (Rafalski 2002). When a large amount of genotyping needs to be done in a timely fashion, automation is essential to dramatically reduce the costs per data point and improve the repeatability of the result among experiments and laboratories. Another major advantage of SNPs is that they can often be identified in candidate genes with potentially important functions (Kuhn et al. 2003, 2004; Borrone et al. 2004). Therefore, it is an ideal system for the association of SNP haplotypes at candidate gene loci with phenotypes. SNP data are also unambiguous so that merge data between platforms or laboratories is straightforward. Because of these advantages, SNPs will likely become the marker-of-choice in the future. Development of SNP markers is progressing rapidly in cocoa (Cryer N, Kuhn D, and Lanaud C; personal communication). The large-scale application of SNPs in cacao will significantly increase our ability to address more complex questions in cocoa germplasm characterization.

6.2.2 Linkage Mapping

The first cacao linkage map was developed for the 'UPA402' x 'UF676' progeny, containing 193 marker loci (mainly RFLPs and RAPDs), covering 759 cM in 10 linkage groups, corresponding to the haploid chromosome number (Lanaud et al. 1995) (Table 6.1). This map was later saturated with additional markers (mainly AFLP and SSRs) to a total of 424 loci (Risterucci et al. 2000), and it became the consensus linkage map and reference for linkage group or chromosome numbering for *T. cacao* (Clément et al. 2001. The latest published version of this high-density map contained 465 markers, 268 of which were SSRs and 16 resistance gene analogs (RGA), covering 782.8 cM, and it was based on 135 individuals (Pugh et al. 2004) (Table 6.1).

Crouzillat et al. (1996) developed a second linkage map, using a backcross population of 131 plants derived from a cross between a single F1 tree from 'Catongo' x 'Pound12' to a recurrent 'Catongo'. The final backcross map contained 140 markers (RFLPs and RAPDs) and included two morphologic loci (self-compatibility and anthocyanin synthesis), and covered 944 cM (Crouzillat et al. 2000b). An additional genetic map was developed for the related F1 population from the cross 'Catongo' x 'Pound12', with 162 markers, covering 772 cM (Crouzillat et al. 2000a).

Subsequently, another eight linkage maps were developed to search for genomic regions associated with resistance to various *Phytophthora* species and strains, evaluated by diverse methods of inoculation (Table 6.1; Flament et al. 2001; Motilal et al. 2002; Clement et al. 2003b; Risterucci et al. 2003). Simultaneously, -QTL- associated with yield components and plant vigor were also assessed in some of these families.

An F2 population, derived from the 'ICS1' x 'Scavina6' cross, has been specifically developed and used to identify genomic regions associated with witches' broom disease resistance, caused by *Moniliophthora perniciosa* (Queiroz et al. 2003; Brown et al. 2005; Faleiro et al. 2006). This F2 high-density genetic map currently contains nearly 500 markers, including 270 SSR loci, and is based on 146 individuals. Recently, additional mapping populations derived from novel sources of resistance to witches' broom were developed in Brazil to identify QTLs associated

Table 6.1 Details about currently published cacao genetic maps with emphasis on identification of QTLs associated with disease resistance

Families	No. of Plants	No. of Markers	No. of Linkage Groups	Map Length (cM)	Evaluation method for resistance	Reference
					Phytophthora	
UPA 402 × UF 676	144	193	10	759	Artificial inoculation of leaf disks	Lanaud et al., 1995;
	181	473		887	and % pod rot rate under field	Risterucci et al., 2000;
	135	465		783	conditions with *P. palmivora*	Pugh et al., 2004;
T60/887 × IFC 2	(59)		11	793	Artificial inoculation of leaf disks	Flament et al., 2001;
T60/887 × IFC 5	(56)				and wounded pods, and % pod	
	112	198			rot rate under field conditions with *P. palmivora*	
Catongo × Pound12	55	162	12	772	Attached pod inoculation without wounding	Crouzillat et al., 2000b
Catongo × (Catongo × Pound12)	131	140	10	944		Crouzillat et al., 2000b
ICS 84 × UPA 134	62	224	15	548	Artificial inoculation of leaf disks	Flament, 1998
SNK 10 × UPA 134	78	151	16	617	and wounded pods, and % pod	Flament, 1998
SNK 413 × IMC 67	58	119	15	419	rot rate under field conditions with *P. megakarya*	Flament, 1998
DR 1 × Catongo	107	192	9	653	% pod rot rate under field	Clement et al., 2003b
S 52 × Catongo	101	138	11	589	conditions with *P. palmivora*	Clement et al., 2003b

Table 6.1 (continued)

Families	No. of Plants	No. of Markers	No. of Linkage Groups	Map Length (cM)	Evaluation method for resistance	Reference
IMC 78 × Catongo	128	223	10	721		Clement et al., 2003b
(Scavina 6 × H) × IFC 1	151	213	10	682	Artificial inoculation of leaf disks with two strains of *P. megakarya*, *P. palmivora*, *P capsicii*	Risterucci et al., 2003
IMC 57 × Catongo	155	235	12	1427	Artificial inoculation of leaf disks with *P. palmivora*	Motilal et al., 2002
					Moniliophthora perniciosa	
F_2 (ICS 1 × Scavina 6)	146	178	10	672	Field resistance	Brown et al., 2005;
	82	342	16	670		Faleiro et al., 2006
ICS 39 × CAB 214	116	57	9	501	Artificial inoculation under field conditions	Figueira et al., 2006
ICS 39 × CAB 208	168	57	9	541		
					Moniliophtora roreri	
Pound7 × UF273	256	180	10	878	Artificial inoculation of attached pods, scored for internal and external lesion	Brown et al., 2006

with resistance (Figueira et al. 2006). Preliminary QTLs have been associated with resistance to frosty pod disease caused by *Moniliophthora roreri*, in the 'Pound7' x 'UF273' progeny (Brown et al. 2006). Another F2 population was developed to identify QTLs associated with flavor and seed quality (Crouzillat et al. 2001).

6.2.3 Quantitative Trait Loci Mapping

Theobroma cacao genetic maps have been used to detect QTLs for various agronomically important traits, including resistance to the three major fungal diseases (Table 6.1), yield components, plant vigor, and quality traits (Lanaud et al. 1996; Motilal et al. 2002; Flament et al. 2001; Clement et al. 2003a; Clement et al. 2003b; Guimarães et al. 2003; Queiroz et al. 2003; Risterucci et al. 2003; Brown et al. 2005).

A. Disease resistance: Black pod rot, caused by various species of *Phytophthora*, is the most important disease of cacao worldwide. *Phytophthora palmivora* occurs globally, while *P. megakarya* is restricted to West Africa, and *P. capsicii* and *P. citrophthora* occur in the Americas. To identify genomic regions associated with resistance to *Phytophthora* species and isolates, 12 linkage maps have been published for populations in Ivory Coast, Costa Rica, Cameroon, Trinidad, and France (Table 6.1). Evaluation of *Phytophthora* resistance for QTL analyses have been performed by either using natural pod rot losses, by artificial inoculation of attached pods (wounded or not), or by inoculation of leaf disks (Table 6.1).

Based on natural infection rates under field conditions in Ivory Coast, QTLs for resistance against *P. palmivora* were detected for both parents of the 'UPA402' x 'UF676' progeny on chromosome 1, explaining 15% to 19% of the variation of the trait, and a minor one on chromosome 9 (Lanaud et al. 2000). When a similar evaluation was conducted for a two-year harvest period using the 'T60/887' x 'IFC2/IFC5' progenies, one major QTL was identified on chromosome 10, explaining 17% of the phenotypic variance (Flament et al. 2001). Based on data between years eight and 13 after planting, Clement et al. (2003b) detected a significant QTL for resistance against *P. palmivora* for two parents 'DR1' and 'IMC78' on the same region of chromosome 4, explaining 10.1% and 22.6%, respectively of the trait variation.

When the resistance against *P. palmivora* of the 'Catongo' x 'Pound12' F1 and the BC1 progenies was evaluated based on artificial pod inoculation under field conditions, six QTLs were detected on five linkage groups (Crouzillat et al. 2000a). Only one QTL (on chromosome 9) was common to both populations, with a major effect on the F1, explaining nearly 48% of the variance for the trait. Flament et al. (2001) also used artificial inoculation of attached pods (nonetheless wounded) in the search for resistance, identifying two QTLs on chromosomes 2 and 6 of 'T60/887'. This minor QTL identified on chromosome 2 was co-localized with one identified by Crouzillat et al. (2000a).

A screening method for resistance against *Phytophthora* based on inoculation of leaf disks has been used to identify QTLs (Lanaud et al. 2000; Flament et al. 2001; Risterucci et al. 2003). However, correlations between resistance evaluated by leaf disk inoculation with field pod rot rate or pod inoculations with *P. palmivora* have been weak and non-significant, possibly because leaf disk inoculations are highly influenced by the environment and have low precision and reproducibility. Flament et al. (2001) identified two QTLs on chromosomes 6 and 3 that were associated with resistance to *P. palmivora* based on the leaf test. These are at distinct locations to those identified by field pod rot rate or by pod inoculation, suggesting either that each evaluation method accounted for a distinct genetic mechanism of resistance, or the size of the progeny and/or the low reproducibility of the tests were not sufficiently accurate to detect all QTLs involved. Similar results were obtained for tests with *P. megakarya* (Flament 1998) with QTLs for resistance identified on chromosome 9 (evaluated by leaf inoculation), and on chromosome 2 (by pod inoculation) on 'UPA 134'.

Based on inoculation of leaf disks of the ('Scavina6' x Hybrid) x 'IFC1' progeny, using two isolates of each of three *Phytophthora* species (*P. palmivora, P. megakarya, P. capsicii*), 13 QTLs for resistance were detected in six chromosome regions, explaining between 7.5% to 12.4% of the phenotypic variation (Risterucci et al. 2003). A region on chromosome 5 contained QTLs for resistance against five strains belonging to the three *Phytophthora* species, whereas two other regions on chromosomes 1 and 6 enclosed QTLs for resistance against two species, suggesting that some of the resistance factors against these *Phytophthora* species might be shared. A QTL for resistance against *P. palmivora* based on field pod inoculation was also identified on chromosome 5 of the BC1 progeny by Crouzillat et al. (2000a), where a cluster of RGA have been also localized (Lanaud et al. 2004). Additionally, a QTL for resistance against one strain of *P. megakarya* was detected on chromosome 3, near another one identified by leaf test against *P. palmivora* in 'T60/887' (Flament et al. 2001). Motilal et al. (2002) using leaf inoculation of 'IMC57'x 'Catongo' with *P. palmivora* identified three major QTLs on chromosomes 1, 9, and 3 or 8, that co-localized with QTLs detected in other studies.

The difficulty in identifying consistent QTLs for *Phytophthora* resistance might be in part due to lack of precision in phenotype evaluation, or a complex multigenic mechanism of resistance, under a large environmental effect, or even because of the small size of the populations segregating for the traits. Nevertheless, identification of QTLs on similar chromosomal regions offers the possibility for marker-assisted breeding for *Phytophthora* resistance, which would include selection of favorable alleles for pyramiding resistance genes in specific populations and in recurrent selection.

Witches' broom disease, caused by the fungus *Moniliophthora perniciosa*, is a severe constraint to cacao production in the Americas, where it was responsible for the collapse of the industry in Surinam, Trinidad, Ecuador, and more recently in Brazil. To identify markers associated with witches' broom resistance, an F2 population, derived from selfing 'TSH516', a selected hybrid from an 'ICS1' x 'Scavina6' cross was developed (Queiroz et al. 2003). A major QTL was identified on chromosome 9 near the SSR locus mTcCIR35, responsible for up to 51% of the phenotypic variance for resistance, while a secondary minor QTL was detected on chromosome 1, near a RGA locus (Brown et al. 2005; Faleiro et al. 2006). To enable map-based cloning of this major resistance gene, a BAC library was constructed from 'Scavina 6' (Clement et al. 2004), and a larger F2 population is currently being developed in Brazil.

Frosty-pod rot of cacao, caused by *Moniliophtora roreri*, is another extremely destructive disease restricted to the Americas where it is a serious yield-limiting factor in Central America. The population 'Pound7' x 'UF273' was used to construct a linkage map and it was evaluated for frosty-pod rot resistance by scoring attached pods for internal and external lesions after artificial inoculations with *M. roreri* (Brown et al. 2006). Three major QTLs for resistance against internal and external lesions by *M. roreri* were detected at the same region on chromosomes 2 and 8, and an additional one for external resistance was found on linkage group 7. QTLs associated with witches' broom and frosty-pod resistance are potentially useful in preventive breeding.

B. Yield components, plant vigor and quality traits: In general, the yield of dry fermented cacao seeds produced by a tree is significantly correlated with the total number of pods harvested, but usually not correlated with the other yield components, such as pod weight, seed weight, or number of seeds per pod. Seed yield tends to be correlated with mature tree vigor, as estimated by trunk girth and/or canopy size.

Dry seed production from 55 F1 individuals of the 'Catongo' x 'Pound12' collected over 15 years allowed for the detection of 10 QTLs, distributed on eight chromosomes (Crouzillat et al. 2000b). Two genomic regions, on chromosomes 4 and 5, each explaining ca. 20% of the total phenotypic variance were detected as early as four years after planting and were consistently detected for another 12 years. Similarly, six QTLs for mean seed yield based on nine years of harvest were identified in five linkage groups in families derived from three genotypes (DR1, S52, and IMC78), each crossed with 'Catongo' (Clement et al. 2003b). The yield QTL identified on chromosome 5 in the Forastero 'IMC78' was detected at a later stage (nine years after planting), but it was more repeatable than those from the Trinitarios ('DR1' and 'S52'). The two yield QTLs identified on chromosomes 4 and 5 in 'IMC78' were located close to those detected in 'Pound12', probably because both Forastero genotypes share a common genetic origin in Peru. A QTL for yield was detected around the same region of chromosome 1 for the two Trinitario 'S52' and 'UF676' (Clément et al. 2001).

A major QTL for pod weight in 'IMC78', explaining 43.5% of the phenotypic variation, was detected on chromosome 4 (Clement et al. 2003b) near a similar QTL detected in the Trinitario 'DR1' and the Forastero 'T60/887' (Clément et al. 2001). A QTL for pod index, defined as the number of pods required to produce one kilogram of dry cacao seeds (function of pod weight), was also identified on chromosome 4 of 'Pound12' (Crouzillat et al. 2000b). Similarly, QTLs for pod weight have been co-localized on chromosome 1 for the Trinitario genotypes DR1 and UF676 (Clément et al. 2001).

In terms of seed weight, a major QTL was co-localized on the same region of chromosome 4 in 'Pound12', 'S52' and 'IMC78', explaining 23.6%, 16.2%, and 13.6% of phenotypic variation, respectively (Crouzillat et al. 2000b; Clement et al. 2003a). Another example of co-localization of QTLs included the region of chromosome 9, containing a QTL for seed weight identified in 'S52' and in 'UF676' (Clement et al. 2003a).

In cacao, seed yield at maturity appears to be associated with plant vigor, usually measured by trunk diameter or canopy size. The significant QTLs identified for canopy width for 'DR1' and 'IMC78' were co-localized with yield QTLs, while QTLs for stem diameter and trunk circumference are closely located to yield QTLs in 'IMC78' (Clement et al. 2003b).

QTLs for fat content and flavor quality attributes of cacao seeds (cocoa, floral and fruity flavor; astringency; acidity) have been identified in the 'UPA402' x 'UF676' progeny (Lanaud et al. 2005). Additionally, a specific F2 population was established in Ecuador to identify QTLs responsible for superior flavor aspects (Crouzillat et al. 2001). More recently, genomic regions involved in yield components have been identified by association or admixture mapping, instead of conventional family-derived mapping (Schnell et al. 2005; Marcano et al. 2007).

Association mapping is a new approach based on the occurrence of linkage disequilibrium over extensive genetic distances on chromosome segments in a population derived from recent hybridization, with defined founding individuals. Using a low density genome-wide scan with SSR loci, it should be possible to identify significant linkage disequilibrium and test for statistical association between phenotypes and markers, without the requirement of specific populations derived from controlled crosses and segregation (Schnell et al. 2005; Marcano et al. 2007). The first application of association mapping in cacao analyzed 99 productive and 50 unproductive trees from a population at Waialua, Hawai'i, and individuals from three presumed parental populations to identify associations between markers and yield components (Schnell et al. 2005). From the 65 SSR loci analyzed, 17 displayed a significant association with yield components, whereas 13 (76.4%) were located in genomic regions previously assigned to QTLs for yield components on chromosomes 1, 2, 3, 4, and 9. Co-localization of significant associations and QTL for pod number in 'DR1' and 'IMC78' were identified on chromosomes 4 and 9 (Clement et al. 2003b).

Based on admixture mapping, Marcano et al. (2007) also identified 15 genomic regions associated with seed and fruit weight in two populations, one including 150 Criollo/Trinitario accessions from a germplasm collection in Costa Rica, and the other 291 trees from a plantation in Venezuela. Linkage disequilibrium extended to up to 25-35 cM in both populations. From the 15 identified genomic regions associated with the traits, 10 were localized near previously identified QTLs for the same traits based on four distinct families (Crouzillat et al. 2000b; Clement et al. 2003a), while five were novel regions, indicating the usefulness of this new approach.

Association mapping offers great potential for cacao breeding, because it minimizes the requisite of developing specific populations, an impediment due to long juvenile phase. This mapping approach opens new possibilities with existing breeding or commercial populations, and germplasm collections, currently under molecular characterization.

Genomic regions associated with yield components and plant vigor have been identified in some of the linkage maps, and more recently by admixture mapping. Many showed co-localization on chromosomes across different genotypes, especially those sharing a common origin. For example, chromosome 4 displayed various major QTLs for distinct yield components for more than one genotype (Crouzillat et al. 2000b; Clement et al. 2003; Schnell et al. 2005; Marcano et al. 2007). Some of the QTLs or markers have potential for use in indirect selection, but the effective use of these markers will require new approaches in cacao breeding. Developing specific populations will be required to apply genotype-building strategies (e.g., population recurrent selection), based on combining favorable alleles at all loci with phenotype evaluation.

6.3 Genomics Resources for Cacao

6.3.1 Development of BAC Libraries: BAC Physical Mapping

Two BAC libraries for cacao have been created. The first library was constructed by CIRAD, the French federal agricultural development research agency, using the Scavina-6 genotype (Clement et al. 2004). The second library was constructed by the Clemson University Genomics Institute (CUGI) under a commission from the USDA Miami Tropical Research Station (USDA-ARS Subtropical Horticulture Research Station (SHRS), Miami, FL. The CUGI library was constructed using the LCT-EEN 37 genotype and is commercially available to the public. Each of the BAC libraries contains over 36,864 clones with average insert sizes of 120 kb, representing 10 haploid genome equivalents.

The BAC libraries serve as a vital resource for a variety of structural and evolutionary genomic studies. For example, the BAC clones can be constructed into contigs by restriction fingerprinting analysis. The resulting contigs can then be anchored onto existing genetic maps to generate a physical map of the cacao genome. The BAC clones can also be utilized as a resource for map-based cloning of agriculturally important genes, and can serve as a template for genomic sequencing assembly in the future.

6.3.2 ESTs

One of the keys to gaining a better understanding of how cacao plants grow, develop, and respond to their environment lies in the unraveling of the gene expression networks underlying these processes. Toward this end, several groups have undertaken discovery programs aimed at identifying the genes that are expressed in given tissues at specific times and in response to specific biotic and abiotic stimuli.

Several cacao EST sequencing projects have resulted in 6,569 ESTs being deposited to date in dbEST (http://www.ncbi.nlm .nih.gov/dbEST/) (Jones et al. 2002; Verica et al. 2004). The Masterfoods Company generated a collection of ESTs isolated from cacao leaves and beans. These ESTs were assembled into a unigene set consisting of 1,380 sequences. Amplified inserts for each of the unigenes were used to construct a microarray which can be used to study the expression of these genes in different tissues and in response to a variety of biotic and abiotic stimuli (Jones et al. 2002).

A second EST collection was developed with a focus on characterizing genes expressed during defense responses in cacao (Verica et al. 2004). Suppressive-subtractive hybridization (SSH) and macroarray analysis were used to identify cacao ESTs representing genes upregulated by defense signaling molecules. Differential expression analysis using macroarrays identified 475 upregulated clones. These clones, as well as 1,639 randomly chosen cDNAs, were sequenced and assembled into contigs. A total of 1,256 unigenes was obtained, including 330 representing upregulated genes. Eight hundred sixty-five unigenes were assigned to functional classes using BLAST. Eight percent of the sequences up-regulated by the defense inducers were similar to defense proteins in other plants. cDNAs were identified with sequences similar to known defense-related genes including heat shock proteins, NPR1 (a transcriptional regulator that mediates the expression of salicylic acid (SA)- and jasmonic acid (JA)- responsive genes), and several pathogenesis related (PR) genes, including chitinases, (shown to enhance resistance against fungal pathogens). An additional 8% of the up-regulated genes are predicted to play roles in signaling, although their roles in defense are unclear. A database including the ESTs from both of these projects was compiled by The Institute for Genomic Resources (TIGR) and is available online (see section 6.3.5).

A major EST sequencing project is currently underway managed by the French cacao research lab within CIRAD (C. Lanaud personal communication). The goal of the project is to sequence and annotate the ends of 200,000 clones from 28 cDNA libraries constructed from mRNA isolated from different cacao tissues, as well as transcripts induced by a variety of biotic and abiotic stimuli. The libraries were contributed in a cooperative effort by the international cacao molecular biology community. The project is scheduled to be completed in 2007, at which time all sequence data will be made publicly available. CIRAD researchers have created an annotated database to house the data and allow text or DNA based searches within a user-friendly web-based interface. With the completion of this project, the accumulated EST resources will contain representatives of nearly all expressed cacao genes. Genes of low abundance and genes exclusive to tissues or inductive conditions not sampled will remain to be discovered.

6.3.3 Microarrays

Although the EST libraries serve as a valuable resource for understanding gene expression, their utility is limited by the fact that they cannot be used simultaneously to compare differences in expression between multiple tissues or between multiple treatments. Microarrays allow the comparative measurement of gene expression levels in thousands of genes in a single experiment. In this type of analysis, sequences corresponding to specific genes are spotted onto array, and RNA from a tissue under study is labeled then hybridized to the array. The arrays can subsequently be used to study the transcription profiles of all the arrayed genes in a given tissue, or in response to a specific stimulus.

Through the efforts of an international cacao research consortium, a unigene dataset of all publicly available cacao DNA sequences has been compiled (M. Guiltinan, unpublished). Using these sequences, a unigene set consisting of 2,781 sequences was assembled. From these, a set of 50-mer oligonucleotides unique to each unigene sequence was designed and synthesized by the MWG company (Germany) with funding from Mars Inc. Oligonucleotide microarrays were printed with five replicate sets on each slide. The specificity of the arrays was validated by comparison of leaf vs. fruit RNA, demonstrating that the majority of genes detected at higher levels in leaves are involved in photosynthesis, as would be expected. The microarrays have been made available to cacao researchers worldwide and have been used to study pathogen and endophyte interactions by one of the authors (MJG). Although microarray experiments have proved an invaluable research tool, these analyses have not been without their problems. For example, microarray analyses have been shown to have a potentially high rate of both false positives and false negatives. In addition, similar analyses performed in different laboratories have resulted in a limited overlap in gene signatures. To confirm the validity of microarray results, it is necessary to use an independent methodology.

6.3.4 Genetic Transformation System

A genetic transformation system for cacao was developed in the Guiltinan lab (Guiltinan et al. 1998; Antúnez de Mayolo 2003; Antúnez de Mayolo 2003; Maximova et al. 2003, reviewed in Maximova et al. 2007). This system is based on somatic embryogenesis as a regenerative system. Primary somatic embryos are used as explants for co-cultivation with *Agrobacterium*-harboring Ti-plasmids containing the GFP marker gene and the kanamycin selectable marker gene. The use of a dual selection system allowed the optimization of the many variables impinging on transformation frequency and for the visible selection of high expressing embryos. Transformation of cacao, regeneration of plantlets, and their subsequent analysis is a time consuming process requiring about one year from construct to small plantlet. Moreover, although the transformation system was shown to be reproducible, the transformation frequencies are low compared to other species, further increasing the difficulty and cost to produce sufficient numbers of independent transformants for analysis.

Once developed, this system was used to study the function of the cacao chitinase A gene by over-expression of this gene in transgenic cacao plants (Maximova et al. 2005). The *chiA* gene was shown to confer enhanced resistance to a fungal pathogen in the transgenic cacao leaves demonstrating the efficacy of this system as a tool for functional genomics research in cacao. The potential application of this technology to crop improvement in cacao is uncertain, however, as consumer and industry concerns over the GMO issue continue.

6.3.5 Informatics Databases

The molecular, morphological, and pedigree data for cacao germplasm are currently managed in several international databases. One is International Cocoa Germplasm Database (ICGD) (University of Reading/Euronext.liffe; http://www.icgd.rdg.ac.uk/), which mainly contains passport data, characterization data, and conservation information of cocoa germplasm, supplied directly by research institutes and from various publications. Another database is CocoaGenDB (http://cocoagendb.cirad.fr/), which was developed and maintained at CIRAD (in partnership with ICGD and USDA). CocoaGenDB combines the genomic data of cacao held in CIRAD's TropGENE database and the germplasm information from ICGD. In CocoaGen DB, a Java applet tool was developed for the visualization of genealogy and for alleles tracking among cocoa germplasm. The CocoaGen DB has been modeled to integrate sequences, clustering, Blast and GO annotation and libraries description of the increasing ESTs data in cocoa. Microsatellites and SNPs derived from the analysis of cDNA collection will also be added to the database. The cocoa gene index was initially developed at TIGR based on publicly available EST data. The TIGR database contains a searchable annotated EST unigene database of all cacao ESTs submitted to the GenBank database at the National Center for Biotechnology Information (NCBI). Genbank contains all publicly available cacao genomic, cDNA and EST data.

6.4 Perspectives

Cacao is a crop of major economic importance to smallholder farmers in the developing world and to the food industry worldwide. It is also of major ecological importance as it is cultivated as a sustainable shade crop inside rainforests. The development of genomics resources for cacao will facilitate the breeding of improved varieties of cacao, most urgently those with improved disease resistance. Because of the importance of cacao to the chocolate industry, its value in economic advancement of developing countries, and its potential as an ecologically sustainable crop, applied genomics to this crop has tremendous potential for exerting a wide array of major impacts. In the future, genomic sequencing of cacao will provide a comprehensive dataset from which markers and genes of interest can be chosen for guiding marker assisted selection programs to speed cacao improvement and allow the relatively rapid introgression of resistance genes, genes for yield and quality traits into varieties of cacao adapted to local conditions throughout producing countries. In this way, genomics will have a major impact on reducing poverty, stabilizing economies, and protecting the fragile environments in the countries where cacao is a major export crop.

In addition to its intrinsic importance as the source of cocoa for the chocolate industry, cacao genomics will also contribute in a wider sense to the growing knowledge base of plant biology in general. Cacao is one of the 14 crops of major economic importance in the Eurosids II group that includes the model plant system Arabidopsis: the others being Brassica sp. (11 crops), cotton, and citrus (Fig. 6.1). There are at least 21 additional species of economic interest in this group. Cacao is particularly well positioned for genomic comparisons with Arabidopsis, cotton, and other species, and offers an important benefit in having a moderately sized, simple diploid genome (in contrast for example to Brassica, citrus, and cotton). Detailed comparative genomics among Arabidopsis and these crops will enrich our understanding of gene function, genome evolution, and developmental mechanisms in all members of the group.

Furthermore, cacao offers unique opportunities for addressing biological questions of broad interest to plant biologists, for example woody tree development, phase change (juvenile to adult growth habit), disease resistance in a tropical

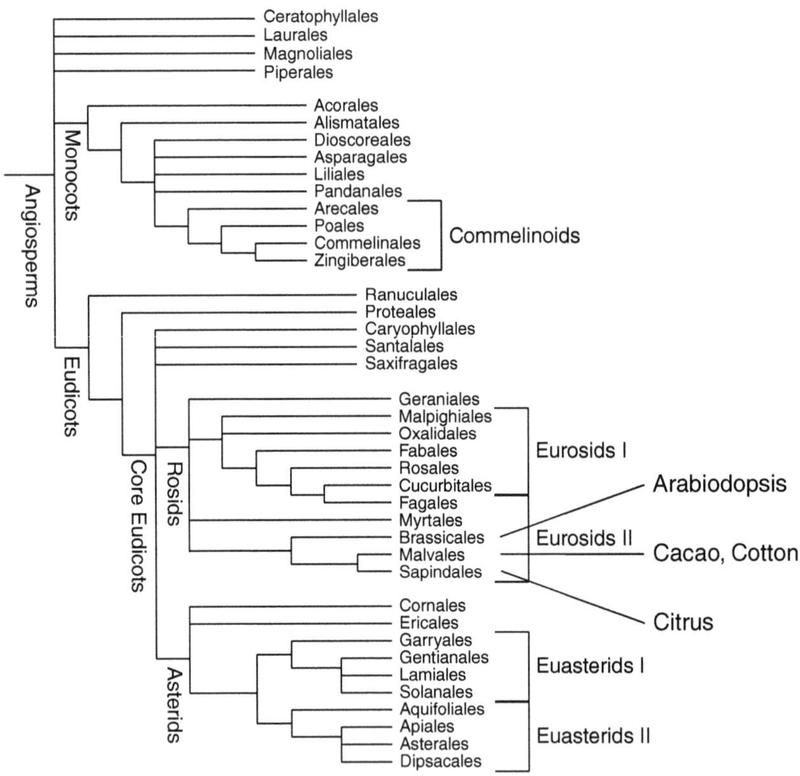

Fig. 6.1 Phylogeny of Flowering Plants Relatedness of Arabidopsis, Brassica, cacao, cotton and citrus

environment, ovarian self-incompatibility, and the phytochemical basis of flavor and aroma. Translation of genomics resources to cacao will make these questions and others accessible to plant biologists in the future. For certain, *Theobroma cacao*, the plant named the Food of the Gods, will hold many interesting and important discoveries for future generations of plant scientists to come.

References

Aime MC, Phillips-Mora W (2005) The causal agents of witches' broom and frosty pod rot of cacao (chocolate, *Theobroma cacao*) form a new lineage of Marasmiaceae. Mycologia 97:1012–1022

Antúnez de Mayolo G (2003) Genetic Engineering Of *Theobroma Cacao* And Molecular Studies On Cacao Defense Responses, A Thesis in Integrative Biosciences. Ph.D. Thesis: The Pennsylvania State University, The Graduate School, Graduate Degree Program in Integrative Biosciences

Antunez de Mayolo G, Maximova SN, Pishak S, Guiltinan MJ (2003) Moxalactam as a counter-selection antibiotic for *Agrobacterium*-mediated transformation and its positive effects on *Theobroma cacao* somatic embryogenesis. Plant Sci 164:607–615

Appiah AA, Flood J, Bridge PD, Archer SA (2003) Inter- and intraspecific morphometric variation and characterization of *Phytophthora* isolates from cocoa. Plant Pathol 52:168–180

Appiah AA, Flood J, Archer SA, Bridge PD (2004) Molecular analysis of the major *Phytophthora* species on cocoa. Plant Pathol 53:209–219

Bartley BGD (1994) A review of cacao improvement: fundamentals, methods and results. http://ingenic.cas.psu.edu/proceedings.htm. Proc Intl Workshop Cocoa Breeding Strategies. pp. 3–16

Bartley BGD (2005) The Genetic Diversity of Cacao and its Utilization. (Wallingford, UK: CABI Publishing)

Bennett AB (2003) Out of the Amazon: *Theobroma cacao* enters the genomic era. Trends Plant Sci 8:561–563

Bhattacharjee R, Aikpokpodion P, Kolesnikova-Allen M, Badaru K, Schnell RJ (2004) West African Cocoa: A pilot study on DNA fingerprinting the germplasm from Cross River State of Nigeria. INGENIC Newslett 9:15–20

Boccara M, Zhang D (2006) Progress in resolving identity issues among the Parinari accessions held in Trinidad: the contribution of the collaborative USDA/CRU project. CRU Ann Rpt 2005, Cacao Research Unit, The University of the West Indies, St. Augustine, Trinidad and Tobago

Borrone JW, Kuhn DN and Schnell RJ. (2004) Isolation, characterization, and development of WRKY genes as useful genetic markers in *Theobroma cacao*. Theor Appl Genet 109: 495–507

Borrone JW, Brown JS, Kuhn DN, Schnell RJ (2007) Microsatellite markers developed from *Theobroma cacao* L. expressed sequence tags. Mol Ecol Notes 7: 236–239

Bowers JH, Bailey BA, Hebbar PK, Sanogo S, Lumsden RD (2001) The impact of plant diseases on worldwide chocolate production (APSNet; Plant Health Progress, http://www.plantmanagementnetwork.org/pub/php /review/ cacao/)

Brown JS, Kuhn DN, Lopez U, Schnell RJ (2005) Resistance gene mapping for witches' broom disease in *Theobroma cacao* L. in an F2 population using SSR markers and candidate genes. J Am Soc Hort Sci 130:366–373

Brown J, Phillips W, Power E, Kroll C, Cervantes-Martinez C, et al. (2006) Preliminary QTL mapping for resistance to frosty pod in cacao (*Theobroma cacao* L.). Proc Intl Cocoa Res Conf 15: (in press)

Buckler ES, Thornsberry J (2002) Plant molecular diversity and applications to genomics. Curr Opin Plant Biol 5:107–111

Chowdappa P, Brayford D, Smith J, Flood J (2003) Molecular discrimination of *Phytophthora* isolates on cocoa and their relationship with coconut, black pepper and bell pepper isolates based on rDNA repeat and AFLP fingerprints. Curr Sci 84:1235–1237

Christopher Y, Mooleedhar V, Bekele F, Hosein F (1999) Verification of accession in the ICG, T using botanical descriptors and RAPD analysis. Ann Rpt 1998. St. Augustine, Trinidad and Tobago: Cocoa Research Unit, The University of the West Indies. pp. 15–18

Clément D, Risterucci AM, Lanaud C (2001) Analysis of QTL studies related to yield and vigour traits carried out with different cocoa genotypes. Proc Intl Workshop New Technol Cocoa Breeding (INGENIC). pp. 127–134

Clement D, Risterucci AM, Motamayor JC, Ngoran J, Lanaud C (2003a) Mapping quantitative trait loci for bean traits and ovule number in *Theobroma cacao* L. Genome 46:103–111

Clement D, Risterucci AM, Motamayor JC, N'Goran J, Lanaud C (2003b) Mapping QTL for yield components, vigor, and resistance to *Phytophthora palmivora* in *Theobroma cacao* L. Genome 46:204–212

Clement D, Lanaud C, Sabau X, Fouet O, Le Cunff L, et al. (2004) Creation of BAC genomic resources for cocoa (*Theobroma cacao* L.) for physical mapping of RGA containing BAC clones. Theor App Genet 108:1627–1634

Couch JA, Zintel HA, Fritz PJ (1993) The genome of the tropical tree *Theobroma cacao* L. Mol Gen Genetics 237:123–128

Crouzillat D, Lerceteau E, Pétiard V, Morera J, Rodriguez H, et al. (1996) *Theobroma cacao* L: A genetic linkage map and quantitative trait loci analysis. Theor Appl Genet 93:205–214

Crouzillat D, Phillips W, Fritz PJ, Petiard V (2000a) Quantitative trait loci analysis in *Theobroma cacao* using molecular markers. Inheritance of polygenic resistance to *Phytophthora palmivora* in two related cacao populations. Euphytica 114:25–36

Crouzillat D, Menard B, Mora A, Phillips W, Petiard V (2000b) Quantitative trait analysis in *Theobroma cacao* using molecular markers. Euphytica 114:13–23

Crouzillat D, Rigoreau M, Cabigliera M, Alvarez M, Bucheli P, Pétiard V (2001) QTL studies carried out for agronomic, technological and quality traits of cocoa in Ecuador. Proc Intl Workshop New Technol Cocoa Breeding (INGENIC). pp. 120–126

Cryer NC, Fenn MGE, Turnbull CJ, Wilkinson MJ (2006a) Allelic size standards and reference genotypes to unify international cocoa (*Theobroma cacao* L.) microsatellite data. Genetic Res Crop Evol 53:1643–1652

Cryer NC, Fenn MGE, Turnbull CJ, Wilkinson MJ (2006b) Allelic size standards and reference genotypes to unify international cocoa (*Theobroma cacao* L.) microsatellite data. Genetic Res Crop Evol, http://dx.doi.org/10.1007 /s10722-005-1286-9

Day RK (1984) Population dynamics of cocoa pod borer *Acrocercops cramerella*: Importance of host plant cropping cycle. Intl Conf Cocoa and Coconuts. pp. 1–9

Dias LAS (2001) Genetic Improvement of Cacao. Editora Folha de Vicosa Ltda. xii + 578 pp. Translation (2005) by Cornelia Elisabeth Abreu-Reichart, Vicosa, M.G., Brazil; aided by the Editor and Peter Griffee of FAO and supported by FAO. Web version: http://ecoport.org/ep?SeachType=earticleView &earticleId=197

Efombagn MIB, Marelli JP, Ducamp M, Cilas C, Nyasse S, et al. (2004) Effects of fruiting traits on the field resistance of cocoa (*Theobroma cacao* L.) clones to *Phytophthora megakarya*. Phytopathology 152:557–562

Efombagn MIB, Sounigo O, Nyass S, Manzanares-Dauleux M, Cilas CB, et al. (2006) Genetic diversity in cocoa germplasm of southern Cameroon revealed by simple sequences repeat (SSRs) markers. African J Biotech 5:1441–1449

Engels J (1981) Genetic resources of cacao, A catalogue of the CATIE collection. (Turrialba, Costa Rica)

Eskes A (2003) Proc Intl Workshop Cocoa Breeding for Improved Production Systems: INGENIC International Workshop on Cocoa Breeding. Bekele F, End M, Eskes A (eds) (Accra, Ghana: INGENIC and Ghana Cocoa Board 2005, http://ingenic.cas. psu.edu/proceedings.htm)

Evans HC, Holmes KA, Reid AP (2003) Phylogeny of the frosty pod rot pathogen of cocoa. Plant Pathol 52:476–485

Faleiro FG, Yamada MM, Lopes UV, Pires JL, Faleiro ASG, et al. (2002) Genetic similarity of *Theobroma cacao* L. accessions mantained in duplicates in the Cacao Research Center germplasm collection, based on RAPD markers. Crop Breed Appl Biotechnol 2:439–444

Faleiro FG, Lopes UV, Yamada MM, Melo GRP, Monteiro WR, et al. (2004) Genetic diversity of cacao accessions selected for resistance to witches' broom disease based on RAPD markers. Crop Breed Appl Biotechnol 4:12–17

Faleiro F, Queiroz V, Lopes U, Guimarães C, Pires J, et al. (2006) Mapping QTLs for witches' broom (*Crinipellis perniciosa*) resistance in cacao (*Theobroma cacao* L.). Euphytica 149:227–235

Figueira A, Janick J, Goldsbrough P (1992) Genome size and DNA polymorphism in *Theobroma cacao*. J Am Soc Hort Sci 117:673–677

Figueira A, Janick J, Levy M, Goldsbrough P (1994) Reexamining the classification of *Theobroma cacao* l. using molecular markers. J Am Soc Hort Sci 119:1073–1082

Figueira A, Albuquerque P, Leal-Jr G (2006) Genetic mapping and differential gene expression of Brazilian alternative resistance sources to witches' broom (causal agent *Crinipellis perniciosa*). Proc Intl Cocoa Res Conf 15: (in press)

Flament MH (1998) Cartographie genetique de facteurs impliques dans la resistance du cacaoyer (Theobroma cacao L.) a *Phytophthora megakarya* et a *Phytophthora palmivora* (Montpellier, France: Ecole Nationale Supérieure Agronomique de Montpellier)

Flament MH, Kebe I, Clement D, Pieretti I, Risterucci AM, et al. (2001) Genetic mapping of resistance factors to *Phytophthora palmivora* in cocoa. Genome 44:79–85

George MLC, Li W, Moju C, Dahlan M, Pabendon M, et al. (2004) Molecular characterization of Asian maize inbred lines by multiple laboratories. Theor Appl Genet. 109:80–91

Guiltinan M (2007) Cacao. In: Pua VE, Davey M (eds) Biotechnology in Agriculture and Forestry - Transgenic Crops. Berlin Heidelbelg: Springer-Verlag

Guiltinan MJ, Li Z, Traore A, Maximova S, Pishak S (1998) High efficiency somatic embryogenesis and genetic transformation of cacao. Ingenic Newsl

Guimarães CT, Queiros VT, Mota JWS, Pereira MG, Daher RF, et al. (2003) A cocoa genetic linkage map and QTL detection for witches' broom resistance. Plant Breeding 122: 268–272

Guyton B, Lumsden R, Matlick BK (2003) Strategic plan for sustainable cocoa production. Manufacturing Confectioner 83:55–60

Hanna AD (1954) Application of a systemic insecticide by trunk implantation to control a mealybug vector of the cacao swollen shoot virus. Nature 173:730–731

Hervé L, Djiekpor E, Jacquemond M (1991) Characterization of the genome of cacao swollen shoot virus. J Gen Virol 72:1735–1739

Hughes JA, Ollennu LAA (1994) Mild strain protection of cocoa in Ghana against cocoa swollen shoot virus-a review. Plant Pathol 43:442–457

Iwaro AD, Bekele FL, Butler DR (2003) Evaluation and utilization of cacao (*Theobroma cacao* L.) germplasm at the International Cocoa Genebank, Trinidad. Euphytica 130:207–221

Johnson L (2003) INGENIC Workshop, Cocoa genomics group. Gro Cocoa 4, 4–5 (http://www.cabi-commodities.org/Acc/ACCrc/PDFFiles/GROC/GROC.htm)

Johnson SE, Mora A, Schnell (2007) Field Guide efficacy in the identification of reallocated clonally propagated accessions of cacao. Genetic Res Crop Evol (in press)

Jones PG, Allaway D, Gilmour DM, Harris C, Rankin D, et al. (2002) Gene discovery and microarray analysis of cacao (*Theobroma cacao* L.) varieties. Planta 216:255–264

Keane PJ (1992) Diseases and pests of cocoa: An overview. Cocoa pest and disease management in Southeast Asia and Australasia, FAO Plant Production and Protection Paper 112:1–12

Kuhn DN, Heath M, Wisser RJ, Meerow A, Brown JS, et al. (2003) Resistance gene homologues in *Theobroma cacao* as useful genetic markers. Theor Appl Genet 107:191–202

Kuhn, DN, Borone J, Meerow A, Motamayor JC, Brown JS, et al. (2004) Single strand conformation polymorphism analysis of candidate genes for reliable identification of alleles by capillary array electrophoresis. Electrophoresis 26:112–115

Lachenaud P, Sounigo O, Oliver G (2004) Genetic structure of Guianan wild cocoa (*Theobroma cacao* L.) described using isozyme electrophoresis. Plant Genetic Res Newsl 139:24–30

Lanaud C (1986) Genetic studies of *Theobroma cacao* L. with the help of enzymatic markers. I. Genetic control and linkage of nine enzymatic markers. Café Cacao Thé 30:259–270
Lanaud C, Risterucci AM, N'Goran AKJ, Clement D, Flament MH, et al. (1995) A genetic linkage map of *Theobroma cacao* L. Theor Appl Genet 91:987–993
Lanaud C, Kebe I, Risterucci AM, Clement D, N'Goran JKA, et al. (1996) Mapping quantitative trait loci (QTL) for resistance to *Phytophthora palmivora* in *T. cacao*. Proc. Intl Cocoa Res Conf 12:99–105
Lanaud C, Risterucci AM, Pieretti I, Falque M, Bouet A, et al. (1999) Isolation and characterization of microsatellites in *Theobroma cacao* L. Mol Ecol 8:2141–2143
Lanaud C, Kébé I, Risterucci A, Clément D, N'Goran J, et al. (2000) Mapping quantitative trait loci (QTL) for resistance to *Phytophthora palmivora* in *T. cacao*. Proc Intl Cocoa Res Conf 12:99–105
Lanaud C, Risterucci AM, Pieretti I, N'goran JAK, Fargeas D (2004) Characterisation and genetic mapping of resistance and defence gene analogs in cocoa (*Theobroma cacao* L.). Mol Breeding 13:211–227
Lanaud C, Boult E, Clapperton J, N'Goran J, Cros E, et al. (2005) Identification of QTLs related to fat content, seed size and sensorial traits of *Theobroma cacao*. Proc Intl Cocoa Res Conf 13:1119–1126
Laurent V, Risterucci AM, Lanaud C (1994) Genetic diversity in cocoa revealed by cDNA probes. Theor Appl Genet 68:193–195
Lerceteau E, Robert T Pétiard V, Crouzillat D (1997) Evaluation of the extent of genetic variability among *Theobroma cacao* accessions using RAPD and RFLP. Theor Appl Genet 95:10–19
Lockwood G, Gyamfi MMO (1979) The CRIG cocoa germplasm collection with notes on codes used in the breeding programme at Tafo and elsewhere (Ghana: Cocoa Research Institute). 62 pp
Marcano M, Pugh T, Cros E, Morales S, Portillo Paez EA, et al. (2007) Adding value to cocoa (*Theobroma cacao* L.) germplasm information with domestication history and admixture mapping. Theor Appl Genet 114: 877–884
Marita JM, Nienhuis J, Pires JL, Aitken WM (2001) Analysis of genetic diversity in *Theobroma cacao* with emphasis on withes' broom disease resistance. Crop Sci 41:1305–1316
Maximova S, Miller C, Antunez de Mayolo G, Pishak S, Young A, et al. (2003) Stable transformation of *Theobroma cacao* L. and influence of matrix attachment regions on GFP expression. Plant Cell Rep 21:872–883
Maximova SN, Marelli JP, Young A, Pishak S, Verica JA, et al. (2005) Over-expression of a cacao class I chitinase gene in *Theobroma cacao* L. enhances resistance against the pathogen, *Colletotrichum gloeosporioides*. Planta 224:740–749
Maximova, SN, Tan CL, Guiltinan MJ (2007) Cocoa In: Kole C, Hall TC (eds) A Compendium of Transgenic Crop Plants. Volume 7. Blackwell Publishing, Malden, MA, USA: In Press
Morais RC (2005) The Genomes of cocoa. In Forbes,(March 14) pp. 110–112
Motamayor JC, Lopez PA, Ortiz CF, Moreno A, Lanaud C (2002) Cacao domestication. I. The origin of the cacao cultivated by the Mayas. Heredity 89:380–386
Motamayor JC, Risterucci AM, Heath M, Lanaud C (2003) Cacao domestication. II. Progenitor germplasm of the Trinitario cacao cultivar. Heredity 91:322–330
Motilal L, Butler D (2003) Verification of identities in global cacao germplasm collections. Genetic Res Crop Evol 50:799–807
Motilal L, Sounigo O, Thévenin J, Risterucci A, Pieretti I, et al. (2002). *Theobroma cacao* L.: genome map and QTLs for *Phytophthora palmivora* resistance. Proc Intl Cocoa Res Conf 13 Kota Kinabalu, Malaysia, 2000. Cocoa Producer's Alliance, Lagos, Nigeria, pp. 111–118
Muller E, Sackey S (2005) Molecular variability analysis of five new complete cacao swollen shoot virus genomic sequences. Arch Virol 150:53–66
N'Goran JAK, Laurent V, Risterucci AM, Lanaud C (1994) Comparative genetic diversity studies of *Theobroma cacao* using RFLP and RAPD markers. Heredity 73:589–597

Nyasse S, Despreaux D, Cilas C (2002) Validity of a leaf inoculation test to assess the resistance to *Phytophthora megakarya* in a cocoa (*Theobroma cacao* L.) diallel mating design. Euphytica 123:395–399

Opoku IY, Akrofi AY, Appiah AA (2002) Shade trees are alternative hosts of the cocoa pathogen *Phytophthora megakarya*. Crop Protection 21:629–634

Opoku IY, Bhattacharjee R, Kolesnikova-Allen M, Enu-Kwesi L, Asante E, et al. (2005) Impact of Breeders' Collection on Cocoa Plantings of Ghana: Assessment by Molecular Marker Analysis and Farmers' Field Survey. A paper presented in First COCOBOD Conference and Promotion of Local Consumption of Cocoa Products and 24 th Biennial Conference of The Ghana Science Association. Accra Ghana . 1–4 August 2005

Pereira JL, Ram A, Defigueiredo JM, Dealmeida LCC (1990) First occurrence of witches' broom disease in the principal cocoa-growing region of Brazil. Trop Agric 67:188–189

Perry MD, Davey MR, Power JB, Lowe KC, Bligh HFJ, et al. (1998) DNA isolation and AFLP genetic fingerprinting of Theobroma cacao (L.). Plant Mol Biol Rep 16:49–59

Presson A, Sobel E, Lange K, Papp J (2006) Merging microsatellite data. J Comput Biol 13:1131–1147

Piasentin F, Klare-Repnik L (2004) Biodiversity conservation and cocoa agroforests. Gro Cocoa 5:7–8

Piry S, Alapetite A, Cornuet J-M, Paetkau D, Baudouin L, et al. (2004) Geneclass2: A Software for Genetic Assignment and First-Generation Migrant Detection. J Heredity 95:536–539

Pound FJ (1940) Witches' broom resistance in cacao. Trop Agric 17:6–8

Pritchard JK, Stephens M, Donnelly PJ (2000) Inference of population structure using multilocus genotype data. Genetics 155:945–959

Pugh T, Fouet O, Risterucci AM, Brottier P, Abouladze M, et al. (2004) A new cacao linkage map based on codominant markers: development and integration of 201 new microsatellite markers. Theor Appl Genet 108:1151–1161

Queiroz VT, Guimaraes CT, Anhert D, Schuster I, Daher RT, et al. (2003) Identification of a major QTL in cocoa (*Theobroma cacao* L.) associated with resistance to witches' broom disease. Plant Breed 122:268–272

Rafalski A (2002) Applications of single nucleotide polymorphisms in crop genetics. Curr Opin Plant Biol 5:94–100

Rice RA, Greenberg R (2003) The Chocolate Tree: Growing cacao in the forest. Natural History 112:36–43

Risterucci AM, Grivet L, Ngoran JA, Pieretti I, Flamen, MH, et al. (2000) A high-density linkage map of *Theobroma cacao* L. Theor Appl Genet 101:948–955

Risterucci AM, Paulin D, Ducamp M, N'Goran JAK, Lanaud C (2003) Identification of QTLs related to cocoa resistance to three species of *Phytophthora*. Theor Appl Genet 108:168–174

Ronning CM, Schnell RJ (1994) Allozyme Diversity in a Germplasm Collection of *Theobroma cacao* L. J. Heredity 85:291–295

Rodriguez LCM, Wetten AC, Wilkinson MJ (2004) Detection and quantification of in vitro-culture induced chimerism using simple sequence repeat (SSR) analysis in *Theobroma cacao* (L.). Theor Appl Genet 110:157–166

Ruf F, Zadi H (2003) Cocoa: From deforestation to reforestation (Migratory Bird Center: Smithsonian National Zoological Park, (http://nationalzoo.si.edu/ConservtionAndScience/MigratoryBirds/Research/Cacao/ruf.cfm))

Russell JR, Hosein F, Johnson E, Waugh R, Powell W (1993) Genetic differentiation of cocoa (*Theobroma cacao* L.) populations revealed by RAPD analysis. Mol Ecol 2:89–97

Santoso D, Chaidamsari T, Wiryadiputra S, de Maagd RA (2004) Activity of *Bacillus thuringiensis* toxins against cocoa pod borer larvae. Pest Management Sci 60:735–738

Sampaio P, Gusmão L, Alves C, Pina-Vaz C, Amorim A, Pais C (2003). Highly polymorphic microsatellite for identification of *Candida albicans* strains. J Clin Microbiol 41:552–557

Saunders JA, Hemeida AA, Mischke S (2001) USDA DNA fingerprinting programme for identification of *Theobroma cacao* accessions. p. 108–114. In F Bekele et al (ed) Proc Intl Workshop New Technol Cocoa Breeding 2000. INGENIC Press, London

Sereno ML Albuquerque PSB, Vencovsky R, Figueira A (2006) Genetic diversity and natural population structure of cacao (*Theobroma cacao* L.) from the Brazilian Amazon evaluated by microsatellite markers. Conservation Genetics 7:13–24

Schnell RJ, Olano CT, Brown JS, Meerow AW, Cervantes-Martinez C, et al. (2005) Retrospective determination of the parental population of superior cacao (*Theobroma cacao* L.) seedlings and association of microsatellite alleles with productivity. J Am Soc Hort Sci 130:181–190

Silva CRS, Figueira A (2005) Phylogenetic analysis of *Theobroma* (Sterculiaceae) based on Kunitz-like trypsin inhibitor sequences. Plant Syst Evol 250:93–104

Sounigo O, Umaharan R, Christopher Y , Sankar A, Ramdahin S (2005) Assessing the genetic diversity in the International Cocoa Genebank, Trinidad (ICG,T) using isozyme electrophoresis and RAPD. Genetic Res Crop Evol 52:1111–1120

Takrama JF, Cervantes-Martinez CT, Phillips-Mora W, Brown JS, Motamayor JC, et al. (2005) Determination of off-types in a cacao breeding program using microsatellites. INGENIC Newsl 10:2–8

Toxopeus H (1969) Cacao (*Theobroma cacao*). pp 79–109. In: Fewerda FP (ed) Outlines of perennial crop breeding in the tropics, (Landbouwhogesschool Wageningen, The Netherlands)

Turnbull CJ, Butler DR, Cryer NC, Zhang D, Lanaud C, et al. (2004) Tackling mislabelling in cocoa germplasm collections. INGENIC Newsl 9:8–11

Verica JA, Maximova SN, Strem MD, Carlson JE, Bailey BA, et al. (2004) Isolation of ESTs from cacao (*Theobroma cacao* L.) leaves treated with inducers of the defense response. Plant Cell Rep 23:404–413

Vos P, Hogers R, Bleeker M, Reijans M, Van de Lee T, et al. (1995) AFLP: a new technique for DNA fingerprinting. Nucleic Acids Res 23:4407–4414

Warren J (1994) Isozyme variation in a number of populations of *Theobroma cacao* obtained through various sampling regimes. Euphytica 72:121–126

Whitkus R, De La Cruz M, Mota-Bravo L, Gomez-Pompa A (1998) Genetic diversity and relationships of cacao (*Theobroma cacao* L.) in southern Mexico. Theor Appl Genet 96:621–627

Wilde J, Waugh R, Powell W (1992) Genetic fingerprinting of *Theobroma* clones using randomly amplified polymorphic DNA markers. Theor Appl Genet 83:871–877

Wood GAR, Lass RA (1985) Cocoa. (New York: Longman Scientific & Technical)

Yamada MM, Faleiro FG, Lopes UV, Bahia RCS, Pires JL, et al. (2001) Genetic variability in cultivated cacao populations in Bahia, Brazil, using isozymes and RAPD markers. Crop Breed Appl Biotechnol 1:377–384

Zhang D-X, Hewitt GM (2003) Nuclear DNA analyses in genetic studies of populations: practice, problems and prospects. Mol Ecol 12:563–584

Zhang D, Mischke S, Goenaga R, Hemeida AA, Saunders JA (2006a) Accuracy and reliability of high-throughput microsatellite genotyping for cacao clone identification. Crop Sci 46:2084–2092

Zhang D, Arevalo-Gardini E, Mischke S, Zúñiga-Cernades L, Barreto-Chavez A, et al. (2006b) Genetic diversity and structure of managed and semi-natural populations of cacao (*Theobroma cacao*) in the Huallaga and Ucayali valleys of Peru. Ann Bot 98:647–655

Chapter 7
Chickpea, a Common Source of Protein and Starch in the Semi-Arid Tropics

Fred J. Muehlbauer and P.N. Rajesh

Abstract Chickpea (*Cicer arietinum* L.) is an important food crop in the semi-arid tropics where it is grown during the cool winter season. Research has concentrated on the development of improved germplasm for resistance to diseases and pests and more recently has focused on the use of genetics and biotechnological tools to enhance the knowledge of the genomics of chickpea. The wild species (8 annuals and 34 perennials) are a potential source of genes for overcoming problems of diseases and pests, and work is underway toward overcoming barriers to interspecific hybridization. Bacterial artificial chromosome libraries are available for genomic research in chickpea and a targeted induced local lesion in genomes, also called TILLING, platform is under development that holds promise for identification of important genes and determination of their function. The chickpea plant is described and the tools for further exploitation of the crop are discussed.

7.1 Introduction

Chickpea had its origin in the Near East Arc, where it was domesticated along with other pulses including lentil (*Lens culinaris* Medik.) and peas (*Pisum sativum* L.) over 7,000 years ago (Ladizinsky 1975). After domestication, chickpea along with other pulses and cereals formed the basis of early agriculture in the Mediterranean and West Asian regions. Chickpea soon spread south to Ethiopia and east to South Asia where it became an important and popular legume food crop and remains so to the present time. Chickpea apparently was taken to the Americas soon after the discovery of the New World and became an important food crop in the Pacific coastal regions of North, Central, and South America.

Chickpea has been divided into two broad groupings, "Kabuli" and "Desi," based on size, shape, and coloration of the seeds. Kabuli types generally have large seeds that weigh in excess of 26 grams per 100 seeds, are rounded, and white or cream colored. Often the Kabuli seed type is referred to as "Ramshead." Plants of Kabuli

F.J. Muehlbauer
USDA-ARS, 303 Johnson Hall, Washington State University, Pullman, WA 99164, USA
e-mail: muehlbau@wsu.edu

© Springer 2008

types are tall and are devoid of purple or violet pigmentation of the leaves, stems, flowers, and pods. Desi types generally have seeds that are less than 26 grams per 100 seeds and have a rough, angular appearance. Coloration can vary from light tan to black with all gradations in between. Seeds are often mottled and speckled. Desi types are typically produced in South Asia where they represent the majority of production; while Kabuli types dominate the production in most other regions and especially in the western hemisphere. Even though Kabuli and Desi types are generally distinguished by seed size and seed coloration, there are some Kabuli types smaller than 26 grams per 100 seeds and some Desi types that are larger than 26 grams per 100 seeds.

Chickpeas are branched, spreading annuals that have considerable variation in form, with some being semi-erect with a main stem and few branches while others can be relatively prostrate with profuse branching. Germination of the seeds is hypogeal and the plants develop a strong, well-developed tap root with numerous lateral roots that are nodulated under normal conditions. Plant height can range from 30 to 60 cm depending on growing conditions. Leaves can be either pinnately compound with 3-8 leaflet pairs along a central rachis or simple (unifoliolate) with a single leaf lamina. The self pollinating flowers, usually 0.6-1.3 cm long, are borne as singles, doubles, or triples on inflorescences that originate from the stem axes. Chickpea is a quantitative "long-day" plant but flowers in all photoperiods (Smithson et al. 1985). Flower color of Kabuli types is usually white; while flower color of Desi types can be white, pink, purple, or blue. Pod shape is rhomboid ellipsoid and about 2.0-5.0 cm long and 1.5-2.0 cm wide with glandular pubescence. Pods have an inflated appearance and usually contain one or two seeds, but rarely more than two. Seedcoat colors vary from white to cream to tan for the Kabuli types, while Desi type seeds can be light tan, green, dark brown or black with many variations of mottling and other forms of pigmentation. The anterior of the seeds of both types are "beaked."

The crop is grown as a cool weather crop in semi-arid regions and as a winter crop in many production regions, particularly in South Asia, the Mediterranean, Australia, and Mexico and other areas where weather permits. Though sensitive to cold, some cultivars can tolerate air temperatures as low as -9.5° C. Very early maturing cultivars will complete their life cycle in as little as 65 days, while later maturing cultivars may require in excess of 120 days. Winter chickpea may require up to 180 days from planting to maturity.

7.1.1 Economic, Agronomic and Societal Importance of Chickpea

Chickpea, also known as "Garbanzo," is the third most important pulse crop in the world after beans(*Phaseolus vulgaris* L.) and peas (*Pisum sativum* L.) and is a major source of protein for human food in semi-arid tropical regions (FAO 2006). By far the majority of the world's chickpea crop is grown in South Asia with India being the largest producer with an estimated production of 6.0 million metric tonnes (mmt) annually and accounting for over half of world production. Other major producing countries include Pakistan (0.87 mmt), Turkey (0.61 mmt), Iran (0.31 mmt), Mexico (0.24 mmt), Myanmar (0.23 mmt), Australia (0.19 mmt), Ethiopia (0.14 mmt), Canada (0.98 mmt), Syria (0.55 mmt), Morocco (0.42 mmt), and numerous countries that produce smaller amounts (FAO 2006). Production in the U.S is currently estimated at 80,000 mt. Of the producing countries, Canada, Australia, Turkey, and the United States are major exporters.

Chickpea is an annual legume that is grown from seeds that are either broadcast on the soil surface and incorporated by tillage or planted using grain drills. Seeding rates vary from 25 to 120 kilograms per hectare depending on soil conditions and method of planting. Chickpea can be cultivated as a sole crop or in mixtures with numerous other crops (Van der Maesen 1972).

Originally, it was considered that strains of *Rhizobium leguminsarum* nodulated the roots of chickpea; however, later evidence indicated that strains of *R. ciceri* were the active bacteria (Nour et al. 1994). Nitrogen fixation through symbiosis can account for a significant portion of the nitrogen needs of the chickpea crop and can be achieved with good nodulation. Estimates of nitrogen fixation by chickpea crops vary considerably with soil type, climatic conditions, and residual soil nitrogen content. It is, therefore, difficult to make predictions for expected nitrogen fixation.

In traditional farming systems chickpea is harvested at maturity or near maturity and allowed to dry completely prior to threshing. Chaff is separated from the seed by winnowing. The crop is generally harvested in developed countries by direct combining when the moisture content of the seeds is less than 15%.

Chickpea is primarily a food crop used in various forms. In India the crop is consumed primarily as dhal, a preparation produced by decorticating the seed and separating the cotyledons. The decorticated and split cotyledons are then used to produce a thick soup that is generally served with rice. Chickpea is also used as a whole pulse and is soaked and boiled. "Chole" is a traditional dish made from whole chickpeas in India. A popular use in the Middle East is as hummus, a dish made from cooked and ground chickpea that is mixed with tahini, olive oil, and various spices and eaten with traditional flat bread. Dry roasted chickpea seasoned with various spices is a popular snack in most countries of the Middle East and North Africa. In India, the young leaves of chickpea are often harvested green and cooked to make a dish similar to spinach.

7.1.2 Chickpea as an Experimental Organism

Chickpea is a self-pollinating diploid annual with $2n = 16$ chromosomes. Natural outcrossing is estimated at less than 1%. Procedures for crossing of chickpea are similar to that used for other cool season legumes and require careful handling of the flowers and good sources of pollen for success. The advantage of chickpea as an experimental organism is its relatively short seed-to-seed reproductive cycle, usually less than 4 months under normal field conditions or in controlled environments. The highly self-pollinating nature of chickpea lends the species to development of recombinant inbred line (RIL) populations for genetic mapping that can be repeatedly and accurately phenotyped for important traits necessary for precise genetic linkage analysis.

The *Cicer* genus comprises nine annual species and 34 perennials. The annuals can be placed into three groupings. The first includes *C. arietinum*, the only cultivated species, *C. reticulatum*, the presumed progenitor, and *C. echinospermum*. Within this group, *C. reticulatum* is completely cross compatible with cultivated *C. arietinum*; while *C. echinospermum* can be crossed with the cultivated species, but with a high degree of sterility in the hybrids and hybrid progenies. The second group includes *C. bijugum*, *C. pinnatifidum*, and *C. judaicum*. There is no verified evidence that species of this group can be crossed to the cultivated species. There have been attempts to facilitate intercrossing of *C. arietinum* with *C. bijugum* through the use of embryo rescue; however, success has yet to be reported and confirmed. A similar approach has been used in attempts to overcome barriers to crossing *C. arietinum* with *C. pinnatifidum*. In both cases, weak albino plants have been obtained, but no seeds have been reported from those crosses. *C. chorassanicum*, *C. yamashitae*, and *C. cuneatum* comprise the third grouping and are considered to be the most distant of the annual species from cultivated *C. arietinum*.

A very limited amount of genetic polymorphism for molecular markers in chickpea has been a distinct handicap in the development of genetic linkage maps in these plants. Initially this handicap was circumvented through the use of interspecific hybrid populations. In populations from those crosses, genetic polymorphism for molecular markers was considerably greater than what was found in intraspecific crosses within cultivated chickpea. While the initial genetic maps of chickpea were developed from interspecific cross populations, more recent mapping efforts have concentrated on the use of intraspecific cross populations and the use of sequence tagged microsatellite site (STMS) markers.

7.2 Genetic Mapping and Tagging in Chickpea

An initial genetic map of chickpea by Simon and Muehlbauer (1997) comprised a limited number of restriction fragment length polymorphism (RFLP) and random amplified polymorphic DNA (RAPD) markers and showed some similarity to the genetic maps of pea and lentil indicating conservation of linkage groups. Later, a map developed by Santra et al. (2000) comprised mostly RAPD markers and had an estimated size of 982 cM.

More recent genetic maps were developed using STMS markers and interspecific crosses that had a greater degree of polymorphism (Huttel et al. 1999; Winter et al. 2000; Cho et al. 2002; Pfaff and Kahl 2003; Flandez-Galvez et al. 2003). These maps have included important genes for resistance to Fusarium wilt, caused by *Fusarium oxysporum* f.sp. *ciceri*. STMS markers have proven useful for mapping within intraspecific crosses and have resulted in more densely populated maps.

The STMS markers and mapping populations were also used to determine the locations of the quantitative trait loci (QTLs) for resistance to Ascochyta blight, caused by *Ascochyta rabiei*, considered to be the most devastating disease of chickpea worldwide. Further refinements of the chickpea map and development of a consensus map of the chickpea genome is currently underway using the numerous genetic maps of chickpea that are available and common markers. At the present time there are more linkage groups than the haploid chromosome number, a problem that will likely be overcome through increased marker density on the chickpea genetic map and through development of a consensus map of the chickpea genome.

Gene tagging for eventual use in marker-assisted selection (MAS) has been a long-term goal of chickpea research programs. To date, the genes for resistance to Fusarium wilt (caused by several races of *Fusarium oxysporum*) have been located on the chickpea linkage map and associated with closely linked molecular markers (Winter et al. 2000). QTLs for Ascochyta blight resistance have been located on the chickpea map (Santra et al. 2000; Tekeoglu et al. 2002) and have formed the basis for further study of resistance mechanisms (Cho and Muehlbauer 2004; Cho et al. 2004).

7.3 Sequence and Marker Diversity

The chickpea genome is considered homogeneous based on the minimal polymorphism for molecular markers. The lack of polymorphism may be a consequence of relatively recent domestication and its self-pollinating reproductive cycle. Hence, the identification of molecular polymorphism is difficult and time consuming. To overcome this problem, researchers have used interspecific crosses that have an increased frequency of molecular polymorphism and highly variable simple sequence repeat (SSR) markers. SSR markers in chickpea were identified as being highly polymorphic compared to other types of markers. There are six published linkage maps based on either intra- and interspecific crosses that are populated with morphological markers and various types of molecular markers including: isozymes, amplified fragment length polymorphism (AFLP), RAPD, inter simple sequence repeat (ISSR), resistant gene analog (RGA), and STMS markers (Santra et al. 2000; Winter et al. 2000; Cho et al. 2002; Collard et al. 2003; Flandez-Galvez et al. 2003; Cho et al. 2004). In spite of the availability of SSR markers, they are insufficient in number to generate a high density linkage map. However, progress in genomic research in chickpea including bacterial artificial chromosome (BAC) end sequencing and expressed sequence tags (ESTs) will provide additional opportunities for developing sequence-based markers that target specific genomic regions.

Single nucleotide polymorphism (SNP) research is relatively new in plant systems but appears to have great potential for marker development. The potential for SNPs was initially shown in *Arabidopsis* (Jander et al. 2002), maize (Ching et al. 2002), and soybean (Zhu et al. 2003), where extensive analyses for SNPs were performed. Recently, Rajesh and Muehlbauer (2006, unpublished data) estimated

SNP frequency at 1 in 94 bp in coding sequences and 1 in 74 bp in genomic regions in two parental chickpea lines (FLIP84-92C [*C. arietinum*] and PI599072 [*C. reticulatum*]) that were previously used to develop an interspecific linkage map. SNP frequency between these two interspecific parental lines was sufficient to allow for additional marker development that was used for increasing marker density in the important *QTL*1 region for Ascochyta blight resistance.

SNP frequencies within intraspecific parental lines of chickpea were lower than within interspecific parents but comparable to that found in sugar beet (1 in 126 bp; Schneider et al. 2001), rice (1 in 89 bp; Nasu et al. 2002), maize (1 in 60-120 bp; Ching et al., 2002), cassava (1 in 62 bp; Lopez et al. 2005), and pearl millet (1 in 59 bp; Bertin et al. 2005).

Other than genetic mapping, SNP-derived markers can be used to genotype chickpea germplasm collections and to associate marker genotype with agronomically important traits with direct implications for breeding programs. This objective can be achieved because technology is available for large scale SNP genotyping with a capacity of 2 million genotypes per day (BeadArray technology, Illumina, Inc, USA), but SNP markers still need to be determined on a large scale.

Genome-wide SNP determination and mapping by high-throughput technology in combination with genetic map integration with physical maps will improve our understanding of the whole genome and will have broad implications for breeding programs.

7.4 BAC Cloning and Utilization

Chickpea genomic research has progressed rapidly over the past five years. Starting with the development of the first BAC library (Rajesh et al. 2002), research has made significant advances in various aspects of genomics including: development of additional BAC libraries (Lichtenzveig et al. 2005), whole BAC sequencing (Rajesh current research), whole genome physical mapping (Zhang personal communication), EST development (Buhariwalla 2003), targeted induced local lesion in genomes (TILLING) mutants (Rajesh personal communication), established protocols for *Agrobacterium*-mediated transformation (AMT) (Sanyal et al. 2003), and SNP discovery (Rajesh et al. 2005)

7.4.1 Development of BAC Libraries

The use of BAC libraries and other large insert libraries has revolutionized genomic research in plants and animals. Although the insert size in BAC libraries is smaller than in yeast artificial chromosome (YAC) libraries, considering the ease of the cloning, long term maintenance of inserts, and lack of chimerism, BAC libraries have gained importance for plant genomic research. BIBAC (binary BAC) libraries have the added advantage of carrying elements in the vector that are necessary for

AMT. This approach eliminates the necessity of sub-cloning the insert from a BAC vector to a binary vector once the gene of interest is identified. As is commonly known, several agronomically important genes, especially those for disease resistance, may be clustered in the genome (Staskawicz et al. 1995). A binary vector that has both a large insert cloning capacity and is amenable to AMT significantly reduces the time requirement when transformation is attempted. The BAC libraries can be used for BAC end sequencing, physical mapping and whole genome sequencing. Since most agronomically important traits are governed by QTLs spread widely across the genome, BAC libraries can enhance the success of map-based cloning by reducing the number of steps in chromosome walking.

The first BAC library for chickpea was constructed using *Hind* III (Rajesh et al. 2002) and has been used to identify clones associated with genes for disease resistance (Rajesh et al. 2004). The BAC library was constructed in the binary vector pCLD04541 (otherwise V41) and had a 3.8X coverage of the genome with a 95% probability of including a genomic fragment with an individual gene of interest. Additional BAC libraries were constructed using *Hind*III and *BamH*I restriction enzymes and vectors pIndigoBAC and pCLD04541, with average insert sizes of 121 and 145 kb, respectively (Lichtenzveig et al. 2005). Two additional *Hind*III BAC libraries have been developed (Doug Cook and Gunther Kahl personal communications) for a total of five libraries from four different cultivars. These resources are available to the chickpea research community and provide combined genome coverage greater than 15X. Since the recognition sites of *Hind*III (AT rich) and *BamH*1(GC rich) are complementary to each other and the libraries are constructed using different vectors, these libraries are expected to represent the entire genome. The ease of AMT, combined with an established protocol for chickpea and the ability to transform inserts larger than 100 kb, enhance direct transformation of large inserts representing agronomically important genomic regions from the BIBAC library without further sub-cloning.

These libraries have been used for various applications in chickpea such as identification of clones containing resistance genes (Rajesh et al. 2004), SNP discovery, whole BAC sequencing (Rajesh and Muehlbauer unpublished data), and developing SSR markers (Lichtenzveig et al. 2005). A limited number of sequence based markers have been placed on the chickpea linkage map and the current number of linkage groups is greater than the haploid number of chickpea chromosomes. BAC end sequencing has enormous potential for developing markers to increase marker density in the linkage maps. Currently, these large insert libraries are in use for whole genome physical mapping by BAC fingerprinting (HB Zhang personal communication).

7.4.2 BAC Physical Mapping

Although development of dense linkage maps facilitates identification of markers tightly linked to the desired trait, it requires establishing physical and genetic correlation to effectively apply the markers in MAS. Hence, developing markers from the genomic region that are physically closer to the gene will have a long term application in breeding programs. Development of genome-wide integrated physical and genetic maps will allow easier chromosome walking, offers an efficient and economical approach for map-based cloning, and will serve to identify tightly linked markers. Physical mapping using BAC clones can be accomplished by fingerprint analysis, fluorescent *in situ* hybridization, optical mapping, iterative hybridization, and chromosome walking (Ren et al. 2005). Of these procedures, fingerprint analysis is the most commonly used method. Successful physical mapping using BAC libraries is enhanced by large insert sizes that minimize the number of clones needed to cover the genome and reduces the number of clones required to span a given genomic region with minimal tiling path. Whole genome representation is improved by using complementary restriction enzymes that target AT-rich and GC-rich nucleotides while making the library (Ren et al. 2005)

The chickpea BAC library of Rajesh et al. (2002) was used to develop markers from the ends of the BACs spanning *QTL*1 for Ascochyta blight resistance and to physically map the *QTL*1 genomic region. Sequencing of BACs from these contigs identified several candidate disease resistance genes. High resolution mapping was performed using the flanking markers (Rajesh et al. unpublished data). Work is in progress to develop and integrate physical and genetic maps by fingerprinting ~50,000 (~10X) BACs from BAC libraries and assemble them into a genome-wide contig map of chickpea. The ultimate goal is to anchor the map contigs to the chickpea genetic maps to which flowering time and Ascochyta blight resistance QTL were mapped. The approach should provide additional markers that can be used to fine map these important QTLs (HB Zhang personal communication) and be used for MAS in breeding programs.

7.4.3 Whole-BAC Sequencing

High throughput technology has made it feasible to sequence entire genomes of different organisms. Genome sequencing of higher organisms is difficult because of their complex nature and large size. However, large scale DNA sequencing is underway and is useful for studying genome organization, comparing genomes of related species, and SNP discovery. Currently, most genome sequencing projects in higher organisms are performed using BAC clones because of their advantages such as containing specific gene-rich genomic regions (euchromatin) that can be sequenced rather than the entire genome. Also, BAC clones allow easy distribution of specific genomic regions to collaborating laboratories. By sequencing BACs representing only euchromatic genomic regions, the *Medicago* Genome Sequencing Consortium predicts that 90% of the gene content will be sequenced by sequencing just 50% of the 500Mb *Medicago truncatula* genome (Town 2005).

In chickpea, although the number of sequences being deposited in the database has gradually increased, whole-BAC sequencing was not attempted until 2003

(Rajesh 2006 unpublished data). Currently, there are 758,388 base pairs that show particular genome organization and composition in chickpea. Based on an estimated gene density of one gene per 9.2 kb and genome size of 740 Mb, approximately 80,000 genes are estimated to be present in chickpea genome. Considering the average gene size and the total number of genes, it is estimated that 200 Mb of the genome (27%) consist of genes. These estimates were based on 11 BAC clones that represented different genomic regions, three of which are located in a genomic region associated with resistance to Ascochyta blight (*QTL*-1). Interestingly, repetitive elements were relatively infrequent (8.68%) in these BAC clones, while in other plants repetitive elements account for more than 50% of the genome. Sequencing of additional BAC clones is underway and should help clarify the organization of the chickpea genome.

7.5 Molecular Cytogenetics

Karyotype analysis has been used to describe the relative size and arm lengths of the eight chromosomes of *C. arietinum* and several of the wild relatives; however, the small size and sticky nature of the chromosomes does not lend itself to more detailed analysis (Ahmad 2000). The relative size and chromosome arm lengths of the chromosomes has made it possible to place some of the species into groups. For example, it has been possible to place the cultivated species, *C. arietinum,* with *C. reticulatum*, *C. echinospermum,* and the perennial species, *C. anatolicum* and *C. songaricum* into the same grouping based on chromosome morphology (Ahmad 2000; Croser et al. 2003). Morphological examination indicates that only one chromosome has a nucleolar organizer region (Ocampo et al. 1992; Ahmad 2000).

Classification of chickpea chromosomes by flow cytometry (Vláčilová et al. 2002) was successful in isolating chromosomes corresponding to five of the eight chickpea chromosomes and associating STMS markers from linkage group 8 to the smallest chromosome. These promising results complement genetic linkage and physical mapping of the chickpea genome by associating specific linkage groups to particular chromosomes.

7.6 Functional Genomics

With the availability of complete genome sequences (*Arabidopsis* and rice) and near complete whole genome DNA sequences of model plants, 25,000 genes have been predicted in *Arabidopsis* and from 32,000 to 50,000 have been predicted in rice. Since the predicted number of genes exceeds the number of genes with known function, it is challenging to determine the function of each gene. Full length cDNA synthesis is a major aspect of functional genomics not only for annotating, but for in vitro and in vivo characterization of each gene. Correlation of functional change by alteration of a gene at the molecular level in mutants will facilitate the determination of the gene function. Emphasis has also been placed on how well these genes are spatially and temporally separated or coordinated by transcriptional profiling of different tissue and at different time intervals. Overall, the study of functional genomics of chickpea is in the very early stages.

7.6.1 EST Development

Full length cDNA synthesis of all the genes is technically laborious and costly; therefore, partial sequencing of the expressed genes, known as expressed sequence tags (ESTs), promulgates nearly as much information as full length cDNA and is less time consuming. Development of an EST database for each crop is necessary for transcriptome profiling and also for gene discovery.

Differentially expressed genes during the *Ascochyta rabiei* -chickpea interaction have been studied using various techniques such as differential cDNA hybridization (Ichinose et al. 2000), differential display reverse transcription (DDRT) (Rajesh et al. 2003), and cDNA-AFLP (Cho et al. 2004). However, global transcriptome analysis representing the entire chickpea genome has not been possible because of the small number of ESTs. The first report of large scale EST development was published in 2003 with 488 entries in Genbank (Buhariwalla et al. 2003). Recently, Coram and Pang (2005) increased this number by depositing an additional 566 ESTs in Genbank and developed a microarray platform to study the *Ascochyta rabiei* – chickpea interaction. At the time of writing this chapter, there were 1,302 chickpea EST entries in Genbank.

ESTs can be used as genetic markers as well. Buhariwalla et al. (2003) showed the utility of ESTs as markers in genetic diversity analysis among chickpea wild species. EST mapping is worthwhile because gene-rich regions in the genome can be defined and gene discovery is possible for traits of interest. SNP discovery in ESTs across different accessions will help in developing SNP markers that can be used in high throughput genotyping. Microsatellites present in ESTs (EST-SSRs) are variable and are likely to be naturally polymorphic. EST-SSR markers designed from *Medicago truncatula* has been attempted in chickpea, lentil and pea. Although the number of polymorphic markers was low, it showed the potential for transferability of markers across the species (Muehlbauer unpublished data). However, these markers were 70% polymorphic among six *Medicago* species (Eujayl et al. 2004). A separate study by Gutierrez et al. (2005) indicated significant transferability of *M. truncatula* microsatellites to three pulse crops (40% to faba bean, 36.3% to chickpea and 37.6% to pea) and that the SSR motifs were variable, although the flanking regions were conserved. It is, therefore, possible to design markers with good transferability and representative of the same genomic regions. This may be achieved by incorporating phylogenetic relationships with molecular analyses to determine genome conservation.

Comparative EST sequence analysis involving two major clades including cool season legumes and warm season legumes identifies the conserved exon sequences and promotes cross-species marker development. Such markers have been utilized for genetic mapping and to determine chromosomal rearrangements among various legumes such as *Medicago truncatula, Medicago sativa, Pisum sativum, Glycine max, Vigna radiata, Phaseolus vulgaris* and *Lotus japonicus* (Choi et al. 2004). These cross-species markers have been analyzed in chickpea and lentil mapping populations and have shown that there is potential for this approach for use in comparative genomics across genera and species of legumes (Muehlbauer unpublished data)

7.6.2 TILLING

TILLING gained popularity among other reverse genetic tools because it can produce an allelic series of point mutations and it is a non-transgenic approach for gene discovery and verification. For example, 246 alleles of the waxy genes were identified by TILLING each homoeolog in 1,920 allohexaploid and allotetraploid wheat individuals (Slade et al. 2005). Traditionally, TILLING mutants are generated by mutagenizing seeds to produce M1 plants. These M1 plants are self-fertilized and the resulting M2 seeds are grown for DNA analysis and M3 seeds. Specific genomic targets in M2 plants are chosen and amplified by polymerase chain reaction using suitable primer pairs. These amplified products are incubated with *Cel*1 enzyme that cleaves to the 3' side of mismatches in the heteroduplex and leaves the duplex intact. High-throughput screening for point mutations is performed by running the digested products on a LI-COR gel analyzer system.

In chickpea, M2 seeds from approximately 9,000 individual M1 plants of chickpea germplasm accession ICC12004, which had been treated with 0.2% ethyl methyl sulfonate (EMS), were obtained in the initial phases for development of a TILLING platform for chickpea. The most important factor in determining the suitability of a population for TILLING production screening is the estimated mutation frequency. The estimated mutation frequency was determined through an analysis of 768 M2 progenies using 20 targets comprising genomic DNA and cDNA sequences. There was a 100% success rate in primer design and screening of the mutants when using genomic sequence, and only a 7% success rate using cDNA sequence. Analysis of 3,763,200 base pairs identified 23 G/C to A/T mutations and determined that the frequency of mutations was 1 per 165 kb (Rajesh and Muehlbauer unpublished data). This frequency is ~1.6x higher than reported in *Arabidopsis*. These TILLING mutants are the only authentic mutant resource available to both the national and international chickpea research community. Availability of mutants has the potential to advance genomics rapidly and has enormous utility in forward and reverse genetics.

The probability of finding one mutation in any G/C pair in the M1 population can be calculated using the formula: $P = 1-(1-F)^N$, where, F= mutation frequency and N= the number of M1 lines. In *Arabidopsis* with a mutation frequency of 1.6×10^{-5}, it has been estimated that 45,000 and 250,000 M1 plants were necessary for 95%

and 98% probability, respectively, to find a mutation in a given G/C base pair (Jander et al. 2003).

However, considering that mutagenesis is a random process and it produces an allelic series of a particular locus, only few mutations will have a deleterious effect on the gene. Such mutants are useful for studying gene function. It has been identified in *Arabidopsis* that treatment of M1 seeds with ethyl methyl sulfonate (EMS) produce G/C to A/T mutations (Till et al. 2004). In chickpea, there were 14 missense, nine silent, and zero truncated mutations among the 23 G/C to A/T mutations analyzed. Based on project aligned related sequences and evaluate SNPs), also called PARSESNP, analysis, five of 14 missense mutations were found to be deleterious in a leucine rich repeat motif (Rajesh et al. 2006 unpublished data).

Ecotilling is performed using natural variation rather than mutating the genomes as in TILLING. This method demands significantly fewer sequencing reactions to study the natural variation because genomes from different samples are mixed with one reference genome in 1:1 ratio in each reaction. Hence, sequencing representative samples will indicate the natural variation at a particular locus that can be expected in all other samples. Ecotilling has been performed to study DNA variation in natural populations. However, this method can be exploited in other studies including genetic mapping and linkage disequilibrium.

7.6.3 Transformation

The first successful efforts toward genetic modification of chickpea were stable integration of the gene*cryIA(c)* by biolistic procedures (Kar et al. 1997) followed by AMT in 2003 (Sanyal et al. 2003). Apart from the *cryIA(c)* gene for Bt toxin production, Sarmah et al. (2004) and Sanyal and Prakash (2006) have reported stable integration and expression of the *Phaseolus* α-amylase inhibitor gene in chickpea via AMT. These genes were used to develop resistance against bruchids and *Helicoverpa armigera*. Currently, there is an effort to exploit proteinase inhibitor genes from non-host and host plants by AMT to impart resistance to *Helicoverpa* infestations (Vidya Gupta personal communication). Although biolistic transformation and AMT have been reported as successful transformation methods in chickpea, recent publications indicate that the latter method is the most commonly used approach (Sanyal et al. 2003; Sarmah et al. 2004; Sanyal 2006).

In addition to genes, there have been efforts to transform large inserts into chickpea. AMT of large DNA inserts directly into plants facilitates the transfer of gene clusters and flanking regulatory elements. Hence, it is recommended that the integrity of large genomic fragments in *Agrobacterium tumefaciens* be verified prior to plant transformation. Rajesh et al. (2007) reported that chickpea genomic DNA fragments up to 100 kb in size can be stably transformed into *A.t.* strain*Agl0* through triparental mating and stably maintained in *A.t.* and *E. coli,* thus improving the prospects for successful transformation.

Selectable marker genes for resistance to antibiotics or herbicides are being used to identify transgenic plants. However, concerns over development of transgenics harboring antibiotic resistance genes and their release for commercial uses have forced implementation of procedures to eliminate selectable marker genes from transgenic plants. Various procedures such as site-specific recombination using Cre-Lox system, co-transformation using two different binary vectors, or a single twin binary vector harboring two T-DNA regions are most commonly used. Attempts have been made to develop transgenic chickpeas using chimeric Bt-*cryIA(c)* genes that were reconstructed using the Cre-Lox excision system to facilitate elimination of the selectable marker gene from segregating progenies. However, considering the possible patent issues related to the use of the Cre-Lox system (owned by DUPONT, USA), emphasis is on developing a twin binary vector system to eliminate marker genes from segregating progeny (BK Sarmah personal communication).

With the rapid advancement of chickpea genomics with BIBAC libraries and cross-species transferability of genomic information and the availability of standardized protocols, AMT has potential in chickpea crop improvement for studies of gene function for agronomically important traits, particularly those for disease resistance.

7.7 Perspective

Chickpea is an important crop in South Asia, the Middle East, and West Asia, where it is produced using traditional farming methods to provide nutritious food to local populations. The crop is often relegated to poor soils without benefit of inputs such as fertilizers, irrigation, or pest control. However, it remains an important food crop and has gained popularity as a "health food" in developed countries. Chickpea has become an important export commodity in Turkey, Canada, Australia, and the United States where it is a valuable nitrogen fixing crop in primarily cereal-based cropping systems. Production hazards such as Ascochyta blight (caused by *Ascochyta rabiei*), Fusarium wilt (caused by *Fusarium oxysporum* f.sp.*ciceris*), and pod borer (*Heliocoverpa armigera*) have prompted geneticists to develop information about the chickpea genome.

Progress has been made in developing genetic maps and the placing genes for resistance to Ascochyta blight and Fusarium wilt as well as genes controlling agronomically important traits such as time to flowering, time to maturity and podding habits. The relatively small genome size of chickpea is an advantage for genomic research and holds promise for eventual comparisons of the genome of chickpea with those of closely related legumes and those of the model legumes, *Medicago truncatula* and *Lotus japonicus*. BAC library construction has furthered the study of the chickpea genome and offers the opportunity to locate and clone important genes. Overall, chickpea is an important world crop and especially so in developing countries of semi-arid regions of South Asia. Continued improvement of chickpea based on emerging genetic information is critical to maintaining its viability in cropping systems worldwide.

Acknowledgments The authors would like to thank the U.S. Department of Agriculture, Agricultural Research Service, for supporting research on chickpea and Washington State University, Pullman, Washington USA, for providing laboratory and field facilities for chickpea research. We also wish to thank the McKnight Foundation for the long support of research that resulted in the first genetic maps of chickpea and some of the initial analyses of species relationships within *Cicer*.

References

Ahmad F (2000) A comparative study of chromosome morphology among the nine annual species of *Cicer* L. Cytobios 101:37–53

Bertin I, Zhu JH, Gale MD (2005) SSCP-SNP in pearl millet—a new marker system for comparative genetics. Theor Appl Genet 110:1467–1472

Buhariwalla HK, Jayashree B, Eshwar K, Crouch JH (2003) Development of ESTs from chickpea roots and their use in diversity analysis of the Cicer genus. BMC Plant Biol 17: 5:16

Ching ADA, Caldwell KS, Jung M, Dolan M, Smith OS, et al. (2002) SNP frequency, haplotype structure and linkage disequilibrium in elite maize inbred lines. BMC Genetics 3:19

Cho S, Muehlbauer FJ (2004) Genetic effect of differentially regulated fungal response genes on resistance to necrotrophic fungal pathogens in chickpea (*Cicer arietinum* L.). Physiol Mol Plant Pathol 64:57–66

Cho S, Kumar J, Shultz JL, Anupama K, Tefera F, et al. (2002) Mapping genes for double podding and other morphological traits in chickpea. Euphytica 128:285–292

Cho S, Chen W, Muehlbauer FJ (2004) Pathotype-specific genetic factors in chickpea (*Cicer arietinum* L.) for quantitative resistance to ascochyta blight. Theor Appl Genet 109:733–739

Choi H-K, Mun JH, Kim DJ, Zhu H, Baek JM, et al. (2004) Estimating genome conservation between crop and model legume species. Proc Natl Acad Sci USA 101:15289–15294

Collard BC, Pang EC, Ades PK, Taylor PW (2003) Preliminary investigation of QTLs associated with seedling resistance to ascochyta blight from *Cicer echinospermum*, a wild relative of chickpea. Theor Appl Genet 107:719–729

Corum T, Pang ECK (2005) Isolation and analysis of candidate ascochyta blight defence genes in chickpea. Part I. Generation and analysis of an expressed sequence tag (EST) library. Physiol Mol Plant Pathol 66:192–200

Croser JS, Ahmad F, Clarke HJ, Siddique KHM (2003) Utilization of wild *Cicer* in chickpea improvement - progress, constraints, and prospects. Austr J Agric Res 54:429–444

Eujayl I, Sledge MK, Wang L, May GD, Chekhovskiy K, et al. (2004) *Medicago truncatula* EST-SSRs reveal cross-species genetic markers for Medicago spp. Theor Appl Genet 2004 108:414–422

FAO (2006) FAO Statistics at http://faostat.fao.org/

Flandez-Galvez H, Ford R, Pang ECK, Taylor PWJ (2003) An intraspecific linkage map of the chickpea (*Cicer arietinum* L.) genome based on sequence tagged microsatellite site and resistance gene analog markers. Theor Appl Genet 106:1447–1456

Gutierrez MV, Vaz Patto MC, Huguet T, Cubero JI, Moreno MT, et al. (2005) Cross-species amplification of *Medicago truncatula* microsatellites across three major pulse crops. Theor Appl Genet 110:1210–1217

Huttel B, Winter P, Weising K, Choumane W, Weigand F et al. (1999) Sequence-tagged microsatellite site markers for chickpea (Cicer arietinum L.). Genome 42:210–217

Ichinose Y, Tiemann K, Schwenger-Erger C, Toyoda K, Hein F, et al. (2000) Genes expressed in *Ascochyta rabiei*-inoculated chickpea plants and elicited cell cultures as detected by differential cDNA-hybridization. Z Naturforsch 55:44–54

Jander G, Baerson SR, Hudak JA, Gonzalez KA, Gruys KJ, et al. (2003) Ethylmethanesulfonate saturation mutagenesis in *Arabidopsis* to determine frequency of herbicide resistance. Plant Physiol 131:139–146

Jander G, Norris SR, Rounsley SD. Bush DF, Levin IM, et al. (2002) *Arabidopsis* map-based cloning in the post-genome era. Plant Physiol 129:440–450

Kar S, Basu D, Das S, Ramkrishnan NA, Mukherjee P, et al. (1997) Expression of *cryIA(c)* gene of *Bacillus thuringiensis* in transgenic chickpea plants inhibits development of pod-borer (*Heliothis armigera*) larvae. Transgenic Res 6:177–185

Ladizinsky G (1975) A new *Cicer* from Turkey. Notes of the Royal Botanic Garden Edinburgh 34:201–202

Lichtenzveig J, Scheuring C, Dodge J, Abbo S, Zhang HB (2005) Construction of BAC and BIBAC libraries and their applications for generation of SSR markers for genome analysis of chickpea, *Cicer arietinum* L. Theor Appl Genet 110:492–510

Lopez C, Pie'gu B, Cooke R, Delseny M, Tohme J, et al. (2005) Using cDNA and genomic sequences as tools to develop SNP strategies in cassava (*Manihot esculenta* Crantz). Theor Appl Genet 110:425–431

Nasu S, Suzuki J, Ohta R, Hasegawa K, Yui R, et al. (2002) Search for and analysis of single nucleotide polymorphism (SNPs) in rice (*Oryza sativa*, *Oryza rufipogon*) and establishment of SNP markers. DNA Res 9:163–171

Nour SM, Fernandez MP, Normand P, Cleyet-Marel JC (1994) Rhizobium ciceri sp. Nov., consisting of strains that nodulate chickpeas (*Cicer arietinum* L.). Intl J Syst Bacteriol 44:511–522

Ocampo B, Venora G, Errico A, Singh KB, Saccardo F (1992) Karyotype analysis in the genus *Cicer*. J Genet Plant Breed 46:229–240

Pfaff T, Kahl G (2003) Mapping of gene-specific markers on the genetic map of chickpea (*Cicer arietinum* L.). Mol Genet Genom 269:243–251

Rajesh PN, Meksem K, Coyne C, Lightfoot D, Muehlbauer FJ (2002). Construction of first BAC library in Chickpea. Intl Chickpea Pigeonpea Newslet 9:29–30

Rajesh PN, Gupta VS, Ranjekar PK, Muehlbauer FJ (2003) Functional genome analysis using DDRT with respect to ascochyta blight disease in chickpea. Intl Chickpea Pigeonpea Newslet 10:35–37

Rajesh PN, Coyne C, Meksem K, Sharma KD, Gupta VS, et al. (2004) Construction of a *Hind*III Bacterial Artificial Chromosome library and its use in identification of clones associated with disease resistance in chickpea. Theor Appl Genet 108:663–669

Rajesh PN, McPhee K, Muehlbauer FJ (2005) Detection of polymorphism using CAPS and dCAPS markers in two chickpea genotypes. Intl Chickpea Pigeonpea Newslet 12:4–6

Rajesh PN, Muehlbauer FJ, McPhee K (2007) Stability of chickpea large genomic DNA inserts in *Agrobacterium*. (In press)

Ren C. Xu Z, Sun S, Lee M, Wu C, et al. (2005) Genomic DNA libraries and physical mapping. The handbook of plant genome mapping pp. 173–214 Wiley-VCH Verlag GmbH &Co. KGaA publishers

Santra DK, Tekeoglu M, Ratnaparkhe MB, Gupta VS, Ranjekar PK, et al. (2000) Identification and mapping of QTLs conferring resistance to Ascochyta blight in chickpea. Crop Sci 40:1606–1612

Sanyal I, Prakash S (2006) *Agrobacterium*-mediated transformation of chickpea with α-amylase inhibitor gene for insect resistance. J. Biosci 31:339–345

Sanyal I, Singh AK, Amla DV (2003) *Agrobacterium tumefaciens* mediated transformation of chickpea (*Cicer arietinum* L.) using mature embryonic axes and cotyledonary nodes. Indian J Biotech 2:524–532

Sarmah BK, Moore A, Tate W, Molvig L, Morton RL, et al. (2004) Transgenic chickpea seeds expressing high levels of a bean α-amylase inhibitor. Mol Breed 14:73–82

Schneider K, Weisshaar B, Borchardt DC, Salamani F (2001) SNP frequency and allele haplotype structure of *Beta vulgaris* expressed genes. Mol Breed 8:63–74

Simon CJ, Muehlbauer FJ (1997) Construction of a chickpea linkage map and comparison with maps of pea and lentil. J Hered 88:115–119

Slade AJ, Fuerstenberg SI, Loeffler D, Steine MN, Facciotti D (2005) A reverse genetic, nontransgenic approach to wheat crop improvement by TILLING. Nat Biotechnol 23:75–81

Smithson JB, Thompson JA, Summerfield RJ (1985) Chickpea (*Cicer arietinum* L.). In: Summerfield RJ, Roberts EH (eds.) Grain Legume Crops. Collins, London, UK pp. 312–390

Staskawicz BJ, Ausubel FM, Baker BJ, Ellis JG, Jones JD (1995) Molecular genetics of plant disease resistance. Science 268:661–667

Tekeoglu M, Rajesh PN, Muehlbauer FJ (2002) Integration of sequence tagged microsattellite sites to the chickpea genetic map. Theor Appl Genet 105:847–854

Till BJ, Reynolds SH, Weil C, Springer N, Burtner C, et al. (2004) Discovery of induced point mutations in maize genes by TILLING. BMC Plant Biol 28:4:12

Town CD (2005) Large-scale DNA sequencing The handbook of plant genome mapping. pp. 337–351 Wiley-VCH Verlag GmbH &Co. KGaA publishers

Van der Maesen LJG (1972) *Cicer* L. a monograph of the genus, with special reference to the chickpea (*Cicer arietinum* L.), its ecology and cultivation. Mededlingen landbouwhogeschool (Communication Agricultural University) Wageningen 72–10. 342 p

Vláčilová K. Ohri D, Vrána J, Číhalíková J, Kubaláková M, et al. (2002) Development of flow cytogenetics and physical genome mapping in chickpea (*Cicer arietinum* L.), Chromosome Res 10:695–706

Winter P, Benko-Iseppon HB, Ratnaparkhe M, Tullu A, Sonnante G, et al. (2000) A linkage map of the chickpea (*Cicer arietinum* L.) genome based on recombinant inbred lines from a *C. arietinum* X *C. reticulatum* cross: localization of resistance genes for fusarium wilt races 4 and 5. Theor Appl Genet 101:1155–1163

Zhu YL, Song QJ, Hyten DL, Van Tassell CP, Matukumalli LK, et al. (2003) Single-nucleotide polymorphisms in soybean. Genetics 163:1123–1134

Chapter 8
Genomics of Citrus, a Major Fruit Crop of Tropical and Subtropical Regions

Mikeal L. Roose and Timothy J. Close

Abstract Genomics in citrus and closely related genera is relatively advanced, with many linkage maps, several bacterial artificial chromosome libraries, a physical map, a fairly large and well-annotated expressed sequence tag collection, several microarrays, and a low coverage (1.2x) genome sequence for sweet orange. This chapter reviews the various genomics resources available for citrus. Integration of these resources, particularly those developed in different laboratories, remains a challenge.

8.1 Introduction

Citrus is a large genus with many different cultivated species, the major ones being *Citrus sinensis* L. (sweet orange), *C. reticulata* Blanco (mandarin), *C. limon* (L.) Burm. (lemon), *C. maxima* (Burm.) Merrill (pummelo), *C. aurantifolia* (Christm.) Swingle (lime), *C. medica* L. (citron), and *C. paradis*i Macf. (grapefruit). Some taxonomic authorities recognize many additional species, particularly within the mandarin group. Here we refer to all mandarin-type fruits (including the important satsuma and Clementine groups) as *C. reticulata.* Some authors refer to pummelo as *C. grandis* Macf., an epithet that was common in the older literature. Cultivated types include many suspected and known hybrids among these "species," and there is good evidence that sweet orange, lemon, lime, and grapefruit are natural hybrids rather than wild species (Federici et al. 1998; Nicolosi et al. 2000; Moore 2001). Several related genera including *Fortunella*, *Poncirus*, *Microcitrus*, and *Eremocitrus* can be hybridized with citrus species, further expanding the germplasm pool of interest to breeders. Citrus is native to China, eastern India, and Southeast Asia, but is now distributed worldwide. The dates of domestication of most commercially important species are unknown, although grapefruit apparently originated in the Carribbean, probably during the eighteenth century (Gmitter 1995).

M.L. Roose
Department of Botany & Plant Sciences, University of California, Riverside, CA 92521 U.S.A.
e-mail; mikeal.roose@ucr.edu

© Springer 2008

All citrus species are trees, about 2–16 m in height, with a relatively thin (<5 mm) bark. Leaves are simple, usually ovate to lanceolate in shape, with petioles that vary in size and shape among species. Flowers are usually single in leaf axils or in short axillary racemes, with 4–8 white petals. Flowers are perfect or staminate due to abortion of the pistil. Fruits are technically called a hesperidium, and consist of juice vesicle-filled segments with seeds near the inner segment angle. The segments are surrounded by a white endocarp (albedo) and then the rind (flavedo) which has many oil glands and is usually colored orange or yellow at maturity. The rind and internal (pulp or flesh) color is quite variable, ranging from green to yellow, orange, and red (see Fig. 8.1). Seed content is variable among and within species.

Many varieties produce polyembryonic seeds that may contain a viable zygotic embryo resulting from fertilization and syngamy plus one or more apomictic embryos that develop from cells of the nucellus and are genetically identical to the mother plant. The developmental pathway that produces these adventitious embryos is usually called "nucellar embryony" and the embryos are termed "nucellar" embryos. Nucellar embryony seems absent from pummelo and citron, two species believed ancestral to several important cultivated groups.

8.1.1 Economic, Agronomic, and Societal Importance of Citrus

Citrus is the one of the most important fruit crops in subtropical and tropical areas. It is produced for use as fresh fruit and for juice, with many additional minor uses including flavorings, aroma, and jams. Citrus fruit also have considerable cultural importance in some societies. The citron, specifically the Etrog variety, is used in certain Jewish ceremonies (Nicolosi et al. 2005). The most familiar constituent of citrus fruit is citric acid (vitamin C), the major acid present in citrus juice. Citrus

Fig. 8.1 Fruit of *Citrus* species and related genera from the Citrus Variety Collection of the University of California, Riverside, one of the world's most diverse citrus germplasm collections. (See color insert)

fruit are rich in many other health-promoting chemicals, including flavonoids, anthocyanins, lycopenes, and antioxidants (Jayaprakasha and Patil 2007). In 2005/06 the largest producers of citrus were Brazil (20.6 mmt), China (15.0 mmt), the USA (10.4 mmt), and Spain and Mexico, with much of the crop from Brazil and Florida being used for orange juice production. In some regions citrus production is the largest single industry when employment in production, packing, shipping, processing, and supporting industries is included.

8.1.2 Citrus as an Experimental Organism

Citrus has several features that make it an excellent model for genomics of trees, but additional characteristics pose limitations. Positive aspects include a small haploid genome size (382 Mb, Arugumanathan and Earle 1991), considerable natural genetic diversity, simple clonal propagation by grafting or production of rooted cuttings, diploidy in nearly all important species, a rich reservoir of natural mutations that are equivalent to isogenic lines, and the availability of efficient transformation systems in a few species. Comparison with *Arabidopsis* should be particularly informative because *Citrus* is in the order Sapindales, a sister order to the Brassicales of the Eurosid II clade, making it closer to *Arabidopsis* than to other plants with major genomics projects (Bausher et al. 2006). However, the degree of similarity in sequence and genome organization have not yet been characterized. Nucellar embryony in citrus is a relatively unusual mechanism of apomixis that is worthy of more detailed investigation. Inheritance of the trait can be studied because many genotypes produce a mixture of sexual and apomitic seedlings. In subtropical climates, most citrus species (with the exception of lemon) flower for about two months during the spring, but in tropical climates flowering can be more continuous if not regulated by water availability.

Less desirable attributes of citrus include relatively long generation time (5–7 years in most species), high heterozygosity in most individuals, difficulty in developing homozygous lines due to inbreeding depression, long seed development time (at least 9 months for most taxa), and difficulty in making some controlled crosses due to apomixis. Production of large hybrid populations by controlled crosses is possible but not easy. Only a portion (5–20%) of pollinated flowers produce fruit, and the number of hybrid seeds per fruit is generally 20 or fewer, except in pummelo where fruit can have 50 or more seeds. A political problem is posed by the differing importance of species in different countries, making it difficult for the international community to agree on a single "model" citrus species on which to focus genomics research.

8.2 Genetic Mapping and Tagging in Citrus

Linkage mapping has been used in citrus since the early 1980s (Durham et al. 1992; Jarrell et al. 1992), but high density maps based on sequence-defined markers have not been reported. Comparative mapping of different citrus species should be particularly interesting because so many species can be hybridized.

8.2.1 Linkage Mapping

Several linkage maps of citrus have been published and many more are under development. Table 8.1 lists several of the reported maps, but see Chen et al. (2007a) for a more complete summary. Several of the maps involve at least one parent that is a *Citrus* x *Poncirus* parent, a strategy that ensures high heterozygosity but creates mapping populations that are of limited use for fruit traits because these hybrids have inedible fruit. Although some of these hybrids are reasonably fertile, which implies a high degree of synteny, such maps may show considerable distortion compared to those developed within species. In general, the various citrus maps share few markers, so comparison of maps is difficult (Roose et al. 2000), but comparative maps of *Citrus* and *Poncirus* (Ruiz and Asins 2003) and sweet orange and mandarin (de Olivera et al. 2005) have been produced. Existing maps are incomplete in that the number of major linkage groups often differs from the number of chromosomes, and too few markers have been mapped to cover the genome. While such maps are still quite useful, more complete, high-density maps would be more useful for both quantitative trait loci (QTL) analysis and for anchoring bacterial artificial chromosome (BAC) and genome sequence. Development of high density maps will also

Table 8.1 Citrus linkage maps

Mapped Parent[z]	Total Markers	Pop. size	Map length cM	Major linkage groups[y]	Percent distorted	Reference
C. maxima x *P. trifoliata*	310	60	874	9	26	Sankar and Moore 2001
C. maxima	34	52	600	4	6	Luro et al. 1995
C. reshni x *P. trifoliata*	97	52	1500	7	37	Luro et al. 1995
C. sunki	63	80	732	5	nd	Cristofani et al. 1999
P. trifoliata	62	80	867	5	nd	Cristofani et al. 1999
C. volkameriana	97	80	460	5	29	Ruiz and Asins 2003
P. trifoliata	73	80	342	4	40	Ruiz and Asins 2003
C. latipes	92	120	433	5	nd	Recupero et al. 2000
C. aurantium	247	120	964	8	nd	Recupero et al. 2000
Citrus x Poncirus (F2)	153	57	701	14	17	Roose et al. 2000
C. sinensis	113	97	776	8	8	Chen et al. 2007a
P. trifoliata	45	97	426	2	10	Chen et al. 2007a

[z] Successive lines with same reference refer to two maps developed from the same hybrid population using a pseudotestcross strategy.
[y] Major linkage groups is the number of groups with more than 4 markers.

require use of larger population sizes to separate closely linked markers. Efforts are in progress to develop high density maps of sweet orange, trifoliate orange, Clementine, and satsuma mandarin using simple sequence repeats (SSRs) and other sequence-based markers (Chen et al. 2007a). These maps should provide a solid foundation for relating physical maps, genome sequences, and traits.

8.2.2 Gene Tagging

Gene tagging has been widely used in citrus. Because most individuals are heterozygous, the typical approach has been to identify individuals differing for a trait, cross them, measure the trait in their progeny, and use bulked-segregant analysis with sequence anonymous RAPDs (Randomly Amplified Polymorphic DNA) and AFLPs (Amplified Fragment Length Polymorphism) to identify segregating markers linked to the genes of interest. This approach has been used for several traits (Table 8.2) including dwarfing (Cheng and Roose 1995), low acid in fruit (Fang et al. 1998a), tolerance to citrus tristeza virus (Fang et al. 1998b; Gmitter et al. 1996; Mestre et al. 1997), citrus nematode resistance (Ling et al. 2000), and nucellar embryony (Garcia et al. 1999; Kepiro 2004). However, none of these genes has been identified at the DNA sequence level with the possible exception of the citrus tristeza virus gene(s) (*Ctv*) from the citrus relative *Poncirus trifoliata*. In this case, a patent for two genes believed to cause resistance has been obtained (US Patent 7,126,004), but the function of the candidate genes has not yet been confirmed in transgenic plants.

QTL analysis has also been used to tag genes affecting various traits in specific crosses. Regions containing genes that influence salinity tolerance (Tozlu et al. 1999b), citrus tristeza virus resistance (Asins et al. 2004), Alternaria brown spot resistance (Dalkilic et al. 2005), and several other traits (Tozlu et al. 1999a) have been identified, but fine mapping of QTLs has not been reported.

Table 8.2 Citrus genes tagged with molecular markers

Trait	Source variety	Marker type	Marker distance	Reference
Dwarfing by rootstock	Flying Dragon	RAPD	13 cM	Cheng and Roose 1995
Citric acid in fruit	Pummelo	RAPD	1.2 cM	Fang et al. 1998a
Citrus tristeza virus resistance	Trifoliate orange	RAPD, SSR	0 cM	Yang et al. 1998b
Citrus nematode tolerance	Trifoliate orange	RAPD	3 cM	Ling et al. 2000
Nucellar embryony	Trifoliate orange	AFLP	1–4 cM	Kepiro 2004
Nucellar embryony	Volkameriana Trifoliate orange	RAPD, SSR	QTLs	Garcia et al. 1999

8.3 BAC Cloning and Utilization

The relatively small genome size of citrus makes preparation and analysis of BAC libraries an attractive approach. For example, an 11x genome coverage library with 120 kb inserts would include only about 35,000 clones, a number that can be spotted in duplicate on only two high-density filters. This makes identification of BAC clones containing specific sequences convenient and inexpensive relative to species with larger genomes. Perhaps because of this, the alternative strategy of screening pools and superpools of BAC clones has not been reported for citrus.

8.3.1 Development of BAC Libraries

Several BAC libraries have been developed for citrus and its near relatives. Specific methods are similar to those used for other organisms: isolation of large DNA fragments by isolating nuclei, embedding them in agarose, partial digestion with a restriction endonuclease, size fractionation by pulsed-field gel electrophoresis, and cloning into a suitable BAC vector.

BAC libraries were developed from *Poncirus* (ca 8x) (Yang et al. 2001) and a *Citrus* x *Poncirus* hybrid (Deng et al. 2001) to identify the citrus tristeza virus resistance gene (*Ctv*) from *P. trifoliata*. Screening high density filters with markers linked to *Ctv* led to development of BAC contigs surrounding the gene (Yang et al. 2001), and eventually to the full sequence of a 282 kb region (Yang et al. 2003).

Libraries recently developed from a number of citrus genotypes, including Ridge Pineapple sweet orange, satsuma, and Clementine mandarin are available from Amplicon Express (http://www.genomex.com/) or the Clemson University Genomics Institute (http://www.genome.clemson.edu/).

8.3.2 BAC Physical Mapping

The heterozygosity present in citrus can complicate BAC fingerprinting to develop physical maps. Completion of the chromosome walk for the *Ctv* gene region was not affected by this problem because the plant that provided the DNA for the library was selected to be homozygous for the targeted region using markers (Yang et al. 2001). An alternative approach is to prepare the library from a wide hybrid such as the *Citrus* x *Poncirus* cross used in Florida (Deng et al. 2001). In this strategy, it is expected that the BAC clones derived from each genome will assemble into two separate contigs. The disadvantage of this approach is that more clones must be analyzed to obtain the same genome coverage. This method facilitates comparison of two different haplotypes which can sometimes be quite valuable.

Physical maps of the 7x Ridge Pineapple library developed by high-throughput, multicolor fingerprinting can be viewed at http://phymap.ucdavis.edu:8080/citrus/. Other groups have not yet released physical maps based on BAC libraries, but additional physical maps are likely to be developed. Future efforts will concentrate on associating physical maps, linkage maps, and genome sequence data to provide an integrated view of the citrus genome.

8.4 Molecular Cytogenetics

Molecular cytogenetics has had limited development in citrus. The chromosomes are small and therefore somewhat difficult to study in detail. Also, meiosis is not easily examined because appropriate flower buds are available only in the spring. Initial work focused on chromosome banding to distinguish the chromosomes of different species and varieties (Guerra 1993; Yamamoto and Tominaga 2003). Fluorescence in situ hybridization (FISH) using repetitive sequence probes such as rDNA has also been useful to detect cytogenetic variation in citrus chromosomes (Matsuyama et al. 1996; Roose et al. 1998; Pedrosa et al. 2000). Combining chromosome banding with rDNA hybridization has revealed distinct patterns for many citrus chromosomes, some of which are species-specific (Moraes et al. 2007). In several cases, the hybrid ancestry of important cultivar groups is clearly supported by molecular cytogenetic analyses showing heterozygosity for karyotypes. There are no reports of localization of individual gene sequences, or even large insert clones such as BACs onto citrus chromosomes, but it seems probable that this will be achieved in the near future.

8.5 Genome Sequencing

The chloroplast genome of sweet orange has recently been fully sequenced (Bausher et al. 2006). Its organization is similar to that of other Eudicots, with a gene order identical to that of the Solanaceae. The chloroplast genome of citrus shows maternal inheritance. Marker and partial sequence data from the chloroplast genomes of other *Citrus* species indicates considerable diversity among the ancestral species so that determining the species that contributed the cytoplasmic genomes to many natural hybrids is relatively easy (Moore 2001).

The only large fragment of citrus genome sequence analyzed in detail is a 282 kb fragment of the *P. trifoliata* genome that should contain a gene for citrus tristeza virus resistance. The sequence was assembled from 8x coverage of four BAC clones, with gap closure by primer walking and directed subclone sequencing. The region contained 22 predicted genes, most of which were confirmed by RT-PCR (reverse transcription polymerase chain reaction), northern blots, or isolation from cDNA libraries. About 22% of the region was composed of retrotransposon-derived sequences with homology to the *gypsy* and *copia* families. Transposable elements related to mutator were also found. Seven of the predicted genes were CC-NBS-LRR type resistance genes, indicating that this region has a high density of such genes (Yang et al. 2003). Limited analyses of the corresponding region from *Citrus* chromosomes indicate considerable rearrangement in certain regions (Mirkov et al. in press). This region had limited synteny with Arabidopsis, only pairs of adjacent genes show clear homology with closely linked Arabidopsis homologs, a result consistent with considerable rearrangement in the region since divergence of these taxa (Yang et al. 2003).

Initial genome sequencing focused on sweet orange because of its commercial importance in many countries. A low-coverage (1.2x) shotgun genome sequence for sweet orange has been released by the US Department of Energy Joint Genome Institute. The sequence files may be downloaded from the National Center for Biotechnology (NCBI) Trace Archives (http://www.ncbi.nlm.nih.gov/Traces/trace.cgi?) and searched using BLAST through http://138.23.191.145/blast/index.html. Relatively little analysis and annotation of this sequence has been completed, but it should yield considerable information on repeat content, sequence organization, and gene structure when analysis is completed. Because of the heterozygosity of sweet orange it is probable that future efforts to obtain and assemble a high quality sequence will focus on a doubled haploid genotype.

8.6 Functional Genomics

Functional genomics in citrus has focused on developing and analyzing EST collections and using this information to develop microarrays for transcript analysis.

8.6.1 EST Development

During the past few years, EST sequences have been produced by several research groups from a number of citrus species (Forment et al. 2005; Terol et al. 2007). Not all of these ESTs have been publicly released and there may be more private than publicly available citrus ESTs as of July 2007. A tally of ESTs available from the NCBI EST database http://www.ncbi.nlm.nih.gov/dbEST/dbEST_summary.html is refreshed weekly and displayed on the International Citrus Genomics Consortium website http://int-citrusgenomics.org/. The June 26 tally included 225,204 ESTs. The largest portion was derived from *C. sinensis* (94,738) and the next largest portion was from mandarin (76,909 total) under the names *C. clementina* (62,250), *C. unshiu* (4,489), *C. reticulata* (3,640), *C. reshni* (2,867), *C. clementina* x *C. temple* (1,823), *C. clementina* x *C. tangerina* (1,766), and *C. clementina* x *C. reticulata* (74). Considering that the genetic composition of sweet orange is probably 75% mandarin, mandarin ESTs comprise nearly 2/3 of these public ESTs. Common rootstocks *Poncirus trifoliata* (28,737) and sour orange *C. aurantium* (5,060) are also represented, along with ESTs from other citrus species and hybrids including rough lemon and citron.

ESTs have been clustered by sequence similarity and may be browsed in the context of these clusters using several publicly available interfaces. NCBI provides the

UniGene web interface at http://www.ncbi.nlm.nih.gov/sites/entrez?db=unigene, from which build #8 of *C. sinensis* sequences displayed 8,811 unigenes and build #1 of *C. clementina* provided 5,872 unigenes. The Institute for Genome Resources (TIGR) provides a web interface to perform BLAST searches and retrieve EST sequences from transcript assemblies for each taxonomically uniquely named group of ESTs within the family Rutaceae (http://plantta.tigr.org/search.shtml). HarvEST: Citrus, available for Windows from http://harvest.ucr.edu or operable online through http://harvest-web.org, provides assemblies of all citrus ESTs, or only those of *C. sinensis* or *P. trifoliata*. Version 1.16 of "HarvEST:Citrus" displayed 85 libraries and 222,457 ESTs from *Citrus* and *Poncirus*. ESTs from 16 libraries produced at University of California Riverside (98,578 ESTs), 10 at University of California Davis (Abhaya Dandekar; 17,446 ESTs), five at USDA/ARS US Horticultural Research Lab in Ft. Pierce, Florida (Robert Shatters, Michael Bausher, Jose Chaparro, Greg McCollum; 16,396 ESTs) and two from Volcani Center, Israel (Avi Sadka; 1,764 ESTs) were derived from chromatograms using the full HarvEST pipeline. These 134,184 ESTs therefore retain their phred quality values and can be viewed more extensively than other sequences in HarvEST:Citrus. All other sequences were downloaded from the GenBank dbEST database by Matt Lyon and Steve Wanamaker at UC Riverside. The latter include about 11,000 additional ESTs from USDA/ARS US Horticultural Research Lab in Ft. Pierce; 70,581 ESTs from Universidad Politecnica de Valencia, Valencia, Spain; 3,104 ESTs from East Tennessee State University (Cecilia McIntosh); 2,496 ESTs from the Laboratory of Biotechnology & Citrus Genome Analysis Team (CGAT), Shizuoka, Japan; 274 cDNA or genomic sequences from the GenBank nr database, including 40 microbial pathogen sequences; and smaller contributions from others. HarvEST:Citrus version 1.16 contained best BLASTX hits from UniProt (January 2007) and the *Arabidopsis* genome (TAIR version 7; April 2007). Another very useful web-accessible citrus EST interface, containing information on 157,608 ESTs as of July 2005, is available from the Genomics Facility at the University of California, Davis (http://cgf.ucdavis.edu/home/).

Citrus EST libraries have been produced from a range of tissues with and without exposure to pests, pathogens, and abiotic stress treatments. A detailed description of each library is available through HarvEST:Citrus, where one may search by tissue or treatment. Briefly, source tissues have included mature and developing fruit (ovary, flavedo, albedo, juice sac, pulp; 45 libraries), scion or seedling vegetative tissue (leaf, stem, bark; 26 libraries), root (10 libraries), flower (two libraries), seed (one library), and callus (one library). Exposure to pests and pathogens has included tristeza virus, several viroids, *Phytophthora* and *Penicillium*, nematodes, thrips, and red scale. Abiotic stress and chemical treatments have included drought, iron deficiency, heat, cold, gibberellic acid, auxin, ethylene and low oxygen. Overall, the spectrum of tissues and treatments providing ESTs encompasses a range of developmentally and economically important aspects of citrus biology, providing what is expected to be a very satisfactory resource for the development of transcriptome and proteome monitoring tools.

8.6.2 Microarrays

As in many species, the first type of array to be used to investigate gene expression was a spotted macroarray in which cDNA clones were spotted on membranes and hybridized with labeled probe molecules (Mozoruk et al. 2006). Arrays composed of cDNA clones spotted on slides have also been developed. For example, Forment et al. (2005) describe a spotted array of 6,875 clones, mainly from Clementine mandarin. This array has been used to investigate gene expression patterns during fruit ripening (Cercós et al. 2006) and other traits. Higher density spotted arrays are being developed. Spotted arrays have also been developed in Japan (Shimada et al. 2005), with an oligonucleotide array under development.

An Affymetrix citrus GeneChip was designed from the HarvEST:Citrus database and became commercially available in January 2006 (Close et al. 2006). This chip contains 30,264 probe sets (22 probes each) for measurement of citrus transcripts and 5,023 probe sets (56 probes each) to serve as single nucleotide polymorphism (SNP) markers for 3,219 genes. The citrus GeneChip also includes tiling of one region of the *Poncirus trifoliata* genome containing a citrus tristeza virus resistance locus, as well as probe sets for detection of several viruses, viroids, *Xylella* species, and commonly used transgenes. Interpretation of data from this chip is supported by gene function annotations and interactive graphical user interfaces in the HarvEST:Citrus software.

It will be valuable to develop databases that interrelate annotation of probes on the various arrays.

8.6.3 Transformation

Citrus transformation systems have been developed for many cultivars because the technology promises to allow improvement of existing varieties without the requirement for a sexual generation in which recombination would disrupt the essential characteristics of the variety. Most research has focused on transformation via *Agrobacterium*. Particularly efficient systems have been developed for the rootstock, Carrizo citrange (Cervera et al. 1998), but systems for commercially important scions including navel orange (Bond and Roose 1998), Rio Red grapefruit (Yang et al. 2000), and others have been developed. Co-cultivation of *Agrobacterium* with seedling epicotyl tissue is the foundation of most transformation protocols but has the undesirable consequence of producing juvenile transgenics that do not flower and fruit for several years, making evaluation of fruit traits expensive and inefficient. Mature tissue of some cultivars has been transformed (Cervera et al. 2005), but other laboratories have been unable to replicate this result. Most selection systems use kanamycin resistance, but this is not always effective and new selection schemes are being developed (Ballester et al. 2007). Regeneration from embryogenic callus can also be used to transform citrus (Duan et al. 2007) but requires production and maintenance of the callus.

There has been relatively little research use of transgenics for functional genomics in citrus. Transformation is relatively difficult and the time and resources required to produce and evaluate each transgenic discourages use of this approach to validate gene function. In some cases, function of citrus genes can be efficiently validated in Arabidopsis as illustrated by transformation of a citrus terminal flower gene into Arabidopsis (Pillitteri et al. 2004), but given the large phenotypic divergence between these species, validation within citrus will often be required. Some work using RNAi to silence resistance gene candidates has been completed (Ye and Roose, in preparation), but little else has been reported. Development of new transformation vectors that permit more efficient screening for transgenics (Chen et al. 2007b) may facilitate the high-throughput transformation that would be valuable for functional genomics.

8.7 Sequence and Marker Diversity

Various types of molecular markers have been used to characterize citrus varieties and germplasm accessions, beginning with isozyme studies in the late 1970s (Torres et al. 1978), followed by restriction fragment length polymorphisms (RFLPs) and RAPDs in the 1980s and early 1990s, and more recently AFLPs, SSRs, ISSRs (inter-simple sequence repeats), IRAPs (inter-retrotransposon amplified polymorphisms), and others (Fang and Roose 1997; Bretó et al. 2001; Sankar and Moore 2001; Ahmad et al. 2003; Barkley et al. 2006; Chen et al. 2006; Pang et al. 2007). The overall picture that has emerged from these studies is high allelic diversity and heterozygosity in certain ancestral species groups, particularly mandarins and pummelos, while citrons and trifoliate orange have moderate-low levels of diversity. Other groups, including sweet orange, lemon, lime, and grapefruit have fairly high heterozygosity, but nearly all cultivars are identical, a pattern consistent with diversification by mutation from a single hybrid ancestor (Federici et al. 1998; Gulsen and Roose 2001; Moore 2001; Barkley et al. 2006).

DNA sequence diversity has not been reported in citrus except for diversity in the chloroplast genome (de Araujo et al. 2003) where pummelo, mandarin, and citron fall into distinct groups, but *Citrus* does not appear monophyletic because citron is isolated from other *Citrus* species. Ongoing projects in several laboratories are likely to reveal much about the origin and type of diversity present within and between groups.

As in other plants, diversity in citrus is driven in part by activity of transposable elements, particularly retrotransposons (Bretó et al. 2001). Retrotransposons compose a substantial portion of the citrus genome, with *copia*-type elements composing about 13% and *gypsy* elements about 10% (Rico-Cabanas and Martínez-Izquierdo 2007). An apparently active *copia*-class element, CIRE1, has been identified in sweet orange (Rico-Cabanas and Martínez-Izquierdo 2007).

8.8 Perspective

Citrus genomics has progressed rapidly in recent years with the development of BAC libraries, an extensive EST collection, microarrays, and dense, sequence-based maps. Prospects for whole genome sequencing and subsequent exploration and exploitation of such data are bright. Because the citrus genomics community is relatively small, increased cooperation and collaboration among groups in different countries will be important to derive practical benefits from such genome analysis.

Acknowledgments We thank Dr. Claire Federici for critically reviewing the manuscript.

References

Ahmad R, Struss D, Southwick SM (2003) Development and characterization of microsatellite markers in *Citrus*. J Am Soc Hort Sci 128:584–590

Arugumanathan K, Earle ED (1991) Nuclear DNA content of some important plant species. Plant Mol Biol Rep 9:208–218

Asins MJ, Bernet GP, Ruiz C, Cambra M, Guerri J, et al. (2004) QTL analysis of citrus tristeza virus-citradia interaction. Theor Appl Genet 108:603–611

Ballester A, Cervera M, Pena, L (2007) Efficient production of transgenic citrus plants using isopentenyl transferase positive selection and removal of the marker gene by site-specific recombination. Plant Cell Rep 26:39–45

Barkley NA, Roose ML, Krueger RR, Federici, CT (2006) Assessing genetic diversity and population structure in a citrus germplasm collection utilizing simple sequence repeat markers (SSRs). Theor Appl Genet 112:1519–1531

Bausher MG, Singh ND, Lee S, Jansen RK, Daniell H (2006) The complete chloroplast genome sequence of *Citrus sinensis* (L.) Osbeck var 'Ridge Pineapple': organization and phylogenetic relationships to other angiosperms. BMC Plant Biol 6:21

Bond JE, Roose ML (1998) *Agrobacterium*-mediated transformation of the commercially important citrus cultivar Washington navel orange. Plant Cell Rep 18:229–234

Bretó MO, Ruiz C, Pina JA, Asins MJ (2001) The diversification of *Citrus clementina* Hort. Ex Tan., a vegetatively propagated crop species. Mol Phylogenet Evol 21:285–293

Cercós M, Soler G, Iglesias DJ, Gadea J, Forment J, et al. (2006) Global analysis of gene expression during development and ripening of citrus fruit flesh. A proposed mechanism for citric acid utilization. Plant Mol Biol 62:513–527

Cervera M, Pina JA, Juarez J, Navarro L, Pena, L (1998) Agrobacterium-mediated transformation of citrange: Factors affecting transformation and regeneration. Plant Cell Rep 18:271–278

Cervera M, Juarez J, Navarro L, Pena L (2005) Genetic transformation of mature citrus plants, In: Pena, L (ed.), Methods in Molecular Biology, Humana Press, pp. 177–187

Chen C, Zhou P, Choi YA, Huang S, Gmitter FG Jr (2006) Mining and characterizing microsatellites from citrus ESTs. Theor Appl Genet 112:1248–1257

Chen C, Bowman KD, Choi YA, Dang PM, Rao MN, et al. (2007a) EST-SSR genetic maps for *Citrus sinensis* and *Poncirus trifoliata*. Tree Genetics and Genomes DOI 10.1007/s11295-007-0083-3

Chen C, Zheng Q, Xiang X, Soneji JR, Huang S, et al. (2007b) Development of pGreen-derived GFP binary vectors for citrus transformation. HortSci 42:7–10

Cheng FS, Roose ML (1995) Origin and inheritance of dwarfing by the citrus rootstock *Poncirus trifoliata* 'Flying Dragon.' J Am Soc Hort Sci 120:286–291

Close TJ, Wanamaker S, Lyon M, Mei G, Davies C, et al. (2006) A GeneChip®for Citrus. Plant & Animal Genomes Conf 14:W82, pg 26

Cristofani M, Machado MA, Grattapaglia D (1999) Genetic linkage maps of *Citrus sunki* Hort. Ex. Tan. and *Poncirus trifoliata* (L.) Raf. and mapping of citrus tristeza virus resistance gene. Euphytica 109:25–32

Dalkilic Z, Timmer LW, Gmitter FG Jr (2005) Linkage of *Alternaria* disease resistance gene in mandarin hybrids with RAPD fragments. J Am Soc Hort Sci 130:191–195

de Araujo EF, de Queiroz LP, Machado MA (2003) What is *Citrus*? Taxonomic implications from a study of cp-DNA evolution in the tribe Citreae (Rutaceae subfamily Aurantioidaeae). Org Divers Evol 3:55–62
de Oliveira RP, Cristofani M, Machado MA (2005) Integrated genetic map of citrus based on RAPD markers. Fruits 60:187–193
Deng Z, Huang S, Ling P, Yu C, Tao Q, et al. (2001) Fine genetic mapping and BAC contig development for the citrus tristeza virus resistance gene locus in *Poncirus trifoliata* (Raf.). Mol Genet Genomics 265:739–747
Duan YX, Guo WW, Meng HJ, Tao NG, Li DD, et al. (2007) High efficient transgenic plant regeneration from embryogenic calluses of *Citrus sinensis*. Biologia Plantarum (Prague) 51: 212–216
Durham RE, Liou PC, Gmitter FG Jr, Moore GA (1992) Linkage of restriction fragment length polymorphisms and isozymes in *Citrus*. Theor Appl Genet 84:39–48
Fang DQ, Roose ML (1997) Identification of closely related citrus cultivars with inter-simple sequence repeat markers. Theor Appl Genet 95:408–417
Fang DQ, Federici CT, Roose ML (1998a) Development of molecular markers linked to a gene controlling fruit acidity in Citrus. Genome 40:841–849
Fang DQ, Federici CT, Roose ML (1998b) A high-resolution linkage map of the citrus tristeza virus resistance gene region in *Poncirus trifoliata* (L.) Raf. Genetics 150:883–890
Federici CT, Fang DQ, Scora RW, Roose ML (1998) Phylogenetic relationships within the genus *Citrus* (Rutaceae) and related genera as revealed by RFLP and RAPD analysis. Theor Appl Genet 96:812–822
Forment J, Gadea J, Huerta L, Abizanda L, Agusti J, et al. (2005) Development of a citrus genome-wide EST collection and cDNA microarray as resources for genomic studies. Plant Mol Biol 57:375–391
Garcia R, Asins MJ, Forner J, Carbonell EA (1999) Genetic analysis of apomixis in *Citrus* and *Poncirus* by molecular markers. Theor Appl Genet 99:511–518
Gmitter FG Jr (1995) Origin, evolution and breeding of the grapefruit. Plant Breeding Rev 13:345–363
Gmitter FG, Xiao SY, Huang S, Hu XL, Garnsey SM, et al. (1996) A localized linkage map of the citrus tristeza virus resistance gene region. Theor Appl Genet 92:688–695
Guerra M (1993) Cytogenetics of Rutaceae. V. High chromosomal variability in *Citrus* species revealed by CMA/DAPI staining. Heredity 71:234–241
Gulsen O, Roose ML (2001) Lemons: diversity and relationships with selected *Citrus* genotypes as measured with nuclear genome markers. J Am Soc Hort Sci 126:309–317
Jarrell DC, Roose ML, Traugh SN, Kupper RS (1992) A genetic map of citrus based on the segregation of isozymes and RFLPs in an intergeneric cross. Theor App Genet 84:49–56
Jayaprakasha GK, Patil BS (2007) In vitro evaluation of the antioxidant activities in fruit extracts from citron and blood orange. Food Chemistry 101:410–418
Kepiro J (2004) Molecular genetic analysis of nucellar embryony (apomixis) in *Citrus maxima* x *Poncirus trifoliata*. pp. 220. Ph.D. Dissertation, University of California, Riverside
Ling P, Duncan LW, Deng Z, Dunn D, Hu X, et al. (2000) Inheritance of citrus nematode resistance and its linkage with molecular markers. Theor Appl Genet 100:1010–1017
Luro F, Lorieux M, Laigret F, Bové JM and Ollitraut P (1995) Cartographie du génome des agrumes á l'aide des marquers moléculaires et distorsions de ségrégation. In: INRA (ed.) Techniques et Utilisations des Marqueurs Moléculaires. INRA, Paris, pp. 69–82
Matsuyama T, Akihama T, Ito Y, Omura M, Fukui K (1996) Characterization of heterochromatic regions in 'Trovita' orange (*Citrus sinensis* Osbeck) chromosomes by the fluorescent staining and FISH methods. Genome 39:941–945
Mestre PF, Asins MJ, Pina JA, Carbonell EA, Navarro L (1997) Molecular markers flanking citrus tristeza virus resistance gene from *Poncirus trifoliata* (L.) Raf. Theor Appl Genet 94:458–464
Mirkov TE, Yang Z, Rai M, Molina JJ, Roose ML, et al. (in press) Toward positional cloning of the *Citrus tristeza virus* resistance gene. In: Karasev, AV and Hilf, ME (ed.), *Citrus tristeza Virus* Complex and Tristeza Diseases, APS Press

Moore G (2001) Oranges and lemons: clues to the taxonomy of Citrus from molecular markers. Trends Genet 17:536–540

Moraes A, Soares Filho WS, Guerra M (2007) Karyotype diversity and the origin of grapefruit. Chromosome Res 15:115–121

Mozoruk J, Hunnicutt LE, Cave RD, Hunter WB, Bausher MG (2006) Profiling transcriptional changes in *Citrus sinensis* (L.) Osbeck challenged by herbivory from the xylem-feeding leafhopper *Homalodisca coagulata* (Say) by cDNA macroarray analysis Plant Sci 170:1068–1080

Nicolosi E, Deng ZN, Gentile A, La Malfa S, Continella G, et al. (2000) *Citrus* phylogeny and genetic origin of important species as investigated by molecular markers. Theor Appl Genet 100:1155–1166

Nicolosi E, La Malfa S, El-Otmani M, Neqbi M, Goldschmidt EE (2005) The search for the authentic citron (*Citrus medica* L.): Historic and genetic analysis. HortSci 40: 1963–1968

Pang X-M, Hu C-G, Deng, X-X (2007) Phylogenetic relationships within *Citrus* and its related genera as inferred from AFLP markers. Genet Resources Crop Evol 54:429–436

Pedrosa A, Schweizer D, Guerra M (2000) Cytological heterozygosity and the hybrid origin of sweet orange *Citrus sinensis* (L.) Osbeck. Theor Appl Genet 100:361–367

Pillitteri LJ, Lovatt CJ, Walling LL (2004) Isolation and characterization of a TERMINAL FLOWER homolog and its correlation with juvenility in Citrus. Plant Physiol 135: 1540–1551

Recupero GR, Russo MP, De Simone M, Natoli A, Marsan PA, Marocco A (2000) Development of molecular marker maps for rootstock breeding in *Citrus*. Acta Hort. 535:33–36

Rico-Cabanas L, Martínez-Izquierdo JA (2007) CIRE1, a novel transcriptionally active *Ty1-copia* retrotransposon from *Citrus sinensis*. Mol Genet Genomics 277:365–377

Roose ML, Schwarzacher T, Heslop-Harrison JS (1998) The chromosomes of *Citrus* and *Poncirus* species and hybrids: identification of characteristic chromosomes and physical mapping of rDNA loci using in situ hybridization and fluorochrome banding. J Hered 89:83–86

Roose ML, Fang D, Cheng FS, Tayyar RA, Federici CT, et al. (2000) Mapping the *Citrus* genome. Acta Horticulturae 535:25–32

Ruiz C, Asins MJ (2003) Comparison between *Poncirus* and *Citrus* genetic linkage maps. Theor Appl Genet 106:826–836

Sankar AA, Moore GA (2001) Evaluation of inter-simple sequence repeat analysis for mapping in *Citrus* and extension of the genetic linkage map. Theor Appl Genet 102:206–214

Shimada T, Fuiii H, Endo T, Yazaki J, Kishimoto N, et al. (2005) Toward comprehensive expression profiling by microarray analysis in citrus: monitoring the expression profiles of 2213 genes during fruit development. Plant Sci 168:1383–1385

Terol J, Conesa A, Colmenero JM, Cercos M, Tadeo FR, et al. (2007) Analyses of 13000 unique *Citrus* clusters associated with fruit quality, production and salinity tolerance. BMC Genomics 8:31

Torres AM, Soost RK, Diedenhofen U (1978) Leaf isozymes as genetic markers in *Citrus*. Am J Bot 65:869–881

Tozlu I, Guy CL, Moore GA (1999a) QTL analysis of morphological traits in an intergeneric BC_1 progeny of *Citrus* and *Poncirus* under saline and nonsaline environments. Genome 42:1020–1029

Tozlu I, Guy CL, Moore GA (1999b) QTL analysis of Na^+ and Cl accumulation related traits in an intergeneric BC_1 progeny of *Citrus* and *Poncirus* under saline and nonsaline environments. Genome 42:692–705

Yamamoto M, Tominaga S (2003) High chromosomal variability of mandarins (*Citrus* spp.) revealed by CMA banding. Euphytica 129:267–274

Yang ZN, Ingelbrecht I, Louzada E, Skaria M, Mirkov, TE (2000) *Agrobacterium* mediated transformation of the commercially important grapefruit cultivar Rio Red. Plant Cell Rep 19:1203–1211

Yang ZN, Ye XR, Choi S, Molina J, Moonan F, et al. (2001) Construction of a 1.2 Mb contig including the *Citrus tristeza virus* resistance gene locus using a bacterial artificial chromosome library of *Poncirus trifoliata*. Genome 44:382–393

Yang ZN, Ye XR, Molina J, Roose ML, Mirkov, TE (2003) Sequence analysis of a 282-kb region surrounding the *Citrus tristeza virus* resistance gene (*Ctv*) locus in *Poncirus trifoliata*. Plant Physiol 131:482–492

Chapter 9
Genomics of Coffee, One of the World's Largest Traded Commodities

Philippe Lashermes, Alan Carvalho Andrade, and Hervé Etienne

Abstract Coffee is one of the world's most valuable agricultural export commodities. In particular, coffee is a key export and cash crop in numerous tropical and subtropical countries having a generally favorable impact on the social and physical environment. While coffee species belong to the Rubiaceae family, one of the largest tropical angiosperm families, commercial production relies mainly on two species, *Coffea arabica* L. and *Coffea canephora* Pierre, known as Robusta. Although a considerable genetic diversity is potentially available, coffee breeding is still a long and difficult process. Nevertheless, genomic approaches offer feasible strategies to decipher the genetic and molecular bases of important biological traits in coffee tree species that are relevant to the growers, processors, and consumers. This knowledge is fundamental to allow efficient use and preservation of coffee genetic resources for the development of improved cultivars in terms of quality and reduced economic and environmental costs. This review focuses on the recent progress of coffee genomics in relation to crop improvement.

9.1 Introduction

Coffee species belong to the Rubiaceae family, one of the largest tropical angiosperm families. Variations in cpDNA classified the Coffeeae tribe into the Ixoroideae monophyletic subfamily, close to Gardenieae, Pavetteae and Vanguerieae (Bremer and Jansen 1991). Two genera, *Coffea* L. and *Psilanthus* Hook. f., were distinguished on the basis of flowering and flower criteria (Bridson 1982). Each genus was divided into two subgenera based on growth habit (monopodial *vs.* sympodial development) and type of inflorescence (axillary *vs.* terminal flowers). More than one hundred coffee species have been identified and new taxa are still being discovered (Bridson and Verdcourt 1988; Davis et al. 2006; Stoffelen et al. 2007). All species are perennial woody bushes or trees in inter-tropical forests of Africa

P. Lashermes
Institut de Recherche pour le Développement, UMR RPB, BP 64501,
34394 MontpellierCedex 5, France
e-mail: Philippe.Lashermes@mpl.ird.fr

P.H. Moore, R. Ming (eds.), *Genomics of Tropical Crop Plants*

© Springer 2008

and Madagascar for the *Coffea* genus, and Africa, Southeast Asia, and Oceania for the *Psilanthus* genus. Plants of the two genera differ greatly in morphology, size, and ecological adaptations. Some species, such as *C. canephora* and *C. liberica* Hiern, are widely distributed from Guinea to Uganda. Other species display specific adaptations, e.g., *C. congensis* Froehner to seasonally flooded areas in the Zaire basin and *C. racemosa* Lour. to very dry areas in the coastal region of Mozambique. Molecular phylogeny of *Coffea* species has been established based on DNA sequence data (Lashermes et al. 1997; Cros et al. 1998). The results suggest a radial mode of speciation and a recent origin in Africa for the genus *Coffea*. Several major clades were identified, which present a strong geographical correspondence (i.e., West Africa, Central Africa, East Africa, and Madagascar).

Commercial production relies mainly on two species, *C. arabica* L. and *C. canephora* Pierre, known as robusta. The cup quality (low caffeine content and fine aroma) of *C. arabica* makes it by far the most important species, representing 65% of the world production. Another species, *C. liberica* (liberica coffee) stands third with a share of less than one per cent of world coffee production.

C. arabica has its primary center of genetic diversity in the highlands of Southwest Ethiopia and the Boma Plateau of Sudan. Wild populations of *C. arabica* also have been reported in Mount Imatong (Sudan) and Mount Marsabit (Kenya) (Thomas 1942; Anthony et al. 1987). Cultivation of *C. arabica* started in Southwest Ethiopia about 1,500 years ago (Wellman 1961). Modern coffee cultivars are derived from two base populations of *C. arabica*, known as Typica and Bourbon, that were spread worldwide in the 18th century (Krug et al. 1939). Historical data indicate that these populations were composed of progenies of very few plants, i.e., only one for the Typica population (Chevalier and Dagron 1928) and the few plants that were introduced to the Bourbon Island (now Réunion) in 1715 and 1718 for the Bourbon population (Haarer 1956).

During the 18th and 19th centuries, only Arabica was produced. However, this species appeared to be very sensitive to parasitic threats, especially orange rust. That is why, in Africa, during the 19th century, the spontaneous forms of other species of coffee, especially *C. canephora*, were cultivated locally. In particular, coffee plants from local forest populations of the Belgian Congo (now the Democratic Republic of Congo) and Uganda were transferred to Java, a major breeding center from 1900 to 1930. At the same time, in Africa, the diversity of material cultivated was extended with the use of local spontaneous forms: Kouilou in Côte d'Ivoire, Niaouli in Togo and Benin, and Nana in the Central African Republic. The material selected in Java was reintroduced in the Belgian Congo around 1916 at INEAC (Institut National pour l'Etude Agronomique au Congo) research center which was the major breeding center of *C. canephora* from 1930 to 1960. Selected plant materials were largely distributed worldwide. Although the overall performance of cultivated trees has increased noticeably after a few breeding cycles, the cultivars nonetheless have remained genetically very close to individuals of the original natural populations (Dussert et al. 2003). Furthermore, the considerable genetic diversity observed in *C. canephora* is still largely unexploited.

Arabica production is constrained by numerous diseases and pests like leaf rust (*Hemileia vastatrix* Berk & Br.), coffee berry disease (*Colletotrichum kahawae*), coffee berry borer (*Hypothenemus hampei*), stem borer (*Xylotrechus quadripes* Chevr.), and nematodes (*Meloidogyne spp.* and *Pratylenchus spp.*). In contrast, Robusta is more tolerant to these diseases and pests. Hence, transfer of desirable genes, in particular for disease resistance, from coffee species into Arabica cultivars without affecting quality traits has been the main objective of Arabica breeding (Carvalho 1988; Van der Vossen 2001). To date, *C. canephora* provides the main source of disease and pest resistance traits not found in *C. arabica*, including coffee leaf rust (*H. vastatrix*), coffee berry disease (*C. kahawae*), and root-knot nematode (*Meloidogyne* spp.). Likewise, other coffee species are of considerable interest in this respect. For instance, *C. liberica* has been used as source of resistance to leaf rust (Srinivasan and Narasimhaswamy 1975), while *C. racemosa* constitutes a promising source of resistance to the coffee leaf miner (Guerreiro Filho et al. 1999). Exploitation of such genetic resources has so far relied on conventional procedures in which a hybrid is produced between an outstanding variety and a donor genotype carrying the trait of interest, and the progeny is backcrossed to the recurrent parent. Undesirable genes from the donor parent are gradually eliminated by selection. In so doing, conventional coffee breeding methodology faces considerable difficulties. In particular, strong limitations are due to the long generation time of a coffee tree (5 years), the high cost of field trials, and the lack of accuracy of current strategy. A minimum of 25 years after hybridization is required to restore the genetic background of the recipient cultivar and thereby ensure good quality of the improved variety.

9.1.1 Economic and Societal Importance of Coffee

Coffee is one of the world's most valuable agricultural export commodities. In particular, coffee represents one of the key export and cash crops in tropical and subtropical countries with generally a favorable impact on the social and physical environment (International Coffee Organization, http://www.ico.org/). About 125 million people depend on coffee for their livelihoods in Latin America, Africa, and Asia. Coffee is produced in more than 68 countries between 22° N and 24° S latitudes, on a total of about 10.6 million ha. Total world production for the year 2005/06 was 6.6 million t green coffee beans, with an estimated market value of $12.2 billion. World coffee production over the last 5 years has fluctuated from 6.3 to 7.4 million t green beans.

Arabica coffee is cultivated in relatively cool mountain climates, at 1,000-2,000 m altitudes in equatorial regions and at 400 -1,200 m altitudes further from the equator. Robusta coffee requires the warm and humid climates of tropical lowlands and foothills. About 63% of the world coffees in 2004/05 were produced in Latin America, 24% in Asia, and 13% in Africa. Brazil (80% arabica) alone produced 35%, Vietnam (97% robusta) 12%, Colombia (100% arabica) 10%, and Ethiopia (100% arabica) 4% of all coffee. Smallholders (< 5 ha) account for about 70% of world coffee production. However, medium to large coffee plantations (30-3,000 ha per estate) can be found in countries like Brazil, El Salvador, Guatemala, India, Indonesia, Vietnam, Kenya, Tanzania, and Ivory Coast.

World coffee consumption was 6.9 million t for 2004/05 (6.7 million t the previous year), of which about 1.8 million t (26%) is consumed domestically in coffee-exporting countries. Brazil, Mexico, and Ethiopia are large coffee exporters with notably high domestic coffee consumption (>40% of production). The European Community, Norway, Switzerland, the United States, and Japan consume about 80% of the coffee exported from producing countries. The demand for coffee has been growing steadily at 1.2% per year, but world production regularly exceeds demand by 0.3-0.6 million t per year. All these factors contribute to a high volatility of price on the main world coffee markets.

9.1.2 Coffea *as an Experimental Organism*

The coffee plant is an evergreen shrub or small tree, which, under free growth, may become 4-6 m tall for *C. arabica* and 8-12 m for *C. canephora*. In cultivation, both species are pruned to manageable heights of less than 2-3 m with one or more stems. The growth of the coffee plant is dimorphic. The main stems (orthotropic axes) grow vertically and the branches (plagiotropic axes) grow horizontally. Horticultural propagation is relatively easy. The plant may flower once or twice a year after a rainfall of at least 10 mm, which follows a period of water stress. Fruits mature in 7-9 (*C. arabica*) to 9-11 (*C. canephora*) months from date of flowering depending on the variety and environment. The coffee seeds do not behave in an orthodox manner when dehydrated or stored at low temperature. For instance, the viability of seeds of *C. arabica* decreases rapidly after 4-6 months at ambient temperature. However, short term storage (up to 3 years) is possible using adapted and controlled storage conditions (Eira et al. 2006).

All coffee species are diploid (2n=2x=22) and generally self-incompatible, except for *C. arabica* Linne which is tetraploid (2n=4x=44) and self-fertile. Molecular analyses (Lashermes et al. 1999) have indicated that *C. arabica* is an amphidiploid formed from the hybridization between two closely related diploid species (i.e., *C. canephora* and *C. eugenioides*). The evidence suggests recent speciation and low divergence between the two constitutive genomes of *C. arabica* and those of its progenitor species. The analysis of segregating molecular markers has confirmed earlier genetic and cytogenetic evidence that *C. arabica* is a functional diploid. Furthermore, homoeologous chromosomes do not pair in *C. arabica*, not as a consequence of structural differentiation, but because of the functioning of pairing regulating factors (Lashermes et al. 2000a).

The nuclear DNA content of several coffee species has been estimated by flow cytometry (Cros et al. 1995). The DNA amount (2C values) varies between diploid coffee species from 0.95 to 1.8 pg. In comparison to other angiosperms (Bennett and Leitch 1995), the genomes of coffee species appear to be of rather low size (i.e.,

810 Mb for *C. canephora* and 1,300 Mb for *C. arabica*). These variations in DNA amount, other than variation due to ploidy level (e.g. *C. arabica*), are probably due almost entirely to variation in the copy number of repeated DNA sequences. Differences may correspond to genomic evolution correlated with an ecological adaptation process. Furthermore, reduced fertility of certain interspecific F_1 hybrids appears to be associated with significant differences in nuclear content of parental species (Barre et al. 1998).

9.2 Genetic Mapping and Tagging in Coffee

9.2.1 Marker Diversity

In coffee, a whole range of techniques has been used to detect polymorphism at the DNA level, including randomly amplified polymorphic DNA (RAPD) (Orozco-Castilho et al. 1994; Lashermes et al. 1996a; Anthony et al. 2001; Aga et al. 2003), cleaved amplified polymorphisms (CAP) (Lashermes et al. 1996b; Orozco-Castilho et al. 1996), restriction fragment length polymorphisms (RFLP) (Paillard et al. 1996; Dussert et al. 2003; Lashermes et al. 1999), amplified fragment length polymorphism (AFLP) (Lashermes et al. 2000b; Anthony et al. 2002; Prakash et al. 2005), inverse sequence-tagged repeat (ISR) (Aga and Bryngelsson 2006), and simple sequence repeats or microsatellites (SSR) (Mettulio et al. 1999; Combes et al. 2000; Baruah et al. 2003; Moncada and McCouch 2004; Poncet et al. 2004). During the last few years, the number of co-dominant markers has been considerably increased by SSR mining in coffee expressed sequence tag (EST) databases (Bhat et al. 2005; Poncet et al. 2006; Aggarwal et al. 2007) offering new possibilities for genetic analysis.

9.2.2 Linkage Mapping

Several genetic maps have been constructed. The low polymorphism has been a major drawback for developing genetic maps of the *C. arabica* genome. Hence, the works reported so far are often restricted to alien DNA introgressed fragments into *C. arabica* (Prakash et al. 2004). Nevertheless, Pearl et al. (2004) recently obtained a genetic map from a cross between Catimor and Mokka cultivars of *C. arabica*. Furthermore, to overcome the limitation of low polymorphism, efforts were directed to the development of genetic maps in *C. canephora* or interspecific crosses.

The earliest attempt to develop a linkage map was based on Canephora doubled haploid (DH) segregating populations (Paillard et al. 1996; Lashermes et al. 2001). *C. canephora* is a strictly allogamous species consisting of polymorphic populations and of strongly heterozygous individuals. Conventional segregating populations are therefore somewhat difficult to generate and analyze. However, the ability to produce DH populations in *C. canephora* offers an attractive alternative approach. The method of DH production is based on the rescue of haploid embryos of maternal origin occurring spontaneously in association with polyembryony (Couturon 1982). Two complementary segregating plant populations of *C. canephora* were produced from the same genotype. One population comprised 92 doubled haploids derived from female gametes, while the other population was a test-cross consisting of 44 individuals derived from male gametes. A genetic linkage map of *C. canephora* was constructed spanning 1,041 cM of the genome (Lashermes et al. 2001). This genetic linkage map comprised more than 40 specific STS markers, either single-copy RFLP probes or SSRs that are distributed on the eleven linkage groups. These markers constituted an initial set of standard landmarks of the coffee genome which have been used as anchor points for map comparison (Herrera et al. 2002) and coverage analysis of bacterial artificial chromosome (BAC) libraries (Leroy et al. 2005, Noir et al. 2004). Furthermore, the recombination frequencies in both populations were found to be almost indistinguishable. These results offer evidence in favor of the lack of significant sex differences in recombination of *C. canephora.*

More recently, Crouzillat et al. (2004) reported the development of a Canephora consensus genetic map based on a segregating population of 93 individuals from the cross of two highly heterozygous parents. Backcross genetic maps were established for each parent (i.e., elite clones BP409 and Q121) and then a consensus map was elaborated. More than 453 molecular markers such as RFLP and SSRs were mapped covering a genome of 1,258 cM. Recently, this map was used to map COS (i.e. Conserved Orthologous sequence) markers and perform comparative mapping between coffee and tomato (Wu et al. 2006).

In parallel several diploid interspecific maps were built. Those maps are based on either an F_1 hybrid population resulting from a cross between coffee diploid species (López and Moncada 2006) or progenies obtained by backcrossing of hybrid plants to one of the parental species (Ky et al. 2000; Coulibaly et al. 2003).

9.3 Molecular Cytogenetics and BAC Cloning

9.3.1 *Fluorescence in situ Hybridization*

The basic chromosome number for the genus *Coffea* is considered to be n = 11, which is typical for most genera of the family Rubiaceae. Coffee somatic chromosomes are relatively small (1.5 to 3 μm) and morphologically similar to each other (Krug and Mendes 1940; Bouharmont 1959). Observations at the meiotic pachytene phase provided a significantly better chromosomal characterization and allowed the identification of most bivalents of *C. arabica* (Pinto-Maglio and Cruz 1998). Characterization of the longitudinal differentiation of the mitotic chromosomes of coffee has progressed using other techniques such as fluorescent banding (Lombello and Pinto-Maglio 2004). In addition, the fluorescence in situ hybridization (FISH) method that opened up new perspectives. In particular, genomic in situ hybridization (GISH) was successfully applied for characterization of genomes and chromosomes in polyploid, hybrid plants, and recombinant breeding lines. The genome organization of *C. arabica* was confirmed by GISH using simultaneously labeled total genomic DNA from the two putative genome donor species as probes (Raina et al. 1998; Lashermes et al. 1999). Furthermore, FISH was used to study the presence of alien chromatin in interspecific hybrids and plants derived from interspecific hybrids between coffee species (Barre et al. 1998; Herrera et al. 2007). More recently, the BAC-FISH procedure was used to rapidly localize a given introgression on a specific chromosome (Herrera et al. 2007).

9.3.2 Development of BAC Libraries

BAC libraries have been reported for both cultivated coffee species, *C. arabica* and *C. canephora*.

An Arabica BAC library using the cultivar IAPAR 59 was successfully constructed and validated (Noir et al. 2004). This introgressed variety, derived from the Timor Hybrid, was selected for resistance to leaf rust and root-knot nematodes, and it is widely distributed in Latin America. The library contains 88,813 clones with an average insert size of 130 kb and represents approximately 8 *C. arabica* dihaploid genome equivalents.

A Canephora BAC library was developed on a relatively good cup quality genotype (i.e., clone 126). The library contains 55,296 clones, with an average insert size of 135 kb, representing almost nine haploid genome equivalents. This resource was used to analyze the genome organization (copy number) of sucrose-metabolizing enzymes (mainly sucrose synthase and invertases) in the *C. canephora* genome (Leroy et al. 2005).

9.3.3 BAC Physical Mapping

Although the integration of physical and genetic mapping information would be particularly useful in coffee genome research, only few activities have been reported so far. Regional physical maps based on BAC contigs corresponding to agronomical important disease resistance genes were developed (Lashermes et al. 2004). Recently, a BAC contig linked to the S_H3 leaf rust resistance gene was used to assess microsynteny between coffee (*C. arabica* L.) and *Arabidopsis thaliana* (Mahé et al. 2007). Microsynteny was revealed and the matching counterparts to *C. arabica* contigs were seen to be scattered throughout four different syntenic segments of *Arabidopsis* on chromosomes (*Ath*) I, III, IV, and V.

9.4 EST Resources and Transcriptome Analysis

Large-scale sequencing of cDNAs to produce ESTs followed by comparisons of the resulting sequences with public databases has become the method of choice for the rapid and cost-effective generation of data, becoming the fastest growing segment of public DNA databases. Hence, ESTs, DNA arrays, large-scale gene expression (transcriptome) profiling and associated bioinformatics are becoming routine in the plant sciences. However, only recently have efforts been dedicated to coffee.

9.4.1 Coffee EST Resources

Currently, there are nearly 43 million ESTs (dbEST release 031607, March 2007) in the NCBI public collection (http://www.ncbi.nlm.nih.gov/dbEST/) and the largest set of plant ESTs comes from the model species *Arabidopsis thaliana* and *Oryza sativa.* Nevertheless, a large variety of EST sequences from other plant species including coffee have been deposited in the dbEST database. Recently, a large set of ESTs (46,914) with a special focus on developing seeds of *C. canephora* has also been released (Lin et al. 2005). In addition, two other *C. canephora* EST sequence sets were developed from mRNA isolated from leaves and fruits at different development and maturation stages with 8,778 valid EST sequences (Poncet et al. 2006). All together, a significant set of 55,692 ESTs, mainly from fruits, is already publicly available for *C. canephora* (Table 9.1). Regarding *C. arabica,* public accessibility to EST collections is limited to 1,226 ESTs, comprising a suppression subtractive hybridization library of 618 EST sequences selected upon infection by *H. vastatrix* (Fernandez et al. 2004), 139 EST sequences identified as a differential response of *C. arabica* leaves and roots to chemically induced systemic acquired resistance (De Nardi et al. 2006), and 469 other ESTs from leaves (Cristancho et al. 2006; Joet et al. 2006). In total, there are fewer than 60 thousand entries of coffee ESTs, publicly available, deposited at the dbEST database.

However, other research groups involved in molecular genetics and genomics of *Coffea* sp. have also generated EST data that will become available to the coffee scientific community in the near future (Table 9.1). The Brazilian Coffee EST Project is such an example, which generated single-pass sequences of a total of 214,964 randomly picked clones from 37 cDNA libraries of *C. arabica*, *C. canephora,* and *C. racemosa*, representing specific stages of cells and plant development that after trimming resulted in 130,792, 12,805, and 10,510 good quality sequences for each species, respectively (Vieira et al. 2006). The ESTs assembled into 17,982 clusters and 32,155 singletons. Blast analysis of these sequences revealed that 22% had no significant matches to sequences in the NCBI database. Manually annotated sequencing results have been stored in two online databases (Vieira et al. 2006). Coffee EST resources have also been developed by the Cenicafe research group in Colombia, which have in their database, to date, 32,000 coffee EST sequences from 22 libraries organized in 9,257 *C. arabica* and 1,239 *C. liberica* unigenes (Cristancho et al. 2006). In addition, the Cenicafe database contains 6,000 *Beauveria bassiana* and 4,000 *H. hampei* (coffee berry borer) EST sequences. Aiming at the development of EST-SSR markers for coffee, an Indian research group reported an interim set of 2,092 ESTs of coffee generated at CCMB (Center for Cellular & Molecular

Biology), Hyderabad, India (Aggarwal et al. 2007). EST sequences of two cDNA libraries from leaves and embryonic roots of *C. arabica* were also produced by an Italian group to develop a cDNA microarray based on 1,587 non-redundant sequences (De Nardi et al. 2006).

With these efforts in progress on EST sequencing, the number of coffee ESTs in the public domain will continue to increase as can be seen from data presented in Table 9.1. In fact, the reports compiled here, from different research groups working worldwide, indicate that around 250,000 good quality ESTs from at least four different coffee species, have already been produced. The novelty and complementary nature of these upcoming data can be estimated from a clustering performed with the available coffee EST data at the dbEST database (NCBI) and the ESTs produced in Brazil (Fig. 9.1). The results indicate that only 13% of the unigenes is redundant. These results are expected in view of the different coffee species and tissues sampled to generate the ESTs.

Table 9.1 Available EST collections of *Coffea* species

Coffea species	Tissue/developmental stage	Number of valid ESTs	Reference
C. canephora	Leaves, young	8, 942	Lin et al. 2005
	Pericarp, all developmental stages	8, 956	Lin et al. 2005
	Whole cherries, 18 and 22 week after pollination	9, 843	Lin et al. 2005
	Endosperm and perisperm of seeds, 30 week after pollination	10, 077	Lin et al. 2005
	Endosperm and perisperm of seeds, 42 and 46 week after pollination	9, 096	Lin et al. 2005
	Embryogenic calli	7, 062	Vieira et al. 2006
	Leaves from water-deficit stressed plants	5, 743	Vieira et al. 2006
	Whole cherries of different developmental stages	5, 086	Poncet et al. 2006
	Leaves, young	3, 692	Poncet et al. 2006
C. arabica	Plantlets and leaves treated with araquidonic acid	4, 098	Vieira et al. 2006
	Suspension cells treated with acibenzolar-S-methyl	5, 605	Vieira et al. 2006
	Suspension cells treated with acibenzolar-S-methyl and brassinoesteroids	8, 252	Vieira et al. 2006
	Hypocotyls treated with acibenzolar-S-methyl	9, 882	Vieira et al. 2006
	Suspension cells treated with NaCl	7, 656	Vieira et al. 2006
	Embryogenic calli	7, 599	Vieira et al. 2006
	Zygotic embryo (immature fruits)	106	Vieira et al. 2006
	Germinating seeds (whole seeds and zygotic embryos)	8, 001	Vieira et al. 2006
	Flower buds in different developmental stages	15, 833	Vieira et al. 2006
	Flower buds + pinhead fruits + fruits at different stages	9, 451	Vieira et al. 2006
	Non embryogenic calli with and without 2,4 D	8, 558	Vieira et al. 2006
	Young leaves from orthotropic branch	9, 939	Vieira et al. 2006
	Mature leaves from plagiotropic branches	10, 319	Vieira et al. 2006
	Roots infected with nematodes	302	Vieira et al. 2006
	Primary embryogenic calli	2, 042	Vieira et al. 2006

Table 9.1 (continued)

Coffea species	Tissue/developmental stage	Number of valid ESTs	Reference
	Leaves infected with leaf miner and coffee leaf rust	3, 072	Vieira et al. 2006
	Roots	149	Vieira et al. 2006
	Roots with acibenzolar-S-methyl	1, 051	Vieira et al. 2006
	Suspension cells stressed with aluminum	4, 981	Vieira et al. 2006
	Stems infected with *Xylella spp.*	8, 045	Vieira et al. 2006
	Water-deficit stressed field plants (pool of tissues)	5, 743	Vieira et al. 2006
	Well-watered field plants (pool of tissues)	798	Vieira et al. 2006
	SSH - young coffee leaves	618	Fernandez et al. 2004
	Young coffee leaves	448	Cristiancho et al. 2006
	SSH-young coffee leaves	16	Joet et al. 2006
	Embryonic roots and leaves	2, 016	De Nardi et al. 2006
	17 cDNA libraries (not described)	30, 000	Cristiancho et al. 2006
C. racemosa	Fruits (*Coffea racemosa*)	5, 041	Vieira et al. 2006
	Fruits, stages 1,2 and 3 (*Coffea racemosa*)	5, 469	Vieira et al. 2006
C. liberica	4 cDNA libraries (not described)	2, 000	Cristiancho et al. 2006
Coffea sp.	cDNA libraries not described	2, 092	Aggarwal et al. 2007
	Total *Coffea sp.* ESTs	246,541	

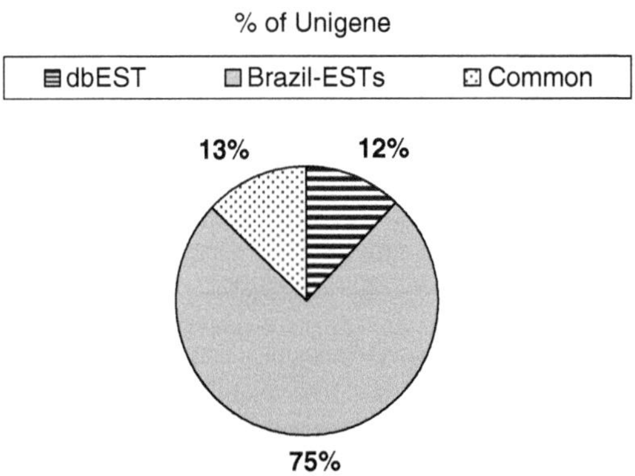

Fig. 9.1 Origin distribution of unigenes from ESTs present on two databases (Brazil-ESTs - 75% of total unigenes are only represented on this database; dbEST – 13 % of total unigenes are only represented on this database; Common- 12 % of total unigenes are represented on both databases)

9.4.2 Coffee EST Exploitation

As sequence and annotation data continue to accumulate, computer-based tools for systematic collection, organization, storage, access, analysis, and visualization of these data will become increasingly valuable to the plant science community. Advances in computational molecular biology and biostatistics will make it possible to analyze large EST datasets more precisely and efficiently producing more reliable digital expression profiling data, taking into account the methodological limitations of constructing cDNA libraries (Alba et al. 2004).

Table 9.2 presents a list of several databases containing coffee data with a diversity of bioinformatics resources. The Solanaceae Genomics Network (SGN; http://sgn.cornell.edu/), serves well as an example of these on-line resources. SGN is dedicated to the biology of Solanaceae species, including tomato, potato, tobacco, eggplant, pepper, petunia, and coffee. This database contains sequence data derived from nearly 300,000 ESTs coming from different plant species, extensive mapping data for the tomato genome, in addition to mapping data for the genomes of potato and eggplant, which can facilitate cross-species homology relationships (via comparative genome analysis) among solanaceus and related species (Wu et al. 2006).

The inventory of the available resources on coffee ESTs provided in this section indicates that considerable information has been generated the last few years. These data are valuable tools for discovering genes (a large number of unknown genes already identified), developing EST-SSR markers, providing the basis for mapping and the establishment of breeding programs based on marker-assisted selection (MAS), and providing the foundation for the development of novel tools such as microarrays. Initiatives for the generation of these arrays have already begun and some results have started to appear (De Nardi et al. 2006). Microarrays and real-time reverse transcription polymerase chain reaction (qRT-PCR) will allow comprehensive transcriptome analysis making it possible to identify and dissect complex genetic networks that underlie important biological processes critical to physiology,

Table 9.2 Bioinformatics resources and databases containing coffee ESTs

Databases	URL	Valid ESTs	Reference
SGN-USA	http://sgn.cornell.edu/	46, 914	Lin et al. 2005
IRD-France	http://www.mpl.ird.fr/bioinfo/	9, 412	Fernandez et al. 2004 Poncet et al. 2006 Joet et al. 2006
CoffeeDNA-Trieste	http://www.coffeedna.net/	5, 391	De Nardi et al. 2006
CCMB-India	http://www.ccmb.res.in/	2, 092	Aggarwal et al. 2007
CBP&D-Café Brazil	http://www.lge.ibi.unicamp.br/cafe/ http://www.cenargen.embrapa.br/biotec/genomacafe/	154, 107	Vieira et al. 2006
Cenicafe-Colombia	http://bioinformatics.cenicafe.org/	32, 000	Cristancho et al. 2006
NCBI	http://www.ncbi.nlm.nih.gov/dbEST/	56, 918	Fernandez et al. 2004 Lin et al. 2005 Poncet et al. 2006 De Nardi et al. 2006

development, quality, and responses to abiotic and biotic stresses in coffee. Recently, the early molecular resistance responses of coffee (*C. arabica* L.) to the rust fungus (*H. vastatrix*) have been monitored using real-time quantitative qRT-PCR (Ganesh et al. 2006). Similarly, several studies of particular biosynthetic pathways in relation with the ripening and development of coffee fruit were reported (Geromel et al. 2006; Hinniger et al. 2006; Simkin et al. 2006; Bustamante et al. 2007; Lepelley et al. 2007). Further characterization of gene networks in coffee plants will help us to identify new targets for manipulation of physiological, biochemical, and developmental processes of this very important crop species.

Moreover, modern biology is facing a new momentum. Recent advances in high-throughput methodologies and equipment allowed researchers from different disciplines to attempt to combine large-scale DNA sequence, gene expression, protein, metabolite, genotype, and/or phenotype data to develop resourceful and integrated databases for a better understanding of biological processes (Alba et al. 2005). These combined tools through integrated efforts of genomics research and breeding will certainly be essential to quickly overcome practical problems faced by the coffee agribusiness such as control of pre- and post-harvest physiological factors involved in quality, disease, and pest control and management of plant responses to environmental changes (e.g., limited water availability and adverse temperatures). Finally, integrated coffee genomic research may result in increased beverage consumption through new value-added products derived from coffee (e.g., nutraceuticals, oils, and flavors), ensuring the sustainability of the coffee production chain.

9.5 Genetic Transformation

Coffee genetic engineering emerged during the last decade as a tool to elucidate the function, regulation, and interaction of agronomically interesting genes through functional genomics approach. Genetic transformation became available only after the establishment of protocols for in vitro regeneration through somatic embryogenesis for the two principal commercial species (Berthouly and Etienne 2000; Etienne 2005). Nowadays, genetic transformation of coffee plants has been successfully achieved by several research groups. However, despite significant advances over the last 15 years, coffee transformation is still very laborious, with bottlenecks in the methodology that make it far from a routine laboratory technique. Up to now, only a few genes have been transferred into coffee genotypes. However, the recent development of coffee genomics has led to the identification of numerous genes involved in agronomically important biological processes and which have potential for transformation.

9.5.1 Direct Gene Transfer

Barton et al. (1991) reported a transformation method of coffee embryos by electroporation using the *nptII* (i.e. kanamycin resistance) gene. This method remained little noticed until the recent works of Fernandez-Da Silva and Menéndez-Yuffá (2003) that described improved conditions to regenerate transformed *C. arabica* somatic embryos expressing the GUS and *bar* genes. The biolistic delivery method has been improved considerably since the first report of GUS transient expression in coffee using a gunpowder driven device by van Boxtel et al. (1995). Using a helium gun device, Rosillo et al. (2003) determined that a short endosperm pre-treatment with two osmotic preconditioning agents (i.e., mannitol and sorbitol) increased the number of cells expressing GUS. Ribas et al. (2005a) described a protocol for transformation of *C. canephora* embryogenic callus and somatic embryos using a helium gun, associated with sub-culturing onto medium containing mannitol before and after bombardment. Their protocol allowed 12.5% of transformed callus expressing GUS-positive reaction to histochemical assay.

9.5.2 Indirect Gene Transfer

Hatanaka et al. (1999) achieved the first successful *A. tumefaciens*-mediated transformation of *C. canephora* plants exhibiting strong GUS stable expression. Leroy et al. (2000) also reported efficient regeneration of *A. tumefaciens*-transformed coffee plants of both *Coffea* sp. expressing the *uidA* and *cry1Ac* genes. The use of embryogenic callus is often preferred to somatic embryos and becomes the most common way for coffee transformation (Fig. 9.2). From this callus, only 30% (*C. canephora*) and 10% (*C. arabica*) developed somatic embryos, and from these, only 50% regenerated into plantlets. Ribas et al. (2006) transformed *C. canephora* explants submitted to sonification during immersion in a suspension of an *A. tumefaciens* strain encoding *uidA* and *bar* genes and regenerated transformed plants. Canche-Moo et al. (2004) transformed leaf explants through

Fig. 9.2 Expression of the 35S-GFP reporter gene on *C. arabica* transformed embryogenic callus three weeks after *Agrobacterium tumefaciens*-mediated transformation

A. tumefaciens-mediated transformation involving a vacuum infiltration protocol in a bacterial suspension. *A. tumefaciens*-mediated transformation has also served to induce stable gene silencing through RNA interference (RNAi) technology of genes encoding theobromine synthase (Ogita et al. 2004) and N-methyl transferase, respectively (Kumar et al. 2004), both genes involved in caffeine biosynthesis, in both cultivated species. Ogita et al. (2003) showed that leaves of 1-year-old transformed trees exhibited reduced theobromine and caffeine content (30 to 50% compared with the control). Ribas et al. (2005b) achieved inhibition of ethylene burst in *C. arabica* by introducing the ACC-oxidase gene in antisense orientation.

Agrobacterium rhizogenes-mediated transformation in both *C. canephora* and *C. arabica* species was first reported by Spiral et al. (1993) and Sugiyama et al. (1995), respectively. However in these studies the regeneration protocol was laborious and plants frequently showed a "hairy" phenotype with short internodes and stunted growth. Such abnormal phenotype is stable as demonstrated by Perthuis et al. (2005), who showed that four out of the nine independently transformed *C. canephora* clones obtained with *A. rhizogenes* were still displaying this phenotype in field conditions; furthermore all these plants died within few months after planting. Kumar et al. (2006) described an adapted method for *A. rhizogenes* sonification-assisted embryos transformation. The percentage of plantlets with aberrant phenotypes was significantly low compared with the results previously described by Sugiyama et al. (1995). Alpizar et al. (2006a) developed an *A. rhizogenes*-mediated transformation protocol (Fig. 9.3) that enables efficient and rapid regeneration of transformed roots from hypocotyls of zygotic embryos and subsequent production of composite plants (i.e., transformed roots induced on non-transformed shoots). This methodology was specifically developed for functional analysis of genes involved

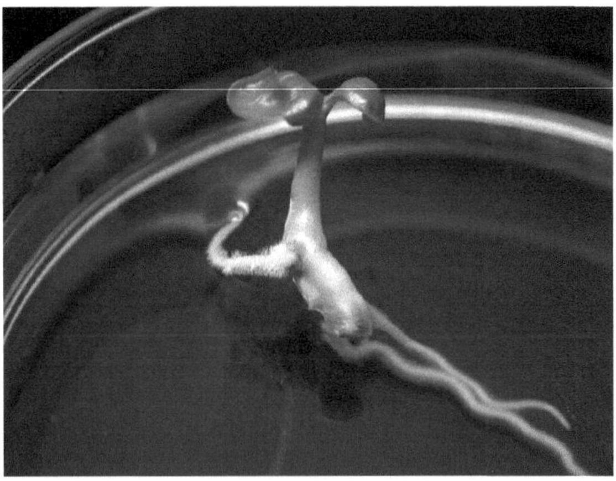

Fig. 9.3 Regeneration of transformed roots on the hypocotyl of zygotic embryo following *Agrobacterium rhizogenes*-mediated transformation protocol. Emergence of a transformed root at the wound site 4-8 weeks after the end of the co-cultivation with A4 RS strain. The transformed zygotic embryo is subcultured every 4 weeks on MS germination medium containing decreasing cefotaxime concentrations (Alpizar et al. 2006a)

Fig. 9.4 Inoculation of nematodes on transformed roots of *Coffea arabica* cv. Caturra composite plants to evaluate the resistance:susceptibility to the root-knot nematode *Meloidogyne exigua*. The Caturra cv. is susceptible to *M. exigua*. Composite plants were generated after 12 weeks by inducing transformed roots on non-transformed shoots. One month after acclimatization to *ex vitro* conditions, they were ready to be inoculated with *M. exigua* nematode individuals. Gall symptoms caused by *M. exigua* on transgenic roots will be visible three months after nematode inoculation (Alpizar et al. 2006a)

in resistance to root specific pathogens like nematodes and/or in plant morphology and development. The authors demonstrated that the *Meloidogyne exigua* root-knot nematode could normally proliferate in transgenic roots and that consequently this transformation methodology could efficiently be applied for the functional analysis of the *Mex-1* resistance gene (Fig. 9.4).

9.5.3 Selection of Transformed Tissue

The first studies in coffee transformation were done using antibiotics and particularly kanamycin. However, kanamycin has exhibited contradictory results as a potential selection agent for transformed coffee embryos. Barton et al. (1991) and Spiral et al. (1993) concluded that kanamycin possesses poor selective capacity since non-transformed somatic embryos could develop even at high concentration (400 mg/L). Nevertheless, kanamycin was recently used successfully at 400 mg/L by Cunha et al. (2004) and at 100 mg/L by Canche-Moo et al. (2006). Hatanaka et al. (1999), Naveen et al. (2002), and Ogita et al. (2004) observed that hygromycin at 50-100 mg/L allowed an acceptable regeneration of *A. tumefaciens*-mediated transformed somatic embryos. Because of the low selection efficiency of antibiotics in coffee, along with biosafety issues, other types of selection markers including herbicide selection or positive selection were used.

Van Boxtel et al. (1997) and Fernandez-Da Silva and Menéndez-Yuffá (2004) showed that low concentration (6 mg/L) of the herbicide ammonium glufosinate was sufficient to inhibit non-transformed callus growth. Subsequently, research by Ribas et al. (2005a, 2006) confirmed the reliability of the *bar* gene as a selection marker in both *C. canephora* and *C. arabica.* Leroy et al. (2000) successfully used selectable marker gene *csr1-1* from *A. thaliana* that allowed selection with the herbicide chlorsulfuron at a concentration of 80 μg/L in the two cultivated species. Cruz et al. (2004) also succeeded in regenerating transformed plantlets containing the *ppt* gene, which confers resistance to phosphinothricine, on selective medium containing 10 μM of the herbicide. Samson et al. (2004) demonstrated the potential use of xylose isomerase (*XylA*) gene as a positive selection marker in coffee transformation.

Visual markers have been used to replace those based on herbicide selection. Ogita et al. (2004) and Canche-Moo et al. (2006) used *gfp*, the gene encoding for green fluorescent protein (Chalfie et al. 1994), and *DsRFP*, the gene encoding for red fluorescent protein (Gallie et al. 1989) as reporter genes for visual selection of somatic embryos of *C. canephora* following *A. tumefaciens*-mediated transformation. The first complete selection of coffee transformed tissues without using any marker gene was achieved by Alpizar et al. (2006a, 2006b), after selection of putative *A. rhizogenes* transformed roots using *uidA* and *gfp* genes through histochemical GUS assay or GFP epifluorescence.

9.5.4 Testing Transgenic Coffee Plants

Ribas et al. (2006) regenerated coffee somatic embryos transformed with the *bar* gene on selective medium containing ammonium glufosinate. Regenerated plants supported up to eight times the herbicide doses recommended for field applications. Leroy et al. (2000) described *C. canephora* plants transformed with *cry1Ac* gene from *Bacillus thuringiensis* expressing resistance against the coffee leaf miner (*Perileucoptera coffeella*) in green-house conditions. Perthuis et al. (2005) reported that this resistance was stable and effective after six releases of a natural population of *P. coffeella* during four years of field assessments. In addition, production of transformed coffee plants with resistance to coffee berry borer (*Hypothenemus hampei*) was attempted by Cruz et al. (2004) using the *α-Al1*gene from common bean encoding for an inhibitor of alfa-amylase (Powers and Culbertson, 1983). The authors achieved transformation with this gene and bioassays with the insect are under way to confirm functional activity in coffee. Following recent achievements in other crops, gene pyramiding can be envisaged in coffee to introduce a larger number of genes for resistance to diverse races of a particular pathogen or a combination of different pathogens.

Ribas et al. (2005b) achieved inhibition of the ethylene burst by introducing the *ACC oxidase* gene in antisense orientation; this technique would permit the understanding of genes involved in fruit maturation and ethylene production.

9.5.5 *Identification of Coffee Promoters*

Availability of tissue-specific and inducible promoters would be helpful for successful development of coffee functional analysis and transgenically improved coffee. With few exceptions, the 35S promoter derived from the cauliflower mosaic virus, has been the most common component of transgenic constructs used. However, Rosillo et al. (2003) comparing pCaMV35S with two coffee promoters (α-tubulin and α-arabicin) showed that they all resulted in similar transient expression of the *uidA* gene. Marraccini et al. (1999) cloned a complete coffee 11S seed storage protein gene and carried out promoter analysis in transgenic tobacco plants. Acuña et al. (1999) also reported the cloning of a bean (endosperm)-specific promoter partially corresponding to the sequence of Marraccini et al. (1999). However, no functional analysis (test in a heterologous plant) of their sequence was performed. In another work, Marraccini et al. (2003) reported the cloning of the rubisco small subunit (*CaRBCS1*) promoter of *C. arabica* that was tested in transgenic tobacco and shown to be leaf-specific. Satyanarayana et al. (2005) reported the cloning of the promoter of N-methyl transferase (NMT) gene involved in caffeine biosynthesis pathway. The promoter region was tested in tobacco, exhibiting expression of the *GUS* reporter gene in leaves. The authors mentioned that current efforts are focused on the use of this promoter sequence for down-regulation of NMT gene through transcriptional gene silencing. This cloning of the first promoter of a gene involved in caffeine biosynthesis together with the near identification of genes involved in sucrose and drought tolerance metabolism opens up the possibility for coffee transformation to validate and study the molecular mechanisms that regulate the production of these important targets for *Coffea* sp. cultivation.

Even without functional analysis in transgenic plants, other coffee promoters were also cloned and described in several publications, some of them recently and directly deduced from the use of EST sequencing: promoter of a protein homologous to the yeast translation initiation factor SUI1 (Gaborit et al. 2003), promoter oleosin *CcOLE-1* (Simkin et al. 2006), promoter dehydrin *CcDH2a* (Hinninger et al. 2006). A number of highly expressed genes that showed high specificity for different stages of seed development, as well as for the pericarp tissue that surrounds the seed, have been identified (Lin et al. 2005). These genes could lead to promoters which can potentially be used to drive gene expression in specific stages/tissues of the coffee plant. Many of these genes are important for insect/pathogen resistance and determining the quality of the coffee bean. It is proposed to create a collection of promoters in the International Coffee Genomics Network (ICGN) as a resource for functional analysis of coffee genes.

9.6 Perspectives

In the last decade, overproduction has resulted in historically low coffee prices that have had devastating effects on coffee producers. Farmers have survived with great difficulties or have abandoned coffee culture and switched to alternative crops.

Consequently, bean quality may be lowered and supply become less stable to adversely affect the coffee market of consumer countries. Furthermore, in the light of climate changes and increasing awareness of the negative impacts of the non-sustainable use of natural resources, coffee production will have to evolve. This is particularly relevant for perennial plants such as coffee whose productive life is very long and for which rapid genetic gains are tedious to obtain.

In the twentieth century, coffee production benefited from the selection of spontaneous Arabica mutants (e.g., the compact cultivar Caturra), natural hybrids between species (e.g., the Timor hybrid), highly productive Canephora clones, and new cultural practices (e.g., open sun monoculture). Coffee germplasm was collected, but little has been done at the molecular level to exploit biodiversity in coffee species as well as their genomic resources.

The recent development of high-capacity methods for analyzing the structure and function of genes, which is collectively termed "genomics," represents a new paradigm with broad implications. The advent of large-scale molecular genomics for plant species such as *C. arabica and C. canephora* will provide access to previously inaccessible sources of genetic variation which could be exploited in breeding programs. Anticipated outcomes in current and future coffee breeding include (1) rapid characterization and managing of germplasm resources, (2) enhanced understanding of the genetic control of priority traits, (3) identification of candidate genes or tightly linked genomic regions underlying important traits, and (4) identification of accessions in genetic collections with variants of genomic regions or alleles of candidate genes having a favorable impact on priority traits. In this way, the recent efforts to set up an international commitment (ICGN, http://www.coffeegenome.org) to work jointly for the development of common sets of genomic tools, plant populations, and concepts would be extremely useful.

References

Acuña R, Bassüner R, Beillinson V, Cortina H, Cadena-Gómez G, et al. (1999) Coffee seeds contain 11S storage proteins. Physiol Plant 105:122–131

Aga E, Bryngelsson T (2006) Inverse sequence-tagged repeat (ISTR) analysis of genetic variability in forest coffee (*Coffea arabica* L.) from Ethiopia. Genetic Res Crop Evol 53:721–728

Aga E, Bryngelsson T, Bekele E, Salomon B (2003) Genetic diversity of forest arabica coffee (*Coffea arabica* L.) in Ethiopia revealed by random amplified polymorphic DNA (RAPD) analysis. Hereditas 138:36–46

Aggarwal RK, Hendre PS, Varshney RK, Bhat PR, Krishnakumar V, et al. (2007) Identification, characterization and utilization of EST-derived genetic microsatellite markers for genome analyses of coffee and related species. Theor Appl Genet 114:359–372

Alba R, Fei Z, Payton P, Liu Y, Moore SL, et al. (2004) ESTs, cDNA microarrays, and gene expression profiling: tools for dissecting plant physiology and development. Plant J 39:697–714

Alba R, Payton P, Fei Z, McQuinn R, Debbie P, et al. (2005) Transcriptome and selected metabolite analyses reveal multiple points of ethylene control during tomato fruit development. Plant Cell 17:2954–2965

Alpizar E, Dechamp E, Espeout S, Royer M, Lecouls A-C, et al. (2006a) Efficient production of *Agrobacterium rhizogenes*-transformed roots and composite plants for studying gene expression in coffee roots. Plant Cell Rep 25:959–967

Alpizar E, Dechamp E, Bertrand B, Lashermes P, Etienne H (2006b) Transgenic roots for functional genomics of coffee resistance genes to root-knot nematodes. Proc Intl Sci Colloquium on Coffee 21:653–659, ASIC (http://www.asic-cafe.org)

Anthony F, Berthaud J, Guillaumet JL, Lourd M (1987) Collecting wild *Coffea* species in Kenya and Tanzania. Plant Genet Res Newsl 69:23–29

Anthony F, Bertrand B, Quiros O, Wilches A, Lashermes P, et al. (2001) Genetic diversity of wild coffee (*Coffea arabica* L.) using molecular markers. Euphytica 118:53–65

Anthony F, Combes MC, Astorga C, Bertrand B, Graziosi G, et al. (2002) The origin of cultivated *Coffea arabica* L. varieties revealed by AFLP and SSR markers. Theor Appl Genet 104:894–900

Barre P, Layssac M, D'Hont A, Louarn J, Charrier A, et al. (1998) Relationship between parental chromosomic contribution and nuclear DNA content in the coffee interspecific hybrid: *C. pseudozanguebariae* x *C. liberica* var. *dewevrei*. Theor Appl Genet 96:301–305

Barton CR, Adams TL, Zarowitz M (1991) Stable transformation of foreign DNA into *Coffea arabica* plants. Proc Intl Conf Coffee Sci 14:460–464

Baruah A, Naik V, Hendre PS, Rajkumar R, Rajendrakumar P, et al. (2003) Isolation and characterization of nine microsatellite markers from *Coffea arabica* L., showing widecross-species amplifications. Mol Ecol Notes 3:647–650

Bennett MD, Leitch IJ (1995) Nuclear DNA amounts in angiosperms. Ann Bot 76:113–176

Berthouly M, Etienne H (2000) Somatic embryogenesis of coffee. In: Coffee biotechnology and quality. Proc Intl Seminar Biotechnology Coffee Agro-industry 13:71–90

Bhat PR, Krishnakumar V, Hendre PS, Rajendrakumar P,Varshney RK, et al. (2005) Identification and characterization of gene-derived EST–SSR markers from robusta coffee variety 'CxR' (an interspecific hybrid of *Coffea canephora* and *Coffea congensis*). Mol Ecol Notes 5:80–83

Bouharmont J (1959) Recherches sur les affinités chromosomiques dans le genre *Coffea*. Publ. INEAC, Série Scientifique 77:94 p

Bremer B, Jansen RK (1991) Comparative restriction site mapping of chloroplast DNA implies new phylogenetic relationships within *Rubiaceae*. Am J Bot 78:198–213

Bridson D (1982) Studies in *Coffea* and *Psilanthus* (*Rubiaceae* subfam. *Cinchonoideae*) for Part 2 of 'Flora of Tropical East Africa': *Rubiaceae*. Kew Bull 36:817–859

Bridson D, Verdcourt B (1988) *Coffea*. In: Polhill RM (ed) Flora of Tropical East Africa. Rubiaceae (Part 2). A.A. Balkema, Rotterdam, pp 703–727

Bustamante J, Campa C, Poncet V, Noirot M, Leroy T, et al. (2007) Molecular characterization of an ethylene receptor gene (*CcETR1*) in coffee trees, its relationship with fruit development and caffein content. Mol Genet Genomics (in press), DOI: 10.1007/s00438-007-0219-z

Canche-Moo RLR, Ku-Gonzales A, Burgeff C, Loyola-Vargas VM, Rodríguez-Zapata LC, et al. (2006) Genetic transformation of *Coffea canephora* by vacuum infiltration. Plant Cell Tiss Org Cult 84:373–387

Carvalho A (1988) Principles and practice of coffee plant breeding for productivity and quality factors: *Coffea arabica*. In: Clarke RJ, Macrae R (eds) Coffee, volume 4: Agronomy. Elsevier Applied Science, London, pp 129–165

Chalfie M, Tu Y, Euskirchen G, Ward WW, Prasher DC (1994). Green florescent protein as a marker for gene expression. Science 263:663–664

Chevalier A, Dagron M (1928) Recherches historiques sur les débuts de la culture du caféier en Amérique. Communications et Actes de l'Académie des Sciences Coloniales (Paris) 5:1–38

Combes MC, Andrzejewski S, Anthony F, Bertrand B, Rovelli P, et al. (2000) Characterisation of microsatellite loci in *Coffea arabica* and related coffee species. Mol Ecol 9:1178–1180

Coulibaly I, Revol B, Noirot M, Poncet V, Lorieux M, et al. (2003) AFLP and SSR polymorphism in a Coffea interspecific backcross progeny [(*C. heterocalyx* x *C. canephora*) x *C. canephora*]. Theor Appl Genet 107:1148–1155

Couturon E (1982) Obtention d'haploïde spontanés de *Coffea canephora* Pierre par l'utilisation du greffage d'embryons. Café Cacao Thé 26(3):155–160

Cristancho MA, Rivera C, Orozco C, Chalarca A, Mueller L (2006) Development of a bioinformatics platform at the Colombia National Coffee Research Center Proc Intl Sci Colloquium Coffee 21: 638–643, ASIC (http://www.asic-cafe.org)

Cros J, Combes MC, Chabrillange N, Duperray C, Monnot des Angles A, et al. (1995) Nuclear DNA content in the subgenus *Coffea* (Rubiaceae): inter- and intra-specific variation in African species. Can J Bot 73:14–20

Cros J, Combes MC, Trouslot P, Anthony F, Hamon S, et al. (1998) Phylogenetic relationships of *Coffea* species: new evidence based on the chloroplast DNA variation analysis. Mol Phylo Evol 9:109–117

Crouzillat D, Rigoreau M, Bellanger L, Priyono S, Mawardi S, Syahrudi, McCarthy J, Tanksley S, Zaenudin I, Petiard V (2004) A Robusta consensus map using RFLP and microsatellites markers for the detection of QTL. Proc Intl Sci Colloquium Coffee 20: 546–553, ASIC (http://www.asic-cafe.org)

Cruz ARR, Paixao ALD, Machado FR, Barbosa MFF, Junqueira CS, et al. (2004) Metodologia para obtenção de plantas transformadas de *Coffea canephora* por co-cultivo e calos embriogênicos com *A. tumefaciens*. Boletim de Pesquisa e Desenvolvimento. Embrapa, Brasilia, Brasil, vol.58:15 p

Cunha WG, Machado FRB, Vianna GR, Texteira JB, Barros EVSA (2004) Obtenção de *Coffea arabica* geneticamente modificado por bombardeio de calos embriogênicos. Boletim de Pesquisa e Desenvolvimento. Embrapa, Brasilia, Brasil, vol. 73:15 p

Davis AP, Maurin O, Chester M, Mvungu EF, Fay MF (2006) Phylogenetic relationship in *Coffea* (Rubiaceae) inferred from sequence data and morphology. Proc Intl Sci Colloquium Coffee 21: 868-875, ASIC (http://www.asic-cafe.org)

De Nardi B, Dreos R, Del Terra L, Martellossi C, Asquini E, et al. (2006) Differential responses of *Coffea arabica* L. leaves and roots to chemically induced systemic acquired resistance. Genome 49:1594–1605

Dussert S, Lashermes P, Anthony F, Montagnon C, Trouslot P, et al. (2003) Coffee (*Coffea canephora*). In: Hamon P, Seguin M, Perrier X, Glaszmann JC (eds) Genetic diversity of cultivated tropical plants. Science Publishers Inc., Plymouth, pp 239–258

Eira MTS, Amaral da Silva EA, de Castro RD, Dussert S, Walters C, et al. (2006) Coffee seed physiology. Brazilian J Plant Physiol 18:149–163

Etienne H (2005) Protocol of somatic embryogenesis: Coffee (*Coffea arabica* L. and *C. canephora* P.). In: Protocols for somatic embryogenesis in woody plants. Series: Forestry Sci Vol. 77, Jain SM, Gupta PK (Eds). Springer, the Netherlands. ISBN: 1-4020-2984-5, pp. 167–179

Fernandez D, Santos P, Agostini C, Bon MC, Petitot AS, et al. (2004) Coffee (*Coffea arabica* L.) genes early expressed during infection by the rust fungus (*Hemileia vastatrix*). Mol Plant Pathol 5:527–536

Fernandez-Da Silva R, Menéndez-Yuffá A (2003) Transient gene expression in secondary somatic embryos from coffee tissues electroporated with genes *gus* and *bar*. Electronic J Biotech 6:29–35

Fernandez-Da Silva R, Menéndez-Yuffá A (2004) Efecto del herbicida glufosinate de amonio en diferentes explantes de *Coffea arabica* cv. Catimor. Acta Científica Venezolana 55:211–217

Gaborit C, Caillet V, Deshayes A, Marraccini P (2003) Molecular cloning of a full-length cDNA and gene from *Coffea arabica* encoding a protein homologous to the yeast translation initiation factor SUI1: expression analysis in plant organs. Braz J Plant Physiol 15: 55–58

Gallie DR, Lucas WJ, Walbot V (1989). Vizualing mRNA expression in plant protoplasts: factors influencing efficient mRNA uptake and translation. Plant Cell 1: 303–311

Ganesh D, Petitot A-S, Silva M, Alary R, Lecouls AC, et al. (2006). Monitoring of the early molecular resistance responses of coffee (*Coffea arabica* L.) to the rust fungus (*Hemileia vastatrix*) using real-time quantitative RT-PCR. Plant Sci 170:1045–1051

Geromel C, Ferreira LP, Guerreiro SMC, Cavalari AA, Pot D, et al. (2006) Biochemical and genomic analysis of sucrose metabolism during coffee (*Coffea arabica*) fruit development. J Exp Bot 57:3243–3258

Guerrero Filho O, Silvarolla MB, Eskes AB (1999) Expression and mode of inheritance of resistance in coffee to leaf miner *Perileucoptera coffeella*. Euphytica 105:7–15

Haarer AE (1956) Modern coffee production. Leonard Hill (books) Limited, London

Hatanaka T, Choi YE, Kusano T, Sano H (1999) Transgenic plants of *Coffea canephora* from embryogenic callus via *Agrobacterium tumefaciens*-mediated transformation. Plant Cell Rep 19:106–110

Herrera JC, D'Hont A, Lashermes P (2007) Use of fluorescence in situ hybridization as a tool for introgression analysis and chromosome identification in coffee (*Coffea arabica* L.). Genome 50:619–626

Herrera JC, Combes MC, Anthony F, Charrier A, Lashermes P (2002) Introgression into the allotetraploid coffee (*Coffea arabica* L.): segregation and recombination of the *C. canephora* genome in the tetraploid interspecific hybrid (*C. arabica* x *C. canephora*). Theor Appl Genet 104:661–668

Hinniger C, Caillet V, Michoux F, Ben Amor M, Tanksley S, et al. (2006) Isolation and characterization of cDNA encoding three dehydrins expressed during *Coffea canephora* (Robusta) grain development. Ann Bot 97:755–765

Joet T, Salmona J, Suzanne W, Descroix F, Dussert S, et al. (2006) Targeted transcriptome profiling during seed development in *C. arabica* cv. Laurina. Proc Intl Sci Colloquium Coffee 21: 687–694, ASIC (http://www.asic-cafe.org)

Krug CA, Mendes AJT (1940) Cytological observations in *Coffea* - IV. J Genet 39:189–203

Krug CA, Mendes JET, Carvalho A (1939) Taxonomia de *Coffea arabica L.* Campinas: Instituto Agronômico do Estado, Bolétim Técnico no 62

Kumar V, Sathyabarayana KV, Indu EP, Sarala Itty S, Giridhar P, et al. (2004) Post transcriptional gene silencing for down regulating caffeine biosynthesis in *Coffea canephora* P. Proc Intl Sci Colloquium Coffee 20:769–774

Kumar V, Satyanarayana KV, Sarala Itty S, Indu EP, Giridhar P, et al. (2006) Stable transformation and direct regeneration in *Coffea canephora* P ex. Fr. *Agrobacterium rhizogenes* mediated transformation without hairy-root phenotype. Plant Cell Rep 25:214–222

Ky CL, Barre P, Lorieux M, Trouslot P, Akaffou S, et al. (2000) Interspecific genetic linkage map, segregation distortion and genetic conversion in coffee Coffea sp. Theor Appl Genet 101:669–676

Lashermes P, Trouslot P, Anthony F, Combes MC, Charrier A (1996a) Genetic diversity for RAPD markers between cultivated and wild accessions of *Coffea arabica*. Euphytica 87:59–64

Lashermes P, Cros J, Combes MC, Trouslot P, Anthony F, et al. (1996b) Inheritance and restriction fragment length polymorphism of chloroplast DNA in the genus *Coffea* L. Theor Appl Genet 93:626–632

Lashermes P, Combes MC, Trouslot P, Charrier A (1997) Phylogenetic relationships of coffee tree species (*Coffea* L.) as inferred from ITS sequences of nuclear ribosomal DNA. Theor Appl Genet 94:947–955

Lashermes P, Combes MC, Robert J, Trouslot P, D'Hont A, et al. (1999) Molecular characterisation and origin of the *Coffea arabica* L. genome. Mol Gen Genet 261:259–266

Lashermes P, Paczek V, Trouslot P, Combes MC, Couturon E, et al. (2000a). Single-locus inheritance in the allotetraploid *Coffea arabica* L. and interspecific hybrid *C. arabica* x *C. canephora*. J Heredity 91:81–85

Lashermes P, Andrzejewski S, Bertrand B, Combes MC, Dussert S, et al. (2000b) Molecular analysis of introgressive breeding in coffee (*Coffea arabica* L.). Theor Appl Genet 100:139–146

Lashermes P, Combes MC, Prakash NS, Trouslot P, Lorieux M, et al. (2001) Genetic linkage map of *Coffea canephora*: effect of segregation distortion and analysis of recombination rate in male and female meioses. Genome 44:589–595

Lashermes P, Noir S, Combes MC, Ansaldi C, Bertrand B, et al. (2004) Toward an Integrated Physical Map of the Coffee Genome. Proc 20th Intl Sci Colloquium Coffee 20: 554–559, ASIC (http://www.asic-cafe.org)

Lepelley M, Cheminade G, Tremillon N, Simkin A, Caillet V, et al. (2007) Chlorogenic acid synthesis in coffee: An analysis of CGA content and real-time RT-PCR expression of HCT, HQT,

C3H1, and CCoAOMT1 genes during grain development in *C. canephora*. Plant Sci (in press), DOI:10.1016/j.plantsci.2007.02.004

Leroy T, Henry A-M, Royer M, Altosaar I, Frutos R, et al. (2000) Genetically modified coffee plants expressing the *Bacillus thuringiensis cry1Ac* gene for resistance to leaf miner. Plant Cell Rep 19:382–389

Leroy T, Marraccini P, Dufour M, Montagnon C, Lashermes P, et al. (2005) Construction and characterization of a *Coffea canephora* BAC library to study the organization of sucrose biosynthesis genes. Theor Appl Genet 111:1032–1041

Lin C, Mueller LA, McCarthy J, Crouzillat D, Pétiard V, et al. (2005). Coffee and tomato share common gene repertoires as revealed by deep sequencing of seed and cherry transcripts. Theor Appl Genet 112:114–130

Lombello RA, Pinto-Maglio CAF (2004) Cytogenetic studies in *Coffea L.* and *Psilanthus* Hook. Using CMA/DAPI and FISH. Cytologia 69:85–91

López G, Moncada MDP (2006) Construction of an interspecific genetic Linkage map from a *Coffea liberica* x *C. eugenioides* F_1 Population. Proc Intl Sci Colloquium Coffee 21: 644–652, ASIC (http://www.asic-cafe.org)

Mahé L, Combes MC, Lashermes P (2007) Comparison between a coffee single copy chromosomal region and *Arabidopsis* duplicated counterparts evidenced high level synteny between the coffee genome and the ancestral *Arabidopsis* genome. Plant Mol Biol 64:699–711

Marraccini P, Deshayes A, Pétiard V, Rogers WJ (1999) Molecular cloning of the complete 11S seed storage protein gene of *Coffea arabica* and promoter analysis in transgenic tobacco plants. Plant Physiol Biochem 37:273–282

Marraccini P, Courjault C, Caillet V, Lausanne F, Lepage B, et al. (2003) Rubisco small subunit of *Coffea arabica*: cDNA sequence, gene cloning and promoter analysis in transgenic tobacco plants. Plant Physiol Biochem 41:17–25

Mettulio R, Rovelli P, Anthony F, Anzueto F, Lashermes P, et al. (1999) Polymorphic microsatellites in *Coffea arabica*. Proc Intl Sci Colloquium Coffee 18:344–347

Moncada P, McCouch S (2004) Simple sequence repeat diversity in diploid and tetraploid Coffea species. Genome 47:501–509

Naveen KS, Sreenath HL, Sreedevi G, Veluthambi K, Naidu R (2002) Transgenic coffee (*Coffea arabica*) plants with markers genes through *Agrobacterium tumefaciens*-mediated transformation. XV Plant Crops Symposium. Placrosym. Mysore, India, pp. 219–225

Noir S, Patheyron S, Combes MC, Lashermes P, Chalhoub B (2004) Construction and characterisation of a BAC library for genome analysis of the allotetraploid coffee species (*Coffea arabica* L.). Theor Appl Genet 109:225–230

Ogita S, Uefuji H, Yamaguchi Y, Sano H (2003) RNA interference: producing decaffeinated coffee plants. Nature 423:823

Ogita S, Uefuji H, Morimoto M, Sano H (2004) Application of RNAi to confirm theobromine as the major intermediate for caffeine biosynthesis in coffee plants with potential for construction of decaffeinated varieties. Plant Mol. Biol 54:931–941

Orozco-Castillo C, Chalmers KJ, Waugh R, Powell W (1994) Detection of genetic diversity and selective gene introgression in coffee using RAPD markers. Theor Appl Genet 87:934–940

Orozco-Castillo C, Chalmers KJ, Powell W, Waugh R (1996) RAPD and organelle specific PCR re-affirms taxonomic relationships within the genus *Coffea*. Plant Cell Rep 15:337–341

Paillard M, Lashermes P, Pétiard V (1996) Construction of a molecular linkage map in coffee. Theor Appl Genet 93:41–47

Pearl HM, Nagai C, Moore PH, Steiger DL, Osgood RV, et al. (2004) Construction of a genetic map for arabica coffee. Theor Appl Genet 108:829–835

Perthuis B, Pradon J, Montagnon C, Dufour M, Leroy T (2005) Stable resistance against the leaf miner *Leucoptera coffeella* expressed by genetically transformed *Coffea canephora* in a pluriannual field experiment in French Guiana. Euphytica 144:321–329

Pinto-Maglio CAF, Da Cruz ND 1998. Pachytene chromosome morphology in *Coffea* L. II. *C. arabica* L. complement. Caryologia, 51:19–35

Poncet V, Hamon P, Minier J, Carasco C, Hamon S, et al. (2004) SSR cross-amplification and variation within coffee trees (*Coffea* spp.). Genome 47:1071–1081

Poncet V, Rondeau M, Tranchant C, Cayrel A, Hamon S, et al. (2006) SSR mining in coffee tree EST databases: potential use of EST-SSRs as markers for the *Coffea* genus. Mol Genet Genomics 276:436–449
Powers J, Culbertson JD (1983). Interaction of a purified bean (*Phaseolus vulgaris*) glycoprotein with an insect amylase. Cereal Chem 60:107–117
Prakash NS, Marques DV, Varzea VMP, Silva MC, Combes MC, et al. (2004) Introgression molecular analysis of a leaf rust resistance gene from *Coffea liberica* into *Coffea arabica* L. Theor Appl Genet 109:1311–1317
Prakash NS, Combes MC, Dussert S, Naveen S, Lashermes P (2005) Analysis of genetic diversity in Indian robusta coffee genepool (*Coffea canephora*) in comparison with a representative core collection using SSRs and AFLPs. Genetic Res Crop Evol 52: 333–343
Raina SN, Mukai Y, Yamamoto M (1998) *In situ* hybridisation identifies the diploid progenitor of *Coffea arabica* (Rubiaceae). Theor Appl Genet 97:1204–1209
Ribas AF, Kobayashi AK, Pereira LFP, Vieira LGE (2005a) Genetic transformation of *Coffea canephora* by particle bombardment. Biol Plant 49:493–97
Ribas AF, Galvão RM, Pereira LFP, Vieira LGE (2005b) Transformação de *Coffea arabica* com o gene da ACC-oxidase em orientação antisenso. 50th Congreso Brasileiro de Genética p. 492
Ribas AF, Kobayashi AK, Pereira LFP, Vieira LGE (2006) Production of herbicide-resistance coffee plants (*Coffea canephora* L.) via *Agrobacterium tumefaciens*-mediated transformation. Braz Arc Biol Technol 49:11–19
Rosillo AG, Acuna JR, Gaitan AL, de Pena M (2003) Optimized DNA delivery into *Coffea arabica* suspension culture cells by particle bombardment. Plant Cell Tiss Org Cult 74:75–79
Samson NP, Campa C, Noirot M, De Kochko A (2004) Potential use of D-Xylose for coffee plant transformation. Proc Intl Conf Coffee Sci 20: 707-713 (ASIC) (http://www.asic-cafe.org)
Satyanarayana KV, Kumar V, Chandrashekar A, Ravishankar GA (2005) Isolation of promoter for *N*-methyltransferase gene associated with caffeine biosynthesis in *Coffea canephora*. J Biotechnol 119:20–25
Simkin AJ, Qian T, Caillet V, Michoux F, Ben Amor M, et al. (2006) Oleosin gene family of *Coffea canephora*: quantitative expression analysis of five oleosin genes in developing and germinating coffee grain. J Plant Physiol 163:691–708
Spiral J, Thierry C, Paillard M, Pétiard V (1993) Obtention de plantules de *Coffea canephora* Pierre (Robusta) transformées par *Agrobacterium rhizogenes*. C. R. Acad Sci Paris 3:1–6
Srinivasan KH, Narasimhaswamy RL (1975) A review of coffee breeding work done at the Government coffee experiment station, Balehonnur. Indian coffee 34:311–321
Stoffelen P, Noirot M, Couturon E, Anthony F (2007) A new caffeine-free coffee species in the deep rain forest of Cameroon. Taxon (in press)
Sugiyama M, Matsuoka C, Takagi T (1995) Transformation of *Coffea* with *Agrobacterium rhizogenes*. Proc Intl Conf Coffee Sci 16:853–859
Thomas AS (1942) The wild arabica coffee on the Boma Plateau, Anglo-Egyptian Sudan. Empire J Expt Agric 10:207–212
van Boxtel J, Berthouly M, Carasco M, Dufour M, Eskes A (1995) Transient expression of ÿ-glucuronidase following biolistic delivery of foreign DNA into coffee tissue. Plant Cell Rep 14:748–752
van Boxtel J, Eskes A, Berthouly M (1997) Glufosinate as an efficient inhibitor of callus proliferation in coffee tissue. In Vitro Cell Dev Biol Plant 33:6–12
Van der Vossen HAM (2001) Agronomy I: Coffee Breeding Practices. In: Clarke RJ, Vitzthum OG (eds) Coffee: recent developments. Blackwell Science, United Kingdom, pp 184–201
Vieira LGE, Andrade AC, Colombo CA, Araujo AH, Metha A et al. (2006) Brazilian coffee genome project: an EST-based genomic resource. Braz J Plant Physiol 18:95–108
Wellman FL (1961) Coffee: botany, cultivation and utilization. London: Leonard Hill Books
Wu F, Mueller LA, Crouzillat D, Pétiard V, Tanksley SD (2006) Combining bioinformatics and phylogenetics to identify large sets of single-copy orthologous genes (COSII) for comparative, evolutionary and systematic studies: A test case in the euasterid plant clade. Genetics 174:1407–1420

Chapter 10
Cowpea, a Multifunctional Legume

Michael P. Timko and B.B. Singh

Abstract Cowpea [*Vigna unguiculata* (L.) Walp.] is an important warm-season legume grown primarily in the semi-arid tropics. The majority of cowpea is grown by subsistence farmers in west and central sub-Saharan Africa, where its grain and stover are highly valued for food and forage. Despite its economic and social importance in developing parts of the world, cowpea has received relatively little attention from a research standpoint. To a large extent it is an underexploited crop where relatively large genetic gains can likely be made with only modest investments in both applied plant breeding and molecular genetics. A major goal of many cowpea breeding and improvement programs is combining resistance to numerous pests and diseases and other desirable traits, such as those governing maturity, photoperiod sensitivity, plant type, and seed quality. New opportunities for improving cowpea exist by leveraging the emerging genomic tools and knowledge gained through research on other major legume crops and model species. The use of marker-assisted selection and other molecular breeding systems for tracking single gene traits and quantitatively inherited characteristics will likely increase the overall efficiency and effectiveness of cowpea improvement programs in the foreseeable future and provide new opportunities for development of cowpea as a food staple and economic resource.

10.1 Introduction

Cowpea [*Vigna unguiculata* (L.) Walp.] is one of the most important food and forage legumes in the semi-arid tropics that includes parts of Asia, Africa, Southern Europe, Southern United States, and Central and South America (Singh 2005; Timko et al. 2007a). It is truly a multifunctional crop, providing food for man and livestock and serving as a valuable and dependable revenue-generating commodity for farmers and grain traders (Singh 2002; Langyintuo et al. 2003). The cowpea plant is a herbaceous, warm-season annual requiring temperatures of at least 18 °C throughout all stages of its development and having an optimal growing

M.P. Timko
Department of Biology, University of Virginia, Charlottesville, VA 22904 USA.
e-mail: mpt9g@virginia.edu

© Springer 2008

temperature of about 28 °C (Craufurd et al. 1997). Seeds of cultivated cowpea types weigh between 80 mg and 320 mg and range in shape from round to kidney-shaped. The seed pods contain between eight and 18 seeds per pod and are cylindrical and curved or straight. The seed coat varies in texture (e.g., smooth, rough, or wrinkled), color (e.g., white, cream, green, buff, red, brown, black), and uniformity (e.g., solid, speckled, or patterned). Seeds of the most well-known cowpea types, such as "blackeye pea" and "pinkeye," are white with a round irregularly shaped black or red pigmented area encircling the hilum that gives the seed the appearance of an eye.

Following germination, emergence of the cowpea seedling from the soil is considered epigeal. This type of emergence makes the seedling more susceptible to injury since the plant cannot regenerate buds below the cotyledonary node. The first two true leaves are opposite, sessile, and entire, whereas the remaining leaves are alternate, petiolate, and trifoliate. Structure of the mature plant varies depending on genotype, growth temperature, and the photoperiod in which the plant grows. The major plant growth habits are erect, semi-erect, prostrate (trailing), or climbing. Most cowpea plants are indeterminate in growth habit. However, some of the newly developed early maturing varieties have a determinate growth phenotype. Early flowering cowpea genotypes can produce a crop of dry grain in 60 days, while longer season genotypes may require more than 150 days to mature, depending on photoperiod.

According to Fery (1985), the inflorescence is axillary and formed of a peduncle 10 to 30 cm long, at the end of which there is a rachis with each node bearing a pair of flowers and a cushion of extrafloral nectaries that contribute to the attraction of insects. Cowpea primarily is self-pollinating. In cultivated forms, the flowers open at the end of the night and close in late morning, with the dehiscence of the anthers taking place several hours before the flower opens. Although considered autogamous, outcrossing rates as high as 5% have been recorded, and therefore some care needs to be taken to avoid outcrossing during the production of breeder and foundation seeds. Two or three pods per peduncle are common, and often four or more pods are carried on a single peduncle if growing conditions are very favorable. The presence of these long peduncles is a distinguishing feature of cowpea, and this characteristic also facilitates hand harvesting.

Cowpea is primarily a short day plant or, in some instances, day-neutral (Ehlers and Hall 1996; Craufurd et al. 1997). Floral bud initiation and development is sensitive to photoperiod in many cowpea accessions, and in some genotypes the degree of photoperiod sensitivity (i.e., the extent of delay in flowering) is influenced by temperature (Wein and Summerfield 1980; Ehlers and Hall 1996). In West Africa, selection for differing degrees of photosensitivity or differences in extent of juvenile growth has occurred in different climatic zones resulting in genotypes where pod ripening occurs at the end of the rainy season in a given locale, regardless of planting date that often varies due to the variable onset of wet seasons (Steele and Mehra 1980). This attribute allows pods to escape damage from excessive moisture and pathogens.

10.2 Economic, Agronomic, and Social Importance

V. unguiculata is known by a variety of names world-wide, with cowpea being among the most prevalent in the literature. In the English speaking parts of Africa it is known as cowpea whereas in the Francophone regions of Africa, the name "niébé" is most often used. Local names for cowpea also include "seub" and "niao" in Senegal, "wake" in Nigeria, and "luba hilu" in the Sudan. In the United States, it is typically referred to as blackeye beans, blackeye peas, and southern peas. On the Indian subcontinent it is called "lobia" and in Brazil it is "caupi."

The seed, or grain as it is sometimes referred to, is the most important part of the cowpea plant for human consumption. The seeds are most often harvested and dried for storage and consumption at a later time, either after cooking whole or after being milled like a flour product and used in various recipes (Nielsen et al. 1997; Ahenkora et al. 1998). As such, cowpea plays a critical role in the lives of millions of people in the developing world, providing them a major source of dietary protein that nutritionally complements low-protein cereal and tuber crop staples. The nutritional profile of cowpea grain is similar to that of other pulses with a relatively low fat content and a total protein content that is two- to fourfold higher than cereal and tuber crops. Similar to other pulses, the storage proteins in cowpea seeds are rich in the amino acids lysine and tryptophan when compared to cereal grains, but low in methionine and cystiene when compared to animal proteins. Total seed protein content ranges from 23% to 32% of seed weight (Nielson et al. 1993; Hall et al. 2003). Cowpea seeds are also a rich source of minerals and vitamins (Hall et al. 2003) and among plants have one of the highest contents of folic acid, a B vitamin necessary during pregnancy to prevent birth defects in the brain and spine (http://www.cdc.gov/ncbddd/folicacid/).

In the southeastern parts of the United States, portions of West Africa, Asia, and in the Caribbean, consuming fresh seeds and green pods is preferred to the cooked dry seeds (Nielsen et al. 1997; Ahenkora et al. 1998). In many parts of Africa and Asia, in addition to the seeds, the fresh or dried leaves are also consumed as a side dish or as part of a stew and provide significant nutritional value. In addition to human consumption, cowpea leaves and stems (stover) are also an important source of high-quality hay for livestock feed (Tarawali et al. 1997, 2002). Cowpea fodder plays a particularly critical role in feeding animals during the dry season in many parts of West Africa (Singh and Tarawali 1997; Tarawali et al. 1997, 2002). Although protease inhibitors have been found in the seed, the use of cowpea grain does not apparently present any serious nutritional problems in animal nutrition and has been used an alternative to other more costly grain protein sources of animal feed (Singh et al. 2006).

Dry grain production is the only commodity of cowpea for which production estimates are generated on a worldwide basis. According to the United Nations Food and Agricultural Organization (FAO), approximately 4 million metric tons (mmt) of dry cowpea grain are produced annually on about 10 million ha worldwide (www.faostat.fao.org/faostat). Worldwide cowpea grain production has gone from an annual average of about 1.2 mmt in the 1970s to approximately 3.6 mmt per annum (during the five-year period spanning 1998 to 2003). This increase in production is partly tied to long-term drought in the Sahelian zone of West Africa that has resulted in many farmers in this part of Africa shifting their production to cowpea because of its drought tolerance (Duivenbooden et al. 2002). Singh et al. (2002) suggest that cowpea production and acreage are actually higher than FAO estimates, with worldwide production of 4.5 mmt on 12 to 14 million ha, because the FAO estimates do not include the acreage and production figures in Brazil, India, and some other countries.

About 70% of cowpea production occurs in the drier Savanna and Sahelian zones of West and Central Africa, where the crop is usually grown as an intercrop with pearl millet (*Pennisetum glaucum*) or sorghum (*Sorghum bicolor*). In these regions, cowpea is less frequently planted in monoculture or intercropped with maize (*Zea mays*), cassava (*Manihot esculenta*), or cotton (*Gossypium* sp.) (Langyintuo et al. 2003). Other important cowpea production areas include the lower elevation areas of eastern and southern Africa, low elevation areas in South America (particularly in Peru and northeastern Brazil), parts of India, and the southeastern and southwestern regions of North America.

Nigeria is the largest producer and consumer of cowpea grain with approximately 5 million ha under cultivation with an annual yield estimate at 2.0 mmt (Singh et al. 2002). After Nigeria, Niger and Brazil are the next largest producers with annual yields estimated at 650,000 mt and 490,000 mt, respectively (Singh et al. 2002). Cowpea grain production in Central America and in east and southern Africa are likely underestimated since these regions also produce significant quantities of common beans (*Phaseolus vulgaris*) and the two are often not distinguished during collection of production statistics. Commercial trading of dry cowpea grain and hay are particularly important to the local and regional economies of West Africa (Singh 2002, 2005; Langyintuo et al. 2003). Most of the cowpea grain sold at large commercial markets in large urban centers of coastal West Africa is produced further inland where climates are drier and favorable to production of high-quality grain. Cowpea production in the United States is estimated at 80,000 mt, with the majority of the production in Texas, California and the southern states of Alabama, Arkansas, Georgia, Louisiana, Missouri, and Tennessee (Fery 2002; Timko et al. 2007a).

Compared to other legumes, cowpea is known to have good adaptation to high temperatures and resistance to drought stress (Hall et al. 2002; Hall 2004). For example, Hall and Patel (1985) reported cowpea grain yields of as much as 1000 kg ha^{-1} of dry grain in a Sahelian environment with low humidity and only 181 mm of rainfall. At present, few other legume crop species are capable of producing significant quantities of grain under these conditions. Cowpea is also a valuable component of farming systems in areas where soil fertility is limiting. This is because cowpea has a high rate of nitrogen fixation (Elawad and Hall 1987), forms effective symbiosis with mycorrhizae (Kwapata and Hall 1985), and has the ability to better tolerate a wide range of soil pH when compared to other grain legumes (Fery 1990). Cowpea is also well recognized as a key component in crop rotation schemes because of its ability to help restore soil fertility for succeeding cereal crops (Carsky et al. 2002; Tarawali et al. 2002; Sanginga et al. 2003). In addition, well-adapted, early maturing cowpea varieties capable of producing seed in as few as 55 days after planting often provide farmers with the first source of food from the current harvest sooner than any other crop (Hall et al. 2003).

In the developing world where soil infertility is high, rainfall is limiting, and most of the cowpea is grown without the use of fertilizers and plant protection measures (i.e., pesticides or herbicides), a wide variety of biotic and abiotic constraints also limit growth and severely limit yield (Singh 2005; Timko et al. 2007a).

While cowpea is inherently more drought-tolerant than other crops, water availability is still among the most significant abiotic constraints to growth and yield. Erratic rainfall at the beginning and towards the end of the rainy season adversely affects plant growth and flowering resulting in a substantial reduction in grain yield and total biomass production. The use of early maturing cultivars helps farmers escape the effects of a late season drought, but plants exposed to intermittent moisture stress during the vegetative or reproductive stages will perform very poorly.

Cowpea is susceptible to a wide range of bacterial, fungal, and viral diseases and a large variety of insect pests (Singh 2005; Timko et al. 2007a). The major insect pests of cowpea are aphids (*Aphis craccivora*), thrips (*Megalurothrips sjostedti*), Maruca pod borer (*Maruca vitrata*), a complex of pod sucking bugs (*Clavigralla* spp., *Acanthomia* spp., *Riptortus* spp.), and the storage weevil *Callosobruchus maculatus*. Of these, thrips and *Maruca* cause major damage in sub-Saharan Africa. There are some location-specific insect pests such as *Lygus* in the Americas, bean fly in Asia and East Africa, and ootheca beetles in wetter regions of the tropics.

Nematodes are important constraints in some areas (Roberts et al. 1996, 1997) and parasitic weeds such as *Striga gesnerioides* and *Alectra vogelii* are a major limitation to cowpea production in Africa (Timko et al. 2007b). *Striga* causes severe damage to cowpeas in the Sudan savanna and Sahel of West Africa, whereas *Alectra* is more prevalent in the Guinea and Sudan savannas of West and Central Africa and in portions of eastern and southern Africa. *Striga* infection in cowpea is more devastating in areas with sandy soils, low fertility, and low rainfall. Both parasites are difficult to control because they produce a large number of seeds and up to 75% of the crop damage is done before they emerge from the ground.

Major opportunities exist for breeders to develop cowpea cultivars with tolerance to a wide range of abiotic factors (e.g., drought, low soil fertility, high salinity), resistance to a variety of diseases, pests, and parasites, and agronomic characteristics (e.g., plant growth habits, flowering times, maturity dates) specifically adapted to agroecological production zones and crop product utilizations (i.e., dual-purpose grain and hay production).

10.3 Taxonomic Relationships

Cowpea [*Vigna unguiculata* (L) Walp.] is a dicotyledonous crop in the order Fabaceae, subfamily Faboideae (Syn. Papillionoideae), tribe Phaseoleae, subtribe Phaseolinae, genus Vigna, and section Catiang (Verdcourt 1970; Maréchal et al.

1978). It contains 22 chromosomes (2n = 2x = 22). The genus *Vigna* is pantropical and highly variable. In addition to cowpea, other members include mungbean (*V. radiata*), adzuki bean (*V. angularis*), blackgram (*V. mugo*), and the bambara groundnut (*V. subterranea*). The genus was initially divided into several subgenera based upon morphological characteristics, extent of genetic hybridization/reproductive isolation, and geographic distribution of species (Maréchal et al. 1978). The major groupings consist of the African subgenera *Vigna* and *Haydonia*, the Asian subgenus *Ceratotropis*, and the American subgenera *Sigmoidotropis* and *Lasiopron*. Under the scheme proposed by Maréchal et al. (1978) cultivated cowpea was placed in the subgenus *Vigna*, whereas mungbean and blackgram were placed in the Asian subgenera.

V. unguiculata subspecies *unguiculata* includes four cultigroups: *unguiculata*, *biflora* (or *cylindrica*), *sesquipedalis*, and *textilis* (Ng and Maréchal 1985). *V. unguiculata* subspecies *dekindiana, stenophylla,* and *tenuis* are the immediate wild progenitors of cultivated cowpea and form the major portion of the primary gene pool of cowpea. Members of subspecies *dekindiana, stenophylla,* and *tenuis* are also considered part of this gene pool. A secondary gene pool is constituted by other wild subspecies like *pubescence* that do not readily hybridize and show some degree of pollen sterility and require embryo rescue (Fatokun and Singh 1987). Observations from recent attempts to cross *V. vexillata* and *V. radiata* with *V. unguiculata* (Barone et al. 1992; Gomathinayagam et al. 1998) indicate that these may constitute a tertiary gene pool for cowpea.

10.3.1 Origin and Diversity of Cultivated Forms

The precise origin of cultivated cowpea has been a matter of speculation and discussion for many years. Early observations showed that the cowpeas present in Asia are very diverse and morphologically different from those growing in Africa, suggesting that both Asia and Africa could be independent centers of origins for the crop. However, the absence of wild cowpeas in Asia as possible progenitors has led some to question whether the Asian center of origin is valid. All of the current evidence suggests that cowpea originated in southern Africa, although, it should be noted that it is difficult to ascertain where on the continent the crop was first domesticated. Based on the distribution of diverse wild cowpeas along the entire length of eastern Africa, from Ethiopia to Southern Africa, Baudoin and Maréchal (1985) proposed east and southern Africa to be the primary region of diversity, and west and central Africa to be the secondary center of diversity. These researchers also proposed Asia as a third center of diversity. More recent studies strongly indicate that the highest genetic diversity of primitive wild forms of cowpea can be found in the region of the African continent currently encompassed by Namibia, Botswana, Zambia, Zimbabwe, Mozambique, Swaziland, and South Africa, with among the most primitive species observed in the Transvaal, Cape Town, and Swaziland (Padulosi 1987, 1993; Padulosi et al. 1990, 1991). Based on this latter observation, Padulosi and Ng (1997) suggested that southern Africa may be site of origin of cowpea with subsequent radiations of the primitive forms to other parts of southern and eastern Africa, and subsequently to West Africa and Asia. The small seed size of wild cowpeas likely facilitated their dispersal by birds throughout East and West Africa contributing to the diversity and development of secondary wild forms. Human selection for larger seeds and better growth habits from natural variants in wild cowpeas likely led to diverse cultigroups and their domestication in Asia and in Africa (Steele 1976; Ng and Padulosi 1988; Ba et al. 2004; Ng 1995).

Based on analysis of chloroplast DNA polymorphisms, Vaillancourt and Weeden (1992) discovered that a loss of a *Bam*HI restriction site in chloroplast DNA (haplotype 0) characterized all domesticated accessions and a few wild (*Vigna unguiculata* ssp. *unguiculata* var. *spontanea*) accessions. Based on these data, they suggested that Nigeria was the center of domestication in West Africa. In contrast, studies based on analysis of amplified fragment length polymorphism (AFLP) profiles led Coulibaly et al. (2002) to propose domestication in northeastern Africa. Currently, the wild cowpea, *Vigna unguiculata* ssp. *unguiculata* var. *spontanea,* is thought to be the likely progenitor of cultivated cowpea (Pasquet 1999; Pasquet and Baudoin 2001). Using a new set of chloroplast DNA primers, Feleke et al. (2006) evaluated 54 domesticated cowpea accessions and 130 accessions from the wild progenitor. They confirmed the earlier observation of Vaillancourt and Weeden (1992) that domesticated accessions, including primitive landraces from cultivar groups biflora and textilis, are missing the *Bam*HI restriction site in chloroplast DNA, suggesting that this mutation occurred prior to domestication. However, 40 var. *spontanea* accessions distributed from Senegal to Tanzania and South Africa showed the alternative haplotype 1. Whereas this marker could not be used to identify a precise center of origin, its very high frequency in West Africa was interpreted as a result of either genetic swamping of the wild/weedy gene pool by the domesticated cowpea gene pool or as the result of domestication by ethnic groups focusing primarily on cowpea as fodder.

It is likely that the cowpea was first introduced to India during the Neolithic period (Pant et al. 1982) and was certainly there before the Christian era, since it has a Sanskrit name in writings dated to 150 BC (Steel and Mehra 1980; Ng and Maréchal 1985). It is at that point that human selection led to it being modified to a form different from that present in Africa. Cowpea probably moved to West Asia and parts of Europe between 800 and 300 BC (Ng and Maréchal 1985; Tosti and Negri 2002). Cowpea is well adapted to parts of southern Europe, including Italy, Spain, Portugal, and Turkey but less adapted to the western parts of Asia and continental Europe (Tosti and Negri 2002). Little variability and selection has taken place relative to South Asia and South East Asia, where small seeded and vegetable cowpeas were developed. Asia is often considered a secondary domestication site for the crop. "Yardlong beans," a unique cultivar group (Sesquipedialis) of cowpea that produces very long pods widely consumed in Asia as a fresh green or "snap" bean, apparently evolved in Asia and is rare in African landrace germplasm.

Spanish explorers are likely responsible for introducing cowpea into the New World, bringing seed to the West Indies in the 16th century (Purseglove 1968). The plant presumably was introduced into Central and South America at about the same time and made its way to the continental United States by 1700 (Purseglove 1968).

10.3.2 Molecular Phylogeny and Genome Organization

The development and use of biochemical-based analytical techniques and molecular-marker technologies, such as restriction fragment length polymorphisms (RFLPs), random amplified polymorphic DNAs (RAPDs) , amplified fragment length polymorphisms (AFLPs), and microsatellites or simple sequence repeats (SSRs), have greatly facilitated the analysis of the structure of plant genomes and their evolution, including relationships among the Legumioseae (Choi et al. 2004; Yan et al. 2004; Gepts et al. 2005) This in turn has contributed significantly to our current understanding of the cowpea genome organization and evolution.

Using RFLP analysis, Fatokun et al. (1993a) analyzed 18 *Vigna* species including five of the subgenus *Ceratotropis* to determine the taxonomic relationship between the subgenus *Ceratotropis* and other subgenera. These investigators showed that a high level of genetic variation exists within the genus, with a remarkably higher amount of variation associated with *Vigna* species from Africa relative to those from Asia. Their data supported the taxonomic separation of the Asian and Africa genera as proposed by Maréchal et al. (1978) and underscored the previously held viewpoint that Africa is the likely center of diversity for *Vigna*. In general, the placement of species and subspecies based upon molecular taxonomic procedures by Fatokun et al. (1993a) substantiated prior classifications based on classical taxonomic criteria, such as morphological and reproductive traits.

Genetic variation in 23 accessions of five species within the subgenus *Ceratotropis* was subsequently reinvestigated by using RAPD analysis by Kaga et al. (1996a). Based on the degree of polymorphism at 404 informative loci, these investigators were able to separate the accessions into two main groups differing by approximately 70% at the molecular level. Within each of the main groups, the accessions could be further divided into five subgroups whose composition were in complete agreement with their taxonomic species classifications.

Sonnante et al. (1996) examined isozyme variation between *V. unguiculata* and other species in the subgenus *Vigna* and showed that *V. unguiculata* was more closely related to *V. vexillata*, a member of the subgenus *Plectotropis*, than to any other species belonging to section *Vigna*. This is not surprising since *V. vexillata* is thought to be the intermediate species between African and Asian *Vigna* species. Vaillancourt and Weeden (1996) reached a similar conclusion about the relatedness of these species. Based on an analysis of variation in chloroplast DNA structure (Vaillancourt and Weeden 1992) and isozyme polymorphisms (Vaillancourt et al. 1993), it was suggested that *V. vexillata* and *V. reticulata* were the closest relatives of *V. unguiculata*. While the close relationship between *V. unguiculata* and *V. vexillata* proposed by Vaillancourt and Weeden (1996) is consistent with previous observations (Maréchal et al. 1978), *V. reticulata* was placed in a different cluster based upon RFLP analysis (Fatokun et al. 1993a).

Polymorphisms in 21 different enzyme systems were used by Pasquet (1999) to evaluate the relationship between 199 accessions of wild and cultivated cowpea differing in breeding system and growth characteristic (i.e., annual vs. perennial growth habit). Based on these allozyme data, perennial subspecies of cowpea (spp. *unguiculata* var. *unguiculata*) were shown to form a coherent group closely related to annual forms (ssp. *unguiculata* var. *spontanea*). Among the 10 subspecies studied, *V. unguiculata* var. *spontanea* and ssp. *pubescens* were the closest taxa to cultivated cowpea. Most recently, Ajibade et al. (2000) used inter simple sequence repeat (ISSR) DNA polymorphism analysis to study the genetic relationships among 18 *Vigna* species. They showed that closely related species within each subgenera clustered together [e.g., *V. umbellata* and *V. angularis* (subgenus *Ceratotropis*), *V. adenantha* and *V. caracalla* (subgenus *Sigmoidotropis*), and *V. luteola* and *V. ambacensis* (subgenus *Vigna*)]. Cultivated cowpea grouped closely with the wild subspecies of *V. unguiculata*, and the entire species was separated from its most closely allied species *V. triphylla* and *V. reticulata*. ISSR polymorphism analysis split *Vigna* into groupings that differed in their composition from previous classifications. For example, the subgenus *Vigna* was split into three lineages, with *V. unguiculata/reticulata/friesorum* forming one group, *V. luteola/ambacensis* forming a second, and *V. subterranea* being far from the other two. *Ceratotropis* split into two sections, with three species (*V. radiata*, *V. mungo*, and *V. acontifolia*) in one section and two species (*V. angularis* and *V. umbellata*) in a second section. While such groupings had been suggested previously (Maréchal et al. 1978; Fatokun et al. 1993a; Vaillancourt and Weeden 1996), it should be noted that ISSR analysis was not as effective at resolving genetic distance relationships at the subgeneric level as it was at resolving relationships at the species level and below. Therefore, the authors note that their conclusions regarding subgeneric classifications should be taken with some caution. There is still considerable need to develop appropriate strategies and molecular techniques to resolve exact taxonomic relationships among members of this important genus.

Repetitive DNA sequences have been shown to represent a substantial fraction of the nuclear genome of all higher plant species and to account for much of the variation in genomic DNA content observed among species (Flavell et al. 1994). Many of the repeat sequences found in plant genomes appear to have originated through the activity of transposable elements (transposons) that move either by first forming an RNA intermediate (i.e., retrotransposons [Boeke et al. 1985]) or by direct DNA transposition intermediates (i.e., transposons [Federoff 1989]). To gain insight into the genomic organization and evolution of species within *Vigna*, Galasso et al. (1997) examined the genomic organization and distribution of Ty1-*copia* type retrotransposons in seven different species and subspecies of *Vigna* and several related leguminous plants. Gel blot analysis of genomic DNA from *V. unguiculata*, *V. luteola*, *V. oblongifolia*, *V. ambacensis*, and *V. vexillata* probed with radioactively labeled probes to the reverse transcriptase gene amplified from *V. unguiculata* ssp. *unguiculata*, *V. unguiculata* ssp. *dekindtania*, *V. luteola*, and *V. vexillata* showed variable hybridization patterns and intensities generally correlating with their previously defined taxonomic position. Fluorescence in situ hybridization analysis of the distribution of the Ty1-*copia* type sequences showed that these elements represented a major fraction of the cowpea genome and were dispersed relatively uniformly over all of the chromosomes. Little or no hybridization was found associated with centromeric, subtelomeric, and nucleolar organizing regions of the chromosomes, indicating that these portions of the genome may not be suitable sites for transposition. Comparisons of retrotransposon structural similarity between *Vigna* and other genera of legumes generally supported the subdivision of the tribes Phaseoleae and Vicieae, with greater homology being seen between members of the Cicereae and Phaseoleae than *Cicer* species and those from the Vicieae (Galasso et al. 1997).

Ba et al. (2004) used RAPD analysis to characterize genetic variation in domesticated cowpea and its wild progenitor, and their relationships. Twenty-six domesticated accessions representing the five cultivar groups and 30 wild/weedy accessions, including accessions from West, East, and southern Africa, were evaluated. Twenty-eight primers generated 202 RAPD bands. One hundred and eight bands were polymorphic among the domesticated compared to 181 among wild/weedy cowpea accessions. Wild accessions were more diverse in East Africa, which is the likely area of origin of *V. unguiculata* var. *spontanea*. *V. unguiculata* var. *spontanea* is thought to have spread westward and southward, with a loss of variability that is counterbalanced in southern Africa by introgressions with local perennial subspecies. Although the variability of domesticated cowpea was the highest ever recorded, cultivar groups were poorly resolved, and several results obtained with isozyme data were not confirmed here. However, primitive cultivars were more diverse than evolved cultivars, suggesting two consecutive bottlenecks within domesticated cowpea evolution. These data support the single domestication hypothesis and further underscore the gap between wild and domesticated cowpea and the widespread introgression phenomena between wild and domesticated cowpea. Furthermore, the findings demonstrated that there is a widely distributed cowpea crop-weed complex all over Africa consistent with previous studies using other molecular marker tools (Pasquet 1999; Coulibaly et al. 2002). Taking into account that there appears to have been a single domestication event, the genetic similarity of some of these wild accessions to the domesticated group would be the result of post-domestication gene flow between wild and domesticated forms due to their sympatric distribution.

10.4 Classical Genetics and Breeding

Significant long-term genetic improvement efforts of cowpea have taken place within national laboratories and universities in several West African countries, India, Brazil, and the USA. Within the Consultative Group on International Agricultural Research (CGIAR), the International Institute of Tropical Agriculture (IITA) based in Ibadan, Nigeria, has the global mandate for improving cowpea cultivars. IITA develops and distributes a range of improved cowpea breeding lines to

65 countries. The accomplishments of some of these programs have been described recently (Ehlers et al. 2002a; Singh et al. 2002; Hall et al. 2003; Singh 2005; Timko et al. 2007a).

10.4.1 Germplasm Collections

Cowpea germplasm is maintained in collections around the world with varying levels of accessibility and documentation. The largest collections are held by the IITA with more than 14,000 accessions. The collection can be accessed via an electronic database maintained through the CGIAR-SINGER system (http://singer.cgiar.org). The United States Department of Agriculture (USDA) maintains a collection with ca. 8,000 accessions. Access to this collection is through the USDA Germplasm Resources Information Network or GRIN system (www.ars-grin.gov). The University of California-Riverside has a collection with ca. 5,000 accessions accessible on a Microsoft Access database. There is also a large collection of Mediterranean and African landraces (ca. 600 accessions) held at the Istituto di Genetica Vegetale at Bari, Italy (www.ba.cnr.it). Other centers maintaining seeds of wild and cultivated cowpeas include the following: Agricultural University-Wageningen (Wageningen, The Netherlands), Botanical Research Institute (Pretoria, South Africa), Le Jardin Botanique National de Belgique (Meise, Belgium), International Plant Genetic Resources Institute (IPGRI) in Harare (Zimbabwe), Institut Français de la Recherché Scientifique pour le Dévelopepement en Coopération (ORSTOM; now IRD) in Montpellier (France), Empresa Brasileira de Pesquisa Agropecuária (EMBRAPA) in Goiana (Brazil), Zentralinstitut fur Genetik und Kultupflanzenforschung (GAT) in Gatersleben (Germany), and the National Bureau of Plant Genetic Resources in New Delhi (India).

In addition to the centers and facilities mentioned above, many national cowpea breeding programs in Africa (including programs in Botswana, Burkina Faso, Ghana, Kenya, Nigeria, and Senegal) also have substantial germplasm collections. The condition of some of these collections, which are important reserves of local diversity, could be improved with funding for germplasm maintenance and facility repair.

10.4.2 General Breeding Strategies

Most cowpea breeders employ backcross, pedigree, or bulk breeding methods to handle segregating populations because cowpea is a self-pollinating species and varieties are pure lines. Higher grain yields and improved grain quality are the primary breeding objectives of nearly all programs. In addition, most breeders seek to incorporate a wide range of abiotic and biotic stress resistance/tolerance characters. The constraints that direct individual breeding programs at the local and national program levels depend on the major diseases and pests encountered in their target environments. Several comprehensive reviews of cowpea breeding have been published, among which the recent efforts by Hall et al. (1997), Singh (2005), and Timko et al. (2007a) are recommended.

The general strategy of most breeding programs is to develop a range of high yielding cowpea varieties adapted to different agroecological zones that possess regionally preferred traits for plant type, growth habit, days to maturity, and seed type. Some of the major breeding objectives for cowpea are summarized in Table 10.1. In general, the focus is on the development of extra-early maturing (60–70 days) and medium maturing (75–90 days), non-photosensitive lines with good grain quality and potential for dual-purpose use (i.e., food and fodder), either for use as a sole crop and as an intercrop in multiple cropping systems. Other traits targeted include resistance to major diseases, insect pests, and parasitic plants (*S. gesnerioides* and *A. vogelii*), tolerance to drought, heat, acidity and low fertility, and seed types with high protein content and low cooking time. For example, new extra-early cowpea varieties have been developed that have erect plant type, early maturity and resistance to major pests, and are capable of yields up to 2.5 tons ha^{-1} within 60 days compared to less than 1 ton/ha of the local varieties, which mature in 100 to 140 days. Similarly,

Table 10.1 The Major Breeding Objectives for Cowpea[1]

Breeding Objective	Selection/Improvement Criteria
High seed yield	Without inputs under intercropping conditions from 100 to 400 kg ha^{-1}
	With inputs under sole cropping conditions from 900 to 3000 kg ha^{-1}
Diverse types	Extra-early maturing (60–70 days) photo-insensitive grain type, for use as sole crop in multiple cropping systems and short rainy seasons
	Medium-maturing (75–90 days) photo-insensitive grain type, for use as a sole crop and intercrop
	Late-maturing (85–120 days) photo-insensitive dual-purpose (grain + leaf) types, for use as a sole crop and intercrop
	Photosensitive early-maturing (70–80 days) grain types, for intercropping
	Photosensitive and photo-insensitive medium-maturing (75–90 days) dual purpose (grain + fodder) types, for intercropping
	Photosensitive late-maturing (85–120 days) fodder type, for intercropping
	High-yielding, bush-type vegetable varieties
Resistance to biotic stresses	Insects: Aphid (*Aphis cracivorra*), Thrips ( ), leaf hoppers (*Empoasca* sp.), podborer (Maruca vitrata), *Clavigralla* spp., *Anoploenemis* spp., *Riptortus* sp., *Nezara viridula*
	Parasitic plants: *Striga gesnerioides* and *Alectra vogelii*
	Diseases: *Colletotrichum* sp., *Xanthomonas* sp., viral mosaics and mottling
Tolerance to abiotic stresses	Drought, high temperatures, low phosphorus, high BNF, and soil acidity; root architecture
Quality and acceptability of the seed	Size, color and texture of seed coat
	Protein content
	Mineral levels (Fe, Zn, Ca, K)
	Low cooking time

[1] Partly adapted and modified from Pasquet and Baudoin (2001)

a number of medium-maturing, dual-purpose cowpea varieties have been developed which yield over 2.5 tons ha^{-1} grain and over 3.0 tons ha^{-1} fodder in 75–80 days. In recent years, over 40 improved cowpea varieties have been released in 60 countries covering Africa, Asia, and Central and South America. Table 10.2 lists a few of the notable improved varieties released in different agroecological regions.

10.4.3 Breeding for Resistance to Biotic Stresses

For many bacterial, fungal and viral diseases, effective screening techniques have been developed that allow researchers to identify cultivars with potential sources of resistance (Ehlers and Hall 1997). In general, good progress has been made using conventional breeding techniques to move resistance to various bacterial, fungal, and viral diseases, parasitic weeds (*S. gesnerioides* and *A. vogelii*), and root-knot nematodes into farmer-acceptable germplasm. Resistance to these pathogens and parasites is usually governed by single genes that are often effective only in a restricted region due to pathogen/parasite variability and may be overcome in a relatively short period of time. Marker-assisted selection can be helpful in assembling more durable resistance by incorporating an array of resistance genes from other regions as discussed below.

Insect pests are a major problem in cowpea in cultivation (Singh and van Emden 1979; Daoust et al. 1985). Therefore, developing cultivars with sustainable resistance to insects is a key objective of many breeding programs worldwide. While in the developed world the problem of insect infestation and damage is easily controlled by treatment with insecticides, in many parts of the developing world access

Table 10.2 Improved Cowpea Varieties Released for Use in Africa, Asia and the Americas

Region	Variety/Breeding Line/Cultivar
Asia and Oceania	IT81D-897, IT82D-752, IT82D-789, IT82D-889, IT82E-18, IT93K-452-1, IT97K-1042-3, IT98K-1111-1, VITA-4, Victory, Breeze, Light,, Sky, Big Buff
East and Southern Africa	IT82E-16, IT82E-18, IT82D-889, IT85F-2020, IT86D-1010, IT87D-611-3, IT89KD-245, IT90K-59, IT90K-76, IT93K-2046-2, IT97K-568-18, IT97K-499, Hope, Pride, Gold from the Sand
West and Central Africa	TVx 3236, IT81D-985, IT81D-994, IT83S-818, IT83S-728-13, IT84S-2246-4, IT86D-719, IT86D-721, IT87D-453-2, IT89KD-245-1, IT89KD-288, IT88D-867-11, IT89KD-374-57, IT90K-76, IT90K-82-2, IT90K-277-2, IT90K-372-1-2, IT93K-452-1, IT97K-499-35, Melakh, Ein El Gazal, Mouride, Son of IITA, Korobalen, Ayiyti, Asontem, Bengpla, CRSP Niebe, Lori Niebe
North, Central, and South America	VITA-1, VITA-3, VITA-6, VITA-7, IT82E-18, IT82D-716, IT82D-789, IT82D-889, IT83D-442, IT83S-841, IT84D-449, IT84D-666, IT84S-2246-4, IT86D-314, IT86D-368, IT86D-782, IT86D-792, IT86D-1010, IT87D-697-2, IT87D-885, IT88S-574-3, TVx1836-01J, IT87D-1627, IT89KD-288, IT90K-284-2, IT91K-118-2, Titan, Cubinata, California Blackeye No.27, Bettergreen, Charleston Greenpack

to the insecticides themselves or the financial resources required to purchase the insecticides and the equipment required for proper application are not available. In addition, the use of insecticides is an environmental and human safety concern. The imposition of new and significantly more stringent restrictions on the use of some popular insecticides is likely forthcoming and therefore alternative approaches to insect control are needed, especially for cowpea, where the number of registered products for use is low.

The development of insect-resistant cowpea cultivars would have a significant impact on yield and food availability and nutritional status in many regions. Achieving this goal will not be easy since cowpea is attacked by a large number and diversity of insect pests throughout its life-cycle and attack by any one of the major pests can be devastating. Therefore, resistance to multiple pests would have to be developed to positively influence seed production/ yield without the use of insecticides. For example, if cultivars were developed with a high level of resistance to flower thrips, capable of protecting their floral buds from damage, any resulting flowers and pods on these plants would likely be destroyed by pod bugs and pod borers. However, resistance to individual pests can reduce the number of sprays needed to obtain optimal yields and would generally increase yields without insect protection in regions where pest pressure is moderate, as in the case of the Sahel.

Screening methods have been developed for several major insect pests of cowpea (Ehlers and Hall 1997). However, despite the evaluation of hundreds to thousands of cowpea accessions, plants with high levels of resistance to most of the most significant pests have not been identified. Among the pest for which good sources of resistance have been identified are the cowpea aphid (*Aphis cracivorra*) and leaf hoppers (*Empoasca* sp.). Low to moderate levels of resistance have been identified in several genotypes for flower thrips, pod bugs, and Maruca pod borer (Singh et al. 2002; Singh 2005). Recurrent selection is being used to combine these resistances, but progress in this area is hampered by the low heritability of the traits based on the field screening methods currently available. The identification of molecular markers for insect resistance would greatly facilitate the transfer and pyramiding of the resistance genes in preferred backgrounds.

Using a combination of field and laboratory screening, a number of cowpea breeding lines have been developed with combined resistance to cowpea yellow mosaic, blackeye cowpea mosaic and many strains of cowpea aphid borne mosaic, *Cercospora*, smut, rust, *Septoria*, scab, *Ascochyta* blight, bacterial blight, anthracnose, nematodes, *Striga, Alectra*, ahpid, thrips and bruchid. Among these, IT82D-889, IT83S-818, IT86D-880, IT86D-1010, IT84S-2246-4, IT89KD-889, IT90K-59, IT90K-76, IT90K-277-2, IT90K-284-2, IT97K-207-15, IT97K-499-35, and IT98K-205-8 are very promising (Van Boxtel et al. 2000; Singh et al. 2002; Lale and Kolo 2007).

10.4.4 Breeding for Tolerance to Abiotic Stresses

Using simple screening methods for heat and drought tolerance and root architecture, major varietal differences for all three traits have been identified and incorporated into improved lines (Matsui and Singh 2003). The best drought-tolerant varieties are IT89KD-374-57, IT88DM-867-11, IT98D-1399, IT98K-131-1, IT97K-568-19, IT98K-452-1, and IT98K-241-2, and the best heat-tolerant lines are IT93K-452-1, IT98K-1111-1, IT93K-693-2, IT97K-472-12, IT97K-472-25, IT97K-819-43 and IT97K-499-38. Significant progress has also been made in developing cowpea breeding lines with enhanced nitrogen fixation and tolerance to low phosphorus. Some of the more promising lines are IT89KD-374-57, IT90K-372-1-2, IT98D-1399, IT99K-1060, IT97K-568-19, IT97K-568-11, IT00K-1148, IT97K-1069-6, IT03K-314-1 and IT03K-351-2.

10.4.5 Breeding for Improved Nutritional Quality

Under the Harvest Plus initiative funded by the Bill & Melinda Gates Foundation (http://www.gatesfoundation.org/default.htm) and others, a systematic breeding program to develop improved cowpea varieties with enhanced levels of protein and micronutrient contents was initiated in 2003. Since its inception, considerable progress has been made and approximately 2,000 genotypes (cultivars and breeding lines) have been evaluated revealing significant genetic variability in protein and micronutrient contents. Typical values are as follows: protein 21% – 30.7%; calcium 545 ppm – 1,300 ppm; iron 48 ppm – 79 ppm; zinc 23 ppm – 48 ppm; and potassium 12,750 ppm – 16,150 ppm. Among the genotypes tested, IT97K-1042-3, IT99K-216-48-1, and IT97K-556-4 appeared to have good levels of all attributes, whereas IT 97K-131-2 and IT86D-724 had the lowest concentration of most of the attributes. These data suggested that cowpea already has fairly high levels of these micronutrients compared to other crops, and there is also a good opportunity to further improve the nutritional attributes of new cowpea varieties.

In developed countries, cowpea is also being considered as a healthy alternative to soyabean as consumers look to more traditional food sources that are low in fat and high in fiber and that have other health benefits. Fat contents of cowpea seeds range from 1.4 to 2.7% (Nielson et al. 1993), while fiber content is about 6% (Bressani 1985). Protein isolates from cowpea grains have good functional properties, including solubility and emulsifying and foaming activities (Rangel et al. 2004), and could be a substitute for soy protein isolates for persons (especially infants) with soy protein allergies. Processed-food products using dry cowpea grain, such as cowpea-fortified baked goods, extruded snack foods, and weaning foods, have been developed (Phillips et al. 2003).

10.4.6 Breeding for Regional Preference in Seed Type

Diverse regional preferences make the breeding objectives very challenging. For example, only white- and brown-seeded varieties with rough seed coats are preferred in West and Central Africa because of the ease of removing the seed coats for local food preparations. On the other hand, red or brown seeded varieties with smooth seed coats are preferred in East and Southern Africa and parts of Central and South America where cowpea is used as boiled beans for which removal of seed coat is not desirable. In Cuba and some of the other countries in Central America, black-seeded cowpea varieties are used as a substitute of black beans for local delicacies. The relative density of cowpea seeds ranges from 1.01 to 1.09, while hardness (crushing weight) ranges from 3.96 kg to 8.4 kg for IT89KD-288 and Aloka local, respectively. The seed coat content ranged from 5.7 % to 13.8 % in IT95K-207-15 and TVu 12349, respectively, and cooking time ranged from 27.5 minutes for IT90K-277-2 to 57.5 minutes for Aloka local. The seed hardness was positively correlated with cooking time (Singh 2005).

Varieties of cowpea with a "persistent-green" grain have been developed by breeding programs in the USA that are a versatile product for frozen vegetable applications (Ehlers et al. 2002a). Persistent-green cowpea grains are green colored when dry but when soaked in water for several hours closely resemble fresh-shelled cowpea that can be used in frozen vegetable products to add color and variety. Because persistent-green cowpea grain can be harvested and stored dry until rehydration and freezing, it is a quite convenient and economical frozen vegetable compared to other frozen vegetable crops that require highly coordinated harvesting and processing operations and expensive long-term frozen storage.

There is a need for late maturing dual purpose cowpea varieties in East and Southern Africa where cowpea leaves are an important vegetable and in West Africa where cowpea stovers are important fodder for the livestock, but most countries would like to have early and medium maturing varieties because cowpea is grown in low rainfall areas. Most of the Asian countries grow cowpea for green pods for vegetables and some grow exclusively for fodder.

10.5 Genetic Maps

Numerous attempts have been made to develop a comprehensive genetic map of cowpea (Fatokun et al. 1992; Fatokun et al. 1993b; Menancio-Hautea et al. 1993b; Menéndez et al. 1997; Li et al. 1999; Ubi et al. 2000). The most complete genetic map currently available was developed by Ouédraogo et al. (2002a) using a recombinant inbred population derived from a cross between IT84S-2049 and 524B (see Menéndez et al. 1997). IT84S-2049 is an advanced breeding line that was developed at IITA in Nigeria for multiple disease and pest resistance and has resistance to several races of Blackeye cowpea mosaic virus (B1CMV) and to virulent root-knot nematodes in California. Line 524B is a blackeye cowpea that shows resistance to *Fusarium* wilt and was derived from a cross between cultivars CB5 and CB3, which encompasses the genetic variability that was available in cowpea cultivars in California.

The map contains a total of 441 markers of which 432 were assigned to one of 11 linkage groups (LGs) spanning a total of 2,670 cM, with an average distance of ca. 6 cM between markers. The markers comprise 242 AFLPs and 18 disease- or pest-resistance-related markers developed by Ouédraogo et al. (2002a) integrated with 133 RAPD, 39 RFLP, and 25 AFLP markers from the map of Menéndez et al. (1997). Among these marker loci, genes for a number of biochemical and phenotypic traits have been located on this map (see Table 10.3). Candidate resistance genes (termed resistance gene analogs or RGAs) were also placed by RFLP analysis in various locations on the integrated cowpea map, including LG2, LG3, LG5, and LG9. However, none of the RGA loci have yet to be associated with specific disease or pest resistance trait underscoring the need for additional disease and pest resistance phenotyping and mapping in cowpea.

V. vexillata (L.) A. Rich is a perennial wild relative of cowpea and has significant potential as a repository of genes for resistance to pests and diseases to which cowpea plants succumb. In fact, many *V. vexillata* lines have been identified as having high levels of resistance to several cowpea insect pests including the pod-sucking bug *Clavigralla tomentosicollis*, the bruchid *Callosobruchus maculatus*, and the pod borer *Maruca vitrata*, and it possesses high resistance to cowpea mottle carmovirus (CPMoV) (Thottappilly et al. 1994, Ogundiwin et al. 2002). However, the usefulness of this species in traditional breeding approaches for cowpea improvement is limited because there is strong cross incompatibility between these two species. It might be possible using molecular cloning approaches to identify and transfer these desirable genes. To facilitate accessibility of desirable genes in *V. vexillata* for cowpea improvement, maps of the wild cowpea *V. vexillata* have also been generated (Ogundiwin et al. 2000; Ogundiwin et al. 2005). The most recent version comprises 120 markers, including 70 RAPDs, 47 AFLPs, one SSR, and two morphological traits, namely, the CPMoV resistance locus and leaf shape (La), utilizing an F2 generation of the intra-specific cross Tvnu 1443 x Tvnu 73 (Ogundiwin et al. 2005). The map has 14 linkage groups, with 11 of the LGs containing at least three markers, ranging in size between 15.0 and 454.9 cM while the remaining three contained two markers each. The map covered 1,564.1 cM of the *V. vexillata* genome. The average distance between markers was 14.75 cM, ranging from 1.0 to 49.0 cM. Of 106 intervals between loci, 38 were below 10 cM. Thirty-nine quantitative trait loci (QTL) associated with nine morphological and agronomic traits (leaf length, leaf width, petiole length, peduncle length, pod length, internode length, number of seed per pod, 100 seed weight, seed/pod ratio) distinguishing both parents were resolved by composite interval mapping (CIM). The QTL detected on the linkage map accounted for between 15.62 and 66.58% of their respective phenotypic variation. Seven chromosomal intervals contained QTL with effects on multiple traits. Further efforts must be made to generate additional markers, thus leading to the development of a linkage map of *V. vexillata* that would assist breeders to improve cowpea to reach its full potential.

Several early studies involving comparative mapping in legumes showed high levels of conservation between the genomes of cowpea and mungbean (*V. radiata*) and mungbean and common bean (*Phaseolus vulgaris*) (Menancio-Hautea et al. 1993a; 1993b; Boutin et al. 1995). The genetic map of mungbean constructed by Menancio-Hautea et al. (1993a) consisted of 172 markers placed into

Table 10.3 Agronomic, growth habit, and disease and pest resistance trait loci currently placed on the cowpea genetic map of Ouédraogo et al. (2002a) and other traits mapped to probable non-analogous linkage groups[1]

Trait	Locus designation	Linkage group/reference map
Pod pigmentation	P	LG1; (LG1-Menéndez et al. 1997)
Resistance to *Striga gesnerioides*-Race 1	*Rsg2-1*	LG1
Resistance to *Striga gesnerioides*-Race 3	*Rsg4-3, Rsg1-1*	LG1
Root-knot nematode (*Meloidogyne incognita*) resistance	Rk	LG1
Nodes to 1st Flower (D1301a)	NTF	LG2; (LG2-Menéndez et al. 1997)
Dehydrin protein	Dhy	LG2; (LG7-Menéndez et al. 1997)
Resistance to cowpea mosaic virus	CPMV	LG2
Resistance gene analog (pathogen unknown)	RGA-438	LG2
Resistance gene analog (pathogen unknown)	RGA-468	LG2
Resistance gene analog (pathogen unknown)	RGA-490	LG2
Resistance to *Fusarium oxysporum*	*Fus*R	LG3
Cowpea severe mosaic virus resistance	CPSMV (*ims*)	LG3
Cowpea mosaic virus resistance	CPMV	LG3
Resistance gene analog (pathogen unknown)	RLRR3-4B	LG3
General flower color factor	C	LG4; (LG1-Menéndez et al. 1997)
Seed weight (OB6a)	SW	LG5; (LG5-Menéndez et al. 1997)
Resistance gene analogs (pathogen unknown)	RGA-434	LG5
Resistance to southern bean mosaic virus	SBMV(*sbc-1,2*)	LG6
Resistance to *Striga gesnerioides*-Race 1	*Rsg3-1, Rsg-994*	LG6
Resistance to blackeye cowpea mosaic virus	BlCMV	LG8
Resistance gene analogs (pathogen unknown)	RLRR3-4T	LG9
Traits mapped in other populations (likely nonanalogous linkage groups to map of Ouédraogo et al. 2002a)		
Resistance to cowpea aphid (*Aphid craccivora*)	*Rac1*	(LG1-Myers et al. 1996)
50% Flowering	50%FL	(LG7-Fatokun et al. 1993)
Seed weight	SW	(LG7-Fatokun et al. 1993)
Plant height	HT	(LG8-Fatokun et al. 1993)
Pod number per plant	PodN	(LG9-Fatokun et al. 1993)

[1]Adapted from genetic maps and data of Ouédraogo et al. (2002a) and Menéndez et al. (1997) that used the same genetic population. There is insufficient marker data to integrate LGs of the maps of Fatokun et al. (1993) and data from Myers et al. (1996) with the map of Ouédraogo et al. (2002a)

11 linkage groups and provided 1,570 cM coverage with an average distance of 9 cM between loci. Significant colinearity was recognized to exist between the cowpea and mungbean genomes (Menancio-Hautea et al. 1993b). Similarily, Kaga et al. (1996b) reported significant blocks of synteny when comparing the linkage map of azuki bean with those of mungbean and cowpea. Choi et al. (2004) combined genetic, phylogenetic, and DNA sequence comparison to examine the degree of conservation of genome microstructure between model legumes such as *M. truncatula* and *L. japonicus* and crop legumes including *G. max* (soybean), *P. sativum* (pea), *V. radiata* (mungbean), and *P. vulgaris* (common bean). These studies revealed extensive conservation of gene order and orthology between the crop and model legumes and also identified features of structural divergence between these genomes.

10.6 Molecular Markers and Marker-Assisted Selection in Cowpea Breeding

There is a clear need for leveraging modern biotechnological tools to complement conventional breeding in cowpea. Such efforts should focus on the development of molecular markers and protocols for use in marker-assisted selection (MAS) and marker-assisted breeding. Support for such endeavors should come from a cooperation of both public sources and private foundations and must integrate national and regional breeding programs (Timko et al. 2007b).

MAS relies on the identification of DNA sequences within or near genes controlling traits of interest that can then be used to track those genes in breeding populations where the phenotypes are difficult or time-consuming to observe. In practice, MAS allows a more efficient means of assembling alleles of interest in an improved cultivar, thereby increasing the overall efficiency and effectiveness of crop improvement programs (Moreau et al. 2000; Charcosset and Moreau 2004). The application of MAS can be relatively straightforward for genes conditioning large and easily scored phenotypic effects. Most important traits are governed by multiple genes, each having relatively small effects. These "quantitative traits" have been difficult to understand and to manipulate in conventional crop breeding programs. The term QTL, quantitative trait loci, refers to the chromosomal regions of genes that control quantitative traits.

Prior to applying MAS, a realistic assessment of the cost-benefit ratio in comparison with phenotypic assays performed in the field, greenhouse, or laboratory needs to be conducted (Dekkers and Hospital 2002; Dreher et al. 2003). In general, traits that are difficult or expensive to measure using phenotypic assays are good candidates for MAS. In some cases, MAS can allow smaller populations to be used, reduce the number of generations needed to reach a goal, or increase the accuracy of evaluations (Sharma et al. 2002). MAS offers the only practical method to combine multiple resistance genes into one cultivar when the genes mask the expression of one another, yet when together provide more durable resistance

(Kelly et al. 2003). Other advantages of MAS are that a single technology can handle selection of diverse types of traits (e.g., pest resistance and grain quality parameters) and that cultivars developed through the use of MAS are not subjected to negative stereotyping as transgenic cultivars (Dubcovsky 2004). Also, selection of traits conferring resistance to quarantined pests can be conducted using MAS, eliminating the need for transfer of quarantined pests and assessment of resistance in expensive quarantine facilities.

The use of MAS has yet to be implemented in cowpea, but some of the groundwork for its application is in place (Kelly et al. 2003). As noted above, a genetic map has been constructed (Ouédraogo et al. 2002a) and loci controlling important pest and disease resistance genes and agronomic traits have been placed on the map. In addition, markers closely linked to some resistance factors whose function has yet to be fully defined have been identified (Gowda et al. 2002; Timko et al. 2007b). Many of these traits are controlled by single genes and therefore are potentially good candidates for MAS. Currently, no QTL with linked markers have been identified for use in selecting for more complex traits such as grain yield.

Based on host differential response of various cowpea genotypes (cultivars and breeding lines) and genetic diversity analysis, at least seven distinct races of *S. gesnerioides* have been identified within the cowpea-growing regions of West Africa (Lane et al. 1996; 1997; Botanga and Timko 2006). Similarly, "resistance-breaking" strains of the root-knot nematode *Meloidogyne incognita*, cowpea aphid (*Aphis craccivora*), cowpea weevil (*Callosobruchis maculatus*), and Fusarium wilt (*Fusarium oxysporum* f. sp. *tracheiphilum*) have been recognized in specific cowpea production areas. Markers for genes conferring resistance to the various strains of these pests would allow efficient development of varieties with resistance that is more broadly effective using MAS.

Ouédraogo et al. (2001) found three AFLP markers linked to *Rsg2-1*, a gene that confers resistance to *Striga* Race 1 (SG1) present in Burkina Faso, and six AFLP markers linked to gene *Rsg4-3*, a gene that provides resistance to *Striga* Race 3 (SG3) from Nigeria (Fig. 10.1). Two of the AFLPs were associated with both *Rsg2-1* and *Rsg4-3*. Boukar et al. (2004) also reported two AFLP markers that are closely linked to *Rsg1-1*, a gene that also confers resistance to SG3 in Nigeria. Five markers were subsequently found linked to the *994-Rsg* gene on LG6 that also confers resistance to SG1 (Ouédraogo et al. 2002b).

Currently, two sequence confirmed amplified region (SCAR) markers suitable for use in MAS for *Striga* resistance have been developed (Timko et al. 2007b). One SCAR marker, designated 61R(E-ACT/M-CAA), was generated from an AFLP marker associated with resistance to SG1 on LG1 (Ouédraogo et al. 2002a). The second SCAR marker, designated as SEACTMCAC83/85, is linked to SG3 on LG1 (Boukar et al. 2004). Analysis has shown that both 61R and a modified version of it termed MahSE2 (Ouédraogo J, Ouédraogo M, and Timko MP, unpublished data) are effective in identifying resistance to *Striga* races SG1 and SG3, but are less well linked to race SG5. At present, these two markers are available for germplasm evaluation and efficacy testing on populations in the field. Work is also currently underway aimed at identifying markers lined to SG2 and SG4z.

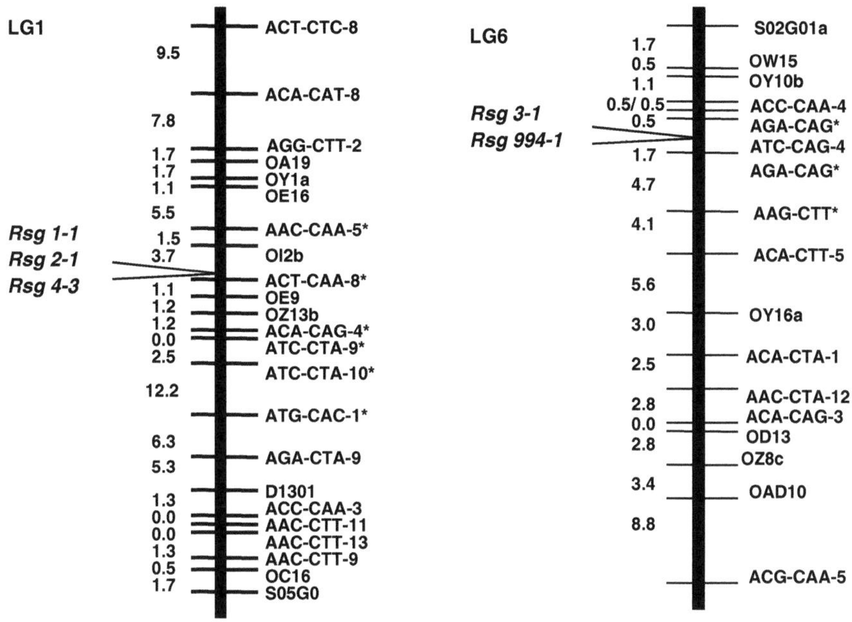

Fig. 10.1 Linkage of molecular markers to S. gesnerioides race-specific resistance genes in cowpea. Shown are the partial linkage maps of linkage groups LG1 (left) and LG6 (right) indicating the position of various AFLP, RAPD, and other molecular markers linked to S. gesnerioides race 1 and race 3 resistance genes. S. gesnerioides race 1 on LG1: Rsg2-1 and Rsg1-1; S. gesnerioides race 3 on LG1: Rsg4-3; S. gesnerioides race 1 on LG6: Rsg3-1 and Rsg994-1. Map distances are shown in cM. Adapted from Ouédraogo et al. (2001)

Breeding lines resistant to root-knot nematodes (*Meloidogyne* spp.) are well characterized in U.S. germplasm, and laboratory and field bioassays to assess resistance to root-knot nematodes in cowpea are effective and reasonably cost effective (Roberts et al. 1997; Ehlers et al. 2002b). Work underway to develop polymerase chain reaction (PCR)-based markers tightly linked to the *Rk* locus (that has multiple resistance specificities to *Meloidogyne* populations) should lead to more effective breeding for nematode resistance in cowpea (Roberts et al. 1996, 1997; Ehlers et al. 2002b).

Chida et al. (2000) obtained three RAPD markers flanking a gene conferring resistance to cucumber mosaic cucumovirus (*Cry* gene) that could be useful in MAS. Linkage analyses of these molecular markers showed that genetic distances of the markers CRGA5, D13/E14-350, WA3-850, and OPE3-500 to the *Cry* locus were 0.7, 5.2, 11.5, and 24.5 cM, respectively.

Insect resistance is a good candidate for MAS in cowpea because assessments of host plant resistance to insects are often difficult to conduct in the field or greenhouse. Most insect resistance factors in cowpea do not provide immunity to the pest and often have low heritability under field conditions. Field screenings that rely on natural insect infestations are subject to natural fluctuations in pest pressure.

When such variability is combined with incomplete resistance, field screens can lead to misclassification and selection of lines lacking the strongest resistance. For example, this has been the case with screening cowpea breeding lines and accessions for resistance to aphids, Lygus bug (*Lygus hesperus*), and pod-sucking bugs (such as *Nezara viridula, Clavigralla tomentosicollis, Riptortus dentipes*). In addition, colonies of insects may be difficult to rear without specialized facilities and trained entomologists to monitor the growth and uses. Such resources may not be available to cowpea breeding programs.

Resistance to the pod bug *Clavigralla tomentosicollis* has been identified in the wild cowpea (ssp. *dekindtiana*) germplasm line TVNu 151 (Koona et al. 2002). The development of effective markers for this trait would allow breeders to use MAS to introgress resistance into cultivated forms using a rapid backcrossing approach, based on the simultaneous selection for the resistance genes (markers) and against markers associated with unwanted wild germplasm characteristics such as small seed size and seed shattering. Clearly, such an approach would require a substantial increase in the number of markers available in cowpea and the development of high-throughput markers such as SSR and Single Nucleotide Polymorphism (SNP) markers.

The application of MAS for improvement of agronomic traits controlled by QTL is much more difficult. Expression of many quantitative traits (such as yield) reflects the influence of many (often interacting) developmental processes over a substantial period of time such as a full growing season. As noted earlier there has been little progress toward the development of markers linked to QTL useful in the selection of agronomic characteristics in cowpea. Progress has been faster in other related legumes (such as *Phaseolus*), and it is possible that some of this information may be leveraged since there is a significant degree of synteny between the bean and cowpea genomes (Kelly et al. 2003).

10.7 Current Genomic Resources

The development of tools for genomics-based research has proceeded rapidly in some legumes, whereas in others the development of such resources has lagged. Among those at the forefront are two species considered to be model legumes, Medicago (*M. truncatula*) and soybean (*G. max*), the latter of major economic importance in the United States. Other species such as alfalfa (*Medicago sativa*), common bean (*P. vulgaris*), pea (*P. sativum*), lentil (*Lens culinaris* Med.), chickpea (*Cicer arietinum*), and peanut (*Arachis hypogaea*), have also received considerable attention (Van den Bosch and Stacey 2003). Until recently, few genomic resources were available for researcher working with cowpea. However, this is likely to change as improvements in technology and reduced costs allow a broader examination of the plant kingdom.

One of the difficulties in developing genomic resources for some plant species stems from the fact that the genomes of many higher plants are relatively large and contain significant amounts of repetitive DNA surrounding the low-copy number expressed regions of the genome (Rabinowicz et al. 1999). A number of experimental approaches have been developed that focus on targeted sequencing of gene-rich regions as an alternative to whole-genome sequencing (Palmer et al. 2003; Whitelaw et al. 2003; Bedell et al. 2005; Rabinowicz et al. 2005). Because of its relatively small genome size (estimated at 620 Mbp), cowpea is a good candidate for reduced representation cloning. To test this, a pilot study was carried out to determine whether methylation filtering technology (http://www.oriongenomics.com/) could be positively applied to analyzing the genespace of cowpea. The results of this pilot study showed that methylation filtering produced a 4.1-fold enrichment of gene-rich clones from cowpea genomic DNA libraries and estimated the size of the hypomethylated, gene-rich space of cowpea to be approximately 151 Mb (Chen et al. 2007). Based on these findings, a large scale analysis of the genespace was undertaken in which the nucleotide sequences of approximately 145,000 clones were determined from the forward and reverse directions, yielding a total of 268, 950 successful gene-space sequence reads (GSRs) with an average read length of 610 bp and an estimated raw coverage of approximately 160 Mb (Chen et al. 2007; Timko, MP unpublished results). A homology-based approach was applied for annotations of the GSRs, mainly using BLASTX against four public FASTA formatted protein databases (NCBI GenBank Proteins, UniProtKB-Swiss-Prot, UniprotKB-PIR [Protein Information Resource], and UniProtKB-TrEMBL). Comparative genome analysis was done by BLASTX searches of the cowpea GSRs against four plant proteomes from *Arabidopsis thaliana, Oryza sativa, Medicago truncatula,* and *Populus trichocarpa.* The results of the analysis, and information on the annotation of individual sequences, can be viewed at (http://cowpeagenomics.med.virginia.edu/). The data provide an excellent starting point for both marker development and comparative genomics.

In addition to the GSRs, other genomics tools are now becoming available. As part of a Generation Challenge Program grant, expressed sequence tags (ESTs) from drought-stressed and non-stressed drought-sensitive and tolerant cowpea lines are being generated (http://www.generationcp.org). A deep-coverage (14X) bacterial artificial chromosome (BAC) library and combinatorial pools of BACs are available from various cowpea cultivars and whole BAC and BAC end sequencing is underway (http://www.medicago.org/genome/BACregistry.php). A 6X BAC library has also been constructed from IT97K-499-35, the advanced line used for the genespace sequencing (M.P. Timko, unpublished). These new initiatives will certainly help in the further development of resources for both marker development and gene expression analysis. Given the rapidity at which sequence data, gene expression information, and other resources are being generated, it is clear that cowpea genomics is poised to begin making significant contributions to crop improvement.

10.8 Transformation Systems for Generating Transgenic Cowpea

Over the last two decades, a substantial number of research laboratories have worked diligently on the development of a reliable genetic transformation and in vitro plant regeneration system for cowpea (Anand et al. 2001; Van Le et al. 2002;

Machuka et al. 2002; Ikea et al. 2003; Avenido et al. 2004). Garcia and his colleagues (Garcia et al., 1986; 1987) were among the first to demonstrate successful transformation of cowpea, obtaining kanamycin-resistant callus, but were unable to achieve plant regeneration. Penza et al. (1991) attempted *Agrobacterium* co-cultivation using longitudinal sections derived from mature embryo slices but could not show evidence of stable integration of either selectable marker or reporter genes. Muthukumar et al. (1995) obtained four cowpea plants after co-cultivation of mature de-embryonated cotyledons and selection on hygromycin-containing media. However, DNA gel blot analysis could demonstrate integration of the *hpt marker* gene in only one of the presumptive transgenic plants, and transference of the marker could not be shown in subsequent generation. Ikea et al. (2003) also observed transformation in cowpea, but the transgenes were transmitted to only a small proportion of the progeny and there was no evidence for stable integration.

The results of the studies described above were at best inconclusive and, unfortunately, cowpea remained one of the last major grain legume species for which an efficient genetic transformation and regeneration system had yet to be developed. This changed in 2006 with the announcement by T.J. Higgins and his colleagues at the CSIRO in Australia that by adapting features of legume transformation systems, they have developed a protocol for *Agrobacterium*-mediated genetic transformation of cowpea that was reliable and modestly efficient in its recovery of transgenic cowpea plants (Popelka et al. 2006). More importantly, these researchers demonstrate for the first time stable transmission and expression of two co-integrated genes in the progeny of transgenic plants. Among the critical parameters in this transformation system are the choice of cotyledonary nodes from developing or mature seeds as explants and a tissue culture medium devoid of auxins in the early stages, but including the cytokinin BAP at low levels during shoot initiation and elongation. Addition of thiol-compounds during infection and co-culture with *Agrobacterium* and the choice of the bar gene for selection with phosphinothricin were also important. Transgenic cowpeas that transmit the transgenes to their progeny can be recovered at a rate of one fertile plant per thousand explants.

These results pave the way for the introduction of new traits into cowpea. Which traits will be selected for initial genetic manipulation will require some critical analysis and should be done in a manner complementary to existing breeding programs. Among the leading candidates are genes conferring strong resistance to insect pests which are a major constraint to productivity and affect post-harvest seed security. These include the use of *Bacillus thurengensis* (Bt) toxin (e.g., Cry1Ab, Cry1C, and CryIIA proteins) against the *Maruca* pod borer (*Maruca vitrata*), the alpha-amylase inhibitor gene from common bean for control of cowpea weevil (*Callosobruchus maculatus),* the soybean cysteine protease inhibitor soyacystatin N (scN) and alpha-amylase inhibitor (alphaAI) from wheat with synergistic effects against the cowpea weevil (Amirhusin et al. 2004), and genes for various plant lectins and plant proteinaceous inhibitors (PIs) of insect proteinases (serine, cysteine, aspartic, and metalloproteinases) (Machuka et al. 2002; Machuka et al. 2002). The development and successful deployment of transgenic cultivars with genes conferring resistance to insects will be a major achievement.

10.9 Conclusions and Perspective

One of the major goals of cowpea programs is to combine resistances to numerous pests and diseases and other desirable traits such as those governing maturity, photoperiod sensitivity, plant type, and seed quality. Parental lines with many desirable traits, such as resistance to cowpea weevil, cowpea aphid, and the parasitic weeds *A. vogelii* and *S. gesnerioides*, along with resistances to bacterial blight, CABMV, and other pathogens, exist in different advanced breeding lines developed by cowpea breeding programs around the world. The release of new improved cowpea varieties in over 60 countries has led to a quiet revolution in cowpea cultivation throughout the tropics. From about 6.3 million ha and 1.1 mmt production in 1974, the global area and production under cowpea in 2004 were about 14.5 million ha and 4.5 mmt, respectively. The new cowpea varieties developed have been given special names like 'Victory' and 'Breeze' in Sri Lanka, 'Light' and 'Sky' in Nepal, 'Big Buff' in Australia, 'Hope' and 'Pride' in Tanzania, 'Gold from the Sand' in Sudan, 'Son of IITA' in Nigeria, 'Korobalen' in Mali, 'Ayiyti', 'Asontem' and 'Bengpla' in Ghana, and 'Titan' and 'Cubinata' in Cuba, etc. Millions of small holder farmers in the tropics are benefiting from the new improved cowpea varieties. The major impact has been in Nigeria where cowpea production has increased from 580,000 mt in 1981 to over 2.3 mmt in 2004 (Singh 2005).

Cowpea remains to a large extent an underexploited crop where relatively large genetic gains can be made with only modest investments in both applied plant breeding and molecular genetics. Because it is grown mostly by poor farmers in developing countries it has received relatively little attention from a research standpoint. Indeed, cowpea has been identified as an "orphan crop" that is recommended for increased public/donor support for biotechnology research (Naylor et al. 2004). The development of new genomics-based resources for cowpea will certainly assist in the future expansion of both marker-assisted selection and marker assisted-breeding. It will also contribute to the development of transgenic plants that can be used in the developing world in a safe, rational, and controlled manner. Future development of cowpea will also benefit from the application of knowledge being gained from basic genomics research on other legume crops and "model species".

Acknowledgments We would like to thank the many friends and colleagues who made helpful suggestions during the preparation of this manuscript especially Drs. Bhavana S. Gowda, Jeremy Ouedraogo, Boukar Ousmane, Jianxiong Li, and Mohammad Ishiyaku. This work was supported in part by funds from the Generation Challenge Program (MPT & BBS), Kirkhouse Trust (MPT) and National Science Foundation (MPT).

References

Ahenkora K, Adu-Dapaah HK, Agyemang A (1998) Selected nutritional components and sensory attributes of cowpea (*Vigna unguiculata* [L.] Walp.) leaves. Plant Foods Hum Nutr 52:221–229

Ajibade SR, Weeden NF, Chite SM (2000) Inter simple sequence repeat analysis of genetic relationships in the genus *Vigna*. Euphytica 111:47–55

Amirhusin B, Shade RE, Koiwa H, Hasegawa PM, Bressan RA, et al. (2004) Soyacystatin N inhibits proteolysis of wheat alpha-amylase inhibitor and potentiates toxicity against cowpea weevil. J Econ Entomol 97:2095–2100

Anand RP, Ganapathi A, Vengadesan G, Selvaraj N, Anbazhagan VR, et al. (2001) Plant regeneration from immature cotyledon-derived callus of *Vigna unguiculata* (L). Walp (cowpea). Curr Sci 80:671–674

Avenido RA, Dimaculangan JG, Welgas JN, Del Rosario EE (2004) Plant regeneration via direct shoot organogenesis from cotyledons and cotyledonary node explants of pole sitao (*Vigna unguiculata* [L] Walp. var *sesquipedalis* [L.] Koern.). Philippine Agric Sci 87:457–462

Ba FS, Pasquet RE, Gepts P (2004) Genetic diversity in cowpea [*Vigna unguiculata* (L.) Walp.] as revealed by RAPD markers. Genet Resource Crop Evol 51:539–550

Barone A, del Guidice A, Ng NQ (1992) Barriers to interspecific hybridization in *V. unguiculata* and *V. vexillata*. Sexual Plant Reproduction 5:195–200

Baudoin JP, Maréchal R (1985) Genetic diversity in *Vigna*. In: Singh SR, Rachie KO (eds) Cowpea Research, Production and Utilization. John Wiley and Sons, Ltd., Chichester, NY, pp. 3–9

Bedell JA, Budiman MA, Nunberg A, Citek RW, Robbins D, et al. (2005) Sorghum genome sequencing by methylation filtration. PLoS Biol 3:e13

Boeke JD, Garfinkel DJ, Styles CA, Fink GR (1985) *Ty* elements transpose through an RNA intermediate. Cell 40:491–500

Botanga CJ and Timko MP (2006) Phenetic relationships among different races of *Striga gesnerioides* (Willd.) Vatke from West Africa. Genome 49: 1351–1365

Boukar O, Kong L, Singh BB, Murdock L, Ohm HW (2004) AFLP and AFLP-derived SCAR markers associated with *Striga gesnerioides* resistance in cowpea. Crop Sci 44:1259–1264

Boutin SR, Young ND, Olson TC, Yu ZH, Shoemaker RC, et al. (1995) Genome conservation among three legume genera detected with DNA markers. Genome 38:928–937

Bressani R (1985) Nutritive value of cowpea. In: Singh SR, Rachie KO (eds) Cowpea Research, Production and Utilization. John Wiley and Sons, Ltd., Chichester, NY, pp. 353–359

Carsky RJ, Vanlauwe B, Lyasse O (2002) Cowpea rotation as a resource management technology for cereal-based systems in the savannas of West Africa. In: Fatokun CA, Tarawali SA, Singh BB, Kormawa PM, M Tamo (eds) Challenges and Opportunities for Enhancing Sustainable Cowpea Production. International Institute of Tropical Agriculture, Ibadan, Nigeria, pp. 252–266

Charcosset A, Moreau L (2004) Use of molecular markers for the development of new cultivars and the evaluation of genetic diversity. Euphytica 137:81–94

Chen X Laudeman TW, Rushton PJ, Spraggins TA, Timko MP (2007) CGKB: an annotation knowledge base for cowpea (*Vigna unguiculata* L.) methylation filtered genomic genespace sequences. BMC Bioinformatics 8:129.

Chida Y, Okazaki K, Karasawa A, Akashi K, Nakazawa-Nasu Y, et al. (2000) Isolation of molecular markers linked to the *Cry* locus conferring resistance to cucumber mosaic cucumovirus infection in cowpea. J Gen Plant Pathol 66:242–250

Choi H-K, Mun J-H, Kim D-J, Zhu H, Baek J-M, et al. (2004) Estimating genome conservation between crop and model legume species. Proc Natl Acad Sci USA 101:15289–15294

Coulibaly S, Pasquet RS, Papa R, Gepts P (2002) AFLP analysis of the phenetic organization and genetic diversity of cowpea [*Vigna unguiculata* (L.) Walp.] reveals extensive gene flow between wild and domesticated types. Theor Appl Genet 104:258–266

Craufurd PQ, Summerfield RJ, Ell RH, Roberts EH (1997) Photoperiod, temperature and the growth and development of cowpea (*Vigna unguiculata*). In: Singh BB, Mohan Raj DR, Dashiell KE, Jackai LEN (eds) Advances in Cowpea Research. Copublication Intl Inst Tropical Agric (IITA) and Japan Intl Res Center Agric Sci (JIRCAS). Sayce, Devon, UK, pp. 75–86

Daoust RA, Roberts DW, Das Neves BP (1985) Distribution, biology and control of cowpea pests in Latin America. In: Singh SR, Rachie KO (eds) Cowpea Research, Production and Utilization. John Wiley and Sons, Ltd., Chichester, NY, pp. 249–264

Dekkers JCM, Hospital F (2002) The use of molecular genetics in the improvement of agricultural populations. Nat Rev Genet 3:22–32

Dreher K, Khairallah M, Ribaut JM, Morris M (2003) Money matters. (I) Costs of field and laboratory procedures associated with conventional and marker-assisted maize breeding at CIMMYT. Mol Breed 11:221–234

Dubcovsky J (2004) Marker-assisted selection in public breeding programs: the wheat experience. Crop Sci 44:1895–1898

Duivenbooden Van H, Abdoussalam S, Mohamed AB (2002) Impact of climate change on agricultural production in the Sahel-Part 2. Case study for groundnut and cowpea in Niger. Climate Change 54:349–368

Ehlers JD, Hall AE (1996) Genotypic classification of cowpea based on responses to heat and photoperiod. Crop Sci 36:673–679

Ehlers JD, Hall AE (1997) Cowpea (*Vigna unguiculata* L. Walp). Field Crops Res 53:187–204

Ehlers JD, Fery RL, Hall AE (2002a) Cowpea breeding in the USA: new varieties and improved germplasm. In: Fatokun CA, Tarawali SA, Singh BB, Kormawa PM, Tamo M (eds) Challenges and Opportunities for Enhancing Sustainable Cowpea Production. Intl Inst Tropical Agric, Ibadan, Nigeria, pp 62–77

Ehlers JD, Matthews WC, Hall AE, Roberts PA (2002b) Breeding and evaluation of cowpeas with high levels of broad-based resistance to root-knot nematodes. In: Fatokun CA, Tarawali SA, Singh BB, Kormawa PM, M Tamo (eds) Challenges and Opportunities for Enhancing Sustainable Cowpea Production. Intl Inst Tropical Agric, Ibadan, Nigeria, pp. 41–51

Elawad HOA, Hall AE (1987) Influences of early and late nitrogen fertilization on yield and nitrogen fixation of cowpea under well-watered and dry field conditions. Field Crops Res 15:229–244

Fatokun CA, Singh BB (1987) Interspecific hybridization between *V. pubescence* and *V. unguiculata* through embryo rescue. Plant Cell Tissue Organ Cult 9:229–233

Fatokun CA, Menancio-Hautea DI, Danesh D, Young ND (1992) Evidence for orthologus seed weight genes in cowpea and mung bean based on RFLP mapping. Genetics 132: 841–846

Fatokun CA, Danesh D, Young ND, Stewart EL (1993a) Molecular taxonomic relationships in the genus *Vigna* based on RFLP analysis. Theor Appl Genet 86:97–104

Fatokun CA, Danesh D, Menancio-Hautea D, Young ND (1993b) A linkage map for cowpea [*Vigna unguiculata* (L.) Walp.] based on DNA markers. In: O'Brien JS (ed) A compilation of linkage and restriction maps of genetically studied organisms, Genetic maps 1992, Cold Spring Harbor Laboratory Press, Cold Spring Harbor, NY, pp. 6.256–6.258

Federoff NV (1989) About maize transposable elements and development. Cell 56:181–191

Feleke Y, Pasquet RS, Gepts P (2006) Development of PCR-based chloroplast DNA markers that characterize domesticated cowpea (*Vigna unguiculata* ssp *unguiculata* var *unguicalata*) and highlight its crop-weed complex. Plant Syst Evol 262:75–87

Fery RL (1985) The genetics of cowpea: a review of the world literature. In: Singh SR, Rachie KO (eds) Cowpea Research, Production and Utilization. John Wiley and Sons, Ltd., Chichester, NY, pp. 25–62

Fery RL (1990) The cowpea: production, utilization, and research in the United States. Hort Rev 12:197–222

Fery RL (2002) New opportunities in *Vigna*. In: Janick J, Whipkey A (eds) Trends in New Crops and New Uses. ASHS, Alexandria, VA, pp. 424–428.

Flavell AJ, Pearce S, Kumar A (1994) Plant transposable elements and the genome. Curr Opin Genet Dev 4:838–844

Galasso I, Harrison GE, Pignone D, Brandes A, Heslop-Harrison JS (1997) The distribution and organization of *Ty1-copia*-like retrotransposable elements in the genome of *Vigna unguiculata* (L.) Walp. (cowpea) and its relatives. Ann Bot 80:327–333

Garcia JA, Hillie J, Goldbach R (1986) Transformation of cowpea *Vigna unguiculata* cells with an antibiotic resistance gene using a Ti-plasmid-derived vector. Plant Sci 44:37–46

Garcia JA, Hillie J, Goldbach R (1987) Transformation of cowpea *Vigna unguiculata* cells with a full length DNA copy of cowpea mosaic virus m-RNA. Plant Sci 44:89–98

Gepts P, Beavis WD, Brummer EC, Shoemaker RC, Stalker HT, Weeden NF, Young ND (2005) Legumes as a model plant family. Genomics for Food nd Feed Report of the Cross-Legume Advances through Genomics Conference. Plant Physiol 137: 1228–1235

Gomathinayagam P, Ram SG, Rathnaswanmy R, Ramaswamy NM (1998) Interspecific hybridization between *Vigna unguiculata* (L.). Walp and *V. vexillata* (L.). A. Rich, through in vitro embryo culture. Euphytica 102:203–209

Gowda BS, Miller JL, Rubin SS, Sharma DR, Timko MP (2002) Isolation, sequence analysis, and linkage mapping of resistance-gene analogs in cowpea (*Vigna unguiculata* L. Walp.). Euphytica 126:365–377

Hall AE (2004) Breeding for adaptation to drought and heat in cowpea. Eur J Agron 21:447–454

Hall AE, Patel PN (1985) Breeding for resistance to drought and heat. In: Singh SR, Rachie KO (eds) Cowpea Research, Production and Utilization. John Wiley and Sons, Ltd., Chichester, NY, pp. 137–151

Hall AE, Singh BB, Ehlers JD (1997) Cowpea breeding. Plant Breed Rev 15:215–274

Hall AE, Ismail AM, Ehlers JD, Marfo KO, Cisse N, et al. (2002) Breeding cowpeas for tolerance to temperature extremes and adaptation to drought. In: Fatokun CA, Tarawali SA, Singh BB, Kormawa PM, M Tamo (eds) Challenges and Opportunities for Enhancing Sustainable Cowpea Production. Intl Inst Tropical Agric, Ibadan, Nigeria, pp. 14–21

Hall AE, Cisse N, Thiaw S, Elawad HOA, Ehlers JD, et al. (2003) Development of cowpea cultivars and germplasm by the Bean/Cowpea CRSP. Field Crops Res 82:103–134

Ikea J, Ingelbrecht I, Uwaifo A, Thottappilly G (2003) Stable gene transformation in cowpea (*Vigna unguiculata* L. Walp.) using particle gun method. Afr J Biotechnol 2:211–218

Kaga A, Tomooka N, Egawa Y, Hosaka K, Kamijima O (1996a) Species relationships in the subgenus Ceratotropis (genus *Vigna*) as revealed by RAPD analysis. Euphytica 88:17–24

Kaga A, Ohnishi M, Ishii T, Kamijima O (1996b) A genetic linkage map of azuki bean constructed with molecular and morphological markers using an interspecific population (*Vigna angularis* x*V. nakashimae*). Theor Appl Genet 93:658–663

Kelly JD, Gepts P, Miklas PN, Coyne DP (2003) Tagging and mapping of genes and QTL and molecular marker-assisted selection for traits of economic importance in bean and cowpea. Field Crops Res 82:135–154

Koona P, Osisanya EO, Jackai LEN, Tamo M, Markham RH (2002) Resistance in accessions of cowpea to the Coreid Pod-Bug *Clavigralla tomentosicollis* (Hemiptera: Coreidae). J Econ Entomol 95:1281–1288

Kwapata MB, Hall AE (1985) Effects of moisture regieme and phosphorus on mycorrhizal infection, nutrient uptake, and growth of cowpeas [*Vigna unguiculata* (L.) Walp.]. Field Crops Res 12:241–250

Lale NES, Kolo AA (2007) Susceptibility of eight genetically improved local cultivars of cowpea to Callosobruchus maculatus (F.) (Coleoptera: Bruchidae) in Nigeria. Intl J Pest Management 44:25–27

Lane JA, Moore THM, Child DV, Cardwell KF (1996) Characterization of virulence and geographic distribution of *Striga gesnerioides* on cowpea in West Africa. Plant Dis 80:299–301

Lane JA, Child DV, Reiss GC, Entcheva V, Bailey JA (1997) Crop resistance to parasitic plants. In: Crute IR, et al. (eds) The Gene-for-Gene Relationship in Plant-Parasite Interactions. CAB, Wallingford, UK, pp. 81–97

Langyintuo AS, Lowenberg-DeBoer J, Faye M, Lamber D, Ibro G, et al. (2003) Cowpea supply and demand in West Africa. Field Crops Res 82:215–231

Li J, He G, Gepts P, Prakash CS (1999) Development of a genetic map for cowpea (*Vigna unguiculata*) using DNA markers. Plant & Animal Genome Conf VII:P327

Machuka J (2002) Potential role of transgenic approaches in the control of cowpea insect pests. In: Fatokun CA, Tarawali SA, Singh BB, Kormawa PM, M Tamo (eds) Challenges and Opportunities for Enhancing Sustainable Cowpea Production. Intl Inst Tropical Agric, Ibadan, Nigeria, pp. 213–232

Machuka J, Adesoye A, Obembe OO (2002) Regeneration and genetic transformation in cowpea. In: Fatokun CA, Tarawali SA, Singh BB, Kormawa PM, M Tamo (eds) Challenges and Opportunities for Enhancing Sustainable Cowpea Production. Intl Inst Tropical Agric, Ibadan, Nigeria, pp. 185–196

Maréchal R, Mascherpa JM, Stainer F (1978) Etude taxonomique d'un group complexe d'especes des genres *Phaseolus* et *Vigna* (Papillionaceae) sur la base de donnees morphologiques et polliniques traitees par l'analyse informatique. Boissiera 28:1–273

Matsui T and Singh BB (2003) Root characteristics in cowpea realted to drought tolerance at the seedling stage. Experimental Agriculture 39:29–38

Menancio-Hautea D, Kumar L, Danesh D, Young ND (1993a) A genome map for mungbean [*Vigna radiata* (L.) Wilczek] based on DNA genetic markers (2N=2X=22). In: O'Brien JS (ed) A compilation of linkage and restriction maps of genetically studied organisms, Genetic maps 1992, Cold Spring Harbor Laboratory Press, Cold Spring Harbor, NY, pp. 6.259–6.261

Menancio-Hautea D, Fatokun CA, Kumar L, Danesh D, Young ND (1993b) Comparative genome analysis of mung bean (*Vigna radiata* L. Wilczek) and cowpea (*V unguiculata* L. Walpers) using RFLP mapping data. Theor Appl Genet 86:797–810

Menéndez CM, Hall AE, Gepts P (1997) A genetic linkage map of cowpea (*Vigna unguiculata*) developed from a cross between two inbred, domesticated lines. Theor Appl Genet 95:1210–1217

Moreau L, Lemarie S, Charcosset A, Gallais A (2000) Economic efficiency of one cycle of marker-assisted selection. Crop Sci 40:329–337

Muthukumar B, Mariamma M, Gnanam A (1995) Regeneration of plants from primary leaves of cowpea. Plant Cell Tissue Organ Cult 42:153–155

Myers GO, Fatokun CA, Young ND (1996) RFLP mapping of an aphid resistance gene in cowpea (*Vigna unguiculata* L. Walp.). Euphytica 91:181–187

Naylor RL, Falcon WP, Goodman RM, Jahn MM, Sengooba T, et al. (2004) Biotechnology in the developing world: a case for increased investments in orphan crops. Food Policy 29:15–44

Ng NQ (1995) Cowpea. In: Smart J, Simonds NW (eds) Evolution of Crop Plants (2nd Edition), Longman, London, UK, pp. 326–332

Ng NQ, Marechal R (1985) Cowpea taxonomy, origin and germplasm. In: Singh SR, Rachie KO (eds) Cowpea Research, Production and Utilization. John Wiley and Sons, Ltd., Chichester, NY, pp. 11–21

Ng NQ, Padulosi S (1988) Cowpea genepool distribution and crop improvement. In: Ng NQ, Perrino P, Attere F, Zedan H (eds.), Crop Genetic Resources of Africa, Vol II. IBPGR, Rome, pp. 161–174

Nielson SS, Brandt WE, Singh BB (1993) Genetic variability for nutritional composition and cooking time of improved cowpea lines. Crop Sci 33:469–472

Nielson SS, Ohler TA, Mitchell CA (1997) Cowpea leaves for human consumption: production, utilization, and nutrient composition. In: Singh BB, Mohan Raj DR, Dashiell KE, Jackai LEN (eds) Advances in Cowpea Research. Copublication Intl Inst Tropical Agric (IITA) and Japan Intl Res Center Agric Sci (JIRCAS). Sayce, Devon, UK, pp. 326–332

Ogundiwin EA, Fatokun CA, Thottappilly G, Aken'Ova ME, Pillay M (2000) Genetic linkage map of *Vigna vexillata* based on DNA markers and its potential usefulness in cowpea improvement. (abstr) World Cowpea Res Conf III, p. 19

Ogundiwin EA, Thottappilly G, Aken'Ova ME, Ekpo EJA, Fatokun CA (2002) Resistance to cowpea mottle carmovirus in *Vigna vexillata*. Plant Breed 121:517–520

Ogundiwin EA, Thottappilly G, Aken'Ova ME, Pillay M, Fatokun CA (2005) A genetic linkage map for *Vigna vexillata*. Plant Breed 124:392–398

Ouédraogo JT, Maheshwari V, Berner D, St-Pierre C-A, Belzile F, et al. (2001) Identification of AFLP markers linked to resistance of cowpea (*Vigna unguiculata* L.) to parasitism by *Striga gesnerioides*. Theor Appl Genet 102:1029–1036

Ouédraogo JT, Gowda BS, Jean M, Close TJ, Ehlers JD, et al.(2002a) An improved genetic linkage map for cowpea (*Vigna unguiculata* L.) combining AFLP, RFLP, RAPD, biochemical markers and biological resistance traits. Genome 45:175–188

Ouédraogo JT, Tignegre J-B, Timko MP, Belzile FJ (2002b) AFLP markers linked to resistance against *Striga gesnerioides* race 1 in cowpea (*Vigna unguiculata*). Genome 45:787–793

Padulosi S (1987) Plant exporation and germplsam collection in Zimbabwe. IITA Genetic Resources Unit Exploration Report. IITA, Ibadan, Nigeria

Padulosi S (1993) Genetic diversity, taxonomy and ecogeographic survey of the wild relatives of cowpea (*V. unguicullata*). Ph.D. Thesis. University Catholique Lovain-la-Neuve, Belgique

Padulosi S, Ng NQ (1997) Origin, taxonomy, and morphology of *Vigna unguiculata* (L.) Walp. In: Singh BB, Mohan Raj DR, Dashiell KE, Jackai LEN (eds) Advances in Cowpea Research. Copublication Intl Inst Tropical Agric (IITA) and Japan Intl Res Center Agric Sci (JIRCAS). Sayce, Devon, UK, pp. 1–12

Padulosi S, Laghetti G, Ng NQ, Perrino P (1990) Collecting in Swaziland and Zimbabwe. FAO/IBPGR Plant Genetic Resources Newsl 78/79, pp. 38

Padulosi S, Laghetti G, Pienaar B, Ng NQ, Perrino P (1991) Survey of wild *Vigna* in southern Africa. FAO/IBPGR Plant Genetic Resources Newsl 83/84, pp. 4–8

Palmer LE, Rabinowicz PD, O'Shaughnessy AL, Balija VS, Nascimento LU, et. al. (2003) Maize genome sequencing by methylation filteration. Science 302:2115–2117

Pant KC, Chandel KPS, Joshi BS (1982) Analysis of diversity in Indian cowpea genetic resources. SABRO J 14:103–111

Pasquet RS (1999) Genetic relationships among subspecies of *Vigna unguiculata* (L.) Walp. based on allozyme variation. Theor Appl Genet 98:1104–1119

Pasquet RS, Baudoin J-P (2001) Cowpea. In: Charrier A, Jacquot M, Harmon S, Nicolas D (eds) Tropical Plant Breeding, Science Publishers, Enfield. pp. 177–198

Phillips RD, McWatters KH, Chinannan MS, Hung Y, Beuchat LR, et al. (2003) Utilization of cowpeas for human food. Field Crops Res 82:193–213

Penza R, Lurquin PF, Filippone E (1991) Gene transfer by cocultivation of mature embryos with *Agrobacterium tumefaciens*: application to cowpea (*Vigna unguiculata*Walp). J Plant Physiol 138:39–43

Popelka JC, Gollasch S, Moore A, Molvig L, Higgins TJ (2006) Genetic transformation of cowpea (Vigna unguiculata L.) and stable transmission of the transgenes to progeny. Plant Cell Rep 25:304–312

Purseglove JW (1968) Tropical Crops - Dicotyledons. Longman, London, UK

Rabinowicz PD, Schutz K, Dedhia N, Yordan C, Parnell LD, et al. (1999) Differential methylation of genes and retrotransposons facilitates shotgun sequencing of the maize genome. Nature Genetics 23:305–308

Rabinowicz PD, Citek R, Budiman MA, Nunberg A, Bedell JA, et al. (2005) Differential methylation of genes and repeats in land plants. Genome Res 15:1431–1440

Rangel A, Saraiva K, Schwengber P, Narciso MS, Domont GB, et al. (2004) Biological evaluation of a protein isolate from cowpea (*Vigna unguiculata*) seeds. Food Chem 87:491–499

Roberts PA, Matthews WC, Ehlers JD (1996) New resistance to virulent root-know nematodes linked to the *Rk* locus of cowpea. Crop Sci 36:889–894

Roberts PA, Ehlers JD, Hall AE, Matthews WC (1997) Characterization of new resistance to root-knot nematodes in cowpea. In: Singh BB, Mohan Raj DR, Dashiell KE, Jackai LEN (eds) Advances in Cowpea Research. Copublication Intl Inst Tropical Agric (IITA) and Japan Intl Res Center Agric Sci (JIRCAS). Sayce, Devon, UK, pp. 207–214

Sanginga N, Dashiell KE, Diels J, Vanlauwe B, Lyasse O, et al. (2003) Sustainable resource management coupled to resilient germplasm to provide new intensive cereal–grain–legume–livestock systems in the dry savanna. Agric Ecosyst Environ 100:305–314

Sharma HC, Crouch JH, Sharma KK, Seetharama N, Hash CT (2002) Applications of biotechnology for crop improvement: prospects and constraints. Plant Sci 163:381–395

Singh BB (2002) Recent genetic studies in cowpea. In: Fatokun CA, Tarawali SA, Singh BB, Kormawa PM, Tamo M (eds) Challenges and Opportunities for Enhancing Sustainable Cowpea Production. Intl Inst Tropical Agric, Ibadan, Nigeria, pp. 3–13

Singh BB (2005) Cowpea [*Vigna unguiculata* (L.) Walp. In: Singh RJ, Jauhar PP (eds) Genetic Resources, Chromosome Engineering and Crop Improvement. Volume 1, CRC Press, Boca Raton, FL, USA, pp. 117–162

Singh BB, Tarawali SA (1997) Cowpea and its improvement: key to sustainable mixed crop/livestock farming systems in West Africa. In: Renard C (ed) Crop Residues in Sustainable Mixed Crop/Livestock Farming Systems, CAB in Association with ICRISAT and ILRI, Wallingford, UK, pp. 79–100

Singh BB, Ehlers JD, Sharma B, Freire Filho FR (2002) Recent progress in cowpea breeding. In: : Fatokun CA, Tarawali SA, Singh BB, Kormawa PM, M Tamo (eds) Challenges and Opportunities for Enhancing Sustainable Cowpea Production. Intl Inst Tropical Agric, Ibadan, Nigeria, pp. 22–40

Singh S, Kundu SS, Negi AS, Singh PN (2006) Cowpea (Vigna unguiculata) legume grains as protein source in the ration of growing sheep. Small Ruminant Res 64:247–254

Singh SR, van Emden HF (1979) Insect pests of grain legumes. Annu Rev Entomol 24:255–278

Sonnante G, Piergiovanni AR Ng NQ, Perrino P (1996) Relationships of *Vigna unguiculata* (L.) Walp., *V. vexillata* (L.) A. Rich., and species of section *Vigna* based on isozyme variation. Genet. Resource Crop Evol 43:157–165

Steele WM (1976) Cowpea, *Vigna unguiculata* (Leguminosae-Papillionatae). In: Simmonds NW (ed) Evolution of Crop Plants., Longman, London, pp. 183–185

Steele WM, Mehra KL (1980) Structure, evolution and adaptation to farming systems and environments in *Vigna*. In: Summerfield RJ, Bunting AH (eds) Advances in Legume Science. Royal Botanic Gardens, Kew, UK, pp. 393–404

Tarawali SA, Singh BB, Peters M, Blade SF (1997) Cowpea haulms as fodder. In: Singh BB, Mohan Raj DR, Dashiell KE, Jackai LEN (eds) Advances in Cowpea Research. Copublication Intl Inst Tropical Agric (IITA) and Japan Intl Res Center Agric Sci (JIRCAS). Sayce, Devon, UK, pp. 313–325

Tarawali SA, Singh BB, Gupta SC, Tabo R, Harris F, et al. (2002) Cowpea as a key factor for a new approach to integrated crop–livestock systems research in the dry savannas of West Africa. In: Fatokun CA, Tarawali SA, Singh BB, Kormawa PM, M Tamo (eds) Challenges and Opportunities for Enhancing Sustainable Cowpea Production. Intl Inst Tropical Agric, Ibadan, Nigeria, pp. 233–251

Thottappilly G, Ng NQ, Rossel HW (1994) Screening germplasm of *Vigna vexillata* for resistance to cowpea mottle carmovirus. Int J Trop Plant Dis 12:75–80

Timko MP, Ehlers JD, Roberts PA (2007a) Cowpea. In: Kole C (ed) Genome Mapping and Molecular Breeding in Plants, Volume 3, Pulses, Sugar and Tuber Crops, Springer Verlag, Berlin Heidelberg. pp. 49–67

Timko MP, Gowda BS, Ouedraogo J, Ousmane B (2007b) Molecular markers for analysis of resistance to *Striga gesnerioides* in cowpea. In: Ejeta G, Gressell J (eds) Integrating New Technologies for Striga Control: Towards Ending the Witch-hunt, World Scientific Publishing Co. Pte Ltd, Singapore, pp. In Press

Tosti N, Negri V (2002) Efficiency of three PCR-based markers in assessing genetic variation among cowpea (*Vigna unguiculata* ssp. *unguiculata*) landraces. Genome 45:656–660

Ubi BE, Mignouna H, Thottappilly G (2000) Construction of a genetic linkage map and QTL analysis using a recombinant inbred population derived from an intersubspecific cross of cowpea (*Vigna unguiculata* (L.) Walp.). Breed Sci 50:161–172

Vaillancourt RE, Weeden NF (1992) Chloroplast DNA polymorphism suggests a Nigerian center of domestication for the cowpea, *Vigna unguiculata* (Leguminosae). Am J Bot 79: 1194–1199

Vaillancourt RE, Weeden NF (1996) *Vigna unguiculata* and its position within the genus *Vigna*. In: Pickersgill B, Lock JM (eds) Advances in Legume Systematics, 8: Legumes of Economic Importance. Royal Botanic Gardens, Kew, UK, pp. 89–93

Vaillancourt RE, Weeden NF, Barnard JD (1993) Isozyme diversity in the cowpea species complex. Crop Sci 33:606–613

Van Boxtel J, Singh BB, Thottappilly G, Maule AJ (2000) Resistance of (*Vigna unguiculata* (L.) Walp.) breeding lines to blackeye cowpea mosaic and cowpea aphid borne mosaic potyvirus isolates under experimental conditions. J Plant Dis Protect 107:197–204

VandenBosch KA, Stacey G (2003) Summaries of legume genomics projects from around the globe. Community resources for crops and models. Plant Physiol 131: 840–865

Van Le B, de Carvalho MHC, Zully-Fodil Y, Thi ATP, Van KTT (2002) Direct whole plant regeneration of cowpea [*Vigna unguiculata* (L.) Walp] from cotyledonary node thin layer explants. J Plant Physiol 159:1255–1258

Verdcourt B (1970) Studies of the *Leguminosae-Papilionoideae* for 'Flora of Tropical East Africa': IV. Kew Bull pp. 507–569

Whitelaw CA, Barbazuk WB, Pertea G, Chan AP, Cheung, F., et al. (2003) Enrichment of gene-coding sequences in maize by geneome filteration. Science 302:2118–2120

Wein HC, Summerfield RJ (1980) Adaptation of cowpeas in West Africa: Effects of photoperiod and temperature responses in cultivars of diverse origin. In: Summerfield RJ, Bunting AH (eds) Advances in Legume Science. Royal Botanic Gardens, Kew, UK, pp. 405–417

Yan HH, Mudge J, Kim DJ, Shoemaker RC, Cook DR, Young ND (2004) Comparative physical mapping reveals features of microsynteny between *Glycine max*, *Medicago truncatula*, and *Arabidopsis thaliana*. Genome 47:141–155

Chapter 11
Genomics of *Eucalyptus*, a Global Tree for Energy, Paper, and Wood

Dario Grattapaglia

Abstract Planted Eucalyptus forests occupy more than 18 million hectares globally and have become the most widely planted hardwood tree in the world, supplying high quality woody biomass for several industrial applications. This chapter attempts to link current eucalypt breeding practice and the genomic tools available or in development. A brief introduction is presented on the main features of modern eucalypt breeding and clonal forestry to provide a better understanding of the challenges and opportunities that lie ahead. Some current low technological input applications of molecular markers in support of operational breeding and clonal deployment are introduced. After reviewing the status of QTL mapping and gene discovery by EST sequencing, the prospects for physical mapping and association genetics in Eucalyptus are discussed. Challenges and opportunities for the application of genomic information to improve relevant traits are described within the framework of molecular breeding for trait improvement. Finally, with the expectation of a draft of a *Eucalyptus grandis* genome within the next three years, a discussion is included on the prospects of gene identification and subsequent applications in breeding.

11.1 Introduction

Intensive production forestry based on exotics began in the southern hemisphere about 50 years ago. Since then, the world forest industry has experienced a slow, steady, but now increasing shift of plantation forestry from the northern hemisphere to the tropics and subtropics on either side of the equator, and to the warmer, temperate climates of New Zealand, Chile, and South Africa. *Eucalyptus* species have played a significant role in this process. High productivity eucalypt forests have supplied high quality raw material for pulp, paper, wood, and energy. Planted forests

D. Grattapaglia
Plant Genetics Laboratory, Embrapa - Recursos Genéticos e Biotecnologia, Parque Estação Biológica, Brasília 70770-970 DF, and Graduate Program in Genomic Sciences and Biotechnology, Universidade Catolica de Brasília – SGAN 916 modulo B, Brasília 70790-160 DF, Brazil
e-mail: dario@cenargen.embrapa.br

© Springer 2008

have provided woody biomass that would otherwise have come from native tropical forests. The expansion of these "fiber farms" will likely be limited by the growth of food and biofuels crops and, in some cases, by pressure of public opinion opposed to plantation forestry. Increased forest productivities and refinements in the quality of wood products by genome assisted breeding and transgenic technologies will become increasingly strategic to the forest industry (Fig. 11.1).

While a number of genes affecting lignin content and composition have been intensively investigated and manipulated in recent years (Boerjan 2005; Bhalerao et al. 2003) significant applications of transgenics in eucalypt production forestry are still to come. At the same time, biosafety challenges persist. Challenges are also faced by molecular breeding applications of genomics. Seventeen years have passed since the first experiments in genetic mapping and molecular breeding of forest trees were described (Neale and Williams 1991; Grattapaglia et al. 1992). From the outset, many expectations of fast and accurate methods for early marker-based selection for growth and wood properties in trees were generated. Significant progress has been made and knowledge gained has prompted some short term opportunities for the incorporation of genomic analysis into tree genetics and breeding. Several challenges are still ahead before high impact applications can be implemented.

Both reverse and forward genomics approaches have been attempted in *Eucalyptus* research. Reverse genomics operates to generate a phenotype from the manipulation of the expression of a given gene through transgenic technology. The forward genomics approach, i.e., going from the analysis of existing phenotypic variation to the causal genetic variants, is based on the wide natural intra and interspecific variation that exists in *Eucalyptus*. The technologies involved in this later

Fig. 11.1 Mosaic of clonal *Eucalyptus* forests and native Atlantic forest in southern Bahia, Brazil. (Photograph courtesy of Veracel – Brazil, by Lasse Arvidson) (See color insert)

approach include genetic mapping, quantitative trait loci (QTL) discovery, physical mapping and genome sequencing (Fig. 11.2). This chapter reviews the status of genomics in *Eucalyptus* with an emphasis on forward genomics and its prospects for integrating genomics into breeding. Recent reviews have described *Eucalyptus* genome research including gene discovery, candidate gene mapping, functional genomics, and physical mapping (Moran et al. 2002; Grattapaglia et al. 2004; Poke et al. 2005; Shepherd and Jones 2005; Myburg et al. 2007). This chapter attempts to link current eucalypt breeding practice and the genomic tools available or in development. A brief introduction is presented of the main features of modern eucalypt breeding and clonal forestry to provide a better understanding of the challenges and opportunities that lie ahead. Some current low technological input applications of molecular markers in support of operational breeding and clonal deployment are also presented. After reviewing the status of QTL mapping in *Eucalyptus*, the challenges and some realistic prospects for the application of genomic information to improve relevant traits are described within the framework of molecular breeding for trait improvement. Finally, with the expectation of a draft of a *Eucalyptus grandis* genome within the next three years, a summary is presented of the prospects for gene identification and subsequent application in breeding.

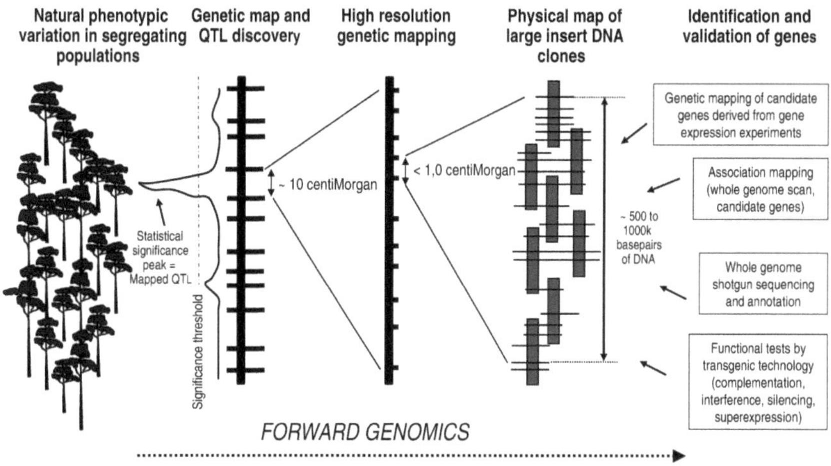

Fig. 11.2 Schematic representation of the forward genomics approach. Genetic mapping and QTL discovery is carried out in segregating populations specifically generated to display large amounts of natural phenotypic variation for wood properties, growth and disease resistance. Major effects QTLs are detected with a resolution of 10 centiMorgans, defining target genomic segments involved in the control of the measured traits. Narrower genomic windows of $<$ 1 cM are defined that can be anchored to physical maps of large insert clones such as BACs covering ~500 – 1000 kbp of DNA by high resolution mapping. Several reverse genomics approaches (boxes) can be used to identify, test and validate specific genes that underlie the target QTL

11.2 Eucalyptus Biology, Domestication, and Breeding

11.2.1 Biology and Domestication

The genus *Eucalyptus* includes some 700 species, some of which are the most widely planted hardwoods in the world. They are long-lived, evergreen trees belonging to the angiosperm family Myrtaceae, which occurs predominantly in the southern hemisphere (Ladiges et al. 2003). They are native to Australia and islands to its north from sea level to the alpine tree line, from high rainfall to semi-arid zones, and from the tropics to 43° south latitude (Ladiges et al. 2003).

Eucalypts were rapidly adopted for plantation forestry around the world following their discovery by Europeans in the 18th century (Eldridge et al. 1993). They were introduced into India, France, Chile, Brazil, South Africa, and Portugal in the first quarter of the 1800s (Doughty 2000) and were quickly chosen for plantations as their fast growth and good adaptability was realized. Following their introduction, several seed collection expeditions were undertaken by government organizations and private forestry companies throughout the world so that large amounts of seed were collected and distributed directly from Australia. Several countries maintain large and diverse germplasm collections of *Eucalyptus,* including several provenances of the most widely planted species.

Species of *Eucalyptus* typically follow a mixed mating system, but are predominantly outcrossers and animal pollinated. Protandry and several incomplete barriers to selfing contribute to the maintenance of high outcrossing rates, together with strong selection against inbreeding (Pryor 1976). Although the major eucalypt subgenera do not hybridize in nature, hybridization among species within the same subgenus has been reported (Pryor and Johnson 1971). Hybridization is more frequent in exotic conditions outside the species' natural range. This property has been exploited by eucalypt breeders in tropical countries who take advantage of the naturally occurring genetic variation for growth and wood properties among species (de Assis 2000). Several artificial hybrid combinations have been produced. Hybrid inviability tends to increase with increasing taxonomic distance between the parents (Griffin et al. 1988; Potts and Dungey 2004).

In several countries, continued planting from local seed sources gave rise to landraces adapted to the specific environment of the seed source (Eldridge et al. 1993). Seed collections from such local exotic plantings became common, and where multiple species plantings occurred, F_1 hybrids were derived (Potts 2004). While several of these F_1 hybrids performed well when deployed as clones, seed collection from hybrid stands often resulted in plantations that were extremely variable and performed poorly in subsequent generations. A textbook case is the Rio Claro hybrid swarm in Brazil (Campinhos and Ikemori 1977; Brune and Zobel 1981). It is a eucalypt arboretum where Navarro de Andrade, the "father of eucalypts" in Brazil first planted 144 different *Eucalyptus* species between 1904 and 1909. Several of these species hybridized once the natural barriers to introgression were removed in the exotic habitat so that seeds collected from these stands were largely interspecific hybrids. Large commercial plantations were established with seeds from this arboretum following Brazilian government fiscal incentives for reforestation starting in 1966. Although some of the resulting forests of these very variable stands were economically inferior, some outstanding trees were found from chance events of recombination for growth, form, and disease resistance. Operational cloning techniques begun in the 1980s captured the superiority of hybrids, which are used today in some of the most productive eucalypt clonal plantations of the world.

The history of eucalypt breeding was detailed by Eldridge et al. (1993) and more recently by Potts (2004). Initial breeding efforts were undertaken by French foresters in Morocco in 1954-55 (Eldridge et al. 1993). The advent of industrially oriented eucalypt stands in the 1960s led to a more formal approach to breeding as exemplified by the establishment of the Florida *E. grandis* breeding program in 1961 (Franklin 1986), *E. globulus* breeding in Portugal in 1965-66 (Potts et al. 2004), and large provenance tests of *E. camaldulensis* in many countries (Eldridge et al. 1993). However, a major breakthrough in eucalypt plantation technology occurred in the 1970s with the establishment of the first commercial stands of selected clones derived from hardwood cuttings in the Democratic Republic of the Congo (Martin and Quillet 1974) followed by Aracruz in Brazil (Campinhos and Ikemori 1977). At the same time, in many tropical countries such as Brazil and South Africa, efforts were intensified to establish extensive provenance/progeny trials of *E. urophylla*, *E. grandis,* and others belonging to the same subgenus *Symphyomyrtus* (Eldridge et al. 1993). These trials were established from open pollinated seed lots collected from selected trees in the wild and constituted the base populations for subsequent selective breeding in many countries. This initial effort was carried out typically by government forestry research institutions and was followed during the 1980s by more intensive collections by private organizations targeting the elite provenances identified in earlier collections as being more adapted for species such as *E. grandis, E. tereticornis,* and *E. viminalis* (Eldridge et al. 1993).

11.2.2 Eucalyptus *Breeding*

Eucalyptus plantation forestry species are well known for their fast growth, straight form, valuable wood properties, wide adaptability to soils and climates, and ease of management through coppicing (Eldridge et al. 1993; Potts 2004). They are now planted in more than 90 countries where the various species are grown for products as diverse as sawn timber, poles, firewood, pulp, charcoal, essential oils, honey, and tannin, as well as for shade and shelter (Doughty 2000). They are an important source of fuel and building material in rural communities of countries such as India, China, Ethiopia, Peru, and Vietnam. The increasing global demand for short fiber pulp has driven the massive expansion of eucalypt plantations and accompanying breeding practices throughout the world during the 20th century (Turnbull 1999).

Their high fiber content relative to other wood components, coupled with the uniformity of fibers relative to other angiosperm species, has led to high demand for eucalypt pulp for coated and uncoated free-sheet paper, bleach board, and sanitary products (fluff pulp), and to a lesser extent, for top liners on cardboard boxes, corrugating medium, and as a filler in long fiber conifer products such as newsprint and containerboard (Kellison 2001). In the last 10 years, the development of new wood drying and sawing technologies have also increased interest in using plantation eucalypts for sawnwood, veneer, and medium density fiberboard and as extenders in plastic and moulded timber (Kellison 2001).

FAO (2000) estimated a total of 17.9 million hectares (ha) of planted *Eucalyptus* worldwide. India was the largest planter with over 8 million ha followed by Brazil with 3 million. The majority of plantations consist of only a few eucalypt species and hybrids. The most important are *E. grandis, E. globulus, E. urophylla,* and *E. camaldulensis*, which together with their hybrids account for about 80% of the plantation area; followed by *E. nitens, E. saligna, E. deglupta, E. pilularis, Corymbia citriodora,* and *E. teriticornis* (Eldridge et al. 1993; Waugh 2004). Market favorites for pulpwood are *E. grandis and E. urophylla* and their hybrids in tropical and sub-tropical regions and *E. globulus* in temperate regions.

Although eucalypt breeding is currently a very dynamic and technically advanced operation, carried out mainly by several private companies, eucalypts are still in domestication infancy when compared with crop species, as most eucalypt breeding programs are only one or two generations removed from the wild. With the combination of ample genetic variation both at the intra and interspecific levels and the ability to clone elite genotypes, eucalypts have quickly become among the most advanced genetic material in forestry. Breeding of eucalypts has moved faster in countries like Brazil, South Africa, Portugal, and Chile that adopted *Eucalyptus* for industrial plantation forestry. Most eucalypts breeding programs worldwide are focused on genetically improving trees for industrial pulp wood production (Kanowski and Borralho 2004). The target traits of most breeding programs include volume growth per ha, wood density, and pulp yield (Borralho et al. 1993 Raymond 2000). Traits such as pest and disease resistance and adaptability to abiotic stresses such as frost, drought, or wind are usually secondary targets that become more important when they impact the main traits. Following the standard concepts in tree breeding, large genetic gains have been obtained in the early stages of eucalypt domestication through species and provenance selection followed by individual selection and establishment of clonal or seedling seed orchards or clonal propagation of elite selections for deployment (Eldridge et al. 1993; Kanowski and Borralho 2004; Potts 2004). Subsequent population improvement has also demonstrated genetic gain through recurrent selection in an open-pollinated breeding population coupled with open or controlled pollinated populations of the most elite selections or specialized breeds (Potts 2004). For species that are easily propagated vegetatively, such as *E. grandis, E. urophylla* and several of their hybrids, clonally propagated breeding populations have enhanced gains by allowing the capture of additive and non-additive genetic effects (Fig. 11.3).

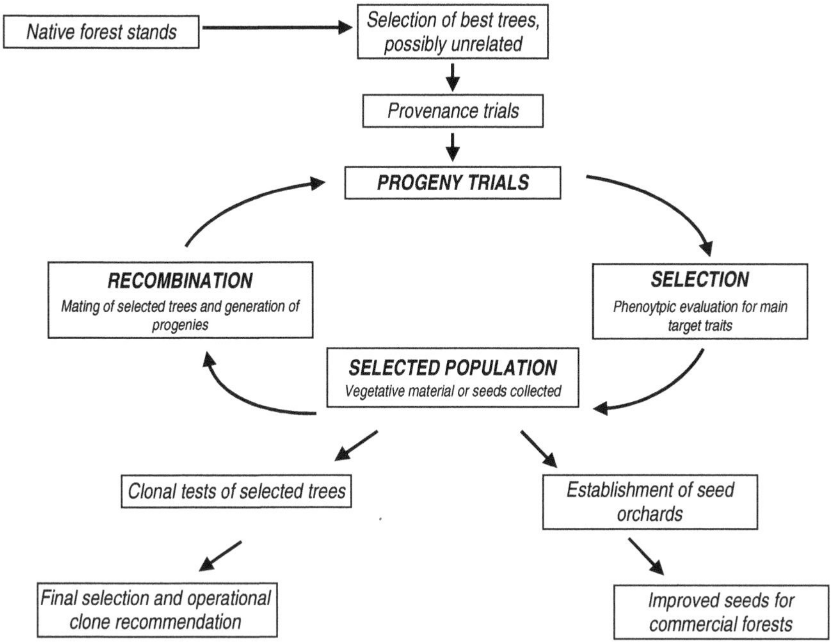

Fig. 11.3 Basic breeding scheme used in *Eucalyptus* breeding involving recurrent selection cycles from which elite parents, for the establishment of seed orchards, and individual trees, to be tested as potential commercial clones, are derived

11.2.3 Clonal Forestry

After more than 25 years following the introduction of clonal forestry of *Eucalyptus* (Campinhos 1980; Brandão et al. 1984), this forest production system is integrated into the strategies of advanced generation breeding programs. Clonal propagation and hybrid breeding are a powerful combination of tools for the improvement of wood and wood product quality. The first hybrid clones were selected by large-scale screening of high yielding spontaneous hybrids, resistant to diseases (such as the eucalypt canker). Today, clones are being derived increasingly from inter-specific hybrid production strategies that involve indoor orchards and highly efficient methods of controlled pollination (de Assis et al. 2005a) (Fig. 11.4). Eucalypt hybrids, involving two or more species and deployed as clones, currently make up a significant proportion of eucalypt plantation forestry, particularly in the tropics and sub-tropics. In Brazil, considering all the large and medium sized companies, the area planted with clones corresponded to more than 1,008,000 ha, involving 362 different clones at a rate of 2 to 40 clones per company, and a range of 10 to 34,000 ha per clone (mean 4,150 ha). The annual introduction of new clonal plantations to support expansion of forest-based industrial production is in the order of 238,000 ha/year, with a mean of 1,820 ha per clone (de Assis et al. 2005b).

Fig. 11.4 Controlled pollination of *Eucalyptus*. (A) Elite parent trees, kept as grafts in indoor insect-proof orchards, are induced with growth regulators for precocious flowering (12 to 15 months). (B) Pollen from the male parent is deposited directly at the base of the style and no bag protection is needed as the greenhouse is kept free of insects. (Photographs courtesy of Teotônio F. de Assis) (See color insert)

An important paradigm shift in eucalypt breeding for pulp and paper began in the 1990s with the increasing realization that the actual "pulp factory" is the tree. Particularly in vertically integrated pulp production systems, as highly productive clonal forests with over $40\,m^3\,ha^{-1}\,yr^{-1}$ became the standard (Binkley and Stape 2004), the focus shifted quickly from volume growth to wood quality, with the objective of improving pulp yield per hectare by reducing wood specific consumption (WSC), i.e., the amount of wood in cubic meters necessary to produce one ton of pulp. Trees that yield more cellulose generate savings all the way from tree harvesting, and transportation to chipping and pulping, while mitigating an accelerated expansion of the commercial forest land base.

Clonal forestry of *E. grandis* x *E. urophylla* selected clones in the 1980s was able to reduce WSC from 4.9 to 4.0 m^3/ton of pulp (Ikemori et al. 1994). *E. globulus* has the best combination of wood properties for pulp and paper among the commercially planted *Eucalyptus* species, resulting in a high pulp yield requiring approximately 25% less wood to produce the same ton of cellulose. *E. globulus* has a very adequate wood density in the range of 550 kg/m^3, the longest fiber length and the largest content of holocellulose and pentosans of any other intensively planted species (Sanchez 2002). While only 3.0 cubic meters of *E. globulus* wood are required per ton of pulp, 4 cubic meters are needed from selected *E. grandis*. *E. globulus*, however, it is much more demanding of soil fertility, it is not adapted to tropical temperatures, it is slower growing and more difficult to clonally propagate than *E. grandis*. In the last 10 years, based on the pioneering experiments in Brazil led by Teotonio de Assis, several breeding programs in tropical countries have started an intensive effort to introgress superior *E. globulus* pulp traits into the tropical and subtropical high yielding genetic backgrounds of *E. grandis* and *E. urophylla*. Given the very high genetic diversity that segregates in such crosses together with intensive within-family selection and clonal propagation, this effort has resulted in exceptional trees that combine superior growth and adaptability to tropical conditions, higher pulp yielding wood and are easily propagated using minicutting/hydroponics technology (de Assis 2000, 2001) (Fig. 11.5). A new wave of

Fig. 11.5 Selection and clonal propagation of elite trees by the minicutting technology. (A) Elite trees are selected, cut and induced for juvenile sprouting. (B) Operational mother plants in hydroponic sand beds from which apical minicuttings are harvested for propagation. (C) Minicuttings are rooted without any growth regulators in controlled environment greenhouses. (D) Clonal plants acclimatizing outdoors. (Photographs B, C, and D courtesy of Teotônio F. de Assis) (See color insert)

clonal forestry is therefore starting that will most likely result in another significant jump in the quality of *Eucalyptus* forests. It is therefore in the context of a highly specialized industrially oriented breeding program that fully exploits the power of hybrid breeding and clonal forestry that one needs to discuss the prospects of genomics and molecular breeding of *Eucalyptus*.

11.3 Marker-Assisted Management of Genetic Variation in Breeding Populations

The use of genome information for the practice of directional selection of superior genotypes still represents a challenge that depends on further and more refined experimental work. Molecular markers can be used immediately to solve several questions related to the management of genetic variation in breeding and production populations. These applications can be useful to essentially any breeding program at any stage of development. Although isozyme markers were initially used for these purposes (Moran and Bell 1983), DNA polymorphisms provide an enhanced level of resolution both at the locus level, with much higher expected heterozygosity values, and at the genome level with greater coverage. Molecular markers can be used to estimate the extent of genetic divergence between individuals selected to compose breeding populations and resolve several issues of individual identity even at high levels of relatedness, including variety protection and the verification of alleged parentage in open pollinated breeding systems. Some operational applications of molecular markers for management of genetic variation in *Eucalyptus* are discussed below.

11.3.1 Identification of Elite Clones

The correct identification of clones is currently the most common application of molecular markers in *Eucalyptus* operational breeding and production forestry. This application is routinely used by several forest companies in Brazil, South Africa, Portugal, Spain, Chile, and Australia. Quality control of large-scale clonal plantation operations is crucial, especially in vertically integrated production systems where the pulp mill depends on the availability of wood from specific clones with specific wood properties at specific times. Given the scale of propagation operations that have to feed plantation programs of several thousand ha per year, (i.e., several million seedlings), mislabeling can seriously affect the expected product. Correct clonal identity also has important implications in breeding procedures such as seed orchard management or controlled pollination programs affecting the expected gains of breeding cycles.

Several technologies are available today to resolve questions of clonal identity in *Eucalyptus*. Dominant markers such as randomly amplified polymorphic DNA (RAPD) or amplified fragment length polymorphisms (AFLP) have been used for clonal fingerprinting of eucalypts (Grattapaglia et al. 1992; Keil and Griffin 1994; Nesbitt et al. 1997). Dominant markers can be used to establish that two individuals are not the same, but the statement that two individuals are identical is usually only approximate and no formal test statistics can be attached to this assertion. The high degree of multi-allelism and the clear and simple Mendelian inheritance of microsatellites provide an extremely powerful system for the unique identification of individuals by fingerprinting and parentage testing particularly when the individuals are expected to be related (Fig. 11.6). Kirst et al. (2005a) demonstrated the high resolving power of this class of markers in *Eucalyptus*. Using only three loci, all 192 trees of a breeding population could be discriminated and a combined probability of identity (i.e., the probability of two individuals having the same multilocus genotype) with six loci was less than 1 in 2 thousand million.

In common with human forensic DNA analysis, the standard method for clonal identification in eucalypts today is based on multiplexed, multicolor fluorescent analysis of microsatellite markers sized in an automatic sequencer. The identity of samples is declared based on a maximum likelihood ratio between two hypotheses, i.e. that the two samples are derived from the same clone and the alternative one, i.e. that the two samples are derived from different clones. With a battery of 10 to 15 microsatellites, likelihoods on the order of 10^{16} are typically reached for indistinguishable samples, given an adequate database of allele frequencies is available.

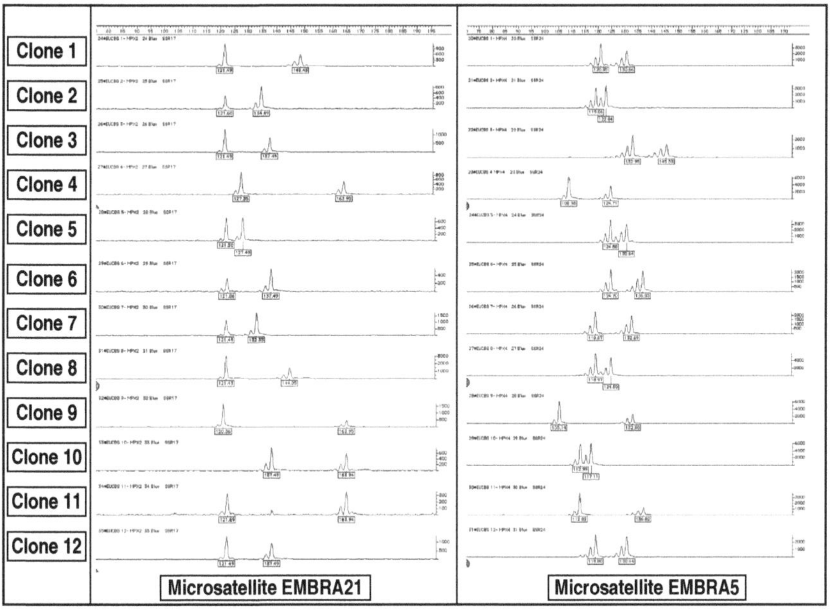

Fig. 11.6 Electropherograms of two microsatellite markers in 12 Eucalyptus clones. Peaks correspond to alleles. In this case, heterozygosity is 100% at both markers giving each clone a unique genetic identity. However, 10 to 15 markers are generally used to assure definitive identification for varietal protection

Most of the microsatellites currently used for clonal identification are derived from dinucleotide and trinucleotide repeats. These markers, while providing powerful discrimination, still present challenges when multilocus profiles need to be compared between labs or automatic sequencers or at different times in the same lab. This is partly due to the small basepair differences among alleles and to the well known phenomenon of stuttering during polymerase chain reaction (PCR) that renders allele declaration challenging (Litt et al. 1993). In human forensic DNA analysis a consensus was reached several years ago that for individual identification purposes only tetra and pentanucleotide repeat markers should be used (Butler 2005). Recently, we have developed a set of novel microsatellites for Eucalyptus based on tetra and pentanucleotide repeats derived from bacterial artificial chromosome (BAC) end sequencing. Although these markers are generally less polymorphic than dinucleotide repeat based microsatellites, allele size differences are clearly interpreted providing very robust multilocus genotype profiles (Sansaloni et al. 2007).

11.3.2 Varietal Protection

Varietal protection of forest trees is still uncommon. Due to the outbreeding nature of forest trees, orchards seed lots are genetically homogeneous. Only specific clones could be targeted for protection. In the case of *Eucalyptus*, due to the high value of some elite clones derived from breeding programs and the upcoming possibility of deploying transgenic clones, protection will likely be an increasing practice. In Brazil, following publication of the varietal protection law, specific instructions for protecting *Eucalyptus* clones were published in 2002 by the Brazilian Ministry of Agriculture based on a set of validated morphological descriptors. To the best of the author's knowledge, this is the only country today that has formalized such descriptors, which include 36 morphological characteristics of leaves, flowers, bark, fruit, as well as wood density. Although these characters generally satisfy the requirements of stability and low environmental influence, they are difficult to evaluate, especially those related to mature traits in flowers and fruits. Furthermore, clones may be related by common ancestry making discrimination difficult. The high power of discrimination coupled with the general acceptance of DNA technology by eucalypt breeders in Brazil resulted in the inclusion of molecular markers as additional descriptors (Grattapaglia et al. 2003). The inclusion of DNA markers represented a remarkable advance in the international landscape of varietal protection of forest trees. Currently all requests for clonal protection are accompanied by a multilocus DNA profile (DNA fingerprint) of 10 to 15 microsatellite markers that were recommended based on robustness, polymorphic information content and availability in the public domain. The perspective for the following years points to an increased number of applications for clone protection by forest companies in view of the value of elite eucalypt clones for competitiveness of the forestry-based industry. DNA markers will add significant power of resolution for distinctness, uniqueness, and stability (DUS) tests in varietal protection of eucalypt clones especially when closely related individuals are under scrutiny in legal disputes.

11.3.3 Characterization of Breeding Populations

Breeding populations can be characterized by quantifying the levels and organization of genetic variation within and between breeding groups, sublines and progenies. These data can be used to improve the structure of breeding populations, infuse new material and decide on selection, enrichment or elimination of germplasm. In incomplete pedigree systems, frequently used in eucalypts, marker-based systems can monitor the levels of random genetic variation throughout the different cycles of a breeding program thus allowing flexibility and control over the rate of reduction of genetic variability. For example, RAPD markers were used to characterize the wide range of genetic variation in a germplasm bank of *Eucalyptus globulus* and thereby assist in the designing further seed collections (Nesbitt et al. 1995). Gaiotto and Grattapaglia (1997) estimated the distribution of genetic variability within and between open pollinated families of a long-term breeding population of *E. urophylla* and proposed a selection strategy within and between families for incomplete pedigreed populations based on genetic diversity. Marcucci-Poltri et al. (2003) used AFLP and microsatellite markers to design a clonal seed orchard using the nine most divergent pairs of genotypes. In a subsequent study, Zelener et al. (2005) used trait selection index and genetic marker information to design an *E. dunni* seed orchard that captured more genetic variation based on individual and family selection rather than on provenance selection.

11.3.4 Mating and Deployment Designs Based on Genetic Distance

Given the wide genetic diversity and multiple sources of germplasm for eucalypt breeding, choices have to be made as to which elite parents should be mated. Some selection based on the individual's performance or pedigree is used before including it in a mating design. Any means of predicting tree performance would be valuable for the breeder. One "holy grail" sought by molecular breeders has been the ability to predict progeny performance based on distance estimates amongst parents from genetic marker data. Vaillancourt et al. (1995a) showed that the ability of RAPD based genetic distance to predict heterosis was significant but accounted only for less than 5% of the variation in specific combing ability in *E. globulus* progenies. Baril et al. (1997) used the structure of RAPD genetic diversity within and between *E. grandis* and *E. urophylla* to develop prediction equations for the tree trunk volume of individual hybrids at 38 months. Surprisingly, this study showed that a genetic distance based on RAPD markers with similar frequencies in the two species successfully predicted the value of a cross with a global coefficient of determination of 81.6%. RAPD markers were used to recommend more divergent crosses in a reciprocal recurrent selection program for hybrid breeding in Brazil based on the premise that crossing more divergent individuals would maximize the appearance of transgressive segregants to be used as clones (Ribeiro et al. 1997).

RAPD data were used to quantify relatedness among elite eucalypt clones for deployment purposes. Little, if any, pedigree information is typically available, even for elite clones. Clonal plantations of *Eucalyptus* generally involve only a few superior genotypes of unknown origin. Costa e Silva and Grattapaglia (1997) used RAPD markers to quantify the genetic relatedness among a group of 15 elite clones. Comparative similarity analyses showed that there was significantly more genomic variation in the group of clones than both within and between unrelated half-sib families from a single species. Data on genetic similarity among clones was also used to propose a deployment strategy in a "genetic mosaic," i.e., avoiding planting more genetically related clones side by side in contiguous forest blocks. This proposed strategy was based on the premise that related clones share a common origin and ancestry, have been subject to similar evolutionary selective pressures, and therefore share common susceptibility/tolerance alleles at pest and pathogen defense loci.

11.3.5 Mating System and Paternity in Breeding Populations

Open pollinated breeding by controlling exclusively the maternal progenitor and half-sib progeny testing is still common practice in some eucalypt breeding programs. Knowledge of outcrossing versus selfing rates is essential for maintaining adequate levels of genetic variability for continuous gains over generations. Eucalypts are preferentially outcrossed both in natural populations as well as seed orchards. Isozyme markers were originally used (Moran et al. 1989), but other types of markers now provide a much higher level of resolution. The outcrossing rate in an open pollinated breeding population of *Eucalyptus urophylla* was estimated at 93% using RAPD markers, indicating predominant outcrossing and maintenance of adequate genetic variability within families (Gaiotto et al. 1997). A complex pattern of mating was described in an *E. regnans* seed orchard in Australia where gene dispersal was influenced by crop fecundity and orchard position of mother trees with approximately 50% of effective pollen gametes coming from males more than 40 meters away from mother trees (Burczyk et al. 2002). In a detailed mating system study in a *E. grandis* orchard in Madagascar, the outcrossing rate was found to be 96.7%, but a pollination rate from outside the seed orchard of 39.2% was estimated based on six microsatellite markers (Chaix et al. 2003).

The ability to determine paternity precisely using DNA markers was recently proposed as a short term breeding tactic for *Eucalyptus*. The conventional way to drive modifications in old forest tree seed orchards is to establish progeny trials involving each parent tree and then evaluate its contribution to the performance of the progeny by estimating its general and specific combining ability (GCA and SCA). Grattapaglia et al. (2004) applied retrospective parent selection based on paternity testing of superior offspring. After identifying seed mixtures, selfed individuals, and offspring sired by pollen parents outside the orchard, one particular pollen parent was found to have sired significantly more high yielding progeny trees. Based on

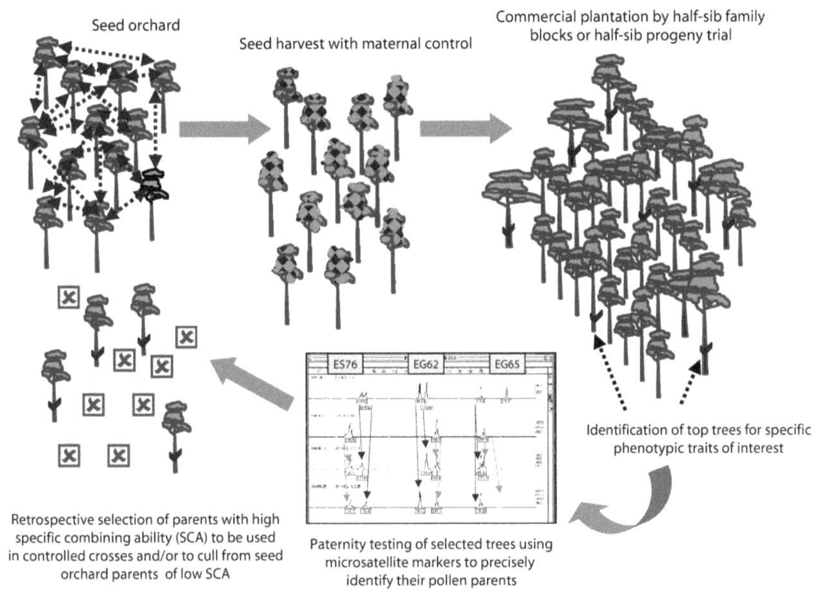

Fig. 11.7 Retrospective selection of elite parents based on paternity testing using microsatellite markers of progeny individuals displaying superior performance (See color insert)

these results, low-reproductive-success parents were culled from the orchard and management procedures were adopted to minimize external pollen contamination. A significant difference ($p < 0.01$) in mean annual increment was observed between forest stands produced with seed from the orchard before and after selection of parents and revitalization of the orchard. An average realized gain of 24.3% in volume growth was obtained from the selection of parents as measured in forest stands at age two to four years. Marker-assisted tree breeding efficiently identified top parents in a seed orchard and resulted in an improved seed variety. It should be applicable for rapidly improving the quality of output from seed orchards especially when the breeder is faced with an emergency demand for improved seeds (Fig. 11.7).

11.4 Molecular Breeding and Genomics

11.4.1 Genetic Resources for Eucalyptus *Genomics*

While public *Eucalyptus* genomic resources including a draft genome sequence will become available in the very near future (http://www.jgi.doe .gov/News/news_6_8_07.html) (http://www.jgi.doe.gov/sequencing/why/CSP2008/eucalyptus.html) biological resources and precise phenotyping represent the real limitation of many genomic projects. Especially in forest trees, where generation times and phenotype assessment can take years, the availability of ideal experimental populations should be one of the main targets in any genomic project. For example, the driving principle adopted in the GENOLYPTUS project in Brazil is that there is ample genetic variation within the genus *Eucalyptus* and more specifically within the subgenus *Symphyomyrtus* to allow profound genetic modification of the current planting stock (Grattapaglia et al. 2004). *Eucalyptus globulus* contrasts with commonly used tropical species such as *E. grandis, E. urophylla,* and *E. camaldulensis*, for it displays a number of wood properties extremely interesting to industry. *Eucalyptus globulus* germplasm supply stands out as a rich source of genetic variation for all the target wood traits and therefore a key resource for eucalypt genomic research especially for the pulp and paper industries.

A number of experimental data from hybridization experiments in Brazil are already available to clearly demonstrate that the introgression of temperate *E. globulus* alleles into tropical hybrid breeding programs coupled to clonal propagation of selected individuals will result in significant reductions in wood specific consumption (de Assis 2000; de Assis et al. 2005b). Most Brazilian breeders are investing heavily on the potential impact of *E. globulus* in their programs. This same view was also adopted in the construction of the biological resources for genomic research in the GENOLYPTUS project. Over 20 intra- and interspecific families involving different *Eucalyptus* species were generated and are currently being used for QTL mapping and validation, gene expression, and proteomics. By establishing a rich resource of genetic variation resulting from hybridization, we may uncover the genetic causes that make the *E. globulus* wood superior to the wood of *E. grandis*.

11.4.2 Molecular Markers and Maps for Eucalyptus

In the last 10 years a number of studies have reported genetic maps for *Eucalyptus* built from combinations of several hundred RAPD and AFLP markers (Grattapaglia and Sederoff 1994; Verhaegen and Plomion 1996; Marques et al. 1998; Myburg et al. 2003). together with RFLP, isozymes, expressed sequence tags (EST), genes and some microsatellites (e.g., Byrne et al. 1995; Bundock et al. 2000; Gion et al. 2000; Brondani et al. 2002; Thamarus et al. 2002). In contrast to crop species where mapping populations are designed based on contrasting inbred lines, map construction in eucalypts has relied on available pedigrees drawn from operational breeding programs. These pedigrees generally involve only the highly heterozygous parents and their F1 progeny, either full-sibs or half-sibs. Genetic mapping has therefore been carried out using a pseudo-testcross strategy, analyzing dominant markers present in one parent and absent in the other (Grattapaglia and Sederoff 1994) (Fig. 11.8). Maps are individual-specific and cannot be aligned or integrated as such unless other markers common to both maps are also used. Consequently, although many genome maps of eucalypts have been constructed, the use of the linkage information tends to remain restricted to the pedigree employed as the mapping population, limiting the inter-experimental sharing of linkage mapping and QTL data generated.

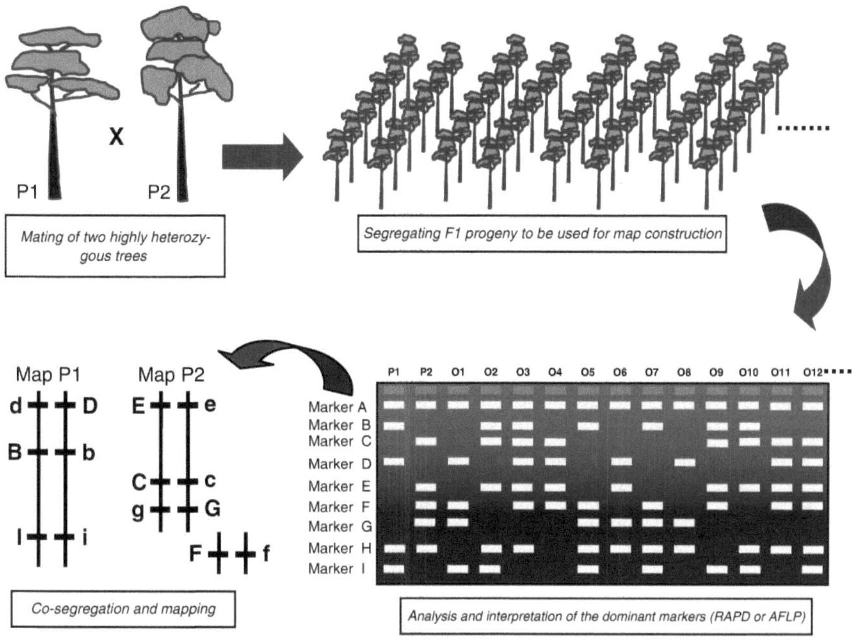

Fig. 11.8 Pseudo-testcross mapping strategy. Analysis and interpretation of the dominant markers: markers B, D, and I that are heterozygous in parent P1 and absent in parent P2, thus segregating in the F1 in a 1:1 ratio (presence: absence); markers C, E, F, and G are heterozygous in parent P2 and absent in parent P1, thus segregating in the F1 in a 1:1 ratio; marker H is heterozygous in both parents, thus segragting in a 3:1 ratio (presence:absence); marker A is homozygous in both parents, thus monomorphic with no segregation. Co-segregation and mapping: Markers B, D, and I co-segregate and are linked in the P1 parental map; markers E, C, and G co-segragate and are linked in the P2 parental map; marker F is not linked to the other markers in the P1 parental map; marker H is not shown but could be used to link both parental maps

Advancements in QTL validation across pedigrees for marker-assisted selection (MAS) in *Eucalyptus* will strongly depend on the availability of higher throughput, higher polymorphism typing systems such as microsatellites, organized in dense genetic maps (Brondani et al. 1998; Thamarus et al. 2002). Until mid 2006, only 137 autosomal microsatellite markers had been published for species of *Eucalyptus*, including 67 from *E. globulus, E. nitens, E. sieberi*, and *E. leucoxyon* (Byrne et al. 1996; Steane et al. 2001; Glaubitz et al. 2001; http://www.ffp.csiro.au/tigr/molecular/eucmsps.html; Ottewell et al. 2005) and 70 from *E. grandis* and *E. urophylla* (Brondani et al. 1998; Brondani et al. 2002). A set of 35 chloroplast DNA microsatellites was developed based on the full cp-DNA sequence of *E. globulus* (Steane et al. 2005). Microsatellite transferability across species of the subgenus *Symphyomyrtus* that includes all the most widely planted species, varies between 80 and 100% depending on the section to which the species belongs. It is around 50 to 60% for species of different subgenera such as *Idiogenes* and *Monocalyptus* and goes down to 25% for the related genus *Corymbia* (Kirst et al. 1997). Microsatellite comparative mapping data have also shown that genome homology across species of the same subgenus *Symphyomyrtus* is very high not only in terms of microsatellite flanking sequence conservation but also marker order along linkage maps (Marques et al. 2002). Although some 10s of microsatellites have been mapped on existing RAPD and AFLP framework maps (Brondani et al. 2002; Marques et al. 2002; Thamarus et al. 2002), the genus *Eucalyptus* still lacked a comprehensive genetic map widely useful for molecular breeding. To fill this gap, a novel set of 230 new microsatellites and a consensus map covering at least 90% of the estimated recombining genome of *Eucalyptus* have recently been assembled and published. This map has 234 mapped loci on 11 linkage groups, an observed length of 1,568, cM and a mean distance between markers of 8.4 cM (Brondani et al. 2006). The generalized use of an increasingly larger set of interspecific transferable markers and consensus mapping information will allow faster and more detailed investigation of QTL synteny among species, validation of QTL and expression-QTL across variable genetic backgrounds, and positioning of a growing number of candidate genes co-localized with QTL, to be tested in association mapping experiments.

11.4.3 QTL Mapping in Eucalyptus

Following the construction of linkage maps, several groups reported the identification of genomic regions that have a significant effect on the expression of economically important traits in *Eucalyptus*. QTL mapping experiments have, without exception, found a few major-effect QTL for all traits considered, in spite of the limited experimental precision, the lack of pre-designed pedigrees to maximize phenotypic segregation, and the relatively small segregating populations evaluated. This can be explained by the undomesticated nature and wide genetic heterogeneity of eucalypts added to the fact that most QTL mapping experiments were carried out in interspecific populations thus taking advantage of contrasting gene pools. QTL for juvenile traits such as seedling height, leaf area, and seedling frost tolerance have been mapped (Vaillancourt et al. 1995b; Byrne et al. 1997a, 1997b), while traits related to vegetative propagation ability such as adventitious rooting, stump sprouting, and in vitro shoot multiplication have also been detected (Grattapaglia et al. 1995; Marques et al. 1999) as has a major QTL for early flowering (Missiaggia et al. 2005a). In addition, QTL for insect resistance and essential oil traits were mapped (Shepherd et al. 1999) and a major QTL for *Puccinia psidii* rust resistance was found and mapped in *E. grandis* (Junghans et al. 2003). Major QTL were also found for crop age traits such as volume growth, wood specific gravity, bark thickness and stem form (Grattapaglia et al. 1996; Verhaegen et al. 1997; Kirst et al. 2004; Thamarus et al. 2004; Kirst et al. 2005b).

Although QTL of relatively large effects have been detected for growth traits, the best opportunities for QTL mapping for MAS are related to specialized wood properties that impact industrial processes. These traits are usually difficult to measure both because they require destructive whole stem sampling and because they are traits that are expressed late. Myburg (2001) demonstrated the application of indirect, high throughput phenotyping of wood quality traits in *Eucalyptus* by near-infrared spectroscopy (NIR) for QTL mapping in a hybrid *E. grandis* x *E. globulus* backcross population. Approximately 300 individuals that had been previously genotyped with AFLP markers were analyzed by NIR and predictions were made for pulp yield, alkali consumption, basic density, fiber length and coarseness, and several wood chemical properties (lignin, cellulose, and extractives). A variety of molecular marker classes and pedigree types were used in these experiments. QTLs were detected in F1, inbred or outbred F2, and half-sib families with or without clonal replicates. Also looking at wood quality traits, Thamarus et al. (2004) used novel high throughput and traditional methods to quantify wood density, fiber length, pulp yield, and microfibril angle (MFA), in two full-sib families of *E. globulus* that shared a common parent. Pulp yield and cellulose content were determined by NIR, and MFA was quantified by SilviScan^{TM}(Evans 1994; Evans et al. 2001). Except for fiber length, QTL for all traits could be detected in both populations, including three QTL in common genetic regions on both crosses for wood density, one for pulp yield and one for MFA. The proportion of phenotypic variation explained by the QTL identified in both crosses ranged from 3.2% to 15.8%.

Although the number of reports detecting QTL in *Eucalyptus* has grown and these have become increasingly sophisticated, the large majority of mapped QTL have been localized on RAPD or AFLP maps. Consequently, it is impossible to compare positions of QTL for the same or correlated traits, seriously limiting the long term value for MAS. Exceptions are QTL studies where transferable markers such as a few microsatellites (Marques et al. 2002; Thamarus et al. 2004) or candidate genes (Gion et al. 2000; Thamarus et al. 2004) were also mapped so that it is at least possible to make a rough preliminary comparison of QTL locations at the linkage group level. Especially in the genus *Eucalyptus* where breeders worldwide take advantage of interspecific genetic variation for wood properties and disease resistance through hybridization, the recent availability of a robust, genus-wide genetic map with highly transferable microsatellite markers (Brondani et al. 2006) should stimulate improved genomic undertakings including QTL validation across pedigrees, co-localization of QTL, and candidate genes for guiding association mapping experiments

Recent QTL validation efforts have started to pinpoint candidate genomic regions controlling traits of interest. The construction and analysis of multiple QTL maps for wood properties has shown the possibility of performing comparative QTL mapping analysis. Genetic maps were constructed for three genetically unrelated families. The first involved a cross between two elite Rio Claro natural hybrid trees involving predominantly *E. grandis* and *E. urophylla*. The other two maps were derived from crosses between pure *E. grandis* and *E. urophylla* select trees, and the F1 progeny was cloned and planted in replicated trials in five environments throughout Brazil. Map construction used as a reference the integrated map involving 234 markers developed by Brondani et al. (2006) and added new markers developed from genomic shotgun as well as EST sequences. In the cross involving hybrid parents, 10 QTLs were detected for parent clone 235, with LOD (logarithmic odds) scores varying between 2.9 (lignin content) and 4.2 (specific wood consumption). For hybrid parent 221, five QTLs were detected with LODs varying from 2.9 (cellulose yield) to 4.8 (basic wood density). Comparative QTL mapping across the three pedigrees, as well as to QTLs and candidate gene mapping carried out in *E. globulus* by other research groups, revealed a number of syntenic QTLs for cellulose yield and lignin content and for different but correlated fiber traits as well as candidate genes for lignification. These are exciting results for *Eucalyptus,* as they revealed the first QTL validation data and demonstrated the power of using microsatellites across multiple pedigrees to allow precise determination of target genomic regions for gene discovery, association mapping and eventually marker assisted selection (Missiaggia et al. 2005b).

11.4.4 Marker-Assisted Selection in Eucalyptus

Twenty-seven years have passed since the first demonstration that QTL for major effects could be identified with molecular markers in plants (Stuber et al. 1980; Paterson et al. 1988; Lander and Botstein 1989). Several reviews have described the potential benefits and caveats of MAS in the plant genetics literature (e.g., Tanksley 1993; Beavis 1998; Young 1999; Mauricio 2001; Dekkers and Hospital 2002). Yet, large scale operational MAS is still largely restricted to very few crops and for very specific applications. Maize is probably the best example, where the financial returns on hybrid seed development coupled to the ability to fully control germplasm has prompted large scale investments in MAS by the private sector based on high throughput single nucleotide polymorphism (SNP) genotyping platforms. Based on a detailed understanding of the molecular architecture of quantitative traits, current applications include yield oriented AB-QTL (advanced backcross QTL) systems, whole-genome selection strategies (Meuwissen et al. 2001), as well as accelerated line conversion following trait introgression by marker assisted backcrossing. In *Eucalyptus* and forest tree breeding in general, the application of molecular markers for directional selection is still an unfulfilled promise. This is largely due to (a) the recent domestication of tree crops and hence the high genetic heterogeneity and linkage equilibrium of breeding populations; (b) the inability to develop inbred lines to allow a more precise understanding of genetic architecture of quantitative traits; (c) the absence of simply inherited traits that could be immediately and more easily targeted; and finally (d) the very limited number of scientists actually working on forest trees.

Eucalyptus breeding programs vary broadly according to the target species or hybrid, the possibility of deploying clones and the amount of resources available to the breeder. However, from the standpoint of integrating MAS, a reasonable premise is that MAS will only be justifiable when the breeding program has already reached a relatively high level of sophistication, fully exploiting all the accessible breeding and propagation tools. Advanced breeding programs that aim at elite clone selection involve a significant amount of time and effort being devoted to clonal testing before effective recommendations can be made concerning new clones for operational plantations. Sub-line breeding for hybrid performance combined with clonal propagation of selected individuals is being used increasingly for extracting new elite clones (Potts 2004). Progeny trials together with expanded single family plots where larger numbers of full-sibs per family are deployed are used for very intensive within-family selection. This selection is generally carried out at half-rotation age based on growth performance and on a preliminary assessment of wood specific gravity using indirect non-destructive techniques (Fig. 11.9). Vegetative propagules are then rescued from selected trees either by coppicing, sequential grafting, or in vitro techniques, multiplied, and then used for the establishment of clonal tests, see Fig. 11.5.

This breeding scheme generates large amounts of linkage disequilibrium by hybridization and substantial amounts of non-additive genetic variation can be captured by vegetative propagation. These are favorable conditions for MAS in forest trees (Strauss et al. 1992). Favorable alleles at QTL segregating within-families

Fig. 11.9 Clonal trial of selected Eucalyptus hybrid clones at age 3 years in Minas Gerais state, Brazil (See color insert)

could be efficiently tagged with microsatellite markers in linkage with the actual functional polymorphisms and used for marker assisted within-family selection for superior individuals (O'Malley et al. 1994). QTL linked markers could be used to carry out early selection thus reducing the time to carry out the first selection especially for traits related to wood properties, and reducing the number of trees to be selected, propagated, and advanced to clonal trials (Fig. 11.10). Given their relatively short rotations and the ability to capture non-additive genetic variation, eucalypts may be the first forest tree crop to broadly apply MAS.

Even though the costs of genotyping have fallen in recent years, DNA analysis is still costly. The most likely cost-beneficial application of MAS in *Eucalyptus* will be for traits that provide significant added value to the final product such as branching habit (for solid wood) or wood chemical traits or that allow clonal deployment such as adventitious rooting or somatic embryogenesis response. Within all possible quality traits, preference would be for those that display medium to high heritabilities but where phenotype assessment is difficult, expensive, or requires waiting until the tree reaches maturity. Wood quality traits typically require the tree to start accumulating mature wood and involve relatively lengthy procedures for phenotypic evaluation in the laboratory. These kinds of traits could be interesting targets for

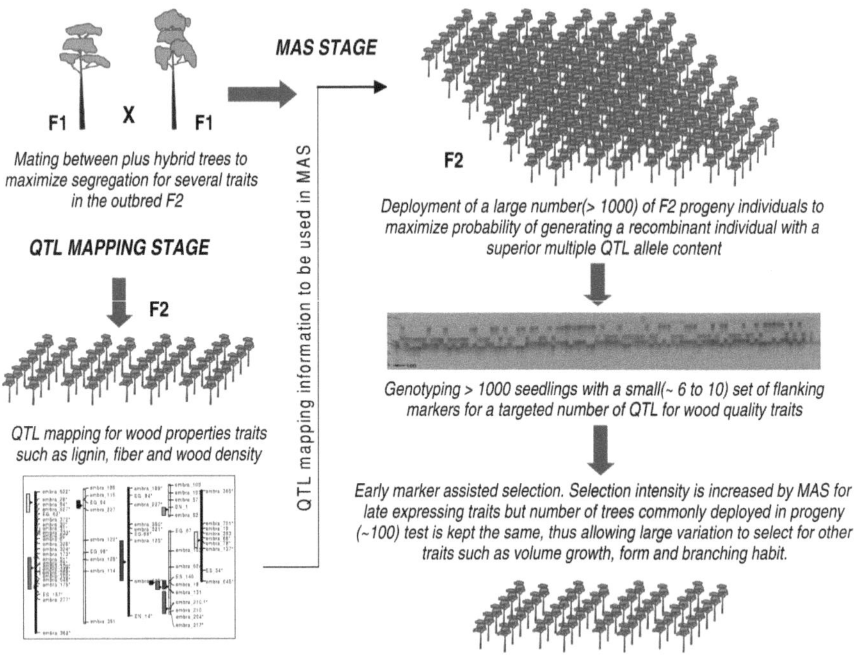

Fig. 11.10 Proposed scheme for early marker assisted selection (MAS) of superior (plus) trees to be used as clones. QTL mapping is carried out and flanking markers in linkage disequilibrium with favorable alleles at major effect QTL alleles are identified. These markers are used for early within-family selection in expanded progeny sizes at very early age to select for mature wood expression traits such as density and lignin content (See color insert)

MAS in *Eucalyptus*, given that the costs of genotyping are sufficiently competitive and precision is high when compared with direct phenotype measurements. It is important to point out, however, that with the recent developments of fast sampling and indirect wood chemistry measurements based on NIR (Schimleck et al. 1996), the potential gain will be realized only on the basis of the time savings provided by very early selection.

11.4.5 Association Mapping in Eucalyptus

With the rapid advancement of genome projects generating a large amount of sequence information and SNPs (one-letter variations in the DNA sequence that contribute to differences among individuals) data, plant genomics has experienced a growing interest in an alternative approach for the identification of genes underlying quantitative traits. The new model is based on the possibility of investigating phenotypes caused directly by sequence variation in genes and not by association with linked markers. This association mapping approach identifies candidate gene sequence variation and relies on the existence of linkage disequilibrium (LD) (non-random association between alleles at linked loci) between detectable sequence polymorphism SNPs and QTN (quantitative trait nucleotide) that ultimately determine phenotypic variation (Neale and Savolainen 2004).

In considering MAS for forest trees, more will likely be learned from experiences in livestock (Dekkers 2004) than from annual crop plants, with the added advantage, however, that gains in forest trees can be quickly realized by large-scale cloning of selected individuals. Three different levels of marker-trait relationships (Dekkers 2004) are relevant to trees: (a) direct markers, i.e., loci that encode functional variants; (b) linkage disequilibrium (LD) markers: loci that are in population-wide linkage disequilibrium with the functional mutation; (c) linkage equillibrum (LE) markers: loci that are in population-wide linkage equilibrium with the functional mutation in outbred populations but do display LD in specific segregating pedigrees (Fig. 11.11).

Other than the recent encouraging discovery of an LD marker for MFA in *Eucalyptus* (Thumma et al. 2005), only LE marker-trait associations have been described in forest trees. LE markers have been readily detected on a genome-wide basis by analyzing large full-sib families with sparse marker maps allowing the detection of most QTL of moderate to large effects as discussed above. For the other two types of marker-trait association, only now the first association mapping experiments are being started to uncover LD markers, i.e., polymorphisms that are sufficiently close to the functional mutation (Neale and Savolainen 2004; González-Martínez et al. 2007). The challenge however is considerable, as LD in outcrossing forest trees such as pines decays very rapidly, in general within 1000 bp (Neale and Savolainen 2004; Krutovsky and Neale 2005) and similar behavior has been seen in the few *Eucalyptus* genes analyzed to date with significant LD extending for only a few hunderd basepairs (Thumma et al. 2005; Kirst et al. 2005c; Faria et al. 2006)

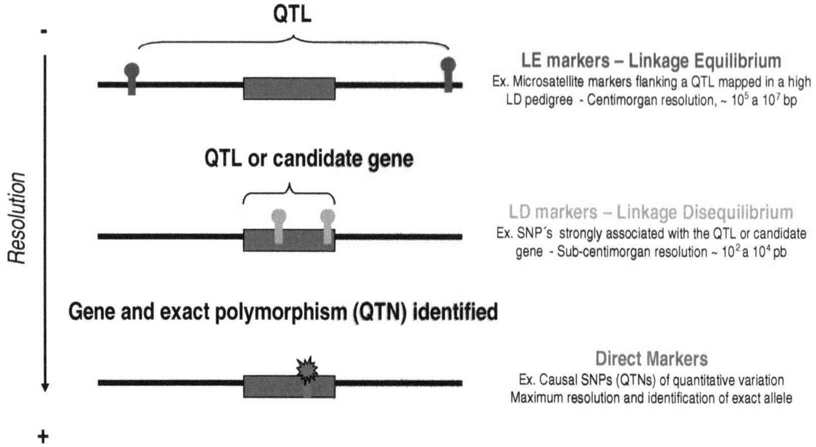

Fig. 11.11 Classification of three different types of marker-trait associations relevant to Eucalyptus MAS (see text for details) (See color insert)

(Fig. 11.12). Genome-wide association studies for LD marker-trait discovery in tress will require very high SNP marker densities that are currently impractical (but see below in Conclusion and Perspectives), so that a candidate gene approach remains attractive. Finally direct markers (i.e., polymorphisms that code for the functional mutations) would be the most valuable and directly applicable in breeding. However, they are the most difficult to detect because causality is very difficult to prove unless the traits of interest have very high penetrance Mendelian inheritance.

The candidate gene approach has the advantage that once a major effect gene is determined and validated, MAS could then be practiced directly on the gene and therefore would not rely on the need for strong association (linkage disequilibrium) between the marker allele and the favorable allele at the gene of interest. The challenge, however, is the correct identification of candidate genes. This is not an easy task and every effort should be made to maximize the probability of choosing the proper genes. The choice of candidate genes is an elusive target for the majority of phenotypes relevant to forest trees. It requires knowledge of biochemistry, physiology, and development that is generally not available even for well-defined phenotypes and/or known metabolic pathways. Testing the role of a candidate gene can be carried out by a conventional co-segregation analysis in structured segregating populations such as outbred F2's, half sib families or diallel designs in small sublines of a few trees where the gene is used as a marker in the attempt to relate the sequence polymorphism in the gene with variation in the quantitative trait. Allelic variation at the gene is defined by haplotypes, comprising a number of SNPs. The majority of SNPs have no effect, or some cause subtle differences in the final effect of the gene, and hence the phenotype. Significant differences in phenotypic means among candidate gene haplotype classes should identify candidate gene alleles with the greatest effect on the trait of interest. Another approach to test and validate candidate genes is to look for SNP-phenotype associations in germplasm collections or natural

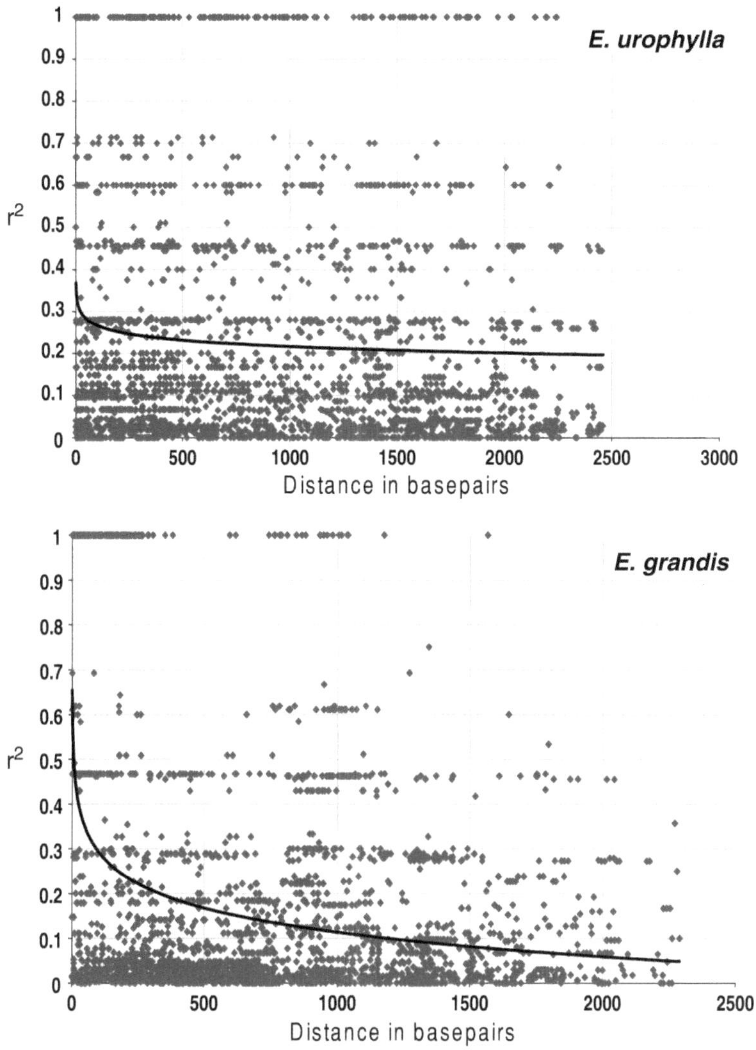

Fig. 11.12 Rapid decay of intragenic linkage disequilibrium estimated as r^2 in the cinnamyl alcohol dehydrogenase (*cad*) gene in two different *Eucalyptus* species (Faria et al. 2006)

populations involving contrasting phenotypes. The objective again is to correlate the distribution of candidate gene genotypes (DNA sequences) and relevant phenotypes.

11.4.6 Gene Discovery by EST Sequencing

Until the completion of the draft sequence of the Poplar genome (Tuskan et al. 2006), gene discovery in trees was based on sequencing expressed portions of the genome, called expressed sequence tags (ESTs). EST sequencing quickly generates many partial gene sequences for the organism of interest making them available in organized collections of clusterized sequences for further molecular investigation. Partial gene sequences have mutiple applications including: (a) the identification of genes and gene families involved in the control of target traits; (b) the identification of new molecular markers, such as microsatellites and SNPs for mapping; (c) supplying sequence information or biological reagents to build microarrays for large scale gene expression studies and (d) supplying sequences or genes for transgenic experiments.

Since the first *Eucalyptus* gene sequence was deposited in Genbank some 14 years ago (Feuillet et al. 1993), the number of public sequences has increased very slowly compared to other crops or major tree species. The reason has been the industrially oriented, proprietary nature of the genomic research done in *Eucalyptus*. About 10 years ago, several tens of thousand sequences were generated by Genesis in New Zealand and later incorporated by the US-based company Arborgen (M. Hinchee personal communication). Dupont in collaboration with the former Australian company Forbio generated a database of approximately 14,000 ESTs from *E. grandis* around 1997 (Scott Tingey pers. comm.) that were later annotated and used in the pioneering work of expression-QTL (Kirst et al. 2004). Other mostly private ESTs databases for *Eucalyptus* were built between 2002 and 2004 by different groups in Brazil (Grattapaglia et al. 2004) and in Japan (Sato et al. 2005). The first cDNA microarray experiments carried out in *Eucalyptus* required the publication of a few thousand sequences (Kirst et al. 2004; Paux et al. 2004). Until 2006, there were only around 2,700 ESTs in Genbank, derived from several different tissues, the large majority deposited by Dupont. Recently the French group led by J. Grima-Pettenati of the Transcriptional Regulation and Wood Formation team of CNRS deposited a larger set of ~13,000 sequences mostly derived from *E. globulus*. As of May 2007 there were 15,950 EST sequences available in Genbank out of the 33,984 deposited nucleotide sequences for *Eucalyptus*.

Small scale targeted EST sequencing efforts have been successful in identifying relevant sets of genes involved in wood formation. Paux et al. (2004) used xylem subtractive libraries enriched for xylem specific sequences resulting in 224 unique sequences with a high proportion of sequences classified either as "no hits" or as "proteins of unknown function," revealing the complexity of secondary xylem in woody angiosperms. Subsequently, a xylem vs. phloem subtractive library generated 263 unique sequences and a total of 87 unigenes preferentially expressed in xylem and where they are involved in hormone signaling and metabolism, secondary cell wall thickening and proteolysis. Some of these genes, including unknown genes, may be xylem-specific and likely to control important functions in xylogenesis (Foucart et al. 2006).

The different species planted around the world and the hybrid breeding system adopted in *Eucalyptus* has driven a distinctive multi-species approach of the EST sequencing efforts. In the GENOLYPTUS project, the initial database of over 125,000 EST sequences is derived from 20 different cDNA libraries from four *Eucalyptus* species with a focus on xylem transcripts. Several thousand cDNA clones were sequenced from *E. globulus*, *E. grandis*, *E. pellita*, and *E. urophylla* xylem and phloem libraries derived from a number of individuals for each species for both gene and SNP discovery at both the inter- and intraspecific levels. This database contains all the known genes for lignin and cellulose metabolism as well as several other genes for cell wall structural proteins that may be involved in the control of chemical wood properties (Pasquali et al. 2005).

With the speed and dramatic cost reductions brought about by pyro-sequencing technologies (Margulies et al. 2005) the size of public *Eucalyptus* EST databases will soon become larger and more diverse. Recently, Novaes et al. (2007) reported on the generation of more than 1 million *Eucalyptus grandis* reads with average lengths of 100-200 bp from only three runs of a Genome Sequencer 20 and FLX Systems (454 Life Sciences Corporation), generating a preliminary assembly with approximately 29,000 contigs. As such large numbers of sequences are made public it will certainly drive the publication of all the existing private EST databases. Such a radical jump in the number of available sequences will consolidate a very valuable source of multi species sequence information for eucalypt genomic research.

11.4.7 Analysis of Gene Expression and Detection of Expression-QTLs

The main focus of gene expression studies in trees has been the elucidation of the metabolic pathways that determine wood formation. In the last few years, several studies have been carried out in Poplar using microarrays that represent the fully transcribed genetic complement of the species. Gene expression has been analyzed in different stages of the lignification process, from meristematic cells all the way to maturation and programmed cell death (e.g., Hertzberg et al. 2001). Genes that encode enzymes involved in lignin and cellulose biosynthesis as well as a number of transcription factors and other genes that regulate wood formation, operate in a well-defined, rigorous, stage-specific way. This approach has revealed genes that are over or under expressed in specific moments of wood development.

In *Eucalyptus*, the lack of a publicly available whole transcriptome microarray platform has hindered public larger scale studies. A transcript profiling study of 231 genes preferentially expressed in differentiating *Eucalyptus* xylem was used to investigate the early changes in gene expression during tension wood formation. Cluster correlation analysis revealed differential regulation of lignin biosynthetic genes and highlighted candidate genes responsible for the genetic and environmentally induced variation of wood quality traits (Paux et al. 2005). A cDNA-AFLP approach was used by Ranik et al. (2006) to profile gene expression in a 6-year-old, fast-growing *Eucalyptus* tree. The expression profiles of 6,385 transcript-derived fragments (TDFs) were analyzed across four major woody tissues (mature xylem, immature xylem, phloem, and cork) to provide a global view of transcript abundance and variability in the *Eucalyptus* stem. About 21% of the TDFs were differentially expressed and could be grouped into clusters representing co-expressed genes. Seventy-one TDFs representing different gene clusters were isolated and characterized. These included genes implicated in cell fate, signal transduction, and cell wall biosynthesis, processes closely associated with xylogenesis. Recently, the GENOLYPTUS project successfully tested a transcriptome-wide oligoarray of ~398,000 probes representing all the 21,400 unique genes discovered by sequencing (Pasquali G, Pappas GJ and Grattapaglia D unpublished). The pilot experiment showed that the oligoarray platform selected (Nimblegen Systems) is extremely robust providing consistency in signal across probe replicates as well as between biological replicates (i.e., different trees of the same clone). Gene expression in differentiating xylem was compared intra- and interspecifically between *E. grandis* and *E. globulus*. The preliminary analyses showed a number of genes differentially expressed both within and between species. These results point to a high complexity of the genetic control of mature wood formation. Further experiments are now planned using this first version of the GENOLYPTUS array where an expression-QTL mapping analysis in segregating populations will be carried out to identify interesting candidate genes for association mapping experiments.

Proving the cause-effect relationship between genes identified in microarray experiments and the phenotypic variation observed is a complex task that requires additional experiments. Microarrays are a very effective but only exploratory way to identify key genes to understand and manipulate wood formation. A more functional and forward genomics approach to the study of gene expression has been the integration between genetics and genomics based on the analysis of gene expression in parallel to an underlying Mendelian framework of genetic mapping and QTL discovery. The so called "genetical genomics" approach (Jansen and Nap 2001) combines gene mapping and gene expression data to identify loci controlling gene expression (eQTLs) that may underlie functional trait variation. This approach allows the co-localization between (a) expression QTLs, i.e., QTLs identified that explain observed differences in expression levels of specific genes on the array; (b) QTLs identified for wood quality traits, and (c) the actual position of the gene on the genetic map. This approach was applied in *Eucalyptus* by monitoring the expression of ~2,700 genes putatively involved in cell wall formation, lignin and cellulose metabolism, cell growth, and protein targeting. The key role of some lignin biosynthesis genes was confirmed and some other new unexpected genes with major effect were discovered highly correlated with volume growth as well (Kirst et al. 2004). In a subsequent study, Kirst et al. (2005b) showed that one identified expression QTL explained up to 70% of the transcript level variation for over 800 genes, and hotspots with co-localized expression QTL were identified on single tree AFLP maps typically containing genes associated with specific metabolic and regulatory pathways, suggesting coordinated genetic regulation. However the expression data also showed that the lack of conservation of the genetic architecture of transcript abundance regulation in different genetic backgrounds indicates that many different loci could be involved in modulation of transcription of these genes and that there is a complex and variable network of gene expression control. A current limitation of this genetical genomics approach in *Eucalyptus* is the lack of a full genome sequence or at least a genetic map including large numbers of mapped genes. This information is required to test whether the genetic control of gene expression occurs in cis- or trans-. In the context of MAS in *Eucalyptus*, a better understanding of such "master expression QTL" that apparently control cascades of gene expression of important biochemical pathways may be very promising targets for detailed characterization in association mapping experiments to uncover relevant polymorphisms to be used in molecular breeding practice.

11.4.8 Physical Mapping and Genome Structure

While genetic mapping of *Eucalyptus* has evolved quite rapidly, efforts have been timid in generating physical mapping resources for species of the genus. A complete physical map for *Eucalyptus* will certainly represent a great experimental resource for years to come as it provides a physical organized equivalent of the genome to access genes and regulatory regions for multiple applications both in transgenics and molecular breeding. To this end, we have constructed a *Eucalyptus grandis* BAC library with an initial genome coverage of ~4x with over 70% of the inserts >150 kb long. Using this BAC library we have been isolating and shotgun sequencing a number of candidate genes involved in wood chemical composition. This strategy has allowed us to identify BACs containing important genes that encode enzymes involved in the lignin or cellulose biosynthetic pathway as well as some transcription factors. For two key genes, 4-CL and CAD, the smaller single BAC (30 kbp for CAD and 120 kbp for 4-CL) as evaluated by Pulse Field Gel Electrophoresis (PFGE) was selected for constructing shotgun libraries with average insert size of 2 kbp. Assembly of 1,052 reads (3.5x) for 4-CL resulted in a 5,477 bp contig covering the whole gene but part of the 5-UTR. For CAD, 768 reads (10x) allowed the assembly of a 9,785 bp contig of what was found to be CAD2 (Brommonschenkel et al. 2005).

With the full genomic sequence of specific genes, it will be possible to identify regulatory regions and carry out a detailed analysis of polymorphism in a set of individuals by resequencing specific upstream regions in an association mapping approach. Our immediate plan is to build in collaboration with the Arizona Genomics Institute a complete physical map for *E. grandis* by fluorescent fingerprinting as a contribution to an international *Eucalyptus* Genome Network recently established. This physical genomic resource from *E. grandis* will facilitate the identification of specific genomic regions in *E. globulus*. It should be faster, for example, to clone the full homolog gene from *E. globulus* and thus compare in detail potential regulatory regions responsible for differential patterns of gene expression and resulting phenotypic variation. Such information would not be available using only translated sequences.

To obtain a general overview of the structure and composition of the *E. grandis* genome, we sample sequenced 7,395 randomly sheared genomic DNA clones representing roughly 3 Mbp of sequence (Lourenço et al. 2005). The analysis revealed that on average the Eucalyptus genome has a GC content of 40.2 % with 39% in introns, 45.5% in exons and the remainder in repetitive regions. From the total bases sequenced approximately 1.4% was located in transposons, distributed in 310 interspersed repetitive genetic elements. We also identified 1636 low complexity sequences and 987 microsatellites. In total we estimated that 5.8% of the

Eucalyptus genome is represented by repetitive elements, possibly due to the existence of elements not yet described and/or exclusive to *Eucalyptus*. To identify putative genes we used an alternative approach by comparing the genomic sequences with the Genolyptus ESTs database using the GenESTate software. The sequences were clustered using the CAP3 software, resulting in 766 contigs and 5428 singlets, the former showing an average of 1200 bp. These 766 contigs were compared with a relatively reduced set of available ESTs at the time (~5,000 *E. grandis* ESTs from mature leaf tissue and ~6,000 *E. urophylla* ESTs from xylem). From the 766 contigs we found 44 that showed high similarity to some ESTs. The coding portion of the sequences accounted for around 2% of the total sequences. Other 166 possible genes were identified, 76 of them classified as housekeeping,

11.4.9 Breeding by Transgenic Technology

Transgenic technology is undoubtedly a powerful complementary tool available to the molecular breeder. Considering that industrial *Eucalyptus* forests are almost exclusively clonal, transgenics will most likely have an increasing role not only in wood quality improvement but in resolving problems related to pest and pathogens susceptibility and/or abiotic stress tolerance (e.g. frost, drought) that might limit the expansion or survival of existing plantations, as in the case of annual crops. The introduction of genes that confer traits that do not display variation within the *Eucalyptus* gene pool or impossible to be attained by the natural recombination processes might radically modify the ways that forests are planted or that forest products are derived.

In spite of the recognized economic importance of Eucalyptus in world forestry, very little has been published on transgenic experiments in species of the genus. While Eucalyptus tissue can be transformed by *Agrobacterium tumefasciens*, major difficulties are faced in the regeneration step. Several reports have documented the production of transformed callus, tissue, and root organs; however, reports on transformed plants are scarce (Machado et al. 1997; MacRae and van Staden 1999). A marked genotype effect has been observed on the efficiency of regeneration and consequently, stable transformation. This fact has prompted several groups to first identify Eucalyptus "lab rats," i.e., easily regenerable genotypes, and only after that develop improved protocols to generate large numbers of independent transformation events. These "lab rats" are not yet available in the public domain and so it is for the transformation protocols. Again, this research is carried out primarily by private companies. The most representative and complete work published on *Eucalyptus* transformation was carried out by a French-Brazilian group where an easily regenerable plant was selected after screening around 300 plantlets of an *E. grandis x E. urophylla* hybrid. This plant exhibited the best compromise between short- and long-term GUS expression level, regeneration, and micropropagation efficiency under selection after transformation. Following this selection step, stably transformed plants were obtained for some reporter genes, and a cinnamyl alcohol dehydrogenase (CAD) antisense cDNA from *Eucalyptus gunnii* at an efficiency of 10% (120 transformed and confirmed plants from 1,200 explants), thus demonstrating the efficiency of the protocol (Tournier et al. 2003). Recently, Chen et al. (2006) also described a basic *Agrobacterium*-mediated genetic transformation protocol through organogenesis for the production of transgenic plants using *Eucalyptus camaldulensis*. More importantly, modifications of the protocol for mature tissues derived from elite trees and other Eucalyptus species were also described. Efficient transformation protocols have also been developed in Japan, where an *E. camaldulensis* "lab rat" has been used (Kawazu et al. 1996) as well as different labs in the USA with *E. grandis* and *E. urophylla* (M. Hinchee and V. Chiang personal communication). However, these have not yet been published although they constitute important components of patent applications.

The current information on *Eucalyptus* transgenesis points to a very promising future as far as the technical possibility of generating stably transformed *Eucalyptus* plants is concerned. However, some strategic issues in the adoption of transgenic technology for wood quality manipulation have been a matter of recent debate, including: (a) What is the magnitude of the attainable gain and cost/benefit relationship by manipulating lignification or cellulose genes when compared to the exploitation of the genetic variation in *Eucalyptus* by hybridization and intensive selection? (b) What are the specific biosafety and intellectual property issues relevant to transgenic eucalypts and the time and investment necessary to solve them to actually be able to plant transgenic trees on a large scale? (c) What is the speed by which breeding programs generate new and better clones for several adaptability traits (growth, pest resistance, clonability, etc.) compared to the time needed for regulatory approval of every new transgenic clone? (d) What is the lifespan of a patent in the local regulation as compared to the time needed to effectively make returns on the patent from the planted forest before the patent goes into public domain? (e) What are the market issues that the company has to consider in adopting transgenics both in relation to public perception and forest certification processes? All these and other issues will have to be carefully considered without overlooking that, just as occurred in annual crops such as soybean, maize, and cotton, the use of transgenics could become a major technology divide and represent the necessary condition for a forest based industry to continue competitive in the world scenario.

11.5 Conclusions and Perspectives

The successful application of molecular breeding in *Eucalyptus* will depend heavily on demonstrating and validating the clear-cut association between a DNA polymorphism and a quantitatively inherited phenotypic trait. In highly heterogeneous eucalypts, while conventional QTL mapping can reveal useful markers to be exploited in within-family selection practices, only a more direct linkage disequilibrium mapping approach will uncover population wide applicable marker-trait associations. Such studies based on candidate genes have begun and the first candidate gene association for MFA was detected. This association explains, however, only a small proportion (3.4%) of the variation needed to be really exciting news to breeders (Thumma et al. 2005). One of the key issues when embarking on an association mapping experiment is the selection of candidate genes. Maximizing the probability of choosing the proper genes requires levels of knowledge of biochemistry, physiology, and development that are generally not available yet even for well defined phenotypes and/or known metabolic pathways.

Co-localization of candidate genes and QTLs for relevant traits on linkage maps together with integrative expression-QTL mapping (Kirst et al. 2004) to identify loci controlling gene expression (eQTLs) that may underlie functional trait variations could be a powerful way forward, although choosing the correct positional candidates will depend heavily on the precision of the QTL localization by high resolution mapping. At the moment, there are two possibilities for circumventing the dilemma of choosing positional candidate genes correctly. The first is microarray-based genotyping with ultra-dense arrays of short (25 nt) oligonucleotides (Borevitz et al. 2003; Hazen and Kay 2003; West et al. 2006) coupled to methods to reduce genome complexity of DNA samples to be hybridized either by cDNA synthesis or methyl filtration. This approach would theoretically allow sufficient throughput for association genetic analysis of thousands of genes at a time. Such an array format could later turn out to be a useful instrument for MAS once validated marker-trait associations have been established. The second would be to have access to a whole genome sequence so that candidate genes in a fine mapping interval delimited by markers flanking a QTL with centimorgan resolution could be mined, reannotated, and then analyzed in association mapping experiments.

A major advance was recently announced (June 8, 2007): that the proposal to sequence the*Eucalyptus grandis* genome, submitted by the International Eucalyptus Genome Network (EUCAGEN) (www.ieugc.up.ac.za; Myburg 2004) was selected as a target eukaryotic genome for FY 2008 by the Joint Genome Institute of the US Department of Energy (http://www.jgi.doe.gov/News/news_6_8_07.html). As published in the press release: "*The biomass production and carbon sequestration capacities of eucalyptus trees match DOE's and the nation's interests in alternative energy production and global carbon cycling. The consortium of eucalyptus draws upon the expertise from dozens of institutions and hundreds of researchers worldwide.*" This public collaborative effort will contribute greatly to the advancement of *Eucalyptus* genetics, genomics, and molecular breeding by bringing together existing private databases and genomic resources and thereby expanding the value of such genome sequences. Key contributions will come from some participating groups including the USA with a large EST collection to be donated by Arborgen Inc. (M. Hinchee personal communication) and our GENOLYPTUS network in Brazil. We will make available our collection of ESTs, provide a high density microsatellite map and will collaborate with the Arizona Genomics Institute to build a public BAC library resource together with the assembly of a physical map of the sequenced genome to aid the future genome assembly. Furthermore, prospects exist that a low coverage draft genome of *E. camaldulensis*, currently being sequenced at the Kazusa DNA Research Institute in Japan (T. Hibino personal communication.) will eventually also be made public through the EUCAGEN initiative.

As this genome project advances and new and more powerful analytical tools become accessible, the true challenge to dissecting the complexity of adaptive and economically important traits in *Eucalyptus* and identifying key genes to be manipulated by molecular breeding or transgenic technologies, will depend to a large extent on our ability to phenotype trees accurately, analyze the overwhelming amount of genomic data available, and translate this into relevant information for breeding. The actual use of genomic information in molecular breeding should be considered on a case-by-case basis. Expectations should not be overemphasized until experimental data on realized gains are validated in industrial forests beyond those attained with comparable investments in conventional breeding by exploiting the extraordinary genetic variation that exists in the genus *Eucalyptus*.

Acknowledgments I am very thankful to the Brazilian Ministry of Science and Technology and the participating forestry companies in the GENOLYPTUS project for the continued support for research and the Brazilian National Research Council (CNPq) for its support through a research fellowship. I am also very grateful to all my undergraduate and graduate students and research collaborators from several Universities in Brazil and around the world as well as breeders from the forest companies for the continued discussion and scientific input.

References

Baril CP, Verhaegen D, Vigneron P, Bouvet JM, Kremer A (1997) Structure of the specific combining ability between two species of *Eucalyptus*. I. RAPD data. Theor Appl Genet 94:796–803

Beavis WD (1998) QTL analyses: power, precision, and accuracy. pp. 145–162. In Paterson AH (ed) Molecular Dissection of Complex Traits. CRC Press, Boca Raton, Florida

Bhalerao R, Nilsson O, Sandberg G (2003) Out of the woods: forest biotechnology enters the genomic era. Curr Opin Biotechnol.14(2):206–213

Binkley D, Stape JL (2004) Sustainable management of eucalypt plantations in a changing world. pp. 11–15. In: Tomé M (ed) IUFRO Conf. *Eucalyptus* in a Changing World, RAIZ, Instituto Investigação de Floresta e Papel, Aveiro, Portugal

Boerjan W (2005) Biotechnology and the domestication of forest trees. Curr Opin Biotechnol 16(2):159–166

Borevitz JO, Liang D, Plouffe D, Chang HS, Zhu T, et al. (2003) Large-scale identification of single-feature polymorphisms in complex genomes. Genome Res 13:513–523

Borralho NMG, Cotterill PP, Kanowski, PJ (1993) Breeding objectives for pulp production of *Eucalyptus globulus* under different industrial cost structures. Can J For Res 23:648–656

Brandão LG, Campinhos E, Ikemori YK (1984) Brazil's new forest soars to success. Pulp Pap Int 26:38–40

Brommonschenkel SH; Brondani RPV, Bucelli RF, Lourenço RT, Novaes E, et al. (2005) A BAC library of *Eucalyptus grandis*: characterization, fingerprinting, bac-end sequencing and shotgun assembly of lignification genes. IUFRO Tree Biotechnology S3.08. http://www.eyevisual.co.za/biotree/viewAbstract.asp

Brondani RPV, Brondani C, Tarchini R, Grattapaglia D (1998) Development, characterization and mapping of microsatellite markers in *Eucalyptus grandis* and *E. urophylla*. Theor Appl Genet 97:816–827

Brondani RPV, Brondani C, Grattapaglia D (2002) Towards a genus-wide reference linkage map for *Eucalyptus* based exclusively on highly informative microsatellite markers. Mol Genet Genomics 267:338–347

Brondani RPV, Williams ER, Brondani C, Grattapaglia D (2006) A microsatellite-based consensus linkage map for species of *Eucalyptus* and a novel set of 230 microsatellite markers for the genus. BMC Plant Biol 6:20

Brune A, Zobel BJ (1981) Genetic base populations, gene pools and breeding populations for *Eucalyptus* in Brazil. Silvae Genetica 30:146–149

Bundock PC, Hayden M, Vaillancourt RE (2000) Linkage maps of *Eucalyptus globulus* using RAPD and microsatellite markers. Silvae Genetica 49:223–232

Burczyk J, Adams WT, Moran GF, Griffin AR (2002) Complex patterns of mating revealed in a *Eucalyptus* regnans seed orchard using allozyme markers and the neighbourhood model. Mol Ecol 11:2379–91

Butler JM (2005) Forensic DNA Typing: Biology, Technology, and Genetics of STR Markers. (2nd Edition). Elsevier Academic Press, New York, 688 pp

Byrne M, Murrell JC, Allen B, Moran GF (1995). An integrated genetic linkage map for *Eucalyptus* using RFLP, RAPD and isozyme markers. Theor Appl Genet 91:869–875

Byrne M, Marquezgarcia MI, Uren T, Smith DS, Moran GF (1996) Conservation and genetic diversity of microsatellite loci in the genus *Eucalyptus*. Aust J Bot 44:331–341

Byrne M, Murrell JC, Owen JV, Kriedemann P, Williams ER, et al. (1997a) Identification and mode of action of quantitative trait loci affecting seedling height and leaf area in *Eucalyptus nitens*. Theor Appl Genet 94:674–681

Byrne M, Murrell JC, Owen JV, Williams ER, Moran GF (1997b) Mapping of quantitative trait loci influencing frost tolerance in *Eucalyptus* nitens. Theor Appl Genet 95:975–979

Campinhos E (1980) More wood of better quality through intensive silviculture with rapid growth improved Brazilian *Eucalyptus*. Tappi 63:145–147

Campinhos E, Ikemori YK (1977) Tree improvement program of *Eucalyptus* spp.: preliminary results, pp. 717–738. In: Third World Consultation on Forest Tree Breeding. CSIRO, Canberra, Australia

Chaix G, Gerber S, Razafimaharo V, Vigneron P, Verhaegen D, et al. (2003) Gene flow estimation with microsatellites in a Malagasy seed orchard of *Eucalyptus grandis*. Theor Appl Genet 107:705–712

Chen ZZ, Ho CK, Ahn IS, Chiang VL (2006) *Eucalyptus*. Methods Mol Biol 344:125–34

Costa e Silva C, Grattapaglia D (1997) RAPD relatedness of elite clones, applications in breeding and operational clonal forestry. pp. 161–166 Proc. International IUFRO Conference on *Eucalyptus* Genetics and Silviculture, Salvador, Brazil

de Assis TF (2000) Production and use of *Eucalyptus* hybrids for industrial purposes. pp. 63–74. In Nikles DG (ed) Proc QFRI/CRC Workshop on Hybrid Breeding and Genetics of Forest Trees. Department of Primary Industries, Brisbane, Australia

de Assis TF (2001) The evolution of technology for cloning *Eucalyptus* in a large scale. Proc IUFRO Conference on Developing the Eucalypt of the Future. Valdivia, Chile, INFOR. 16 pp (CDROM)

de Assis TF, Warburton P, Harwood C (2005a) Artificially induced protogyny: an advance in the controlled pollination of *Eucalyptus*. Austr Forestry 68:27–33

de Assis TF, Rezende GDSP, Aguiar AM (2005b) Current status of breeding and deployment for clonal forestry with tropical eucalypt hybrids in Brazil. Intl Forestry Rev 7:61. XXII IUFRO World Congress. Forests in the Balance: Linking Tradition and Technology, Brisbane, Australia

Dekkers JC (2004) Commercial application of marker-and gene-assisted selection in livestock: strategies and lessons. J Anim Sci. 82:313–328

Dekkers JC, Hospital F (2002) The use of molecular genetics in the improvement of agricultural populations. Nat Genet Rev 3:22–32

Doughty RW (2000) The *Eucalyptus*: A natural and commercial history of the gum tree. The Johns Hopkins University Press, Baltimore and London

Eldridge K, Davidson J, Harwood C, van Wyk G (1993) Eucalypt domestication and breeding. Clarendon Press, Oxford 288 pp

Evans R (1994) Rapid measurement of the transverse dimensions of tracheids in radial woodsections from Pinus radiata. Holzforschung 48: 168–173

Evans R, Kibblewhite RP, Stringer SL (2001) Variation in microfibril angle, density and fibre orientation in twenty-nine Eucalyptus nitens trees. Appita J 53(5):450–457.

FAO (2000) Global forest resources assessment 2000: main report. FAO Forestry paper http://www.fao.org /forestry/fo/fra/main/index.jsp

Faria DA, Alves TPM, Pereira RW, Grattapaglia D (2006). Frequência de SNPs e extensão do desequilíbrio de ligação ao longo dos genes CCR e CAD em *E. grandis*, *E. globulus* e *E. urophylla*. abstract GP251. 52nd Brazilian Genetics Congress

Feuillet C, Boudet AM, Grima-Pettenati J (1993) Nucleotide sequence of a cDNA encoding cinnamyl alcohol dehydrogenase from *Eucalyptus*. Plant Physiol 103:1447

Foucart C, Paux E, Ladouce N, San-Clemente H, Grima-Pettenati J, Sivadon P (2006) Transcript profiling of a xylem vs phloem cDNA subtractive library identifies new genes expressed during xylogenesis in Eucalyptus. New Phytol 170(4):739–752

Franklin EC (1986) Estimation of genetic parameters through four generations of selection in *Eucalyptus grandis*. pp. 12–17. Proc IUFRO Joint Meeting of Working Parties on Breeding Theory, Progeny Testing and Seed Orchards

Gaiotto FA, Grattapaglia D (1997) Estimation of genetic variability in a breeding population of *Eucalyptus urophylla* using AFLP (amplified fragment length polymorphism) markers. pp. 46–52. Proc Intl IUFRO Conf *Eucalyptus* Genetics and Silviculture

Gaiotto FA, Bramucci M, Grattapaglia, D (1997) Estimation of outcrossing rate in a breeding population of *Eucalyptus urophylla* s.t. Blake with dominant RAPD and AFLP markers. Theor Appl Genet 95:842–849

Gion J-M, Rech P, Grima-Pettenati J, Verhaegen D, Plomion C (2000) Mapping candidate genes in *Eucalyptus* with emphasis on lignification genes. Mol Breed 6:441–449

Glaubitz JC, Emebiri LC, Moran GF (2001) Dinucleotide microsatellite from *Eucalyptus* sieberi: Inheritance, diversity, and improved scoring of single-base differences. Genome 44: 1014–1045

González-Martínez SC, Wheeler NC, Ersoz E, Nelson CD, Neale DB (2007) Association genetics in Pinus taeda L. I. Wood property traits. Genetics 175(1):399–409

Grattapaglia D (2000) Molecular breeding of *Eucalyptus* - State of the art, operational applications and technical challenges. pp. 451–474. In: Jain SM, Minocha SC (eds) Molecular biology of woody plants. Kluwer Academic Publishers, The Netherlands

Grattapaglia D (2004) Integrating genomics into *Eucalyptus* breeding. Gen Mol Res 3:369–379

Grattapaglia D, Sederoff R (1994) Genetic linkage maps of *Eucalyptus grandis* and *Eucalyptus urophylla* using a pseudo-testcross: mapping strategy and RAPD markers. Genetics 137:1121–1137

Grattapaglia D, Chaparro J, Wilcox P, Mccord S, Werner D, Amerson H, Mckeand S, Bridgwater F, Whetten R, O'malley D, Sederoff R R (1992) Mapping in woody plants with RAPD markers: applications to breeding in forestry and horticulture. Proceedings of the Symposium "Applications of RAPD Technology to Plant Breeding". Crop Science Society of America, American Society of Horticultural Science, American Genetic Association, pp. 37–40

Grattapaglia D, O'Malley DM, Sederoff RR (1992) Multiple applications of RAPD markers to genetic analysis of *Eucalyptus* sp. pp. 436–450. Proc IUFRO Intl Conf "Breeding tropical trees" Section 2.02–08 Cali, Colombia

Grattapaglia D, Bertolucci FLG, Sederoff R (1995) Genetic mapping of QTLs controlling vegetative propagation in *Eucalyptus grandis* and *E. urophylla* using a pseudo-testcross mapping strategy and RAPD markers. Theor Appl Genet 90:933–947

Grattapaglia D, Bertolucci FLG, Penchel R, Sederoff R (1996) Genetic mapping of quantitative trait loci controlling growth and wood quality traits in *Eucalyptus grandis* using a maternal half-sib family and RAPD markers. Genetics 144:1205–1214

Grattapaglia D, Pimenta D, Campinhos EN, Rezende GDS, Assis TF (2003) Marcadores moleculares na proteção varietal de *Eucalyptus*. pp. 1–13. Proc 8th Brazilian Forestry Congress. Published in CD, SBS, Brazilian Soc Silviculture

Grattapaglia D, Ribeiro VJ, Rezende, GD (2004) Retrospective selection of elite parent trees using paternity testing with microsatellite markers: an alternative short term breeding tactic for *Eucalyptus*. Theor Appl Genet 109:192–199

Griffin AR, Burgess IP, Wolf L (1988) Patterns of natural and manipulated hybridisation in the genus *Eucalyptus* L'Herit: a review. Austr J Bot 36:41–66

Hazen SP, Kay SA (2003) Gene arrays are not just for measuring gene expression. Trends Plant Sci 8:413–416

Hertzberg M, Aspeborg H, Schrader J, Andersson A, Erlandsson R, et al. (2001) A transcriptional roadmap to wood formation. Proc Natl Acad Sci USA 98:14732–14737

Ikemori YK, Penchel RM, Bertolucci FLG (1994) Integrating biotechnology into *Eucalyptus* breeding, pp. 79–84. Proc Intl Symp Wood Biotechnol, TAPPI, Japan Wood Research Society and Nippon Paper Industries

Jansen RC, Nap JP (2001) Genetical genomics: the added value from segregation. Trends Genet 177:388–391

Junghans DT, Alfenas AC, Brommonschenkel SH, Oda S, Mello EJ, et al. (2003) Resistance to rust (*Puccinia psidii* Winter) in *Eucalyptus*: mode of inheritance and mapping of a major gene with RAPD markers. Theor Appl Genet 108:175–80

Kanowski PJ, Borralho NMG (2004) Economics of tree improvement. pp. 1561–1568. In: Youngquist JA (ed) Encyclopedia of Forest Science. Elsevier Science, Oxford

Kawazu T, Dol K, Tatemichi Y, Ito K, Shibata M (1996) Regenaration of transgenic plants by nodule culture systems in *Eucalyptus camaldulensis*. pp. 492–497. In: Proc IUFRO Conf: "Tree improvement for sustainable tropical forestry

Keil M, Griffin AR (1994) Use of random amplified polymorphic DNA (RAPD) markers in the discrimination and verification of genotypes in *Eucalyptus*. Theor Appl Genet 89: 442–450

Kellison RC (2001) Present and future uses of eucalypts wood in the world. In: Barros S (ed) Developing the Eucalypt of the Future. IUFRO Intl Symp INFOR, Chile (published in CDROM)

Kirst M, Brondani RPV, Brondani C, Grattapaglia D (1997) Screening of designed primer pairs for recovery of microsatellite markers and their transferability among species of *Eucalyptus*, pp. 167–171. Proc IUFRO Conf *Eucalyptus* Genetics and Silviculture

Kirst M, Myburg AA, De Leon JP, Kirst ME, Scott J, et al. (2004) Coordinated genetic regulation of growth and lignin revealed by quantitative trait locus analysis of cDNA microarray data in an interspecific backcross of *Eucalyptus*. Plant Physiol 135:2368–2378

Kirst M, Cordeiro CM, Rezende GD, Grattapaglia, D (2005a) Power of microsatellite markers for fingerprinting and parentage analysis in *Eucalyptus grandis* breeding populations. J Hered 96:161–166

Kirst M, Basten CJ, Myburg A, Zeng Z-B, Sederoff, R (2005b) Genetic architecture of transcript level variation in differentiating xylem of *Eucalyptus* hybrids. Genetics 169:2295–2303

Kirst M, Marques CM, Sederoff R (2005c) Nucleotide diversity and linkage disequilibrium in three *Eucalyptus globulus* genes. Section 5, P 28. (abs) IUFRO Tree Biotechnol Conf

Krutovsky KV, Neale DB (2005) Nucleotide diversity and linkage disequilibrium in cold-hardiness- and wood quality-related candidate genes in Douglas fir. Genetics. 171:2029–2041

Ladiges PY, Udovicic F, Nelson, G (2003) Australian biogeographical connections and the phylogeny of large genera in the plant family Myrtaceae. J Biogeogr 30:989–998

Lander ES, Botstein D (1989) Mapping mendelian factors underlying quantitative traits using RFLP linkage maps. Genetics 121:185–199

Litt M, Hauge X, Sharma V (1993) Shadow bands seen when typing polymorphic dinucleotide repeats: some causes and cures. BioTechniques 15:280–284

Lourenço RT, Grattapaglia D, Pappas GJ Jr, Pereira GA (2005) Sample sequencing of 3 megabases of shotgun DNA of *Eucalyptus grandis* : genome structure, repetitive elements and genes. IUFRO Tree Biotechnology S1.08 http://www.eyevisual.co.za/biotree/viewAbstract.asp

MacRae S, van Staden J (1999) Transgenic eucalyptus. In: Bajaj YPS (ed) Biotechnology in Agriculture and Forestry. 44:88–114. Springer, Heidelberg

Machado LO, de Andrade GM, Cid LPB, Penchel RM, Brasileiro ACM (1997) Agrobacterium strain specificity and shooty tumour formation in eucalypt *Eucalyptus grandis* × *E. urophylla*. Plant Cell Rep 16:299–303
Marcucci-Poltri SN, Zelener N, Rodriguez Traverso J, Gelid P, Hopp H (2003) Selection of a seed orchard of *Eucalyptus dunnii* based on genetic diversity criteria calculated using molecular markers. Tree Physiol 23:625–632
Margulies M, Egholm M, Altman WE, Attiya S, Bader JS, et al. (2005) Genome sequencing in microfabricated high-density picolitre reactors. Nature 437:376–380
Marques CM, Araujo JA, Ferreira JG, Whetten R, O'Malley DM, et al. (1998) AFLP genetic maps of *Eucalyptus globulus* and *E. tereticornis*. Theor Appl Genet 96:727–737
Marques CM, Vasquez-Kool J, Carocha VJ, Ferreira JG, O'Malley DM, et al. (1999) Genetic dissection of vegetative propagation traits in *Eucalyptus tereticornis* and *E. globulus*. Theor Appl Genet 99:936–946
Marques CM, Brondani RPV, Grattapaglia D, Sederoff R (2002) Conservation and synteny of SSR loci and QTLs for vegetative propagation in four *Eucalyptus* species. Theor Appl Genet 105: 474–478
Martin B, Quillet J (1974) The propogation by cuttings of forest trees in the Congo. Bois et Forets des Tropiques 155:15–33
Mauricio R (2001) Mapping quantitative trait loci in plants: uses and caveats for evolutionary biology. Nat Rev Genet 2:370–381
Meuwissen TH, Hayes BJ, Goddard ME (2001) Prediction of total genetic value using genome-wide dense marker maps. Genetics 157:1819–1829
Missiaggia AA, Piacezzi AL, Grattapaglia D (2005a) Genetic mapping of Eef1, a major effect QTL for early flowering in *Eucalyptus grandis*. Tree Genet Gen 1:79–84
Missiaggia AA, Mamani EM, Novaes E, Pappas MCR, Padua JG, et al. (2005b) Microsatellite based QTL mapping and validation across multiple pedigrees of *Eucalyptus*. IUFRO Tree Biotechnology (abstr) s5.34 http://www.eyevisual. co.za/biotree/viewAbstract.asp
Moran GF, Bell JC (1983) *Eucalyptus*. pp 423–441. In: Tanksley SD, Orton TJ (eds) Isozymes in plant genetics and breeding. Elsevier, Amsterdam
Moran G, Bell JC, Griffin AR (1989) Reduction in levels of inbreeding in a seed orchard of *Eucalyptus* regnans F. Muell, compared with natural populations. Silvae Genetica 38:32–36
Moran GF, Thamarus KA, Raymond CA, Qiu D, Uren T, et al. (2002) Genomics of *Eucalyptus* wood traits. Ann For Sci 59:645–650
Myburg AA (2001) Genetic architecture of hybrid fitness and wood quality traits in a wide inter-specific cross of *Eucalyptus* tree species. PhD thesis. North Carolina State University, Raleigh, NC (http://www.lib.ncsu.edu/theses/ available/etd-20010723–175234)
Myburg AA (2004) The International *Eucalyptus* Genome Consortium (IEuGC): Opportunities and Resources for Collaborative Genome Research in *Eucalyptus*. In: Li B, McKeand S (eds) Forest Genetics and Tree Breeding in the Age of Genomics: Progress and Future, pp. 154–155. IUFRO Joint Conference Division 2, Conf Proc http://www.ncsu.edu/feop/iufro_genetics 2004/proceedings.pdf
Myburg AA, Griffin AR, Sederoff RR, Whetten RW (2003) Comparative genetic linkage maps of *Eucalyptus grandis*, *Eucalyptus globulus* and their F1 hybrid based on a double pseudo-backcross mapping approach. Theor Appl Genet 107:1028–1042
Myburg AA, Potts B, Marques CM, Kirst M, Gion JM, et al. (2007) *Eucalyptus*, pp. 115–160; In Kole C (ed) Genome Mapping & Molecular Breeding in Plants Vol 7: Forest Trees. Springer, Heidelberg, Berlin, New York & Tokyo
Neale D.B., Williams C.G. (1991) Restriction fragment length polymorphism mapping in conifers and applications to forest genetics and tree improvement. Can J For Res. 21:545–554.
Neale DB, Savolainen O (2004) Association genetics of complex traits in conifers. Trends Plant Sci 9:325–330
Nesbitt KA, Potts BM, Vaillancourt RE, West AK, Reid JB (1995) Partitioning and distribution of RAPD variation in a forest tree species, *Eucalyptus globulus* (Myrtaceae). Heredity 74:628–637

Nesbitt KA, Potts BM, Vaillancourt RE, Reid JB (1997) Fingerprinting and pedigree analysis in Eualyptus *globulus* using RAPDs. Silvae Genetica 46:6–11

Novaes E, Drost D, Farmerie B, Kirst M (2007) Rapid, high-throughput gene discovery in *Eucalyptus* by massive parallel pyrosequencing. IUFRO Tree Biotechnol Conf (abstr) SIII.10p. http://www.itqb.unl.pt/iufro2007/SciProg .html

O'Malley D, Sederoff R, Grattapaglia D (1994) Methods For Within Family Selection In Woody Perennials Using Genetic Markers United States Patent and Trademark Office - Pat #6,054,634 (www.uspto.gov)

Ottewell KM, Donnellan SC, Moran GF, Paton DC (2005) Multiplexed microsatellite markers for the genetic analysis of *Eucalyptus* leucoxylon, Myrtaceae and their utility for ecological and breeding studies in other *Eucalyptus* species. J Hered 96:445–451

Pasquali G, Bastolla FM, Kirch RP, Brondani RPV, Coelho ASG, et al. (2005) Sequencing of the *Eucalyptus* transcriptome in the Genolyptus project. IUFRO Tree Biotechnol (abstr) S1.21 http://www.eyevisual.co.za/biotree /viewAbstract .asp

Paterson AH, Lander ES, Hewitt JD, Peterson S, Lincoln SE, et al. (1988) Resolution of quantitative traits into Mendelian factors by using a complete linkage map of restriction fragment length polymorphisms. Nature 335:721–726

Paux E, Tamasloukht M, Ladouce N, Sivadon P, Grima-Pettenati J (2004) Identification of genes preferentially expressed during wood formation in *Eucalyptus*. Plant Mol Biol 55:263–80

Paux E, Carocha V, Marques C, Mendes de Sousa A, Borralho N, et al. (2005) Transcript profiling of *Eucalyptus* xylem genes during tension wood formation. New Phytol 167(1):89–100

Poke FS, Vaillancourt RE, Potts BM, Reid JB (2005) Genomic research in *Eucalyptus*. Genetica 125:79–101

Potts BM (2004) Genetic improvement of eucalypts, pp. 1480–1490. In: Burley J, Evans J, Youngquist JA (eds) Encyclopedia of Forest Science. Elsevier Science, Oxford

Potts BM, Dungey HS (2004) Hybridisation of *Eucalyptus*: key issues for breeders and geneticists. New Forests 27:115–138

Potts BM, Vaillancourt RE, Jordan GJ, Dutkowski GW, Costa e Silva J, et al. (2004) Exploration of the *Eucalyptus globulus* gene pool. pp. 46–61. In: Tomé M (ed) *Eucalyptus* in a changing world. Aveiro, Portugal RAIZ, Instituto Investigação de Floresta e Papel

Pryor LD (1976) The biology of eucalypts. Edward Arnold, London

Pryor LD, Johnson LAS (1971) A classification of the eucalypts. Australian National University Press, Canberra

Ranik M, Creux NM, Myburg AA. (2006) Within-tree transcriptome profiling in wood-forming tissues of a fast-growing Eucalyptus tree. Tree Physiol 26(3):365–75

Raymond, CA (2000) Genetics of *Eucalyptus* wood properties. Ann For Sci 59:525–531

Ribeiro VJ, Bertolucci FLG, Grattapaglia D (1997) RAPD marker - guided matings in a reciprocal recurrent selection program of *Eucalyptus*. pp. 156–160. Proc. Intl IUFRO Conf *Eucalyptus* Genetics and Silviculture, Salvador, Brazil

Sansaloni C, Pappas GJ, Grattapaglia D (2007) Desenvolvimento de sistemas otimizados de fingerprinting de *Eucalyptus* baseados em microssatélites de tetra e pentanucleotídeos. 53rd Brazilian Genetics Congr, São Lourenço (abstr) 15465

Sato S, Yamada N, Nakamoto S, Hibino T (2005) Expression profiling of the *Eucalyptus* transcription factor in differentiating xylem tissues. Plant Animal Genome Conf 13:P520, pg 201

Schimleck LR, Michell AJ, Vinden P (1996) Eucalypt wood classification by NIR spectroscopy and principal components analysis. Appita J 49:319–324

Shepherd M, Jones M (2005) Molecular markers in tree improvement: Characterisation and use in *Eucalyptus*. pp. 399–409. In: Lörz H, Wenzel G (eds) Molecular marker systems in plant breeding and crop improvement. Springer-Verlag, Heidelberg

Shepherd M, Chaparro JX, Teasdale R (1999) Genetic mapping of monoterpene composition in an interspecific eucalypt hybrid. Theor Appl Genet 99:1207–1215

Steane DA, Vaillancourt RE, Russell J, Powell W, Marshall D, et al. (2001) Development and characterization microsatellite loci in *Eucalyptus globulus* (Myrtacea). Silvae Genet. 50:89–91

Steane DA, Jones RC, Vaillancourt RE (2005) A set of chloroplast microsatellite primers for *Eucalyptus*, Myrtaceae. Mol Ecol Notes 5:538–541

Strauss SH, Lande R, Namkoong G (1992) Obstacles to molecular-marker-aided selection in forest trees. Can J For Res 22:1050–1061

Stuber CW, Moll RH, Goodman MM, Schaffer HE, Weir BS (1980) Allozyme frequency changes associated with selection for increased grain yield in maize. Genetics 95: 225–236

Tanksley SD (1993) Mapping polygenes. Ann Rev Genet 27:205–233

Thamarus K, Groom K, Murrell J, Byrne M, Moran G (2002) A genetic linkage map for *Eucalyptus globulus* with candidate loci for wood, fibre and floral traits. Theor Appl Genet 104:379–387

Thamarus KA, Groom K, Bradley A, Raymond CA, Schimleck LR, et al. (2004) Identification of quantitative trait loci for wood and fibre properties in two full-sib pedigrees of *Eucalyptus globulus*. Theor Appl Genet 109:856–864

Thumma RF, Nolan MF, Evans R, Moran GF (2005) Polymorphisms in cinnamoyl CoA reductase (CCR) are associated with variation in microfibril angle in *Eucalyptus* spp. Genetics 171:1257–1265

Tournier V, Grat S, Marque C, El Kayal W, Penchel R, et al. (2003) An efficient procedure to stably introduce genes into an economically important pulp tree (*Eucalyptus grandis* x *Eucalyptus urophylla*). Transgenic Res 12:403–411

Turnbull JW (1999) Eucalypt plantations. New Forests 17: 37–52

Tuskan GA, Difazio S, Jansson S, Bohlmann J, Grigoriev I, et al. 2006 The genome of black cottonwood, Populus trichocarpa (Torr. & Gray). Science 313:1596–1604

Vaillancourt RE, Potts BM, Watson M, Volker PW, Hodge GR, et al. (1995a) Prediction of heterosis using RAPD generated distances between *Eucalyptus globulus* trees. pp. 455–456. Proc CRC/IUFRO Conf Eucalypt plantations, improving fibre yield and quality. Hobart, Australia

Vaillancourt RE, Potts BM, Manson A, Eldridge T, Reid JB (1995b) Using RAPD's to detect QTLs in an interspecific F2 hybrid of *Eucalyptus*. pp. 430–433. Proc. CRC/IUFRO Conf Eucalypt plantations, improving fibre yield and quality, Hobart, Australia

Verhaegen D, Plomion C (1996) Genetic mapping in *Eucalyptus urophylla* and E *grandis*. RAPD markers. Genome 39:1051–1061

Verhaegen D, Plomion C, Gion JM, Poitel M, Costa P, et al. (1997) Quantitative trait dissection analysis in *Eucalyptus* using RAPD markers. 1. Detection of QTL in interspecific hybrid progeny, stability of QTL expression across different ages. Theor Appl Genet 95:597–608

Waugh G (2004) Growing *Eucalyptus globulus* for high-quality sawn products. pp. 79–84. In: Tomé M (ed) *Eucalyptus* in a changing world. Aveiro, Portugal, Instituto Investigação de Floresta e Papel

West MA, van Leeuwen H, Kozik A, Kliebenstein DJ, Doerge RW, et al. (2006) High-density haplotyping with microarray-based expression and single feature polymorphism markers in Arabidopsis. Genome Res 16:787–795

Young ND (1999) A cautiously optimistic view for marker-assisted selection. Mol Breed 5:505–510

Zelener N, Poltri SN, Bartoloni N, Lopez CR, Hopp HE (2005) Selection strategy for a seedling seed orchard design based on trait selection index and genomic analysis by molecular markers: a case study for *Eucalyptus dunnii*. Tree Physiol 25:1457–1467

Chapter 12
Ginger and Turmeric, Ancient Spices and Modern Medicines

David R. Gang and Xiao-Qiang Ma

Abstract Ginger and turmeric have been used in human cuisine and in traditional medicinal practice for thousands of years. They are widely popular spices used extensively in Asian cuisine and growing in use in western cuisine. They have been used as medicinal plants, due to their anti-inflammatory properties, to treat a wide array of illnesses and conditions, such as arthritis (osteo and rheumatoid), inflammatory bowel disease, cancer, Alzheimer's disease, the common cold, etc. Two groups of compounds, the diarylheptanoids (including the curcuminoids) and the gingerol-related compounds, are potent anti-inflammatory compounds and contribute to, or are responsible for, many of the medicinal properties in these plants. They also contribute to the color of turmeric used in curries and to the pungency of ginger. Several of these compounds, most notably curcumin and [6]-gingerol, are now the targets of drug development. Despite their great medicinal and culinary importance, very little basic scientific research has been done on these plants. This is now changing. Belonging to the Zingiberaceae, they are members of the Zingiberales. The closest relative to this large group of plants that has been studied in some detail is banana (Musa spp, see chapter in this book), although that plant is not that closely related. The relationships of these plants to other plant groups, their diversity, their production, and recent and proposed efforts to understand the genetic and genomic makeup of these plants are discussed.

12.1 Introduction

12.1.1 Importance and Uses of Ginger and Turmeric

Ginger, *Zingiber officinale* Rosc., and turmeric, *Curcuma longa* L., are perennial sub-tropical and tropical herbs that have been used by people for thousands of years. Originating apparently in southern and southeastern Asia, these plants were adopted

D.R. Gang
University of Arizona, Department of Plant Sciences and BIO5 Institute, 1657 E. Helen Street, Tucson, AZ 85719
e-mail: gang@ag.arizona.edu

P.H. Moore, R. Ming (eds.), *Genomics of Tropical Crop Plants*

© Springer 2008

into the pantheons of Indian and Chinese traditional medicines, as well as into their cuisines. These plants are now loved the world over for their medicinal properties as well as for their roles as important spices.

The world production of ginger is about 100,000 tons annually, with about 80% grown in China (Langner et al., 1998). The rhizome of the plant is the tissue used for most purposes, although it is often called "ginger root." The Caribbean and Hawaiian islands are also significant producers of ginger for the culinary market in the United States. Australia, too, is a growing producer of ginger for the world market. Although it is well known in food products such as ginger ale, crystallized ginger, and ginger confections and as a spice for Chinese, Thai, and other Asian dishes (Langner et al., 1998), ginger also has a long history as a traditional Chinese herbal remedy to treat a number of health ailments including: motion sickness, nausea, vomiting, headache, arthritis, rheumatism, muscular aches and pains, coughs, sinusitis, sore throats, diarrhea, colic, cramps, indigestion, loss of appetite, fever, and chills and infectious diseases such as influenza and the common cold (College, 1985; Grant and Lutz, 2000; Vutyavanich et al., 2001). In addition, ginger is used in Indian Ayurvedic medicine as an anti-inflammatory agent for the treatment of muscular discomfort, rheumatic disorders, and arthritis, among other conditions (Srivastava and Mustafa, 1992). Due to its use as a traditional medicinal herb, ginger has drawn the attention of both the general population and the medical community. It is now one of the top 20 selling herbal supplements in the United States, and it is used to treat rheumatoid and osteoarthritis, motion sickness, and nausea and vomiting in pregnancy. Many of these properties are attributed to the presence of compounds such as [6]-gingerol, one of the major pungent principals in ginger that has been shown to have great potential in treating chronic inflammation, such as occurs in asthma and rheumatoid arthritis (Jolad et al., 2004; Jiang et al., 2005a).

Due to these medicinal uses, ginger and [6]-gingerol have recently been the subject of a growing number of clinical studies (Langner et al., 1998; Lacroix et al., 2000; Lien et al., 2003). Well planned studies, with appropriate randomized, double-blind, placebo-controlled, and parallel-group experimental designs have indicated that ginger and its extracts (including [6]-gingerol) are effective in treating the symptoms of osteoarthritis (Altman and Marcussen, 2001), of nausea and vomiting in pregnant women (Lacroix et al., 2000; Vutyavanich et al., 2001; Keating and Chez, 2002; Willetts et al., 2003; Smith et al., 2004a), and of motion sickness (Stewart et al., 1991; Lien et al., 2003). In these studies, few gastrointestinal-related side effects were observed (Altman and Marcussen, 2001), and the beneficial effects were significantly superior to the placebo groups (Wigler et al., 2003). All of these studies concluded that ginger and its extracts are effective treatments for inflammation, nausea or vomiting. Moreover, ginger was at least as effective in alleviating nausea symptoms in early pregnancy as vitamin B6, well known for its benefits in this regard (Smith et al., 2004a).

As is the case for ginger, turmeric has historically been grown mainly in Asia. Its cultivation has spread in recent years from India to China, Southeast Asia, Northern Australia, the West Indies, South and Central America, and Hawaii. India is still the world's the largest producer of turmeric, producing annually about 485,000 metric tons for export and domestic use (Joe et al., 2004). Ayurvedic Indian medicine uses turmeric to treat a number of conditions, including: arthritis, anorexia, coryza, cough, diabetic wounds, rheumatism, and sinusitis, as well as muscular, hepatic, and biliary disorders (Ammon et al., 1992). Turmeric has found use in traditional Chinese medicine as well, where it is used as a topical analgesic and for conditions ranging from flatulence, colic, arthralgia, psychataxia, dysmenorrhea, and ringworm to hepatitis and chest pain (Grant and Schneider, 2000; Sasaki et al., 2002). In the United States, turmeric is used to color foods such as pickles and curries, and it is also gaining rapid ground in the herbal supplements/remedies field as an alternative treatment for numerous diseases and for disease prevention (Grant and Schneider, 2000). Recent evidence suggests that many of its beneficial effects can be attributed to the presence of compounds such as curcumin and the related curcuminoids and other diarylheptanoids, which are a major class of natural products in turmeric rhizomes (Jiang et al., 2006b; Jiang et al., 2006a; Ma and Gang, 2006). Curcumin is the most abundant of these compounds in most turmeric samples. As has been found for whole turmeric rhizome, curcumin has been found to possess potent anti-inflammatory activity, giving it great promise in preventing and treating a wide variety of ailments that have ties to inflammation, including: arthritis (osteo- and rheumatoid), inflammatory bowel disease, cystic fibrosis, cancer, and Alzheimer's disease, among others. (Aggarwal et al., 2003; Chainani-Wu, 2003; Egan et al., 2004; Ringman et al., 2005; Yang et al., 2005; Atamna and Boyle, 2006; Balasubramanian and Eckert, 2006; Dikshit et al., 2006; Ono et al., 2006; Rapaka and Coates, 2006; Yang et al., 2006).

12.1.2 Taxonomy

The genus *Zingiber* contains between 100 and 150 species, which are distributed throughout tropical to warm-temperate Asia (http://www.efloras.org). The origin of ginger (*Z. officinale*) is unknown, and it is widely cultivated in the tropics and subtropics. Three varieties which are especially popular in the United States are Chinese white ginger, Japanese yellow ginger, and blue ring ginger, as they are called by growers in Hawaii. Numerous additional varieties are grown in China, Australia, and other parts of the world. The genus *Curcuma* contains about 50 species. These are naturally distributed across India, and in southeast Asia and Australia (http://www.efloras.org). Hundreds of varieties of turmeric are grown throughout south and southeastern Asia. Ginger and turmeric belong to the Zingiberaceae family of the Zingiberales, which is a large and diverse plant order, and which, along with sister group Commelinales, resides sister to the Poales in the monocot class of higher plants (Davis, 1995; Chase, 2004; Davis et al., 2004). Commelinales is a minor order, with no members that are economically significant. The most notable members of that order are a few common ornamental garden and house pot plants, such as wandering Jew and spiderwort. Poales, on the other hand, is the most economically significant plant order on the planet, providing the majority of the world's staple crops. A significant amount of resources are now being brought to bear on deciphering the genomes of the three most economically important staple plants: rice, maize, and wheat, all members of Poales. Although not as economically important as their distant cousins the grasses, members of the Zingiberales, including banana (*Musa acuminata*, Musaceae), ginger (*Zingiber officinale* Rosc., Zingiberaceae), and turmeric (*Curcuma longa* L., Zingiberaceae) are well known worldwide and are of great importance to many local economies. Musa spp. are discussed in detail in another chapter of this book. Not only would more detailed genomics information about these three latter species be important for efforts to improve these crops, both for food and for medicine, but one may well argue that research that seeks to improve the cereal crops could benefit from detailed information about the biology and genome structure of these related plants. Several recent investigations have sought to more clearly define the relationships of members of these groups of plants.

12.1.3 Phylogenetic Analysis of Ginger, Turmeric and Related Species

DNA sequence or chemically based phylogenetic investigation can be used to authenticate the identity of plant material and distinguish the relationships between different species or genera. A phylogenetic study of 104 species in 41 genera, representing all four tribes of the Zingiberaceae has been reported (Kress et al., 2002). That study, which was based on DNA sequences of the nuclear internal transcribed spacer (ITS) and plastid matK regions, did not include *Z. officinale* (ginger) or *Curcuma longa* (tumeric) but did include other members of the genera *Zingiber* (*Z. corallinum, Z. wrayi, Z. sulphureum, Z. gramineum, Z. ellipticum,* and *Z. species*) and *Curcuma* (*C. comosa, C. aeruginosa, C. roscoeana, C. thorelii, C. bicolor,* and *C. species*).

In addition, a DNA sequence based phylogenetic study was coupled to a chemotaxonomical study of ginger and related species (Jiang et al., 2006c). In that study, 10 varieties of ginger and samples of related species within the genus *Zingiber* were compared to *Alpinia galanga* (as outgroup), and two unrooted phylogenetic trees were generated. The first was based on combined *TrnL* and *rps16* sequences, whereas the second was based on volatile compound composition of the rhizomes. Both trees gave essentially the same topology even though many of these sample lines had been obtained from very different geographical origins and suggested that ginger is monophyletic (Jiang et al., 2005b). In addition, anti-inflammatory activity of the various samples was evaluated in this study and the results of this investigation suggested that this activity did not necessarily follow phylogeny, suggesting that large amounts of variation within and between species exists in the genus *Zingiber*. A comparable study was performed with turmeric (*Curcuma longa*) and related *Curcuma* lines (Deeb et al., 2005). Like ginger, turmeric was found to possess high variability both at the sequence and especially at the chemical level. The results of this investigation suggested that the turmeric lines investigated may not be monophyletic, and that instead, "turmeric" may have originated multiple times, probably from hybridization events.

12.1.4 Ploidy Levels and Genetic Diversity in Ginger and Turmeric

One area of research close at hand to genomics-based approaches, an area in which knowledge is absolutely critical, is the ploidy level of the species. In this regard, some research has been carried out on Zingiberaceae plants, including ginger and turmeric. Most turmeric varieties are believed to be sterile triploids, derived from the hybridization of two closely related species. The results described directly above support this conclusion (Deeb et al., 2005). One of these species was likely to be *Curcuma petiolata* or *Curcuma aromatica*. *C. petiolata* is recognized to be one of the closest relatives of turmeric (Deeb et al., 2005). What the other species was that hybridized to *C. petiolata/aromatica*, is not perfectly clear. It is clear that several different hybridization events have occurred and produced what is now called turmeric. This is likely the cause of the great diversity seen in the varieties of turmeric that have been described. Therefore, as suggested above, what we now call turmeric is likely to be a paraphyletic species, if it can be called a species at all.

Ginger, on the other hand, is likely to be a diverse monophyletic species. The results outlined above support this conclusion (Jiang et al., 2005b). Several reports have addressed the issue of ploidy level within this species. Although some of these have claimed that ginger exists at higher ploidy levels, the ginger that is available worldwide is mostly likely to be diploid. Tetraploid clones of ginger have been produced artificially by colchicine treatment (Adaniya and Shirai, 2001; Smith et al., 2004b; Wohlmuth et al., 2005; Wohlmuth et al., 2006). Since some of these tetraploids had enhanced levels of important medicinal compounds, such as 6-gingerol; this area of investigation may lead to significant improvements in select ginger varieties (Smith et al., 2004b; Wohlmuth et al., 2005; Wohlmuth et al., 2006).

12.2 Progress in Genomics

Despite their importance in cuisine and especially in medicine, very little information has been available about the genomes of ginger and turmeric. Although several groups are working to improve ginger and turmeric quality through traditional breeding programs, these plants have not been the target of major modern breeding efforts, and no tools such as bacteria artificial chromosome (BAC) libraries or molecular marker-based genetic maps have been produced, or, at least, no such resources have been released or even mentioned in the literature. There are several reasons why these resources have not been made available. First of all, these plants are not very amenable to production of genetic maps. Turmeric is a sterile triploid, and production of genetic maps that rely on recombination cannot be produced with this plant. Traditional outcrossing-based breeding is not possible with this species. Instead, new turmeric varieties must either be collected anew from the wild, as a result of new hybridization events that have likely occurred many times through history, or new mutations in existing clonal lines must be identified and then further propagated. Ginger is not a triploid, but it also does not outcross readily. Ginger plants do not usually produce inflorescences in their first year of growth, but instead produce these in their second year. When they do produce inflorescences, they often produce only one flower at a time per plant, with flowers remaining viable for only one or two days. In the five years that we have been working with these plants in our greenhouses, we have never been able to collect a single seed from either species. Perhaps others have been successful in this area, but no reports of such success have been made in the literature. Both ginger and turmeric are produced commercially via vegetative propagation, not by seed. Thus, both ginger and turmeric must be treated as clonally propagated species. Although presenting problems for efforts to produce genetic maps, this property does have benefits in efforts to perform experiments based on the genome. The only mapping that has been done with any species related to ginger or turmeric has been done with banana (*Musa acuminata*), in the Musaceae, see the *Musa* chapter for more information on this topic. Nothing in this area has been reported with any species within the Zingiberaceae.

Prior to our investigations involving expressed sequence tag (EST) production (see below), the only genomic information regarding these plants came from fragments of a few genes from ginger and turmeric and related species in the Zingiberaceae. In all cases, these gene fragments were produced and sequenced as part of phylogenetic studies and were partial gene sequences from variable regions of nuclear *rDNA* ITS, *matK*, *TrnL*, and *rps16* (Andersson and Chase, 2001; Specht et al., 2001). Many of these genes are chloroplastic and are thus not expected to be expressed in the rhizome. Within the last two years, several additional genes have been identified from ginger, including: a cysteine protease, a chalcone synthase, a polyphenol oxidase, a germacrene D synthase, violaxanthin de-epoxidase, an NBS-LRR disease resistance protein, among a few others. No other genes from turmeric have been characterized. Clearly, much work is yet to be done regarding the genome structure of these species, and our investigations and those of others completed to date are just a beginning of what should be done.

12.3 EST Database

We have developed an EST database from two ginger lines (white and yellow ginger) and from the major turmeric line (Hawaiian red) grown in Hawaii. The purpose of this database was to begin to develop genomics tools to address two important questions that are ideally answered using ginger/turmeric as model plants: what genes are involved in rhizome development and how is complex metabolism controlled and regulated in important medicinal plants? To address these questions, RNAs for EST production were isolated from rhizomes, roots, and leaves of these plants. In total, 50,409 ESTs were produced (12,535 from turmeric, 37,874 from ginger) and assembled into a database of 20,599 contigs/unigenes (see Table 12.1 and http://ag.arizona.edu/research/ganglab/ArREST.htm). A small number of ESTs (~120) from leaves of three other ginger cultivars have been deposited in GenBank by other researchers as of July 1, 2007, but were not included in our database.

Table 12.1 Turmeric and ginger EST database composition

Species	line	Tissue	LibID	Contigs	Singletons	ESTs
Turmeric	Red	Rhizome	CL_Ea	2275	772	5703
		Leaf	CL_Eb	2371	879	6832
Ginger	GY	Rhizome	ZO_Ed	2261	1038	6483
		Leaf	ZO_Ea	2046	564	6006
		Root	ZO_Ee	2292	551	5672
	GW	Rhizome	ZO_Ec	2374	1226	6239
		Leaf	ZO_Eg	2520	944	7141
		Root	ZO_Ef	2480	908	6333
Total				13717	6882	50409

A simplified gene expression profile of the ESTs database was developed based on the Gene Ontology (GO, http://www.geneontology.org) system (Fig. 12.1). Genes involved in transcription, metabolism (including metabolism, carbohydrate metabolism, lipid metabolism, amino acid and derivative metabolism, and nucleobase and nucleic acid metabolism), cellular process, and transport are expressed at a high level in ginger and turmeric tissues. In contrast, secondary metabolism and biological process categories are underrepresented as a whole, based on GO categorization. Actually, the biosynthesis of most "secondary metabolites" in ginger and turmeric has not been characterized, and we expect that many genes from other metabolism categories will be involved in production of compounds such as the gingerols and curcuminoids, so these categorizations are not that surprising and suggest that metabolic processes in general are highly active in developing ginger and turmeric tissues.

Based on the EST data, the network of metabolic pathways leading from sucrose to the phenylpropanoids and terpenoids formed in turmeric and ginger could be evaluated at the transcriptional level. Transcripts encoding known components of these pathways were highly represented in the EST database, suggesting that these metabolic pathways are highly expressed in ginger and turmeric tissues (Fig. 12.2). One interesting finding in this regard was that the genes of the mevalonate independent 2-C-methyl-D-erythritol 4-phosphate (MEP) pathway, leading to production of IPP, the precursor of terpenoids of different classes, were expressed at significantly higher levels than were genes of the mevalonate pathway. Indeed, several genes in the latter pathway were not represented at all in the EST database, whereas all genes of the MEP pathway were. This suggests that production of precursors for different terpenoid classes, including the sesquiterpenoids, are likely derived from the MEP pathway and not the mevalonate pathway, as has been shown for other species (Dudareva et al., 2005).

Another interesting finding in the EST database was that many signal transduction genes were expressed at higher levels in ginger and turmeric rhizomes than in leaves. The exact significance of this observation is yet to be determined, but it suggests differential regulation of metabolism and development between these tissue types. Several such genes that are upregulated in the rhizome or expressed exclusively in the rhizome are good candidates for genes that may be involved

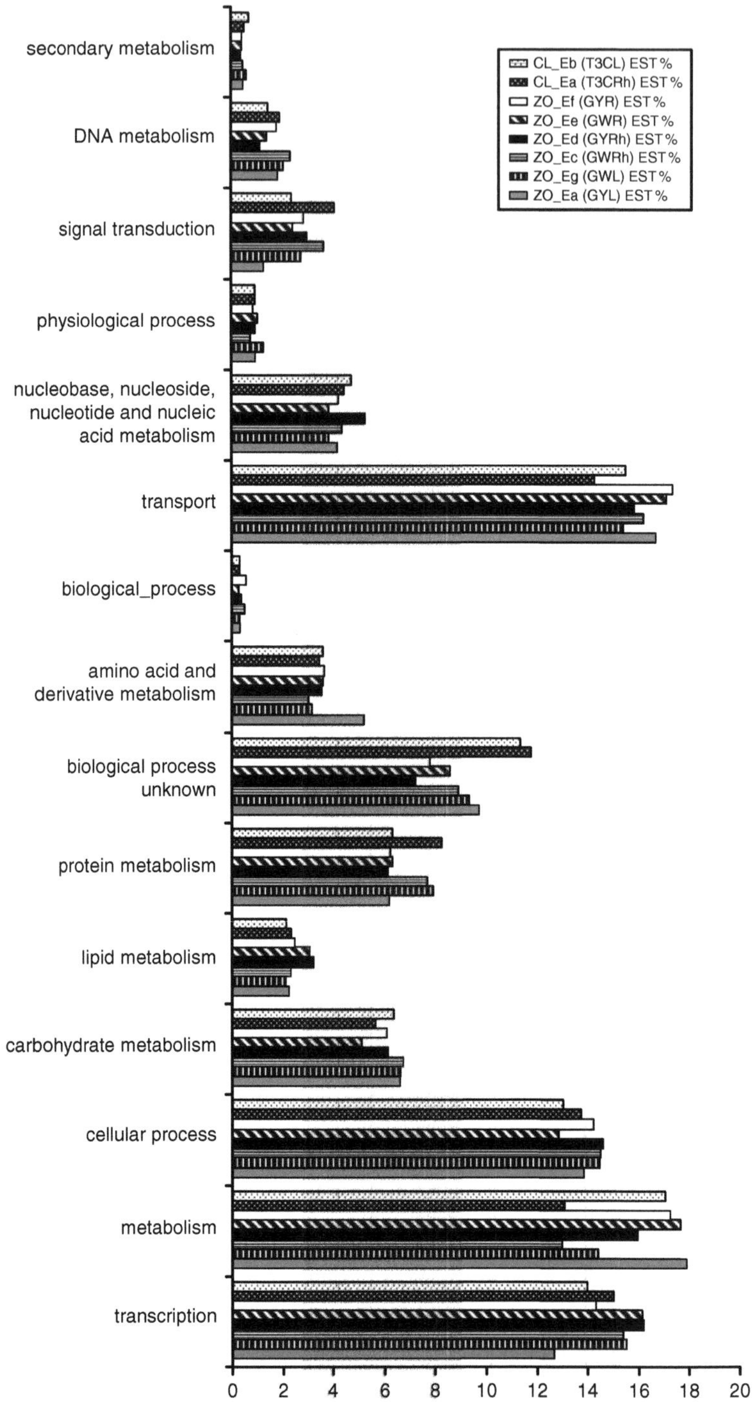

Fig. 12.1 A simplified gene expression profile of the ESTs database based on the GeneOntology system (GO, http://www.geneontology.org)

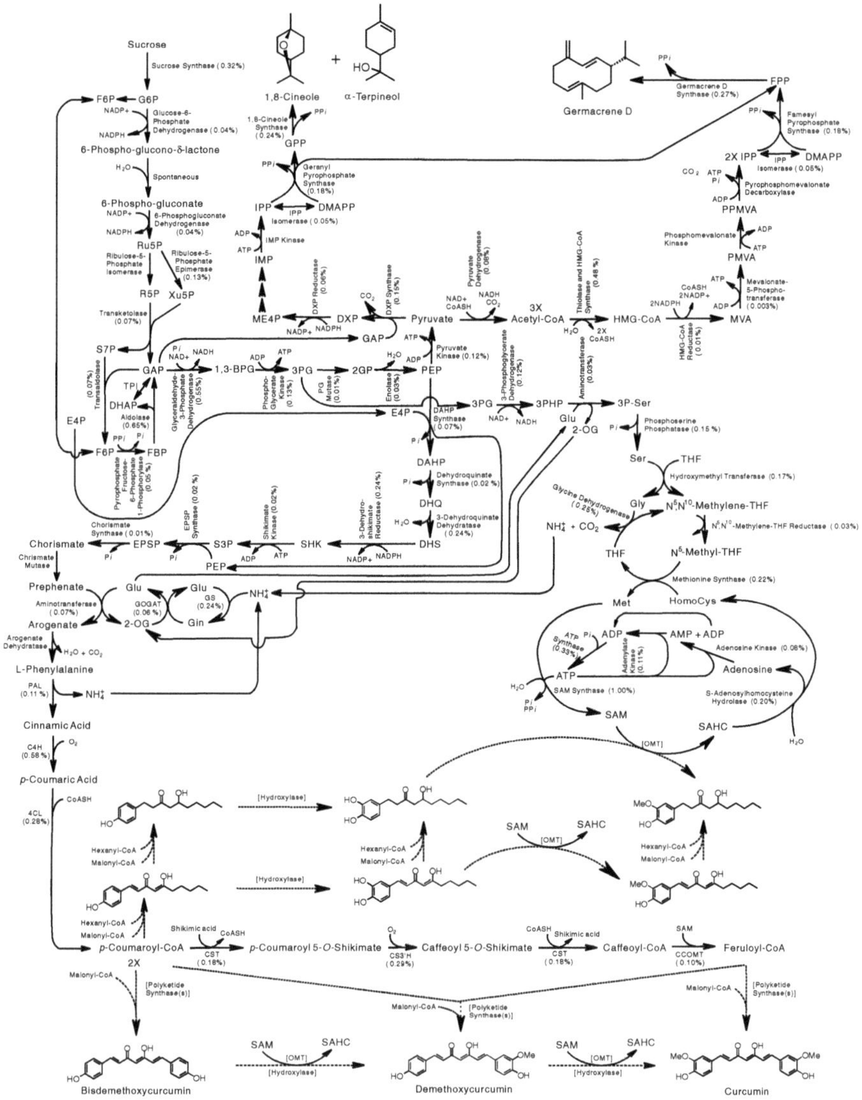

Fig. 12.2 Network of metabolic pathways evaluated at the transcriptional level leading from sucrose to the phenylpropanoids and terpenoids formed in turmeric and ginger

in determining cell fate and tissue differentiation leading to rhizome development and maintenance.

Furthermore, to identify possible *trans*-acting transcriptional regulators in ginger and turmeric, the EST database was queried for sequences with the associated gene ontology identifier for DNA binding: GO0003677. This led to the identification of 1,372 nonredundant contigs with this identifier. These sequences were then further analyzed using the protein motif identification program INTERPROSCAN

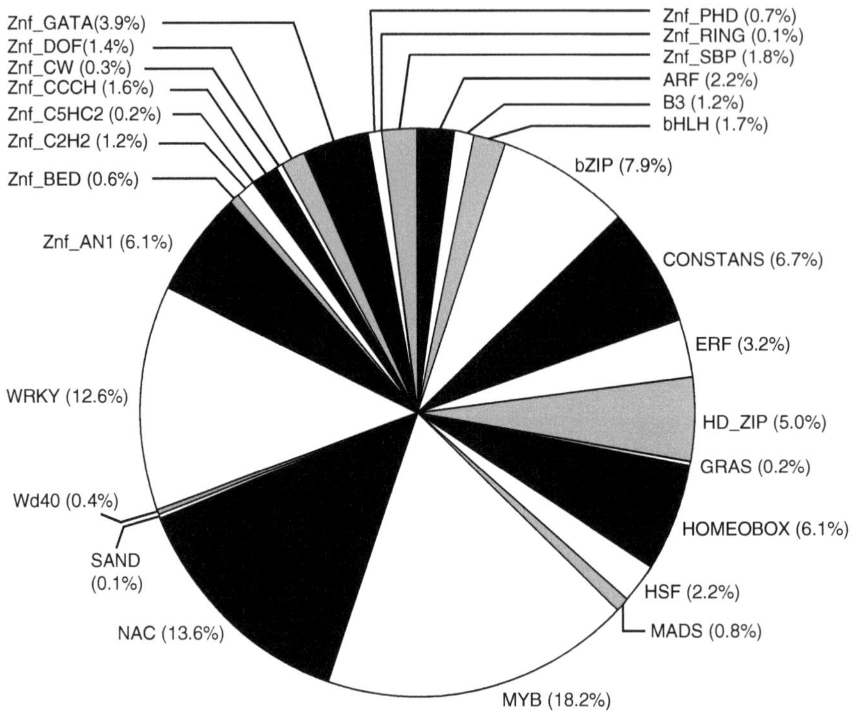

Fig. 12.3 Categorization of 818 contigs according to DNA binding domains and sorted according to contig and EST number

(Syed and Upton, 2006) to identify contigs that contained only generalized DNA binding domains, which were then eliminated from further analysis. After manual curation and categorization by DNA binding domain, 818 contigs were categorized according to DNA binding domain and sorted according to contig and EST number (Fig. 12.3). The most highly expressed classes of transcription factors are MYB (18.2% of total transcription factors), NAC (13.6%), WRKY (12.6%), BZIP (7.9%), CONSTANS (6.7%), HOMEOBOX (6.1%), Znf_AN1 (6.1%), and HD_ZIP (5%).

12.4 Perspective

As discussed above, ginger and turmeric, two important medicinal and culinary plants have been neglected by the genomics community, but this is changing. Development of new sequencing technologies (454 Lifesciences/Roche FLX sequencer, Solexa sequencer, etc.) will enable us to obtain at very reasonable costs large amounts of gene space sequence for these and other species. Production of BAC libraries for these species, a logical next step, will be required for this large scale sequence production and will enable production of a physical map of the genomes of these species as well. These technologies will also allow for the rapid and large volume production of EST data to expand and complement the work that we have already done using older technologies.

Although there is great interest in ginger and turmeric worldwide, especially in Asia, Australia, and the United States, a cohesive group of researchers working on these plants has not yet been developed, as has occurred for species such as rice, Arabidopsis, wheat, maize, tomato, etc. This is in spite of efforts by several individuals to make contacts and to try to develop collaborations across national borders and is due in part to governmental restrictions on the part of some countries, which have justifiable intellectual property concerns. This situation was not helped when two medical researchers applied for a US patent for the use of turmeric to treat common ailments (and were astonishingly granted the patent by the USA patent office, even though there are thousands of years of historically documented use of this plant in India for this purpose!). That patent was eventually overturned, but governmental institutional memories can be long. Scientists from the world over who are working on ginger and turmeric need to be brought together to develop a plan to promote the development of more extensive genomics resources for these important medicinal plants.

Acknowledgments The authors would like to thank Hyun Jo Koo, Eric T. McDowell, Dr. Zhengzhi Xie, and Dr. Jeremy Kapteyn for assistance with data analysis; Dr. Dave Kudrna, Dr. HyeRan Kim, Dr. Yeisoo Yu, and Dr. Rod A. Wing at the Arizona Genomics Institute for help with DNA sequencing; and Dr. Karl Haller and Dr. Carol A. Soderlund at the Arizona Genomics Computational Laboratory for help with production of the ginger and turmeric EST database. This research was funded by grant DBI-0227618 to DRG from the National Science Foundation Plant Genome Program. The contents of this publication are solely the responsibility of the authors and do not necessarily represent the official views of the National Science Foundation.

References

Adaniya S, Shirai D (2001) In vitro induction of tetraploid ginger (Zingiber officinale Roscoe) and its pollen fertility and germinability. Sci Hortic 88:277–287

Aggarwal BB, Kumar A, Bharti AC (2003) Anticancer potential of curcumin: preclinical and clinical studies. Anticancer Res 23:363–398

Altman RD, Marcussen KC (2001) Effects of a ginger extract on knee pain in patients with osteoarthritis. Arthritis Rheum 44:2531–2538

Ammon HP, Anazodo MI, Safayhi H, Dhawan BN, Srimal RC (1992) Curcumin: a potent inhibitor of leukotriene B4 formation in rat peritoneal polymorphonuclear neutrophils (PMNL). Planta Med 58:226

Andersson L, Chase MW (2001) Phylogeny and classification of Marantaceae. Botanical Linnean Society 135:275–287

Atamna H, Boyle K (2006) Amyloid-beta peptide binds with heme to form a peroxidase: relationship to the cytopathologies of Alzheimer's disease. Proc Natl Acad Sci USA 103:3381–3386

Balasubramanian S, Eckert RL (2006) Curcumin suppresses AP1 transcription factor-dependent differentiation and activates apoptosis in human epidermal keratinocytes. J Biol Chem 282:6707–6715

Chainani-Wu N (2003) Safety and anti-inflammatory activity of curcumin: a component of tumeric (*Curcuma longa*). J Altern Complement Med 9:161–168

Chase MW (2004) Monocot relationships: An overview Am J Bot 91:1645–1655

College JNM (1985) The Dictionary of Traditional Chinese Medicine. (Shanghai: Shanghai Sci-Tech Press)

Davis JI (1995) A phylogenetic structure for the Monocotyledons, as inferred from chloroplast DNA restriction site variation, and a comparison of measures of clade support. Syst Bot 20:503–527

Davis JI, Stevenson DW, Petersen G, Seberg O, Campbell LM, et al. (2004) A Phylogeny of the monocots, as inferred from rbcL and atpA sequence variation, and a comparison of methods for calculating jackknife and bootstrap values. Syst Bot 29:467–510

Deeb DD, Jiang H, Gao X, Divine G, Dulchavsky SA, et al. (2005) Chemosensitization of hormone-refractory prostate cancer cells by curcumin to TRAIL-induced apoptosis. J Exp Ther Oncol 5:81–91

Dikshit P, Goswami A, Mishra A, Chatterjee M, Jana NR (2006) Curcumin induces stress response, neurite outgrowth and prevent NF-kappaB activation by inhibiting the proteasome function. Neurotox Res 9:29–37

Dudareva N, Andersson S, Orlova I, Gatto N, Reichelt M, et al. (2005) The nonmevalonate pathway supports both monoterpene and sesquiterpene formation in snapdragon flowers. Proc Natl Acad Sci USA 102:933–938

Egan ME, Pearson M, Weiner SA, Rajendran V, Rubin D, et al. (2004) Curcumin, a major constituent of turmeric, corrects cystic fibrosis defects. Science 304:600–602

Grant KL, Lutz R (2000) Ginger Am J Health Syst Pharm 57:945–947

Grant KL, Schneider CD (2000) Turmeric Am J Health Syst Pharm 57:1121–1122

Jiang H, Timmermann BN, Gang DR (2006a) Use of liquid chromatography-electrospray ionization tandem mass spectrometry to identify diarylheptanoids in turmeric (*Curcuma longa* L.) rhizome. J Chromatogr A 1111:21–31

Jiang H, Solyom AM, Timmermann BN, Gang DR (2005a) Characterization of gingerol-related compounds in ginger rhizome (*Zingiber officinale* Rosc.) by high-performance liquid chromatography/electrospray ionization mass spectrometry. Rapid Commun Mass Spectrom 19:2957–2964

Jiang H, Somogyi A, Jacobsen NE, Timmermann BN, Gang DR (2006b) Analysis of curcuminoids by positive and negative electrospray ionization and tandem mass spectrometry. Rapid Commun Mass Spectrom 20:1001–1012

Jiang H, Xie Z, Koo H, McLaughlin SP, Timmermann BN, Gang DR (2005b) Metabolic profiling, phylogenetic analysis and anti-inflammatory investigation of *Zingiber* species: tools for authentication of ginger (*Zingiber officinale* Rosc.). Phytochem 67:232–244 doi:210.1016/j.phytochem.2005.1008.1001

Jiang H, Xie Z, Koo HJ, McLaughlin SP, Timmermann BN, Gang DR (2006c) Metabolic profiling and phylogenetic analysis of medicinal *Zingiber* species: Tools for authentication of ginger (*Zingiber officinale* Rosc). Phytochemistry 67:1673–1685

Joe B, Vijaykumar M, Lokesh BR (2004) Biological properties of curcumin-cellular and molecular mechanisms of action. Crit Rev Food Sci Nutr 44:97–111

Jolad SD, Lantz RC, Solyom AM, Chen GJ, Bates RB, et al. (2004) Fresh organically grown ginger (*Zingiber officinale*): composition and effects on LPS-induced PGE2 production. Phytochemistry 65:1937–1954

Keating A, Chez RA (2002) Ginger syrup as an antiemetic in early pregnancy. Altern Ther Health Med 8:89–91

Kress WJ, Prince LM, Williams KJ (2002) The phylogeny and a new classification of the gingers (Zingiberaceae): evidence from molecular data. Am J Bot 89:1682–1696

Lacroix R, Eason E, Melzack R (2000) Nausea and vomiting during pregnancy: A prospective study of its frequency, intensity, and patterns of change. Am J Obstet Gynecol 182:931–937

Langner E, Greifenberg S, Gruenwald O (1998) Ginger: history and use. Adv Ther 15:25–44

Lien HC, Sun WM, Chen YH, Kim H, Hasler W, et al. (2003) Effects of ginger on motion sickness and gastric slow-wave dysrhythmias induced by circular vection. Am J Physiol Gastrointest Liver Physiol 284:481–489

Ma X, Gang DR (2006) Metabolic profiling of turmeric (*Curcuma longa* L.) plants derived from in vitro micropropagation and conventional greenhouse cultivation. J Agric Food Chem 54:9573–9583

Ono K, Naiki H, Yamada M (2006) The development of preventives and therapeutics for Alzheimer's disease that inhibit the formation of beta-amyloid fibrils (fAbeta), as well as destabilize preformed fAbeta. Curr Pharm Des 12:4357–4375

Rapaka RS, Coates PM (2006) Dietary supplements and related products: a brief summary. Life Sci 78:2026–2032

Ringman JM, Frautschy SA, Cole GM, Masterman DL, Cummings JL (2005) A potential role of the curry spice curcumin in Alzheimer's disease. Curr Alzheimer Res 2:131–136

Sasaki Y, Fushimi H, Cao H, Cai SQ, Komatsu K (2002) Sequence analysis of Chinese and Japanese *Curcuma* drugs on the 18S rRNA gene and trnK gene and the application of amplification-refractory mutation system analysis for their authentication. Biol Pharm Bull 25:1593–1599

Smith C, Crowther C, Willson K, Hotham N, McMillian V (2004a) A randomized controlled trial of ginger to treat nausea and vomiting in pregnancy. Obstet Gynecol 103:639–645

Smith MK, Hamill SD, Gogel BJ, Severn-Ellis AA (2004b) Ginger (Zingiber officinale) autotetraploids with improved processing quality produced by an in vitro colchicine treatment. Aust J Exp Agric 44:1065–1072

Specht CD, Kress WJ, Sevenson DW, Rob D (2001) A molecular phylogeny of Costaceae (Zingiberales). Molec Phyl Evol 21:333–345

Srivastava KC, Mustafa T (1992) Ginger (*Zingiber officinale*) in rheumatism and musculoskeletal disorders. Med Hypotheses 39:342–348

Stewart JJ, Wood MJ, Wood CD, Mims ME (1991) Effects of ginger on motion sickness susceptibility and gastric function. Pharmacology 42:111–120

Syed A, Upton C (2006) Java GUI for InterProScan (JIPS): a tool to help process multiple InterProScans and perform ortholog analysis. BMC Bioinformatics 7:462

Vutyavanich T, Kraisarin T, Ruangsri RA (2001) Ginger for nausea and vomiting in pregnancy: Randomized, double-masked, placebo-controlled trial. Obstet Gynecol 97:577–582

Wigler I, Grotto I, Caspi D, Yaron M (2003) The effects of Zintona EC (a ginger extract) on symptomatic gonarthritis. Osteoarthritis Cartilage 11:783–789

Willetts KE, Ekangaki A, Eden JA (2003) Effect of a ginger extract on pregnancy-induced nausea: a randomised controlled trial. Aust N Z J Obstet Gynaecol 43:139–144

Wohlmuth H, Leach DN, Smith MK, Myers SP (2005) Gingerol content of diploid and tetraploid clones of ginger (Zingiber officinale Roscoe). J Agric Food Chem 53:5772–5778

Wohlmuth H, Smith MK, Brooks LO, Myers SP, Leach DN (2006) Essential oil composition of diploid and tetraploid clones of ginger (Zingiber officinale roscoe) grown in Australia. J Agric Food Chem 54:1414–1419

Yang F, Lim GP, Begum AN, Ubeda OJ, Simmons MR, et al. (2005) Curcumin inhibits formation of amyloid beta oligomers and fibrils, binds plaques, and reduces amyloid in vivo. J Biol Chem 280:5892–5901

Yang X, Thomas DP, Zhang X, Culver BW, Alexander BM, et al. (2006) Curcumin inhibits platelet-derived growth factor-stimulated vascular smooth muscle cell function and injury-induced neointima formation. Arterioscler Thromb Vasc Biol 26:85–90

Chapter 13
Genomics of Macadamia, a Recently Domesticated Tree Nut Crop

Cameron Peace, Ray Ming, Adele Schmidt, John Manners, and Vasanthe Vithanage

Abstract The tree nut crop known as macadamia includes two cultivated species that readily hybridize. This Australian native from subtropical rainforests was domesticated recently, and cultivated trees are very few generations from their wild progenitors. A genomic understanding of the crop has the potential to deliver massive genetic improvements to a worldwide industry, and reveal the genetic changes that have occurred through the domestication process. The bulk of research efforts in this field have focused on the development of molecular marker technology and its various applications in assessing both cultivated and wild germplasm. Markers were also used as the basis of genetic linkage mapping for macadamia's chromosomes, but regions controlling important traits have not yet been localized. Gene sequence information for macadamia is very limited, although two genes encoding proteins with antimicrobial properties have been described. Macadamia is the most economically valuable member of the ancient Proteaceae family, and has few cultivated relatives. This crop is the obvious target within the family for developing further genomics resources. Comparing the structure of the macadamia genome and its functional components with those of crops from other plant families, particularly nut, fruit, and tree species, should provide insights for understanding the evolution and genomic regulation of many important biological and agronomic traits.

13.1 Introduction

Macadamia is a tree nut crop representing the first Australian native plant to be cultivated as a major food crop. Worldwide, macadamia was valued in 2005 at US$290 million at the farm gate, and US$330 million after processing to raw kernel (Jones 2006; K. Jones personal communication). The largest producer is Australia (10750 tonnes in 2005), followed by Hawaii (6200 tonnes), and South Africa (4205 tonnes) (Piza 2006). Although macadamia has subtropical natural origins, it appears to have wide climatic adaptability and is grown in many tropical regions of the

C. Peace
Department of Horticulture and Landscape Architecture, Washington State University, Pullman, WA 99164, USA
e-mail: cpeace@wsu.edu

P.H. Moore, R. Ming (eds.), *Genomics of Tropical Crop Plants*

© Springer 2008

world, including Hawaii, Kenya, Malawi, Guatemala, Brazil, and Thailand, as well as subtropical and temperate climes of Australia, South Africa, California, China, and New Zealand (Hardner et al. 2007). Limitations to crop production that may be addressed through breeding, and thus benefit from genomics assistance, run the full gamut from propagation, through tree growth and stress resistance, to yield and kernel quality (Hardner et al. 2007).

The macadamia as a crop is based on two members of the Proteaceae family, *Macadamia integrifolia* Maiden & Betche and *M. tetraphylla* L.A.S. Johnson (Stanley and Ross 1986). Historically, *M. integrifolia* is the preferred species, although hybrid cultivars are common, and the industry in some countries relies heavily on *M. tetraphylla* (Hardner et al. 2007). These species are found naturally in isolated populations in rainforests along the east coast of Australia, from approximately 26° to 29° south of the equator, *M. integrifolia* in the northern parts and *M. tetraphylla* in the south, with a natural hybrid zone in between (Peace 2002). Two closely related species, *M. ternifolia* F. Muell. and *M. jansenii* C.L. Gross & P.H. Weston, grow within or nearby the native range of *M. integrifolia* but are not cultivated because of their small, bitter, inedible nuts (Gross 1995). Macadamia is preferentially outcrossing, with a gametophytic partial self-incompatibility system (Sedgley et al. 1990) and pollination relying on introduced bees and native insects including native bees (Vithanage and Ironside 1986). Natural seed dispersal is thought to be via animals and water, based on observations of native populations of macadamia often concentrated along waterways and in nearby open rocky areas (McConachie 1980).

Domestication of macadamia began only in the last one and a half centuries. The first selections were made from Australian rainforests in the mid 1800s, and the first orchard plantings, of seedling *M. tetraphylla* trees, were in Australia towards the end of that century (McConachie 1980). However, a viable macadamia industry developed first in Hawaii, based on three small introductions – one of *M. integrifolia* by W.H. Purvis in 1881, another of *M. integrifolia* by R.A. Jordan in 1892, and a government introduction in 1892–1894 of *M. tetraphylla* (Wagner-Wright 1995). It was in Hawaii that critical advances in commercialization (clonal propagation, use of superior selections, processing, and marketing) started from the 1930s (Wagner-Wright 1995). Cultivars bred in Hawaii from this time constitute the bulk of the macadamia industry worldwide, consisting entirely of *M. integrifolia*, though certain *M. integrifolia* and hybrid cultivars developed in Australia are also widely planted, and many other regions have breeding programs for local adaptation (Peace 2002; Hardner et al. 2007). *M. integrifolia* appears to perform better in tropical regions, and hybrids and *M. tetraphylla* are preferred in cooler regions (Hardner et al. 2007). Cultivated hybrids appear to have arisen entirely in cultivation through artificial species combinations rather than from sourcing of natural hybrid zones (Peace 2002). Best estimates indicate that modern cultivars are at most six generations removed from their wild progenitors, but typically less, with some cultivars being direct seedlings of wild trees (Fig. 13.1). Strong selection pressures for certain traits have been applied only in the last one to two generations (Hardner et al. 2007).

Much can be learned from the study of such a relatively undomesticated genome. Because cultivars are virtually wild themselves, systematic evaluation of wild

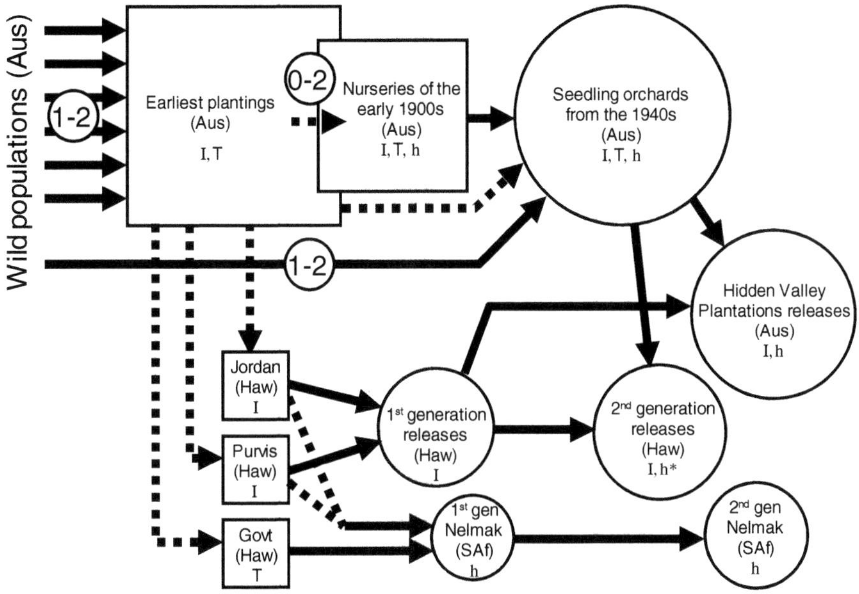

Fig. 13.1 Macadamia domestication – germplasm sources and generations in cultivation leading to current cultivars grown in the three largest macadamia-producing regions of the world: Australia (Aus), Hawaii (Haw), and South Africa (SAf). Circles represent cultivated germplasm groups that include currently-grown cultivars in at least one of the three regions; boxes represent past germplasm groups without current cultivars; sizes of these shapes roughly correspond to degree of genetic diversity within groups. Arrows represent one generation unless noted otherwise; presumed germplasm flow is indicated by a dashed line. I = *M. integrifolia*. T = *M. tetraphylla*. h = interspecific hybrids. * One Hawaiian second generation release has approximately one-quarter *M. ternifolia* heritage, and is probably a direct selection or seedling from an unspecified Australian orchard/planting existing in the 1950s. Adapted from Peace (2002) and Hardner et al. (2007)

populations may reveal trees that are directly useable in orchard production. Hawaiian-bred cultivars are derived from a very limited gene pool of only a small number of *M. integrifolia* seeds taken from Australia. Wild populations represent rich sources of genetic diversity for various traits of interest to commercial production. Interspecific hybridization over multiple generations in cultivation has confused the actual species status of many cultivars, though hybrids could combine the best features of each species and exhibit hybrid vigor. The identification of pure species from wild populations can set the baseline for genome organization of each species and allow examination of the commercial value of each species, of intraspecific genepools, and of interspecific hybridization. Wild populations are under serious threat from human activities, so the opportunity to investigate genomic changes during the first stages of domestication of this tree nut crop is quickly passing.

Macadamia has been the subject of limited genomic research to date. Cytogenetic studies from the 1960s revealed the basic physical genome organization. Various genetic marker systems have been developed, from isozymes in the early 1990s through multi-locus dominant markers to simple sequence repeats (SSRs) at present.

These markers have been used to determine germplasm organization and evaluate genetic diversity to aid conservation and breeding decisions, establish cultivar identity and verify or deduce pedigrees, monitor pollen flow to optimize orchard design, and seek marker-trait associations to enable marker-assisted selection in breeding. Two macadamia genes of value have been isolated and characterized. A genetic map of macadamia, the first for the Proteaceae family, was constructed that is giving preliminary insights into the genetic structure of macadamia chromosomes. No physical genomic resources such as bacterial artifical chromosome (BAC) or expressed sequence tag (EST) libraries have been developed for macadamia.

13.2 Cytogenetics

All macadamia species are diploid, with a haploid chromosome number of 14 (Ramsay 1963; Storey 1965; Storey and Saleeb 1970). Hybridization (*M. tetraphylla* x *M. tntegrifolia)* does not appear to disrupt normal chromosome pairing and disjunction and the chromosome number of the F_1 progeny remains at n = 14 (Storey and Saleeb 1970). Stace et al. (1998) reviewed cytological data for 188 species in 65 genera of Proteaceae. Genera of subfamily Grevilleoideae are almost entirely diploid, with chromosome base numbers of n = x = 10, 11, 12, 13, and 14, with two observed instances of triploidy. Stace et al. (1998) argued against an earlier hypothesis that these chromosome base numbers represented "paleo-polyploidy" from an ancestral genome of x = 5 or 7 (e.g., Venkata Rao 1970; Johnson and Briggs 1975). Instead, Stace et al. (1998) suggested that members of Proteaceae are derived from an ancestral genome of x = 12 or 21, with 24 chromosome arms. The ancestral grevilleoid genome of x = 14 is probably of old Gondwanan origin and consists of ten metacentric and four short telocentric chromosomes (Stace et al. 1998). Many members of tribe Macadamieae appear to have retained this original genome: n = x = 14 for *Macadamia*, *Brabeium*, *Floydia*, and others, with n = x = 13 (fusion of a telocentric and a metacentric chromosome) for*Hicksbeachia* and *Gevuina* (Stace et al. 1998). The number of isozyme loci in *Macadamia* reported by Vithanage and Winks (1992) and Aradhya et al. (1998), and in other Australian genera of Proteaceae (Stace et al. 1998), is consistent with an ancestral diploid rather than tetraploid origin. Although information on chromosomal size in *Macadamia* is not available, five other genera surveyed from subfamily Grevilleoideae have relatively small chromosomes, further evidence against a paleo-polyploid origin (Stace et al. 1998). The genome size of *Brabeium stellatifolium*, in the same subtribe as *Macadamia*, was the smallest reported (mean chromosome length of 1.0 μm) for those five genera (Stace et al. 1998), suggesting that *Macadamia,* too, has a relatively small genome size.

13.3 Genetic Markers

Many marker systems have been developed and applied to the genetic analysis of macadamia, usually for a specific initial purpose. Each marker system has specific technical requirements, and its advantages or disadvantages depend on the purpose to which it is applied. However, differences also reflect advances in technologies, so that in some cases, more efficient marker systems have superseded others for the same applications. In any case, the various marker systems developed for macadamia represent a collective toolbox that future researchers can readily choose from, depending on particular research needs and available resources.

The ideal marker system is one that, with minimal technical requirement, cheaply and rapidly generates many informative markers. Information is maximized with markers that are polymorphic between any two individuals, and therefore, codominant loci with multiple alleles are the most informative. However, marker systems that generate these types of markers usually amplify only one locus per assay, and so marker systems that generate many, though less informative, dominant markers per assay can offer an analytical advantage. Another important consideration is the initial costs involved in developing the specific marker assays, such as primers for sequence-specific markers or probes for hybridization-based markers, and the preliminary testing of the assays on germplasm subsets. Whether markers are gene/sequence-specific or random apply primarily to the purpose to which they are to be used and therefore do not necessarily reflect inherent benefits. Visualized genetic markers are proxies for underlying DNA sequence variation, so the ultimate DNA marker system would be DNA sequencing of as many parts of the genome as desired, targeted and gene-specific, or otherwise. As sequencing technology becomes cheaper and more accessible, such "markers" will replace all others, for macadamia and every other organism. In the meantime, a multitude of marker systems are available, many have been developed for macadamia, and direct comparisons between most of them have been made to aid decisions for future genetic analyses of this crop.

13.3.1 Isozyme

Among the cultivated crops, the utility of isozymes in plant breeding and improvement were appreciated because of the commonly codominant nature of these markers. At a time when perennial crop breeding, including for macadamia, was largely dependent on open-pollinated progeny selection (Hamilton and Ito 1976), the advent of isozyme technology was a promising breakthrough. In macadamia, the immediate benefit was in being able to identify the numerous cultivars that had the potential to be used as breeding parents. The industry also benefited indirectly as isozyme technology enabled the easy identification of cultivars and cleared doubts about mis-identifications, which were rampant at the time. Nine isozyme systems – acid phosphatase, glutamate oxaloacetate transaminase, isocitric dehydrogenase, leucine amino peptidase, malate dehydrogenase, malic enzyme, 6-phosphogluconate dehydrogenase, phosphoglucoisomerase, and peroxidase – established for macadamia exhibited sufficient polymorphism to discriminate 73 macadamia genotypes (Vithanage and Winks 1992). Phosphoglucoisomerase was the most polymorphic isozyme locus observed. Of the nine loci identified, seven were useful in the unequivocal identification of all cultivars. This knowledge was further extended (Aradhya et al. 1998) to include 16 isozyme loci, screened across 45 macadamia genotypes including several crop relatives. Considerable genetic diversity was detected, and two distinct groups of Hawaiian cultivars were observed. The isozyme results in general supported the known origin of some of the Hawaiian cultivars but cast doubts on some cultivars of Australian origin.

13.3.2 RAPD and STMS

Limitations in isozymes and advances in polymerase chain reaction (PCR) technology led to the development of two more marker systems for macadamia, random amplified polymorphic DNA (RAPD) and sequence-tagged microsatellite sites (STMS). These marker systems represent two major types based on primer design: (i) use of primers based on sequence information, STMS and (ii) arbitrary primers, RAPD. In contrast to STMS, the RAPD marker system is relatively cheap to develop and typically produces 5–20 dominant markers per assay, with a lower degree of polymorphism per marker, but as for many other multilocus dominant marker techniques, markers can be difficult to transfer between labs.

Eight STMS loci were developed for macadamia after isolating microsatellite DNA from a small insert library (Vithanage et al. 1999). From 74 colonies that tested positive for GA repeats, 22 were sequenced to produce 11 suitable for primer design. In one case, a TA repeat was also revealed lying very close to a GA repeat. Seven pairs of primers were designed for GA repeats and another pair designed for the TA repeat microsatellite (Vithanage et al. 1999). These primers amplified loci in all of the southern clade macadamia species, which were separated on agarose gels. Three to 13 putative alleles were detected per locus; however, allelism could not be verified and so amplified bands, while highly reproducible, were scored as binary data (Vithanage et al. 1998, 1999). This scoring nullified the advantage associated with codominant multi-allelic loci. Mendelian inheritance in an F_1 population was demonstrated for three of the STMSs, and segregation data for one locus provided an STMS marker for linkage mapping (Vithanage et al. 1998; Peace et al. 2003). Amplification products for all eight STMS primer pairs were later separated on large polyacrylamide gels, which confirmed codominant inheritance and multi-allelism, and in several cases, multiple loci per primer pair (Peace et al. 2004).

For RAPD, 40 random primers were screened and the best ten chosen to survey a large collection of mostly cultivated macadamias (Vithanage et al. 1998, 1999). Agarose gels were used to separate RAPD markers. The ten primers generated 126 markers, and together with the SSR (agarose) data, the RAPD markers were used for cultivar identification and to estimate genetic diversity in the germplasm set (Vithanage et al. 1998, 1999). The ten primers were also used on an F_1 population to generate 124 markers, of which 48 were polymorphic, with 40 of these (83%) displaying Mendelian inheritance (Vithanage et al. 1998, 1999). Five of these RAPD markers were placed on the framework genetic map of macadamia (Peace et al. 2003).

13.3.3 AFLP

The amplified fragment length polymorphism (AFLP) marker system was applied to macadamia to provide a more abundant source of markers than previous techniques. Most AFLP markers are dominant and scored as present/absent, though co-dominant markers are occasionally generated in a mapping population. The major advantages of AFLP markers are: (i) no previous sequence information is needed; and (ii) distribution is random in both coding and non-coding regions of the genomes. These properties make AFLP markers useful for genetic and genomic analysis of minor crops.

AFLP markers were used for fingerprinting 26 macadamia accessions representing four *Macadamia* species and one *Hicksbeachia* species (Steiger et al. 2003). Polyacrylamide gels were used to separate the AFLP markers. Fifteen *Eco*R I – *Mse* I primer combinations with three selective nucleotides were used to survey three accessions of *M. integrifolia* and one of *M. tetraphylla* to assess variation detected by different primer sets. This allowed for selection of those primer pairs that generated the highest level of polymorphism for full-scale analysis. High levels of variation between the two species were found. The average number of polymorphic markers per primer pair within the *M. integrifolia* accessions was 14.6, with a range of 9 to 21. All of the primers screened generated many polymorphic markers within the samples surveyed. Based on the survey results, six primer pairs were chosen for the final analysis, and 105 polymorphic AFLP markers were used to evaluate the 26 accessions (Steiger et al. 2003).

13.3.4 RAF and RAMiFi

Similar to AFLP, the randomly amplified DNA fingerprinting (RAF) marker system is an efficient means of generating dominant markers for species such as macadamia where gene sequence information is not readily available. RAF assays are as robust and identify as many markers as AFLP (that depends on the means of marker visualization, e.g., 80–100 or more with radioactive labeling, or typically half as many with silver staining), and yet require fewer laboratory steps including only a single PCR step, tolerate lower quality DNA, and are cheaper per assay. Similarly to RAPD, RAF typically uses 10-mer primers and amplifies equivalent DNA fragments that usually have dominant inheritance.

The initial need of RAF for macadamia was in genetic map construction, which had begun with RAPD and AFLP. When RAF assays were attempted on macadamia, efforts were met with immediate success (unpublished results). The power of RAF compared to RAPD was obvious, and RAF soon became a useful marker system for adding many markers to the macadamia genetic map. The first 16 primers screened across a progeny population each generated approximately 100 markers, averaging 22 polymorphic RAF DNA fragments each (Peace et al. 2003). Of these polymorphic fragments identified, 317 (90%) exhibited dominant inheritance, 30 were suspected to represent alleles of 15 size-variant codominant loci, while another RAF marker contained a microsatellite with four distinct alleles (Waldron et al. 2002; Peace et al. 2003).

The discovery of microsatellites within some RAF markers led to the advent of the randomly amplified microsatellite fingerprinting (RAMiFi) marker system (Peace et al. 2004). The stuttered appearance of putative microsatellite-containing markers, which typically have codominant inheritance except when "null" alleles are common, helps distinguish them from nearby dominant markers in a RAF profile and is critical in the RAMiFi primer screening process (Peace et al. 2004). RAMiFi markers are defined as codominant or dominant markers amplified by RAMiFi primers, i.e., those primers chosen specifically for their ability to amplify microsatellite loci via the RAF protocol. Six new RAMiFi primers were developed, which identified an average of 3.2 codominant markers each (putatively containing microsatellites, with 2–10 alleles per marker) and averaging 13.5 accompanying polymorphic dominant markers each when screened across a set of 30 macadamia cultivars (Peace et al. 2002). When the best five of these same RAMiFi primers were applied to expanding the genetic map of macadamia, an average of three codominant and 15 dominant markers were identified for each (Peace 2002).

Besides providing the basis for genetic map construction, RAF/RAMiFi markers have also been applied to examining heterozygosity in cultivars and segregation distortion in marker inheritance (Peace et al. 2003), tracing pollen flow in orchards (Vithanage et al. 2002), deducing parentage of elite cultivars and verifying pedigrees (Peace et al. 2001a, b, 2002, 2003, 2005; Peace 2002), tracing natural origins (Peace et al. 2001b; Peace 2002), investigating genetic relationships among cultivars (Peace et al. 2001b, 2002, 2005; Peace 2002), identifying hybrids and quantifying species compositions in cultivation (Peace et al. 2000, 2001a, b, 2002, 2005) and in wild populations (Peace et al. 2001b; Peace 2002), studying demographics of wild populations (Neal 2006), determining relationships within and between natural genepools of *Macadamia* species (Peace 2002), and delimiting natural distributions of species (Peace 2002).

Peace et al. (2004) compared development costs and data generated for several marker systems across a common set of 14 macadamia genotypes. At that time, the RAMiFi approach was identified as the most efficient and economical. Overall, for genetic analysis of macadamia, RAMiFi is a very versatile tool, due to its ability to amplify codominant markers (anonymous microsatellites) efficiently, together with abundant dominant markers. The different marker types generated by RAMiFi make it very suitable for cultivar identification, pedigree analysis, diversity analysis, and linkage mapping – the only marker system identified to have such versatility (Peace et al. 2004).

13.3.5 SSR

Although they have proven useful in previous studies of macadamia genetics, dominant markers such as RAPD, RAF, and AFLP, and sequence-tagged markers scored in a dominant manner such as STMS, have relatively limited utility for studying genetic diversity. For studies of proximal relatedness and population genetic variation, locus-specific codominant markers such as simple sequence repeats (SSRs), which are based on microsatellites (regions of DNA containing tandem repeats of a core 2–6 bp nucleotide sequence), are considerably more powerful.

Previous attempts to isolate and characterize sequence-tagged microsatellite loci generated small numbers of dominant loci with limited reliability (Vithanage et al. 2002). Peace et al. (2004) suggested that RAMiFi represented a cheaper alternative to standard codominant microsatellite marker development, particularly for under-researched and under-funded crops. However, the RAMiFi procedure identified a relatively small number of loci, and locus-specific primers were not generated.

A new suite of SSR markers has recently been developed for use in macadamia (Schmidt et al. 2006). Unlike the RAMiFi markers, these were isolated directly from genomic DNA, generating 100 locus-specific primer pairs (Schmidt et al. 2006; Schmidt et al. manuscript in preparation). The markers, representing a range of different repeat motifs, have been tested and used in *M. integrifolia*, *M. tetraphylla*, and *M. jansenii* (Schmidt et al. 2006; Schmidt et al. manuscript in preparation) and, following initial investigation of the nature and extent of polymorphism within and between cultivars/species (Schmidt et al. 2005; Schmidt et al. 2006; Schmidt et al. manuscript in preparation), subsets have been employed in parentage analysis (Neal 2006) and investigation of diversity and gene flow in natural populations (Neal 2006). The long-term goal in developing the markers, however, was to facilitate marker-assisted selection and breeding, for which efforts are currently underway.

While these SSRs were costly to develop for macadamia, they are now the most powerful markers available for the purposes of cultivar identification, pedigree analysis, and diversity analysis in this crop. Development and application of SSRs in the model fruit tree genus, *Prunus* (e.g., Dirlewanger et al. 2002; Howad et al. 2005), indicates that while as few as 20 well-chosen SSR loci can provide comprehensive data on germplasm organization, hundreds of SSRs are required to be widely useful for genetic linkage mapping. Once the macadamia SSR loci are mapped and polymorphism across various cultivars and genepools have been established, these and further SSRs will be the most informative and widely applicable markers for macadamia genetic studies, particularly if efforts are also made to combine several SSRs into single assays (i.e., multiplexing) and polymorphic loci are detectable on a range of platforms (e.g., agarose gels, "sequencing" gels, and high-throughput semi-automated systems). However, deep phylogeny will require gene or organelle sequence data.

13.4 Linkage Mapping

As a clonally-propagated tree crop with a long juvenile phase, macadamia offers the opportunity for great gains to be made from marker-assisted selection (MAS). Such a breeding strategy allows selection at the seedling stage for traits not expressed until reproductive maturity, including the many components of nut yield and quality.

Genetic linkage map construction is a useful first step towards identifying markers linked to genes controlling trait expression, and genetic maps are particularly advantageous in elucidating the genetic architecture of quantitative traits. Due to long generation times and varying degrees of self-incompatibility for macadamia (Meyers 1997), it is difficult to construct the F_2 or backcross populations typically used for linkage mapping research in annual species. The "two-way pseudo-testcross strategy" is useful for heterozygous tree crops as it enables efficient map construction with F_1 populations (Grattapaglia and Sederoff 1994). This mapping strategy was adopted for macadamia.

The first efforts towards macadamia linkage mapping began with the creation of useful F_1 populations. A cross-compatibility study in macadamia (Meyers 1997) led to the formation of several F_1 progeny populations of important cultivars in the Australian industry. Reciprocal crosses between two parents formed progeny populations. The populations were field-planted in the late 1990s and became a useful base for a new breeding program at CSIRO Plant Industry, Brisbane. One of these populations, containing 56 F_1 progeny of cultivars 246 ('Keauhou' in Hawaii) and A16, was chosen as the first to be used for linkage mapping. Cultivars 246 and A16 differ in field performance for several important agronomic traits including kernel recovery and tree shape. Cultivar 246 is *M. integrifolia* while A16 is a hybrid with a quarter of *M. tetraphylla* in its pedigree (Peace et al. 2002; 2005). This population resulted from a reciprocal cross, where for 21 of the progeny, '246' was the female parent, and for the remaining 35 progeny, 'A16' was the female parent. Preliminary inheritance and linkage analysis for 30 progeny of the '246' x 'A16' population was performed with RAPD markers (unpublished results). That study identified 48 markers from ten RAPD primers, which were assembled into just six linkage groups of 2–3 markers each. Attempts were also made to develop the AFLP marker system for macadamia mapping, but no markers resulted (unpublished results).

The RAF marker system proved immediately useful for linkage mapping in macadamia, and quickly replaced RAPD for this purpose. Using 56 progeny of the '246' x 'A16' population, RAF markers formed the basis of a genetic map for macadamia. This map, the first reported for the Proteaceae family, had 24 linkage groups that covered 70–80% of the genome of macadamia (Peace et al. 2003). The number of linkage groups was greater than the haploid chromosome number of macadamia, and some regions of the macadamia genome remained uncovered. The map contained 265 framework markers (259 RAF, 16 of which being codominant, five RAPD, and one SSR) from analysis of 382 polymorphic markers. RAF and RAPD markers appeared to be well distributed across the map, although some clustering was evident in several linkage groups. The average spacing between markers on this consensus map was 4.7 cM, with no gaps larger than 20 cM. Individual parent maps were initially constructed, with the 'A16' map including 190 framework markers in 18 linkage groups and spanning 920 cM (61–69% of the estimated genome size for this cultivar). The '246' map contained 119 framework markers in 19 linkage groups, spanning 530 cM (48–53% of the estimated genome size for this cultivar). Ninety bridging loci allowed merging of these two maps to produce the consensus map (Peace et al. 2003).

Examination of marker inheritance in these macadamia cultivars confirmed that inheritance was typical for Mendelian loci in a diploid species. Markers with segregation distortion accounted for 16% of the 382 markers available for inheritance analysis, and 36 markers on the framework map exhibited distorted segregation patterns, with most occurring in four linkage blocks of other such markers (Peace et al. 2003). The occurrence of blocks of markers with skewed segregation suggests linkage to loci under selection pressure. Most of the distorted markers were from 'A16', and further analysis of these markers considering the reciprocal nature of the progeny population identified that the most prominent linkage block of distorted segregation occurred when 'A16' was the maternal parent (Peace 2002). This finding indicates pre- or post-zygotic selection against a deleterious locus, and may be associated with the interspecific ancestry of the A16 cultivar.

Efforts to expand the genetic map of macadamia have continued. Five RAMiFi primers were used to generate 90 new markers for the '246' x 'A16' population. The extended map included 328 framework markers (240 dominant RAF, 52 dominant RAMiFi, five dominant RAPD, 17 size-variant codominant RAF/RAMiFi, 13 RAMiFi microsatellites, and one codominant SSR marker) (Peace 2002) (Fig. 13.2). Map coverage was increased to 1,195 cM (an estimated 75–85% of the genome), decreasing the average distance between markers to 4.0 cM. The microsatellite markers (13 RAMiFi and one SSR) were distributed across 12 linkage groups, and because each microsatellite locus was multi-allelic across macadamia cultivars, any cross was likely to be segregating for at least one allele for each locus. The presence of such markers on the '246' x 'A16' linkage map should therefore facilitate comparative mapping when linkage maps for other cultivars are constructed for this crop. Linkage analysis of multiple inter-related populations is currently being conducted (C. Hardner, personal communication). The next steps are to use the genetic map(s) to identify genomic regions controlling traits of interest, obtain closely linked markers, and implement marker-assisted selection for macadamia. This strategy would greatly benefit from the application of statistics that allow genotypic and phenotypic data from several populations linked through pedigrees to be combined (Hardner et al. 2005).

13.5 Gene Sequencing

Although macadamia is commercially the most important group of species in the botanically and evolutionarily intriguing Proteaceae family, very little is known at the gene level. Partial gene sequences from *M. integrifolia* and*M. ternifolia* exist in the National Centre for Biotechnology Information (NCBI) databases for several reference genes that have been used to study phylogenetic relationships involving the Proteaceae. These include the 18s and 26s ribosomal genes, RNA polymerase II large sub-unit, ÿ-subunit of ATP synthase, large subunit of ribulose-1,5-bisphosphate carboxylase/oxygenase, and the maturase K genes. To date, these macadamia gene sequences have been used as outgroups for studies of other taxa

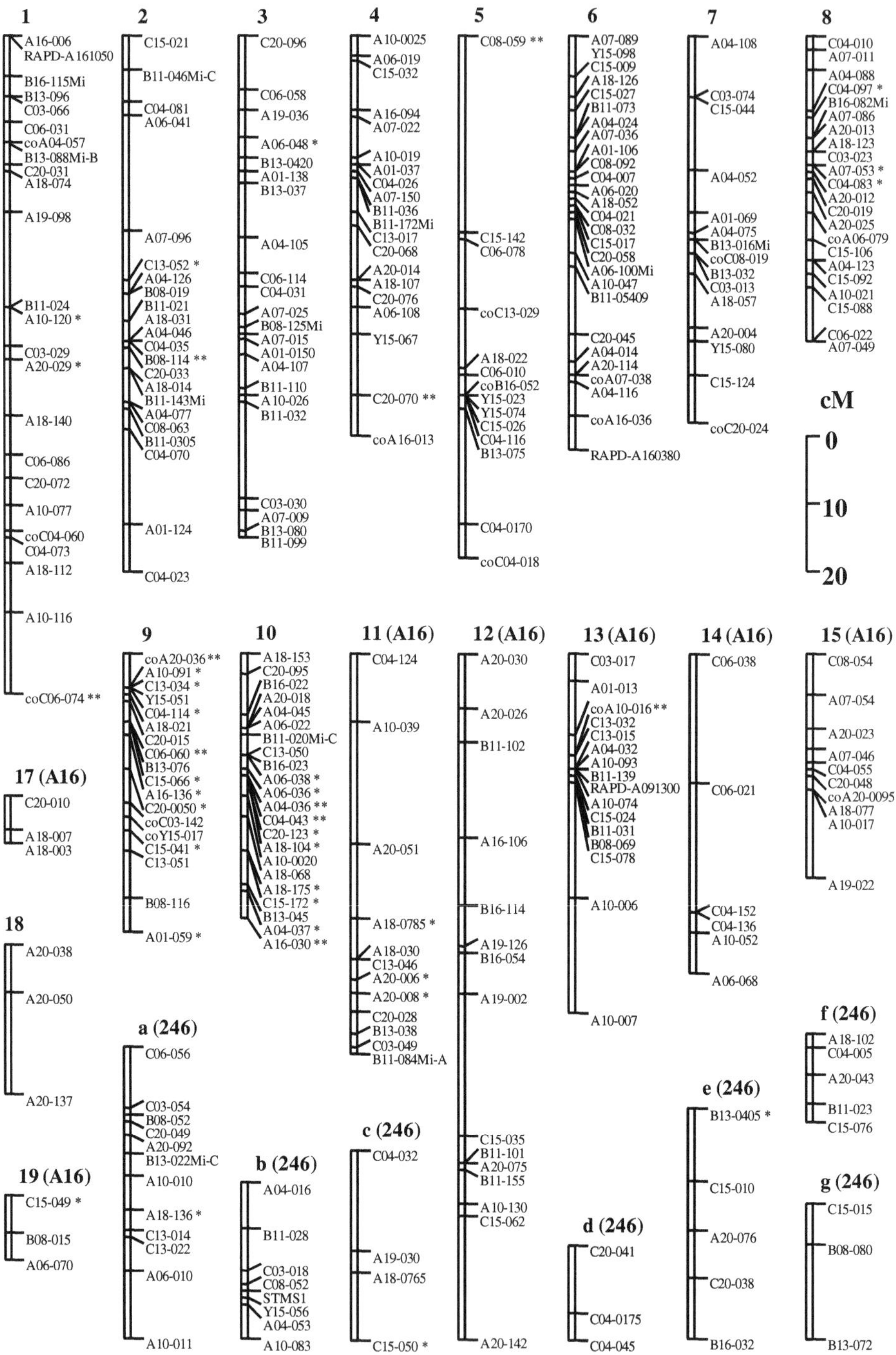
1
A16-006
RAPD-A161050
B16-115Mi
B13-096
C03-066
C06-031
coA04-057
B13-088Mi-B
C20-031
A18-074
A19-098
B11-024
A10-120 *
C03-029
A20-029 *
A18-140
C06-086
C20-072
A10-077
coC04-060
C04-073
A18-112
A10-116
coC06-074 **
2
C15-021
B11-046Mi-C
C04-081
A06-041
A07-096
C13-052 *
A04-126
B08-019
B11-021
A18-031
A04-046
C04-035
B08-114 **
C20-033
A18-014
B11-143Mi
A04-077
C08-063
B11-0305
C04-070
A01-124
C04-023
3
C20-096
C06-058
A19-036
A06-048 *
B13-0420
A01-138
B13-037
A04-105
C06-114
C04-031
A07-025
B08-125Mi
A07-015
A01-0150
A04-107
B11-110
A10-026
B11-032
C03-030
A07-009
B13-080
B11-099
4
A10-0025
A06-019
C15-032
A16-094
A07-022
A10-019
A01-037
C04-026
A07-150
B11-036
B11-172Mi
C13-017
C20-068
A20-014
A18-107
C20-076
A06-108
Y15-067
C20-070 **
coA16-013
5
C08-059 **
C15-142
C06-078
coC13-029
A18-022
C06-010
coB16-052
Y15-023
Y15-074
C15-026
C04-116
B13-075
C04-0170
coC04-018
6
A07-089
Y15-098
C15-009
A18-126
C15-027
B11-073
A04-024
A07-036
A01-106
C08-092
C04-007
A06-020
A18-052
C04-021
C08-032
C15-017
C20-058
A06-100Mi
A10-047
B11-05409
C20-045
A04-014
A20-114
coA07-038
A04-116
coA16-036
RAPD-A160380
7
A04-108
C03-074
C15-044
A04-052
A01-069
A04-075
B13-016Mi
coC08-019
B13-032
C03-013
A18-057
A20-004
Y15-080
C15-124
coC20-024
8
C04-010
A07-011
A04-088
C04-097 *
B16-082Mi
A07-086
A20-013
A18-123
C03-023
A07-053 *
C04-083 *
A20-012
C20-019
A20-025
coA06-079
C15-106
A04-123
C15-092
A10-021
C15-088
C06-022
A07-049
cM
0
10
20
9
coA20-036 **
A10-091 *
C13-034 *
Y15-051
C04-114 *
A18-021
C20-015
C06-060 **
B13-076
C15-066 *
A16-136 *
C20-0050 *
coC03-142
coY15-017
C15-041 *
C13-051
B08-116
A01-059 *
10
A18-153
C20-095
B16-022
A20-018
A04-045
A06-022
B11-020Mi-C
C13-050
B16-023
A06-038 *
A06-036 *
A04-036 **
C04-043 **
C20-123 *
A18-104 *
A10-0020
A18-068
A18-175 *
C15-172 *
B13-045
A04-037 *
A16-030 **
11 (A16)
C04-124
A10-039
A20-051
A18-0785 *
A18-030
C13-046
A20-006 *
A20-008 *
C20-028
B13-038
C03-049
B11-084Mi-A
12 (A16)
A20-030
A20-026
B11-102
A16-106
B16-114
A19-126
B16-054
A19-002
C15-035
B11-101
A20-075
B11-155
A10-130
C15-062
A20-142
13 (A16)
C03-017
A01-013
coA10-016 **
C13-032
C13-015
A04-032
A10-093
B11-139
RAPD-A091300
A10-074
C15-024
B11-031
B08-069
C15-078
A10-006
A10-007
14 (A16)
C06-038
C06-021
C04-152
C04-136
A10-052
A06-068
15 (A16)
C08-054
A07-054
A20-023
A07-046
C04-055
C20-048
coA20-0095
A18-077
A10-017
A19-022
17 (A16)
C20-010
A18-007
A18-003
18
A20-038
A20-050
A20-137
19 (A16)
C15-049 *
B08-015
A06-070
a (246)
C06-056
C03-054
B08-052
C20-049
A20-092
B13-022Mi-C
A10-010
A18-136 *
C13-014
C13-022
A06-010
A10-011
b (246)
A04-016
B11-028
C03-018
C08-052
STMS1
Y15-056
A04-053
A10-083
c (246)
C04-032
A19-030
A18-0765
C15-050 *
d (246)
C20-041
C04-0175
C04-045
e (246)
B13-0405 *
C15-010
A20-076
C20-038
B16-032
f (246)
A18-102
C04-005
A20-043
B11-023
C15-076
g (246)
C15-015
B08-080
B13-072

Fig. 13.2

(Mast et al. 2005) or as family representatives in higher order phylogenetic analyses (Martin and Dowd 1991). However, these macadamia sequences have not been applied to rigorously address questions of species boundaries in the *Macadamia* genus. A broader survey of sequence diversity in the genus would provide clarification of macadamia species and their evolutionary origins to complement the neutral molecular marker systems used so far.

The most comprehensively studied genes in macadamia are those encoding abundant proteins of the kernel, which also have in vitro anti-microbial activity. In a survey of protein extracts of seeds of Australian native flora (Manners et al. 1998), it was observed that extracts from mature macadamia kernels had substantial in vitro activity against a range of filamentous fungal phytopathogens (Marcus et al. 1997; 1999). The original hypothesis for seeking such proteins in the Australian flora was that because of their generally primitive nature and isolated evolutionary history, proteins with novel sequences may be revealed that were lost or highly diverged in more recent angiosperms. This appears to have been borne out by detailed analysis of two families of antimicrobial protein genes from *M. integrifolia*.

The first antimicrobial protein to be purified and cloned from macadamia kernels was termed MiAMP1 (Marcus et al. 1997). The MiAMP1 coding region encodes a protein that has anti-microbial activity against fungi and gram positive bacteria in vitro. The three-dimensional structure for MiAMP1 was determined using NMR techniques, with a unique structure for plant proteins that was proposed to represent a new class of plant defense proteins termed ÿ-barrelins (McManus et al. 1999). A potential role in plant defense has been demonstrated by expressing MiAMP1 in transgenic canola plants, where it provided protection against attack by the fungal phytopathogen *Leptosphaeria maculans* (Kazan et al. 2002). The second type of antimicrobial protein purified from macadamia, termed MiAMP2, consists of a family of small homologous antimicrobial peptides, probably all derived from a single large precursor protein with four homologous subunits (a – d). Regions of this protein have substantial homology to vicilin seed storage proteins that are common in many plants.

Intriguingly, the four subunits observed in the macadamia protein were the largest number recorded among plants. Macadamias represent the first plants where an

Fig. 13.2 (continued) An extended genetic linkage map of macadamia. Minimum non-redundant genome coverage (75–85%) comprises 19 linkage groups (1–19) spanning 1,020 cM with 288 markers. Possible additional coverage is provided by seven linkage groups (a-g) spanning 175 cM with 40 markers. Marker labels: "co" = size-variant codominant RAF/RAMiFi marker; "Mi" = microsatellite RAMiFi marker; "STMS" = STMS marker; "RAPD" = RAPD marker; RAMiFi markers include in their labels the primers A06, B08, B11, B13, B16, or Y15; all others are standard RAF markers. Markers with distorted segregation are indicated by asterisks ($^*p < 0.05$, $^{**}p < 0.01$). Linkage groups 1 to 10 are combinations of 'A16' and '246' markers, groups 11 to 17 and 19 are 'A16'-only (a previous group 16 joined group 12 through two RAMiFi markers from primer B16), group 18 contains only markers that were heterozygous in both cultivars, and groups a to g were '246-only' with no known anchor points to 'A16' groups. Adapted from Peace (2002) and following labeling conventions of Peace et al. (2003)

antimicrobial function for processed vicilin proteins has been demonstrated. However, the diversity of structures observed for these proteins across the plant kingdom has so far failed to reveal a logical evolutionary progression for their assembly. The MiAMP1 protein and the MiAMP2 b, c, and d peptides have been shown to be released from macadamia nuts after imbibing water and during germination (Marcus et al. 1999). It is therefore believed that they may have a function in providing a zone of inhibition of microbial growth around the germinating seed preventing attack by potential pathogens that would be abundant in the soil and litter of the rainforest floor.

Suggested targets of further gene isolation in macadamias are genes responsible for the unique oil composition of macadamia nuts. The oil from macadamia nuts has an unusual composition (Badami and Patil 1980) and contains about 30% palmitic acid ($16{:}1^{\Delta 9}$). It is predicted that this unusual fatty acid contribution is mediated by a specific acyl-acyl carrier protein (ACP) desaturase enzyme. Gummeson et al. (2000) have reported on the isolation and sequencing of an ACP desaturase from macadamia, but functional analysis indicated that it was not the enzyme specifically responsible for the palmitic acid ($16{:}1^{\Delta 9}$) profile.

Gene sequence information for macadamia is clearly very fragmentary and one could extend this conclusion to all members of the Proteaceae. However, the sequences of antimicrobial protein encoding genes have shown that studies of macadamia genes can reveal novel functional features of plant development and raise interesting questions about evolution of proteins in the plant kingdom. Clearly, a systematic analysis of the macadamia genome at the sequence level would provide many more insights into the evolution of plants, particularly the important but poorly studied Proteaceae family, as well as providing a basis for gene-based improvement and manipulation through molecular breeding of macadamia for improved nut yield and quality.

13.6 The Macadamia Genome

Macadamia genomics research to date has provided a broad sense of what is a "typical" macadamia genome in cultivation. Cytogenetics and genetic linkage mapping have revealed macadamia to be a medium-sized diploid crop genome. Macadamia has high heterozygosity, and correspondingly, each individual is genetically distinct and genetic diversity among individuals is typically high. Genomes can be classified primarily by species, then natural region of origin within species. Breeding has occurred for so few generations that it rarely has a noticeable effect on germplasm organization. For the southern clade macadamias, interspecific hybridization readily occurs in nature, and a mixture of species genomes is common in cultivation. While most of the macadamia genome has been mapped with a framework of markers, the specific functional loci and gene networks that separate macadamias from other species and contribute to agronomic production are not yet known other than the two isolated antimicrobial protein genes.

Analyses of marker inheritance and heterozygosity (Peace 2002; Peace et al. 2003) have shown macadamia cultivars to be quite heterozygous – typically around 40–50% for any given locus, and even higher for hybrids. Such values reflect the predominantly outcrossing nature of macadamia. Vithanage et al. (2002) found that even in the middle of a solid block planting of a single cultivar, 27 rows wide, cross-pollination across rows from outside the block was greatly favored over selfing. Neal (2006) determined that despite severe fragmentation of native stands of macadamia trees, heterozygosity of individual trees remains high.

The application of genetic markers has enabled, for any particular genotype of macadamia, distinction from all others. Multi-locus profiles of AFLP and RAF readily achieve cultivar distinction (Peace et al. 2001a, 2002; Steiger et al. 2003). Profiles of multiple codominant loci are also useful for cultivar identification and determining parentage, and single loci of some codominant loci can display many alleles that achieve the same objective (Vithanage and Winks 1992; Aradhya et al. 1998; Peace et al. 2000; Peace 2002; Steiger et al., 2003;).

When differences in marker profiles of macadamia genotypes are quantified, studies find that genetic diversity is abundant in cultivars. This diversity is not randomly distributed, as genotypes can be arranged into germplasm groups (Aradhya et al. 1998; Peace et al. 2001a, b, 2002; Peace 2002; Steiger et al. 2003). Cultivar genetic groupings were once thought to reflect mainly the breeding and selection history of each cultivar. However, the most comprehensive genetic diversity study to date, surveying 85 cultivars and 372 accessions of a germplasm collection of mostly sourced from wild trees, concluded that the primary determinant of cultivated germplasm organization is species status and composition, followed by native region of origin, while breeding and selection origin has had the least effect (Peace 2001b; Peace 2002). At the genomic level, cultivars are mostly indistinguishable from wild *M. integrifolia* and *M. tetraphylla*. However, one group of Hawaiian cultivars cluster together and separately to wild accessions and appear to represent a gene pool of pure *M. integrifolia* that has undergone 2–3 generations of open-pollinated inbreeding within the same gene pool in Hawaii (Vithanage and Winks 1992; Aradhya et al. 1998; Peace et al. 2001b; Peace 2002; Steiger et al. 2003). A second, smaller group of Hawaiian cultivars tends to genetically cluster with certain other Australian cultivars and wild accessions, despite an identical breeding and selection history to the first Hawaiian group (Peace 2002). Tracing natural origins of cultivars by marker comparisons to wild accessions has also revealed that the two Hawaiian groups and many Australian-bred *M. integrifolia* cultivars probably originated from wild populations growing in the northern-most parts of that species' range; pure *M. integrifolia* cultivars and wild accessions from the southern parts are more genetically distinct (Peace et al. 2001b; Peace 2002). Australian cultivars often have some degree of *M. tetraphylla* in their ancestry, and it was revealed that species mixing has occurred much more than is apparent from morphology (Peace et al. 2001b, 2002, 2005; Peace 2002). Studies that include mostly Hawaiian cultivars and Australian hybrid cultivars (which are often misclassified as pure *M. integrifolia*) therefore detect strong breeding origin effects that are instead probably attributed to natural origin and species status.

The genomes of many cultivars and wild trees are mixtures of *M. integrifolia* and *M. tetraphylla*. Hybridization occurs so readily when species are in proximity (Hardner et al. 2000; Peace 2002) that the four species of the southern clade macadamias can be considered a single species complex. According to genetic marker studies and grafting efforts, no further species are likely to be cross-compatible with the southern clade (Storey and Frolich 1964, Hardner et al. 2000; Peace 2002; Steiger et al. 2003). Cultivars and wild trees with species compositions along the entire scale from pure *M. integrifolia* to pure *M. tetraphylla* were detected when screened with 165 RAMiFi markers that had been initially assessed for species-specificity, indicating that hybridization has occurred for multiple generations (Peace et al. 2001a, b, 2002, 2005; Peace 2000). Tracing natural origins with RAMiFi markers suggests that hybrid cultivars are the result of species mixing in cultivation, with no evidence for cultivars being sourced from the natural hybrid zone of *M. integrifolia* and *M. tetraphylla* (Peace et al. 2001b; Peace 2002). One tri-species hybrid, combining *M. integrifolia*, *M. tetraphylla*, and *M. ternifolia*, was detected with RAMiFi markers, representing a popular cultivar in South Africa that was originally selected in Hawaii (Peace et al. 2001a, 2005; Peace 2002). This cultivar has distinct morphological features that are probably derived from *M. ternifolia*, though this heritage was not previously suspected because *M. ternifolia* is otherwise not used in cultivation whether as a pure species or in hybrid form (Peace 2002; Peace et al. 2005). Despite the ease of hybridization, there is some evidence for interspecific incompatibility, as described in the linkage mapping section.

13.7 Prospects

Genomics research on macadamia has thus far provided significant insights into the genetic makeup of a typical macadamia tree, but has only scratched the surface of what is possible for manipulating crop production. Unlike the vast majority of field and plantation crops, and most fruit and nut orchard crops, macadamia has barely diverged from its wild progenitors, if at all. Genetic marker investigations have begun to describe macadamia's chromosomal architecture, enabled the detection of genetic differences between cultivars including their species composition, and given a broad view of cultivated and wild germplasm organization. The development of further genomics tools would undoubtedly have a tremendous impact in these and other areas. However, comprehensive genomics resources appear far off for this crop. In the meantime, new statistical approaches may enable the application of markers to enhance breeding efforts.

Neutral marker studies have established the genetic baseline of the macadamia crop and its available germplasm resources, to catalog genomic changes during the first stages of domestication. Such research has been valuable before the crop's genetic origins are lost through unrecorded mixing of cultivated species, widespread germplasm exchange, and genetic erosion of wild populations by habitat destruction and pollen contamination from nearby orchard trees. Based on marker information, growers, germplasm curators, and breeders can now make genetically informed decisions on orchard design, germplasm conservation, and parent selection in crossing to avoid inbreeding, to take advantage of particular cultivar germplasm groups, to utilize wild germplasm, and to exploit interspecific hybridization.

An even greater impact on crop production and genetic improvement is expected when further genomic tools can be developed. Macadamia needs a reference genetic linkage map, saturated with markers that are readily transferable to other cultivars/populations. A logical step towards this end would be to map the available SSR marker set. The previously-mapped '246' x 'A16' population may suffice, but a *M. integrifolia* x *M. tetraphylla* F_1 population would serve better, having a greater degree of polymorphism for ease of marker and gene mapping and greater relevance for current and future cultivated germplasm. Hardner et al. (2005) reviewed the potential gains and anticipated obstacles to developing and implementing a marker-assisted selection scheme in macadamia breeding. Given limited industry size and correspondingly limited research support available for macadamia breeding, traditional approaches to identifying marker-trait linkages that rely on separate experimental populations are not feasible. Instead, strategies with a greater likelihood of success are those that directly use breeding populations and phenotypic data. This can be achieved with association mapping, whether via a pedigree genotyping approach or linkage disequilibrium mapping if pedigree relationships are unknown, and such statistics have been developed for tree fruit breeding (e.g., van de Weg et al. 2004). These approaches require genome-spanning markers, much more so when pedigree information is not available, and ideally as part of a saturated genetic map. For most macadamia cultivars, pedigree information is unknown, and given the wide diversity and few generations of cultivation, there are many founder genotypes, both of which hamper the effectiveness of pedigree genotyping. However, macadamia breeding populations, with their known pedigree structures of small to large full-sib families linked through shared parents, are very amenable to pedigree genotyping (Hardner et al. 2005).

Although seemingly unaffordable at present, more comprehensive genomics resources such as BAC libraries, EST libraries, and a whole genome sequence should eventually be developed to provide a sound understanding of the macadamia genome for the crop and as a model for the Proteaceae family. A large insert library (e.g., BACs) would allow researchers to readily study DNA sequences of specific loci of interest. A catalogue of single nucleotide polymorphisms based on the genome sequence, a comprehensive EST database, or some other genome-wide array would enable whole genome surveys to readily identify DNA sequences and gene pathways associated with traits of horticultural importance. Once such functional markers are identified and verified, they must be put into practice on a large scale to be most effective for crop improvement. Fortunately, advances in high-throughput technology being developed for higher-value crops that should be readily adaptable to macadamia.

Macadamia is a minor crop but with a product of high economic and nutritional value, apparent wide climatic adaptability, and a quickly expanding industry. This crop has much to gain from the application of approaches and technologies developed for other plants. It is remarkable that this essentially wild crop already has a strong worldwide industry base. With the application of advances in genomics, macadamia production and the nut itself are likely to undergo impressive changes in the next few decades.

References

Aradhya MK, Yee KL, Zee FT, Manshardt RM (1998) Genetic variability in *Macadamia*. Genet Resour Crop Evol 44:19–32

Badami RC, Patil KB (1980) Structure and occurrence of unusual fatty acids in minor seed oils. Progr Lipid Res 19:119–153

Dirlewanger E, Cosson P, Tavaud M, Aranzana MJ, Poizat C, et al. (2002) Development of microsatellite markers in peach [*Prunus persica* (L.) Batsch] and their use in genetic diversity analysis in peach and sweet cherry (*Prunus avium* L.). Theor Appl Genet 105:127–138

Grattapaglia D, Sederoff R (1994) Genetic linkage maps of *Eucalyptus grandis* and *Eucalyptus urophylla* using a pseudotestcross mapping strategy and RAPD markers. Genetics 137:1121–1137

Gross CL (1995) Macadamia. Flora of Australia 16:419–425

Gummeson PO, Lenman M, Lee M, Singh S, Stymne S (2000) Characterisation of acyl-ACP desaturases from *Macadamia integrifolia* Maiden & Betch and *Nerium oleander* L. Plant Sci 154:53–60

Hamilton RA, Ito PJ (1976) Development of macadamia nut cultivars in Hawaii. Cal Macadamia Soc Yearbook 22:94–100

Hardner CM, McConchie CA, Vivian-Smith A, Boyton S (2000) Hybrids in macadamia improvement. In: Dungey HS, Dieters MJ, Nikles DG (eds) Proc QFRI/CRC-SPF Symp Hybrid Breeding Genetics Forest Trees. Department of Primary Industries, Brisbane, Australia, pp. 472–476

Hardner C, Peace C, Henshall J, Manners J (2005) Opportunities and constraints for marker assisted selection in macadamia breeding. Acta Hort 694:85–90

Hardner C, Peace C, Neal J, Pisanu P, Powell M, et al. (2007) Genetic resources of macadamia. Hort. Rev. in press

Howad W, Yamamoto T, Dirlewanger E, Testolin R, Cosson P, et al. (2005) Mapping with a few plants: using selective mapping for microsatellite saturation of the *Prunus* reference map. Genetics 171:1305–1309

Johnson LAS, Briggs BG (1975) On the Proteaceae - the evolution and classification of a southern family. Bot J Linn Soc 70:83–182

Jones K (2006) Industry development manager's report. Aust Macadamia Soc News Bull 33:44

Kazan K, Rusu AG, Marcus JP, Goulter KC, Manners JM (2002) Enhanced quantitative resistance to *Leptosphaeria maculans* conferred by expression of a novel antimicrobial peptide in canola (*Brassica napus* L.) Mol Breed 10:63–70

Manners JM, Goulter KC, Marcus JP, Kazan K, Harrison S, et al. (1998) Application of genes encoding anti-microbial peptides from Australian native plants for the control of fungal pathogens of crop plants. In: Larkin PJ (ed) Agricultural Biotechnology: Laboratory, Field and Market, Proc 4th Asia-Pacific Conf Agri Biotech. UTC Publishing, Canberra, pp. 13–15

Marcus JP, Goulter KC, Green JL, Harrison SJ, Manners JM (1997) Purification of an antimicrobial peptide from *Macadamia integrifolia*. Eur J Biochem 244:743–749

Marcus JP, Green JL, Goulter KC, Manners JM (1999) A family of antimicrobial peptides is produced by processing of a 7S globulin protein in *Macadamia integrifolia* kernels. Plant J 19:699–677

Martin PG, Dowd JM (1991) A comparison of 18s ribosomal RNA and rubisco large subunit sequences for studying angiosperm phylogeny. J Mol Evol 33:274–282

Mast AR, Jones EH, Havery SP (2005) An assessment of old and new DNA sequence evidence for the paraphyly of *Banksia* with respect to *Dryandra* (Proteaceae). Aust Syst Bot 18:75–88
McConachie I (1980) The macadamia story. Calif Macadamia Soc Yearbook 26:41–74
McManus AM, Nielsen KJ, Marcus JP, Harrison SJ, Green JL, et al. (1999) MiAMP1, a novel protein from *Macadamia integrifolia* adopts a Greek key beta-barrel fold unique amongst plant antimicrobial proteins. J Mol Biol 293:629–38
Meyers N (1997) Pollen parent effects on macadamia yield. PhD dissertation, University of Queensland, Australia
Neal J (2006) The demographic and genetic consequences of fragmentation in *Macadamia integrifolia* populations. PhD dissertation, University of New England, Australia
Peace CP (2002) Genetic characterisation of macadamia with DNA markers. PhD dissertation. University of Queensland, Australia
Peace C, Hardner C, Vithanage V, Carroll BJ, Turnbull C (2000) Resolving hybrid status in macadamia. In: Dungey HS, Dieters MJ, Nikles DG (eds) Proc QFRI/CRC-SPF Symp Hybrid Breeding Genetics Forest Trees. Department of Primary Industries, Brisbane, Australia, pp. 472–476
Peace C, Allan P, Vithanage V, Turnbull C, Carroll BJ (2001a) Identifying relationships between macadamia varieties in South Africa by DNA fingerprinting. S Afr Macadamia Growers Assoc Yearbook. pp. 64–71
Peace C, Hardner C, Brown AHD, O'Connor K, Vithanage V, et al. (2001b). Diversity and origins of macadamia cultivars. Proc Austr Macadamia Soc Tech Conf, Australia, pp. 34–37
Peace C, Vithanage V, Turnbull C, Carroll BJ (2002) Characterising macadamia germplasm with codominant radiolabelled DNA amplification fingerprinting (RAF) markers. Acta Hort 575:481–490
Peace CP, Vithanage V, Turnbull CGN, Carroll BJ (2003) A genetic map of macadamia based on randomly amplified DNA fingerprinting (RAF) markers. Euphytica 134:17–26
Peace CP, Vithanage V, Neal J, Turnbull CGN, Carroll BJ (2004) A comparison of molecular markers for the genetic analysis of macadamia. J Hort Sci Biotech 79:965–970
Peace CP, Allan P, Vithanage V, Turnbull, CN, Carroll BJ (2005) Genetic relationships amongst macadamia varieties grown in South Africa as assessed by RAF markers. S Afr J Plant Soil 22:71–75
Piza PT (2006) Macadamia world resume. Proc III Intl Macadamia Symp, Brazil, pp. 15–16
Ramsay HP (1963) Chromosome numbers in the Proteaceae. Aust J Bot 11:1–20
Schmidt AL, O'Connor K, Vithanage V (2005) Development and testing of methods for extraction of DNA from roasted macadamia kernel and genotyping using DNA markers. In: Vithanage V, Schmidt AL (eds) Determining the origin of macadamia kernels from market samples, HAL Final Report MC01002. Austr Macadamia Soc Horticulture Australia, Australia, pp. 13–27
Schmidt AL, Scott L, Lowe AJ (2006) Isolation and characterisation of microsatellite loci from *Macadamia*. Mol Ecol Notes 6:1060–1063
Sedgley M, Bell FDH, Bell D, Winks CW, Pattison SJ, et al. (1990) Self- and cross-compatibility of macadamia cultivars. J Hort Sci 65:205–213
Stace HM, Douglas AW, Sampson JF (1998) Did 'paleo-polyploidy' really occur in Proteaceae? Aust J Bot 11:613–629
Stanley TD, Ross EM (1986) Flora of South-eastern Queensland Volume 2. QDPI, Brisbane, Australia
Steiger DL, Moore PH, Zee F, Liu Z, Ming R (2003) Genetic relationships of macadamia cultivars and species revealed by AFLP markers. Euphytica 132:269–277
Storey WB (1965) The ternifolia group of *Macadamia* species. Pacif Sci 19:507–513
Storey WB, Frolich EF (1964) Graft compatibility in *Macadamia*. Calif Macadamia Soc Yearbook 10:54–58
Storey WB, Saleeb WF (1970) Interspecific hybridization in macadamia. Calif Macadamia Soc Yearbook 16:75–89
van de Weg WE, Voorrips RE, Finkers HJ, Kodde LP, Jansen J, et al. (2004) Pedigree genotyping: a new pedigree-based approach of QTL identification and allele mining. Acta Hort 663:45–50

Venkata Rao C (1970) Studies in the Proteaceae XIV. Tribe Macadamieae. Proc Nat Ins. Sci India 36B:345–363
Vithanage HIMV, Ironside DA (1986) The insect pollinators of macadamia and their relative importance. J Aust Inst Agric Sci 52:155–160
Vithanage V, Winks CW (1992) Isozymes as genetic markers for *Macadamia*. Scientia Hort 49:103–115
Vithanage V, Hardner C, Peace C, Anderson KL, Meyers N, et al. (1998) Progress made with molecular markers for genetic improvement of macadamia. Acta Hort 461:199–207
Vithanage V, Peace C, Thomas MR, Anderson KL (1999) Development and utility of molecular markers of macadamia. In: McConchie CA, Hardner C, Vithanage V, Campbell A, Mayers P (eds) Macadamia Improvement By Breeding, Final Report MC 96002. Horticultural Research and Development Corporation, Australia, pp. 107–123
Vithanage V, Peace C, O'Connor K, Meyers N, McConchie CA (2002) Pollen flow in macadamia orchards can be followed by codominant RAF markers. In: Vithanage V, Meyers N, McConchie C (eds), Maximising the benefits from cross-pollination in macadamia orchards, HRDC Final Report MC98027. Horticultural Research and Development Corporation, Australia, pp. 20–32
Wagner-Wright S (1995) History of the macadamia nut industry in Hawai'i, 1881–1981. The Edwin Mellen Press, Canada
Waldron J, Peace CP, Searle IR, Furtado A, Wade N, et al. (2002) Randomly Amplified DNA Fingerprinting (RAF): a culmination of DNA marker technologies based on arbitrarily-primed PCR amplification. J Biomed Biotech 2:141–150

Chapter 14
Genomics of Tropical Maize, a Staple Food and Feed across the World

Yunbi Xu and Jonathan H. Crouch

Abstract Tropical maize is a major staple crop providing food and feed across the developing world. Genomics of maize is very well advanced but heavily focused on temperate germplasm. Tropical maize germplasm is substantially more diverse than temperate maize with a wide range of landraces and types of varieties. Thus, diversity analysis at genetic, molecular, and functional levels is important for underpinning translational genomics from temperate to tropical maize. Virtually all types of markers have been used for molecular linkage mapping in maize over the past decade. However, single nucleotide polymorphic markers are now very well developed in maize and are becoming the marker of choice for most applications. Both linkage and association-based mapping has been used for identifying marker-trait associations. Maize genome sequencing is now well advanced but focused on gene-rich regions due to its high density of repetitive elements. Functional genomics activities have made use of insertional mutation-based cloning as well as expressed sequence tags and map-based cloning. A wide range of genomic databases and tools have been developed, of which MaizeGDB features a wealth of data and resources facilitating the scientific study of maize. Genomics-assisted breeding is at an advanced stage in temperate, especially in private sector breeding programs, and applications in tropical maize are also common. Marker-assisted selection has been used in maize for yield, grain quality, abiotic and biotic stresses. Using these approaches, commercial maize breeding programs have reported twice the rate of genetic gain compared with phenotypic selection. However, reports in the literature from public breeding programs are inconsistent and generally less promising. Applied maize genomics in the tropics should in the future focus on tropical maize fingerprinting, haplotype establishment, allele mining, gene discovery, understanding genotype-by-environment interactions, and development of decision support tools and networks for developing countries to facilitate effective applications of genomics in maize breeding.

Y. Xu
Genetic Resources and Enhancement Unit, International Maize and Wheat Improvement Center (CIMMYT), Carretera Méx-Veracruz, El Batán, Texcoco, México CP56130
e-mail: y.xu@cgiar.org

© Springer 2008

14.1 Introduction

Maize is primarily a cross-pollinating species, a feature that has contributed to its broad morphological variability and geographic adaptability. Maize is also a C4 plant with high photosynthetic efficiency and high biomass production potential. Maize is classified into two distinct types depending on the latitude and the environment in which it is grown. Maize growing in warmer environments located between 30°N and 30°S latitudes is referred to as tropical maize, while that grown in cooler climates beyond 34°N and 34°S is classified as temperate maize. An intermediate type, subtropical maize, is grown between the 30° and 34° latitudes. Here we will only focus on tropical maize which is further classified into three subclasses: lowland (sea level to $\leq$ 1000 masl [meters above sea level]), mid-altitude (1000 to 1600 masl) and highland ($\geq$ 1600 masl).

Tropical maize occupies about 60% of the area harvested and represents 40% of the world production. It is grown in over 60 countries. The average grain yield of a maize crop in the tropics is 1.8 tons/ha, as compared to the global average of 4.2 tons/ha. Although the average crop yield under temperate conditions is over twice that under tropical conditions, temperate maize varieties have a longer crop cycle than do most tropical maize varieties. Thus, the relative daily yield is not so different between tropical and temperate maize (Paliwal 2000). Maize grain is an important cereal for human consumption, particularly in tropical Africa and Latin America. The FAO estimates that an additional 60 Mt of maize production will be needed by 2030. In addition, the demand for maize as animal feed will continue to grow at a very fast rate, particularly in Asian countries, which are estimated to increase production from 165 Mt today to almost 400 Mt by 2030 (Paliwal 2000). Finally, it is likely that the demand for maize by the biofuel industry will also continue to dramatically increase, in some cases to the detriment of food and feed needs (Ortiz et al. 2006; Rosegrant et al. 2006).

Genetic resources of tropical maize include wild relatives (teosinte, *Tripsicum*), landraces, open-pollinated varieties, synthetic varieties, inbreds, hybrids, germplasm complexes, pools and populations, and various genetic stocks (mutant, permanent populations, near-isogenic lines, introgression lines, etc). Landraces of maize may also be referred to as races, super-races, sub-races, primitive maize types, racial or geographical groups and racial complex. About 300 tropical maize landraces, comprising thousands of different varieties, have been identified worldwide. (see Figs. 14.1 and 14.2)

Maize breeding in the tropics targets a great diversity of environments, a wide range of cropping systems, and varieties that range from high-tech and highly tailored single cross hybrids to open-pollinated improved varieties to farmers' local varieties and landrace selections (Smith and Paliwal 1996). Maize breeding research in developing countries started in agricultural colleges and during the 1960s and 1970s national research institutions became important centers of maize research and improvement. Yield improvement was the most important breeding target across all tropical countries while the importance of other traits varied from continent to

Fig. 14.1 Native races of maize tend to have lower harvest indexes, producing larger stalks and more foliage than improved varieties, an advantageous quality when they are grown to produce fodder. (photo by CIMMYT, used with permission) (See color insert)

continent (Pandey and Gardner 1992): diseases and drought in Africa, diseases and maturity in Asia, plant height and diseases in Latin America, and maturity and grain type in the Middle East. A comprehensive review of progress in breeding for hybrid maize, biotic and abiotic stress resistance, and special purpose maize has been provided by Paliwal et al. (2000). Hybrid maize is now well established across the tropics with average yields of five to six tonnes/ha over large areas.

The pioneering work of Barbara McClintock revealed maize as a highly important model plant for genetic research. A large body of genetic knowledge and huge amounts of experimental and molecular data are already available, although most of this is derived from temperate maize-based studies. The genetic divergence required for adaptation to significantly different tropical and temperate environments

Fig. 14.2 The Jala maize landrace from the western coastal state of Nayarit, Mexico, is known for its gigantic ears. This and other native maize races are in danger of extinction, as the smallholder farmers who grow them leave rural areas to seek better fortunes in cities or other countries, and those who remain increasingly sow improved varieties. (photo by CIMMYT, used with permission) (See color insert)

contributes to diverse target traits associated with tropical and temperate maize breeding programs. Maize has a number of characteristics that make it suitable as an experimental model for crop plants, including (1) an intermediate genome size compared to rice and wheat; (2) typical outbreeding system with flexibility for inbreeding; (3) existence of multiple breeding products (inbreds, hybrids, synthetic varieties, open pollinated varieties and improved landraces); (4) wide adaptability, especially for stressed environments; and (5) a multiple-purpose crop that can be used as food, feed, fuel and many industrial products.

This article will present an overview of various aspects in maize genomics, including genetic and molecular diversity, genetic mapping and trait tagging, physical mapping and genome sequencing, functional genomics, genomic databases and tools, and genomics-assisted plant breeding. Most advances in maize genomics have resulted from researchers working on temperate maize, but tropical maize is only significantly different in terms of its adaptation to different growing environments. Thus, this chapter will review genomic research in both temperate and tropical germplasm but with emphasis on traits important for tropical production systems and highlighting the gaps that need to be addressed in the future by tropical maize improvement programs.

14.2 Molecular Diversity Analysis and Mining

Genetic variation and diversity among germplasm accessions provides a pool of novel alleles and genes for plant breeding. Conventional plant breeding has been dependent on the evaluation of genetic diversity at the phenotypic level. Developments in genomics and molecular biology have provided various tools to identify and manipulate genetic variation at the molecular level. As it becomes increasingly feasible to do this in a functional and trait targeted way, genomics will have an increasing impact on plant genetic research and plant breeding.

14.2.1 Molecular Cytogenetics

The maize cytogenetic map is based on the pachytene stage chromosome karyotype and on the chromosome's fractional arm length unit referred to as centiMcClintocks (cMC). The cytogenetic map has many translocation breakpoint loci, but relatively few genetic marker loci (restriction fragment length polymorphisms, RFLPs) to define those breakpoints. The rapid and cost-efficient method that can be employed for construction of a cytogenetic map is to use segments of sorghum DNA (maize-marker-selected sorghum bacteria artificial chromosome [BAC] clones) as probes to stain the corresponding regions of maize chromosomes by fluorescence *in situ* hybridization (FISH) (Koumbaris and Bass 2003). A pachytene cytogenetic FISH map of the maize genome is under development using sorghum BACs corresponding to the 90 maize Core Bin Marker (CBM) loci (Figueroa et al. 2006).

The RFLP Full Length Insert Sequencing (FLIS) Project aims to determine and submit to GenBank a high-quality (bidirectional), full length insert sequence for maize RFLP markers including ~ 90 CBMs and potentially as many as 500 other markers from the UMC 98 map (http://gremlin3dev.gdcb.iastate.edu/prj/RFLP_FLIS/). In addition to being useful for the Cytogenetic Map of Maize Project (http://www. maizegdb.org/CMM protocols.php), these sequences are useful in that they can anchor assemblies of the maize genome to the genetic and cytogenetic maps. A cytogenetic map of the maize genome using RFLP marker-selected sorghum BACs as FISH probes is under construction (www.cytomaize.org). The results can be integrated with other genome maps and released immediately into GenBank and MaizeGDB for public access.

More recently, PCR-based FISH probes for identification of maize mitotic chromosomes were developed (Danilova et al. 2006). Maize centromeres were mapped using telosomes and isochromosomes produced by spontaneous chromosome breaks, radiation-induced chromosome breaks recovered in T-B translocation lines or in oat-maize radiation hybrid lines, single-locus FISH, and half-tetrad analysis (Okagaki et al. 2006a).

14.2.2 Genetic Diversity Analysis

Over the past 10,000 years, man has used the rich genetic diversity of the maize genome as the raw material for domestication and subsequent crop improvement. Structural diversity appears to be largely mediated by *helitron transposable elements*. Patterns of diversity are yielding insights into the number and type of genes involved in maize domestication and improvement, and functional diversity experiments are helping develop allele mining protocols that may identify novel genetic variation for use in future crop improvement (Buckler et al. 2006). Molecular markers have been used in genetic diversity studies of tropical maize for diverse purposes including:

1. Examination of genotype frequencies for deviations from Hardy-Weinberg equilibrium at individual loci (Reif et al. 2004)
2. Test for linkage disequilibrium (LD) between pairs of loci (Reif et al. 2004)
3. Construction of "phylogenetic" trees or classification of germplasm accessions based on genetic distance (Warburton et al. 2002; Betrán et al. 2003; Liu et al. 2003; Reif et al. 2004; Xia et al. 2004, 2005)
4. Characterization of molecular variation within populations and/or between populations (Warburton et al. 2002; Reif et al. 2004)
5. Determination of heterotic groups (Warburton et al. 2002; Xia et al. 2004, 2005)
6. Analysis of correlation between the genetic distance and hybrid performance, heterosis, and special combining ability (Betrán et al. 2003)
7. Comparison of genetic diversity among different groups of maize germplasm including those from temperate and tropical areas (Liu et al. 2003; Tarter et al. 2004; Xia et al. 2005)

It has been shown that tropical and subtropical inbreds possess a greater number of alleles and greater genetic diversity than their temperate counterparts. Comparison of diversity in equivalent samples of inbreds and open-pollinated landraces revealed that maize inbreds capture less than 80% of the alleles in the landraces, suggesting that landraces are likely to provide important additional genetic diversity for maize breeding. In addition, tropical highland germplasm is poorly represented in maize inbreds (Liu et al. 2003). The incorporation of a substantial percentage of tropical germplasm in an inbred line does not necessarily negatively impact its combining ability for grain yield or other agronomic traits (Tarter et al. 2004).

Thus, tropical maize accessions represent a valuable source of exotic germplasm to broaden the genetic base of temperate maize without hindering agronomic performance.

Maize has been shown to possess 88% of the gene diversity found in teosinte and 76% of the number of alleles, indicating a modest genome-wide deficit of diversity in maize relative to teosinte (Vigouroux et al. 2005). The pattern of genetic diversity at maize microsatellite or simple sequence repeat (SSR) and single nucleotide polymorphism (SNP) loci can be explained largely by a bottleneck effect with a variable effect from artificial selection depending on the germplasm (Wright et al. 2005).

Genetic, molecular and functional diversity analysis will play a key role in translating temperate germplasm-based genomics outputs into valuable information for tropical germplasm research and enhancement. Tropical maize improvement will benefit from translational genomics through the methodologies and technologies developed in temperate germplasm-based genomics, including molecular markers, genotyping system, transformation protocols, marker-trait association, etc. On the other hand, novel genes and alleles, gene and linkage blocks, distinct haplotypes and heterotic patterns can be identified from tropical germplasm and characterized for improvement of both temperate and tropical maize. As the tropical germplasm hosts a large number of open-pollinated varieties, landraces and wild relatives, which are adapted to more diverse environments and serve as a reservoir of genes for wide adaptation, genomics-assisted breeding systems have to be modified for more effective application in tropical maize genetic resources. For example, molecular and functional diversity of the maize genome can be characterized through allele mining, identification of distinct haplotypes for different inbred lines, single feature polymorphism analysis, and discovery of nearly identical paralogs and their evolutionary implications. Novel or unique alleles may not ever have been found via simple phenotypic screens, either because it is not possible to grow and measure every plant in a large germplasm collection under all possible environmental conditions, because its effect may be masked in an unsuitable genetic background, or because its effect may be so small that it will not be found unless specifically sought in carefully controlled phenotypic screens (not generally possible on a very large scale). Therefore, a combination of various methods is required to identify those alleles to ascertain their function.

14.2.3 Allele Mining

In general, there are two approaches that have been elaborated for allele mining: re-sequencing and eco-tilling (Comai et al. 2004). However, eco-tilling is not being widely used in maize at this time, due to the very high numbers of sequence differences found between different maize accessions, which confound interpretation. Whole genome genotyping using sequence-based markers can be carried out for the re-sequencing method. Allele discovery from germplasm collections is in its infancy, currently constrained by the difficulty of establishing which of the various alleles present is functionally different from the wild type in an agronomically beneficial way. Methods to ascertain allele function include marker-assisted back-crossing, transformation, transient expression assays, and association analysis using an independent association mapping set from the one that was used to identify the original allele (Sheen 2001).

There is a growing awareness that levels and patterns of allelic diversity contribute to the chromosomal context of a locus. "Diversity maps" showing the distribution of allelic diversity across the chromosomes and genomes of a variety of organisms suggest that in certain well-defined genepools there is an association between chromosome structural features such as centromeres and telomeres and with selection in particular well-defined gene pools (Dvorak et al. 1998; Hamblin and Aquadro 1999; Gaut et al. 2000). Diversity analysis of individual genes promises to shed new light on crop productivity and evolutionary processes underlying plant domestication (Wang et al. 1999). As one of the crops with high resolution of genetic maps, maize is an ideal choice to develop a diversity map that promises a whole host of new information about the consequences of natural selection, domestication, and polyploidy formation, and perhaps even heterosis. Clearly, such approaches of relating molecular level variation to phenotypic diversity are an essential backdrop for future studies of diversity in large populations of candidate genes. Using quantitative trait locus (QTL) information together with association approaches can focus the list of candidates to a manageable number that can be directly related to a specific phenotype. Mapping QTL (see section 14.3.2) to the level of individual genes will provide new insights into the molecular and biochemical basis of genetic diversity and allelic variation for maize improvement. Linkage disequilibrium (LD)-based association mapping will help identify alleles associated with wide adaptation. A range of new alleles of previously identified genes can be identified based on molecular analysis of germplasm subsets and characterized to determine their relative value (see section 14.3.3).

14.3 Genetic Mapping and Trait Tagging

Understanding genes and their relationship to traits and the influence of their location on chromosomes is highly important for plant breeding. It will facilitate manipulation of genes through more efficient identification, introgression, and selection. There has been great emphasis on elucidating the genetic basis of agronomically important traits. Genetic maps are constructed using molecular markers and then the maps are used to locate genes for the traits of agronomic importance using biometrical tools.

14.3.1 Linkage Mapping

The first generation of maize molecular maps was constructed using RFLP markers (Coe et al. 1987; Burr et al. 1988), which have been subsequently saturated with various types of PCR-based markers (Lee et al. 2002; MaizeGDB database at http://www.maizegdb.org/). More recently, linkage mapping has been revolutionized by high density SNP-based maps while candidate gene maps have also become possible providing the opportunity of linking forward and reverse genetics approaches. Several types of mapping panels have been used in maize, including F_2 (Coe et al. 1987; Beavis and Grant 1991), immortalized F_2 (Gardiner et al. 1993; Davis et al. 1999), and recombinant inbred line (RIL) populations (Burr et al. 1988; Causse et al. 1996; Taramino and Tingey 1996). These mapping efforts were based on populations with 48 to 214 individuals screened with 92 to 1736 markers. In addition, composite maps have been constructed from multiple crosses (Causse et al. 1996). The Maize Mapping Project (MMP) has assembled a high-resolution genetic map for the intermated B73 × Mo17 (IBM; I-intermated; B-B73; M-Mo17) population (Lee et al. 2002), consisting of ~1000 RFLP and ~1000 SSR markers (MaizeDGB, www.maize gdb.org). With the second panel of intermated recombinant inbred lines (IRILs) developed from F2 × F252, the two IRILs were recently used for linkage mapping of 1454 maize candidate genes (Falque et al. 2005).

SNPs are the most abundant type of sequence variation encountered in most genomes and are ideally suited to the generation of high density genetic maps (Cho et al. 1999). A total of 14,832 SNPs have been identified from 102,551 maize ESTs (Batley et al. 2003) and 169 SNPs and indels from 36 maize inbreds (Ching et al. 2002). Most recently efforts have been made to combine many physically and genetically mapped probes and genes onto a single consensus map (Schaeffer et al. 2006) and integrate this into MaizeGDB. This includes: (1) the IDP maps of Pat Schnable (maizemapping.plantgenomics.iastate.edu); (2) the Genoplante cDNA maps (Falque et al. 2005); (3) Genetic 2005 maps (Ed Coe); (4) SNP maps, incorporated into the community IBM94 maps (Mike McMullen).

14.3.2 Gene Tagging/QTL Mapping

Molecular marker-facilitated gene mapping in maize started in the early 1990s (Edwards et al. 1992; Stuber et al. 1992). Since then, there have been large numbers of studies for identifying associated major genes (gene tagging) and quantitative traits (QTL) mapping. Although much of this has been focused on temperate maize germplasm for major genes and QTL with large effects, many alleles may be shared between template and tropical germplasm as revealed by molecular markers (Tarter et al. 2004; Xia et al. 2005). Thus, the temperate germplasm-based gene tagging and QTL mapping is likely to be valuable for the genetic improvement of tropical maize.

The most intensive mapping study to date was based on a population of 1000 individuals derived from two elite inbred lines, which was phenotyped in 19 environments (Schon et al. 2004) for grain yield, grain moisture, and plant height. In another recent study, the high and low oil and protein content lines derived from 70 generations of long-term selection were crossed, intermated, and used for mapping (Laurie et al. 2004). Both studies identified numerous QTL of very small effect by using high statistical power and robust innovative analyses. These results support the hypothesis that traits with quantitative phenotypic variation are often the product of numerous QTL of small effect. Conventional QTL mapping of drought tolerance in maize over the past decade has also detected many QTL but all of only minor effect (Ribaut et al. 2004). However, the parental lines used in these studies are probably capturing only a small proportion of maize functional variation (Buckler et al. 2006). The founders of the elite inbred lines used as parental genotypes were the products of intensive breeding selection for more than 50 years, a process that might have eliminated all large-effect QTL. Thus, mapping studies using more diverse maize germplasm should be carried out in order to determine whether QTL of major effect in fact exist. In rice, such a large-effect QTL for grain yield under reproductive-stage drought stress has been identified from upland rice, explaining 51% of genetic variance (Bernier et al. 2007).

Tropical maize populations have a broad genetic base with greater variability than temperate germplasm (Lanza et al. 1997). Also tropical growing areas are more prone to environmental stresses caused by precipitation and temperature variability and by different types of soils than are temperate areas (Ribaut et al. 1997). Thus, QTL mapping in tropical maize germplasm is likely to identify QTL not present in temperate germplasm. The traits that have been tagged with tropical germplasm include insect resistance, with focus on sugarcane borer (Bohn et al. 1996, 1997; Grohn et al. 1998; Khairallah et al. 1998); plant height (Khairallah et al. 1998); kernel oil content (Mangolin et al. 2004); flowering parameters (Ribaut et al. 1996; Khairallah et al. 1998); drought tolerance and its secondary traits including anthesis-silking-interval (ASI; Ribaut et al. 1996, 2004; Vargas et al. 2006); and yield components (Ribaut et al. 1997; Lima et al. 2006). Because of the high genotype-by-environment interaction in tropical areas, gene tagging/QTL mapping has focused not only on mapping QTL but also on analysis of QTL-by-environment interaction (Crossa et al. 1999; Lima et al. 2006; Vagas et al. 2006).

Gene tagging for quality traits has received great attention. High essential amino acids and vitamin A content have received particular attention in tropical germplasm. *Opaque 2,* which confers high lysine, is a recessive single gene trait, but a number of modifier genes cause it to behave in a quantitative manner. By bulked segregant analysis of vitreous and opaque seed from a cross of a South African QPM line (K0326Y) and a soft *o*2 inbred (W64Ao2), Lizarraga Guerra et al. (2006) found that there are two loci clearly linked to the modified phenotype. The first was found in bin 7.02 and is near the 27-kD gamma-zein locus, which is consistent with prior results from RFLP mapping. The second was found in bin 9.02 and may be associated with starch synthesis genes. Sequencing of starch synthesis genes in isogenic backgrounds showed that four of these genes have sequence differences in a *mo2* compared to *o*2 or normal. The mapping of provitamins A and total carotenoids in maize grain has been reported by Stevens et al. (2006), where near-isogenic populations for a mutation in the phytoene synthase (*y1*) gene show a huge range in concentrations of beta-carotene, alpha-carotene, beta-cryptoxanthin, lutein, zeaxanthin, and total carotenoids. QTL mapping has also been reported for starch, protein, and oil concentrations using high-oil maize parental genotypes (Zhang et al. 2006).

Domesticated maize (*Zea mays* spp. *mays*) and its wild progenitor, teosinte (*Z. mays* spp. *parviglumis*), differ dramatically in their overall plant architecture and the morphology of their female inflorescences. However, the major morphological differences between maize and teosinte are conferred by two QTL that have been dissected into single Mendelian loci: *teosinte branched 1* (*tb1*) that suppresses lateral branching leading to apical dominance (Doebley et al. 1995, 1997; Wang et al. 1999) and *teosinte glume architecture* (*tga1*) that affects the hardness of the seed coat in teosinte (Dorweiler et al. 1993; Wang et al. 2005). Other key loci controlling the differences between maize and teosinte have also been identified (Briggs et al. 2006).

14.3.3 Association or Linkage Disequilibrium Mapping

Association mapping, also known as linkage disequilibrium (LD) mapping, is a method that relies on LD to study the relationship between phenotypic variation and genetic polymorphism (Flint-Garcia et al. 2003a). The high resolution of association mapping depends on the structure of LD or the correlation between polymorphic loci within the test population. LD decays particularly rapidly in maize, so association studies in landraces and a broad sample of tropical and temperate inbreds will be especially powerful as LD often declines to nominal levels within 1.5 kb (Tenaillon et al. 2001; Remington et al. 2001). In contrast, elite breeding materials have less rapid LD decay and are therefore less valuable for association mapping studies (Ching et al. 2002; Jung et al. 2004). Association mapping in maize has focused on candidate gene markers from known pathways and genes. This has led to the identification of SNP markers for genes affecting starch (Wilson et al. 2004), carotenoid (Palaisa et al. 2003), and maysin contents (Szalma et al. 2005), as well as aluminum tolerance (Krill et al. 2006).

With the availability of dense genomewide genotyping in maize, a novel integrated association mapping strategy has been developed for both qualitative and quantitative traits. This has been named nested association mapping (NAM), and has been developed based on 25 maize populations, each of which comprises 200 RILs derived by crossing 25 diverse inbred lines to a common inbred line, B73. With a dense coverage (2.6 cM) of common-parent-specific (CPS) markers, the genome information for the 5000 RILs can be inferred based on the parental genome information leading to genome-wide high-resolution mapping. The power of NAM with 5000 RIL allowed 30% to 79% of the simulated QTL to be precisely identified (Stich et al. 2007; J. Yu et al. Cornell, personal communication). With the ongoing genome sequencing projects, NAM will greatly facilitate the dissection of complex traits in many species in which a similar strategy can be readily applied.

Both linkage analyses and LD mapping have limitations when used alone. Association mapping will not replace linkage mapping for determining marker-gene associations, as it suffers from false positive results (NCI-NHGRI 2007), but it will provide a valuable first step in many cases and a method to validate associations found through independent analyses or germplasm. In addition, a joint linkage and

LD mapping strategy has been devised for genetic mapping (Wu and Zeng 2001; Wu et al. 2002). This strategy has power to simultaneously capture the information about the linkage of the markers (as measured by recombination fraction) and the degree of LD created during historic time. The NAM populations developed by the Molecular and Functional Diversity of the Maize Genome Project (http://www.panzea.org/) provide an opportunity for a joint linkage and LD mapping.

14.3.4 Marker Validation

Marker-trait associations need to be validated before entering into large-scale marker-assisted selection (MAS) applications, whatever methodology was used to identify the association (Nicholas 2006). Several factors contribute to the inconsistency of QTL mapping results including population structure and size, genetic background and epistasis effects, QTL-by-environment interaction, and level of LOD threshold (Beavis 1998; Moreau et al. 1998). Additionally, inaccurate phenotyping of the mapping populations further reduces the power and precision of QTL detection. Cross validation of QTL in independent populations, different genetic backgrounds and environments is necessary to obtain unbiased estimates of QTL position and effects.

The availability of thousands of SNP markers, rather than several hundred SSR markers, makes it practical to validate marker-trait association through high-precision genotyping using the same set of markers to screen different parental lines and breeding populations. Alternatively, marker validation can also be carried out at the same time by mapping multiple independent populations, selective genotyping and pooled DNA analysis, and development of gene-based, closely linked markers. Validation requirements can be minimized by focusing on large-effect QTL (Price 2006); precision phenotyping; identification of context independent QTL; mapping as you go approaches (Podlich et al. 2004); association mapping using large numbers of inbreds; genomewide association scans (as shown in human genomics, e.g., Meaburn et al. 2006); using breeding materials for mapping; and utilization of haplotype-based selection rather than single-marker based selection. Another useful strategy used for confirmation of candidate QTL or fine mapping is the application of NIL (near-isogenic lines). This approach reduces much of the "noise" caused by genetic background effects, thus mapping with NIL offers more accurate QTL effect estimates than RILs if multiple QTL are segregating in the populations, although power to detect a single QTL may be greater in RILs than NILs (Kaeppler 1997; Szalma et al. 2007).

14.4 Physical Mapping and Genome Sequencing

Genetic markers, genes, and other genetic elements in the maize genome can be characterized by their physical positions and biochemical composition. The finest physical map is the nucleotide sequence from which it is possible to determine gene structure and function. Physical mapping and genome sequencing are essential prerequisites for gene isolation and functional characterization of genes. Based on technical advances in the related fields, physical mapping and genome sequencing has beome much easier in recent years for large-genome plants including maize.

14.4.1 Development of BAC Libraries

The construction of a physical framework for the maize genome has been based on a large insert genomic library generated from a temperate maize genotype. Assuming an average insert size of 100 kb, almost 300,000 colonies would be required for an 11-fold representation of the haploid maize genome. In order to develop a comprehensive physical framework for maize using fingerprinting and BAC end sequencing technologies, two deep coverage BAC libraries were developed from the maize inbred lines LH132 Dekalb and B73. The LH132 library consists of 427,392 clones stored in over one thousand 384-well microtiter plates. Based on a haploid genome size of 2,500 Mb, the coverage of the library is about 20 maize genome equivalents (Tomkins et al. 2000a). The coverage of the B73 library is about 13.5 maize genome equivalents (Tomkins et al. 2000b). The two libraries are well suited to construct a comprehensive physical framework of the maize genome due to their high overlap and large average insert size. A third BAC library was constructed using CHORI-201 (http://bacpac.chori.org/maize201.htm). It is this library that is being used for physical mapping and sequencing of the maize genome. The BAC clones have been arrayed into nearly three hundred 384-well microtiter plates and gridded onto nylon high-density filters for screening by probe hybridization. The Maize Mapping Project (MMP, http://www.maize map.org/) has also constructed two BAC libraries from the inbred B73 (*Hin*dIII + *Eco*RI). The BAC libraries from both sources consist of nearly half a million clones which cover the B73 genome of 2,365 Mb nearly 30 times. The MMP is also generating a physical map by DNA fingerprinting of the BAC libraries and linking them to the IBM map.

14.4.2 BAC Physical Mapping

Several genome-wide physical maps are reported in maize (Yim et al. 2002; Coe et al. 2002; Cone et al. 2002) that are useful in genome sequencing, targeted marker development, efficient positional cloning, and high throughput EST mapping. The IBM markers have been hybridized to fingerprinted BACs and the BAC-marker associations have been used to create fingerprint contigs and thereby integrate genetic and physical maps (Coe et al. 2002).

Anchoring the physical and genetic maps requires many high-quality single-locus markers placed on both maps, and the unavailability of such data is one of the significant bottlenecks of the map integration process. In addition to classical RFLP- or SNP-based projects, numerous other approaches have been used to generate denser maps, including ESTs (Wen et al. 2002), radiation hybrid panels (Okagaki et al. 2001, Davis et al. 2000) and physically sheared DNA (Dear and Cook 1993).

Oat-maize addition (OMA) and radiation hybrid lines have been developed for physical and genetic mapping (Okagaki et al. 2006b). OMA lines are available for maize chromosomes 1–10, and also where a maize B chromosome has been individually added to the oat genome by wide crossing. Previously, maize chromosomes 1–10 from Seneca 60 had been recovered individually in OMA lines. Gamma irradiation of the OMA lines has yielded several hundred Radiation Hybrid (RH) lines each having just a fragment of a maize chromosome either after partial deletion of the rest of the maize chromosome or after translocation of a maize chromosome fragment into an oat chromosome.

These genetic resources, BAC libraries, and physical maps in temperate maize provide an important foundation for genomics of tropical maize. However, DNA libraries from tropical maize lines will now be important in order to exploit the genetic diversity and novel alleles that do not exist in temperate maize germplasm.

14.4.3 BAC DNA Sequencing

Sequencing the maize genome will greatly improve our potential to understand the molecular basis of important agronomic traits, gene regulation, genome evolution, plant development and biology. However, the large size of the maize genome and the high frequency of repetitive elements have prompted the examination of sequencing technologies expected to target gene rich regions as an alternative to whole genome sequencing. Based on the one-eighth of the maize genome already sequenced, (307 Mb) repeat sequences represent 58–66% while the predicted 42–59,000 genes represent just 7.5% of the genome (Messing et al. 2004; Haberer et al. 2005).

A large-scale effort to sequence the maize genome was commenced in 2006 through the NSF-funded Maize Genome Project (a collaboration between the Washington University Genome Sequencing Center, the Arizona Genomics Institute, Iowa State University, and Cold Spring Harbor Laboratory), aiming to sequence the maize genespace of a maize inbred B73 using a BAC-based approach (Wilson et al. 2006). This effort will utilize a minimal tiling path of approximately 19,000 mapped BAC clones, and will focus on producing high-quality sequence coverage of all identifiable gene-containing regions of the maize genome. These regions will be ordered, oriented, and, along with all of the intergenic sequences, anchored to the extant physical and genetic maps of the maize genome.

To further prepare for sequencing the entire genome, the Sequencing the Maize Genome Project (STMG, PI, J. Messing, Rutgers University) sought to improve on the B73 physical map by high information content fingerprinting (HICF) and by BAC end sequencing (BES). The Consortium for Maize Genomics (CMG), consisting of The Donald Danforth Plant Science Center, The Institute for Genomic Research (TIGR), Purdue University, and Orion Genomics, sought to test fractionation techniques of genomic regions containing only genes and not repetitive DNA elements. Two strategies, methyl-filtration and high- C_0t selection, have been proposed to enrich for gene rich regions. Both STMG and CMG are collaborating to sequence two 5 Mb homoeologous regions of the maize genome and to analyze the maize genome structure and shotgun sequence assemblies of a larger interval. Using a methylation filtration strategy, Bedell et al. (2005) sequenced 96% of the genes with an average coverage of 65% across their length in sorghum. This strategy filtered away a high proportion of repetitive elements when sequencing the genome of sorghum that reduced the amount of sorghum DNA to be sequenced by two thirds, from 735 Mb to approximately 250 Mb. Both methylation filtration and high C_0t have already been used for efficient characterization of the maize gene space (Palmer et al. 2003; Whitlaw et al. 2003). These methods are used for highly repetitive genomes such as maize and sorghum (see above).

A recent report (July 2007) from the Center for Research and Advanced Studies of the National Polytechnic Institute (CINVESTAV), Mexico, indicates that a major sequencing project is being carried out using bulked plants from a landrace Palomero accession (Mexican popcorn maize). This maize accession has 30% less DNA and is phylogenetically closer to teosinte than to B73. The project focuses on important gene rich regions. The sequence generated so far has about 3–7 fold coverage of the popcorn genome. Assembly and annotation are still on-going (http://www.niherst.gov.tt/s-and-t/s-and-t-news/; Dr. Alfredo Herrera Estrella, CINVESTAV, personal communications).

14.5 Functional Genomics

Functional genomics can be defined as a field of molecular biology that makes use of the vast wealth of data and information produced in genomics to define gene (and protein) functions and interactions. It includes function-related aspects of the genome itself such as mutation and polymorphism analysis, as well as measurement of molecular activities and characterization of the phenotype. The latter comprise a number of "-omics" such as transcriptomics (gene expression), proteomics (protein expression), phosphoproteomics and metabolomics by quantifying the various biological processes to drive increased understanding of gene and protein functions and interactions.

14.5.1 Insertional Mutation

The insertion of a T-DNA element into a gene can lead to the loss or gain of a function. The use of this phenomenon has led to the identification of many genes and regulatory elements in *Arabidopsis* and many other plant species including maize (Cowperthwaite et al. 2002; Singh et al. 2003; Ma and Dooner 2004; Kolkman et al. 2005); for a review on maize insertional mutations, see Lisch (2002). With classical genetic techniques and non-transgenic materials, Ahern et al. (2006) are generating a collection of 10,000 families, each harboring a unique *Ds* insertion distributed throughout the genome. DNA sequences flanking the *Ds* elements are cloned and sequenced providing a precise physical location for each insertion in the maize genome (Liu et al. 2006). Importantly, each *Ds* insertion is stable in the absence of *Ac*, but can be remobilized using a stabilized transposase source. As *Ds* tends to move to closely linked, gene-rich regions of the genome, each insertion will also serve as a platform for additional rounds of mutagenesis targeting linked genes. In addition, *Ac/Ds* transposons can be used for generation of an allelic series within a single gene (Bai et al. 2007).

Another type of DNA transposon, the *Mu* elements, accumulates to high-copy numbers within maize lines, which allows a relatively small population of ~40,000 plants to have a high chance of mutating most genes within the genome. Consequently, multiple groups have developed *Mu* transposon-tagging populations that can be used for both forward and reverse genetics (as reviewed by Settles 2005).

Gene knockouts are an essential resource for functional genomics. Many groups have developed reverse genetics populations in order to identify knockouts in any gene within the maize genome. These include: the Trait Utility System for Corn (Pioneer Hi-bred International; http://www. pioneer.com), the *RescueMu* population (Maize Gene Discovery Project; http://www.maizegdb.org/rescuemu-phenotype.php), and the *Mu*AFLP-based resources (BBSRC Gene Function Initiative). In addition, flanking sequence tags (FSTs) have been generated using DNA transposons to anchor each mutant to a specific locus within the genome. Settles et al. (2006) have shown that FSTs from the UniformMu population create easy to use knockout resources. The genetic markers in the UniformMu population allow for the selection of stable transposition events.

14.5.2 EST Development

Expressed sequence tags (ESTs) are currently the most abundant group of sequence resources. ESTs provide a robust sequence resource that can be exploited for gene discovery, genome annotation and comparative genomics. However, a large proportion of ESTs in public databases are unedited, automatically processed, single read sequences produced from cDNAs that provide only a very preliminary indication of nature and potential function of candidate genes. There are over a million maize ESTs in Genebank (http://www.ncbi.nlm.nih.gov/dbEST_summary.html) including a large number from tissue and growth stage specific libraries or even limited populations of plant cells (e.g. Fernandes et al. 2002; Lê et al. 2005).

A genomics initiative to sequence full length cDNA's for 30,000 genes is underway through the Arizona Genomics Institute (http://www.maize cdna.org/). This effort is based on EST alignment assemblies followed by primer walking. A full length cDNA library from maize seedlings contributed 2073 full length cDNA sequences, of which over 80% represented new genes in the databases (Jia et al. 2006).

Such cDNAs form an excellent research resource as well as aiding efforts to annotate the maize genome.

Recently, a new sequence technology (454 Life Sciences; Emrich et al. 2007) was used to generate more than 261,000 ESTs from laser capture microdissection (LCM) of the shoot apical meristem from a single sequencing run. From this data $> 25,000$ maize genomic sequences were annotated, and several novel EST sequences were discovered (Emrich et al. 2007).

One of major uses of ESTs is to develop gene-based markers that are perfectly associated with the trait of interest, more conserved among related species, and may lead to elucidation of the function of genes influencing the target trait. ESTs have been localized on high resolution maps using different methods such as conventional mapping of the IBM RILs (Chen et al. 2007; http://maize-mapping.plantgenomics.iastate.edu/) and maize single feature polymorphism Genechips (Zhu et al. 2006). Integration of these gene markers with other types of genetic and physical markers will provide an important resource for the identification and marker-assisted selection of genes that control complex traits. The best functional and positional candidates should first be subjected to a functional validation process, such as reverse genetics mutant collection screening, overexpression or knockout transgenic approaches, or association studies using germplasm with allelic polymorphism of the gene and phenotypic data.

14.5.3 Gene cloning

Progress in maize genetics and gene discovery is confounded by the following: (1) the maize genome is big (thus not amendable to whole genome sequencing as has been successful in Arabidopsis and rice); (2) subspecies genome size and even gene order varies greatly; (3) the maize genomes contain multiple copies of most genes; and (4) jumping genes or transposons make up a large portion of the genome.

Traditionally, gene discovery looks for genes in two complementary ways. One method searches for ESTs. These represent genes that are turned "on" in a specific tissue, at a specific development stage, or in response to a specific biotic or abiotic stress. The second method, takes advantage of transposons that insert copies of themselves inside maize genes. Researchers evaluate the phenotype of maize plants that contained a specific, engineered transposon tag called *RescueMu*. The transposon, whose sequence is known and easily traceable, inserts itself in new chromosome locations, but always within a gene. The researchers then find genes by sequencing the DNA on both sides of *RescueMu*. The latter approach has the added advantage of being able to directly compare the sequence of the interfered gene with changes in phenotype.

Positional cloning in maize has been considered near impossible because of the vast amounts of repetitive DNA. However, conservation of synteny across the cereal genomes, in combination with new maize resources, has made chromosomal walking much faster than the more traditional methods of gene isolation (for a review, see Bortiri et al. 2006a). The first gene isolated through positional cloning was the *teosinte glume architecture* (*tga1*) locus, which encodes a transcriptional regulator (Wang et al. 2005). A recent report on cloning of *indeterminate gametophyte1* used a combination of positional cloning and transposon insertion to test candidate genes (Evans 2007). Buckler et al. (2006) have recently reviewed the advent of positional cloning and association approaches that allow for the dissection of complex trait down to the gene and nucleotide level. These are clearly highly relevant to both temperate and tropical maize.

Reproductive Biology: It is well established that small gene sequence changes can have dramatic effects on flowering time (Thornsberry et al. 2001). Recently, several genes that regulate inflorescence architecture in maize have been cloned. The gene responsible for the mutated phenotype of a highly branched tassel and a branched ear, *ra1*, was cloned by transposon tagging rather than by using synteny with rice (Vollbrecht et al. 2005). Two other maize inflorescence genes have been cloned using the map position of the mutation in combination with synteny with a candidate gene in rice. Mutants at the *barren stalk 1* (*ba1*) locus lack tassel branches, spikelets, and ears. The positional cloning of *lax panicle* in rice (Komatsu et al. 2003) provided a candidate gene for *ba1*. In the second case, a maize *clavatal 1* (*clv 1*) ortholog was mapped to chromosome 5 in the same region as *thick tassel dwarf 1* (*td1*). The phenotype of *td1* mimics that of *Arabidopsis clv* mutants, which have larger inflorescence meristems and more floral organs. Proof that *td1* was the *clv1* ortholog came from analysis of a large number of *Mu*-induced alleles (Bommert et al. 2005). More recently, *ra2* (Bortiri et al. 2006b), *ra3*, and *tasselseed4* (*ts4*) have also been isolated through positional cloning.

Forage Quality: Silage maize is a major source of forage for dairy cattle due to its high energy content and good digestibility. Lignin structure and cross-linking between cell wall components influence digestibility (Barrière et al. 2003). Analysis of allelic diversity in relation to cell wall digestibility revealed *ZmPox3* peroxidase as a candidate gene for improvement of silage maize digestibility (Guillet-Claude et al. 2004) as it is co-localized with a cell wall digestibility and lignification QTL (Barrière et al. 2003). Brown midrib (*bm*) mutants in maize have an increased digestibility but inferior agronomic performance (Barrière and Argillier 1993). Two of the four *bm* genes (*bm1* and *bm3*) have been shown to be involved in monolignol biosynthesis (Barrière et al. 2003). These and other lignin biosynthesis genes have been isolated based on sequence homology. Candidate genes, putatively affecting forage quality, have been identified in a collection of maize inbred lines by expression profiling using isogenic *bm* lines, leading to the detection of association between a polymorphism at the caffeic acid *O*-methyl transferase locus and the digestible neutral detergent fiber locus (Lübberstedt et al. 2005).

Pest and Disease Resistance: Conserved domains or motifs shared amongst known resistance genes have been extensively exploited to identify resistance gene analogs (RGAs). In an attempt to isolate and map all potential RGAs from the maize genome, three approaches were adopted by Xiao et al. (2006), including modified AFLP, modified rapid amplification of cDNA ends (RACE), and data-mining.

In response to attack by herbivorous insects, plants synthesize and release volatile chemical signals that attracts the natural enemies of the herbivore. Lin et al. (2006) reported the isolation and characterization of the maize *sesquiterpene cyclase2* gene (*stc2*) that is an ortholog of *stc1*, a gene induced in response to attack by beet armyworm larvae.

14.5.4 Transcription Profiling

Comprehensive, low-cost, public sector long-oligonucleotide (70mer) microarrays have been developed for gene expression analysis in maize based on single and assembled ESTs plus some non-redundant repeat elements, organelle genes, and other community favorites (Iniguez et al. 2006). Transcriptional profiling has already been applied to the study of a range of traits in temperate maize and most of these are highly relevant to tropical production systems.

eQTL Mapping: Transcript abundance levels that differ between the parents of a mapping populations and segregate amongst the progeny can be mapped and characterized as quantitative traits (Cheung and Spielman 2002). Using microarrays the expression levels of large numbers of genes can be determined and compared with variation in the phenotype of target traits. The genomic regions of these gene expression QTLs (eQTL) can then be determined using statistical tools developed for conventional QTL analysis (Jansen and Nap 2001).

Heterosis: Recent data suggest that regulation of gene expression might play an important role in determining hybrid vigor in maize. Between 800 to 1000 genes were identified as being significantly over expressed in the F_1 hybrid as compared to the parental genotypes (Swanson-Wagner et al. 2006; Scheuring et al. 2006). A parallel study concluded that cis-transcriptional variation between genes from the different parents led to additive expression patterns in the F_1 hybrid (Stupar and Springer 2006). A further study found both dominance and over-dominance components were involved in non-additive gene expression variation encompassing a wide variety of biological processes (Pea et al. 2006).

Grain Quality: Transcript profiling has also been carried out on maize material associated with the longest continuous genetic selection experiment in higher plants (Moose et al. 2006). Microarray comparisons of developing seeds revealed significant expression differences in many genes, with the seed storage protein genes exhibiting the most dramatic changes.

Abiotic Stress Tolerance: Gene expression profiling has been widely used for studying abiotic stress tolerance. Flowering is the developmental stage that is most vulnerable to abiotic stress leading to significant yield loss associated with the resultant aberrant floral development, and impaired ear and kernel growth. Genes within the starch biosynthetic pathway are collectively down-regulated during drought stress, resulting in reduced starch content. Many other genes are consistently up-regulated or down-regulated by drought stress (Zinselmeier et al. 2002; Yu and Setter 2003). There are similar reports for cold stress during germination and desiccation tolerance (Kollipara et al. 2002). Finally, transcriptome analysis of the low-phosphorus responses in roots and shoots of a phosphorus-efficient *Zea mays* line identified alterations of several metabolical and physiological processes (Calderon-Vazquez et al. 2006).

14.5.5 Transformation

Transformation is an important tool for maize genetic research and germplasm improvement, in addition to its extensive use in the development of pest- and herbicide-resistant new varieties. *Agrobacterium tumerfaciens*-mediated transformation is the preferred method for genetic transformation because it generates a high proportion of independent events with single, or low, transgene copy numbers, which is considered to favor consistent transgene expression in progeny generations (Meyer and Saedler 1996).

Maize varieties resistant to glufosinate (Liberty) and Roundup herbicides have been produced in the USA. Maize varieties have also been transformed to express the *Bt* toxin by inserting a gene from the soil-dwelling bacteria *Bacillus thuringiensis*. This gene codes for a toxin that will crystallize in the digestive tract of insect larvae, leading to its starvation. This has been particularly effective against the European corn borer *Ostrinia nubilalis* that destroys corn crops by burrowing into the stem, causing the plant to lodge. It can be expected that transgenic maize will have a significant influence on tropical maize production once the public concern issues related to genetically modified organisms have been resolved.

Biswas et al. (2006) demonstrated the feasibility of producing a bacterial cellulose within maize biomass for possible biomass conversion into fermentable sugars by introducing the catalytic domain of an endo-1,4-p-D-glucanase gene from the eubacterium, *Acidothermus cellulolyticus*. For the food industry, the availability of foods that are low in sugar content, yet high in flavor, is critically important to millions of individuals conscious of carbohydrate intake in relation to diabetic or dietetic concerns.

The reported discovery of transgenes in maize landraces of small-scale Mexican farmers (Quist and Chapela 2001) raised questions about whether the commercial introduction of transgenic maize varieties might have a deleterious effect on the diversity of maize landraces and on traditional small-scale agricultural systems. An important concern in assessing the risk of growing a genetically modified crop in its center of domestication is gene flow between the transgenic crop and its landraces and wild relatives. However, a more recent study suggests that it is unlikely that the presence of transgenes per se will automatically reduce the diversity of alleles in local maize populations or the level of diversity of morphological variants managed by small-scale farmers (Bellon and Berthaud 2004).

14.5.6 TILLING

A non-transgenic method for reverse genetics called Targeting Induced Local Lesions In Genomes (TILLING) has been developed as a method for inducing and identifying novel genetic variation. TILLING is a targeted version of conventional mutation breeding with the added advantage of mutation detection in the gene of interest. TILLING employs a mismatch-specific endonuclease to detect single-base-pair (bp) allelic variation in a target gene using a high-throughput assay. Its advantages over other reverse genetic techniques include its applicability to virtually any organism, its high throughput nature, and its independence of genome size, reproductive system or generation time (Gilchrist and Haughn 2005). A public TILLING service has been established through the Maize TILLING Project (MTP) at Purdue University (http://genome.purdue.edu/maizetilling) (Till et al. 2004). The current TILLING population contains ~2,900 mutant lines with ~165,000 mutations in exons (Monde et al. 2006). In addition, the mtmDB web site (http://mtm.cshl.org) contains a knockout resource of a population of 43,776 plants containing stabilized *Mu* insertions that are available to the global scientific community (May et al. 2003).

14.6 Genomic databases and tools

As more and more information and data have been generated in various fields of genomics, databases and bioinformatic tools are needed to store, integrate, and manage the data plus extract and analyze useful information for use in genetic improvement. One of the most important genomic databases and tools for maize is MaizeGDB (http://www.maizegdb.org). The site features a wealth of data and resources facilitating the scientific study of maize:

- Sequence databases including integration with various contig assemblies
- Detailed genetic, physical, and cytogenetic maps
- Molecular marker primer databases
- Integrated tools for map comparisons, sequence similarity searches, and comparisons with and links to other databases, such as Gramene (http://www.gramene.org/) and NCBI (http://www.ncbi.nlm.nih.gov/)
- Web-based community curation tools that enable researchers to edit and annotate their own data and to enter new data into MaizeGDB directly
- Informatics support for maize community initiatives such as the annual Maize Genetics Conference and community-wide workshops, and maintains data for maize community research projects
- QTL information in the literature from the mid-1990's coupled to other information about germplasm, nearby loci and sequence information

To permit this work to continue at MaizeGDB, a new, web accessible curation interface has been designed and implemented. The new design accommodates a legacy trait hierarchy developed at MaizeGDB and recently harmonized with the rice Trait Ontology at Gramene, and trait descriptors used by GRIN (the Germplasm Resources Information Network).

The Maize Assembled Genomic Island (MAGI, http://magi.Plant genomics.iastate.edu) is a resource for maize genome assembly, annotation and mapping. ~3,100,000 maize genomic sequences primarily composed of gene-enriched GSSs, random whole genome shotgun (WGS) sequences, and BAC shotgun reads were assembled into MAGI (Emrich et al. 2004). Similarly ~550,000 methyl filtered (MF) sequence reads from *Sorghum bicolor* (BT × 623) were assembled into Sorghum Assembled genoMic Islands (SAMIs). To identify genomic contigs associated with particular genes, MAGIs and SAMIs can be searched using the BLAST tool. GBrowse, a component of GMOD, is used to display annotated assemblies. Segregation data in the IBM RIs have been generated for ~5,000 MAGIs and ESTs. A new genetic map based on these data and generated using MultiMap, including linkages to AGI's physical map, can be viewed via CMap. The MAGI website serves as a community resource for map-based cloning projects as well as for analyses of genome structure and comparative genomics.

There are many computer software and decision support tools developed by the plant genomics community, including those for germplasm evaluation, breeding population management, genetic map construction, marker-trait association analysis, marker-assisted selection, genotype-by-environment interaction analysis, breeding design and simulation, and information management. Since these programs have been reviewed else-where (Dwivedi et al. 2007), only two software packages that are more specific to maize will be discussed here. The first is TE Nest that was developed to facilitate the annotation of the 1.5 Mb chromosome 3 centromeric *rf1*-spanning sequence constructed from 19 contiguous BAC clones (Kronmiller et al. 2006). Considering that 85% of the maize genome consists of transposable elements (TEs), with more than 70% of TEs found nested within one another, an accurate nested TE identification tool for complete annotation of the maize genome was needed. TE Nest contains an up-to-date database of maize canonical TEs and their associated long terminal repeats (LTRs), if applicable. The second software is an integrated program developed by Schroeder et al. (2006) to aid researchers in the SNP discovery process across several maize, teosinte, and *Tripsacum* lines. An integrated set of tools consisting of a relational database and applications for data loading, editing and reporting has been developed. All stages of SNP discovery from tracking sequences, generating alignments, editing alignments, and reporting are covered. Central to this system is an intuitive, quality score based alignment editing tool designed to simplify manual editing of the highly polymorphic and complex *Zea* alignments.

14.7 Genomics-assisted breeding of tropical maize

One of the most important uses of plant genomics is the application of molecular biology information and tools to improve the efficiency and scope of plant breeding. Genomics-assisted plant breeding includes using molecular markers associated with traits of agronomic importance to help improve selection efficiency, or using genetic transformation with functionally characterized genes. Here we are focusing on the use of markers to improve the identification, introgression, and manipulation of genetic variation. Molecular markers can increase the accuracy and speed of all three of these aspects of the breeding process compared to conventional phenotypic selection. Contrary to the situation in developed countries, where almost all maize grown by farmers are hybrid varieties, developing countries (almost all of which are in the tropics) are growing various types of maize including hybrid, synthetic and open-pollinated varieties, and landraces. Irrespective of the nature of the target breeding product, breeding objectives must focus on traits required for tropical environments, which are very different from those in temperate cropping regions, particularly regarding abiotic and biotic stresses.

Large multi-national seed companies are now routinely using applied genomic tools to (i) dissect the genetic structure of their germplasm to understand gene pools and germplasm (heterotic) groups, (ii) provide insights into allelic content of potential germplasm for use in breeding, (iii) screen early generation breeding populations to select segregants with desired combinations of marker alleles associated with beneficial traits (in order to avoid costly phenotypic evaluations), and (iv) establish genetic identity (fingerprinting) of their products (Fu and Dooner 2002; Xu 2003; Niebur et al. 2004; Cooper et al. 2004; Crosbie et al. 2006). MAS has been successfully applied in the private sector for maize variety development for recovering an ideal genotype, defined as a mosaic of favorable chromosomal segments from the parental genotypes. More specifically MAS has been used to simultaneously select for multiple traits (selection based on marker information only) such as yield, biotic and abiotic stress resistance, and quality attributes (Ragot et al. 2000; Eathington 2005), several of which are polygenic in nature. Using these approaches, commercial breeding programs have reported twice the rate of genetic gain over phenotypic selection in maize (Eathington 2005; Crosbie et al. 2006). The first commercial products of holistic molecular breeding (rather than unilateral MAS interventions) are expected from all the multinational breeding companies very soon. The first molecular breeding hybrids developed by Monsanto entered the U.S.A. commercial portfolio in the 2006 cropping season, and it is estimated that by 2010 over 12% of the commercial crop in the U.S.A. will be derived from molecular breeding (Fraley 2006).

One of the most important applications of MAS is for gene pyramiding to maximize utilization of existing gene resources. Genes controlling resistance to different races or biotypes of pests and pathogens can be pyramided together with agronomic and/or seed quality traits to ensure simultaneous introgression of several traits into an improved genetic background. Traditionally, wild relatives are considered as good sources of resistance to many pests and diseases not found in cultivated species, thus making them a valuable resource for genes to transfer to cultivated species. Both conventional crossing and selection, and molecular biology techniques (MAS and transgenic approaches) have been used to transfer pest- and disease-resistance from wild relatives to cultivated crop species. Resistant gene(s) from wild relatives have enabled large-scale cultivation of crops in disease/pest endemic regions of the world. Many major genes (recessive or dominant) and/or

QTL conferring resistance to pests and diseases have been reported in maize. Using MAS coupled with field evaluation, researchers have been able to combine multiple resistances to these pests and diseases (Widstrom et al. 2003; Quint et al. 2002). More recently, modeling and simulation analysis has been able to define the most efficient breeding strategies for generating such marker-assisted pyramided products (Wang et al. 2007).

Development of exotic genetic libraries, also known as chromosome segment substitution line (CSSL), introgression lines (IL), and contig lines is another approach to enhance utilization of wild relatives to expand crop gene pools. These genetic stocks consist of marker-defined genomic regions taken from wild species and introgressed into the background of elite crop lines, thus providing a potential resource for overcoming the yield barriers through pyramiding of beneficial loci and fixing positive heterosis.

14.7.1 Yield and Heterosis

Dominance, overdominance and epistasis have all been proposed to have a role in the genetic control of superior hybrid performance. The dominance model attributes increased vigor to the action of favorable dominant alleles from both parents combined in the hybrid, whereas the overdominance model postulates the existence of loci at which the heterozygous state is superior to either homozygote (Xiao et al. 1995; Yu et al. 1997; reviewed by Xu 2003). Evidence for the role of epistasis (interaction of the favorable alleles at different loci contributed by the two parents) in hybrid vigor has also been reported (Stuber et al. 1992; Li et al. 2001; Luo et al. 2001; reviewed by Xu 2003). Further, detailed description of the genetic basis of heterosis, heterotic groups, hybrid prediction and hybrid performance, relationships between heterozygosity and genetic distance with hybrid performance and heterosis, and use of MAS in hybrid breeding has been discussed elsewhere (Xu 2003).

The establishment of heterotic groups and heterotic patterns is an empirical task in hybrid maize breeding that has, in temperate maize germplasm, contributed to large increases in yield. Reciprocal recurrent selection (RRS) programs have proven to be effective in the improvement of heterotic groups through maximizing selection gains within a heterotic group and differences between heterotic groups. In temperate maize, such as the U.S. Corn Belt germplasm, the Reid Yellow Dent × Lancaster Sure Crop, a heterotic pattern was recognized by Sprague over 60 years ago (from Iowa State Corn Breeding Annual Report 1939, 1940) (referred to in Troyer and Rocheford 2002). However, the first mention of the term "heterotic pattern" (or heterotic group) was in 1972 by B. Tsotsis (1972); the concept was further developed through the 1970s (Tracy and Chandler 2006). As an example, inbred lines such as B73 and Mo17, which are from two different heterotic groups, were chosen as testers for the selection of new maize inbreds.

Following successful deployment of hundreds of OPVs in the 1970s and early 1980s, the International Maize and Wheat Improvement Center (CIMMYT) maize program began the development of hybrid maize to meet the needs of hybrid-oriented farms and markets in the developing world. In the 1990s, 10 pairs of subtropical, midaltitude, and highland populations were developed as heterotic partners. These were subsequently used in the RRS programs at CIMMYT to enlarge genetic distance between partner groups and maximize the heterosis between inbred lines selected from complementary populations. The heterotic patterns include: Pop33 x Pop45, Pop42 x Pop44, Pop501 x Pop502, Pop401 x Pop402, Pop445 x Pop446, INT-A x INT-B, LAT-A x LAT-B, DR-A x DR-B, Z97EWA x Z97EWB, and Pop902 x Pop903. More recently, testers from each population have been used to test the hybrid performance of inbreds from the partner populations and to help assign new inbred lines to an appropriate heterotic group (Xia et al. 2005).

Molecular markers are a powerful complement to help define heterotic groups and to examine the relationships among inbred lines at the DNA level. Various molecular marker types have been used to investigate relationships among inbred maize lines from different heterotic groups (for example in tropical maize, see Xia et al. 2004). Markers can also be used to assign lines to new or currently existing heterotic groups (Dubreuil et al. 1996; Smith et al. 1997; Yuan et al. 2001).

14.7.2 Quality

Micronutrient deficiencies affect millions of people worldwide, particularly in tropical countries. Although maize can supply the minimum daily caloric requirement for humans, it is a poor source of the essential amino acids lysine and tryptophan. A diet in which maize predominates can lead to serious deficiency disorders such as pellagra and kwashiorkor. Mertz et al. (1964) discovered the mutant *opaque 2* (*o2*) that increases the lysine content in maize endosperm. Unfortunately, this gene is associated with inferior agronomic traits, including brittleness and insect susceptibility. However, with the discovery of "modifier genes" (*mo2*) that alter the soft, starchy texture of the endosperm, maize breeders developed hard endosperm *o2* mutants designated as "Quality Protein Maize" (Prasanna et al. 2001). These have the phenotypes and yield potential of normal maize while maintaining the increased lysine content of *o2*. As summarized by Bjarnason and Vasal (1992) and Krivanek et al. (2007), breeding of QPM varieties requires manipulation of three genetic systems: 1) the *opaque-2* (*o2*) gene must be in its homozygous recessive form, thereby reducing the rate of transcription of genes encoding zein proteins, which contain very small quantities of lysine and tryptophan; 2) modifier genes of the *o2* gene must be selected, to modify the undesirable soft and chalky (opaque) kernel features that are typical of *opaque-2* maize; and 3) additional (non-*o2*) genes affecting lysine and tryptophan concentration in grain must be selected to ensure that concentrations of these amino acids are within the high range of variation observed for maize.

Using SSRs and backcross breeding, Babu et al. (2004) developed maize lines that had twice the amount of lysine and tryptophan than the native lines and recovered up to 95% of the recurrent parent genome. Yang et al. (2005) reported a new lysine mutant, *o16*, which contained a similar level of lysine content to *o2*, but was located on a different chromosome. The genetic effect of o16 needs to be confirmed under different genetic backgrounds. Then MAS for combining both *o2* and *o16* alelles will help develop new high lysine maize varieties.

14.7.3 Abiotic stresses

The most important abiotic stress in many tropical countries, particularly in Africa, is drought, which affects agricultural production in about 60% of the land area in the tropics. Other abiotic stresses in the tropics include low soil fertility stress, soil acidity and high aluminum saturation, extreme temperatures, waterlogging, and salinity. Maintenance of root elongation is an important adaptive response to drought conditions. In addition, abscisic acid accumulation is required for root growth maintenance under water deficits (Leach et al. 2006). Short anthesis-silking interval (ASI) has been used as an important criterion for drought tolerance in maize. Thus, CIMMYT initiated a major marker-assisted breeding program to transfer five genomic regions involved in the expression of short ASI from Ac7643, a drought tolerant line, to CML247, an elite tropical breeding line (Ribaut et al. 1996, 1997). As a result, the best five marker assisted backcrossing-derived hybrids yielded, on average, at least 50% more than the control hybrids under water stress conditions (Ribaut et al. 2002; Ribaut and Ragot 2007). However, drought tolerance is genetically so complex that success stories from public sector MAS program are limited, or have so far been negligible in maize (Tuberosa and Salvi 2006). With all the recent technological breakthroughs (identifying and pyramiding QTL of minor effects from diverse germplasm resources) it is likely that genomics-assisted breeding will soon lead to the release of cultivars improved for quantitative traits.

14.7.4 Biotic stresses

Diseases that are of a global nature and occur in most maize growing environments include leaf blights, leaf rusts, leaf spots, stalk rots and ear rots. Diseases that are of regional economic importance in the tropics include:

- Asia - downy mildews, which are also spreading to some parts of Africa and the Americas
- Africa - maize streak virus and the parasitic weed Striga
- Latin America - maize stunt and tar spot

There are several MAS reports regarding various biotic stresses but we will focus here on just three. The first example is MAS for European corn borer (ECB)

(*Ostrinia nubilalis* Hubner). Two separate experiments were conducted to assess the efficiency of both phenotype-based selection and MAS for second generation ECB tolerance and for stalk strength (Flint-Garcia et al. 2003b). In some populations MAS was more effective than phenotypic selection, although in some other populations the opposite was true. In some cases MAS was effective for selection of resistance and susceptibility, whereas in others MAS was only effective for selection of susceptibility. Finally, MAS for QTL from some sources was much more effective than MAS for QTL from other sources. In a similar study, no significant difference was observed between the products of MAS and the products of phenotypic selection (Willcox et al. 2002). In a third study, selection using MAS data only was less efficient than phenotypic selection, except when combining marker and phenotypic data which increased the relative efficiency, but only by 4% (Bohn et al. 2001).

In summary, MAS has been widely used in large breeding companies for both major gene and QTL controlled traits, while in the public sector MAS applications have generally focused on simply inherited disease and insect resistances. In tropical maize, major efforts have been devoted to genetic mapping for drought tolerance, although as yet without any resultant MAS successes. From several reports comparing MAS and phenotypic selection, MAS does not always provide a better selection response than conventional phenotypic selection. For example, Moreau et al. (2004) characterized 300 $F_{3:4}$ families derived from inbred an early European flint inbred and an early dent inbred from USA using 93 markers and phenotypic evaluation in multiple environments. Three methods of selection were applied – (i) two cycles of conventional phenotypic selection, (ii) two cycles of MAS based on an index combining phenotypic values and QTL genetic values, and (iii) one cycle of combined MAS followed by two cycles of selection based only on the QTL effects estimated in the first generation. In this study, the allele frequencies showed that selection using markers was very efficient only for fixing favorable QTL in the initial population. Genetic gain was significant for each method of selection. However, the differences between phenotypic selection and combined MAS were not significant. Two additional cycles of MAS using only marker data did not improve significantly the genetic value of the population, indicating that QTL effects estimated in the initial population were not stable due to epistasis and/or QTL × E interactions. In many cases, however, MAS does not need to be superior to phenotypic selection to still have a significant impact on the overall breeding efficiency, especially when the cumulative effects of multiple traits are considered.

14.8 Future Perspectives

Despite the abundance of recent technological breakthroughs, the overall contribution of genomics-assisted breeding to the release of maize cultivars improved for quantitative traits such as drought tolerance has so far been negligible (Tuberosa and Salvi 2006). This may be because the majority of achievements in maize genomics have been based on temperate maize germplasm where drought tolerance is less important than it is in tropical maize germplasm. However, markers and gene sequences can still be valuable for tropical maize, although clearly they must first be carefully validated. There are significant differences in applied genomics between temperate and tropical maize. Technology transfer from temperate maize to tropical maize and capacity building in tropical countries are needed for improvement of tropical maize. Comparative genomics across tropical maize germplasm and temperate maize will help identify novel genes and alleles required for improvement of both temperate and tropical maize. Introgression of genes between temperate and tropical maize should be emphasized in order to further improve maize in both regions. North-south collaborations in maize genomics should be strengthened through scientists in both theoretical and applied genomics.

As for the future of genetics and genomics studies in maize, we would like to add our support to the 15 priority areas identified by the 48th Maize Genetics Conference:

1. Turning the analysis of phenotypes and traits into a high-throughput endeavor without sacrificing agronomic relevance
2. Surveying diverse maize (teosintes and landraces) for novel genes, genetic polymorphisms and phenotypic variation
3. Simultaneously manipulating multiple alleles for scores of genes (QTL) across diverse genetic backgrounds
4. Using proteomics/mass spectrometry as a tool for the analysis of maize mutants/QTL
5. Approaches for mapping all the mutants to the gene space so one can quickly move from phenotype to gene candidate
6. A complete collection of all ESTs from all tissues and developmental stages to make more complete microarrays (for example, meiotic genes are greatly underrepresented in GSS and EST collections)
7. Sequence-indexed collections for all the tools related to gene discovery including TILLING, MTMdB, PML, Rescue*Mu*, Uniform*Mu*, and *Ac/Ds*
8. More reverse genetics resources, i.e., more mutants in more genes (TUSC, Tilling etc)-targeted gene disruption strategies (Zn finger nucleases, etc)
9. Additional expression profiling tools or informational resources (e.g., detailed analysis of antisense expression, active promoter mapping, and alternative splicing)
10. Sequencing-based expression profiling platforms and their advantages and disadvantages over current hybridization based approaches
11. The B73 genome sequence and its annotation (and even ESTs) for proteomics/mass spectrometry analysis of other inbred lines and (distant) landraces (e.g. tropical highland varieties)
12. Better events (single or low transgene copy, site specific insertion to minimize the transgene expression variation due to transgene random integration on the chromosome)
13. A few transformable inbred lines amendable for Agrobacterium-mediated transformation

14. Cost effective plant transformation: introduction of GFP, tap-tagged fusion proteins of choice under native or conditional promotors
15. Strategies, tools, and policies that can be developed to increase data submission into the various public databases

We can expect that the rapid developments across maize genetics and genomics, although currently based mainly on temperate maize germplasm, will be transferable and increasingly valuable for tropical maize improvement. Applied maize genomics in tropical countries should focus on the areas specific to tropical maize including: (1) fingerprinting of tropical maize germplasm including adapted landraces; (2) establishing distinct “haplotypes” for tropical maize germplasm; (3) allele mining and gene discovery from tropical germplasm; (4) understanding of genotype-by-environment interactions that are specific to the tropics; (5) developing decision support tools that are more suitable for developing countries and institutions in the tropics; (6) developing information and data management systems that facilitate North-South collaborations; and (7) establishing networks and supporting systems that promote applications of genomics in maize breeding. With the resolution of many practical, logistical and genetical bottlenecks in MAS (review by Xu and Crouch 2008) and the ongoing development of powerful decision support tools for molecular plant breeding (Wang et al. 2007), it can be expected that genomics-assisted breeding will increasingly become a routine component of breeding programs focused on the development of tropical maize varieties.

Acknowledgments The authors would like to gratefully acknowledge the contribution of Marilyn Warburton, Debra Skinner and Manilal William (CIMMYT) through discussions on several areas relevant to this chapter.

References

Ahern K, Deewatthanawong P, Conrad L, Schnable J, Dong Q, et al. (2006) A two component *Activator/Dissociation* platform for reverse and forward genetic analysis in maize. Maize Genet Conf (abstr) 48, P155

Babu ER, Man VP, Gupta HS (2004) Combining high quality protein and hard endosperm traits through phenotypic and marker assisted selection. In: Fhisher T (ed) New Directions for a Diverse Planet. Proc 4th Intl Crop Sci Congress, Published on CDROM. Webside www.cropscience.org.au.

Bai L, Singh M, Lauren Pitt L, Sweeney M, Brutnell TP (2007) Generating novel allelic variation through activator (Ac) insertional mutagenesis in maize. Genetics 175:981–992

Barrière Y, Argillier O (1993) Brown-midrib genes of maize: A review. Agronomie 13:865–876

Barrière Y, Guillet C, Goffner D, Pichon M. (2003) Genetic variation and breeding strategies for improved cell digestibility in annual forage crops: a review. Animal Res 52:193–228

Batley J, Barker G, O'Sullivan H, Edwards KJ, Edwards D (2003) Mining for single nucleotide polymorphisms and insertions/deletions in maize expressed sequence tag data. Plant Physiol 132:84–91

Beavis WD (1998) QTL analysis: power, precision and accuracy. pp 145–162. In: Paterson AH (ed) Molecular Dissection of Complex Traits. CRC Press, Boca Raton, FL

Beavis WD, Grant D (1991) A linkage map based on information from 4 F_2 populations of Maize (Zea mays L.). Theor Appl Genet 82:636–644

Bedell JA, Budiman MA, Nunberg A, Citek RW, Robbins D, et al. (2005) Sorghum genome sequencing by methylation filtration. PLoS Biol 3:0103–0115

Bellon MR, Berthaud J (2004) Transgenic maize and the evolution of landrace diversity in Mexico. The importance of farmers' behavior. Plant Physiol 134:883–888

Bernier, J., Kumar, A., Venuprasad, R., Spaner, D., and Atlin, G. (2007) A large-effect QTL for grain yield under reproductive-stage drought stress in upland rice. Crop Sci 47, 505–517.

Betrán FJ, Ribaut JM, Beck D, Gonzalez de León D (2003) Genetic diversity, specific combining ability, and heterosis in tropical maize under stress and nonstress environments. Crop Sci 43:797–806

Biswas GCG, Ransom C, Sticklen M (2006) Expression of biologically active Acidothermus cellulolyticus endoglucanase in transgenic maize plants. Plant Sci 171:617–623

Bjarnason M, Vasal SK (1992) Breeding of quality protein maize (QPM). Plant Breed Rev 9:181–216

Bohn M, Khairallah MM, González-de-León D, Hoisington DA, Utz HF, et al. (1996) QTLs mapping in tropical maize: I. Genomic regions affecting leaf feeding resistance to sugarcane borer and other traits. Crop Sci 36:1352–1361

Bohn M, Khairallah MM Jiang C, González-de-León D, Hoisington DA, et al. (1997) QTL mapping in tropical maize: II. Comparison of genomic regions for resistance to *Diatraea* spp. Crop Sci 37:1892–1902

Bohn M, Groh S, Khairallah MM, Hoisington DA, Utz HF, et al. (2001) Re-evaluation of the prospects of marker-assisted selection for improving insect resistance against *Diatraea* spp. in tropical maize by cross validation and independent validation. Theor Appl Genet 103:1059–1067

Bommert PB, Lunde C, Nardmann J, Vollbrecht E, Running PM, et al. (2005) *thick tassel dwarf1* encodes a putative maize orthologue of the *Arabidopsis* CLAVATA1 leucine-rich receptor-like kinase. Development 132:1235–1245

Bortiri E, Jackson D, Hake S (2006a) Advances in maize genomics: the emergence of positional cloning. Curr Opin Plant Biol 9:164–171

Bortiri E, Chuck G, Vollbrecht E, Rochefort TF, Martienssen R, et al. (2006b) *ramosa2* ancodes a LOB domain protein that determines the fate of stem cells in branch meristems of maize. Plant Cell 18:574–585

Briggs WH, McMullen M, Gaut BS, Doebley J (2006) QTL analysis of morphological traits in a large maize-teosinte backcross population. Maize Genet Conf (abstr) 48:T24

Buckler ES, Gaut BS, McMullen MD (2006) Molecular and functional diversity of maize. Curr. Opin. Plant Biol 9:172–176

Burr B, Burr F, Thompson KH, Albersten M, Stuber CW (1988) Gene mapping with recombinant inbreds in maize. Genetics 118:519–526

Calderon-Vazquez C, Ibarra-Laclette E, Caballero-Perez J, Herrera-Estrella A, Martinez de la Vega O, et al. (2006) Transcriptome analysis of the low-phosphorus responses in roots and shoots of a phosphorus-efficient *Zea mays* line identifies alterations of several metabolical and physiological processes. Maize Genet Conf (abstr) 48:P203

Causse M, Santoni S, Damerval C, Maurice A, Charcosset A, et al. (1996) A composite map of expressed sequences in maize. Genome 39:418–432

Chen HD, Guo L, Fu Y, Emrich SJ, Ronin YI, et al. (2007) High-density genetic map of maize genes. Maize Genet Conf (abstr) 49, P 141

Cheung VG, Spielman RS (2002) The genetics of variation in gene expression. Nat Genet 32 (suppl.):522–525

Ching A, Caldwell KS, Jung M, Dolan M, Smith OS, et al. (2002) SNP frequency, haplotype structure and linkage disequilibrium in elite maize inbred lines. BMC Genet 3:19

Cho RJ, Mindrinos M, Richards DR, Sapolsky RJ, Anderson M, et al. (1999) Genome-wide mapping with biallelic markers in *Arabidopsis thaliana*. Nat Genet 23:203–207

Coe EH, Hoisington DA, Neuffer MG (1987) Linkage map of corn (maize) (*Zea mays* L.). Maize Genet Coop Newsl 61:116–147

Coe E, Cone K, McMullen M, Chen S-S, Davis G, et al. (2002) Access to the maize genome: An integrated physical and genetic map. Plant Physiol 128:9–12

Comai L, Young K, Till BJ, Reynolds SH, Greene EA, et al. (2004) Efficient discovery of DNA polymorphisms in natural populations by Ecotilling. Plant J 37:778–786

Cone KC, McMullen MD, Bi IV, Davis GL, Yim Y-S, et al. (2002) Genetic, physical, and informatics resources for maize. On the road to integrated map. Plant Physiol 130:1598–1605

Cooper M, Smith OS, Graham G, Arthur L, Feng L, et al. (2004) Genomics, genetics, and plant breeding: A private sector perspective. Crop Sci 44:1907–1913

Cowperthwaite M, Park W, Xu Z, Yan X, Maurais SC, et al. (2002) Use of the transposon *Ac* as a gene-searching engine in the maize genome. Plant Cell 14:713–726

Crosbie TM, Eathington SR, Johnson GR, Edwards M, Reiter R, et al. (2006) Plant breeding: past, present, and future. pp. 3–50. In: Lamkey KR, Lee M (eds) Plant Breeding: The Arnel R. Hallauer International Symposium. Blackwell Publishing, Ames, Iowa

Crossa J, Vargas M, Van Eeuwijk FA, Jiang C, Edmeades GO, et al. (1999) Interpreting genotype × environment interaction in tropical maize using linked molecular markers and environmental covariables. Theor Appl Genet 99:611–625

Danilova T, Lamb J, Bauer M, Meyer J, Birchler J (2006) Development of PCR based FISH probes for identification of maize mitotic chromosomes. Maize Genet Conf (abstr) 48:P74

Davis DW, Cone KC, Chomet P, Cox D, Brady S, et al. (2000) Maize whole-genome radiation hybrids: a progress report. Plant Animal Genome Conf 8:P255

Davis GL, McMullen MD, Baysdorfer C, Musket T, Grant D, et al. (1999) A maize map standard with sequenced core markers, grass genome reference points and 932 expressed sequence tagged sites (ESTs) in a 1736-locus map. Genetics 152:1137–1172

Dear PH, Cook PR (1993) Happy mapping: linkage mapping using a physical analogue of meiosis. Nucleic Acids Res 21:13–20

Doebley J, Stec A, Gustus C (1995) *Teosinte branched 1* and the origin of maize: Evidence for epistasis and the evolution of dominance. Genetics 141:333–346

Doebley J, Stec A, Hubbard L (1997) The evolution of apical dominance in maize. Nature 386:485–488

Dorweiler J, Stec A, Kermicle J, Doebley J (1993) *Teosinte glume architecture 1*: a genetic locus controlling a key step in maize evolution. Science 262:233–235

Dubreuil P, Dufour P, Drejci E, Causse M, de Vienne D, et al. (1996) Organization of RFLP diversity among inbred lines of maize representing the most significant heterotic groups. Crop Sci 36:790–799

Dvorak J, Luo MC, Yang ZL (1998) Restriction fragment length polymorphism and divergence in the genomic regions of high and low recombination in self-fertilizing and cross-fertilizing *Aegilops* species. Genetics 148:423–434.

Dwivedi SL, Crouch JH, Mackill DJ, Xu Y, Blair MW, et al. (2007) The molecularization of public sector crop breeding: progress, problems and prospects. Adv Agron 95:163–318.

Eathington SR (2005) Practical applications of molecular technology in the development of commercial maize hybrids. In: Proc 60th Ann Corn and Sorghum Seed Res Conf. American Seed Trade Association, Washington, D.C.

Edwards MD, Helentjaris T, Wright S, Stuber CW (1992) Molecular-marker-facilitated investigations of quantitative trait loci in maize. 4. Analysis based on genome saturation with isozyme and restriction fragment length polymorphism markers. Theor Appl Genet 83 :765–774

Emrich SJ, Aluru S, Fu Y, Wen TJ, Narayanan M, et al. (2004) A strategy for assembling the maize (*Zea mays* L.) genome. Bioinformatics 20:140–147

Emrich SJ, Barbazuk WB, Li L, Schnable PS (2007) Gene discovery and annotation using LCM-454 transcriptome sequencing. Genome Res 17:69–73

Evans MMS (2007) The *indeterminate gametophyte1* gene of maize encodes a LOB domain protein required for embryo sac and leaf development. Plant Cell 19:46–62

Falque M, Décousset L, Dervins D, Jacob AM, Joets J, et al. (2005) Linkage mapping of 1454 new maize candidate gene loci. Genetics 170:1957–1966

Fernandes J, Brendel V, Gai X, Lal S, Chandler VL, et al. (2002) Comparison of RNA profiles based on maize expressed sequence tag frequency analysis and micro-array hybridization. Plant Physiol 128:896–910

Figueroa D, Amarillo I, Ring B, Strobel C, Lawrence C, et al. (2006) Constructing a cytogenetic map of maize core bin markers in oat addition lines using sorghum BACs as FISH probes. Maize Genet Conf (abstr) 48: P71

Flint-Garcia SA, Thornsberry JM, Buckler IV ES (2003a) Structure of linkage disequilibrium in plants. Ann Rev Plant Biol 54:357–374

Flint-Garcia SA, Darrah LL, McMullen MD, Hibbard BE (2003b) Phenotypic versus marker-assisted selection for stalk strength and second-generation European corn borer resistance in maize. Theor Appl Genet 107:1331–1336

Fraley R (2006) Presentation at Monsanto European Investor Day, 10 November 2006. www.monsanto.com

Fu H, Dooner HK (2002) Intraspecific violation of genetic colinearity and its implications in maize. Proc Natl Acad Sci USA 99:9573–9578

Gardiner JM, Coe EH, Melia-Hancock S, Hoisington DA, Chao S (1993) Development of a core RFLP map in maize using an immortalized F2 population. Genetics 134:917–930

Gaut BS, Le Thierry I' EM, Peek AS, Saukins MC (2000) Maize as a model for the evolution of plant nuclear genomes. Proc Natl Acad Sci USA 97:7008–7015

Gilchrist EJ, Haughn GW (2005) TILLING without a plough: a new method with applications for reverse genetics. Curr Opin Plant Biol 8:1–5

Grohn S, González-de-León D, Khairallah MM, Jiang C, Bergvinson M, et al. (1998) QTL mapping in tropical maize: III. Genomic regions for resistance to *Diatraea* spp. and associated traits in two RIL populations. Crop Sci 38:1062–1072

Guillet-Claude C, Birolleau-Touchard C, Manicacci D, Rogowsky PM, Rigau J, et al. (2004) Nucleotide diversity of the *ZmPox3* maize peroxidase gene: Relationships between a MITE insertion in exon 2 and variation in forage maize digestibility. BMC Genet 5:19

Haberer G, Young S, Bharati AK, Gundlach H, Raymond C, et al. (2005) Structure and architecture of the maize genome. Plant Physiol 139:1612–1624

Hamblin MT, Aquadro CF (1999) DNA sequence variation and the recombinational landscape in *Drosophila pseudoobscura*: a study of the second chromosome. Genetics 153:859–869

Iniguez AL, Gardiner J, Hogan M, Smith A, Buell R, et al. (2006) Antisense expression analysis in the maize transcriptome and microarray crossplatform comparisons. Maize Genet Conf (abstr) 48:P2

Jansen RC, Nap J-P (2001) Genetical genomics: the added value from segregation. Trends Genetics 17:388–391

Jia J, Fu J, Zheng J, Zhou X, Huai J, et al. (2006) Annotation and expression profile analysis of 2073 full-length cDNAs from stress-induced maize (*Zea mays* L.) seedlings. Plant J 48:710–727

Jung M, Ching A, Bhattramakki D, Dolan M, Tingey S, et al. (2004) Linkage disequilibrium and sequence diversity in a 500-kbp region around the *adh1* locus in elite maize germplasm. Theor Appl Genet 109:681–689

Kaeppler SM (1997) Quantitative trait locus mapping using sets of near-isogenic lines: relative power comparisons and technical considerations. Theor Appl Genet 95:384–192

Khairallah MM, Bohn M, Jiang C, Deutsch JA, Jewell DC, et al. (1998) Molecular mapping of QTL for southwestern corn borer resistance, plant height and flowering in tropical maize. Plant Breed 117:309–318

Kolkman JM, Conrad LJ, Farmer PR, Hardeman K, Ahern KR, et al. (2005) Distribution of *Activator* (*Ac*) throughout the maize genome for use in regional mutagenesis. Genetics 169:981–995

Kollipara KP, Saab IN, Wych RD, Lauer MJ, Singletary GW (2002) Expression profiling of reciprocal maize hybrids divergent for cold germination and desiccation tolerance. Plant Physiol 129:974–992

Komatsu K, Maekawa M, Ujiie S, Satake Y, Furutani I, et al. (2003) LAX and SPA: major regulations of shoot branching in rice. Proc Natl Acad Sci USA 100:11765–11770
Koumbaris G, Bass HW (2003) A new single-locus cytogenetic mapping system for maize (*Zea mays* L.): overcoming FISH detection limits with marker-selected sorghum (*S. propinquum* L.) BAC clones. Plant J 35:647–659
Krill A, Hoenkenga O, Kirst M, Kochian L, Buckler E (2006) Association analysis of candidate genes for aluminum tolerance in maize. Maize Genet Conf (abstr) 48:P210
Krivanek AF, De Groote H, Gunaratna NS, Diallo AO, Friesen D (2007) Breeding and disseminating quality protein maize (QPM) for Africa. African J Biotechnol 6:312–324
Kronmiller B, Werner K, Wise R (2006) TE Nest: Automated chronological annotation and visualization of maize nested transposable elements. Maize Genet Conf (abstr) 48:P58
Lanza LLB, Souza Jr CL, Ottoboni LLM, Vieira MLC, Souza AP (1997). Genetic distance of inbred lines and prediction of maize single cross performance using RAPD markers. Theor Appl Genet 94:1023–1030
Laurie CC, Chasalow SD, LeDeaux JR, McCarrol R, Rush D, et al. (2004) The genetic architecture of response to long-term artificial selection for oil concentration in the maize kernel. Genetics 168:2141–2155
Lê Q, Gutièrrez-Marcos,JF Costa LM, Meyer S, Dickinson HG, et al. (2005) Construction and screening of subtracted cDNA libraries from limited populations of plant cells: a comparative analysis of gene expression between maize egg cells and central cells. Plant J 44:167–178
Leach K, Davis D, Maltman R, Hejlek L, Nguyen H, et al. (2006) Genetic diversity of maize primary root growth and abscisic acid content to water stress. Maize Genet Conf (abstr) 48:P219
Lee M, Sharopova N, Beavis WD, Grant D, Katt M, et al. (2002) Expanding the genetic map of maize with the intermated B73 x Mo17 (IBM) population. Plant Mol Biol 48:453–461
Li ZK, Luo LJ, Mei HW, Wang DL, Shu QY, et al. (2001) Overdominant epistatic loci are the primary genetic basis of inbreeding depression and heterosis in rice. I. Biomass and grain yield. Genetics 158:1737–1753
Lima MLA, de Souza CL, Bento DAV, de Souza AP, Carlini-Garcia LA (2006) Mapping QTL for grain yield and plant traits in a tropical maize population. Mol Breed 17:227–239
Lin C, Shen B, Xu Z, Dooner H (2006) Isolation and characterization of maize sesquiterpene cyclase2 (stc2) gene involved in insect resistance. Maize Genet Conf (abstr) 48:P21
Lisch D (2002) Mutator transposons. Trends Plant Sci 7:498–504
Liu K, Goodman M, Muse S, Smith JS, Buckler ED, et al. (2003) Genetic structure and diversity among maize inbred lines as inferred from DNA microsatellites. Genetics 165:2117–2128
Liu W, Gao Y, Teng F, Shi Q, and Zheng Y (2006) Construction and genetic analysis of mutator insertion mutant population in maize Construction and Genetic analysis of the maize mutator-transposon insertional mutant pool. Chinese Sci Bull 51:2604–2610
Lizarraga Guerra R, Gibbon B, Larkins B (2006) Genetic analysis of opaque2 modifier genes. Maize Genet Conf (abstr) 48:P15
Lübberstedt TL, Zein I, Andersen JR, Wenzel G, Krützfeldt B, et al. (2005) Development and application of functional markers in maize. Euphytica 146:101–108
Luo LJ, Li ZK, Mei HW, Shu QY, Tabein R, et al. (2001) Overdominant epistatic loci are the primary genetic basis of inbreeding depression and heterosis in rice. II. Grain yield components. Genetics 158:1755–1771
Ma Z, Dooner HK (2004) A mutation in the nuclear-encoded plastid ribosomal protein S9 leads to early embryo lethality in maize. Plant J 37:92–103
Mangolin CA, Souza Jr CL, Garcia AAF, Garcia AF, Sibov ST, et al. (2004) Mapping QTLs for kernel oil content in a tropical maize population. Euphytica 137:251–259
May BP, Liu H, Vollbrecht E, Senior L, Rabinowicz PD, et al. (2003) Maize-targeted mutagenesis: A knockout resource for maize . Proc Natl Acad Sci USA 100:11541–11546
Meaburn E, Butcher LM, Schalkwyk LC, Plomin R (2006) Genoptyping pooled DNA using 100K SNP microarrays: a step towards genomewide association scans. Nucleic Acids Res 34:No.4, e28

Mertz ET, Bates LS, Nelson OE (1964) Mutant gene that changes protein composition and increases lysine content of maize endosperm. Science 145:279–280
Messing J, Bharti AK, Karlowski WM, Gundlach H, Kim HR, et al. (2004) Sequence composition and genome organization of maize. Proc Natl Acad Sci USA 101:14349–14354
Meyer P, Saedler H (1996) Homology-dependent gene silencing in plants. Ann Rev. Plant Physiol Plant Mol Biol 47:23–48
Monde R-A, Till BJ, Sahm H, Laport R, Haywood N, et al. (2006) The Maize TILLING Project: progress report for year 3. Maize Genet Conf (abstr) 48:P31
Moose S, Schneerman M, Zhang M, Zhang K, Schneeberger R, et al. (2006) Transcript profiling of the Illinois protein strains and derived germplasm. Maize Genet Conf (abstr) 48:P201
Moreau L, Charcosset A, Hospital F, Gallais A (1998) Marker-assisted selection efficiency in populations of finite size. Genetics 148:1353–1365
Moreau L, Charcosset A, Gallais A (2004) Experimental evaluation of several cycles of marker-assisted selection in maize. Euphytica 137:111–118
NCI-NHGRI Working Group on Replication in Association Studies (2007) Replicating genotype-phenotype associations. Nature 447:655–660
Nicholas FW (2006) Discovery, validation and delivery of DNA markers. Aus J Exp Agric 46:155–158
Niebur WS, Rafalski JA, Smith OS, Cooper M (2004) Applications of genomics technologies to enhance rate of genetic progress for yield of maize within a commercial breeding program. In: Fhisher T (ed) New Directions for a Diverse Planet. Proc 4th Intl Crop Sci Congr, www.cropscience.org.au
Okagaki RJ, Kynast RG, Livingston SM, Russell CD, Rines HW, et al. (2001) Mapping maize sequences to chromosomes using oat-maize chromosome addition materials. Plant Physiol 125:1228–1235
Okagaki R, Jacobs M, Schneerman M, Kynast R, Buescher E, et al. (2006a) A comparison of centromere mapping techniques. Maize Genet Conf (abstr) 48:P68
Okagaki R, Kynast R, Stec A, Schmidt C, Jacobs M, et al. (2006b) Oat-maize addition and radiation hybrid lines for the physical and genetic mapping of the maize genome. Maize Genet Conf (abstr) 48:P149
Ortiz R, Crouch JH, Iwanaga M, Sayre K, Warburton M, et al. (2006) Agriculture and energy in developing countries: Bio-energy and Agricultural Research-for-Development. IFPRI "2020 Focus" Policy Brief #14 (www.ifpri.org/pubs/catalog. htm#focus)
Rosegrant MW, Msangi S, Sulser T, Valmonte-Santos R (2006) Biofuels and the Global food Balance. IFPRI "2020 Focus" Policy Brief #14 (www.ifpri.org/pubs/catalog. htm#focus)
Palaisa KA, Morgante M, Williams M, Rafalski A (2003) Contrasting effects of selection on sequence diversity and linkage disequilibrium at two phytoene synthase loci. Plant Cell 15:795–1806
Paliwal RL (2000) Introduction to maize and its importance. pp. 1–3. In: Paliwal RL, Granados G, Lafitte HR, Violic AD, Marathee JP (eds) Tropical Maize: Improvement and Production. FAO, Rome
Paliwal RL, Granados G, Lafitte HR, Violic AD, Marathee JP (2000) Tropical Maize: Improvement and Production. FAO, Rome. 363 pp
Palmer LE, Rabinowicz PD, O'Shaughnessy AL, Balija VS, Nascimento LU, et al. (2003) Maize genome sequencing by methylation filtration. Science 302:2115–2117
Pandey S, Gardner CO (1992) Recurrent selection for population, variety, and hybrid improvement in tropical maize. Adv Agron 48:1–87
Pea G, Ferron S, Gianfranceschi L, Krajewski P, Pe ME (2006) Wide-scale survey of transcriptional heterosis in F_1 maize immature ear. Maize Genet Conf (abstr) 48:P207
Podlich DW, Winkler CR, Cooper M (2004) Mapping as you go: an effective approach for marker-assisted selection of complex traits. Crop Sci 44:1560–1571
Prasanna BM, Vasal SK, Kassahun B, Singh NN (2001) Quality protein maize. Current Sci 81:1308–1319

Price AH (2006) Believe it or not, QTLs are accurate! Trends Plant Sci 11:213–216
Quint M, Mihaljevic R, Dussle C, Xu ML, Melchinger A, et al. (2002) Development of RGA-CAPS markers and genetic mapping of candidate genes for sugarcane mosaic virus resistance in maize. Theor Appl Genet 105:355–363
Quist D, Chapela IH (2001) Transgenic DNA introgressed into traditional maize landraces in Oaxaca, Mexico. Nature 414:41–543
Ragot M, Gay G, Muller J-P, Durovray J (2000) Efficient selection for the adaptation to the environment through QTL mapping and manipulation in maize. pp. 128–130. In: Ribaut J-M, Poland D (eds) Molecular Approaches for the Genetic Improvement of Cereals for Stable Production in Water-Limited Environments. CIMMYT, México, D.F.
Reif JC, Xia XC, Melchinger AE, Warburton ML, Hoisington DA, et al. (2004) Genetic diversity determined within and among CIMMYT maize populations of tropical, subtropical, and temperate germplasm by SSR markers. Crop Sci 44:326–334
Remington DL, Thornsberry JM, Matsuoka Y, Wilson LM, Whitt SR, et al. (2001) Structure of linkage disequilibrium and phenotypic associations in the maize genome. Proc Natl Acad Sci USA 98:11479–11484
Ribaut, JM, Ragot M (2007) Marker-assisted selection to improve drought adaptations in maize: the backcross approach, perspectives, limitations, and alternatives. J Exp Bot 58:351–360.
Ribaut J-M, Hoisington DA, Deutsch JA, Jiang C, Gonzalez-de-Leon D (1996) Identification of quantitative trait loci under drought conditions in tropical maize 1. Flowering parameters and the anthesis-silking-interval. Theor Appl Genet 92:905–914
Ribaut J-M, Jiang C, Gonzalez-de-Leon D (1997) Identification of quantitative trait loci under drought condition in tropical maize. 2. Yield components and marker-assisted selection strategies. Theor Appl Genet 94:887–896
Ribaut JM, Bänziger M, Betran J, Jiang C, Edmeades GO, et al. (2002) Use of molecular markers in plant breeding: drought tolerance improvement in tropical maize. In: Kang MS (ed) Quantitative Genetics, Genomics, and Plant Breeding. CABI Publishing, pp. 85–99
Ribaut J-M, Bänziger M, Setter T, Edmeades G, Hoisington D (2004) Genetic dissection of drought tolerance in maize: a case study. pp. 571–611. In: Nguyen H, Blum A (eds) Physiology and Biotechnology Integration for Plant Breeding. Marcel Dekker Inc., New York
Rosegrant MW, Msangi S, Sulser T, Valmonte-Santos R (2006) Biofuels and the Global food Balance. IFPRI "2020 Focus" Policy Brief #14 (www.ifpri.org/pubs/catalog. htm#focus)
Schaeffer M, Sanchez-Villeda H, Gerau M, McMullen M, Coe E (2006) The New IBM Neighbors: genetic and physical probed sites. Maize Genet Conf (abstr) 48:P151
Scheuring C, Barthelson R, Galbraith D, Betran J, Cothren JT, et al. (2006) Preliminary analysis of differential gene expression between a maize superior hybrid and its parents using the 57K maize gene-specific long-oligonucleotide microarray. Maize Genet Conf (abstr) 48:P193
Schon CC, Utz HF, Groh S, Truberg B, Openshaw S, et al. (2004) Quantitative trait locus mapping based on resampling in a vast maize testcross experiment and its relevance to quantitative genetics for complex traits. Genetics 167:485–498
Schroeder S, Sanchez-Villeda H, Flint-Garcia S, Houchins K, Yamasaki M, et al. (2006) Integrated software for SNP discovery in maize. Maize Genet Conf (abstr) 48:P50
Settles AM (2005) Maize community resources for forward and reverse genetics. Maydica 50:405–414
Settles M, Holding D, Tan B-C, Latshaw S, Suzuki M, et al. (2006) Maize sequence indexed knockouts using the UniformMu transposon-tagging population. Maize Genet Conf (abstr) 48:P180
Sheen J (2001) Signal transduction in maize and Arabidopsis mesophyll protoplasts. Plant Physiol 127:1466–1475
Singh M, Lewis PE, Hardeman K, Bai L, Rose JK, et al. (2003) *Activator* mutagenesis of the *pink scutellum 1/vivparous 7* locus of maize. Plant Cell 15:874–884
Smith ME, Paliwal RL (1996) Contributions of genetic resources and biotechnology to sustainable productivity increases in maize. In: Watanabe K, Pebu E (eds.) Plant Biotechnology and Plant Genetic Resources for Sustainability and Productivity. Lande and Academic Press, Austin, TX

Smith JSC, Chin ECL, Shu H, Smith OS, Wall SJ, et al. (1997) An evaluation of the utility of SSR loci as molecular markers in maize (*Zea mays* L.): Comparisons with data from RFLPs and pedigree. Theor Appl Genet 95:163–173

Stevens R, Paul C, Islam S, Wong J, Harjes C, et al. (2006) Genetic approaches to enhance provitamins A and total carotenoids in maize grain. Maize Genet Conf (abstr) 48:P217

Stich B, Yu J, Melchinger AE, Piepho H, Utz HF, et a. (2007) Power to detect higher-order epistatic interactions in a metabolic pathway using a new mapping strategy. Genetics 176:563–570

Stuber CW, Lincoln SE, Wolff DW, Helentjaris T, Lander ES (1992) Identification of genetic factors contributing to heterosis in a hybrid from two elite maize inbred lines using molecular markers. Genetics 132:823–839

Stupar RM, Springer NM (2006) *Cis*-transcriptional variation in maize inbred lines B73 and Mo17 leads to additive expression patterns in the F_1 hybrids. Genetics 173:2199–2210

Swanson-Wagner R, Jia Y, Borsuk L, DeCook R, Nettleton D, Schnable P (2006) All possible modes of gene action are observed in a global comparison of gene expression in a maize F1 hybrid and its inbred parents. Proc Natl Acad Sci USA 103:6805–6810

Szalma SJ, Buckler ES, Snook ME, McMullen MD (2005) Association analysis of candidate genes for maysin and chlorogenic acid accumulation in maize silks. Theor Appl Genet 110:1324–1333

Szalma SJ, Hostert BM, LeDeaux JR, Stuber CW, Holland JB (2007) QTL mapping with near-isogenic lines in maize. Theor Appl Genet 114:1211–1228

Taramino G, Tingey S (1996) Simple sequence repeats for germplasm analysis and mapping in maize. Genome 39:277–287

Tarter JA, Goodman MM, Holland JB (2004) Recovery of exotic alleles in semiexotic maize inbreds derived from crosses between Latin American accessions and a temperate line. Theor Appl Genet 109:609–617

Tenaillon MI, Sawkins MC, Long AD, Gaut RL, Doebley JF, et al. (2001) Patterns of DNA sequence polymorphism along chromosome 1 of maize (*Zea mays* ssp. *mays* L.). Proc Natl Acad Sci USA 98:9161–9166

Thornsberry JM, Goodman MM, Doebley J, Kresovich S, Nielsen D, et al. (2001) *Dwarf8* polymorphisms associate with variation in flowering time. Nat Genet 28:286–289

Till BJ, Reynolds SH, Weil C, Springer N, Burtner C, et al. (2004) Discovery of induced point mutations in maize by TILLING. BMC Plant Biol 4:12

Tomkins JP, Frisch DA, Byrum JR, Jenkins MR, Barnett LJ, et al. (2000a) Construction and characterization of a maize bacterial artificial chromosome (BAC) library for the inbred line LH132. Maize Genet Coop Newsl 74:18

Tomkins JP, Frisch DA, Jenkins MR, Barnett LJ, Luo M, et al. (2000b) Construction and characterization of a maize bacterial artificial chromosome (BAC) library for the inbred line B73. Maize Genet Coop Newsl 74:18–19

Tracy WF, Chandler MA (2006) The historical and biological basis of the concept of heterotic patterns in corn belt dent maize. pp. 219–233. In: Lamkey KR, Lee M (eds) Plant Breeding: the Arnel R. Hallauer International Symposium. Blackwell Publishing, Ames, IA

Troyer AF, Rocheford TR (2002) Germplasm ownership: related corn inbreds. Crop Sci 42:3–11.

Tsotsis B (1972) Objectives of industry breeders to make efficient and significant advances in the future. pp. 93–107. In: Wilkinson D (ed) Proc 27th Ann Corn and Sorghum Res Conf. American Seed Trade Association, Washington D.C.

Tuberosa R, Salvi S (2006) Genomics-based approaches to improve drought tolerance of crops. Trend Plant Sci 11: 405–412

Vargas M, van Eeuwijk FA, Crossa J, Ribaut J-M (2006) Mapping QTLs and QTL × environment interaction for CIMMYT maize drought stress program using factorial regression and partial least squares methods. Theor Appl Genet 112:1009–1023

Vigouroux Y, Mitchell S, Matsuoka Y, Hamblin M, Kresovich S, et al. (2005) An analysis of genetic diversity across the maize genome using microsatellites. Genetics 169:1617–1630

Vollbrecht E, Springer PS, Gol L, Buckler ES, Martienssen R (2005) Architecture of floral branch systems in maize and related grasses. Nature 436:1119–1125

Wang J, Chapman SC, Bonnett DB, Rebetzke GJ and Crouch JH (2007) Application of population genetic theory and simulation models to efficiently pyramid multiple genes via marker-assisted selection. Crop Science 47:582–588.

Wang H, Nussbaum-Wagler T, Li B, Zhao Q, Vigouroux Y, et al. (2005) The origin of naked grains of maize. Nature 436:714–719

Wang RL, Stec A, Hey J, Lukens L, Doebley J (1999) The limits of selection during maize domestication. Nature 398:236–239

Warburton ML, Xia X, Crossa J, Franco J, Melchinger AE, et al. (2002) Genetic characterization of CIMMYT inbred maize lines and open pollinated populations using large scale fingerprinting methods. Crop Sci 42:1832–1840

Wen T, Qiu F, Guo L, Lee M, Russell K, et al. (2002) High-throughput mapping tools for maize genomics. Maize Genet Conf (abstr) 44:8

Whitelaw CA, Barbazuk WB, Pertea G, Chan AP, Cheung F, et al. (2003) Enrichment of gene-coding sequences in maize by genome filtration. Science 302:2118–2120

Widstrom NW, Butron A, Guo BZ, Wilson DM, Snook ME, et al. (2003) Control of preharvest aflatoxin contamination in maize by pyramiding QTL involved in resistance to ear-feeding insects and invasion by Aaperigillus spp. Eur J Agron 19:563–572

Willcox MC, Khairallah MM, Bergvinson D, Crossa J, Deutsch JA, et al. (2002) Selection for resistance to Southwestern corn borer using marker-assisted and conventional backcrossing. Crop Sci 42:1516–1528

Wilson LM, Whitt SR, Ibanez-Carranza AM, Goodman MM, Rocheford TR, et al. (2004) Dissection of maize kernel composition and starch production by candidate gene association. Plant Cell 16:2719–2733

Wilson R, Wing R, McCombie WR, Martienssen R, Ware D, et al. (2006) Sequencing the maize genome. Maize Genet Conf (abstr) 48:T11

Wright SI, Bi IV, Schroeder SG, Yamasaki M, Doebley JF, et al. (2005) The effects of artificial selection on the maize genome. Science 308:1310–1314

Wu R, Zeng Z-B (2001) Joint linkage and linkage disequilibrium mapping in natural populations. Genetics 157:899–909

Wu R, Ma C-S, Casella G (2002) Joint linkage and linkage disequilibrium mapping of qualtitative trait loci in natural mapping populations. Genetics 160:779–792

Xia XC, Reif JC, Hoisington DA, Melchinger AE, Frisch M, (2004) Genetic diversity among CIMMYT maize inbred lines investigated with SSR markers: I. Lowland tropical maize. Crop Sci 44:2230–2237

Xia XC, Reif JC, Melchinger AE, Frisch M, Hoisington DA, et al. (2005) Genetic diversity among CIMMYT maize inbred lines investigated with SSR markers: II. Subtropical, tropical midaltitude, and highland maize inbred lines and their relationships with elite U.S. and European maize. Crop Sci 45:2573–2582

Xiao J, Li J, Yuan L, Tanksley SD (1995) Dominance is the major genetic basis of heterosis in rice as revealed by QTL analysis using molecular markers. Genetics 140:745–754

Xiao W, Xu M, Zhao J, Wang F, Li J, Dai J (2006) Genome-wide isolation and mapping of resistance gene analogs. Theor Appl Genet 113:63–72

Xu Y (2003) Developing marker-assisted selection strategies for breeding hybrid rice. Plant Breed Rev 23:73–174

Xu Y, Crouch JH (2008) Marker-assisted plant breeding: from publications to practice. Crop Sci (in press)

Yang W, Zheng Y, Zheng W, Feng R (2005) Molecular genetic mapping of a high-lysine mutant gene (*opaque-16*) and the double recessive effects with *opaque-2* in maize. Mol Breed 15:257–269

Yim Y-S, Davis GL, Duru NA, Musket TA, Linton EW, et al. (2002). Characterization of three maize bacterial artificial chromosome libraries toward anchoring of the physical map to the genetic map using high-density bacterial artificial chromosome filter hybridization. Plant Physiol 130:1686–1696

Yu L-X, Setter TL (2003) Comparative transcriptional profiling of placenta and endosperm in developing maize kernels in response to water deficit. Plant Physiol 131:568–582

Yu SB, Li JX, Xu CG, Tan YF, Gao YJ, et al. (1997) Importance of epistasis as the genetic basis of heterosis in an elite rice hybrid. Proc Natl Acad Sci USA 94:9226–9231

Yuan LX, Fu JH, Zhang SH, Liu X, Peng Z, et al. (2001) Heterotic grouping of maize inbred lines using RFLP and SSR markers. Acta Agron Sinica 27:149–156

Zhang J, Lv X, Song X, Yan J, Song T, et al. (2006) Quantitative trait loci mapping for starch, protein, and oil concentrations with high-oil maize by SSR markers. Maize Genet Conf (abstr) 48:P233

Zheng Y, Gao Y, Liu W, Yang W, Shi Q, et al. (2006) Construction and Genetic analysis of the maize mutator-transposon insertional mutant pool. Maize Genet Conf (abstr) 48:P165

Zinselmeier C, Sun Y, Helentjaris T, Beatty M, Yang S, et al. (2002) The use of gene expression profiling to dissect the stress sensitivity of reproductive development in maize. Field Crops Res 73:111–121

Zhu T, Xia Y, Chilcott C, Dunn M, Dace G, et al. (2006) Maize ultra high-density gene map for genome-assisted breeding. Maize Genet Conf (abstr) 48:P181

Chapter 15
Molecular Research in Oil Palm, the Key Oil Crop for the Future

Sean Mayes, Farah Hafeez, Zuzana Price, Don MacDonald, Norbert Billotte, and Jeremy Roberts

Abstract African oil palm (*Elaeis guineensis* Jacq) produces more than five times the yield of oil/year/hectare of any annual oil crop. In consequence, it represents a key species for meeting future vegetable oil needs (both for food and for industry) against the background of a rising world population. As it is a tree crop and naturally out-crossing current planting material is, in contrast to most annual crops, generally heterogeneous. This complicates breeding for future needs. Recent developments in molecular biology have the potential to substantially alter approaches for the genetic improvement of oil palm. Some of these biotechnological approaches have already made an impact, for example, somatic embryogenesis for clonal propagation and routine genetic fingerprinting for quality control. The recent development of plant sequence-based approaches, supported by bioinformatics and broadly classified into genomics (DNA) and functional genomics (mRNA, protein and metabolites) could lead to a step-wise change in our understanding of the genetic basis of agronomic traits and the development of practical tools and trait information for plant breeding. These high-throughput developments add significant new potential to the two broad approaches generally adopted in crop molecular research. The "bottom-up" approach involves investigating individual genes and the pathways in which they operate with a long-term aim to develop a complete understanding of these networks and their importance in trait specification. The alternative "top-down" approach starts with the trait in the species of interest and uses inheritance studies, anonymous molecular markers, and physiological techniques to begin to dissect the trait and its interactions with the environment. Genomics and functional genomics represent a suite of techniques which can help to bridge the gap between the bottom-up and the top-down approaches. In this article we review recent progress in developing molecular resources for oil palm and assess the potential impact that specific cutting edge techniques, already developed and implemented in other plant and crop species, could have in oil palm. The article gives background information on the various technologies, but is not intended to be exhaustive. A number of good text books and

S. Mayes
School of Biosciences, Sutton Bonington Campus, Nottingham University, Loughborough, Leicestershire, LE12 5RD, UK
e-mail: sean.mayes@nottingham.ac.uk

P.H. Moore, R. Ming (eds.), *Genomics of Tropical Crop Plants*

© Springer 2008

articles are available which go into greater detail. It also does not go into depth on discoveries in other species, except to illustrate their potential utility in oil palm.

15.1 Introduction

The oil palm (*Elaeis guineensis* Jacq.) is a tropical perennial crop originating in Central and Western Africa in the humid lowland tropics, within $\pm 10^\circ$ latitude of the equator (Corley and Tinker 2003). Oil palm is a monocotyledonous tree belonging to the Family Arecacea, tribe *Cocoseae,* and subtribe *Elaeidinae*. This subtribe consists of the genus *Elaeis*. The genus name, *Elaeis*, is from the Greek elaia for olive. The African oil palm's specific name, *guineensis*, refers to the tree's discovery by Jacquin in the Gulf of Guinea area (Jacquemard 1998). The chromosome number is $2n = 32$ (Maria et al. 1995) and the genome size is $1C = 980$ Mbp (Bennett and Smith 1991; Bennett and Leitch 2005). The genus also contains *E. oleifera* HBK (the South American oil palm). The major interest in oil palm is the high oil content of the mesocarp and kernel of its fruits that are produced in heavy bunches. The plant is monoecious with out-crossing maintained through successive cycles of male and female inflorescence production on a single palm. The length of the male and female cycles varies widely according to genotype and environment (Corley and Donough 1995). The fruit is a sessile drupe produced in bunches of up to 3,000 fruits on mature palms, with an average of around 1,500 fruit/bunch. The fruits vary in size and shape and may weigh from 3 g to 30 g. At maturity, the fruit is red-brown and consists of the pulp, the shell, and the kernel. The pulp (mesocarp; 60% to 90% of fruit weight, 35% to 55% of bunch weight) yields an edible, orange-red oil commonly known as palm oil. The endosperm or kernel when crushed produces a clear yellowish oil that is known as palm kernel oil and is similar to coconut oil. Palm oil and its refined derivatives, palm olein and palm stearin, are the major commercial products of oil palm (Corley and Tinker 2003).

The plant generation time is prolonged, with seeds taking around 100–120 days to germinate, including a heat-treatment, followed by 10–12 months in the nursery before the young seedlings are ready for field planting. The oil palm starts to bear fruit after 2–3 years of field planting and approaches maturity at around 10 years. The economic life of plantings varies from 20–30 years, depending on local conditions, with excessive palm height being the major factor for replanting (Corley and Tinker 2003).

Of all oil-bearing plants, the oil palm produces the most oil per hectare per year, with reported oil yields exceeding 12 t/ha in some experimental plots in Malaysia and 3 to 7 t/ha more typically in commercial fields, although average Malaysian yield is still currently below 4 t/ha. Thirty-five percent oil-to-bunch content has also been reported (Soh et al. 2003a) with 22–28% being more common. Palm kernel oil has been estimated at 2–3% of bunch weight but can reach 5%. Corley (1985) predicted a potential oil yield based on physiological modelling of 17 t/ha/year, suggesting that significant breeding progress can still be made. However, current yield potential, i.e., achievable yield under ideal conditions, still has a significant way to go before approaching this theoretical potential. A mature palm tree initiates one inflorescence in the axil of each leaf and this develops over a period of 2–3 years. Palms usually produce one or two fruit bunches every month (in female phase) or one or two male inflorescences per month (in male phase) (Adam et al. 2005). A mature female bunch may easily weigh 50 kg and is composed of over 3,000 fruitlets each attached by an abscission zone (Henderson and Osborne 1990; Henderson et al. 2001) to the fibrous bunch. Interestingly, one of the main domestication adaptations of most crops, i.e., non-shattering fruit, here more appropriately termed non-shedding, has not yet been developed in oil palm. The current signal for harvesting a ripe bunch is the accumulation of loose fruit at the base of the tree. A single male inflorescence on a mature palm can weigh 2–3 kg and yield significant quantities of pollen. The pollen can be preserved by freeze-drying for future controlled crosses, potentially enabling a significant genetic contribution to commercial palms from pollen parents.

15.1.1 Economic, Agronomic, and Societal Importance of Oil Palm

While oil palm cultivation arose in Africa, where it still makes an important contribution to diet and industry, the countries most dependent upon oil palm economically are in Southeast Asia, particularly Malaysia and Indonesia, with Thailand and Papua New Guinea also having significant plantations. The rapid increase in plantation area in Malaysia from 300,000 ha in 1970 to 4.1 million ha in 2006, and a total area of over 5 million ha in Indonesia in 2006, indicates the economic importance of this plantation crop and the growing world demand for palm oil. Palm oil has recently overtaken soybean oil as the world's leading vegetable oil, with Europe and China at present the major markets. Production in African countries is now beginning to increase, although Southeast Asian production dominates world trade. Plantations in South American countries have also been established and are increasing output (Corley and Tinker 2003).

Palm oil's unique composition makes it versatile for applications in food manufacturing and in the chemical, cosmetic, and pharmaceutical industries, with palm oil often being separated into olein and stearin fractions before export (Pantzaris 1997).

About 90% of the world's palm oil is used for edible purposes (Sambanthamurthi et al. 2000). Palm oil generally has a 1:1 ratio of saturated to unsaturated fatty acids. Despite a concerted effort by US soybean farmers to label palm oil as a specific risk factor, nutritional studies show that a diet with a high proportion of palm oil as the fat component is as healthy as any other, particularly with regard to atherosclerosis risk factors and coronary heart disease (Corley and Tinker 2003). Palm oil is semi-solid at normal room temperature (22 °C), which favors its use as the solid-fat component for margarines. Many temperate vegetable oils require catalyst-based hydrogenation to increase the level of saturation and melting temperature before they can be used as a solid-fat component. In biological systems, single double-bonds are always in the

CIS configuration. Catalyst-based hydrogenation leads to trans-fatty acids, which have been implicated in cardiovascular disease. The recent decision of a number of major supermarkets in the UK to remove all trans-fatty acids from their products and the imposition of legislation in the United States requiring the labelling of products exceeding a threshold (0.5–1.0%) of trans fatty acids (US FDA, 2006) have provided an additional boost for the use of palm oil. It is also highly suitable for deep frying because of the low content of polyunsaturated linoleic acid and a higher level of saturated fatty acids (Sambanthamurthi et al. 2000), which are less susceptible to oxidation. Palm oil contains high levels of natural antioxidants such as tocopherols, tocotrienols, and carotenoids, although it is generally bleached during processing. Industrial extraction of these antioxidants for the manufacture of health supplements has started.

About 10% of palm oil is used for non-food products such as oleochemicals (e.g., Kuntom and Hamirin 2000), cosmetics, and, increasingly, biofuels (Chuah et al. 2006). It is also used in the metal and leather industries. Fatty acid methyl esters from palm oil can be used as substitutes for diesel (Choo and Cheah 2000), or alcohol can be produced by fermentation of carbohydrates (Corley and Tinker 2003) to provide fuel for burning. Demand from the biofuel sector has recently led the Malaysian and Indonesian governments to limit the amount of palm oil going into biofuels to 12 mt annually (http://www.oilworld.biz; Oil World Weekly, July 21st 2006) to protect the food-use supply of palm oil. Biodegradable plastics such as polyhydroxyburate (PHB) could also be produced from oil palm and to develop this application is one of the objectives of the Malaysia/Massachusetts Institute of Technology Biotechnology Partnership Programme (http://minihelix.mit.edu/malaysia/), by manipulating the flux of acyl groups from acetyl coenzyme A to polyhydroxyalkonates (Houmiel et al. 1999; Masani Mat Yunus et al. 2001).

Palm kernel is also widely used in luxury soaps and as a direct substitute for coconut in confectionary fats, ice cream, and coffee whiteners. Palm kernel meal, which is left over after kernel oil extraction, is used as livestock feed (Corley and Tinker 2003) although it is nutritionally poor and is often mixed with other cattle feeds (Soh et al. 2003a). In recent years there has also been considerable interest in the potential for developing palm material into additional by-products such as chipboard (Hassan and Sukaimi 1993), aggregate for concrete (Mannan et al. 2005), an additive for recycled paper production (Ibrahim 2003), and many others. On many plantations, the empty bunches remaining after fruit harvest are used as a source of nutrients and as a mulch for the plantation. The fibers and shells remaining after oil extraction of the fruit are used as a fuel to generate electrical energy for the plantation. After the oil has been extracted, the remaining palm oil mill effluent (POME) is potentially a rich fertilizer source and can be used to make biogas or a livestock feed. Recently, POME was shown to contain substantial quantities of phenolics and flavonoids that have potent antioxidant properties (Sundram et al. 2003). Technology currently being developed by the Malaysian Palm Oil Board (MPOB; http://www.mpob.gov.my/) is geared towards turning this material into a nutraceutical product that could have substantial health benefits.

15.1.2 Oil Palm as an Experimental Organism

As an experimental organism, oil palm has a number of features/traits to recommend it. It is long-lived (e.g., a study in 1970 to evaluate breeding progress was able to collect seed from unselected material planted in 1878; Corley and Lee 1992), has high individual value justifying the expense of molecular work, and a selected individual can produce significant numbers of progeny (around 1,500 per mature female bunch and millions of offspring for a male pollen contribution). Genetically, it has an intermediate genome size and behaves as a diploid. This makes molecular work reasonably straightforward. A further advantage of oil palm is the ability to generate clonal material through somatic embryogenesis (Soh et al. 2003a). Breeding populations of restricted origin (BPROs; Rosenquist 1985) form the basis of many breeding programs. These often show trait variation between origins, so introgression of different origins remains one way to make new breeding progress without completely sacrificing progress already made. Despite some 60 million years since the geographical speciation of *E. guineensis* (African oil palm) from *E. oleifera* (South American oil palm), probably with the break-up of Gondwanaland, fertile F_1 hybrids can be formed (Hardon 1969; Hardon 1969). This has been important in South America, where disease problems have had an effect on pure *E. guineensis* material (de Franqueville 2003) with the *E. oleifera* and *E. oleifera* x *E. guineensis* hybrids appearing to be less susceptible. Scientifically, material from hybrids is also of interest, as it is clear from molecular analysis that there has been significant genome divergence between the two species, particularly in repetitive sequences (Price et al. 2002, 2004; Kubis et al. 2003)

There are, however, many disadvantages to the use of oil palm as a model organism. The selection cycle is long and breeding trials take considerable areas of land, with 143 plants/ha being a common planting density. Genetically, oil palm exists as heterogeneous material. Few fully inbred lines are available and oil palm usually exhibits significant inbreeding depression if self-pollinated (Luyindula et al. 2005a). The lack of genetically homozygous inbred lines hampers genetic analysis and prevents stable propagation of elite genotypes by seed. Somatic embryogenesis does have the potential to fix elite genotypes (which will be heterozygous, rather than homozygous) and a number of groups are also working on the techniques to generate doubled haploid material (Madon et al. 2005b). This approach has been widely adopted in cereal genetics.

15.2 Genetics, Breeding, and Biotechnology

15.2.1 The Genetics and Breeding of Oil Palm

Oil palm breeding and research are dominated by two major factors –shell-thickness and long selection cycles.

The thickness of the fruit shell, or endocarp, is primarily controlled by two alleles of the shell thickness gene *Sh* (Beirnaert and Vanderweyen 1941). The homozygous, thick-shelled *dura* fruit type typically produces 30% less palm oil than does the heterozygous, thin-shelled, *tenera* fruit type. The other homozygote, *pisifera*, has no shell and in most germplasm the fruit bunches abort during development, resulting in no yield. It has been proposed that *pisifera* is the result of a mutation that fails to lignify the region in which the shell would normally form (Sparnaaji 1969; Bhasker and Mohankumar 2001). In *dura* fruit, two copies of the wild-type gene lead to a replacement of 30% of the mesocarp with shell, reducing oil yield by 30%, compared to *tenera* (heterozygous) fruit. There is an overlap of the range of shell-thickness for *dura* and *tenera*, with the definitive feature for classifying fruit forms into *dura* or *tenera* being the presence of a fiber ring around the shell in the *tenera* form. This overlap of shell-thickness ranges may suggest that different sources of alleles for *Sh* are not identical. It has also been suggested that other loci may carry modifiers of shell-thickness, including possibly maternally inherited genes (Okwuagwu and Okolo 1992, 1994). Sources of female-fertile *pisifera* exist and it has been postulated that a fertility gene is linked closely to the shell-thickness gene (Wonkyi-Appiah 1987).

Selection cycles are typically 10–12 years for *dura* (female parent) palms and 16 years for *pisifera* (male parent) palms, where sterility of *pisifera* requires sib-breeding and extensive progeny testing. This means that since the 1910s to 1920s, when palm breeding began in a systematic way, there have been perhaps only eight generations of breeding and selection. Despite the limited number of generations for improvement, oil yields quadrupled up to the 1990s in Malaysia, with half of this increase attributed to genetic improvement, with 30% of the gain from replacing the *dura* thick-shelled fruit form with the *tenera* thin-shelled form and the other half due to improved agronomy (Corley and Lee 1992). The majority of traits of agronomic importance are thought to be polygenic; however, a limited number of mono- or possibly oligogenic traits have been identified. The most important monogenic trait, as already mentioned, is shell-thickness. Other candidate traits include:

Fruit traits: *Nigrescens* (fruit black when unripe, ripening to reddish brown, the wild-type state); *Virescens* (fruit green when unripe, ripening to bright orange due to absence of carotenoids in the exocarp); *albescens* (mesocarp lacks carotenoids and remains pale when ripe); *genetic mantling* (*poissoni*; development of the rudimentary androecium in female fruits to produce supplementary carpels).

Vegetative traits: *idolatrica* (fusion of pinnae), *dumpy* (short height mutant), although simple inheritance is still controversial (Soh et al. 2003a; Luyindula et al. 2005b) and *crown disease* (twisting of the rachis in juvenile palms) (Blaak 1970; Breure and Soebagjo 1991).

Disease traits: Resistance to *Fusarium oxysporum* fsp elaedis has been postulated as being controlled by two genes, but also remains controversial (de Franqueville and de Greef 1987; Flood 2005).

Most other traits, including oil yield, are thought to be polygenic. Breaking down a complex trait, such as oil yield, into components, e.g., oil-to-mesocarp, mesocarp-to-fruit, fruit-to-bunch etc. (termed "bunch analysis"), is one common approach to try to reduce the oil yield trait to simpler, perhaps more heritable, components for breeding and selection (Blaak et al. 1963; Corley and Tinker 2003). Physiological approaches to oil palm improvement have been used to try to understand the components that make up final oil yield, examining the balance of the source and sink traits (Corley and Tinker 2003).

A major threat to yield is the recent significant increase in incidence and severity of the disease caused by the fungus *Ganoderma boninensis* which brings about basal and upper stem rot (Flood and Bridge 2000; Pilotti et al. 2003, Sanderson 2005; Susanto et al. 2005; Hasan et al. 2005). Oil palms attacked by this soil-borne basidiomycete must be removed to prevent the spread of the disease. The identification of sources of resistance and markers that segregate with such disease resistance genes would be a valuable tool in future breeding programs (Durand-Gasselin et al. 2005) and the development of markers specific to the disease causing *G. boninensis* is a useful first step for diagnostics (Latiffah et al. 2002; Panchal and Bridge 2005; Utomo et al. 2005).

The total phenotypic variation present for any trait can be partitioned into components, reflecting genetic, environmental, and genetic x environmental variances. A trait will respond to selection in a breeding program only if it is strongly transmitted to the next generation, i.e., the trait is largely under genetic control. Estimates of oil palm heritability and GxE interactions for particular traits provide information to the breeder for increased selection efficiency (Soh and Tan 1983; Rafii et al. 2001; Rafii et al. 2002; Soh et al. 2003a). Soh et al. (2003b) demonstrated various methods of estimating broad sense heritability. These concurred with earlier estimates that yield component trait heritabilities were generally low, probably due to their continual use as selection criteria.

The same approach can be used to derive breeding values for individual parental palms. Estimates of General Combining Abilities and Specific Combining Abilities essentially reflect additive and non-additive effects, respectively, together with an error component (Falconer 1989). Other approaches, such as the Best Linear Unbiased Predictor (BLUP) have potential to assist the evaluation process (Purba et al. 2001).

The domination by the single gene for shell-thickness (*Sh*) of the breeding of oil palm, which requires all commercial material to be of the heterozygous *tenera* (thin-shelled) form, has a significant effect on the approaches adopted. This and the long generation and selection times has led to the development of variations of the Recurrent Reciprocal Selection (RRS) or the Family and Individual Selection (FIS) systems being widely adopted. Many breeding programs involve the separate development of maternal and paternal germplasm, followed by test-crossing to evaluate palm quality and to select good parents from each pool to produce the commercial hybrid. Variations on this approach have been suggested (Durand-Gasselin et al. 1999). As an alternative to the RRS-based approach, an extreme FIS approach can lead to continual crossing to unrelated material, producing polyhybrids, as has been used extensively in Africa (Corley and Tinker 2003; Soh et al. 2003a). In practice, many breeding programs are a blend of approaches, with pure RRS being impossible because of female sterility in many *pisifera* sources. Soh (1999) and Soh et al. (2003a) have reviewed this in detail.

15.2.2 Biotechnology of Oil Palm

The potential of oil palm biotechnology has been reviewed a number of times (e.g. Rival et al. 2001; Rival et al. 2003) and one area where considerable progress (with one serious set-back) has been made is clonal propagation. As oil palm has only a single vegetative meristem, it is not possible to replicate elite genotypes by cuttings or graftings. The absence of fully inbred lines means that it is also not possible to commercially propagate elite individual oil palm genotypes by seed. These two constraints led to the development of a clonal propagation system based on somatic embryogenesis in a number of laboratories in the 1970s. The development of this approach and the predicted benefits of it have been extensively reviewed (see Corley and Tinker 2003). The discovery of abnormal flowering (Corley et al. 1986) in clonal material which led to bunch abortion and sterility was a significant set-back for the adoption of clonal palms, and considerable effort has gone into trying to understand the basis of this (Rival et al. 1998; Jaligot et al. 2000; Matthes et al. 2001; Syed Alwee 2001; Toruan-Mathius et al. 2001; Jaligot et al. 2003; Kubis et al. 2003; Jaligot et al. 2004; Morcillo et al. 2006). Hypomethylation of a specific (as yet unknown) gene or genes currently seems to be the most likely cause (Rival et al. 2000), probably acting through phytohormone levels (Jones 1998). Recent characterization of the oil palm equivalents (orthologues) of genes involved in floral patterning and subsequent inflorescence architecture in other species may accelerate its identification (Adam et al. 2006; Syed Alwee et al. 2006). Recent work has identified the risk factors involved (Eeuwens et al. 2002), and current clonal production with an emphasis on limited production from each embryogenic callus, while not completely eliminating the presence of abnormality, has reduced occurrence to economically viable levels. The development of suspension culture systems (Texeira et al. 1995; Alberlenc Bertossi et al. 1999; Soh et al. 2003a) for the production of somatic embryos and of cryopreservation techniques for their storage also offers the possibility of mass production of artificial seed (Dumet et al. 2000; Chaudhury and Malik 2004 Tarmizi et al. 2004). In the last 10 years there has been something of a "cautious Renaissance" in the commercial planting of clonal oil palm (Soh et al. 2001; Soh et al. 2003a). However, the development of a reliable marker for the abnormality is critical for high throughput oil palm clonal propagation; to provide an early warning system, and to give added confidence to growers.

In addition to somatic embryogenesis enabling the production of large numbers of elite genotypes, both for research and commercial exploitation, tissue culture methods have provided a basis for research into genetic transformation of oil palm. The first report of transient expression in oil palm tissues, using the biolistics approach, was presented at the International Oil Palm Congress (PIPOC) in 1993 (Mayes et al. 1995) and significant progress has been made since. Work on establishing conditions for transformation (Te-chato et al. 2002; Zubaidah and Siti Nor 2003; Rohani et al. 2003; Abdullah et al. 2005; Lee 2006) and evaluating potential for such technology (Parveez 2003; Siti Nor et al. 2001; Murphy 2006) together with the genes needed to achieve these ends (e.g., for oil biosynthesis, Shah and Cha 2000; Asemota and Shah 2004; for carotenoids; Khemvong and Suvachittanont 2005; for kernel expression, Cha and Shah 2001) have also been reported. In many species, transformation approaches based on *Agrobacterium tumefaciens* have been developed. Compared to the biolistics approaches, this method generates simpler and lower copy inserts in transformed plants. This is important for stability of transgene expression (instability of expression often being caused by the construct producing aberrant dsRNA). *Agrobacterium*-mediated transformation also produces unlinked inserts in a reasonable number of transformants, potentially facilitating the segregation of the selective marker away from the gene of interest – so-called clean gene technology (Afalobi et al. 2004). Development of an efficient *Agrobacterium* transformation system for oil palm is a key requirement for commercial deployment of this technology. Many Southeast Asian countries (such as Malaysia) are still in the process of developing biosafety systems and protocols for scientific and commercial use of genetically modified (GM) organisms. The commercial future of such work in oil palm will undoubtedly be influenced by this. Time scales are also a major concern, particularly to develop significant numbers of palms to make niche market transgenic palms viable. With modifications to oil composition, separate processing facilities will also be required to maintain the premium obtained from the modification. From induction of callus to production of field-planted palms producing a yield can take 8-10 years. Multiplication of material would take significant time, although introduction of GM by using the trait expressing palm as a pollen source is one option, as male inflorescences produce considerable amounts of pollen. Transgenic approaches would raise potential containment issues of the modified gene (as well as issues of safeguarding the research investment) and in some countries theft of material from nurseries is common. The targeting of transgenes to the plastid might resolve the issue of GM pollen release and might also improve the efficiency with which expression takes place. However, the potential to produce a high-oleate palm oil and other novel oils for nutraceutical or industrial markets (e.g., Murphy 2006) and the possibility of tackling some currently intractable problems (such as Ganoderma, for which a genetic/breeding solution is still a number of generations off) makes transgenic technology inviting (for current status in a number of crops see FAO, 2005).

For both scientific study and even for potentially fairly direct commercial exploitation, a robust and high-efficiency transformation system for oil palm is desirable.

15.3 Genomics and Gene Mapping

Plant genomes vary significantly in DNA content between species. A simple example would be to compare the genome size of rice (*Oryza sativa*; 430 Mb) with the genome size of barley (*Hordeum vulgare*; 5,000 Mb). Both species are diploids and both are members of the Poaceae, with the major difference being that rice is a tropical and barley a temperate grass species. It is also likely that to survive both species need similar numbers of genes. The difference in genome size between such similar species has been termed the C-Value Paradox (Thomas 1971). Recent work suggests that the majority of the difference in genome size between such species is due to repetitive DNA, particularly a class of DNA element called retrotransposons, which have the ability to make an RNA copy of themselves, convert it into DNA, and insert the new copy into the plant genome elsewhere, thus increasing the genome size (SanMiguel and Bennetzen 1998; Feshotte et al. 2002; Schulman and Kalendar 2005; Vicient et al. 2005). It has been estimated that in maize 70% of the genome consists of such elements and it seems highly likely that such elements represent a significant proportion – if not a majority – of the DNA in all crop species.

Genomics is primarily concerned with the DNA components of the crop (see Benfrey and Protopapas 2005 for an introduction). However, sequencing the entire genome of a species could be very wasteful if the majority of the DNA is not directly involved in specifying gene products, including the subset of the gene products which are involved in agronomic traits. Other techniques, such as molecular genetics, mapping, and QTL analysis (Section 15.3.1–15.3.4) are not significantly affected by genome size, with genetic distance being dependent upon the number of DNA crossovers during meiosis which is more a function of chromosome number than genome size. Polyploidy is common in crop plants and can significantly contribute to genome size and complexity. Luckily, oil palm is a diploid, although earlier genome duplication events cannot be ruled out. Functional genomics focuses on the expressed portions of the genome, such as mRNA, the protein products derived from it, and even the metabolites generated in cellular reactions (usually termed transcriptomics, proteomics, and metabolomics, respectively). This has the advantage that the expressed portions of the genome are being investigated, but the disadvantage that the genomic context in which the genes are expressed, including the sequences controlling gene expression, is lost. Bioinformatics underpins both of these approaches, permitting analysis and screening of extremely large information sets.

Three broad distinctions can be made for marker types; DNA-based versus non DNA-based; multiple banding (multilocus) versus locus-specific banding (single locus), and dominant (partially informative) versus co-dominant markers (fully informative).

Isozymes are an example of a non-DNA based marker system, where there may be tissue/developmental stage specificity (e.g., seed storage proteins). Most markers used are DNA-based and these are generally of two types. Multilocus systems have the advantage that prior knowledge of the genome sequence is not needed to use the marker system, making them generic systems (although this is not necessarily true for the copia-based systems), but the disadvantage is that they do not uniquely identify alleles associated with a single locus (i.e., they are dominant systems). Single locus markers require prior sequence information (or cloned fragments of genomic DNA for markers such as restriction fragment length polymorphisms, RFLPs), so are harder to establish and can be species specific. However, they are able to distinguish heterozygote from homozygote forms (co-dominant systems), making them more informative at each locus. Once co-dominant markers have been developed, it is often possible to multiplex a number of them in a single reaction, which increases the information obtained per reaction to similar levels seen with some dominant marker systems, but still retaining the discriminative power of the co-dominant marker system (see Collard et al. 2005; Masi et al. 2003). The future development of single nucleotide polymorphism (SNP) markers is important (see Section 15.3.5) and other technologies, such as Diversity Array Technology (DArT) markers (Jaccoud et al. 2001; Akbari et al. 2006) also have a potential part to play in oil palm.

Table 15.1 summarizes the current state in oil palm. These will be discussed in turn below, before taking a look at the current constraints and prospects for oil palm genomics in the future.

Table 15.1 Current Status of Genomics and Functional Genomics in Oil Palm

3. Genomics	Current State in Oil Palm	Comments
3.1 Fingerprinting	Multiple marker types available.	Can be used as routine quality control.
3.2 Diversity Analysis	Large numbers of SSRs available.	Large comparable dataset can be generated by using common SSRs.
3.3 Linkage Mapping	Map have been constructed using AFLPs, SSRs, RAPDs and retrotransposons markers.	Populations are generally based on heterozygous individuals, which complicates the genetic analysis.
3.4 QTL Analysis	QTLs have been identified for a number of traits, including kernel-to-fruit and height.	The lack of recombinant inbred lines (RILs) or doubled haploid (DH) populations means that a new map is required for each analysis.
3.5 SNP and 'Perfect' Markers	None.	Cloning and conversion of the shell-thickness gene to give a perfect marker is feasible
3.6 Mutation Stocks	None.	Spontaneous mutations could be screened for in commercial plantings (eg., non-abscissing fruit) or eco-TILLING is possible now.
3.7 BAC Libraries	Complete 'non-selective' BAC libraries have been developed and 'targeted' BACs tested.	This resource allows the use of map-based cloning to identify genes of agronomic interest, such as shell-thickness.
3.8 Physical Mapping and Genome Sequencing	A number of linkage maps exist and initial ESTs/sequence/BAC make this feasible. Methylation-filtration sequence has been generated.	Development of a physical map assists map-based cloning and (eventually) efficient genome sequencing. Methy-filtration sequence-should represent >80% of the coding DNA.
3.9 Genome Structure and Conserved Synteny	Molecular cytogenetics, characterization of repeat sequences and gene targeting of 'gene-rich islands' through methylation patterns. Some evidence for conserved synteny through the use of rice RFLPs.	Individual chromosomes have been identified, studied in meiosis and a retrotransposon-based marker system developed Conserved orthologous set (COS) markers offer a way to test this rigorously and in silico approaches may be possible.

Table 15.1 (continued)

4. Functional Genomics	**Current State in Oil Palm**	**Comments**
4.1 Expressed Sequence Tags (ESTs)	>30,000 ESTs have been reported.	As a comparison, wheat has > 1 million.
4.2 Transcriptomics	First 3,805 slide spot array reported.	For highest throughput oligo-based methods (Affymetrix, Agilent, Illumina) significantly more EST data is needed.
4.3 Proteomics	Isozymes have been used in oil palm for many years.	2D and high thoughput methods are possible, although without more sequence data, spot identification is difficult.
4.4 Metabolomics	Used to examine abnormal flowering and other traits.	Fragment database analysis is possible, as many metabolites are conserved across species.
5. Bioinformatics		
Within Oil Palm	Searchable EST database held at MPOB.	Access to significant sequence data is critical for future progress.
Across Species	Information from model and other crops species could facilitate studies on conserved gene order, cloning of oil palm homologues of known genes and testing their in planta effects in model species before their manipulation in oil palm.	The database and analytical tools already exist in crop and model species and could be applied to oil palm, once there is sufficient data. These approaches will be critical for work in oil palm and could alleviate the need to repeat what has already been done for other species.

15.3.1 Genetic Fingerprinting

There are very limited numbers of simple morphological markers present in commercial breeding material. Of the oligogenic traits listed above, shell-thickness is probably the only one which can be used to confirm the identity of planting material in a crossing program (Fig. 15.1). This is most effective when a single shell-type is expected, such as in a *dura* x *pisifera* cross, where all offspring should be of the *tenera* shell-type. The presence of *dura* or *pisifera* fruit within such a cross is evidence for some out-crossing or other error. Shell-thickness could be used when there is segregation of shell-type, e.g., *tenera* x *dura* which should give a 1:1 *tenera* to *dura* ratio, and deviations from this can be tested with a Chi-square test. However, as many breeding test crosses have low numbers of palms or small sample sizes, only major deviations are likely to be identified.

Initial genetic fingerprinting was carried out using isozymes and RFLP markers (Ghesquiére 1984, 1985; Baudouin 1992; Jack and Mayes 1993; Jack et al. 1995; Mayes et al. 1996). Genetic fingerprinting has been used successfully in a number of programs, as in the Unilever clonal propagation program at Unifield, Bedford, UK (Corley and Tinker 2003). Another example of the value of fingerprinting is

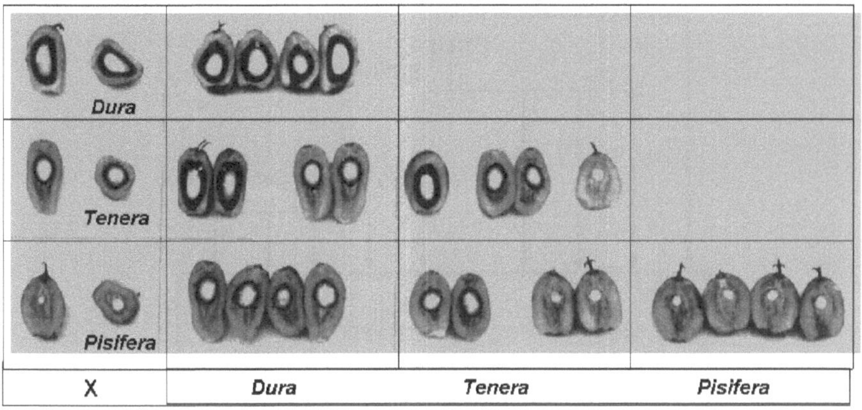

Fig. 15.1 Segregation for fruit shell-thickness. Parental palm types are given bottom and left. The offspring classes produced from these crosses are given in the intersection. Clear differences can be seen for certain crosses (e.g., in the *tenera* x *tenera* cross three shell classes can be found; these segregate 1:2:1, as expected for a single Mendelian gene controlling the trait. '*dura*' = thick-shelled, '*tenera*' = thin-shelled, '*pisifera*' = no shell.) All commercial planting material is of the hybrid '*tenera*' type, which has more mesocarp at final harvest than other fruit-types (See color insert)

provided from the Unilever inbreeding trial planted at Binga, Democratic Republic of Congo, in 1973. The palm Bg. 312/3 was self-pollinated to generate a population with an inbreeding coefficient of Fx = 0.56, reflecting the chance of two loci being identical by descent. The progeny and subsequent crosses were planted in numerous countries as part of a combined breeding program. Palms within this particular progeny (Bg. 143) were further selected due to their low levels of apparent inbreeding depression. Genetic fingerprints confirmed that a majority of the palms tested were actually out-crosses, including the parent of the first mapping cross (Mayes et al. 1996; Mayes et al. 1997; Corley 2005; Fig. 15.2).

The development of significant numbers of publicly available microsatellite, simple sequence repeat (SSRs) markers (Billotte et al. 1999; Billotte et al. 2001; Billotte et al. 2005) means that it is now possible to apply genetic fingerprinting at key points in both breeding and commercial palm production to monitor systems and to ensure quality control. The importance of such measures was recently considered by Corley (2005) and by Lim and Rao (2004). SSRs are probably the best available tool currently for routine fingerprinting via multiplexed pools, making the approach cheap and highly discriminative, although concerns were expressed by Lim and Rao (2005).

15.3.2 Diversity Analysis

Diversity analysis is a potentially powerful approach, particularly in species where early breeding records are missing or unclear (or in the case of oil palm, only started

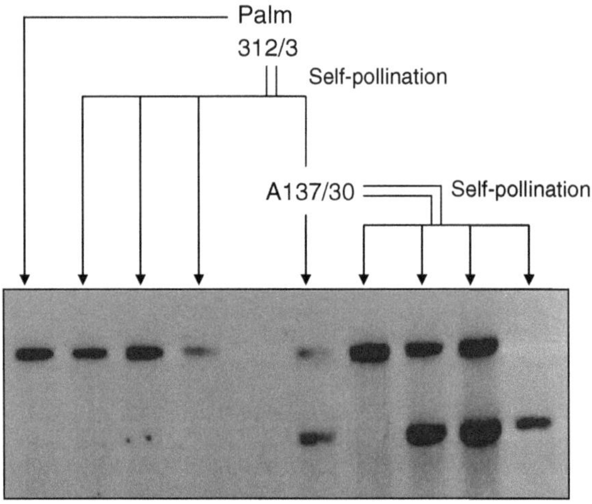

Fig. 15.2 Genetic fingerprinting in an inbreeding trial. The figure shows an RFLP analysis of palms derived from the self-pollination of palm 312/3. Many of the offspring were selected on the basis of their low apparent inbreeding depression. RFLP analysis confirmed that many of these were actually out-crosses, rather than true self-pollinations. A137/30, derived from the expected self-pollination of 312/3 is clearly shown to be an out-cross, although the subsequent self-pollination appears to be correct. A137/30 self-pollinated forms the basis for the genetic map constructed in Mayes et al. 1997.The probe is pOPg95 and restriction enzyme *Hin*dIII

systematically in 1920s) and in species which are natural out-breeders, where controlled crossing is not always completely effective.

Understanding the diversity within breeding material developed at different research stations and from different BPROs can potentially help to identify new elite material to introduce into the genetic base of commercial material. Extensive collections have been made of wild and semi-wild material, particularly by the Malaysian Palm Oil Board (MPOB, formerly PORIM), and how to use these collections to improve current commercial germplasm is an important concern (Lawrence et al. 1995; Rajanaidu et al. 2000; Mohd Din et al. 2005).While molecular markers do not remove the need to field-evaluate the traits of particular accessions, they do offer an additional level of information which potentially allows the minimum number of desirable accession palms to be introduced to capture the maximum amount of genetic diversity. The narrowness of the genetic base is an on-going concern in Southeast Asia, where the dominant contribution on the Malaysian/Indonesian *dura* side is believed to be derived from four palms planted in Indonesia's Bogor Botanic Garden in 1848. On the paternal side, the situation is significantly worse in many programs, where the palm SP540 and its derivatives are the most significant pollen source (Rosenquist 1985). A number of diversity analyses have been reported in the literature, including those based on (1) isozyme markers (Ghesquiére 1984, 1985; Baudouin 1992; Choong et al. 1996; Hayati et al. 2004), (2) RFLP markers (Mayes et al. 2000; Maizura et al. 2001; Moretzsohn et al. 2002; Maizura et al. 2006),

(3) amplified fragment length polymorphisms (AFLP) plus RFLP markers (Barcelos et al. 2002), (4) randomly amplified polymorphic DNA (RAPD) markers (Shah et al. 1994), (5) AFLP markers (Kularatne et al. 2001; Galeano 2005), and (6) AFLP plus isozyme markers (Purba et al. 2000).

The development of retrotransposon-based (Kumar and Bennetzen, 1999) marker systems, both generic (Rohde 1996; Kalendar and Schulman, 2006) and an oil palm specific copia-based system (Price et al. 2004) have significant potential for phylogenetic reconstruction, with retrotranposon movement being an essentially irreversible process. The recent development of large numbers of public SSRs offers a significant boost to attempts to understand genetic diversity in both wild and BPRO material.

15.3.3 Linkage Mapping

The first reasonably complete genetic linkage map was reported in the literature by Mayes et al. (1997). It was really with the development of the RFLP technique that sufficient markers became available to produce reasonable genome coverage. The study also identified an RFLP marker which was 9 cM from the shell-thickness gene (*Sh*). Further linkage studies have been undertaken using RAPD (Moretzsohn et al., 2000; identifying two further markers to shell-thickness at 23.9 cM and 17.5 cM) and AFLP (Chua et al. 2001; developing an AFLP map of 20 linkage groups.) The nearest marker to shell-thickness was an AFLP marker mapped at 4.7 cM by Billotte et al. (2005). The development of markers close to the shell-thickness gene, preferably flanking markers, would allow shell thickness to be predicted before field planting. This would have relevance for breeding trials, allowing potentially female sterile *pisifera* palms to be planted separately from other shell-types (as they can exhibit excessive vegetative vigor) or even allowing a robust quality control test of *tenera* seed lots for commercial production. The recent publication of a comprehensive genetic map combining SSR and AFLP markers by Billotte et al. (2005) represents a significant step forward for oil palm molecular genetics (Fig. 15.3). A number of the oligogenic traits in oil palm have also been tentatively placed onto the genetic linkage maps in addition to shell-thickness, including *virescens* and a component of crown disease (Jack et al. 1998). The existence of a good genetic map is an important prerequisite for the development of a physical map. The publication by Billotte et al. (2005) and the recent development of a comprehensive map by MPOB, including SNP markers, mean that these resources are now available.

15.3.4 QTL Analysis

The majority of traits of commercial importance are thought to be polygenic and often have significant environmental interactions. Estimating heritabilities in well designed experiments can assist in understanding how significant the genotype (G)

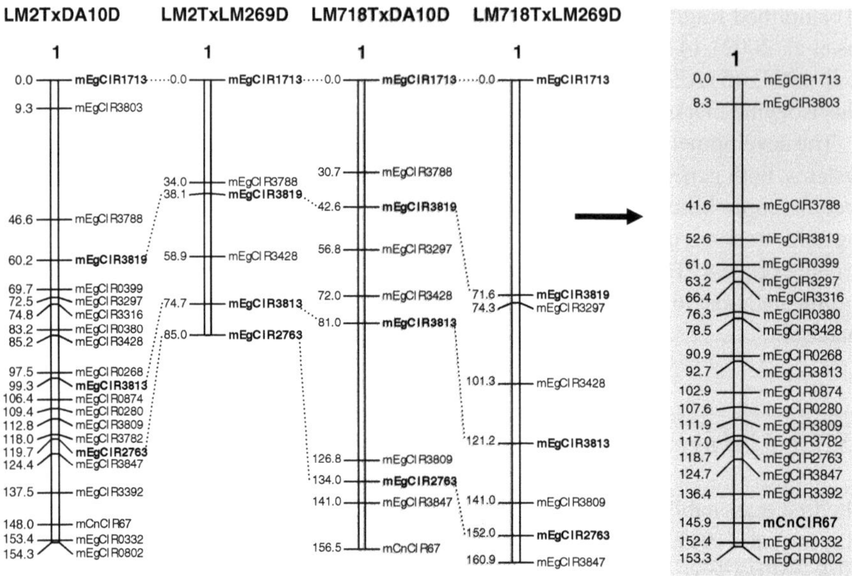

Fig. 15.3 Integration of microsatellite data from the four crosses
From the LINK2PALM program for Linkage Group 1 (Billotte et al. 2006)

effect is compared with the environmental (E) influence. However, even the use of bunch analysis to simplify the individual components of oil yield has still not produced clearly identifiable major gene effects beyond those already identified. This may be because genetic analysis is complicated by significant G x E interactions masking either a relatively simple mode of inheritance, pleiotropic effects, genuinely polygenic inheritance, epistasis, dominance, or disequilibrium effects.

One approach to try to resolve this problem is through quantitative trait locus (QTL) analysis. (Cochard et al. 2005) There is currently one QTL analysis reported in the literature and data available for a second (LINK2PALM - http://www2.mpizkoeln.mpg.de/~rohde/link2palm.html). Rance et al. (2001) used an update of the map developed by Mayes et al. (1997) to carryout a QTL analysis. This analysis identified QTL for components of 11 traits studied, explaining from 8.2% to 44% of the phenotypic variation observed within the cross for that trait (Fig. 15.4). The larger effects may represent major gene effects which could be used for marker-assisted breeding. The LINK2PALM mapping crosses were identified specifically for the LINK2PALM project and initial data on QTL for height increment are available. As the population matures, the linked series of crosses should yield significant QTL data.

The figure illustrates a partial QTL analysis for the bunch analysis component traits mesocarp-to-fruit (MF) and kernel-to-fruit (KF). The y-axis gives likelihood of difference (LOD) score and the x-axis gives centimorgan (cM) position along linkage group 11 (see Rance et al. 2001). The MF and KF lines indicate the probability that a gene(s) explaining part of the variation for that trait exists at each point

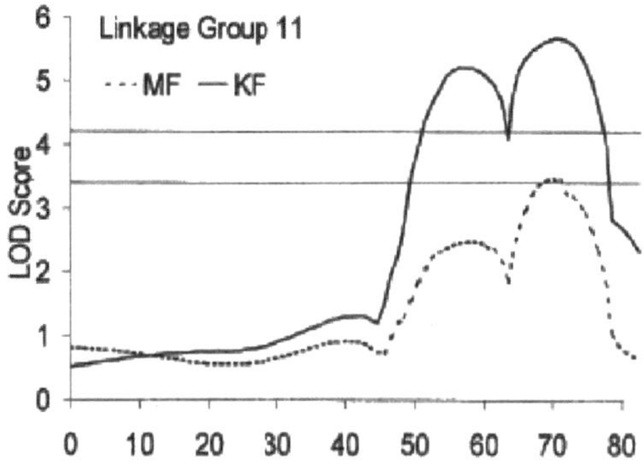

Fig. 15.4 QTL analysis in an oil palm breeding population

along the linkage group (from 0 cM to 80 cM). When the LOD score exceeds a predetermined threshold, the presence of a gene(s) is inferred (genome-wide empirical significant thresholds; $P < 0.01$ LOD 4.2 and $P < 0.05$ LOD 3.4.). Effects can be seen for both MF and KF, with MF reaching significance at $P < 0.01$. The diagram shows the magnitude of effects; in fact the MF and KF effects are in opposite directions, as would be expected for two negatively correlated traits.

The major disadvantage of QTL analysis is that it detects only QTL that are segregating in a particular cross. Whether the genetic markers are fully informative also has an effect on the completeness of the genetic data for comparison to trait data. However, the advantages are also significant. When the QTL analysis is carried out in material of direct commercial or breeding interest, there is the potential to select for the presence of a QTL using flanking DNA markers in subsequent breeding. An important target for future breeding programs could be the generation of a high oleic acid genotype (Murphy 2007). Efforts are currently being focused on the development and exploitation of molecular markers to facilitate the screening of African *E. guineensis* and South American *E. oleifera* germplasm for such a trait. Marker-assisted breeding would be one way to speed up the development of elite germplasm with this desirable trait.

15.3.5 Developing SNP and "Perfect" Markers for Traits

Molecular genetics and, more generally, biotechnology do not offer a panacea for oil palm breeding, or for the breeding of any crop, but they do represent a significant tool kit for improving the efficiency with which breeding progress can be achieved. So far, the limited expressed sequence tag (EST) and genomic sequence resources available for oil palm have made the production of single nucleotide polymorphism (SNP) markers slow. SNP systems, using various methods to detect differences, can potentially distinguish individuals by DNA differences as small as a single base pair and so are potentially extremely powerful, once developed. They can also be very highly multiplexed, either in solution or on a chip format. However, the ideal situation would be to clone the gene known to be responsible for a particular trait and to directly test for the different allelic forms in the offspring, thus predicting the phenotype of the palm before expression of that gene. This is now starting to become a reality in a number of crops. An example would be recent progress in wheat (*Triticum aestivum* L.), perhaps one of the most difficult genomes to work in, given that bread wheat is a hexaploid and has a genome 6x human in size. Over the last five years, genes responsible for vernalisation, (*Vrn1* and *Vrn2*), initially cloned in the diploid ancestor wheat, *T. monococcum* (Yan et al. 2003, 2004; Fu et al. 2005), photoperiod response (*PpdH1*) from barley (Turner et al. 2005) and from wheat (*Ppd-D1*), and plant height (*Rht1* and *Rht2*) – (Ellis et al. 2002) have been cloned. Marker tests for these genes are becoming available. These allow, for example, offspring to be screened for spring/winter habit (vernalisation and photoperiod response). More of these gene tests will become available in the next five years in wheat and such a tool kit will be invaluable for focused genetic dissection of traits. The potential value of gene tests for important agronomic traits in oil palm, particularly shell-thickness, has long been recognized, but the necessary molecular resources to achieve these ends have shown limited progress.

15.3.6 Mutation Stocks

In many species, mutation stocks have been developed in an attempt to correlate genotype with phenotype. In *Arabidopsis thaliana*, for example, lines exist which contain T-DNA insertions in the majority of the known genes (www.arabidopsis.info/). These lines allow the effects of inactivating a single gene on phenotype to be examined. For most crops species, such approaches are less feasible, although Tos17 in rice and Mu elements in maize are examples where this approach does work well (Hanley et al. 2000; Yamazaki et al. 2001), and often mutations are generated by a range of physical and chemical agents. One of the first approaches developed was the use of hard radiation, such as neutron or alpha particle mutagenesis which can generate significant deletions/rearragements (from a few bases to megabases). A more recent development of this approach is the production of Targeting Induced Local Lesions in Genomes (TILLING systems (McCallum et al. 2000; Till et al. 2003) which generate mutations through the use of chemical agents, such as ethylene methyl sulphonate (EMS). This introduces methyl groups into the DNA and leads to largely single base-pair changes which may alter amino acid codons in gene sequences and even produce a stop-codon, truncating the protein. Such lines are screened through a pooling strategy using a gel denaturing system, where specific gene primers are used and any changes in those genes identified through a change in DNA mobility of the PCR product within those pools. Such systems are well established for many cereals, e.g., for barley in the UK (http://germinate.scri.sari.ac.uk/barley/mutants/).

Another way in which mutated lines could be generated would be through tissue culture. Reports in rice using actin RFLPs to monitor changes in the actin genes during tissue culture suggested significant rearrangements and changes in ploidy number. The ability to trace oil palm genotypes from ortet through tissue culture to regenerated ramets (Mayes et al. 1996) would argue for significant stability of the oil palm genome, and considerable effort has gone into trying to identify DNA markers for tissue culture-induced influorescence abnormality without success (Rival et al. 1997, 1998a; Jaligot et al. 2004). That somaclonal variation does exist in oil palm is clear, from the experience of abnormal flowering in clonal material. However, the consensus is that this is an epigenetic phenomenon, i.e., it does not alter the primary DNA sequence.

Any of these approaches for generating random mutations would be highly valuable in oil palm, but are likely to prove difficult to exploit in the near future. At conventional planting densities of 143/ha, such populations would require significant areas of land. Multiple mutations are likely to be present, particularly with EMS, and the long generation cycle for oil palm makes confirming that the phenotype segregates with the identified genome location time consuming, requiring a backcross and a self-pollination for a recessive trait. An alternative is to try to exploit the natural variation that exists from the far lower levels of spontaneous mutation, by asking harvesters on commercial oil palm to report trees with particular phenotypes, such as non-abscising fruit. An Eco-TILLING approach would also be possible (Comai et al. 2004). This uses the TILLING approach with bulks of samples from palms showing natural variation/mutation. Significant numbers of palms could be screened for nucleotide variation in specific genes of interest. This may work well in oil palm, where recessive mutations are likely to be hidden in the heterozygous genotypes and observed inbreeding depression arguing for the presence of mutant alleles.

15.3.7 Developing BAC Libraries for Oil Palm

The ability to maintain large fragments of a plant genome artificially is one of the key steps to developing genomic resources for a species. One system is the cosmid system which has been used for oil palm (Sniady et al. 2003), another is the bacterial artificial chromosome (BAC) system. BAC vectors are capable of cloning and stably maintaining DNA fragments of up to 300 kb in *Escherichia coli*, exhibit low levels of chimerism, and are amenable to high-throughput and economical DNA purification, and thus, large-scale genome sequencing (Woo et al. 1994; Frengen et al. 1999). For oil palm, with a haploid genome size of 980 Mb (nearly 8x arabidopsis and over 2x rice), an average BAC insert size of 120 kb would require 8,000 bacterial clones to represent one haploid equivalent of the genome. Having this basic resource is critical for genomic approaches.

Piffanelli et al. (2002) reported the first BAC library of oil palm constructed using *Hin*dIII endonuclease; nine tropical crop libraries were constructed in the BACTROP program (http://www.intl-pag.org/10/abstracts/ PAGX_P96.html). Preliminary attempts have also been reported to construct an oil palm BAC library in Malaysia (Singh et al. 2003). The potential for using methylation-sensitive restriction endonucleases to construct BACs targeting the lower-copy and coding regions has also been investigated with some success (Hafeez 2005). Initial work with *Mlu*I, which shows methylation sensistivity and will not cut DNA containing methyl-groups on Cytosine residues, suggested that BAC libraries constructed by this approach would be able to span repeat regions, would have ends enriched for coding sequence, and could even completely eliminate some classes of retrotransposons (Hafeez 2005).

Having an arrayed and searchable BAC resource for oil palm is particularly important for map-based cloning approaches. One major concern with such approaches is the lack of consistent relationship between genetic distance (cM) and physical distance (Mb) along chromosomes. In many species, centromeric regions of the chromosome are highly suppressed for recombination, which could make cloning genes in these regions by map-based cloning difficult, if not impossible. Initial map positions for shell-thickness, however, suggest that it is distal on the chromosome (Billotte et al. 2005), making such an attempt possible. Some of the issues with map-based cloning are reviewed in Salvi and Tuberosa (2005).

15.3.8 Physical Mapping and Genome Sequencing

Eventually, the ideal situation would be to have the entire genome of oil palm sequenced. To date, only two plant genomes have been completely sequenced, arabidopsis (*Arabidopsis thaliana)* and rice (*Oryza sativa*). Many others, including barley (*Hordeum vulgare*) and maize (*Zea mays*) are underway and even the sequencing of hexaploid wheat (*Triticum aestivum*) is being seriously planned. Oil palm is a medium-sized genome and the development of the complete sequence is feasible within the next five years, given the political will. One staging post to doing this in an ordered and coherent fashion is the development of a physical map. This essentially involves combining genetic mapping data with the position of physical clones, such as BACs. Once this has been done, it is possible to choose and sequence individual clones in the most efficient way to produce a contiguous genome sequence. BAC or cosmid contigs can be generated using a variety of approaches (Klein et al. 2000; Tao et al. 2001; Sniady et al. 2003) and linked into the genetic map.

A comprehensive program has just been completed to create methylation-filtration sequence data for oil palm. The methodology involves creation of a representative genomic library in a bacterial strain which does not tolerate methylation of DNA bases. This leads to propagation of clones that come from non-methylated regions of the genome, which are expected to be the low-copy and coding regions. This approach has been used very successfully in maize (Palmer et al. 2003; Whitelaw et al. 2003; Rabinowicz 2003), although a similar approach in wheat produced lower levels of enrichment for coding sequences. The only real disadvantage of this technique is that methylation is so common in plant species that the clone sizes, i.e., the continuous sequence obtained, is limited, usually averaging in the 1–2 kbp range.

Methyl filtration sequence data represents an important resource for oil palm genomics and should assist the development of physical maps of the coding regions. Such sequence and clones can be used as one form of "marker" which links together the genetic and physical maps, as well as providing detailed sequence information for the species of interest. Clearly, a complete physical map would also enhance map-based walking approaches, by allowing a series of markers to be developed and tested across the region of interest, without the step-by-step BAC walking. The development of a detailed genetic map, ordered BAC libraries, EST resources, and methylation filtration sequence information for oil palm makes the development of a physical map for the oil palm genome now feasible (http://minihelix.mit.edu/malaysia).

15.3.9 Genome Structure and Conserved Synteny

Another approach which can also assist in this process, while generating further information on how the genome of oil palm is organized, is molecular cytogenetics. Work in Heslop-Harrison's lab in Leicester University (UK) has pioneered recent attempts to visualize the oil palm genome (Castilho et al. 2000; Kubis et al. 2003) using genomic in situ hybridization (GISH) and fluorescent in situ hybridization (FISH) techniques. A number of repeat sequences used as probes, such as copia-like retrotransposon sequences and SSR repeat motifs, have permitted the chromosomes of oil palm to be distinguished and a comparison of repeat sequence distribution made. This has recently been extended to examine oil palm chromosomes during meiosis (Madon et al. 2005a). Such examination would be a particularly useful selection tool for introgression of desirable*E. oleifera* genes into advanced *E. guineensis* germplasm.

Retrotransposons have already been shown to be important components in the genome of oil palm (Price et al. 2002; Kubis et al. 2003) with evidence for divergent evolution of different classes of these sequences since the split between *E. guineensis* and *E. oleifera*. For example, hybridization and sequence analysis with cloned elements suggested that one class of high-copy-number LTRs (long terminal repeats) are retrotransposon, representing approximately 10% of the *E. guineensis* DNA, may even be absent from the *E. oleifera* genome (Price et al. 2002).

Conserved synteny, i.e., conservation of gene order between related species, has been identified in many taxonomic families. At its most advanced, conserved synteny allows related species to be considered as being a complex of related genetic or physical maps, with the presence of a gene for a particular trait, e.g. height, shattering, grain color, in one species being used to imply that an orthologous gene should exist at the syntenic position in another species.

Initial work in oil palm (Hafeez 2005) suggests that there may be evidence for some conservation of gene order between oil palm and rice. However, RFLP approaches are slow and cumbersome and the recent application of conserved orthologous set (COS) (Fulton et al. 2002) approaches offer significantly more potential. In essence, conserved primers are made for a particular gene from a range of species and polymerase chain reaction (PCR) used to isolate the gene and detect polymorphism within a mapping cross. As available sequence increases in all species, much of this initial work may be possible through in silico approaches.

15.4 Functional Genomics

Functional genomics in oil palm has focused on developing and analyzing EST collections and using this information to explore the potential for using microarrays for transcript analysis.

15.4.1 Expressed Sequence Tag (EST) Development

A recent MPOB-MIT collaborative program has developed the first significant numbers of ESTs for oil palm. Six thousand five hundred are currently available on the MPOB site in a searchable database (Rajinder et al. 2001), a further 2,411 have been reported by CIRAD-CP (Jouannic et al. 2005), and about 20,000 have been generated from an oil palm tissue culture cDNA library (Elyana et al. 2005). These have been used to generate a microarray chip containing 3,806 oil palm gene clones (Low et al. 2006). This is an important step on the way to comprehensive transcriptomics resources.

15.4.2 Transcriptomics

The original technology used to examine patterns of gene expression in a tissue was northern analysis, where radiolabelled probes were hybridized to total RNA, or mRNA, which had been size-separated and immobilized onto a filter (Sambrooke and Russell 2000). A few dozen samples could be interrogated with a few genes at a time. Development of a 3,806 clone array spotted onto a slide (Low et al. 2006) allows a thousand times more genes to be evaluated by labeling the mRNA itself, having an immobilized gene probe. Initially, this was used to compare normal and abnormal tissue culture material, but it has important potential for other applications. In the longer term, the development of comprehensive oligo-based chips will be feasible (e.g. microarrays from Affymetrix (www.affymetrix. com/); Agilent (www.agilent.com/); Illumina (www.illumina.com/)).

15.4.3 Proteomics

For oil palm, the basic separation methods have been available for many years; e.g., isozymes (Baudouin 1992), and there is little to prevent the high throughput approaches being optimized. If this can be coupled with using cross-species information, progress can be made in a short space of time.

15.4.4 Metabolomics

While metabolites are not directly genetically determined, they are a reflection of cellular activity and in many families are directly responsible for regulation of developmental responses as well as cellular metabolism. An added advantage of metabolomics is that many metabolites are the same in many species, so identification of metabolites through a system such as HPLC coupled to mass spectrography can use existing databases to screen for structures. While gene sequence, structure and action may all vary in their detail between species, the final products or messengers, in terms of metabolites may be identical. This represets another level at which oil palm can investigated, even in the absence of detailed sequence information. The potential applications of this technology were recently review in Dettmer et al. (2007).

15.5 Bioinformatics and Information from Other Plant Systems

15.5.1 Bioinformatics

Bioinformatics is critical for the successful application of genomics and functional genomics technologies. This is literally true in the sense that it is impossible to analyze the amounts of data generated in a coherent fashion without dedicated tools and in the more general sense that cross species comparisons offer an extremely powerful approach for working in crops where resources and information are more limited. For plants, the most information and resources rich species is the model dicot arabidopsis. The entire sequence is available and held in a particularly intuitive way in AtEnsembl, with other information, such as gene models, ESTs, predicted protein sequences, T-DNA insertion lines, transcriptomics probe positions and Brassica BACs and sequence anchored on the primary sequence (www.arabidopsis.info/). This approach is slowly being developed to include other species data (e.g., Gramene Ensembl, based around the complete rice sequence; http://www.gramene.org/genome_browser/index.html) and it is diffcult to overstate how powerful such a framework is. Whether rice will prove close enough to act as the anchor sequence for oil palm remains to be seen, but the detailed methylation filtration sequence that has been generated in oil palm will allow this to be tested. Some of the ways in which bioinformatics may develop are discussed in Mayes et al. (2005b) and there are a number of excellent reviews and text books available (e.g., Pevsner 2003).

Beyond this, we do not intend to discuss bioinformatics here. It would be true to say that the tools immediately needed for oil palm bioinformatics tools already exist in other crop and model species and could be easily applied to oil palm – the limitation in oil palm is currently the amount of sequence-based data, not the tools to handle it.

15.5.2 Validation and Characterization of Oil Palm Genes in Other Species

While the evolutionary time between the common ancestor of arabidopsis and oil palm makes the use of Arabidopsis as a genome-model for oil palm very weak, both arabidopsis and rice could be used as gene-models for oil palm. The additional step of characterizing oil palm gene function or spatial and temporal expression patterns in transgenic arabidopsis and other model systems is also an attractive option. In particular, the existence of T-DNA insertion-inactivation lines for most arabidopsis genes allows the testing of complementation of arabidopsis gene function by oil palm genes, permitting the effects of the gene to be trialed in a model species, before the significant time and resource investment needed to repeat this in oil palm. By focusing on genes of interest in oil palm and provisionally assessing them in arabidopsis, many false leads could be avoided.

15.6 Perspective

In 2005, oil palm overtook soybean as the major supplier of plant oils. The ability to yield on an annual basis significantly more oil per year per hectare than other oil crop is critical in increasing food oil supply in the coming decades in the absence of more cultivatable land. Increasing yields on land already under cultivation is also important to help to protect native forests. Sustainable production of palm oil is a serious issue and is receiving an even higher profile than previously (e.g., ChanKook 2005; Mohd Basri et al. 2005; Cochard et al. 2005) with the Roundtable on Sustainable Oil Palm Production dedicated to furthering responsible production of palm oil within and beyond the industry (http://www.rspo.org/). This will be an important area of contention for the foreseeable future. The fact that oil palm is a tree crop offers some distinct advantages, but also a number of clear disadvantages. In particular the timescale for a breeding cycle of between 12 and 16 years means that genomics-assisted breeding through the development of direct markers for breeding, an understanding of traits and the production of transgenic material has possibly more potential in oil palm than almost any other crop. It is also significantly more difficult to apply to its full potential, for the same reasons. Integration across disciplines is still required for the full exploitation of genomics as they develop and exactly how this can be done is being tackled in a number of species (e.g., Mayes et al. 2005a; Tuberosa and Salvi 2006).

Oil palm, coconut palm, and date palm are all related species and, as has been shown by the LINK2PALM program, work in at least oil palm and coconut palm is mutually beneficial and may shed light on a number of common traits and genomic regions within palms. Ganoderma, in particular, is believed to have spread from coconut to oil palm and work in both crops may prove synergistic (Abdullah 2000; Flood and Bridge 2000).

More broadly, cross-species information has already been used to assist in the cloning and characterization of a number of genes (e.g., MADS-box; Syed Alwee et al. 2006) and the two way exchange of information between oil palm and model systems is an important possible additional level of characterization. Initial use of the Affymetrix *Arabidopsis* ATH1 chip for work in species where such resources are unlikely ever to exist (e.g., *Thalaspi* species, Hammond et al. 2005, 2006; bambara groundnut (*Vigna subterranea*- May and Mayes, unpublished data)- Xspecies; http://arabidopsis.info/ is promising and a similar approach may be possible in oil palm, until full genome dedicated chips become available. Genomic hybridization of oil palm to available high density array chips also offers the potential to identify primers from the individual chip features which could be used to develop COS markers (Fulton et al. 2002). For highly conserved genes, this could be instrumental in linking conserved regions of the oil palm genome with models such as rice and arabidopsis. With the development of the methylation-filtration sequence for oil palm, a great deal of this process of forming linkages could initially be done in silico with COS, BACs, and other markers being used to physically relate these islands to the chromosomes and genetic maps of oil palm. Such bridges could also be the basis for integrating oil palm into the sequence anchored bioinformatics resources which currently exist, such as AtEnsembl and Gramene Ensembl. This would have major implications for the use of cross-species data in oil palm. The resources available in oil palm, derived from the DNA and the expressed portions of the genome, are now approaching levels where an attempt to clone the shell-thickness gene would have a good chance of success. Given the importance of this gene in crop production and breeding for oil palm the development of a Perfect marker would be worthwhile and should perhaps be considered as the prime target for genomics work in oil palm.

The development of genomics resources in oil palm will provide palm breeders with an additional and very powerful tool kit to improve breeding and to develop alternative breeding avenues to pursue, both conventionally and through genetic modification. Progress over the last decade has been impressive, and if progress is continued at this rate in the next decade, oil palm will reach its potential as the key oil crop for the future.

There are two essential developments needed in oil palm genomics and functional genomics today; first, the development of significant sequence-based resources, second, the use of these sequence resources to integrate oil palm research into the substantial informational and physical resources that already exist in other species. A few decades ago, crop researchers worked in their own species with limited reference to what was happening elsewhere. Bioinformatics, genomics, and functional genomics have developed to the point where most researchers today will at the very least use information from other species as a guide to work in the crop of interest, if not as a surrogate. For oil palm, this could mitigate the difficulties of working in this species, while still allowing the particular strengths of oil palm to be exploited.

Acknowledgments The authors would like to acknowledge the valuable comments made by both reviewers in the development of this article. Thanks, also, to Dr Susan Liddell and Dr Rob Linforth for suggesting initial references for proteomics and metabolomics.

References

Abdullah F (2000) Sequential and spatial mapping of the incidence of basal stem rot of oil palms (*Elaeis guineensis*) on a former coconut (*Cocos nuciifera*) plantation. pp 183–194. In: Flood J, Bridge PD, Holderness M (eds) Ganoderma disease of perenial crops. CABI international, Wallingford, UK

Abdullah R, Zainal A, Heng WY, Li LC, Beng YC, et al. (2005) Immature embryo: A useful tool for oil palm (*Elaeis guineensis* Jacq) genetic transformation studies. Electronic J Biotech 8(1):25–34

Adam H, Jouannic S, Escoute J, Duval Y, Verdeil J-T, et al. (2005) Reproductive developmental complexity in African oil palm (*Elaeis guineensis* Arecaceae). Am J Bot 92(11):1836–1852

Adam H, Jouannic S, Morcillo S, Richard F, Duval Y, et al. (2006) MADS box genes in oil palm (*Elaeis guineensis*): Patterns in the evolution of the SQUAMOSA, DEFICIENS, GLOBOSA, AGAMOUS, and SEPALLATA subfamilies. J Mol Evol 62(1):15–31

Afolabi AS, Worland B, Snape JW, Vain P (2004) A large-scale study of rice plants transformed with different T-DNAs provides new insights into locus composition and T-DNA linkage configurations. Theor Appl Genet 109:815–826

Akbari M, Wenzl P, Caig V, Carling J, Xia L, et al. (2006) Diveristy Arrays Technology (DArT) for high-throughput profiling of the hexaploid wheat genome. Theor Appl Genet 113:1409–1420

Alerlenc Bertossi F, Noirot M, Duval Y (1999) BA enhances the germination of oil palm somatic embryos derived from embryogenic suspension cultures. Plant Cell Tissue Organ Cult 56(1):53–57

Asemota O, Shah FH (2004) Detection of mesocarp oleoyl-thioesterase gene of the South American oil palm *Elaeis oleifera* by reverse transcriptase polymerase chain reaction. Afr J Biotech 3(11):595–598

Barcelos E, Amblard P, Berthaud J, Seguin M (2002) Genetic diversity and relationship in American and African oil palm as revealed by RFLP and AFLP molecular markers. Pesquisa Agropecuaria Brasileira 37(8):1105–1114

Baudouin L (1992) Use of molecular markers for oil palm breeding. I. Protein markers. Oleagineux 47:681–691

Beinaret A, Vanderweyen R (1941) Contribution à l'étude génétique et biométrique des variétés d'*Elaeis guineensis* Jacq. Publs INEAC Sér Sci 27

Benfrey PN, Protopapas AD (2005) Genomics. Pearson Education Ltd, London ISBN 0-13-047019-8

Bennett MD, Leitch IJ (2005) Plant DNA C-values database (release 4.0) http://www.rbgkew.org.uk/cval/homepage.html

Bennett MD, Smith JB (1991) Nuclear DNA amounts in angiosperms. Philo Trans Roy Soc London B 334:309–345

Bhasker S, Mohankumar C (2001) Association of lignifying enzymes in shell synthesis of oil palm fruit (*Elaeis guineensis - dura* variety). Indian J Exp Bio 39(2):160–164

Billotte N, Lagoda PJL, Risterucci A-M, Baurens F-C (1999) Microsatellite-enriched libraries: applied methodology for the development of SSR markers in tropical crops. Fruits 54: 277–288

Billotte N, Risterucci AM, Barcelos E, Noyer JL, Amblard P, et al. (2001) Development, characterisation, and across-taxa utility of oil palm (*Elaeis guineensis* Jacq) microsatellite markers. Genome 44:413–425
Billotte N, Marseillac N, Risterucci AM, Adon B, Brottier P, et al. (2005) Microsatellite-based high density linkage map in oil palm (*Elaeis guineensis* Jacq). Theor Appl Genet 110(4):754–65
Billotte N, Amblard P, Durand-Gasselin T, Flori A, Nouy B, et al. (2006) Oil palm biotechnology at CIRAD. Proc Intl Oil Palm Conf 15:19–22
Blaak G (1970) Epistasis for crown disease in the oil palm (*Elaeis guineensis* Jacq) Euphytica 19(1):22–24
Blaak G, Sparnaaji LD, Menendez T (1963) Breeding and inheritance in oil palm (*E. guineensis* Jacq) Part II. Methods of bunch quality analysis. J W Afr Inst Oil Palm Res 4:146–155
Brenner S, Johnson M, Bridgham J, Golda G, Lloyd DH, et al. (2000) Gene expression analysis by massively parallel signature sequencing (MPSS) on microbead arrays. Nature Biotechnol 18:630–634
Breure CJ, Soebagjo FX (1991) Factors associated with occurrence of crown disease in oil palm (*Elaeis guineensis* Jacq) and its effect on growth and yield. Euphytica 54(1):55–64
Castilho A, Vershinin A, Heslop-Harrison JS (2000) Repetitive DNA and the chromosomes in the genome of oil palm (*Elaeis guineensis*). Ann Bot 85:837–844
Cha TS, Shah FH (2001) Kernel-specific cDNA clones encoding three different isoforms of seed storage protein glutelin from oil palm *Elaeis guineensis*. Plant Sci 160(5):913–923
ChanKook W (2005) Best-developed practices and sustainable development of the oil palm industry. J Oil Palm Res 17:124–135
Chaudhury R, Malik SK (2004) Genetic conservation of plantation crops and spices using cryopreservation. Indian J Biotech 3:348–358
Chen JJ, Rowley JD, Wang SM (2000) Generation of longer cDNA fragments from serial analysis of gene expression tags for gene identification. Proc Natl Acad Sci USA 97(1):349–353
Chen S, Harmon AC (2006) Advances in plant proteomics. Proteomics 6:5504–5516
Choo YM, Cheah KY (2000) Biofuel. In: Basiron Y, Jalani BS, Chan KW (eds) Adv Oil Palm Res 2:1293–1345
Choong CY, Shah FH, Rajanaidu N, Zakri AH (1996) Isoenzyme variation of Zairean oil palm (*Elaies guineensis* Jacq) germplasm collection. Elaeis 8:45–53
Chua KL, Singh R, Cheah SC (2001) Construction of oil palm (*Elaies guineensis* Jacq) linkage maps using AFLP markers. Proc Intl Palm Oil Congr 2001:461–465
Chuah TG, Wan Azlina AGK, Robiah Y, Omar R (2006) Biomass as renewed energy sources in Malaysia: an overview. Int J Green Energy 3(3):323–346
Cochard B, Amblard P, Durand-Gasselin T (2005) Oil palm genetic improvement and sustainable development. (Feature: research, oil palm and sustainable development) OCL - Oleagineux, Corps Gras, Lipides 12(2):141–147
Collard BCY, Jahufer MZZ, Brouwer JB, Pang ECK (2005) An introduction to markers, quantitative trait loci (QTL) mapping and marker-assisted selection for crop improvement: The basic concepts. Euphytica 142:169–196
Comai L, Young K, Till BJ, Reynolds SH, Greene EA, et al. (2004) Efficient discovery of DNA polymorphisms in natural populations by ecotilling. Plant J 37:778–86
Corley RHV (1985) Yield potentials of plantation crops. In: Potassium in the agricultural systems of the humid tropic. Int Potash Inst Bern, Switzerland
Corley RHV (2005) Illegitimacy in oil palm breeding – a review. J Oil Palm Res 17(1):64–69
Corley RHV, Donough CR (1995) Effects of defoliation on sex differentiation in oil palm clones. Exp Agric 31(17):177–189
Corley RHV, Lee CH (1992) The physiological basis for genetic improvement of oil palm in Malaysia. Euphytica 60:179–184
Corley RHV, Tinker PB (2003) The oil palm. 4th Ed. World Agricultural Series. Blackwell Publishers Ltd, Oxford UK
Corley RHV, Lee CH, Law IH, Wong CY (1986) Abnormal flower development in oil palm clones. The Planter 62:233–240

de Franqueville H (2003) Oil palm bud rot in Latin America. Exp Agric 39(3):225–240

de Franqueville H, de Greef W (1987) Hereditary transmission of resistance to vascular wilt of oil palm: facts and hypotheses. In: Halim A, Hassan Y, Chew PS, Wood BJ, Pushparajah E (eds) Proc Intl Oil Palm Conf:118–129

Dettmer K, Aronov PA, Hammock BD (2007) Mass spectrometry-based metabolomics. Mass Spectrometry Rev 26:51–78

Dumet D, Engelmann F, Chabrillange N, Dussert S, Duval Y (2000) Cryopreservation of oil-palm polyembryonic cultures. (JIRCAS International Agriculture Series No8) In: Cryopreservation of tropical plant germplasm: current research progress and application. Proc Intl workshop Tsukuba Japan October 1998 International Plant Genetic Resources Institute (IPGRI) Rome Italy pp. 172–177

Durand-Gasselin T, Baudouin L, Cochard B, Adon B, Cao TV (1999) Oil palm genetic improvement strategies. Plantations, Recherche, Developpement 6(5):344–358

Durand-Gasselin T, Asmady H, Flori A, Jacqeumard JC, Hayun Z, et al. (2005) Possible sources of genetic resistance in oil palm (*Elaeis guineensis* Jacq) to basal stem rot caused by *Ganoderma boninense* – prospects for future breeding. Mycopathologia 159(1):93–100

Ellis MH, Spielmeyer W, Gale GR, Rebetzke GJ, Richards RA (2002) "Perfect" markers for the Rht-B1b and Rht-D1b dwarfing genes in wheat. Theor Appl Genet 105:1038–1042

Elyana MD, Halimah A, Boon SH, Amos TCY, Low ETL, et al. (2005) Expressed sequence tags (ESTs): an approach to gene discovery in oil palm (*Elaeis guineensis* Jacq) tissue culture. Proc Conf Biotechnol Plantation Commodities:344–353

Eeuwens CJ, Lord S, Donough CR, Rao V, Vallejo G, et al. (2002) Effects of tissue culture conditions during embryoid multiplication on the incidence of "mantled" flowering in clonally propagated oil palm. Plant Cell Tiss Organ Cult 70:311–323

Falconer DJ (1989) Introduction to quantitative Genetics. 3rd Edition. Longman Scientific and Technical, Harlow, UK

FAO (2005) Status of research and application of crop biotechnologies in developing countries. FAO, Rome ISBN 92-5-105290-5

Feschotte C, Jiang N, Wessler S (2002) Plant transposable elements: where genetics meets genomics. Nat Rev Genet 3:329–341

Flood J (2005) A review of Fusarium wilt of oil palm caused by *F. oxysporum* f sp. Elaedis. Phytopathol 95(6) suppl

Flood J, Bridge P (2000) Ganoderma diseases of perennial crops. CABI Publishing, Wallingford, UK

Frengen E, Weichenhan D, Zhao B, Osoegawa K, van Geel M, et al. (1999) A modular, positive selection bacterial artificial chromosome vector with multiple cloning sites. Genomics 58:250–253

Fu D, Scuzs P, Yan L, Helguera M, Skinner JS, et al. (2005) Large deletions within the first intron in VRN-1 are associated with spring growth habit in barley and wheat. Mol Gen Genomics 273:54–65

Fulton T, van der Hoeven R, Eannetta N, Tanksley S (2002) Identification, analysis and utilization of a conserved ortholog set (COS) markers for comparative genomics in higher plants. Plant Cell 14:1457–1467

Galeano CH (2005) Standardising amplified fragment-length polymorphisms (AFLP) for *Dura* oil palm (*Elaeis guineensis* Jacq) and preliminary molecular characterization studies. Agronomia Colombiana 23(1):42–49

Ghesquiére M (1984) Enzyme polymorphism in oil palm (*Elaeis guineensis* Jacq) I. Genetic control of 9 enzyme-systems. Oleagineaux 39:561–574

Ghesquiére M (1985) Enzyme polymorphism in oil palm (*Elaies guineensis* Jacq) II Variability and genetic structure of seven origins of oil palm. Oleagineux 40:529–540

Guo Y, Fu Z, Van Eyk JE (2007) A Proteomic primer for the clinician. Proc Am Thoracic Soc 4:9–17

Hafeez F (2005) PhD thesis submitted to the University of Cambridge 'Genome structure and organisation of oil palm (*E. guineensis* Jacq)

Hammond JP, Broadley MR, Craigon DJ, Higgins J, Emmerson ZF, et al. (2005) Using genomic DNA-based probe-selection to improve the sensitivity of high-density oligonucleotide arrays when applied to heterologous species. Plant Methods 1:10

Hammond JP, Bowen HC, White PJ, Mills V, Pyke PA, et al. (2006) A comparison of the *Thlaspi caerulescens* and *Thlaspi arvense* shoot transcriptomes. New Phytologist 170(2): 239–260

Hanley S, Edwards D, Stevenson D, Haines S, Hegarty M, et al. (2000) Identification of transposon-tagged genes by the random sequencing of Mutator-tagged DNA fragments from *Zea mays* Plant J 23 (4):557–566

Hardon JJ (1969) Interspecific hybrids in the genus *Elaeis* II vegetative growth and yield of F_1 hybrids *E. guineensis x E. oleifera.* Euphytica 18(3):380–388

Hardon JJ, Tan GY (1969) Interspecific hybrids in the genus *Elaeis* I crossability, cytogenetics and fertility of F_1 hybrids of *E. guineensis x E. oleifera.* Euphytica 18(3):372–379

Hasan Y, Foster HL, Flood J (2005) Investigations on the causes of upper stem rot (USR) on standing mature oil palms. Mycopathologia 159(1):109–112

Hassan K, Sukaimi J (1993) Industrial moulding of oil palm particles. I. Suitability of oil palm trunk and frond for moulded table-tops. Palm Oil Inst of Malaysia Bulletin 27:1–7

Hayati A, Wickneswari R, Maizura I, Rajanaidu N (2004) Genetic diversity of oil palm (*Elaeis guineensis* Jacq) germplasm collections from Africa: implications from improvement and conservation of genetic resources. Theor Appl Genet 108: 1274–1284

Henderson J, Osborne DJ (1990) Cell separation and anatomy of abscission in the oil palm *Elaeis guineensis* Jacq J Exp Bot 41(2):203–210

Henderson J, Davies HA, Heyes SJ, Osborne DJ (2001) The study of a monocotyledon abscission zone using microscopic, chemical, enzymatic and solid state ^{13}C CP/MAS NMR analyses. Phytochem56 131–139

Houmiel KL, Slater S, Broyles D, Casagrande L, Colburn S, et al. (1999) Poly (beta-hydroxybutyrate) production in oil seed leucoplasts of *Brassica napus.* Planta 20:547–550

Ibrahim R (2003) Structural, mechanical and optical properties of recycled paper blended with oil palm empty bunch pulp. J Oil Palm Res 15(2):28–34

Jaccoud D, Peng K, Feinstein D, Kilian A (2001) Diversity arrays: a solid state technology for sequence information independent genotyping. Nucleic Acids Res 29:e25

Jack PJ, Mayes S (1993) Use of molecular markers for oil palm breeding. II. Use of DNA markers (RFLPs). Oleagineux 48(1):1–8

Jack PL, Dimitrijevic TAF, Mayes S (1995) Assessment of nuclear, mitochondrial and chloroplast RFLP markers in oil palm (*Elaeis guineensis* Jacq). Theor Appl Genet 90: 643–649

Jack PL, James C, Price Z, Groves L, Corley RHV, et al. (1998) Application of DNA markers in oil palm breeding. Proc Intl Oil Palm Conf Indonesian Oil Palm Res Institute, Medan, Indonesia, pp 315–324

Jacquemard J-C (1998) The tropical agriculturist; oil palm. MacMillan Education Ltd, London and Basingstoke, UK

Jaligot E, Rival A, Baule T, Dussert S, Verdeil J-L (2000) Somaclonal variation in oil palm (*Elaies guineensis* Jacq): the DNA methylation hypothesis. Plant Cell Rep 19:684–690

Jaligot E, Beule T, Verdeil J-L, Tregear J, Rival A (2003) DNA methylation vs. somaclonal variation in higher plants: oil palm as a case study. Acta Horticulturae 625:345–353

Jaligot E, Beule T, Baurens F-C, Billotte N, Rival A (2004) Search for methylation-sensitive amplification polymorphisms associated with the "mantled" variant phenotype in oil palm (*Elaeis guineensis* Jacq) Genome 47(1):224–228

Jouannic S, Argout X, Lechauve F, Fizames C, Borgel A, et al. (2005) Analysis of expressed sequence tags from oil palm (*Elaeis guineensis*). FEBS Letters 579(12):2709–2714

Jones LH (1998) Metabolism of cytokinins by tissue culture lines of oil palm (*E. guineensis* Jacq) producing normal and abnormal flowering palms. J Plant Growth Reg 17(4):205–214

Kalendar R, Schulman HA (2006) IRAP and REMAP for retrotransposon-based genotyping and fingerprinting. Nature Protocols 1(5):2478–2484

Khemvong S, Suvachittanont W (2005) Molecular cloning and expression of a cDNA encoding 1-deoxy-D-xylulose-5-phosphate synthase from oil palm *Elaeis guineensis* Jacq. Plant Sci 169(3):571–578

Klein PE, Klein RR, Cartinhour SW, Ulanch PE, Dong J, et al. (2000) A high-throughput AFLP-based method for constructing integrated genetic and physical maps: progress toward a sorghum genome map. Genome Res 10:789–807

Kubis SE, Castilho AMMF, Vershinin AV, Seymour J, Heslop-Harrison JS (2003) Retroelements, transposons and methylation status in the genome of oil palm (*Elaeis guineensis*) and the relationship to somaclonal variation. Plant Mol Biol 52:69–79

Kularatne RS, Shah FH, Rajanaidu N (2001) The evaluation of genetic diversity of Deli *dura* and African oil palm germplasm collection by AFLP technique. Trop Agric Res 13: 1–12

Kumar A, Bennetzen JL (1999) Plant retrotransposons. Ann Rev Genet 33 479–532

Kuntom A, Hamirin K (2000) Soaps from palm products. In: Basiron Y, Jalani BS, Chan KW (eds Advances in Oil Palm Research, Vol 2 Malaysian Palm Oil Board. Kuala Lumpur. Malaysian Palm Oil Board (MPOB). pp 1102–1140

Latiffah Z, Harikrishna K, Tan SG, Tan SH, Abdullah F, Ho YW (2002) Restriction analysis and sequencing of the ITS regions and 5.8S gene of rDNA of Ganoderma isolates from infected oil palm and coconut stumps in Malaysia. Annals Appl Biol 141(2):133–142

Lawrence MR, Marshall DF, Davies P (1995) Genetics of genetic conservation. II. Sample size when collecting seed of cross-pollinating species and the information that can be obtained from the evaluation of material held in gene banks. Euphytica 84(2):101–107

Lee J, Cooper B (2006) Alternative workflows for plant proteomic Analysis. Mol BioSyst 2:621–626

Lee M-P, Yeun LH, Abdullah R (2006) Expression of Bacillus thuringiensis insecticidal protein gene in transgenic oil palm. Electronic J Biotech 9(2):117–126

Lim CC, Rao V (2004) DNA marker technology and private sector oil palm breeding. The Planter 80:611–628

Lim CC, Rao V (2005) DNA fingerprinting of oil palm – choice of tissues. J Oil Palm Res 17:136–144

Low ETL, Tan JS, Chan PL, Boon SH, Wong YL, et al. (2006) Developments toward the application of DNA chip technology in oil palm tissue culture. J Oil Palm Res (Special Issue – April 2006):87–96

Luyindula N, Mantantu N, Dumortier F, Corley RHV (2005a) The effects of inbreeding on growth and yield of oil palm. Euphytica 143(1–2):9–17

Luyindula N, Corley RHV, Mantantu N (2005b) A comparison of the Deli Dumpy and Pobe Dwarf short-stemmed oil palms and their outcrossed progenies. J Oil Palm Res 17:152–159

Madon M, Heslop-Harrison JS, Schwarzacher T, Mohd Rafdi MH, Clyde MM (2005a) Cytological analysis of oil palm pollen mother cells (PMCs). J Oil Palm Res 17:176–180

Madon M, Clyde MM, Hashim H, Mohd Yusuf Y, Mat H, Saratha S (2005b) Polyploid induction of oil palm through colchicine and oryzalin treatments. J Oil Palm Res 17:110–123

Maizura I, Cheah SC, Rajanaidu N (2001) Genetic diversity of oil palm germplasm collections using RFLPs. Proc 2001 PIPOC Intl Palm Oil Congr – Cutting-edge technologies for sustained competitiveness (Agriculture):526–535

Maizura I, Rajanaidu N, Zakri A, Cheah S (2006) Assessment of Genetic Diversity in Oil Palm (*Elaeis guineensis* Jacq) using restriction fragment length polymorphism (RFLP). Genet Resources Crop Evol 53(1):187–195(9)

Mannan MA, Alexander J,Ganapathy C, Teo DCL (2005) Quality improvement of oil palm shell (OPS) as coarse aggregate in lightweight concrete. Building Environ 41:1239–1242

Maria M, Clyde MM, Cheah SC (1995) Cytological analysis of *Elaeis guineensis* (tenera) chromosomes. Elaeis 7:122–134

Masani Mat Yunus A, Ho CL, Parveez GKA (2001) Construction of PHB gene expression vectors for the production of biodegradable plastics in transgenic oil palm. Proc Int Palm oil Congr:694–711

Masi P, Zeuli PLS, Donini P (2003) Development and analysis of multiplex microsatellite markers sets in common bean (Phaseolus vulgaris L). Mol Breed 11(4):303–313
Matthes M, Singh R, Cheah S-C, Karp A (2001) Variation in oil palm (*Elaeis guineensis* Jacq) tissue culture-derived regenerants revealed by AFLPs with methylation-sensitive enzymes. Theor Appl Genet 102:971–979
Mayes S, James CM, Pluhar V, Batty N, Jack PL, Corley RHV (1995) The application of biotechnology to oil palm – prospects and progress. In: Rao V, Henderson IE, Rajanaidu N (eds) Recent developments in oil palm tissue culture and biotechnology. Proc Intl Palm Oil Congr, Palm Oil Res. Inst. Malaysia, Kuala Lumpur, Malaysia: pp 171–189
Mayes S, James CM, Horner SF, Jack PL, Corley RHV (1996) The application of restriction fragment length polymorphism for the genetic fingerprinting of oil palm (*Elaeis guineensis* Jacq). Mol Breed 2:175–180
Mayes S, Jack PL, Marshall DF, Corley RHV (1997) Construction of a RFLP genetic linkage map for oil palm (*Elaeis guineensis* Jacq). Genome 40:116–122
Mayes S, Jack PL, Corley RHV (2000) The use molecular markers to investigate the genetic structure of an oil palm breeding programme. Heredity 85:288–293
Mayes S, Holdsworth MJ, Pellegrineschi A, Reynolds M (2005a) Allying genetic and physiological innovations to improve productivity of wheat and other crops. pp 89–122. In: Sylvester-Bradley R, Wiseman J (eds) Yields of farmed species – constraints and opportunities in the 21st century. Nottingham University Press, Nottingham
Mayes S, Parsley K, Sylvester-Bradley R, May S, Foulkes MJ (2005b) Integrating Genetic information into plant breeding programmes: how will we produce new varieties from molecular variation using bioinformatics? Annals Appl Biol 146:223–237
McCallum CM, Comai L, Greene EA, Henikoff S (2000) Targeting induced local lesions in genomes (TILLING) for plant functional genomics. Plant Physiol 123:439–442
Mohd Basri W, Siti Nor AA, Henson IE (2005) Oil palm – achievements and potential. Oil Palm Bull 50:1–13
Mohd Din A, Rajanaidu N, Kushairi A (2005) Exploitation of genetic variability in oil palm. Proc MOSTA best practices workshops: agronomy and crop management, Malaysian Oil Sci Technol Assoc:19–42
Morcillo F, Gagneur C, Adam H, Richaud F, Rajinder S, et al. (2006) Somaclonal variation in micropropagated oil palm. Characterization of two novel genes with enhanced expression in epigenetically abnormal cell lines and in response to auxin. Tree Physiol 26(5): 585–594
Moretzsohn MC, Nunes CDM, Ferreira ME, Grattapaglia D (2000) RAPD linkage mapping of the shell thickness locus in oil palm (*Elaeis guineensis* Jacq). Theor Appl Genet 100:63–70
Moretzsohn MC, Ferreira MA, Amaral ZPS, Coelho PJA, Grattapaglia D, et al. (2002) Genetic diversity of Brazilian oil palm (EoHBK) germplasm collected in the Amazon rainforest. Euphytica 124:35–45
Murphy DJ (2006) Molecular breeding strategies for the modification of lipid composition. In Vitro Cell Dev Biol 42(2):89–99
Murphy DJ (2007) Future prospects for oil palm in the 21st century: biological and related challenges. Eur J Lipid Sci Technol 109:1–11
Okwuagwu CO, Okolo EC (1992) Maternal inheritance of kernel size in the oil palm (*Elaeis guineensis* Jacq). Elaeis 4:72–73
Okwuagwu CO, Okolo EC (1994) Genetic control of polymorphism for kernel to fruit ratio in oil palm (*Elaeis guineensis* Jacq). Elaeis 6:75
Palmer LE, Rabinowicz PD, O'Shaughnessy AL, Balija VS, Nascimento LU, et al. (2003) Maize genome sequencing by methylation filtration. Science 302:2115–2118
Panchal G, Bridge PD (2005) Following basal stem rot in young oil palm plantings. Mycopathologia 159(1):123–127
Pantzaris TP (1997) Pocketbook of palm oil uses. Palm Oil Res Inst Malaysia, Kuala Lumpur
Parveez GKA (2003) Novel products from transgenic oil palm. AgBiotechNet 5 113:1–8
Pevsner J (2003) Bioinformatics and Functional Genomics. Hoboken, New Jersey: John Wiley and Sons Inc

Piffanelli P, Noa-Carrazana JC, Clement D, Ciampi J, Vilarinhos A, et al. (2002) A Platform of genomic resources to study organization and evolution of tropical crop species. SO4–3. (abstracts) Plant, Animal, Microbial Genome Conf X:58

Pilotti CA, Sanderson FR, Aitken EAB (2003) Genetic structure of a population of *Ganoderma boninense* on oil palm. Plant Pathol 52(4):455–463

Poole R, Barker G, Wilson ID, Coghill JA, Edwards KJ (2007) Measuring global gene expression in polyploidy; a cautionary note from allohexaploid wheat. Funct Integr Genomics 7(3): 207–219

Price Z, Dumortier F, MacDonald DW, Mayes S (2002) Characterization of copia-like retrotransposons in oil palm (*Elaeis guineensis* Jacq). Theor Appl Genet 104:860–867

Price Z, Schulman A, Mayes S (2004) Development of new marker system: oil plam. Plant Genet Resources: Charact and Util 1(2/3):105–115

Purba AR, Noyer JL, Baudouin L, Perrier X, Hamon S, et al. (2000) A new aspect of genetic diversity of Indonesian oil palm (*Elaies guineensis* Jacq) revealed by isoenzyme and AFLP markers and its consequences for breeding. Theor Appl Genet 101:956–961

Purba A, Flori R, Baudouin L, Hamon S (2001) Prediction of oil palm (*Elaeis guineensis* Jacq) agronomic performances using the best linear unbiased predictor (BLUP). Theor Appl Genet 102(5):787–792

Rabinowicz PD (2003) Constructing gene-enriched plant genomic libraries using methylation filtration technology. Methods Mol Biol 236:21–36

Rafii MY, Rajanaidu N, Jalani BS, Zakri AH (2001) Genotype x environment interaction and stability analyses in oil palm (*Elaeis guineensis* Jacq) progenies over six locations. J Oil Palm Res 13(1):11–41

Rafii MY, Rajanaidu N, Jalani BS, Kushairi A (2002) Performance and heritability estimations on oil palm progenies tested in different environments. J Oil Palm Res 14(1):15–24 ,

Rajanaidu N, Kushairi A, Rafii M, Din M, Maizura I, Jalani BS (2000) Oil palm breeding and genetic resources. pp 171–227. In: Basiron Y, Jalani BS, Chan KW (eds) Advances in Oil Palm Research. Malaysian Palm Oil Board, Kuala Lumpur

Rajinder S, Cheah SC, Madon M, Ooi LCL, Rahimah AR (2001) Genomic strategies for enhancing the value of the oil palm. Proc PIPOC Intl Palm Oil Congr:3–17

Rance KA, Mayes S, Price Z, Jack PL, Corley RHV (2001) Quantitative trait loci for yield components in oil palm (*Elaeis guineensis* Jacq). Theor Appl Genet 103(8):1302–1310

Rohde W (1996) Inverse sequence-tagged repeat (ISTR) analysis: a novel and universal PCRbased technique for genome analysis in the plant and animal kingdom J Genet Breeding 50:249–61

Rival A, Beule T, Barre P, Hamon S, Duval Y, et al. (1997) Comparative flow cytometric estimation of nuclear DNA content in oil palm (*Elaeis guineensis* Jacq) tissue cultures and seed derived plants. Plant Cell Rep 16:884–887

Rival A, Tregear J, Verdeil J-L, Richaud F, Beule T, et al. (1998a) Molecular search for mRNA and genomic markers of the oil palm "mantled" somaclonal variation. Acta Horticulturae 461:165–172

Rival A, Bertrand L, Beule T, Combes MC, Trouslot P, et al. (1998b) Suitability of RAPD analysis for the detection of somaclonal variants in oil palm (*Elaeis guineensis* Jacq). Plant Breed 117:73–76

Rival A, Jaligot E, Beule T, Verdeil JL, Tregear J (2000) DNA methylation and somaclonal variation in oil palm. Acta Horticulturae 530:447–454

Rival A, Tregear J, Jaligot E, Morcillo F, Aberlenc F, et al.(2001) Oil palm biotechnology: progress and prospects. OCL - Oleagineux, Corps Gras, Lipides 8(4):295–306

Rival A, Tregear J, Jaligot E, Morcillo F, Aberlenc F, et al. (2003) Biotechnology of the oil palm (*Elaeis guineensis* Jacq).In: Singh RP, Jaiwal PW (eds) Plant genetic engineering. - Houston : Sci Tech Publishing. p 261–318

Rohani O, Zamzuri I, Tarmizi A H (2003) Oil palm cloning: MPOB protocol. MPOB Technology Malaysian Palm Oil Board (MPOB), Kuala Lumpur, Malaysia, 26, ii + 20

Rosenquist EA (1985) The genetic base of oil palm breeding populations. Proc Palm Oil Res Inst Malaysia 10:10–27

Salvi S, Tuberosa R (2005) To clones or not to clone plant QTLs: present and future challenges. Trends Plant Sci 10(6):297–304

Sambanthamurthi R, Sundram K, Tan YA (2000) Chemistry and biochemistry of palm oil. Progr Lipid Res 39:507–558
Sambrooke J, Russell DW (2000) Molecular Cloning: A Laboratory Manual, Cold Spring Harbor Laboratory, Princetown. ISBN-13-9780879695774
Sanderson FR (2005) An insight into spore dispersal of *Ganoderma boninense* on oil palm. Mycopathologia 159(1):139–141
SanMiguel P, Bennetzen JL (1998) Evidence that a recent increase in maize genome size was caused by the massive amplification of intergene retrotranposons. Ann Botany 82 (suppl A):37–44
Schulman AH, Kalendar R (2005) A movable feast: Diverse retrotransposons and their contribution to barley genome dynamics. Cytogenetic Genome Res 110(1–4):598–605
Shah FH, Rashid O, Simons AJ, Dunsdon A (1994) The utility of RAPD markers for the determination of genetic variation in oil palm (*Elaeis guineensis*). Theor Appl Genet 89:713–718
Shah FH, Cha TS (2000) A mesocarp-and species-specific cDNA clone from oil palm encodes for sesquiterpene synthase. Plant Sci 154(2):153–160
Singh R, Lessard PA, Guan TS, Panandam JM, Sinskey A, et al. (2003) Preliminary attempts at the construction of large insert DNA libraries for oil palm (*Elaeis guineensis* Jacq). J Oil Palm Res 15(1):12–20
Siti Nor AA, Sambanthamurthi R, Parveez GKA (2001) Genetic modification of oil palm for producing novel oils. Proc PIPOC Intl Palm Oil Congr:18–30
Sniady V, Becker D, Herrán A, Ritter E, Rohde W(2003) A rapid way of physical mapping in oil palm. www.tropentag.de/2003/proceedings/node255.html
Soh AC (1999) Breeding plans and selection methods in oil palm. p 65–95 In: Rajanaidu N, Jalani BS (eds) Proc Symp "The Science of Oil Palm Breeding", Palm Oil Res Inst Malaysia, Kuala Lumpur
Soh AC, Tan ST (1983) Estimation of genetic variance, variance and combining ability in oil palm breeding. In: Yap TC, Graham KM (eds) Proc 4th Int SABRAO Congr Crop Improvement Res. p 379–388
Soh AC, Wong G, Tan CC, Chew PS, Hor TY, et al. (2001) Recent advances towards commercial production of elite oil palm clones. Proc PIPOC Intl Palm Oil Congr:33–44
Soh AC, Wong G, Hor TY, Tan CC, Chew PS (2003a) Oil palm genetic improvement. Plant Breed Rev 22:165–219
Soh AC, Gan HH, Wong G, Hor TY, Tan CC (2003b) Estimates of within family genetic variability for clonal selection in oil palm. Euphytica 133:147–163
Sparnaaij LD (1969) Oil palm (*Elaeis guineensis* Jacq). In: Outlines of perennial crop breeding in the tropics. Misc Papers Landbouwhogesschool No 4 Wageningen, Veenam, Wageningen pp 339–387
Sundram K, Sambanthamurthi R, Tan Y-A (2003) Palm fruit chemistry and nutrition. Asia Pacific J Clin Nutr 12:355–362
Susanto A, Sudharto PS, Purba PY (2005) Enhancing biological control of basal stem rot disease (*Ganoderma boninense*) in oil palm plantations. Mycopathologia 159(1):153–157
Syed Alwee SSR (2001) Genes controlling flowering: possible roles in oil palm floral abnormality. Oil Palm Bull 43:1–13
Syed Alwee S, van der Linden CG, van der Schoot J, de Folter S, Angenent GC, et al. (2006) Characterization of oil palm MADS box genes in relation to the mantled flower abnormality. Plant Cell, Tissue Organ Cult 85:331–344
Tao Q, Chang YL, Wang J, Chen H, Islam-Faridi MN, et al. (2001) Bacterial artificial chromosome-based physical map of the rice genome constructed by restriction fingerprint analysis. Genetics 158:1711–1724
Tarmizi AH, Norjihan MA, Zaiton R (2004) Multiplication of oil palm suspension cultures in a bench-top (2-litre) bioreactor. J Oil Palm Res 16(2):44–49
Te-chato S, Hilae A, Yeedum I (2002) Improved callus induction and embryogenic callus formation from cultured young leaves of oil palm seedling. Thai J Agric Sci 35(4):407–413

Texeira JB, Sondahl MR, Nakamura T, Kirby EG (1995) Establishment of oil palm cell suspensions and plant regeneration. Plant Cell Tissue Organ Cult 40(2):105–111
Thomas CA (1971) The genetic organization of chromosomes. Ann Rev Genetics 5:237–256
Till BJ, Reynolds SH, Greene EA, Codomo CA, Enns LC, et al. (2003) Large-scale discovery of induced point mutations with high-throughput TILLING. Genome Res 13:524–530
Toruan-Mathius N, Bangun SII, Bintang M (2001) Analysis abnormalities of oil palm (*Elaeis guineensis* Jacq) from tissue culture by random amplified polymorphic DNA (RAPD). Menara Perkebunan 69(2):58–70
Tuberosa R, Salvi S (2006) Genomics approaches to improve drought tolerance in crops. Trends Plant Sci 11:405–412
Turner A, Beales J, Faure S, Dunford RP, Laurie DA (2005) The pseudo-response regulator, *Ppd-H1* provides adaptation to photoperiod in barley. Science 310:1031–1034
US FDA (2006) Questions and answers about *Trans* fat nutrition labeling. http;//www.cfsan.fda.gov/~dms/qatrans2.html#s3q1
Utomo C, Werner S, Neipold F, Deising HB (2005) Identification of Ganoderma, the causal agent of basal stem rot disease in oil palm using a molecular method. Mycopathologia 159(1):159–170
Vicient CM, Kalendar R, Schulman AH (2005) Variability, recombination and mosaic evolution of the barley BARE-1 retrotransposon. J Mol Evol 61(3):275–91
Whitelaw CA, Barbazuk WB, Pertea G, Chan AP, Cheung F, et al. (2003) Enrichment of gene-coding sequences in maize by genome filtration. Science 302:2118–2120
Wonkyi-Appiah JB (1987) Genetic control of fertility in the oil palm (*Elaeis guineensis* Jacq). Euphytica 36:505–511
Woo S-S, Jiang J, Gill BS, Paterson AH, Wing RA (1994) Construction and characterisation of a bacterial artificial chromosome library of *Sorghum bicolor*. Nucleic Acids Res 22:4922–4931
Yamazaki M, Tsugawa H, Miyao A, Yano M, Wu J, et al. (2001) The rice retrotransposon Tos17, prefers low-copy marker sequence as a target. Mol Genetics Genomics 265:336–344
Yan L, Loukoianov A, Tranquilli G, Helguera M, Fahima T, et al. (2003) Positional cloning of the wheat vernalization gene VRN1. Proc Natl Acad Sci USA 100(10):6263–6268
Yan L, Loukolanov A, Blechi A, Tranquilli G, Ramakrishna W, et al. (2004) The wheat VRN2 gene is a flowering repressor downregulated by vernalization. Science 303:1640–1644
Zubaidah R, Siti Nor AA (2003) Development of a transient promoter assay system for oil palm. J Oil Palm Res 15(2):62–69

Chapter 16
Genomics of Papaya, a Common Source of Vitamins in the Tropics

Ray Ming, Qingyi Yu, Andrea Blas, Cuixia Chen, Jong-Kuk Na, Paul H. Moore

Abstract Papaya is a major fruit crop of the tropics and is grown to a lesser extent in the subtropics. The genome is small (372 Mbp) and has evolutionarily primitive sex chromosomes. These characters justify papaya genomics programs. In addition to whole genome sequencing, a second major goal is to completely sequence the male specific region of the Y chromosome (MSY) and its corresponding region of the X chromosome. Genomic resources such as high density genetic maps, a physical map, and an expressed sequence tag database have been generated to support genome sequencing and as tools for papaya improvement. The papaya genome is currently being sequenced by the Hawaii Papaya Genome Consortium. Physical mapping of the MSY is near completion. Sequencing the papaya genome and the MSY will enhance our capacity to explore the origin and evolution of dioecy in the family of Caricaceae, to expand our knowledge on genome evolution by serving as an outgroup for the intensively studied family Brassicaceae, identify candidate genes for target traits, and provide genome-wide DNA markers for papaya improvement.

16.1 Introduction

Papaya, *Carica papaya* L., is a soft-wooded, mostly unbranched, herb-like tree crowned with large, palmately-lobed leaves having stout petioles attached directly on the main stem. The main stem or trunk is straight and tapers from a 20–30 cm base to a 5–7 cm width at the crown. The trunk, with aging, is hollow between nodes and is covered with a thin bark that is smooth with prominent leaf scars. Under optimal conditions, papaya trees are fast growing, producing 2–4 leaves per week alternating spirally at the apex of the trunk. Healthy plants will have 15–30 mature leaves confined to the upper 2 m of the trunk.

Flower-bearing, cymose inflorescences arise in axils of mature leaves. The type of inflorescence produced depends on the sex of the tree. Varieties typically are

R. Ming
Department of Plant Biology, University of Illinois at Urbana-Champaign, 1201 W. Gregory Drive, Urbana, IL 61801, USA
e-mail: rming@life.uiuc.edu

P.H. Moore, R. Ming (eds.), *Genomics of Tropical Crop Plants*

© Springer 2008

either dioecious (with unisexual flowers and exclusively male and female plants) or gynodioecious (with bisexual flowers on hermaphrodite plants and unisexual flowers on female plants). Male trees are characterized by long, pendulous, many-flowered inflorescences bearing slender flowers lacking a pistil. Female trees have short inflorescences with few flowers bearing large functional pistils without stamens. Hermaphroditic trees have short inflorescences bearing bisexual flowers that can be sexually variable.

Fruit, one to three per node and superficially resembling melons, hang from stalks attached to the upper trunk with the youngest fruit nearer the top and mature fruit lower down where leaves have abscised. Depending on the variety and sex of the flowers, fruit shape varies from spherical to pear-shaped or elongate and fruit weight ranges from 0.35 to 12 kg. Ripe papaya fruit have smooth, thin, yellow-orange colored skin. Depending on the variety, the flesh color ranges from pale yellow to red and flesh thickness from 1.5 to 4 cm. Fruit have a five-lodicule central cavity containing numerous grey-black, spherical seeds about 5 mm in diameter.

16.1.1 Economic, Agronomic, and Societal Importance of Papaya

Papaya, one of the most important fruit crops in the tropics, is primarily a fresh-market fruit that is also used in drinks, jams, and as a dried and crystallized fruit candy. Green fruit, leaves, and flowers can be cooked as a vegetable (Watson 1997). Papaya fruit is rich in vitamins A and C and is a good source for the minerals K, Mg, and B. The content of one medium papaya exceeds the adult minimum daily requirements of vitamins A and C established by the U.S. Food and Nutrition Board (USDA 2001). Since papaya grows relatively easily and quickly from seed, it is commonly grown in small gardens for home consumption or local trade in many tropical countries. The commercially reported production of papaya in 56 countries reached 6.8 million metric tons (mmt) harvested from 389,172 ha in 2005 (FAOSTAT 2006). The largest producer was Brazil with 1.65 mmt followed by Mexico (0.96), Nigeria (0.76), India (0.70), and Indonesia (0.65).

In addition to its food value, papaya has industrial uses, primarily for its proteolytic enzyme papain, a major component of the mixture of enzymes extracted from the latex of unripe fruit. Papain (EC: 3.4.22.2) is used directly in applications for protein digestion such as a red meat tenderizer, chill proofing of beer, and the external treatment of hard tissues such as warts and scars. Indirect applications of papain include the development of selective inhibitors to the animal cystein proteases that exhibit abnormal activity in a variety of diseases including muscular dystrophy, osteoporosis, pulmonary emphysema, and tumor growth (Czaplewski et al. 1999).

16.1.2 Papaya as an Experimental Organism

Papaya has a number of characteristics that contribute to its being used as an experimental model for tree crops. Papaya trees are small, generally requiring less than

5 m^2 per plant for a field density of 1,200–2,000 trees per hectare, depending on the variety and where grown. As previously mentioned, papaya exhibits rapid growth and development that results in a short 3–8 month juvenile phase (from germination to flowering) and only a 9–15 month generation time (from seed of one generation to seed of the next generation). Flowering and fruiting are continuous throughout the year with the production of one to three ripe fruit per week and hundreds of fruit over the life of the tree. Although hermaphrodite trees are mostly selfing, flower anthers and stigma are large by time of anthesis thus facilitating controlled crossing. Each fruit matures in about 4–5 months and has about 800 (hermaphrodite) or 1000 (female) seeds to provide an abundance of offspring for genetic studies. Genetic analyses are also facilitated by the fact that papaya is easily cloned from cuttings to allow growing the same individual across multiple environments. Various aspects of genomics are relatively easy with papaya because it is diploid with nine pairs of chromosomes, a comparatively small haploid genome of 372 Mb, 86% as large as the rice genome (Arumuganathan and Earle 1991), and an established transformation system.

Papaya is among the limited number of plant species that are polygamous with three sexes – female, male, and hermaphrodite. In an extensive catalog of sexuality among plant species, Yampolsky and Yampolsky (1922) reported that only 7% of the 120,000 species examined were gynodioecious, which is the condition of the more tropical papaya varieties, while 4% were strictly dioecious, which is the condition of the less tropical papaya varieties. A more recent survey showed that 6% of the estimated 250,000 species were dioecious (Renner and Ricklefs 1995). Sex determination in papaya is controlled by a pair of primitive sex chromosomes showing typical characteristics of chromosome rearrangements and suppression of recombination around the sex determination gene (Liu et al. 2004), which is postulated to have three allelic forms (Hofmeyr 1938; Storey 1938a, 1938b). Papaya, by having multiple sex types in a single species, offers a rare opportunity to use genomics for developing an understanding of the evolutionary history of plant sexuality.

16.2 Genetic Diversity

Cultivated papaya, *Carica papaya* L., belongs to Caricaceae, a family comprised of six genera and 35 species distributed throughout tropical and sub-tropical regions (Ming et al. 2005). Genetic studies over the past decade have resulted in the division of the genus *Carica* into two genera, *Vasconcellea* with 21 species and *Carica* with only one species, *C. papaya*. (Badillo 2000). While few studies have evaluated genetic diversity within cultivated papaya, diversity is generally considered to be low (Aradhya et al. 1999; Jobin-Décor et al. 1997; Kim et al. 2002; Van Droogenbroeck et al. 2002). Cultural preference and geographic isolation have forced selection of cultivated papaya from a relatively narrow genetic base to result in extremely low genetic diversity. Hawaiian "solo" papayas, for instance, were developed from a single introduction from Barbados in 1910 (Storey 1969). While a wide range of morphological characters is visible in the field, Kim et al. (2002) reported only ~12% genetic variation among 63 accessions by amplified fragment length polymorphism (AFLP) analysis. Of the 63 accessions analyzed, 82% of the pair-wise comparisons exhibited genetic similarity greater than 0.85 and fewer than 4% showed less than 0.80 similarity. Genetic variation is attributed to natural outcrossing events and the genetic similarity is attributed to each accession's origin according to a specific breeding or selection program (Kim et al. 2002). Van Droogenbroeck et al. (2002) included six accessions of cultivated papaya in a study of genetic relationships among 95 accessions representing three genera and 11 species of the Caricaceae from Ecuador. AFLP analysis of these accessions showed low genetic variation (0.99 average similarity) and also showed cultivated papaya to be very distinct from the other genera tested, with an average genetic similarity of only 0.23, supporting the idea that *C. papaya* diverged early from its wild relatives and proceeded to evolve in isolation (Aradhya et al. 1999; Van Droogenbroeck et al. 2002).

The low genetic variation within cultivated papaya indicates a need to introgress desirable traits from wild relatives. However, the genetic dissimilarity between papaya and its wild relatives prevents natural hybridization. While many of papaya's wild relatives are intercompatible, spontaneously producing natural hybrids in areas of overlapping distributions, cultivated papaya requires embryo rescue to produce intergeneric hybrids from its wild relatives (Manshardt and Wenslaff 1989; Manshardt and Drew 1998). Isozyme and RAPD analysis revealed 73% and 69% dissimilarity, respectively, between cultivated papaya and highland papayas of the genus *Vasconcellea*, formerly regarded as a section of *Carica* (Jobin-Décor et al. 1997). AFLP analysis of a chloroplast DNA (cpDNA) intergenic spacer region also grouped cultivated papaya in a separate clade away from *Vasconcellea* with a boot-strap analysis confidence level of 64% (Aradhya et al. 1999). Use of an intergeneric hybrid between *C. papaya* and *V. quercifolia* backcrossed to *C. papaya*, in a joint Australia-Philippines project, has thus far produced three lines with resistance to Australian isolates of *Papaya* ringspot virus (PRSV) and six lines with resistance to Philippine PRSV isolates. Infertility and incompatibility problems are no longer a concern by the second backcross generation (Drew et al. 2006). Refinement of these techniques for intergeneric hybridization and embryo-rescue will further facilitate introgression of desirable traits and genetic diversity into *C. papaya* in future studies.

16.3 Cytogenetics

Conventional chromosome karotyping revealed nine pairs of chromosomes in papaya (Heiborn 1921). Selected species from *Vasconcellea* also have the same chromosome numbers $2n = 2x = 18$ (Heiborn 1921; Storey 1976). Papaya chromosomes are small and were thought in early cytogenetic studies to be similar to each other in size (Kumar et al. 1945; Storey 1953). Recent analyses of papaya metaphase chromosomes demonstrated variation in chromosome sizes with the largest chromosome pair twice the size of the smallest pair (C.M. Wai, R. Paull, P. Moore, R. Ming, Q. Yu, unpublished). Seven of the nine pairs of chromosomes appear to be metacentric, whereas the other two pairs are submetacentric. However, despite the improvement in cytological techniques, it is still not possible to identify individual papaya chromosomes based on their karyotype images.

Early attempts to identify the sex chromosomes in papaya failed to discover a heteromorphic chromosome pair among somatic chromosomes or among chromosomes in various stages of meiosis (Meurman 1925; Suguira 1927; Lindsay 1930; Hofmeyr 1938; Storey 1941). Kumar et al. (1945) first reported the observation of precocious separation of a pair of chromosomes at anaphase I of meiosis of pollen mother cells in males and hermaphrodites; this observation was confirmed by Storey (1953). It was suggested that this early separating pair of chromosomes might be the sex chromosomes (Kumar et al. 1945).

16.4 Genome Mapping

16.4.1 Genetic Mapping

A high-density genetic map, essential for the integration of genetic and physical maps, and ultimately for assigning the bacterial artificial chromosome (BAC) DNA sequences to papaya chromosomes, is the first step toward isolating and cloning genes of interest via chromosome walking or chromosome landing (Martin et al. 1993; Tanksley et al. 1995). Good genetic maps are also an important tool for genomic dissection of complex traits, comparative analysis of plant genomes, and marker-assisted selection (Klein et al. 2000; Paterson et al. 2000; Draye et al. 2001).

Several genetic linkage maps have been constructed for papaya. The first genetic map was reported more than 60 years ago and consisted of only three morphological markers: sex form, flower color, and stem color (Hofmeyr 1939). The second map developed was based on 62 randomly amplified polymorphic DNA (RAPD) markers and mapped the sex determination gene *Sex1* on linkage group 1 (Sondur et al. 1996). Two flanking markers, OPT12 and OPT1C, were approximately 7 cM on each side of *Sex1*. Deputy et al. (2002) mapped RAPD markers tightly linked to *Sex1* and cloned three RAPD products to sequence-characterized amplified region (SCAR) markers. SCAR W11 and SCAR T12 showed linkage within 0.3 cM of *Sex1*. The third map was constructed using 1,501 markers, including 1,498 amplified fragment length polymorphism (AFLP) markers, the papaya ringspot virus coat protein marker, morphological sex type, and fruit flesh color (Ma et al. 2004). These markers were mapped into 12 linkage groups and covered a total length of 3,294 cM, with an average distance of 2.2 cM between adjacent markers.

Sequence-based DNA markers, such as microsatellite or simple sequence repeat (SSR) and single nucleotide polymorphism (SNP), are highly informative markers for integration of genetic, physical, and cytomolecular maps. Recently, we constructed a high-density genetic map using SSR markers derived from BAC-end and whole-genome shotgun sequences generated by the Hawaii Papaya Genome Consortium. A total of 713 markers, including 712 SSR markers and one morphological marker, were mapped on nine major linkage groups corresponding to the nine chromosomes and to three minor linkage groups, which with the addition of more markers are expected to merge with the nine major groups (unpublished). This map is being used for molecular cytogenetic mapping of papaya chromosomes.

16.4.2 Physical Mapping

A papaya BAC library was constructed from hermaphrodite plants of the transgenic cultivar SunUp and consists of 39,168 clones from two separate ligation reactions (Ming et al. 2001). The average insert size was 132 kb (86 kb for 18,700 clones from ligation #1 and 174 kb for 20,468 clones from ligation #2). Two chloroplast probes, *ropB* and *trnK*, were hybridized separately to the library, yielding a total of 504 chloroplast clones or 1.3% of the library. A cotton rDNA probe hybridized to 61 BACs (0.16%). This library was estimated to provide 13.7x papaya genome equivalents, excluding the false positive (empty clones) and chloroplast clones. Eleven papaya cDNA and 10 *Arabidopsis* cDNA probes detected an average of 22.8 BACs per probe in the library.

The entire set of 39,168 BAC clones of the papaya BAC library was fingerprinted using the high-information-content fingerprinting system (Luo et al. 2003) to produce high quality fingerprints for physical map construction. One-fifth of these fingerprints were excluded due to empty insert clones, incomplete restriction enzyme digestion, highly repetitive sequences, or failure to size on the capillary sequencer. A total of 30,824 fingerprints, estimated as 11x genome equivalents, were used to construct a papaya physical map. After automated overlap evaluation and manual review, 26,466 papaya BAC clones were assigned to 963 contigs. A total of 4,358 singleton clones could not be assigned to the fingerprint contigs. The three largest contigs included over 200 BACs, whereas 204 contigs contained only two BACs. The remainder of the 756 contigs contained 3 to 199 BACs.

In an exploratory experiment for *Brassica* physical mapping, Dr. Andrew Paterson at the University of Georgia tested 2,277 OVERlapping oliGOnucleotide (overgo) probes, representing anchors to *Arabidopsis* and genetically mapped *Brassica* loci, against 36,864 papaya BACs. These overgos were applied in four multiplexed experiments involving 576 probes applied with triple redundancy ($24 \times 24 \times 24$). The overgo data produced have been incorporated into the current papaya physical map. These data provide a starting point for the comparative analysis of papaya and *Arabidopsis* genomes. A total of 1,259 overgo probes and 16 single copy gene probes were anchored on the fingerprint contigs. Ten of the 16 single copy gene probes were anchored on single contigs. Among the 1,259 overgo probes, 483 of them were anchored on single contigs (Q. Yu, M Luo, A. Paterson, P. Moore, R. Ming, unpublished data).

16.5 Genome Organization

The first assessment of papaya genome organization was through large scale sequencing of papaya BAC ends (BES). A total of 50,661 BAC ends from 26,017 BAC clones was sequenced (Lai et al. 2006). The resulting 17.5 Mb BES represents 4.7% of the papaya genome.

The papaya BES provided the first estimate of the papaya genome gene content. Of the 35,472 BES, 6,769 (19.1%) shared homology with at least one *Arabidopsis thaliana* cDNA. Applying this percentage to the papaya genome of 372 Mb yielded an estimated potential coding sequence of 71.1 Mb. Using an average gene length of 2 kb, similar to that of *Arabidopsis* (*Arabidopsis* Genome Initiative 2000), the gene number in papaya was estimated as approximately 35,000 (Lai et al. 2006).

Repetitive sequences are a major component of plant genomes. Comparison of papaya BESs with the plant repeat database revealed 5,733 (16.2%) of the sequences contained repeat elements. These repeat elements were further classified as class I retrotransposons (4,773 or 83.3%), class II transposon (426 or 7.4%), class II miniature inverted-repeat transposable elements (MITEs) (9 or 0.2%), centromere-related sequences (242 or 4.2%), telomere-related sequences (19 or 0.3%), rDNA (167 or 2.9%), and unclassified repeat elements (97 or 1.7%). The papaya genome appeared to harbor few MITEs, but it does contain abundant retrotransposons. Lai et al. (2006) also identified 10 large clusters of papaya-specific repeat sequences, six of which appeared to be male-specific. A total of 7,456 SSR markers, at least 12 nucleotides in length, were identified from 5,452 (15.4%) BESs. Among them, 1,174 (21.5%) at least 20 nucleotides in length were hypervariable and can be used in genetic mapping and marker-assisted selection.

16.6 EST Resources

Five papaya flower cDNA libraries were constructed, three from pre-meiosis (< 4 mm) flower buds (male, hermaphrodite, and female) and two from mature flower buds (hermaphrodite and female). ESTs from these five libraries were sequenced from the 5’ end to produce 31,652 clean sequences with a minimum length of 200 nucleotides. The average read length of a clean sequence was 486 nucleotides with a minimum quality score of 20. The final clean sequences were used in clustering and assembly using a paracel transcript assembler Contaminant sequences from *E. coli*, mitochondria, chloroplast, cloning vector, and RNA were filtered during the cleanup stage. Repeat sequences were masked and annotated. EST sequences were then clustered based on local similarity scores of pair wise comparison using 88% similarity over 100 nucleotides (nt). Clusters containing only one sequence were grouped as singletons. The EST clusters were assembled into contigs (contiguous sequence) by multiple-sequence alignment that generates a consensus sequence for each of the clusters, with criteria of 95% identity over 30 nt overlap. A unigene set of 8,571 EST contigs and singletons was assembled. Blast analysis indicated that about 82% of the unigenes from these papaya libraries have homologous sequences in the protein database of *Arabidopsis* (Q. Yu, P. Moore, R. Ming, unpublished).

16.7 Sex Chromosomes

Sex determination in papaya is controlled primarily by genetic factors, although environmental conditions can trigger sex reversal, particularly in hermaphrodite and male plants. Hermaphrodites and males are heterogametic, whereas females are homogametic. Seeds from selfed hermaphrodite trees and the occasional male flowers with recovered carpels always segregate into hermaphrodite to female, or male to female, at the ratio 2:1, possibly because the combination of homozygous male or hermaphrodite sex determination factors is lethal. Seeds from female trees segregate hermaphrodite to female, or male to female, at the ratio of 1:1. Hermaphrodites are preferred in many regions of the world for their higher productivity since every hermaphrodite tree will produce fruit, whereas depending on female trees for fruit production involves the loss of 6–10% of field space for growing male trees to pollinate the females. However, in subtropical regions having cool winters, female production is preferred because female flowers are stable at low temperature while hermaphrodite flowers tend to abort the carpels and produce no fruit. The lack of true breeding hermaphrodite varieties results in reduced productivity due to sex segregation among the seedlings. Farmers using hermaphrodites for production need to germinate a minimum of five seedlings per hill to assure there are no more than 3% female trees in the field. The five plants in a hill must be grown for 4–6 months until sexes can be determined. Finally, the plants must be thinned to obtain maximum hermaphrodites conducive to optimal productivity. This process is inefficient of time, labor, water, and nutrients, and also results in delayed production due to competition among the seedlings for the several months until sexing is done.

Advances in genomics over the past two decades led to the development of DNA markers linked to papaya sex determination. Four SCAR markers were developed that have proven reliable and accurate for predicting sex types (Parasnis et al. 2000, Deputy et al. 2002; Urasaki et al. 2002). However, the moderate cost for sexing seedlings and the lack of automation for transplanting selected individuals make it impractical to test and then transplant hectares of seedlings in a brief time for commercial production. Understanding the fundamental mechanism and identification of the sex determination gene could provide the ultimate approach to solving the problems of segregation of sex types in papaya.

16.7.1 Identification of Primitive Sex Chromosomes in Papaya

A high density linkage map of the papaya genome was constructed to further characterize the papaya sex determination locus. AFLP markers (1,498), the PRSV coat protein marker, morphological sex type, and fruit flesh color were used

(Ma et al. 2004). The sex determination locus was mapped to the middle of a large linkage group (LG1) having a large cluster of 225 sex co-segregating markers. This non-recombinant block accounted for 66% of the 342 markers on LG1 and 15% of all markers mapped on the genome. This map clearly demonstrated severe suppression of recombination at or around the sex determination locus indicating that this linkage group is the sex chromosome in papaya.

The sex determination locus was fine-mapped using 4,380 informative chromosomes (two each from 2,190 female and hermaphrodite plants of three F_2 and one F_3 populations) and six DNA markers. Despite the large populations screened, not a single recombination event was detected (Liu et al. 2004). The non-recombining (NR) region was physically mapped using our 13.7x BAC library to produce a 2.5 Mb physical map containing 57% of the random sex co-segregating markers developed from co-segregating AFLP markers. To assess the genomic features of the NR region, random subclones from BACs on the physical map were sequenced. Sequencing results revealed that the NR region consists of a mosaic of conserved, X-degenerated, and ampliconic sequences. Based on 684 reads totaling 517 kb (GenBank Accessions CG026197 to CG026996), the NR region showed 37.7% lower gene density, 27.6% higher retroelement density, and 188.9% higher inverted repeat (IR) density compared to a genome-wide sample of papaya DNA. All of the evidence generated, including suppression of recombination at the sex determination loci, sequence divergence between NR and its homologous region, and degeneration of the NR region as exemplified by the lethal effects of homozygous NR genotypes, is compatible with features of primitive sex chromosomes as envisioned by evolutionary biologists (Charlesworth and Charlesworth 1978; Charlesworth 1991). It was concluded that sex determination in papaya was controlled by a pair of primitive sex chromosomes and that the NR region is the male specific region of the Y chromosome (MSY) (Liu et al. 2004).

Recognizing that papaya has a pair of homomorphic primitive sex chromosomes offers an explanation for the earlier observation of precocious separation of a pair of chromosomes at anaphase I of meiosis of pollen mother cells. The small MSY in the middle of the Y chromosome was not pairing with its X chromosome counterparts. A weak attraction of this pair of chromosomes would make their separation and migration towards each pole of the dividing cell easier and earlier.

Two slightly different Y chromosomes exist in papaya; one controlling males, designated as Y, and the other controlling hermaphrodites, designated as Y^h (Ming et al. 2007). Comparison of 13 male MSY and hermaphrodite MSY DNA sequences showed that they were not identical, but nearly so, suggesting that these two Y chromosomes might have originated from the same ancestral chromosome (Liu et al. 2004). The papaya male Y chromosome probably triggers carpel abortion and the elongation of male flower peduncles, while the hermaphrodite Y chromosome does not. Any combination of these two Y chromosome will cause embryo abortion. The lethal effects of the YY genotype indicate that degeneration of the Y chromosome resulted in the loss of function of critical regulatory genes during early embryo development, apparent about 25–50 days after pollination (Chiu 2000).

16.7.2 Molecular Cytogenetics of Sex Chromosomes

The MSY of the Y^h chromosome was mapped to near the middle of the genetic linkage group (Ma et al. 2004) and because most papaya chromosomes are metacentric, we postulated that the MSY might be in the vicinity of the centromere. To physically locate the MSY on the Y^h chromosome, two MSY BACs, 54H01 and 76M08, were hybridized on interphase, prometaphase, metaphase, and anaphase chromosomes using florescent in situ hybridization (FISH) (Yu et al. 2007a). Both BACs located on or near the centromere. BAC 54H01 hybridized strongly on the Y^h chromosome and weakly on the X chromosome. BAC 76M08 hybridized only on the Y^h chromosomes but not on the X chromosome, suggesting more extensive sequence divergence between the X and Y^h chromosomes in this region. The X and Y sequence divergence was further demonstrated by simultaneous pachytene FISH mapping of BAC 76M08 and the neighboring BAC 79C23, each with a non-MSY BAC on the same slide. The two MSY BACs did not hybridize to their X chromosome counterparts while the non-MSY BACs showed strong signals on two homologous autosomal chromosomes (Yu et al. 2007a).

16.7.3 Physical Mapping of the MSY Region

Physical maps are essential for analysis of genomic structure and organization and complete genome sequencing. Physical mapping of the sex determination locus was initiated from the male specific marker W11 (Deputy et al. 2002). Screening the 13x hermaphrodite BAC library with the sex-linked SCAR marker W11 produced four positive BACs. A contig map was constructed by cloning the BAC ends and hybridizing the ends to the four positive BACs. The two outermost ends were used to screen the BAC library to identify and confirm two groups of positive BACs. One large BAC contig spanning 990 kb was constructed by this stepwise chromosome walking process. In addition, 42 AFLP derived SCAR markers were hybridized to the BAC library and generated four more contig maps. After exhausting the genomic resources available at the time, the first MSY physical map was constructed spanning 2.5 Mb and consisting of two major and three smaller contigs containing four SCAR, 82 *Carica papaya* BAC end (cpbe), and 24 *Carica papaya* sex-specific markers (cpsm) around the sex determination locus.

The second phase of the MSY physical mapping began with fingerprinting all 39,168 clones of the papaya hermaphrodite BAC library (see physical mapping section above). Previously identified MSY BACs were confirmed by FISH. The positive BAC clones were used to detect contigs from the genome wide physical map. Chromosome walking extended the contigs. The relative positions of a set of MSY BACs were verified by fiber FISH and pachytene FISH mapping. Sex co-segregating SSR markers from genetic mapping were used for physical mapping. These SSR markers frequently fell in the already established contigs, but occasionally an SSR marker would provide a new starting point on the MSY or on the corresponding region of the X chromosome. To date, about 7.6 Mbp of the MSY region has been mapped in four ordered contigs.

The physical size of papaya MSY was initially estimated as 4–5 Mbp based on the assumption that papaya chromosomes were similar in size (Storey 1953) and the fact that the 2.5 Mb physical map contained 57% of the random cpsm markers (Liu et al. 2004). However, as stated previously, papaya chromosomes are not similar in size with the largest chromosomes being twice as large as the smallest ones. Genetic mapping indicated that, despite the suppression of recombination at the MSY, the sex chromosomes corresponded to the largest linkage group (C. Chen, Q. Yu, P. Moore, R. Paull, R. Ming, unpublished data). Our revised estimate of the papaya MSY is about 7–8 Mbp, which is about 10–15% of the Y chromosome (Q. Yu, P. Moore, J. Jiang, A. Paterson, R. Ming, unpublished data).

16.7.4 Sequencing of X- and Y-BACs

Five mapped BACs, 54H01, 76M08, 42B05, 41F24, and 94E22, were sequenced to examine the genomic features of the MSY region. None of the five BACs contained known centromere-specific sequences, but they each contained abundant gypsy retroelements and several copia elements, which are features typical of the pericentromeric regions of plant chromosomes. Expression analysis revealed no genes in these five BACs, thus demonstrating the extreme gene paucity in the MSY region. At least 19.7% of the sequences are repetitive based on a search of the *Arabidopsis* repeat database. However, this estimate is likely low since six of the 10 most abundant papaya-specific repeats shared homology with the papaya MSY sequences (Lai et al. 2006), and there is no papaya specific repeat database for comparison. Detailed sequence comparison among these five BACs revealed numerous small scale duplication events. An ancient inverted duplication appears to have occurred in the region covered by MSY BAC 54H01, which was followed by a more recent direct duplication event involving BACs 54H01 and 42B05 as shown by direct sequence alignment between these two BACs (Yu et al. 2007a).

Direct comparison of homologous X and Y^h BAC sequences provided quantitative data for documenting the process of the Y chromosome degeneration and for estimating the time of divergence between the X and Y chromosomes. Two X chromosome specific BACs, 61H02 (168 kb) and 53E18 (252 kb), along with their Y^h chromosome counterparts, 95B12 (150 kb) and 85B24 (294 kb), were sequenced. Direct alignment of the two pairs of X and Y^h BACs revealed three inversion events. One inversion occurred in the matched part of the X and Y^h BAC pair 61H02 and 95B12, and the other two inversions occurred in the matched parts of the X and Y^h BAC pairs 53E18 and 85B24 (Yu et al. 2007b). Further analysis of the aligned sequences of the two X and Y^h BAC pairs showed 9.6% DNA sequence expansion on the Y^h-specific BAC 95B12 and 35.2% DNA sequence expansion on 85B24.

Gene expression analyses indicated seven genes on the two X-BACs and four genes on the two Y-BACs. All four genes on the Y BACs had their X counterparts.

One of the three unmatched genes appeared to have been either deleted or translocated to another part of the MSY or to autosomes, since this gene located within the matching regions of the X and Y BAC pair. The other two unmatched genes located in the unaligned region of the X BACs. Structures of the four gene pairs are well conserved in the X and Y chromosomes. The number of nucleotides in each exon and intron of these four gene pairs was nearly the same between the X and Y^h homologs, except for introns 5 and 7 in Gene 2. However, the nucleotide sequences of these conserved exons and introns on the Y^h chromosome had diverged from the X chromosome within a range of 94.8% to 100% identity in the exons and a range of 78.0% to 99.2% in the introns. The weighted average sequence identity for the 44 exons totaling 7,350 bp was 97.9% between the X and Y^h. The weighted average for the 40 introns totaling 59,354 bp was 80.7% (Yu et al. 2007b).

The coding sequences of the four X-Y^h gene pairs were used to determine the degree of synonymous (K_s) and nonsynonymous (K_a) divergence between them. All four X-Y^h gene pairs have K_a/K_s ratios ranging from 0.04–0.5, considerably less than 1.0, suggesting their divergence has been functionally constrained. The degree of silent site (e.g., synonymous and noncoding sites) nucleotide divergence (K_{sil}) between the four X and Y^h gene pairs ranged from 0.016 to 0.066. Assuming a synonymous substitution rate of 1.5×10^{-8} synonymous substitutions/site/year for dicot nuclear genes (Koch et al. 2000), the time of divergence between the X-Y^h gene pairs was estimated to be between 0.5 to 2.2 million years ago (mya), supporting the concept of recent origin of the sex chromosomes in papaya (Yu et al. 2007b).

16.8 Genetic Engineering for Pest and Pathogen Resistance

Mites, insects, and nematodes can be major pests of papaya depending upon cultivar, environmental conditions, and cultural practices. However, more important than the pests are the pathogens that infect various developmental stages and organs of the plant. As with pests, the severity of pathogen infection depends upon the genotype by environment interaction. Nishijima (1994) lists for Hawaii 12 fungal diseases, two bacterial diseases, three viral diseases, and four miscellaneous diseases caused by suboptimal temperature, mineral nutrient supply, or mycoplasmas. A more recent description of the world's principle papaya diseases, provided by Persley and Ploetz (2003), includes at least eight additional viral diseases. A more extensive listing of papaya diseases by common name and causative pathogen is maintained by the American Phytopathological Society (http://www.apsnet.org/online/common/names/papaya.asp). A majority of papaya diseases affecting plant growth or post-harvest fruit loss can be controlled by cultivating tolerant varieties and by use of pesticide sprays. However, these options are not available for viral diseases that usually impose more serious limitations on papaya productivity.

The most serious viral disease, papaya ringspot virus (PRSV), is a potyvirus responsible for major crop losses world-wide due to the lack of resistance genes in the *Carica papaya* germplasm pool. Genetic resistance to PRSV exists in several species of the genus *Vasconcellea*, close relatives of papaya capable of crossing to produce hybrids when embryo rescue is used. Magdalita and co-workers (1988, 1997) reported PRSV resistance in *C. cauliflora* (revised to *Vasconcella cauliflora*) and hybrids between *C. papaya* and *C. cauliflora*. However, the F1 progenies displayed extensive sterility and extreme hybrid weakness with hundreds of progeny dying in the absence of virus infection.

A research team of scientists from Cornell University, the University of Hawaii, and the Agriculture Research Service of the U.S. Department of Agriculture developed the first transgenic papaya (Fitch et al. 1992) that was resistant to PRSV. Resistance was obtained by expression of the virus coat protein gene (CP) to elicit post-transcriptional gene silencing (PTGS) that blocked virus replication. The newly developed PRSV-resistant cultivars brought about a rapid reversal of the decline in the papaya industry in Hawaii (Gonsalves et al. 2004) and served as the basis for re-engineering the CP construct as a chimera from different PRSV strains for developing broader and more stable resistance (Chiang et al. 2001) against the PRSV strains occurring world wide. Pathogen-derived resistance to several papaya viruses other than PRSV should be possible using the technologies already established, however, this has not yet been reported.

Success in developing papaya as the first transgenic virus-resistant fruit crop led to subsequent engineering of papaya for resistance to the carmine spider mite (McCafferty et al. 2006) and to phytophthora (Zhu et al. 2004), which is responsible for rots of roots, stem, and fruit.

16.9 Perspective

The primitive sex chromosomes in papaya present an unprecedented opportunity to study their initiation and formation in flowering plants and the evolutionary forces driving their evolution. Physical mapping and genomic sequencing of the MSY and the corresponding region of the X chromosomes are well underway. The genomic sequences from the sex chromosomes should reveal candidate genes for sex determination that triggered the sex chromosome evolution and genes for the lethal effect of the YY genotype that degenerated on the Y chromosome. Identification of the sex determination genes could lead to the development of true breeding hermaphrodite varieties to solve problems of over-planting and culling, which have been obstacles in papaya production since the beginning of the papaya industry.

Sequences and annotation of the papaya genome will offer not only key information about the unique reproductive biology of papaya, but also numerous benefits for developing a better understanding of plant evolution. *Carica papaya* is in the same taxonomic order (Brassicales) as are *Arabidopsis* and *Brassica*, and is thus an excellent outgroup for analysis of character states at the molecular level. The recent demonstration that the most recent of three genome-wide duplications in *Arabidopsis* occurred after Brassicales diverged from the nearest plant order (Malvales) raises questions about whether this event affects all, or only a subset, of the Brassicales (Bowers et al. 2003). Papaya is in a taxonomic position to address these questions.

The genomic resources in papaya will have profound impact on papaya improvement through better understanding of relevant biology and direct application of genomic tools in breeding programs. It is possible to investigate on the impact of particle bombardment on genome structure and function in transgenic papaya. More candidate genes can be targeted in selection or by genetic transformation to improve disease resistance and fruit quality. Abundant DNA markers enhanced the capacity for assessment of genetic diversity and germplasm conservation to ensure the continuous improvement and production of this highly nutritious tropical fruit.

Acknowledgments We thank the following agencies and programs for funding relevant parts of our research: NSF Plant Genome Research Program, USDA-ARS Cooperative Agreements with the Hawaii Agriculture Research Center, USDA T-STAR program through the University of Hawaii at Manoa, College of Tropical Agriculture and Human Resources, and the University of Illinois at Urbana-Champaign.

References

Arabidopsis Genome Initiative. (2000). Analysis of the genome sequence of the flowering plant *Arabidopsis thaliana*. Nature 408:796–815

Aradhya, MK, Manshardt RM, Zee F, Morden CW (1999) A phylogenetic analysis of the genus *Carica* L. (Caricaceae) based on restriction fragment length variation in a cpDNA intergenic spacer region. Genet Res Crop Evol 46:579–586

Arumuganathan K, Earle ED (1991) Nuclear DNA content of some important plant species. Plant Mol Biol Rep 93:208–219

Badillo VM (2000) *Carica* L. vs. *Vasconcella* St. Hil. (Caricaceae): con la rehabilitación de este último. Ernstia 10:74–79

Bowers JE, Chapman BA, Rong J-K, Paterson AH (2003). Unravelling angiosperm chromosome evolution by phylogenetic analysis of chromosomal duplication events. Nature 422:433–438

Charlesworth B (1991) The evolution of sex chromosomes. Science 251:1030–1033

Charlesworth B, Charlesworth D (1978) A model for the evolution of dioecy and gynodioecy. Am Nat 112:975–997

Chiang,CH, Wang JJ, Jan FJ, Yeh SD, Gonsalves D (2001). Comparative reactions of recombinant papaya ringspot viruses with chimeric coat protein (CP) genes and wild-type viruses on CT-transgenic papaya. J Gen Virology 82:2827–2836

Chiu C-T (2000). Study on sex inheritance and horticultural characteristics of hermaphrodite papaya. Master Thesis. National Pingtung University of Science and Technology. Republic of China. 60 pp

Czaplewski C, Grzonka Z, Jaskolski M, Kasprzykowski F, Kozk M, et al. (1999) Binding modes of a new epoxysuccinyl-peptide inhibitory of cysteine proteases. Where and how do cysteine proteases express their selectivity? Biochem et Biophysica Acta 1431:290–305.

Deputy JC, Ming R, Ma H, Liu Z, Fitch MMM, et al. (2002) Molecular markers for sex determination in papaya (Carica papaya L.). Theor Appl Genet 106:107–111

Draye X, Lin Y, Qian X, Bowers JE, Burow GB, et al. (2001) Toward integration of comparative genetic, physical, diversity, and cytomolecular maps for grasses and grains, using the sorghum genome as a foundation. Plant Physiol 125:1325–1341

Drew RA, Siar SV, O'Brien CM, Sajise AGC (2006) Progress in backcrossing between *Carica papaya* × *Vasconcellea quercifolia* intergeneric hybrids and *C. papaya*. Austr J Exp Agri 46:419–424

FAOSTAT (2006) Papayas. http://apps.fao.org/page/collections?subset=agriculture last updated April 2005

Fitch MMM, Manshardt RM, Gonsalves D, Slightom JL, Sanford JC (1992) Virus resistant papaya derived from tissues bombarded with the coat protein gene of papaya ringspot virus. Biotechnol 10:1466–1472

Gonsalves D, Gonsalves C, Ferreira S, Pitz K, Fitch M, et al. (2004). Transgenic virus resistant papaya: from hope to reality for controlling papaya ringspot virus in Hawaii. APSnet Feature, Am Phytopathol Soc July 2004

Heilborn O (1921) Taxonomical and cytological studies on cultivated Ecuodorian species of Carica. Ark Bot 17:1–16

Hofmeyr JDJ (1938) Genetical studies of *Carica papaya* L. I. The inheritance and relation of sex and certain plant characteristics. II. Sex reversal and sex forms. So Afr Dept Agri Sci Bul No. 187. 64pp

Hofmeyr JDJ (1939) Sex-linked inheritance in *Carica papaya* L. So Afr J Sci 36:283–285

Jobin-Décor MP, Graham GC, Henry RJ, Drew RA (1997) RAPD and isozyme analysis of genetic relationships between *Carica papaya* and wild relatives. Gene. Res Crop Evol 44:471–477

Kim MS, Moore PH, Zee F, Fitch MMM, Steige, DL, et al. (2002) Genetic diversity of *Carica papaya* as revealed by AFLP markers. Genome 45:503–512

Klein PE, Klein RR, Cartinhour SW, Ulanch PE, Dong J, et al. (2000) A high-throughput AFLP-based method for constructing integrated genetic and physical maps: progress toward a sorghum genome map. Genome Res 10:789–807

Koch MA, Haubold B, Mitchell-Olds T (2000). Comparative evolutionary analysis of chalcone synthase and alcohol dehydrogenase loci in Arabidopsis, Arabis, and related genera (Brassicaceae). Mol Biol Evol 17:1483–1498

Kumar LSS, Abraham A, Srinivasan VK (1945) The cytology of *Carica papaya* Linn. Indian J Agr Sci 15:242–253

Lai CW, Yu Q, Hou S, Skelton RL, Jones MR, et al. (2006) Analysis of papaya BAC end sequences reveals first insights into the organization of a fruit tree genome. Mol Genet Genomics 276:1–12

Lindsay RH (1930) The chromosomes of some dioecious angiosperms. Am J Bot 17:152–174.

Liu A, Moore PH, Ma H, Ackerman CM, Makandar R, et al. (2004). A primitive Y chromosome in papaya marks incipient sex chromosome evolution. Nature 427:348–352

Luo MC, Thomas C, You FM, Hsiao J, Ouyang S, et al. (2003). High-throughput fingerprinting of bacterial artificial chromosomes using the SNaPshot labeling kit and sizing of restriction fragments by capillary electrophoresis. Genomics 82:378–389

Ma H, Moore PH, Liu Z, Kim MS, Yu Q, et al. (2004) High-density linkage mapping revealed suppression of recombination at the sex determination locus in papaya. Genetics 166:419–436

Magdalita PM, Villegas VN, Pimentel RB, Bayot RG (1988) Reaction of papaya (*Carica papaya* L.) and related *Carica* species to ringspot virus. Philippine J Crop Sci 13:129–132

Magdalita PM, Persley DM, Godwin ID Drew RA, Adkins SW (1997) Screening *Carica papaya* x *C. Cauliflora* hybrids for resistance to papaya ringspot virus-type P. Plant Pathol 46:837–841

Manshardt RM, Drew RA (1998) Biotechnology of papaya. Acta Hort 461:65–73

Manshardt RM, Wenslaff TF (1989) Inter-specific hybridization of papaya with other species. J Am Soc Hort Sci 114:689–694

Martin G, Brommonschenkel SH, Chunwongse J, Frary A, Ganal M, et al. (1993) Map-based cloning of a protein kinase gene conferring disease resistance in tomato. Science 262:1432–1436

McCafferty HRK, Moore PH, Zhu YJ (2006) Improved *Carica papaya* tolerance to carmine spider mite by the expression of *Manduca sexta* chitinase transgene. Transgenic Res 15:337–347

Meurman O (1925) The chromosome behavior of some dioecious plants and their relatives with special reference to the sex chromosomes. Soc Sci Fennica comm Biol 2:105p

Ming R, Moore PH, Zee F, Abbey CA, Ma H, et al. (2001). Construction and characterization of a papaya BAC library as a foundation for molecular dissection of a tree-fruit genome. Theor Appl Genet 102:892–899
Ming R, Van Droogenbroeck B, Moore PH, Zee FT, Kynd, T, et al. (2005) Molecular diversity of *Carica papaya* and related species. In: Sharma AK, Sharma A (eds) Plant Genome: Biodiversity and Evolution. Volume 1B: Phanerograms. pp. 229–254. Science Publishers, Enfield, New Hampshire, USA
Ming R, Yu Q, Moore PH (2007) Sex determination in papaya. Semin Cell Dev Biol (in press)
Nishijima W (1994) Papaya. In: Ploetz RC, Zentmyer GA, Nishijima WT, Rohrbach, KG, Ohr HD (eds) Compendium of Tropical Fruit Disease. pp. 54–70. American Phytopath Soc Press, St. Paul, MN
Parasnis AS, Gupta VS, Tamhankar SA, Ranjekar PK (2000) A highly reliable sex diagnostic PCR assay for mass screening of papaya seedlings. Mol Breed 6:337–344
Paterson AH, Bowers JE, Burow MD, Draye X, Elsik CG, et al. (2000) Comparative genomics of plant chromosomes. Plant Cell 12:1523–1539
Persley DM, Ploetz RC (2003) Diseases of papaya. In: Ploetz RC (ed) Diseases of Tropical Fruit Crops. pp. 373–412. CABI Publishing, Wallingford, Oxon, UK
Renner SS, Ricklefs RE (1995) Dioecy and its correlates in the flowering plants. Am J Bot 82:596–606
Sondur SN, Manshardt RM, Stiles JI (1996) A genetic linkage map of papaya based on randomly amplified polymorphic DNA markers. Theor Appl Genet 93:547–553
Storey WB (1938a) The primary flower types of papaya and the fruit types that developed from them. Proc Am Soc Hort Sci 35:80–82
Storey WB (1938b) Segregations of sex types in Solo papaya and their application to the selections of seed. Proc Am Soc Hort Sci 35:83–85
Storey WB (1941) The botany and sex relations of the papaya. Hawaii Agr Exp Sta Bul 87:5–22
Storey WB (1953) Genetics of papaya. J Heredity 44:70–78
Storey WB (1969) Papaya. In: Ferwerda FP, Wit F (eds) Outlines of perennial crop breeding in the tropics. H Veenman & Zonen N.V., Wageningen, The Netherlands, pp. 21–24
Storey WB (1976) Papaya. In: Simmonds NW (ed) The evolution of crop plants.. pp. 21–24. Longman, London
Suguira T (1927) Some observations on the meiosis of the pollen mother cells of *Carica papaya, Myrica rubra, Acuba japonica,* and *Beta vulgaris.* Bot Mag 41:219–224
Tanksley SD, Ganal MW, Martin GB (1995) Chromosome landing: a paradigm for map-based gene cloning in plants with large genomes. Trends Genet 11:63–68
Urasaki N, Tokumoto M, Tarora K, Ban Y, Rayano T, et al. (2002) A male and hermaphrodite specific RAPD marker for papaya (Carica papaya L). Theor Appl Genet 104:281–285
USDA (2001) USDA National Nutrient Database for Standard Reference, Release 17. Papayas, raw: Measure 3 (whole papaya, edible portion). http://www.nal.usda.gov/fnic/foodcomp/Data/SR17/reports/sr17fg09.pdf
Van Droogenbroeck B, Breyne P, Goetghebeur P, Romeijn-Peeters E, Kyndt T, et al. (2002) AFLP analysis of genetic relationships among papaya and its wild relatives (Caricaceae) from Ecuador. Theor Appl Genet 105:289–297
Watson B (1997) Agronomy/Agroclimatology notes for the production of papaya. Min Agric, Forests Fisheries Meterol, Australia
Yampolsky C, Yampolsky H (1922) Distribution of sex forms in the phanerogamic flora. Bibl Genet 3:1–62
Yu Q, Hou S, Feltus FA, Jones MR, Murray JE, et al. (2007a) Recent origin of papaya sex chromosomes.*Submitted*
Yu Q, Hou S, Hobza R, Feltus FA, Wang X, et al. (2007b) Chromosomal Location and Gene paucity of the male specific region on papaya y chromosome. Mol. Genet. Genomics *In press*
Zhu YJ, Agbayani R, Jackson MC, Tang CS, Moore PH (2004) Expression of the grapevine stilbene synthase gene VST1 in papaya provides increased resistance against diseases caused by *Phytophthora palmivora.* Planta 220:241–250

Chapter 17
Genomics of Peanut, a Major Source of Oil and Protein

Mark David Burow, Michael Gomez Selvaraj, Hari Upadhyaya, Peggy Ozias-Akins, Baozhu Guo, David John Bertioli, Soraya Cristina de Macedo Leal-Bertioli, Marcio de Carvalho Moretzsohn, and Patricia Messenberg Guimarães

Abstract Peanut, as a source of oil and protein, is the second-most important grain legume cultivated. The perceived lack of molecular variation in the cultivated species had, until recently, resulted in a focus on characterization and mapping of wild species and on transformation of peanut with genes for improved disease resistance. With development of simple sequence repeats and potentially single nucleotide polymorphism-based markers and improved minicore collections, the focus is shifting towards the molecular characterization of the cultivated species. The development of large-inset libraries, expressed sequence tags, genomic clone libraries, characterized mutant collections, and bioinformatics is expected to advance peanut genomics.

17.1 Introduction

Peanut, or groundnut, is a member of the legume family (Leguminosae or Fabaceae). Plants of the cultivated peanut species, *Arachis hypogaea* L., are annuals, either erect (up to 60 cm tall) or prostrate (usually under 30 cm tall). Inflorescences consist of approximately three perfect flowers that usually appear several days apart; the inflorescences occur in the axils of foliage leaves. Peanut differs from most angiosperms in that the fertilized ovary is carried on a peg (gynophore) into the ground, where pod development begins and the fruits are produced. Depending on genotype and environment, fruits mature 2.5 to 6 months after sowing.

Considerable morphological diversity exists among wild peanut species. For example, although many species produce flowers similarly to the cultigen, some

M.D. Burow
Texas A&M University, Texas Agricultural Experiment Station; 1102 East FM 1294, Lubbock, TX 79403 USA; and Texas Tech University, Department of Plant and Soil Science, 15th and Detroit, Lubbock, TX 79409 USA
e-mail: mburow@tamu.edu

© Springer 2008

produce subterranean flowers and some produce rhizomes in addition to flowers. Few wild species are raised commercially, but some are grown for forage. Wild species may possess valuable alleles apparently lacking in the cultigen.

17.1.1 Economic, Agronomic, and Societal Importance of Peanut

Cultivated peanut is the second-most important grain legume crop worldwide after soybean, with 33 million tons of seed produced in 2003/04 (USDA-FAS 2006). Countries producing more than 1 million tons of seed were China, India, the United States, Nigeria, and Indonesia.

Peanut is important for both its versatility and value. The seed typically contains 36% to 54% oil, 16% to 36% protein, and 10% to 20 carbohydrates (Knauft and Ozias-Akins 1995) as well as high amounts of P, Mg, riboflavin, niacin, folic acid, and vitamin E (The Peanut Institute 2006). Worldwide, the largest use is for oil, with the meal being used as a high-protein dietary supplement for human and animal consumption. In the U.S. and some other countries, the peanut seed is used primarily for food, the seed being roasted or boiled and eaten whole or as part of confections, or ground into peanut butter. Peanut fixes nitrogen symbiotically, requires little fertilizer, and improves the quality of the soil in rotation with other crops. Peanut hay may be used for fodder, and the shells used for fuel or livestock feed.

Peanut cultivation also has significant social consequences. The high protein and energy contents make peanut valuable as a subsistence crop in some countries, and the demand for peanut oil allows peanut to be sold as a cash crop. In some countries, peanut is a significant source of income for women, with significant participation in farming, processing, and sales (Balikrishnan et al. 1998).

17.1.2 Geographic Origin of Peanut

The center of origin of *Arachis* is South America, with wild species having been collected in Brazil, Bolivia, Paraguay, Argentina, and Uruguay (Krapovickas and Gregory 1994; Singh and Simpson 1994; Jarvis et al. 2003). The nine sections within the genus comprise 79 defined wild and one cultivated species (Valls and Simpson 2005). Section Arachis has the widest geographical distribution, with 31 wild species collected in five countries.

The cultigen is considered to have originated in northern Argentina or eastern Bolivia, although an origin in Western Peru has been suggested (Simpson et al. 2001). The archaeological record indicates that peanut was domesticated by indigenous peoples at least 3,500 years ago (Singh and Simpson 1994); older peanut specimens consisted only of wild species. At the time of the explorations by the Spanish and Portuguese in the 16th century, peanut was cultivated in many parts of South America, as well as in the Caribbean and Mexico. As a result of explorations, peanut cultivation spread quickly from the Americas to Africa and Asia.

The cultivated species is divided into two subspecies. Subspecies *hypogaea* is characterized by a spreading growth habit, alternating vegetative and reproductive nodes, lack of flowers on the mainstem, medium-to-large seeds, medium-to-late maturity, and the subspecies include botanical varieties *hypogaea* (Virginia and runner market types) and the less-frequently cultivated *hirsuta.* The *fastigiata* subspecies is typified by erect growth habit, sequential reproductive nodes, flowers on the mainstem, small seeds, early maturity, and the subspecies include botanical varieties *fastigiata* (Valencia), *vulgaris* (Spanish), *peruviana*, and *aequatoriana*. The latter two are not cultivated widely.

17.1.3 Peanut as an Experimental Organism

Most genetic research on peanut is focused on varietal improvement. Major research emphases are resistance to biotic and abiotic stresses and edible seed quality. Peanut is also an excellent system for study of several fundamental biological processes.

Numerous biotic stresses cause significant yield losses (Porter et al. 1990). Early (*Cercospora arachidicola* Hori) and late (*Cercosporidium personatum* (Berk. & Curt.) Deighton) leafspot and rust (*Puccinia arachidis* Speg.) cause substantial losses in Africa and Asia; chemical control is available in the United States but is expensive. Sclerotonia blight (*Sclerotinia minor* Jagger) is a major problem in U.S. areas with cool autumns. Nematodes, especially *Meloidogyne arenaria* (Neal) Chitwood and *M. javanica* (Treub) Chitwood, cause significant losses worldwide. Groundnut rosette and tomato spotted wilt viruses cause major losses where insect vectors are present.

Abiotic stress is a growing concern for peanut cultivation. Many production areas are in semiarid environments or have unreliable rainfall, and global climate changes and growing demand for fresh water pose major challenges. Physiological adaptation and selection for drought tolerance have been studied by many researchers (Cruickshank et al. 2003; Reddy et al. 2003). Contamination of peanut under drought stress by *Aspergillus* spp. produces aflatoxin mycotoxins, which can cause liver cancer and suppress immune response.

Quality issues are also important. The high-oleic varieties developed recently in the United States have monounsaturated fatty acid contents similar to olive oil, improved oxidative stability, and beneficial effects on coronary health (O'Byrne et al. 1997). However, certain seed proteins are associated with allergic response in sensitive individuals, and may cause hypersensitive, potentially fatal, reactions in some people.

Peanut has several important characteristics for basic scientific research. The gravitropic development of peg and pod is present in few other species. Subterranean flower development in some wild species contrasts with the cultivated species (Krapovickas and Gregory 1994). The ability to perform symbiotic nitrogen fixation makes peanut an ideal crop for cross-species comparison. The presence of diploid and tetraploid species can also allow for the study of ploidy on gene function, expression, and crop evolution. It is possible to obtain field production data at many locations, allowing analysis of gene × environment interaction.

Study of peanut genomics has been limited by biological constraints, and many basic tools of genomics have yet to be developed (National Peanut Genome Initiative 2005; Gepts et al. 2005). The peanut genome is large, making insertional mutagenesis and whole-genome sequencing expensive using current technology, and requiring large genomic libraries for physical mapping and positional cloning. The reproductive isolation of *A. hypogaea* from wild species and the limited genetic base of the former made marker analysis difficult until recently. Regeneration of peanut from tissue culture has been characterized by long regeneration times and relative inefficiency. Concern over consumer acceptance and regulatory costs has prevented release of transgenic varieties to date.

17.2 Cytogenetics, Markers, and Species Identification

Some of the first uses of molecular markers have been related to classification of wild species and in identification of the likely diploid ancestors of the cultivated peanut. This origin is of special interest because it could suggest mechanisms for transferring useful wild species alleles to the cultigen.

17.2.1 Cytology and the Origin of Cultivated Peanut

Arachis hypogaea is a tetraploid (2n=4x=40) (Husted 1936), and there is only one other known tetraploid species, *A. monticola*, in section Arachis. The remaining species of section Arachis are diploid and are grouped into three genomes (A, B, and D) each having 20 chromosomes, with the exception of three species with 18 chromosomes (Krapovickas and Gregory 1994; see Valls and Simpson 2005). Hybridization between the cultigen and section Arachis diploids is possible, but no evidence has been found that this has contributed to ongoing gene flow into the cultigen in nature. Cultivated peanut is considered to be an AB tetraploid, arising from hybridization between A and B diploid species (Smartt et al. 1978).

To date, 20 A-genome diploid species have been described (Krapovickas and Gregory 1994); among these are perennials *A. cardenasii, A. diogoi, A. helodes, A. villosa*, and *A. correntina,* and annuals *A. duranensis* and *A. stenosperma*. Based on cytological evidence and cross-hybridization data, *A. cardenasii* was considered originally to be the most-probable A-genome ancestor of *A. hypogaea* (Smartt et al. 1978).

Until recently, only one annual B-genome species had been identified (Smartt et al. 1978); the B genome is always associated with the absence of a specific small pair of A chromosomes (Fernández and Krapovickas 1994). *A. batizocoi* was first proposed as the B genome donor to the cultigen (Smartt et al. 1978). However, cytological measurements by Stalker and Dalmacio (1986) discounted *A. batizocoi* as B-genome donor.

17.2.2 Markers and Insights into the Origin of the Cultivated Peanut

Lack of marker polymorphism in the cultigen using restriction fragment-length polymorphism (RFLP) and random amplified polymorphic (RAPD) markers (Kochert et al. 1991; Halward et al. 1991) contributed to the hypothesis that all varieties and botanical types of *A. hypogaea* share common diploid progenitors (Kochert et al. 1996). RFLP analysis determined that *A. duranensis* had much greater similarity to *A. hypogaea* than did *A. cardenasii* (Kochert et al. 1991, 1996), leading to the conclusion that *A. duranensis* is the most likely A-genome ancestor. However, subsequent marker analyses have also proposed *A. villosa* (Raina and Mukai 1999), *A. helodes*, and *A. simpsonii* (Milla et al. 2005) as potential A-genome donors.

Molecular marker data from Kochert et al. (1991, 1996) supported *A. ipaënsis* instead of *A. batizocoi* as B genome donor. Fluorescent in situ hybridization analysis using rDNA as labeled probe also suggested *A. ipaënsis* as the donor (Raina and Mukai 1999; Seijo et al. 2004.) Only recently has it been possible to make hybrids between *A. ipaënsis* and *A. hypogaea* (Fávero et al. 2006). Additional work has confirmed the existence of up to 10 B-genome diploids (Krapovickas and Gregory 1994; Valls and Simpson 2005; Milla et al. 2005). This provides additional potential introgression routes into the cultigen.

17.3 Genetic Mapping and Tagging

Development of genetic linkage maps is a common method of studying the structure and organization of the genome of a species. Linkage maps also are needed for mapping of traits and marker-assisted selection, anchoring physical maps from large-insert libraries and positional cloning of important genes, comparisons of synteny within and across species, and ordered genome sequencing.

17.3.1 Markers and Genetic Linkage Mapping

In *Arachis*, five major linkage maps have been reported to date, all involving wild species because of limited molecular variability in *A. hypogaea* (Kochert et al. 1991; Halward et al. 1991). The first genetic linkage map was developed using an F_2 population of a cross between A-genome diploids *A. stenosperma* and *A. cardenasii*. The 117 mapped markers were distributed among 11 linkage groups over 1,063 cM (Halward et al. 1993). A second map based on these parents used a BC_1 population, and added 167 RAPD markers to a skeleton of 39 RFLP markers from the first map (Garcia et al. 2005). The largest map of peanut to date, and the only one to involve a parent from *A. hypogaea,* was constructed from a tetraploid cross of the cultivar Florunner × the synthetic amphidiploid TxAG-6 {*A. batizocoi* × [*A. cardenasii* ×

A. diogoi]}^{4x}. A total of 370 RFLP loci were mapped onto 23 linkage groups, for a map distance of 2,210 cM (Burow et al. 2001). The map was characterized by pairing of homoeologous linkage groups, consistent with a disomic nature of the cultigen.

The first microsatellite-based map was based on an F_2 population from a cross between A genome diploids *A. duranensis* and *A. stenosperma*. The published map had 170 microsatellite markers on 11 linkage groups covering 1,231 cM; further work increased the number of markers to 182 and created the expected 10 linkage groups. To create a map of the B genome, a diploid F_2 population was made by crossing *A. ipaënsis* and *A. magna*. This map has 130 mapped microsatellites and 10 linkage groups (Gobbi et al. 2006). The same marker set was used for both maps, and the relationships are mostly co-linear, in agreement with Burow et al. (2001).

The development of simple sequence repeats (SSRs) may allow mapping crosses involving two cultivated parents. The first characterization of peanut SSRs was made by Hopkins et al. (1999); and several papers have reported the majority of the additional markers (Ferguson et al. 2004b; He et al. 2005; Moretzsohn et al. 2005). This totals 702 markers, about 223 polymorphic for cultivated peanut. Approximately 100 additional SSR primers have been developed from genomic libraries (Umesh Reddy, personal communication) and approx. 2,000 have been developed from screening GenBank and unpublished expressed sequence tags (ESTs) (see Luo et al. 2005a). Short motif repeat number (5–15) dinucleotide repeats present in about 3–4% of ESTs are polymorphic among wild species, however, for cultivated peanut most of the polymorphisms have been found only with longer (15 or more) dinucleotide repeats (Moretzsohn et al. 2005). For most populations involving wild species, this is enough markers for map construction, and the increasing number of SSRs expected from sequencing efforts may make mapping cultivated × cultivated crosses possible soon.

Development of linkage maps from cultivated parents has proved difficult. A partial linkage map was constructed using an F_2 population (Herselman et al. 2004). Five linkage groups with 11 markers spanning 139.4 cM of the genome were reported. Reports of a map of cultivated peanut have been made by He et al. (1999) using AFLP markers; this has been extended by addition of AFLP and SSR markers (He, personal communication).

17.3.2 Placing Arachis *in a Unified Genetic Map of the Papilionoids*

Conservation of gene order is a major finding of genomics, with grasses and brassicas being outstanding examples. Comparative alignment has shown that major genes and quantitative trait loci are often conserved between different crops, providing a powerful tool to generate candidate genes, facilitating identification and cloning of genes that determine key traits. For legumes, a unified genetic map is just starting to be developed (Choi et al. 2004), but to date, peanut has been omitted from this effort.

Development of anchor markers for comparison between genomes is the limiting step in the construction of comparative genetic maps. A database of 459 candidate intron-based legume anchor primers pairs has been published (Fredslund et al. 2006). Using a subset of these primer pairs, 66 size polymorphic, cleaved amplified polymorphisms (CAPs) or derived CAPs (dCAPs) anchor markers have been developed on the A genome mapping population (Hougaard et al. 2005). This is allowing the first comparisons of the *Arachis, Lotus,* and *Medicago* genomes. Although analysis is still underway, some clear blocks of macrosynteny are apparent, for instance between *Arachis* LG6 and one end of *L. japonicus* LG1.

17.3.3 Markers and Phenotypic Analysis

The development of trait maps is a useful approach for evaluating the inheritance and feasibility of accelerating gains from selection. One important objective of gene tagging is marker-assisted selection (MAS). MAS is especially useful for traits which are difficult to measure, have low heritability, or are controlled by a few quantitative trait loci (QTLs) with large phenotypic variance.

A small but growing number of markers have been reported in peanut, almost all for biotic stress resistance. The first markers developed were for resistance to root-knot nematode (*Meloidogyne arenaria*). Root-knot nematode resistance was introduced into *A. hypogaea* from *A. cardenasii* (Garcia et al. 1996). RAPD and sequence characterized amplified region (SCAR) markers were identified for genes for reduced galling and egg number. Three RAPD markers were associated with nematode resistance in several backcross breeding populations derived from the interspecific hybrid TxAG-6, [*A. batizocoi* × (*A. cardenasii* × *A. diogoi*)]^{4x} (Burow et al. 1996). RFLP markers were used in breaking the linkage between resistance and low yield in development of the nematode-resistant variety NemaTAM (Church et al. 2000; Simpson et al. 2003).

Markers for additional traits have been developed also. Stalker and Mozingo (2001) identified RAPD markers explaining up to 35% of phenotypic variation for early and late spot resistance in an interspecific peanut population. Milla et al. (2005) reported AFLP-based markers for *A. cardenasii*-derived resistance to aflatoxin contamination, and a PCR-based marker for resistance to *Sclerotinia minor* have been reported by Chenault and Maas (2005). An AFLP marker explaining 76.1% of phenotypic variation for aphid resistance was identified in a cultivated cross (Herselman et al. 2004). Finally, SNP-based markers for high-oleic trait in peanut have been developed (Patel et al. 2004) and are being used to score a segregating F_2 population (Lopez and Burow 2004).

17.3.4 Defining Regions of the Genome that Control Disease Resistance

An important success of genomics has been the identification and mapping of entire classes of resistance genes. Several dozen such resistance genes have been described in the last decade (see Hammond-Kosack and Parker 2003). Many include the

NB-ARC domain, which is thought to act in signal transduction pathways, and which is commonly referred to as nucleotide binding site (NBS) domain genes.

Using degenerate primers that were designed using motifs conserved in these sequences, resistance gene analogs (RGAs) have been identified in peanut. Seventy-eight NBS-encoding regions were characterized (Bertioli et al. 2003). Of nine markers, four were mapped by RFLP and RGA-display on the A genome mapping population described by Moretzsohn et al. (2005) and Leal-Bertioli (unpublished results). An additional 234 sequences were identified and mapped using overgos onto 250 non-redundant BAC clones (Yüksel 2005).

17.4 Large-Insert Libraries and Physical Mapping

Genome-wide physical maps are important tools for ordered genome sequencing, targeted marker development, and positional cloning. Several technologies exist for making large-insert libraries, but bacterial artificial chromosome (BAC) libraries are currently the most useful method.

The first large-insert library for peanut was developed by Yüksel and Paterson (2005) from a cultivated component line of the variety 'Florunner'. The *Hind*III BAC library contained 182,784 clones, with an average insert size of 104 kb, giving an estimated 6.5× genome coverage. Potential problems in physical mapping are ambiguities associated with duplicated segments on homoeologous chromosomes and difficulties in anchoring contigs to duplicated genetic markers. To investigate the practicality of physical mapping in peanut, 117 oligonucleotide-based probes derived from genetically mapped RFLP probes were mapped onto these clones in a multiplex experimental design, and 91.5% of the overgos identified at least one BAC clone.

An alternative approach is to produce BAC libraries from diploid progenitors of peanut to eliminate problems specific to mapping a tetraploid. Recently, BAC libraries for each of the probable diploid ancestors of peanut, *A. duranensis* (A genome) and *A. ipaënsis* (B genome), have been developed (Guimarães, unpublished data). Both *Hind*III BAC libraries represent approximately 6.5 haploid genome equivalents and were constructed at CIRAD, Montpellier, France. The BAC library for *A. duranensis* contains 79,872 clones with an average insert size of 112 kb, and the *A. ipaënsis* library contains 77,184 clones with an average insert size of 100 kb.

17.5 Functional Genomics

Much of the genomics work in peanut to date has involved the development of markers, maps, and the classification of species. However, improvement of methods for gene and genome sequencing, and analysis of gene expression, promise to allow rapid strides in the identification and understanding of gene function and phenotype.

17.5.1 Gene Sequencing

Sequencing is one of the most important sources of information of genomic information. The peanut genome (2,800 Mb/1C) is large in comparison to many model plants such as *Arabidopsis* (128 Mb), rice (420 Mb), and *Medicago truncatula* (500 Mb), and is larger than soybean (1,100 Mb) and maize (2,500 Mb) (Arumuganathan and Earle 1991). The large genome size makes it unlikely that the peanut genome will be sequenced completely in the near future.

Sequencing of large numbers of expressed genes (expressed sequence tags, ESTs) can deliver substantial amounts of genetic information on protein-coding genes for comparative and functional genomics studies. As of November 6, 2006, 12,781 ESTs were deposited in GenBank. The majority were 7,454 and 2,184 sequences from seed libraries of the Chinese accessions Luhua14 and Shanyou 523, respectively, and 1,347 sequences from pod and leaf libraries of the U.S. varieties Tifrunner and C34–24. The latter are part of a larger set of 43,296 sequenced cDNA clones from 10 non-normalized peanut cDNA libraries (Luo et al. 2005a; Chen et al. 2006). From these, ca. 10,000 unique sequences have been identified. In addition, ca. 16,000 additional ESTs have been developed from normalized peanut libraries (H. T. Stalker personal communication), and smaller numbers are being sequenced by other programs.

An alternative to whole-genome sequencing is sequencing of gene-rich islands of hypomethylated DNA. A set of 1,312 genomic sequences was isolated from SSR-enriched libraries (Jayashree et al. 2005). A much larger set of sequences is under development, made by cloning of hypomethylated sequences isolated by methylation-filtration (S. J. Knapp personal communication) A set of 2,989 and 6,528 sequences from methylation-filtered and unfiltered libraries, respectively, of the cultigen, and A and B genome diploids *A. duranensis* and *A. batizocoi* have been deposited. This approach has the potential to identify genes of low expression levels that are under-represented in cDNA libraries, and to compare the A and B peanut genomes. In all, 25,259 *Arachis* sequences of all types were present in GenBank at the time of writing of this manuscript.

17.5.2 Analysis of Gene Expression

The study of gene expression has great power for associating genes with phenotype. Macroarray (nylon-based) and microarray (glass slide-based) screening methods allow for the simultaneous determination of the expression levels of thousands of genes, making it possible to attain a global view of the transcriptional state in a cell or tissue and to associate genes with functions or specific physiological conditions.

Several small scale studies have been performed. The first involved response of peanut to drought stress (Jain et al. 2001), in which 43 differentially regulated transcripts were identified by differential display. More recently, microarray technology has been demonstrated in two publications (Luo et al. 2005b, 2005c). In these,

EST-derived microarrays of ca. 400 unigenes were probed under different conditions. Twenty-five ESTs potentially associated with drought stress and response to *A. parasiticus* were identified. Likewise, 56 up-regulated transcripts were identified and confirmed by real-time PCR upon infection with *Cercospora arachidicola*. A 70-mer oligonucleotide microarray consisting of more than 10,000 gene elements is under development (Guo, unpublished results).

17.5.3 TILLING for Analysis of Gene Function and for Generation of Mutant Phenotypes

In addition to the study of gene expression by analysis of transcript levels, gene function can be studied by various types of mutational analysis or use of gene traps. The size of the peanut genome would require a large number of insertional mutant lines and an efficient transformation system, although an initial effort has been made to develop promoter traps (Anuradha et al. 2006). Targeting induced local lesions in genomes (TILLING), using ethylmethanesulfonate to generate a population of mutants and to use specific gene sequences to identify mutations, does not suffer from the same drawback because mutation frequency (number of substitutions per bp of DNA) is the key factor (Henikoff et al. 2004).

TILLING in peanut is being undertaken with the initial aim of knocking out *Ara*h2 (allergen) genes (Ozias-Akins, unpublished results). High-quality genomic sequences and an assessment of number of gene copies of *Ara*h2 were lacking until recently (Ramos et al. 2006). This information was necessary to allow the design of primers that would amplify only the A- or B-genome copy of A*rah2*, not both. After producing a pilot EMS-mutagenized population in peanut and screening 4-fold pools of 384 individuals for mutations in *Arah2*, the mutation frequency appears to be comparable to that achieved in *Arabidopsis*, i.e., 1/170 kb (Greene et al. 2003). Variations in mutagen treatment currently are being tested to scale up the size of the population. For *Arabidopsis*, a population of ca. 6,900 mutants is available, but a sufficient number of mutations (ca. eight per gene) was found after screening about half that number (Greene et al. 2003).

17.5.4 Transformation

Transformation and regeneration are important components of the success of future genomics work in peanut. A high-throughput system is needed for determination (or confirmation) of function of the rapidly-expanding number of gene sequences. In addition, transformation is instrumental in development of transgenic plants with important value-added traits.

Regeneration of genetically engineered peanuts was accomplished in 1993 (Ozias-Akins et al. 1993). *Agrobacterium tumefaciens*-mediated transformation, typically performed on shoot-forming cultures using neomycin phosphotransferase (*nptII*) as a selectable marker, is characterized by strong genotype specificity for Valencia-type peanuts (Cheng et al. 1996). The advantage of*Agrobacterium* transformation is the higher frequency of low-copy number inserts (Sharma and Anjaiah 2000) and a reduced amount of gene silencing. Microprojectile bombardment of somatic embryos using hygromycin phosphotransferase (*hph*) as a selectable marker is the most broadly applicable method for peanut by virtue of working across genotypes and has been performed by multiple groups. Although high copy number and gene rearrangements remain more common with bombardment, regenerated lines with low-copy number inserts can be selected by Southern blot analysis.

Concerns regarding peanut transformation/regeneration for genomics and varietal development include low efficiency, sterility of some regenerants, and the 12 to 18 months required for the process (Egnin et al. 1998). The use of antibiotic resistance genes is a concern for consumer acceptance, but it is possible to visually select for transformed embryogenic tissues expressing green fluorescent protein (Joshi et al. 2005).

Numerous genes have been introduced into peanut (Ozias-Akins and Gill 2001; Ozias-Akins 2005), and most involve disease resistance. Coat protein-mediated virus resistance was shown first in tobacco (Powell et al. 1986), and subsequent research has shown pathogen-derived resistance to be applicable to many viruses and crop species. Greenhouse and field studies have shown that peanut lines expressing a sense or antisense nucleocapsid protein gene of TSWV can show elevated levels of tolerance to virus infection (Li et al. 1997; Magbanua et al. 2000; Yang et al. 2004). A similar strategy resulted in resistance to peanut stripe potyvirus (Higgins et al. 2004) and peanut clump virus (see Dar et al. 2006).

Peanut production also suffers from several fungal diseases. Chitinase, glucanase, or oxalate oxidase genes have been introduced to target sclerotinia blight (Chenault et al. 2005; Livingstone et al. 2005). The maize *RIP* 1 gene has been reported to reduce aflatoxin in transgenic peanut (Weissinger et al. 2006). Some strategies for reducing problems caused by fungi are indirect. For example, the insect pest the lesser cornstalk borer can cause damage to peanut pods and in the process also inoculate pods with *Aspergillus*. A reduction in pod damage due to the expression of *cryIA*(c) may also reduce pod infection with *Aspergillus* and consequent aflatoxin production (Ozias-Akins et al. 2002). In addition, attempts are underway to introduce the dehydrin-responsive element (DREB) element for drought resistance, which could also reduce aflatoxin contamination (see Dar et al. 2006).

17.6 Proteomics and Allergen Genes

Food allergies are common in the population, and involve various foods, including fish, milk, eggs, soybean, and tree nuts. Some of the major peanut allergens are seed storage proteins. Conarachin is known as *Ara*h1, a vicilin, and genes for expressing this protein have been described (Burks et al. 1995). Similarly, genes for arachin (glycinin, *Ara*h3) (Rabjohn et al. 1999) and *Ara*h2 (conglutin) (Stanley et al. 1997) have been cloned and sequenced. *Ara*h1 and *Ara*h2 are considered to be the major allergens because they are recognized by serum IgE from > 90% of peanut allergic individuals (Burks et al. 1998). *Ara*h1 and *Ara*h3 probably are encoded by multigene families, whereas *Ara*h2 has been shown to derive from two genes in cultivated peanut, one from each genome (Ramos et al. 2006). *Ara*h4 through *Ara*h7 are other seed proteins, present in smaller amounts (Kelber-Jancke et al. 1999). Up to six additional allergens have been identified (De Jong et al. 1998).

High-resolution 2-D gels combined with matrix-assisted laser desorption/ionization time-of-flight (MALDI-TOF) analysis is being applied to the characterization of allergens, including peanut allergens (Law et al. 2005; Boldt et al. 2005). Several accessions have been found to lack one or more allergen subunits (Liang et al. 2006). These methods are essential for the manipulation of peanut seed proteins by mutagenesis or transgenic approaches.

17.7 Biodiversity and Markers

One of the strengths of peanut is the large number of accessions in germplasm collections (Holbrook and Stalker 2002). These harbor sources of alleles that have been underused. Development of core collections is a major emphasis, and application of genomics to these is expected to enhance the use of these resources.

17.7.1 Collections

There are four major peanut germplasm collections in the world. The largest collection consists of 14,966 accessions of cultivated and 453 accessions of 44 wild *Arachis* species from 93 countries; this collection is housed at the International Crops Research Institute for the Semi-Arid Tropics (ICRISAT), Patancheru, India. Other major holders are the USDA Southern Regional Plant Introduction Station, Griffin, GA, USA (9,027 accessions), the National Research Center for Groundnut, Junagadh, India (7,935 accessions), and the Chinese national collection (5,890 accessions).

17.7.2 Assessing Phenotypic Diversity and Enhancing Germplasm Use

Core collections (ca. 10% of the entire collection) of large germplasm collections have been established for greater ease of screening. These include 1,704 (ICRISAT), 831 (US), and 582 accessions (China) (Upadhyaya et al. 2003; Holbrook et al. 1993; Jiang et al. 2004). The phenotypic diversity of the ICRISAT core and important descriptor traits were determined by Upadhyaya et al. (2003) Evaluation was performed for 16 morphological and 15 agronomic characters during the rainy and post-rainy seasons. The average phenotypic diversity index was higher in the *fastigiata* group (0.146) than the *hypogaea* group (0.141), but the maximum phenotypic diversity (0.453) was observed between two *hypogaea* accessions. The two subspecies differed significantly for all traits except leaflet surface and oil content. Principal coordinate and principal component analysis showed that 12 morphological descriptors and 15 agronomic traits were important in explaining multivariate polymorphism.

Evaluation of the U.S. core resulted in identification of resistance to root-knot nematode (*Meloidogyne arenaria* (Neal) Chitwood race 1 (Holbrook et al. 2000), tomato spotted wilt (Anderson et al. 1996), cylindrocladium black rot [*Cylindrocladeum crotalariae* (Loos) Bell and Sobers] and early leafspot (*Cercospora arachidicola* Hori) (Isleib et al. 1995), rhizoctomia limb rot (*Rhizoctonia solani* Kuhn) (Franke et al. 1999), sclerotinia blight [*Sclerotinia minor* Jagger], and reduced preharvest aflatoxin contamination (Holbrook et al. 1998).

To further facilitate the use of germplasm accessions, Upadhyaya, et al. (2002) and Holbrook and Dong (2005) developed peanut minicore collections (1% of entire collection) consisting of only 184 and 111 accessions, respectively. The evaluation of the ICRISAT core and minicore collections led to identification of new sources for drought tolerance (Upadhyaya 2005) and early maturity (Upadhyaya et al. 2006a). Preliminary evaluation of the U. S. minicore has identified new sources of heat tolerance (Kottapalli, unpublished results).

Wild species are also an important potential source of alleles. Alleles for strong resistance to various diseases and pests are present in wild species, including root-knot nematodes (*Meloidogyne arenaria* (Neal) Chitwood), early leafspot (*Cercospora arachidicola* Hori), late leafspot (*Cercosporidium personatum* (Berk. et Curt.) Deighton), rust (*Puccinia arachidis* Speg.), groundnut rosette virus, tomato spotted wilt virus, peanut stunt virus, peanut mottle virus, and lesser cornstalk borer (*Elasmopalpus lignioselIus* Zeller) (Stalker and Moss 1987).

17.7.3 Use of Markers for Measuring and Using Diversity

Efficient use of germplasm collections would be improved by identification of accessions possessing different alleles for specific traits and by development of selectable markers for breeding.

Recent studies have disproved the belief that cultivated peanut lacks variation at the molecular level (He and Prakash, 1997; Subramanian et al. 2000). SSR markers were shown to be useful for detecting diversity in cultivated peanut and can be used for population studies (Moretzsohn et al. 2004; Ferguson et al. 2004a). ICRISAT is using SSR markers to analyze genetic diversity in cultivated germplasm resistant to late leafspot, rust, and bacterial wilt (Mace et al. 2006, 2007). In some cases, more than half of the markers detected polymorphism with polymorphism information content (PIC) values of over 0.5. A more elaborate study involving about 1,000 accessions and 21 SSRs revealed 491 alleles (5–46 alleles/locus) (Upadhyaya et al. 2006b). The mean PIC value was 0.796, and *fastigiata* and *hypogaea* subspecies formed different clusters. The 184 minicore accessions accounted for about 75% alleles of the cultivated accessions in the composite collection. However, wild species had greater diversity: 52 accessions of 14 wild *Arachis* species had more alleles (373) than 333 accessions of *hypogaea* (308 alleles) and 365 *fastigiata* accessions (365 alleles).

Likewise, analysis of the U. S. core subset demonstrated considerable molecular diversity (Kottapalli et al. 2007). Moderate levels of genetic variation were found with genetic distances (D) values among accessions ranging from 0.088 to 0.254. Seventy-two primers amplified 528 bands. PIC values ranged from 0.027 to 0.375, with an average value of 0.15, and from two to 28 polymorphic bands per primer were observed. Distinct groupings of the accessions were observed based on subspecies, and runner/Virginia, Spanish, and Valencia market types were clearly distinguished for approximately 90% of the accessions tested. Twelve of the markers, mapped previously to the A genome, were found sufficient to identify subspecies and botanical types and gave a clustering pattern very similar to the entire 67 SSR marker set.

17.8 Bioinformatics

One of the important needs for genomics has been the informatics resources for analysis of the large amounts of data produced and for comparison of data among species. A map database for peanut, called PeanutMap, has been published recently (Jesubatham and Burow 2006). Peanut Map contains the published maps of the peanut genome, plus smaller map sets of markers associated with traits. The database software allows comparison among linkage groups in a map, showing marker correspondences among homoeologous chromosomes, as well as among maps from different publications.

As awareness of the significance of comparative genomics has increased; databases encompassing data from multiple species are being created. A cross-legume database, called the Legume Information System, has been released to incorporate species-specific data and permit cross-legume comparisons (Gonzales et al. 2005). This will permit data held in different legume databases to be used for comparison of synteny among different species. Legume Information System also incorporates sequence data, making it possible to search for genes expressed in different tissues and at different physiological conditions.

17.9 Perspective

Peanut genomics is beginning to make the progress needed for greater utility in genetic improvement of the species. Advances in peanut have been hindered by a large genome, apparent lack of polymorphism in the cultigen, difficulty in interspecific gene transfer, and a slow transformation/regeneration system the successes of which have been limited by the lack of marketability of transgenic peanut.

These and other limitations are changing for several reasons. There is increased realization of the genetic variability present in the genus, both in the cultigen and wild species. Development of SSR markers has been successful in identifying this variation, making marker work in both the cultigen and wild species feasible. The development of tools in parallel for the cultigen and A and B genome diploids will help in simplifying the work of understanding the organization and evolution of peanut. Mapping of simple polymerase chain reaction-based markers will assist with paving a way for QTL fine mapping and efficient MAS. There is still an unmet need for SNP-based maps and markers. Combined with recent theoretical developments in linkage disequilibrium and genetic association mapping, additional options may be available for identification of biotic and abiotic stress tolerance markers.

Construction of multiple cDNA and genomic libraries and the beginnings of significant sequencing are making data available that can be used for multiple projects. Identification of EST unigene sets and genomic sequences will be useful, especially in comparison with sequencing of the *Medicago* and soybean genomes. Sequence matching of peanut genes with other genomes will identify putative orthologous loci and facilitate transfer of information on gene function, and because of colinearity among legume genomes, mapping of genes can be extended to other species.

Work on TILLING, transformation, and gene expression analysis will assist with the understanding of gene function. TILLING has the potential to overcome some of the disadvantages of the large peanut genome size, and development of transformation-competent BAC libraries and more-efficient transformation systems would assist with studies of gene function. Identification of critical gene pathways and specific genes will be useful for conventional and transgenic improvement programs.

References

Anuradha TS, Jami SK, Datla RS, Kirti PB (2006) Genetic transformation of peanut (*Arachis hypogaea* L.) using cotyledonary node as explant and a promoterless *gus::npt*II fusion gene based promoter. J Biosci 31:235–246

Anderson WG, Holbrook CC, Culbreath AK (1996) Screening the core collection for resistance to tomato spotted wilt virus. Peanut Sci 23:57–61

Arumuganathan K, Earle ED (1991) Nuclear DNA content of some important plant species. Plant Mol Biol Rept 9:208–218

Balikrishnan R, Chombreeda P, Rimkere H (1998) Gender roles in peanut sector for household food security. Bangkok: Kesetsart Agricultural and Agro-Industrial Product Improvement Institute

Bertioli DJ, Leal-Bertioli SCM, Lion MB, Santos VL, Pappas JR, et al. (2003) A large scale analysis of resistance gene homologues in *Arachis*. Mol Gen Genomics 270:34–45

Boldt A, Fortunato D, Conti A, Peterson A, Ballmer-Weber B, et al. (2005) Analysis of the composition of an immunoglobulin E reactive high molecular weight protein complex of peanut extract containing Ara h 1 and Ara h 3/4. Proteomics 5:675–686

Burks W, Sampson HA, Bannon GA (1998) Peanut allergens. Allergy 53:725–730

Burks AW, Cockrell G, Stanley JS, Helm RM, Bannon GA (1995) Isolation, identification, and characterization of clones encoding antigens responsible for peanut hypersensitivity. Internatl Arch Allergy Immunol 107:248–250

Burow MD, Simpson CE, Paterson AH, Starr JL (1996) Identification of peanut (*Arachis hypogaea* L.) RAPD markers diagnostic of root – knot nematode (*Meloidogyne arenaria* (Neal) Chitwood) resistance. Mol Breed 2:369–379

Burow MD, Simpson CE, Starr JL, Paterson AH (2001) Transmission genetics of chromatin from a synthetic amphidiploid to cultivated peanut (*Arachis hypogaea* L.): Broadening the gene pool of a monophyletic polyploidy species. Genetics 159:823–837

Chen H, Holbrook CC, Guo BZ (2006) Peanut seed transcriptome: construction of six peanut seed cDNA libraries from two peanut cultivars. Am Peanut Res Educ Soc

Chenault KD, Maas A (2005) Identification of a simple sequence repeat (SSR) marker in cultivated peanut (*Arachis hypogaea* L.) potentially associated with sclerotinia blight resistance. Proc Am Peanut Res Educ Soc pp 24–25

Chenault KD, Melouk HA, Payton ME (2005) Field reaction to sclerotinia blight among transgenic peanut lines containing antifungal genes. Crop Sci 45:511–515

Cheng M, Jarret R, Li Z, Xing A, Demski J (1996) Production of fertile transgenic peanut (*Arachis hypogaea* L.) plants using *Agrobacterium tumefaciens*. Plant Cell Rep 15:653–657

Choi H-K, Mun J-H, Kim D-J, Zhu H, Baek J-M, et al. (2004) Estimating genome conservation between crop and model legume species. Proc Natl Acad Sci USA 101:15289–15294

Church GT, Simpson CE, Burow MD, Paterson AH, Starr JL (2000) Use of RFLP markers for identification of individuals homozygous for resistance to *Meloidogyne arenaria* in peanut. Nematology 2:575–580

Cruickshank AW, Rachaputi NC, Wright GC, Nigam SN (2003) Breeding of drought-resistant peanuts. Canberra: ACIAR

Dar WD, Reddy BVS, Gowda CLL, Ramesh S (2006) Genetic resources enhancement of ICRISAT-mandate crops. Curr Sci 91:880–884

De Jong EC, van Zijverden M, Spanhaak S, Koppelman SJ, Pellegrom H (1998) Identification and partial characterization of multiple major allergens in peanut proteins. Clin Exp Allergy 28:743–751

Egnin M, Mora A, Prakash CS (1998) Factors enhancing *A. tumefaciens*-mediated gene transfer in peanut (*Arachis hypogaea* L.) In Vitro Cell Dev Biol Plant 34:310–318

Fávero AP, Simpson CE, Valls JFM, Vello NA (2006) Study of the evolution of cultivated peanut through crossability studies among *Arachis ipaënsis*, *A. duranensis*, and *A. hypogaea*. Crop Sci 46:1546–1622

Ferguson M, Bramel P, Chandra S (2004a) Gene diversity among botanical varieties in peanut (*Arachis hypogaea* L.) Crop Sci 44:1847–1855

Ferguson ME, Burow MD, Schulze SR, Bramel PJ, Paterson AH, et al. (2004b) Microsatellite identification and characterization in peanut (*A. hypogaea* L.). Theor Appl Genet 108: 1064–1070

Fernández A, Krapovickas A (1994) Cromosomas y evolucion en *Arachis (Leguminosae)*. Bonplandia 8:187–220

Franke MD, Brenneman TB, Holbrook CC (1999) Identification of resistance to rhizoctonia limb rot in a core collection of peanut germplasm. Plant Dis 83:944–948

Fredslund J, Madsen LH, Hougaard BK, Nielsen AM, Bertioli D, et al. (2006) A general pipeline for the development of anchor markers for comparative genomics in plants. BMC Genomics 7:207

Garcia GM, Stalker HT, Shroeder E, Kochert G (1996) Identification of RAPD, SCAR and RFLP markers tightly linked to nematode resistance genes introgressed from *Arachis cardenasii* to *A. hypogaea*. Genome 39:836–845

Garcia GM, Stalker HT, Shroeder E, Lyerly JH, Kochert G (2005) A RAPD-based linkage map of peanut based on a backcross population between the two diploid species *Arachis stenosperma* and *A. cardenasii*. Peanut Sci 32:1–8

Gepts P, Beavis WD, Brummer EC, Shoemaker R, Stalker HT, et al. (2005) Legumes as a model plant family. Genomics for food and feed report of the cross-legume advances through genomics conference. Plant Physiol 137:1228–1235

Gobbi A, Texeira C, Moretzsohn M, Guimarães P, Bertioli SL, et al. (2006) Development of a linkage map to species of B genome related to the peanut (*Arachis hypogaea* – AABB). Plant Animal Genome P679

Gonzales MD, Archuleta E, Farmer A, Kajendran K, Grant D, et al. (2005) The Legume Information System: an integrated information resource for comparative legume biology. Nucleic Acids Res 33:D660–D665

Greene EA, Codomo CA, Taylor NE, Henikoff JG, Till BJ, et al. (2003) Spectrum of chemically induced mutations from a large-scale reverse-genetic screen in *Arabidopsis*. Genetics 164:731–740

Halward T, Stalker HT, Kochert G (1993) Development of an RFLP linkage map in diploid peanut species. Theor Appl Genet 87:379–384

Halward T, Stalker HT, Laure EA, Kochert G (1991) Genetic variation detectable with molecular markers among unadapted germplasm resources of cultivated peanut and related wild species. Genome 34:1013–1020

Hammond-Kosack KE, Parker JE (2003) Deciphering plant-pathogen communication - fresh perspectives in molecular resistance breeding. Curr Opin Biotech 14:177–193

He G, Prakash CS (1997) Identification of polymorphic DNA markers in cultivated peanut (*Arachis hypogaea* L.) Euphytica 97:143–149

He GH, Li J, Prakash CS, Smith OD, Lopez Y (1999) AFLP mapping and QTL analysis in cultivated peanut (*Arachis hypogaea* L.). Plant and Animal Genome. P236

He GH, Meng R, Gao H, Guo B, Gao G, et al. (2005) Simple sequence repeat markers for botanical varieties of cultivated peanut (*Arachis hypogaea* L.). Euphytica 142:131–136

Henikoff S, Till BJ, Comai L (2004) TILLING. Traditional mutagenesis meets functional genomics. Plant Physiol 135:630–636

Herselman L, Thwaites R, Kimmins FM, Courtois B, van der Merwe PJA, et al. (2004) Identification and mapping of AFLP markers linked to peanut (*Arachis hypogaea* L.) resistance to the aphid vector of groundnut rosette disease. Theor Appl Genet 109:1426–1433

Higgins C, Hall R, Mitter N, Cruickshank A, Dietzgen R (2004) Peanut stripe potyvirus resistance in peanut (*Arachis hypogaea* L.) plants carrying viral coat protein gene sequences. Transgenic Res 13:59–67

Holbrook CC, Dong W (2005) Development and evaluation of a mini core collection for the US peanut germplasm collection. Crop Sci 45:1540–1544

Holbrook CC, Stalker HT (2002) Peanut Breeding and Genetic Resources. Plant Breed. Rev 22:297–356

Holbrook CC, Wilson DM, Matheron ME (1998) Source of resistance to pre-harvest aflatoxin contamination in peanut. Proc Am Peanut Res Educ Soc 30:54

Holbrook C, Anderson W, Pittman R (1993) Selection of a core collection from the U.S. germplasm collection of peanut. Crop Sci 33:859–861

Holbrook CC, Stephenson MG, Johnson AW (2000). Level and geographical distribution of resistance to *Meloidogyne arenaria* in the U.S. peanut germplasm collection. Crop Sci 40:1168–1171

Hopkins MS, Casa AM, Wang T, Mitchell SE, Dean RE, et al. (1999) Discovery and characterization of polymorphic simple sequence repeats (SSRs) in peanut. Crop Sci 39: 1243–1247

Hougaard BK, Madsen LH, Fredslund J, Schauser L, Nielsen AM, et al. (2005) Development of legume anchor markers: A bioinformatic driven tool for genetic mapping and comparative genomics. Phaseomics IV (abstract)

Husted L (1936) Cytological studies of the peanut Arachis. II. Chromosome number, number morphology and behavior and their application to the origin of cultivated forms. Cytologia 7:396–423

Isleib TG, Beute MK, Rice PW, Hollowell JE (1995) Screening the peanut core collection for resistance to Cylindrocladium black rot and early leaf spot. Proc Am Peanut Res Educ Soc 27:25

Jain AK, Basha SM, Holbrook CC (2001) Identification of drought-responsive transcripts in peanut (*Arachis hypogaea* L.). Elect J Biotech 4:2

Jarvis A, Ferguson ME, Williams DE, Guarino L, Jones PG, et al. (2003) Biogeography of wild *Arachis*: assessing conservation status and setting future priorities. Crop Sci 43:1100–1108

Jesubatham AM, Burow MD (2006) PeanutMap: an online genome database for comparative molecular maps of peanut. BMC Bioinformatics 7:375

Jayashree B, Ferguson M, Ilut D, Doyle J, Crouch JH (2005) Analysis of genomic sequences from peanut (*Arachis hypogaea*) Elec J Biotech 8:226–237

Jiang H, Liao B, Duan N, Holbrook CC, Guo B (2004) Development of a core collection of peanut germplasm in China. Proc Am Peanut Res Educ Soc 36:33

Joshi M, Niu C, Fleming G, Hazra S, Chu Y et al. (2005) Use of green fluorescent protein as a non-destructive marker for peanut genetic transformation. *In Vitro* Cell Dev Biol Plant 41:437–445

Kleber-Jancke T, Crameri R, Appenzeller U, Schlaak M, Becker WM (1999) Selective cloning of peanut allergens, including profiling and 2S albumins, by phage display technology. Internatl Arach Allergy Immunol 119:265–274

Knauft D, Ozias-Akins P (1995) Recent methods for germplasm enhancement and breeding. In: Pattee HE, Stalker HT (eds), Advances in peanut science. Stillwater: APRES pp. 54–94

Kochert G, Halward T, Branch WD, Simpson CE (1991) RFLP variability in peanut (*Arachis hypogaea* L.) cultivars and wild species. Theor Appl Genet 81:565–570

Kochert G, Stalker HT, Gimenes M, Galgaro SL, Lopes CR, et al. (1996) RFLP and cytological evidence on the origin and evolution of allotetraploid domesticated peanut, *Arachis hypogaea* (Leguminosae). Am J Bot 83:1282–1291

Kottapalli KR, Burow MD, Burow G, Burke J, Puppala N (2007) Molecular characterization of the U.S. peanut minicore using microsatellite markers. Crop Sci 47:1718–1727

Krapovickas A, Gregory WC (1994) Taxonomía del género *Arachis* (Leguminosae). Bonplandia 8:1–186

Law A, Gupta N, Louie M, Poddar R, Ray A, et al. (2005) Identification and characterization of plant allergens using proteomic approaches. Curr Proteomics 2:147–164

Li Z, Jarret R, Demski J (1997) Engineered resistance to tomato spotted wilt virus in transgenic peanut expressing the viral nucleocapsid gene. Transgenic Res 6:297–305

Liang XQ, Luo M, Holbrook CC, Guo BZ (2006) Storage protein profiles in Spanish and runner market type peanuts and potential markers. BMC Plant Biol 6:24

Livingstone DM, Hampton JL, Phipps PM, Grabau EA (2005) Enhancing resistance to *Sclerotinia minor* in peanut by expressing a barley oxalate oxidase gene. Plant Physiol 137:1354–1362

Lopez Y, Burow MD (2004) Development and validation of CAPS markers for the high oleate trait in peanuts. Proc Am Peanut Res Educ Soc 36:25–26

Luo M, Dang P, Guo BZ, He G, Holbrook CC, et al. (2005a) Generation of expressed sequence tags (ESTs) for gene discovery and marker development in cultivated peanut. Crop Sci 45:346–353

Luo M, Dang P, Holbrook CC, Bausher MG, Lee RD, et al. (2005b) Identification of transcripts involved in resistance responses to leaf spot disease caused by *C. personatum* in peanut (*A hypogaea* L.). Phytopathol 95:381–387

Luo M, Liang X, Dang P, Holbrook CC, Baushe MG, et al. (2005c) Microarray-based screening of differentially expressed genes in peanut in response to *Aspergillus parasiticus* infection and drought stress. Plant Sci 169:695–703

Mace ES, Phong DT, Upadhyaya HD, Chandra S, Crouch JH (2006) SSR analysis of cultivated groundnut (*Arachis hypogaea* L.) germplasm resistant to rust and late leaf spot diseases. Euphytica 152:317–330

Mace ES, Yuejin W, Boshou L, Upadhyaya HD, Chandra S, et al. (2007) SSR-based diversity analysis of groundnut (*Arachis hypogaea* L.) germplasm resistant to bacterial wilt. Plant Genetic Res 5:27–36

Magbanua Z, Wilde H, Roberts J, Chowdhury K, Abad J, et al. (2000) Field resistance to tomato spotted wilt virus in transgenic peanut (*Arachis hypogaea* L.) expressing an antisense nucleocapsid gene sequence. Mol Breed 6:227–236

Milla SR, Isleib TG, Tallury SP (2005) Identification of AFLP markers linked to reduced aflatoxin accumulation in *A. cardenasii*-derived germplasm lines of peanut. Proc Am Peanut Res Educ Soc 37:90

Moretzsohn MC, Hopkins MS, Mitchell SE, Kresovich S, Valls JFM, et al. (2004) Genetic diversity of peanut (*Arachis hypogaea*) and its wild relatives based on the analysis of hyper variable regions of the genome. BMC Plant Biol 4:11

Moretzsohn MC, Leoi L, Proite K, Guimarães PM, Leal-Bertioli SCM, et al. (2005) A micro satellite–based, gene–rich linkage map for the AA genome of *Arachis* (Fabaceae). Theor Appl Genet 111:1060–1071

National Peanut Genome Initiative (2005) Accomplishment Report. USDA-ARS

O'Byrne DJ, Knauft DA, Shireman RB (1997) Low fat-monounsaturated rich diets containing high-oleic peanuts improve serum lipoprotein profiles. Lipids 32:687–695

Ozias-Akins P (2005) Peanut. In: Put EC, Davey MR (eds) Biotechnology in Agriculture and Forestry – Tropical Crops I. Heidelberg: Springer-Verlag

Ozias-Akins P, Gill R (2001) Progress in the development of tissue culture and transformation methods applicable to the production of transgenic peanut. Peanut Sci 28:123–131

Ozias-Akins P, Schnall JA, Anderson WF, Singsi, C, Clemente TE, et al. (1993) Regeneration of transgenic peanut plants from stably transformed embryogenic callus. Plant Sci 93:185–194

Ozias-Akins P, Yang H, Gill R, Fan H, Lynch RE (2002) Reduction of aflatoxin contamination in peanut: genetic engineering approach. ACS Symp Series 829:151–160

Patel M, Jung S, Moore K, Powell G, Ainsworth C, et al. (2004) High-oleate peanut mutants result from a MITE insertion into the FAD2 gene. Theor Appl Genet 108:1492–1502

Peanut Institute. (2006) http:// www.peanut-institute.org/NutritionBasics.html.

Porter DM, Smith DH, Rodríguez-Kábana R (1990) Compendium of Peanut Diseases. 2nd edn. St. Paul: APS Press

Powell A, Nelson R, De B, Hoffmann N, Rogers S, et al. (1986) Delay of disease development in transgenic plants that express the tobacco mosaic virus coat protein gene. Science 232:738–743

Rabjohn P, Helm EM, Stanley JS, West CM, Sampson HA, et al. (1999) Molecular cloning and epitope analysis of the peanut allergen Arah3. J Clin Invest 103:535–542

Raina SN, Mukai Y (1999) Genomic in situ hybridization in *Arachis* (Fabaceae) identifies the diploid wild progenitors of cultivated (*A. hypogaea*) and related wild (*A. monticola*) peanut species. Plant Sys Evol 214:1–4

Ramos ML, Fleming G, Chu Y, Akiyama Y, Gallo M, et al. (2006) Chromosomal and phylogenetic context for conglutin genes in *Arachis* based on genomic sequence. Mol Gen Genomics 275:578–592

Reddy TY, Reddy VR, Anbumozhi V (2003) Physiological responses to groundnut (*Arachis hypogea* L.) to drought stress and its amelioration: a critical review. Plant Growth Regul 41: 75–88

Seijo JG, Lavia GI, Fernández A, Krapovickas A, Ducasse D, et al. (2004) Physical mapping of the 5S and 18S–25S rRNA genes by FISH as evidence that *A. duranensis* and *A. ipaënsis* are the wild diploid progenitors of *A. hypogaea* (Leguminosae). Am J Bot 91:1294–1303

Sharma KK, Anjaiah V (2000) An efficient method for the production of transgenic plants of peanut (*Arachis hypogaea* L.) through *Agrobacterium tumefaciens*-mediated genetic transformation. Plant Sci 159:7–19

Simpson CE, Krapovickas A, Valls JFM (2001) History of *Arachis* including evidence of *A. hypogaea* L. progenitors. Peanut Sci 28:78–80

Simpson CE, Starr JL, Church GT, Burow MD, Paterson AH (2003) Registration of 'NemaTAM' Peanut. Crop Sci 43:1561

Singh AK, Simpson CE (1994) Biosystematics and genetic resources. In: Smartt J . (ed), The groundnut crop: a scientific basis for improvement. London: Chapman and Hall. pp 96–137

Smartt J, Gregory WC, Gregory MP (1978) The genomes of *Arachis hypogaea*. 1. Cytogenetic studies of putative genome donors. Euphytica 27:665–675

Stalker HT, Dalmacio RD (1986) Karyotype relationships and analysis among varieties of *Arachis hypogaea* L. Cytologia 51:167–629

Stalker HT, Moss JP (1987). Speciation, cytogenetics, and utilization of *Arachis* species. Adv Agron 41:1–40

Stalker HT, Mozingo LG (2001) Molecular markers of *Arachis* and marker-assisted selection. Peanut Sci 28:117–123

Stanley JS, King N, Burks AW, Huang SK, Sampson H, et al. (1997) Identification and mutational analysis of the immunodominant IgE binding epitopes of the major peanut allergen Ara h 2. Arch Biochem Biophys 342:244–253

Subramanian V, Gurtu S, Nageswara Rao RC, Nigam SN (2000) Identification of DNA polymorphism in cultivated groundnut using random amplified polymorphic DNA (RAPD) assay. Genome 43:656–660

Upadhyaya HD (2005) Variability for drought resistance related traits in the mini core collection of peanut. Crop Sci 45:1432–1440

Upadhyaya HD, Bramel PJ, Ortiz R, Singh S (2002) Developing a mini core of peanut for utilization of genetic resources. Crop Sci 42:2150–2156

Upadhyaya HD, Ortiz R, Bramel PJ, Singh S (2003) Development of a groundnut core collection using taxonomical, geographical and morphological descriptors. Genet Res Crop Evol 50:139–148

Upadhyaya HD, Reddy LJ, Gowda CLL, Singh S (2006a) Identification of diverse groundnut germplasm: Sources of early-maturity in a core collection. Field Crop Res 97:261–267

Upadhyaya HD, Bhattacharjee R, Hoisington DA, Chandra S, Varshney R, et al. (2006b) Molecular characterization of groundnut (*Arachis hypogaea* L.) composite collection. Research Meeting of the Generation Challenge Program.

USDA-FAS (2006) USDA Foreign Agricultural Service, Circular, WAP-05–06, May 2006

Valls JFM, Simpson CE (2005) New species of *Arachis* (Leguminosae) from Brazil, Paraguay and Bolivia. Bonplandia 14:35–63

Yang H, Ozias-Akins P, Culbreath A, Gorbet D, Weeks J, et al. (2004) Field evaluation of Tomato spotted wilt virus resistance in transgenic peanut (*Arachis hypogaea*). Plant Dis 88:259–264

Weissinger A, Wu M, Isleib T, Stalker T, Shew B, et al. (2006). Expression of an active form of maize *RIP* 1 in transgenic peanut inhibits fungal infection and aflatoxin contamination. Fungal Genomics Workshop

Yüksel B, Paterson AH (2005) Construction and characterization of peanut *HindIII* BAC library. Theor ApplGenet 111:630–639

Yüksel B, Estill JC, Schulze SR, Paterson AH (2005) Organization and evolution of resistance gene analogs in peanut. Mol Gen Genomics 274:248–263

Chapter 18
Genomics of Pineapple, Crowning The King of Tropical Fruits

Jose Ramon Botella and Mike Smith

Abstract Pineapple [*Ananas comosus* (L.) Merr.] is the third most important tropical fruit in world production after banana and citrus. Nevertheless, and despite its commercial importance, very little genomics research has been performed in this crop. Development of molecular markers has been reported recently to study genetic relationships among the different *Ananas* species and with other members of the Bromeliaceae family. Results from those studies suggest that the existing classification of the seven *Ananas* species needs to be reconsidered. A basic pineapple genetic map is available, although it needs to be developed with the addition of additional markers. Medium scale expressed sequence tag (EST) projects have been undertaken using developing fruits and nematode-infested roots as tissue sources. A bioinformatic resource providing sequence and functional information on all EST clones has been developed. Finally, pineapple microarrays containing in excess of 9,000 EST clones have been produced. Although research in pineapple genomics is taking momentum, much more is needed before the tools developed can be used for the benefit of the industry. An international collaborative effort to develop additional molecular markers and perhaps a genome sequencing initiative is needed.

18.1 Introduction

Pineapple (*Ananas comosus*) is native to South America and was first seen by Europeans when Columbus landed on the inhabited island that he named Guadalupe on 4 November 1493 during his second voyage to the New World. It is generally recognized that the indigenous peoples of South America contributed substantially to the domestication of the pineapple (Leal and d'Eeckenbrugge 1996), probably through the selection of spontaneous mutations expressing desirable traits, e.g. improved palatability, improved fruit size, seedlessness, smooth leaves, and in some cases improved leaf fiber properties which are not commonly found in wild

J.R. Botella
Plant Genetic Engineering Laboratory, School of Integrative Biology, University of Queensland, Brisbane 4072, Australia
e-mail: j.botella@uq.edu.au

P.H. Moore, R. Ming (eds.), *Genomics of Tropical Crop Plants*
© Springer 2008

types (Collins 1951; d'Eeckenbrugge et al. 1997). The pineapple was used not only for fresh fruit consumption, but also for wine making, medicinal purposes, and the rotted fruit for poisoning the tip of arrows (Leal and Amaya 1991; Leal and d'Eeckenbrugge 1996). Crowns, slips. and suckers withstand considerable desiccation and resume growth when planted. Consequently, pineapples have been easily dispersed as the result of mankind's many migrations and are now found throughout the tropics.

Based on the key of Smith and Downs (1979), the genus *Ananas* contains seven species: *A. comosus*, *A. ananassoides*, *A. nanus*, *A. bracteatus*, *A. parguazensis*, *A. fritzmuelleri,* and *A. lucidus*. The closely related genus, *Pseudananas*, contains the monotypic *P. sagenarius*. Molecular studies suggest a revision of the current classification system is needed, which would lead to fewer species within the genus *Ananas*. A new system proposed by d'Eeckenbrugge and Leal (2003) would have the seven valid *Ananas* species downgraded to the level of five botanical varieties of *A. comosus*. *Pseudananas sagenarius* would also become *Ananas macrodontes* under their new classification.

All *A. comosus* have a diploid number of 50 small, spherical chromosomes (2n =2x=50) (Collins and Kerns 1931; Marchant 1967; Brown and Gilmartin 1986; Brown et al. 1997). Within the genus *Ananas* there are triploid, tetraploid, and heteroploid cultivars, while *Pseudananas sagenarius* is a naturally occurring tetraploid with 100 chromosomes (Collins 1960). Most varieties of *A. comosus* are self-incompatible due to the inhibition of pollen tube growth in the upper third of the style (Kerns 1932), which is gametophytically controlled by a single locus with multiple alleles (Brewbaker and Gorrez 1967). Some cultivars exhibit partial incompatibility (Cabral et al. 2000), which may be temperature dependent. The wild types, *A. ananassoides* and *P. sagenarius*, are either partially or completely self-compatible and self-compatibility is common in the other wild pineapples.

Pineapple is highly heterozygous and improvement of many different characters is possible. Breeding programs have made both intraspecific and interspecific crosses and selection has encompassed many aspects of productivity, fruit quality, and pest and disease resistance. In addition, clonal selection has also been utilized with up to 30 different somatic mutations described for the Smooth Cayenne cultivar (Collins and Kerns 1938). Once a desirable cultivar has been bred or selected it is relatively easy to propagate by vegetative means. The pineapple breeding system therefore combines very efficient vegetative reproduction with functional allogamous sexual reproduction.

World production of pineapple is estimated at greater than 14.6 million tonnes annually (FAOSTAT 2005) and more than 70% is consumed locally in the area of production. Although only a third of its output is used for processing (e.g., canned slices, chunks, crush, and juice), pineapple products account for more than two-thirds of the trade in pineapple by value. The processing industry is dominated by a single cultivar, Smooth Cayenne, with export earnings estimated at US$1.2 billion for countries in Asia and parts of Africa and Latin America. A recent trend in the industry has been the development of new hybrids specifically aimed for domestic fresh-fruit markets. A first result of these efforts has been the successful introduction of a low-acid cultivar by Del Monte from Costa Rica into the European and American markets (Rohrbach et al. 2003).

Pineapple is the third most important tropical fruit in world production after banana and citrus, however, very little is known about the molecular genetics of pineapple. No molecular markers have been used in breeding programs to date, although they could be of tremendous use if they could be linked to important agronomic traits or to disease and pest resistance. Only recently have genes been isolated, described, and utilized in genetic transformation programs (Smith et al. 2005).

18.2 Progress in Genomics

Very little progress had been made in pineapple genomics until the last five to six years. The available data on *Ananas* genetic diversity is limited and is mostly based on morphological characters. Most of the initial molecular work focused on the genetic relationships among the seven *Ananas* species and the neighboring monospecific genus *Pseudoananas*, as well as their position within the Bromeliaceae family, to clarify classification and for phylogenetic analysis (Noyer et al. 1995; Terry et al. 1997; Duval et al. 2001; Ruas et al. 2001; Duval et al. 2003). Duval et al. (2001) studied molecular diversity in a set of 301 *Ananas* and *Pseudananas* accessions using restriction fragment length polymorphism (RFLP) and 18 pineapple genomic DNA probes. Factorial analysis differentiated *Pseudananas* from *Ananas*, but nevertheless, the two genera shared 58.7% of all bands, suggesting the existence of intergeneric gene flow. Genetic variation revealed by the set of RFLP markers used by these authors seems continuous with most variation found at the intraspecific level but no clear species partition was evident within *Ananas*. This lack of correspondence between the molecular and the taxonomical data was also observed in previous studies (Noyer 1991; Noyer et al. 1995). A different study by Ruas et al. (2001) was somewhat more successful in grouping different *Ananas* species using a much larger set of 148 RFLP markers but fewer accessions (a total of 16 from four *Ananas* species). Nevertheless, the generated dendogram had a number of abnormalities positioning several accessions in the wrong clusters and splitting species into different branches.

Chloroplast DNA has also been used to study phylogenetic relationships between *Ananas* and related genera (Duval et al. 2003). One hundred fifteen accessions representing the seven *Ananas* species and seven other Bromelioideae were analyzed using polymerase chain reaction-RFLP. Phenetic and cladistic analyses positioned *Ananas* and *Pseudananas* in a monophyletic group, with three distinct sub-groups. Interestingly, these groups do not reflect the different *Ananas* species but the geographical origin of the accessions.

A. comosus varieties cultivated for fruit have been divided into a number of groups based on similarity of morphological characters. Phenotypically, these groups are well differentiated and have been extensively characterized (Samuels 1970; Leal and Soule 1977; Dewald et al. 1988; Duval and d'Eeckenbrugge 1993; Noyer et al. 1995). Nevertheless, and despite their wide morphological variation, RFLP analysis of 168 *Ananas comosus* accessions showed a relatively homogeneous group with low level of polymorphism when compared to wild *Ananas* species (Duval et al. 2001). Sripaoraya et al. (2001b) used random amplified polymorphic DNA (RAPD) to study three commercial cultivar groups, Cayenne, Queen, and Spanish, with the Cayenne and Queen groups appearing as separate clusters in the dendogram but failed to position the Spanish group representative in an independent cluster.

In contrast, amplified fragment length polymorphism (AFLP) markers seem to be more effective than RFLPs for the assessment of genetic diversity. A recent study of 148 *A. comosus* accessions using AFLP markers revealed a high degree of genetic variation within this species (Kato et al. 2004). But even though different DNA patterns could be assigned to each of the commercial cultivars studied, AFLP markers were still unsuccessful in clearly separating major cultivar groups (Kato et al. 2004). In contrast, Paz et al. (2005) also used AFLP markers to characterize the Mexican germplasm collection, mostly composed of *A. comosus* accessions, but reported a low level of diversity.

To explain the apparent conflict between taxonomical and molecular data, it has been suggested that the main phenotypic traits that characterize the different commercial cultivar groups are due to similar mutations that appeared on different genetic backgrounds when the cultivars were selected (Duval et al. 2001; Kato et al. 2004). Therefore, even though there is considerable genetic variation as detected by AFLP markers, this variation does not necessarily lead to the same traditional groupings. A good example is the smooth leaves that characterize the Smooth Cayenne cultivar. Presence of leaf spines is controlled by a single genetic locus with three possible alleles (Kinjo 1993; Cabral et al. 1997; Kato et al. 2004); therefore, the presence or absence of leaf spines (spininess) can arise in very genetically different plants by the mutation of a single gene.

Isozyme and RAPD markers have been used to study the genotypic fidelity of micropropagated pineapple plantlets. Two micropropagation systems, stationary and temporary immersion, were evaluated and even though neither of the two markers was successful in identifying significant differences individually, a combination of the two was able to determine that micropropagation by temporary immersion resulted in the lower frequency of somaclonal variants (Feuser et al. 2003).

The first and only pineapple genetic map available to this date was published by Carlier et al. (2004). The authors used the two-way pseudo-testcross approach to construct two individual maps of *A. comosus* and *A. bracteatus* using a segregating population of 46 F1 individuals from fully fertile crosses between the two species. To construct the map, a combination of three different types of markers, RAPDs, AFLPs, and inter simple sequence repeats (ISSRs), were used. The *A. comosus* map contained 157 markers (33 RAPD, 115 AFLP, eight ISSR, and the piping locus) with 30 linkage groups, 18 of which assembled four markers or more (Carlier et al. 2004). A relatively large percentage (43%) of markers remained unlinked, a fact perhaps reflecting the small size of the mapped population. This map covered approximately 31% of the *A. comosus* genome estimated as 4,146 cM with a calculated ratio of 127 kb/cM for the relationship between physical and genetic distance. In the case of *A. bracteatus*, 50 linkage groups were established containing 335 markers (60 RAPDs, 264 AFLPs, and 11 ISSRs) with 26 linkage groups containing at least four markers. In this case, map coverage was approximately 57.2% of the *A. bracteatus* genome calculated as 3693 cM with a ratio of 120 kb/cM.

Since the publication of the first *A. comosus* linkage map, Dr. Leitao's group has greatly improved the quality and resolution of the map and a new version has been kindly provided for this chapter (Fig. 18.1). The linkage groups shown in this new map gather a total of 651 markers, with 505 AFLP, 124 RAPD, 20 SSRs, one expressed sequence tag (EST) and one morphological trait (piping).

Despite the economic importance of the crop, very little sequence information is still available. In fact, only 51 pineapple sequences had been deposited in the GenBank nucleotide sequence database as of 2004. Twenty-four of those sequences were reported by Neuteboom et al. (2002), who used differential screening to isolate genes preferentially expressed in root tissues. Northern analysis using RNA isolated from roots, fruits, and aerial tissues revealed that eight of the clones were predominantly expressed in roots with the rest being present in two or more tissues. The most important contribution of pineapple sequences has been provided by Moyle et al. (2005a), who reported the cloning and sequencing of 1,548 EST clones isolated from cDNA libraries constructed from green, mature fruits (408 clones) and yellow, fully ripe fruits (1,140 clones). Relative EST clone abundance in green and yellow libraries correlated well with mRNA abundance in their respective tissues as shown by northern analysis. A number of genes strongly upregulated during fruit ripening were identified; among the most interesting were two metallothionein genes and a MADS box gene. One of the metallothionein clones was extremely abundant with over 40% of all library colonies hybridizing to a radio-labeled probe. The metallothionein expression level was calculated by quantitative real time PCR to be over 50 fold higher than the β-actin control in ripening fruit tissues. The MADS box gene was highly upregulated during fruit ripening and was not detected in any other tissue. MADS box proteins are transcription factors involved in regulating various aspects of plant development (Parenicova et al. 2003). Interestingly, the recessive ripening-inhibitor (*rin*) mutation in tomato that inhibits ripening even in the presence of exogenous ethylene has been identified as a MADS box gene (LeMADS-RIN) (Vrebalov et al. 2002). It has been suggested that LeMADS-RIN acts upstream of ethylene during ripening and could provide a common mechanistic link between climacteric and non-climateric fruit ripening. It is possible that the pineapple MADS box gene is the orthologue of the tomato LeMADS-RIN and complementation studies in *rin* tomato mutants are underway.

In a further EST project devised to study gene expression in roots after nematode infestation, 4,102 EST sequences were obtained, including 1,298 early infection clones, 2,461 late infection clones, and 343 non-infected root tip clones (Moyle and Botella unpublished results). Northern analysis and quantitative real time PCR have identified a variety of genes differentially expressed during gall formation. Analysis of clone distribution by functional classification reveals that the late infection library contains a named "high proportion of clones associated with oxidative

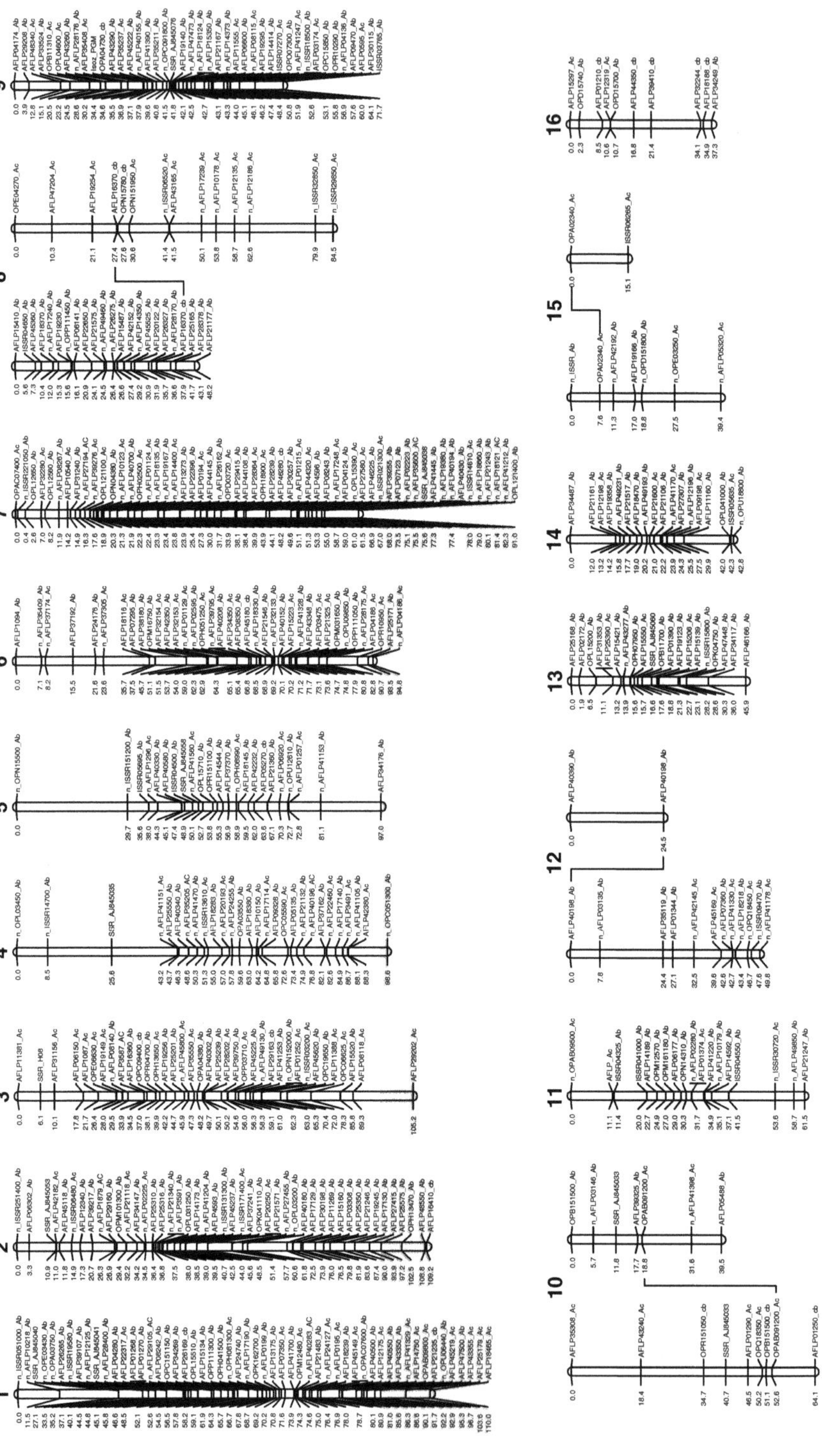

Fig. 18.1 Integrated genetic map of *Ananas comosus* (pineapple). (1–16) Linkage groups that integrate molecular markers of var. *bracteatus* and var. *comosus*; (17–25) Linkage groups that integrate markers only of var. *bracteatus*; (26–32) Linkage groups that integrate markers only of var. *comosus*. Forty-seven very small linkage groups are not shown

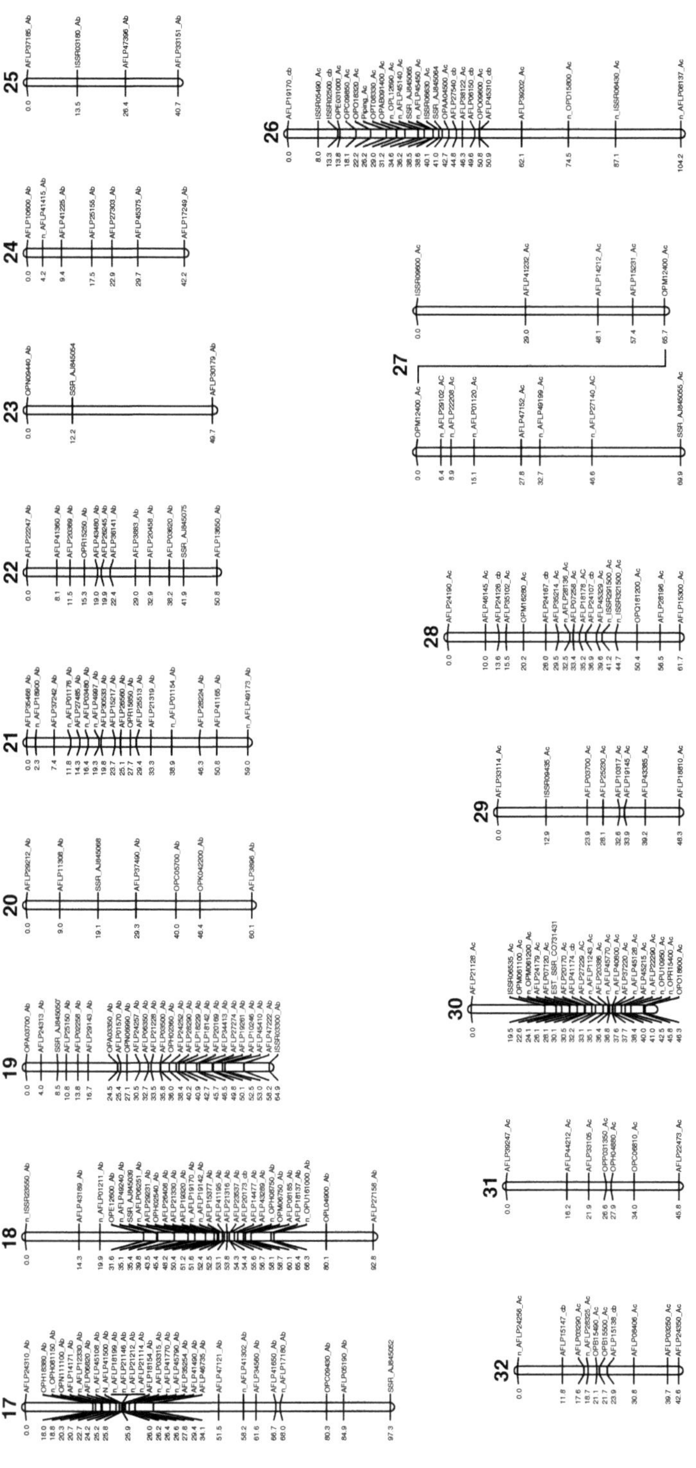

Fig. 18.1 (continued)

Single search | Text search | Multiple search | Homology search | Multiple Homology search

PineappleDB search results

Full Search for 146

All clones in Contig 146
JBW019C06, JBW019C07, JBW019C08, JBW020A12, JBW020E01, JBW020E04, JBW020H04, JBW022A10, JBW023B12, JBW023F05, JBW023F06, JBW023H04, JBW023H05, JBW024A11, JBW024D05, JBW024D07, JBW024E03, JBW024E11, JBW025A07, JBW025D04, JBW026D03, JBW026D04, JBW026F04, JBW027F05, JBW027H07, JBW028B10, JBW028D05, JBW028D07, JBW028F09, JBW028H11, JBW029C10, JBW029F12, JBW029G05, JBW030A12, JBW030B03, JBW030B11, JBW030E10, JBW030H01, JBW031F04, JBW031G11, JBW032A08, JBW032D03, JBW033D02, JBW033E06, JBW033H03

Contig sequence	CGTCCGATATCCTCTCTACCTCGAAGCTCCTCCTCTCCATAATTATATTCCCCTCCCCCGCTACCCACTTCCTCCGAGA AGTTGAAGAGTAACACAACTACTCTACTCGGCACCACTTGAAGAGTGTGTGTTTGTGTGTGTGTGAGAGAGAGAGAG AGAGAGAGCTCGATAAATTTGAGGGATCTTAATTGGAGGAGGAGGATCTGAGTAGAGGTAGTAGAGATGGGGAGAGGGA GAGTTGAGCTGAAGAGGATCGAGAACAAAATCAACCGGCAAGTGACGTTCTCGAAGCGCCGCAACGGGCTCCTCAAGAA GGCCTACGAGCTCTCCGTCCTCTGCGACGCCGAGGTCGCCCTCATCATCTTCTCCAGCCGCGGCAAGCTCTACGAGTTC GGAAGCGTTGGCACAAGCAAGACATTAGAACGATATCAACGCTGCTGTTACAATTCTCAGGATTCAAGTGCTGGTGATC GTGAGACTCAGAGCTGGTACCAGGAAATGTCCAAGTTGAAGGCAAAATTTGAGTCTCTGCAGCGCTCTCAGAGGCATTT GCTCGGGGAGGATCTTGGACCGCTGAGTGTGAAAGAATTGCAGCAACTGGAGCGACAGCTTGAGTCTGCTCTTTCACAA GCCAGACAGAGAAAGACTCAAATAATGATGGATCAGATGGAAGAACTTCGCAAAAAAGAACGTCAACTGGGAGAAATAA ATAAGCAGCTTAGAAACAAGCTTGAGGCCGAAGGCGGCGCCTTTAGGGCCATTCAAGGATCATGGGCTTCTGATGCTAT TGTGAGTGGAAATGCATTCAATATGCAGCACGCCCAATCGAGCAGCATGGAATGCGAACCCACTCTGCAAATAGGGTAT CACCAATTTGTTCCTCCTGAGGCAACCATTCCAAGAACCGCCGGTGGGGAGAACAATTTCATGCTTGGTTGGGTTCTGT GAACCATTCTGGAACCTACCAGAAGCCCATATTATCGGTAATGTTTGTACTGAAATTATTATTATTATTATCATTAATG TTATTATTATTATGTTCTGCTATTGATCGTGCCTCTGAGAACTTATGCTGTTTAATCCGTTTTGCGGACCAARTGTTGC TGCTTTCTAGTGTGTGTTATCCTTTTGAACAATGTAAATTCTGTATGGCAATGCTACTAGCTTCTTGTGGGAACAATAT TTACTTAATATAAATGCTATTAAGTGTTGTCATTTAAC
Name	MADS box protein
Contig length	1241
Full match description	gb|AAQ83835.1| MADS box protein [Asparagus officinalis]
Matching accession number	AAQ83835
Length of match	234
Match homology	92%
Functional class	TRANSCRIPTION
Functional class number	04.05.01.04
Alternative Functional subclass	transcriptional control
Functional subclass number	0
Arabidopsis homolog	At2g45650
Gene Ontology Based on Arabidopsis homolog	Molecular function: GO:0003677 - DNA binding (ISS) GO:0003700 - transcription factor activity (ISS) Cellular component: GO:0005634 - nucleus (IEA) Biological process: GO:0006355 - regulation of transcription, DNA-dependent (IEA)
Splice variants	None

Distribution of cDNAs

	Total	Fruit	Green mature fruit	Yellow ripe fruit	Root	Root tips	Gall VC 1-4 weeks	Gall VC 5-10 weeks
Number of cDNAs	45	45	0	45	0	0	0	0
% for that tissue	0.797	2.907	0.000	3.947	0.000	0.000	0.000	0.000

Return to main search page

Please send comments etc to Mark Crowe

Fig. 18.2 PineappleDB, the online pineapple bioinformatics resource (www.pgel.com.au). View of a typical entry containing sequence, functional, homology and database information

stress responses and detoxification of free radicals (Moyle and Botella unpublished results).

The entire collection of ESTs generated in the Botella lab has been made publicly available by an online pineapple bioinformatics resource PineappleDB" (www.pgel.com.au) (Moyle et al. 2005b). PineappleDB is a curated database providing integrated access to annotated EST data for cDNA clones isolated from pineapple fruit, root, and nematode-infected root gall vascular cylinder tissues. The database contains in excess of 5,600 EST sequences and 3,383 contig consensus sequences. All entries contain associated bioinformatic data including splice variants, *Arabidopsis* homologues, clone distributions, MIPS (Munich Information Center for Protein Sequence) based and gene ontology functional classifications (Fig. 18.2). In addition, the online resource provides extensive search capabilities by text or sequence homology (BLAST).

Finally, microarrays have been produced with 9,312 pineapple cDNAs printed in duplicate including the entire EST collection generated by the Botella group plus a number of unknown clones. These microarrays have been used in two independent studies to (a) study gene expression in roots and gall tissues at different times after nematode infestation, and (b) perform gene expression profiling of pineapple fruits during ripening, and the results will soon be available (Moyle and Botella unpublished results). These microarrays are available to the research community.

18.3 Prospects

It is clear that pineapple genomics is at its infancy and that many more resources need to be developed. Although some molecular markers are now available and a basic genetic map has been constructed, more polymorphic markers and a denser map are needed. In this respect, Leitao's group is improving the existing pineapple map and complementing it with additional molecular markers (J.M. Leitao personal communication).

Expanded EST projects are also needed to enrich the pool of pineapple cDNAs available to the research community and the development of bioinformatic based analyses to identify interesting candidate genes for the genetic improvement of this crop. A full genome sequencing initiative has not been explored yet, but it could prove invaluable for the advancement of classical breeding as well as the development of biotechnological solutions to the most important agronomic problems. Consumer oriented fruit quality improvement, such as sugar, vitamin and acid content, can also be targeted if more genomic resources are developed.

Biotechnology will undoubtedly play an important role in pineapple improvement in the not too distant future and could take advantage of new developments in genomics. Herbicide-tolerant transgenic varieties have already been produced (Sripaoraya et al. 2001a). Another important agronomic problem such as natural flowering has also been tackled and transgenic pineapples have been produced with delayed natural flowering (Trusov and Botella 2006). Fruit quality issues have also been addressed with Ko et al. (2006) introducing transgenes to control blackheart, an internal browning of fruit flesh as a result of chilling injury.

Acknowledgments The authors wish to thank Dr. Carlier and Dr. Leitão for providing unpublished information shown in Figure 18.1 of this manuscript.

References

Brewbaker JL, Gorrez DD (1967) Genetics of self-incompatibility in the monocot genera, *Ananas* (pineapple) and *Gasteria*. Am J Bot 54:611–616

Brown GK, Gilmartin AJ (1986) Chromosomes of the *Bromeliaceae*. Selbyanna 9:85–93

Brown GK, Palaci CA, Luther HE (1997) Chromosome numbers in *Bromeliaceae*. Selbyanna 18:85–88

Cabral JRS, de Matos AP, d'Eeckenbrugge GC (1997) Segregation for resistance to fusariose, leaf margin type and leaf colour from the EMBRAPA Pineapple Hybridization Programme. Acta Hort 425:193–200

Cabral JRS, d'Eeckenbrugge GC, de Matos AP (2000) Introduction of selfing in pineapple breeding. Acta Hort 529:165–168

Carlier JD, Reis A, Duval MF, d'Eeckenbrugge GC, Leitao JM (2004) Genetic maps of RAPD, AFLP and ISSR markers in *Ananas bracteatus* and *A. comosus* using the pseudo-testcross strategy. Plant Breeding 123:186–192

Collins JL (1951) Antiquity of the pineapple in America. Southwestern J Anthrop 7:145–155

Collins JL (1960) The pineapple. (New York: Interscience Publishers)

Collins JL, Kerns KR (1931) Genetic studies of the pineapple. I. A preliminary report on the chromosome number and meiosis in seven pineapple varieties (*Ananas sativus* Lindl) and in *Bromelia pinguin* L. J Hered 22:139–142

Collins JL, Kerns KR (1938) Mutations in pineapples. A study of thirty inherited abnormalities in the Cayenne variety. J Hered 29:169–173

d'Eeckenbrugge CG, Leal F (2003) Morphology, anatomy and taxonomy. In: Bartholomew DP, Paull RE, Rohrbach KG, (eds), The Pineapple: Botany, Production and Uses. CAB International, Wallingford, UK, pp 13–32

d'Eeckenbrugge CG, Leal F, Duval MF (1997) Germplasm resources of pineapple. Hort Reviews 21:133–175

Dewald MG, Moore GA, Sherman WB (1988) Identification of pineapple cultivars by isozyme genotypes. Am Soc Hort Sc 113:935–938

Duval MF, d'Eeckenbrugge GC (1993) Genetic variability in the genus *Ananas*. Acta Hort 27–32

Duval MF, Noyer JL, Perrier X, d'Eeckenbrugge GC, Hamon P (2001) Molecular diversity in pineapple assessed by RFLP markers. Theor Appl Genet 102:83–90

Duval MF, Buso GSC, Ferreira FR, Noyer JL, d'Eeckenbrugge GC, et al. (2003) Relationships in *Ananas* and other related genera using chloroplast DNA restriction site variation. Genome 46:990–1004

FAOSTAT (2005) http://apps.fao.org

Feuser S, Meler K, Daquinta M, Guerra MP, Nodari RO (2003) Genotypic fidelity of micropropagated pineapple (*Ananas comosus*) plantlets assessed by isozyme and RAPD markers. Plant Cell Tissue Organ Cult 72:221–227

Kato CY, Nagai C, Moore PH, Zee F, Kim MS, et al. (2004) Intra-specific DNA polymorphism in pineapple (*Ananas comosus* (L) Merr) assessed by AFLP markers. Gen Res Crop Evol 51:815–825

Kerns KR (1932) Concerning the growth of pollen tubes in pistils of Cayenne flowers. Pineapple Quarterly 1:133–137

Kinjo K (1993) Inheritance of leaf margin spine in pineapple. Acta Hort 334, 59–66

Ko HL, Campbell PR, Jobin-Décor MP, Eccleston KL, Graham MW, et al. (2006) The introduction of transgenes to control blackheart in pineapple (*Ananas comosus* L) cv Smooth Cayenne by microprojectile bombardment. Euphytica 150:387–395
Leal F, Amaya L (1991) The curaga (*Ananas lucidus*, Bromeliaceae) crop in Venzuela. Econ Bot 45:216–217
Leal FJ, d'Eeckenbrugge GC (1996) Pineapple. In: Janick J, Moore JN (Eds) Fruit Breeding, Vol.I. Tree and Tropical Fruits. John Wiley & Sons, New York, pp 565–606
Leal FJ, Soule J (1977) Maipure, a new spineless group of pineapple cultivars. HortSci 12:301–305
Marchant CJ (1967) Chromosome evolution in the Bromeliaceae. Kew Bull 21:161–168
Moyle R, Fairbairn DJ, Ripi J, Crowe ML, Botella JR (2005a) Developing pineapple fruit has a small transcriptome dominated by metallothionein. J Exp Bot 56:101–112
Moyle R, Crowe ML, Ripi-Koja J, Fairbairn DJ, Botella JR (2005b) PineappleDB: An online pineapple bioinformatics resource. BMC Plant Biol 5:21
Neuteboom LW, Kunimitsu WY, Webb D, Christopher DA (2002) Characterization and tissue-regulated expression of genes involved in pineapple (*Ananas comosus* L) root development. Plant Sci 163:1021–1035
Noyer JL (1991) Etude preliminaire de la diversite genetique du genre *Ananas* par les RFLPs. Fruits (numero especial *Ananas*), 372–375
Noyer JL, Lanaud C, d'Eeckenbrugge GC, Duval MF (1995) RFLP study on rDNA variability in *Ananas* genus. Acta Hort 425:153–159
Parenicova L, de Folter S, Kieffer M, Horner DS, Favalli C, et al. (2003) Molecular and phylogenetic analyses of the complete MADS-box transcription factor family in Arabidopsis: New openings to the MADS world. Plant Cell 15:1538–1551
Paz EY, Gil K, Rebolledo L, Rebolledo A, Uriza D, et al. (2005) AFLP characterization of the Mexican pineapple germplasm collection. J Am Soc Hort Sci 130:575–579
Rohrbach KG, Leal F, d'Eeckenbrugge CG (2003) History, distribution and world production. In: Bartholomew DP, Paull RE, Rohrbach KG (eds) The Pineapple: Botany, Production and Uses. CAB International, Wallingford, UK, pp 1–12
Ruas CD, Ruas PM, Cabral JRS (2001) Assessment of genetic relatedness of the genera *Ananas* and *Pseudananas* confirmed by RAPD markers. Euphytica 119, 245–252
Samuels G (1970) Pineapple cultivars. Am Soc Hort Sci Proc:13–24
Smith LB, Downs RJ (1979) Bromelioides (*Bromeliaceae*). Flora Neotropica Mono 14:1493–2142
Smith MK, Ko HL, Sanewski GM, Botella JR (2005) *Ananas comosus,* Pineapple. In: RE Litz, (ed), *Biotechnology of Fruit and Nut Crops*. CAB International, Wallingford, UK, pp 157–172
Sripaoraya S, Marchant R, Power JB, Davey MR (2001a) Herbicide-tolerant transgenic pineapple (*Ananas comosus*) produced by microprojectile bombardment. Ann Bot 88:597–603
Sripaoraya S, Blackhall NW, Marchant R, Power JB, Lowe KC, et al. (2001b) Relationships in pineapple by random amplified polymorphic DNA (RAPD) analysis. Plant Breeding 120:265–267
Terry RG, Brown GK, Olmstead RG (1997) Examination of subfamilial phylogeny in *Bromeliaceae* using comparative sequencing of the plastid locus ndhF. Am J Bot 84:664–670
Trusov Y, Botella JR (2006) Silencing of the ACC synthase gene ACACS2 causes delayed flowering in pineapple (*Ananas comosus* (L) Merr). J Exp Bot 57: 3953–3960
Vrebalov J, Ruezinsky D, Padmanabhan V, White R, Medrano D, et al. (2002) A MADS-box gene necessary for fruit ripening at the tomato ripening-inhibitor (Rin) locus. Science 296:343–346

Chapter 19
Genomics of Tropical Solanaceous Species: Established and Emerging Crops

Richard C. Pratt, David M. Francis, and Luz S. Barrero Meneses

Abstract The primary solonaceous food crops are found within several genera (*Solanum*, *Capsicum*, and *Physalis*). The genus Solanum contains three of the four leading crops (potato, tomato, and eggplant) in the Solanaceae family, and is arguably the most economically important plant genus. Peppers (*Capsicum* spp.) are also one of the leading vegetable crops cultivated throughout the tropics. The centers of diversity of *Solanum*, *Capsicum* and *Physalis* are located in the tropics. Many local varieties and both cultivated and wild relatives are often found in close proximity to these centers. Locally important varieties from all of these genera are cultivated in diverse tropical agro-ecosystems worldwide and some have considerable promise for further development. Additional progress in ascertaining phylogenetic relationships will foster improved utilization of these genetic resources. Among the solanaceous crops, tomato and potato have emerged as "model" scientific research organisms. In 2003, tomato was the first diploid crop among the asterids to be chosen for genome sequencing. It is intended that a high quality sequence of the tomato euchromatin will serve as a reference for the Solanaceae. In addition, the Potato Genome Sequencing Consortium aims to complete the DNA sequence of the entire potato genome by 2008. High quality, annotated, genome sequence information from these two species is expected to benefit improvement of multiple solanaceous crops. Sequence and functional genomics resources are being generated for the Solanaceae through international cooperation. Tomato cDNA, genomic DNA, conserved orthologous set, and restriction fragment length polymorphism markers have been mapped to allow comparative genome analysis between eggplant, tomato, potato and pepper. Additional comparative analyses and their implications for development of two Andean *Solanum* species are presented.

19.1 Introduction

Cultivated solanaceous species contribute incalculable value to human well-being throughout the tropical regions of the world. The leading economic species in the

R.C. Pratt
The Ohio State University, Department of Horticulture and Crop Science, Ohio Agricultural Research and Development Center, 1680 Madison Ave., Wooster, Ohio 44691-4096
e-mail: pratt.3@osu.edu

© Springer 2008

genus *Solanum*, potato (*S. tuberosum* L.) and tomato (*S. lycopersicum* L., formerly *Lycopersicon esculentum* Mill.), and eggplant (*S. melongena* L.) provide an energy staple (tubers) and highly valued fruit used as vegetables in the diet. Another important solanaceous crop, pepper (*Capsicum* spp.) is comprised collectively of five domesticated species (*C. annuum* L., *C. frutescens* L., *C. chinense* L. Jacq., *C. baccatum* L., and *C. pubescens* Ruiz and Pavon) (Bosland 1996). Additional *Solanum* species, and those in other related genera such as *Physalis*, comprise minor crops found throughout diverse tropical agro-ecosystems.

The development and application of translational genomics tools and strategies across taxonomic boundaries in the Solanaceae is particularly attractive because of the abundance of genetic diversity, high economic value of the crops, and the rich genetic resources available from numerous relatives – some wild, and some that have achieved minor crop status. In this chapter we will examine taxonomy, germplasm, genetic, and genomics resources available for improvement of the leading food crops mentioned above, and also for the development of emerging crops. We also present current progress and initiatives at the community level that will foster enhanced integration of genomics, and genetic resources for improvement of both major and minor solanaceous crops.

19.1.1 Economic Importance of Solanceous Food Crops

Potato and tomato are the world's leading vegetable crops. Their average annual production exceeded 300, and 100, million metric tons during the period 2000 to 2004, respectively (ERS/USDA 2007). Peppers (*Capsicum* spp. collectively) and eggplant also are consistently among the top 10 vegetable crops produced globally. *C. annuum* is the leading pepper species in overall global production, but it is not well adapted to the humid lowland tropics. The other *Capsicum* species predominate in tropical regions. The total area devoted to production of these four leading solanaceous crops is difficult to estimate, but it likely approaches 30 million hectares (m ha) worldwide. (The area dedicated to potato production alone is about 19 m ha.) Although potato production in Europe has fallen since the early 1960s, this decline has been more than offset by growth in Asia, Africa and Latin America (CIP 2007). Both China and India consistently are top-five global producers of all four leading solanaceous crops.

19.1.2 Genetics of Leading Solanaceous Food Crops

Tomato has played a key role in genetics research for decades. It has served as an important crop species utilized in molecular research. Extensive expressed sequence tag (EST) databases and a genome sequencing effort are resulting in high quality sequence data that will serve as a reference for the Solanaceae. Annotated genome sequence information of tomato will be expected to benefit improvement programs of tomato and other solanceous crops. While both tomato and potato are increasingly becoming "model" research organisms, other minor crops with considerable potential have received little attention from the international community. The genomes of tomato, potato, eggplant and pepper are fairly well conserved, so the prospects for broader application of genomics information from tomato and potato to pepper and eggplant are good. There also exist little known, but extremely interesting, cultivated *Solanum* species related to the leading crops. The extent to which genomic information from tomato and potato will benefit broad phylogenetic studies and improvement of minor solanaceous crops is still unknown, but considered promising.

19.1.3 Taxonomy of the Genus Solanum

Vast genetic resources are available within the genus *Solanum*. It has been termed a "hyperdiverse" genus comprising 1500 or more species, the largest in the Solonaceae family (Weese and Bohs 2007). The magnitude and diversity of genetic variation is exceptional, but the range of that diversity also presents a formidable challenge for those who would aspire to utilize it in an efficient manner. The taxonomy of the *Solanum* genus is still in debate and the resources dedicated to genomic studies of the leading solanaceous crops are highly variable. A major revision of the genus *Lycopersicon*, formerly considered to be a small genus in the family Solanaceae, was proposed in the 1990s (Spooner et al. 1993). The genus *Lycopersicon* was transferred into Solanum and a section called *Lycopersicum* was named. Scientific investigations to characterize the genome organization and evolutionary relationships within Solanum continue (Weese and Bohs 2007) and a refined phylogeny is expected to facilitate the development of more efficient breeding strategies for utilization of germplasm.

19.2 Promising "Minor" Crop Relatives

Wild relatives have long been important germplasm donors for improvement of the leading solanaceous food crops. Closely related wild species have also been used as parents in mapping populations when there does not exist high levels of polymorphism for molecular markers within the cultivated species. There exists a growing interest in further utilizing emerging genomics tools to improve our understanding of phylogenetic relationships with the wild relatives, but also with related species that already are fully or partially domesticated. Transfer of desirable genes between domesticated donors should result in fewer problems associated with the undesirable horticultural traits of the wild relatives. Some of the less known cultivated crops may also offer opportunities for their further development or introduction into other regions. We will examine several species in both the *Solanum* and *Physalis* genera that are noteworthy, and refrain from reviewing the wild species, which would be beyond the scope of this chapter. Tomato does not display high levels of reproductive affinity with any related cultigens.

19.2.1 "Minor" Solanum Crops of the Andes

The tree tomato, or tamarillo, (*Solanum betaceum*) bears egg-shaped fruit prized for their distinctive juice, and *Solanum quitoense* L. (lulo or naranjilla) fruit is especially popular in Colombia and Ecuador (NRC 1990) where it is also used for juice. (Fig. 19.1). *Solanum muricatum* (pepino dulce) produces succulent fruit reminiscent

Fig. 19.1 A) The lulo fruit (*Solanum quitoense*) B) The tree tomato fruit (*Solanum betaceum*). (Taken from: www.ocati.com)

of honey-dew melons, and like lulo products, may have rising commercial prospects. These species are native of the Andean region circumscribed by Colombia, Ecuador, and Peru (Heiser and Anderson 1999). They are increasingly entering commercial production on a limited scale, but remain largely unknown outside South America.

19.2.2 Diversity of Cultivated Tuber-bearing Potatoes

Within the genus *Solanum*, the section Petota (Huamán and Spooner 2002), includes the tuber-bearing species, of which the cultivated potato (*S. tuberosum*) is best known worldwide. There are approximately 180 wild species in the Petota section, and they are present primarily in the Peruvian and Bolivian Andes. Additional species names have been utilized to describe cultivated (tuber-bearing) land races (Hawkes 1990). Huamán and Spooner (2002) recently proposed a single species, *S. tuberosum*, with eight cultivar-groups: Ajanhuiri Group, Andigenum Group, Chaucha Group, Chilotanum Group, Curtilobum Group, Juzepczukii Group, Phureja Group, and Stenotomum Group. These diverse types are adapted to a broad range of environments, and constitute a rich source of genetic diversity for improvement of leading cultivars and for introduction into other agro-ecosystems (NRC 1990, Ghislain et al. 2006).

The Phureja Group consists of potato landraces referred to as golden potatoes in the broad region of the Andes from western Venezuela to central Bolivia. This group forms an important germplasm resource due to its excellent culinary properties and other traits that are potentially useful in the development of modern varieties. The Phureja group is characterized by short-day adaptation, diploidy ($2n = 2x = 24$), and lack of tuber dormancy. All of these factors make utilization of this germplasm difficult through application of conventional breeding techniques. Additional tools from genomics research could have substantial bearing on facilitating gene flow between the Phureja group and *S. tuberosum*.

The tubers of the tetraploid Andigenum Group species are larger, rounder, and more uniform than are those of other related groups. They are cultivated extensively from northern Argentina to Venezuela and on the mountainsides of Central America and the *cordilleras* of Mexico (NRC 1990). Members of the Stenotomum group, include the diploid Limena, or yellow potato, a favorite in Peru because of its exceptional flavor. Triploid Chaucha potatoes from Peru (Chaucha group) are thought to have arisen through interspecific hybridization of Andigena and Stenotomum groups (NRC 1990).

19.2.3 Domesticated Relatives of Eggplant

Solanum aethiopicum (African or Ethiopian eggplant, nakati), is used primarily as a green leafy vegetable (Facciola 1990, Lester 1986). It is resistant to many pests and diseases and it is of increasing importance in Africa. Plants bear bitter, but slightly sweet, edible fruit varying in shape (oval to round) and color (white, striped and green - turning red or orange when ripe). The small, immature fruit are used in Thai cooking. Because of the fruit's lycopene content, *S. aethiopicum* fruit also are allowed to ripen so that their ornamental value may be appreciated. Interspecific hybrids arising from crosses between *S. melongena* and *S. aethiopicum* are infertile. Isshiki and Taura (2003) have demonstrated fertility restoration of interspecific hybrids following chromosome doubling.

19.2.3 Domesticated Relatives of Pepper (C. annuum)

C. chinense and C. frutescens are closely related to *C. annuum* whereas *C. baccatum*, and *C. pubescens* are more distantly related (Bosland 1996, Pickersgill 1997). Non-pungent *C. annuum* peppers are often referred to as sweet, bell, or paprika peppers, whereas more pungent members of all species are generally referred to as chiles or hot peppers (aji in Andean South America). The best known *C. frutescens* cultivar is 'Tabasco' and this species is also well adapted in Africa, India, & Polynesia (USDA 2007). *C. chinense* (habanero or bonnet) is widely cultivated in the neotropics, and is particularly popular in Brazil. *C. baccatum* is the most common hot pepper in the Andean region of South America and it too is widely adapted. Its fruit are brilliantly colored and it offers diverse aromatics and flavors. The black-seeded *C. pubescens* (peron or rocoto) also is widely grown in the Andes (NRC 1990) and is sometimes called tree chile. It provides a range of colors and shapes and may grow for a decade in cooler regions that other chiles cannot tolerate. Reproductive barriers to interspecific gene transfer occur among these species, and they are similar to those found in other genera of Solanaceae: unilateral incompatibility, post-fertilization abortion, and nucleo-cytoplasmic interactions leading to male sterility or other abnormalities (Pickersgill 1997). The genetic diversity available across these domesticated species has not yet been extensively utilized.

19.2.4 "Minor" Crops in the Genus Physalis

All but one of the more than 70 species of the genus *Physalis* are indigenous to Meso-America or North America, and they all share distinctive fruit morphology. As the fruit develops, the outer calyx of the flower also enlarges into an inflated, papery, bladder-like sheath ("husk") which then encloses the fleshy, tomato-like berry. The most widely grown and distributed species in the genus is *P. philadelphica* Lam. syn. *Physalis ixocarpa* Auct., commonly called husk-tomato or tomatillo in English-speaking countries. It is known as tomatillo, or tomate near its Center of Origin in Meso-America (Gómez-Pompa and Sosa 1978). This species is often confused with *Physalis pubescens* L (ground-cherry). It too may be called husk-tomato in English-speaking areas because of the similar fruit morphology. *P. pubescens* is known in Meso-America as tomato verde (Facciola 1990).

Cape-gooseberry, also referred to as goldenberry (*P. peruviana* L.), has long been a minor crop in the Andes and it is now produced to a limited extent in New Zealand, South Africa, India and Hawaii (NRC 1990). It is known as poha in Hawaii and uchuva in Colombia (NRC 1990). The small, spherical, golden berries are eaten fresh and used in jams and jellies.

19.3 Translational Genomics for Solanaceous Crops

The term "translational genomics" has been borrowed from medical sciences and it describes an intended process whereby information derived from genome technologies is adapted and utilized for applied outcomes (Minna and Gazdar 1996). Thinking about agricultural research from the broader perspective of taxonomic groups and DNA sequence homology, rather than within traditional commodity boundaries, will enable translational genomics research to more fully exploit genome sequence information. Models for organizing translational research in crop plants are now emerging. One example, the USDA/NRI Coordinated Agricultural Project (CAP) program (http://www.csrees.usda.gov/fo/fundview.cfm?fonum= 1601) currently offers competitive funding to coordinate research and tool development while maximizing cooperation and minimizing redundancy. The applied plant genomics CAPs were initiated to bring together scientists and stakeholders with a shared vision and plan to facilitate translation of basic discoveries and technology. The goal of CAP is to create an inclusive community consisting of applied and basic, private and public researchers combined with participation of commodity groups, growers, and end users.

The Solanaceae Coordinated Agricultural Project (SolCap) has been formed to organize the applied research community in order to exploit emerging genomics resources (Van Deynze et al. 2007). Planning has been undertaken from the outset so that translational outcomes will facilitate both discovery and application.

19.4 Functional Genomics Resources

19.4.1 Tomato

Numerous types of genetic markers with good genome coverage have been combined with excellent web-resources for tomato researchers. Current genetic maps for tomato include 2,200 restriction fragment length polymorphisms (RFLPs), cleaved amplified polymorphic sequences (CAPs), and microsatellites or simple sequence repeats (SSRs), as well as emerging genetic resources which include a comparative map with *Arabidopsis* of over 500 conserved ortholog set (COS) markers (Solanaceae Genomics Network, http://www.sgn.cornell.edu/; Tanksley et al. 1992). These maps were derived from populations that were developed between wild relatives (various *Solanum* species) and cultivated varieties. That approach maximized genetic variation and led to the discovery and introgression of useful alleles into cultivated germplasm.

Comparative mapping should be useful for the rapid identification of genes similar to those already mapped in related genera. The tomato map, comparative analyses, and their implications for genome evolution in the Solanaceae have been presented and discussed (Doganlar et al. 2002). Grube et al. (2000) examined genomic positions of phenotypically defined resistance "R" genes (conferring resistance to pathogen infection) in tomato, potato, and pepper using a direct comparative approach. The authors determined that positional correspondence across genomes did not necessarily imply homology. They concluded that mapping the specificity of host R genes may be imprecise because many R genes appear to be evolving rapidly. However, the use of DNA sequence information from R genes has been useful in numerous studies to define the genetic position and function of loci that confer resistance to pathogens and insects (Kang et al. 2005, Ori et al. 1997, Stewart et al. 2005, Zhang et al. 2002, 2004).

Large collections of expressed sequence tags (ESTs) are available for tomato (200,248 ESTs; 45,585 unique; http://plantta.tigr.org/cgi-bin/ plantta_release.pl). These EST resources are international as efforts in North America have been complemented by the on-going tomato, full-length cDNA sequencing effort in Japan (Tsugane et al. 2005, http://www.kazusa.or.jp/jsol/microtom/index.html). Approximately 106 validated and mapped single nucleotide polymorphisms (SNPs) have been developed by mining EST resources (Yang et al. 2004, Labate and Baldo 2005). A collaborative USDA project will release 1505 SNPs, 593 of which are in breeding germplasm and represent an additional 161 loci (Van Denyze et al. 2007). The tomato genome sequencing initiative started in 2003 and is focused on sequencing the euchromatic region of cultivar 'Heinz 1706' using a BAC-by-BAC approach. To date, this initiative is 27% complete (http://sgn.cornell.edu/about/tomato_sequencing.pl).

19.4.2 Potato

Approximately 219,485 ESTs (Ronning et al. 2003, Flinn et al. 2005) representing 81,072 unique sequences are available (http://plantta.tigr.org/cgi-bin/plantta_release.pl) for potato. These ESTs were primarily generated from three cultivars: 'Kennebec', 'Shepody', and 'Bintje'. In addition, an international consortium has begun to sequence the entire genome of the diploid potato RH89-039-16 (van Os et al. 2006). The Institute for Genomic Research (TIGR) has constructed spotted microarrays that contain approximately 12,000 potato clones from ESTs that have been re-sequenced and validated. Samples from other Solanaceae species can be successfully hybridized to the microarrays due to the high level of sequence similarity within the genus. Resources to aid in the application of diverse resources generated for potato are freely available (http://www. tigr.org/tdb/sol/).

19.4.3 Eggplant

Public genomic resources have lagged for eggplant in comparison with tomato and potato. A map based on an F_2 population from an interspecific cross between *S. linnaeanum* (MM195) and *S. melongena* (MM738) contains 233 restriction fragment-length polymorphisms (RFLP) markers, 22 of which are COS markers (http://www.sgn.cornell.edu/cview/map.pl? map_id=6). *S. linnaeanum* is a spiny wild relative of cultivated eggplant whose fruit characteristics contrast with those of *S. melongena* MM738, a non-spiny commercial type. Comparative mapping efforts have been undertaken (Doganlar et al. 2002). Transcript assemblies are currently unavailable from TIGR.

19.4.4 Pepper

Like eggplant, pepper resources lag those of tomato and potato but pepper has benefited from leveraging of resources developed for the primary crops. Livingstone et al. (1999) created a genetic map of *Capsicum* from an interspecific F_2 population. To date 15,404 unique sequences from 30,830 ESTs are available (http://plantta.tigr.org/cgi-bin/plantta_release.pl). A second set of 122,503 ESTs representing 29,580 unique sequences from mainly a single F_1 cultivar, 'Bukang', was developed by Dr. D. Choi (http://210.International International 218.199.240/ SOL/index.php?menu= intro&species=3).

19.5 Marker Discoveries for Breeding and Genetic Studies

Genome projects in the Solanaceae family now encompass basic studies to elucidate genome sequences, and translational research for marker development, germplasm curation, and breeding. The resources described above can guide marker discovery with application to breeding programs and genetic studies. TIGR has identified SSRs from 12 solanaceous species including tomato, potato and pepper and designed primers to these sequences (http://www.tigr.org/tdb/sol/sol_SSR.shtml). TIGR has also identified SNPs in the current potato EST collection; 4,798 high confidence candidate SNPs could be identified in 2,667 contigs (http://www.tigr.org/tdb/sol/sol_SNP.shtml).

This type of data-mining can be productive, but a problem arises because the majority of tomato and pepper EST sequences are derived from single genotypes. In addition, the polymorphisms discovered remain "putative" until experimental verification, and resources for such verification have been scarce. For example, tomato EST resources are biased toward the line TA496 (with 116,736 ESTs); the next most abundant cultivar is 'Rio Grande' (or progeny) with 21,382 ESTs. This limits polymorphism discovery because sequencing ESTs from only a few cultivars is unlikely to reveal much of the genetic variation in a cultivated species. The MicroTom project (Shibata 2005) offers new sequence data for SNP discovery, but this "lab-strain" dwarf variety contains a large portion of the *S. pimpinellifolium* genome. Thus, a high percentage of SNPs discovered using MicroTom sequences are not polymorphic in breeding germplasm.

19.6 Application of Translational Genomics Tools to Improvement of Emerging Andean Solanum Species

The focus of this section will be on how genomics tools developed for tomato and potato can be used to enhance research on minor *Solanum* crops for which little or no genomic information exists. Two exotic fruit species, known as lulo, or naranjilla, (*S. quitoense*), and tree tomato or tamarillo (*S. betaceum*) are indigenous to the Andes. Lulo has been described as "the golden fruit of the Andes." Both species yield fruit that provides an excellent source of vitamins C, B1, B2, B3, B6, E, and pro vitamin A, minerals (iron, calcium, phosphorus, potassium, and nitrogen), proteins, and carbohydrates. Their fruit can be eaten raw or cooked or used to make juice, pies, jellies, jams, ice cream and have been also used for medicinal purposes (NRC 1990; International Plant Genetic Resources Institute 2006). In areas where illicit crops are grown, these indigenous species can serve as alternative crops whose cultivation can conserve and regenerate Andean agro-ecosystems.

Despite their increasing market value, major constraints for their adoption by local farmers include little technological support and little availability of breeding materials to address a need to improve fruit yield, quality and resistance to biotic and abiotic stresses (Lobo M, personal communication). Production constraints may be alleviated with the development of high yielding, sustainable cultivars, which should start with a broad genetic base.

One of the largest germplasm collections for these species is maintained at the Corporation for Agricultural Research (CORPOICA), Colombia. This collection has been partially characterized using phenotypic and genotypic (AFLP marker) information (Garcia et al. 2002, Fory et al. 2004, Lobo M., personal communication). However, additional tools are necessary for a deeper knowledge and proper use of these genetic resources.

A consortium of about 30 countries embarked on a new project called the International Solanaceae Genomics Project (http://sgn.cornell.edu /solanaceae-project) with the mission of developing genomic tools for this family. Here we show an example of how genomic information derived from this project in well developed species such as tomato can be used in less developed species such as lulo and tree tomato for their future improvement.

A strategy using COS markers has been developed by several groups to characterize variation within cultivated germplasm by using genomic resources of Arabidopsis and other crops to identify conserved single-copy genes. A COS is defined as a set of genes that are conserved throughout plant evolution in both sequence and copy number (Fulton et al. 2002, Lyons et al. 1997). Sequences from a single-copy

COS of Arabidopsis are used as a reference so that relationships can be "bridged" across genes identified through ESTs in crops. A COS in tomato (SGN COS I) was identified by comparing the set of unique gene sequences in tomato to all translated proteins in the fully sequenced genome of Arabidopsis (Fulton et al. 2002, Wu et al. 2006). Approximately 10% of the 27,000 tomato unigene sequences meet the defined COS criteria. This set was further refined to include a wider range of species, as had been tried in the Compositae (http://cgpdb.ucdavis.edu/compositae/ncbi_compositae_resources.php), Comparison of tomato ESTs to pepper, potato, eggplant and coffee resulted in a combined COS (SGN COS II) of 2,869 unigenes across five species (Wu et al. 2006). There are multiple methods to identify a COS (e.g. http://cgpdb.ucdavis.edu/COS_Arabidopsis/), and most are suitable for evolutionary, phylogenetic, and comparative genomic studies. Primers from COS genes can be designed to amplify intronic or exonic regions, thus providing greater flexibility for identification of polymorphisms.

More than 400 COSII primers covering the 12 tomato chromosomes have been evaluated in lulo and tree tomato (Olarte A, Tanksley S, Barrero L, unpublished results). Of these, about 85% in lulo, and 70% in tree tomato, have produced PCR amplification products. This result suggests that COSII can be used as a framework for marker use in comparative genomics research on Solanaceae and related species where genomic information is unavailable. Lulo and tree tomato parental lines, and wild relatives, have been evaluated for COSII polymorphisms (Table 19.1). The lulo wild relative *S. hirtum* and the tree tomato wild relative *S. uniloba* are able to produce fertile hybrids with their cultivated counterparts (Lobo M., personal communication). DNA from accessions listed in Table 19.1 has been amplified with 84 COSII primers dispersed in the tomato genome and single band amplification PCR products have been sequenced. Pair-wise comparisons between possible parental combinations have been made using the program CAPS designer to find polymorphisms (http://www.sgn. cornell.edu /tools/caps_designer)

Table 19.1 Accessions of lulo, tree tomato and wild relatives analyzed with COSII. Provided by M. Lobo (CORPOICA, Colombia)

Accession Number	Species Name	Traits of interest
120039	*S. quitoense var. septeptrionale*	Big fruit and foliar area.
120044	*S. quitoense var. quitoense*	Susceptible to
120052	*S. quitoense var. quitoense*	*Phytophtora*
120101	*S. quitoense var. quitoense*	
120103	*S. quitoense var. septeptrionale*	
120060	*S. hirtum*	Small fruit. Resistant to
120061	*S. hirtum*	*Phytophtora* and
120062	*S. hirtum*	nematodes
120071	*S. hirtum*	
285028	*S. betaceum*	Big fruit. Susceptible to
6002056	*S. betaceum*	*Colletotrichum*
6002061	*S. betaceum*	(anthracnose)
6002078	*S. uniloba*	Small fruit. Resistant to anthracnose

The results gathered so far indicate that the cultivated counterpart *S. quitoense* produces a higher percentage of single band amplifications and a lower percentage of multiple band amplifications as compared to its wild relative (about 80% of single band amplifications of the total COSII amplified for cultivated vs. 70% for wild) suggesting that *S. hirtum* is more heterozygous and/or has been subjected to more duplication events at COSII loci. Per locus, pair-wise comparisons between accessions showed that SNPs and insertions and deletions (INDELs) are the predominant types of polymorphisms and indicated that any parental combination is suitable for mapping. Several of these *in silico* polymorphisms have been visualized in agarose gels (either directly or by using CAPs) and are being used for diversity and mapping studies (Enciso F, Cardenas Z, Barrero L, unpublished results). Currently, the evaluation of about 460 COSII markers for selected parents is being expanded.

The sequence information developed so far (about 1400 sequences) is published in the Solanaceae genome database (http://www.sgn.cornell.edu) and will assist studies of homology and diversity, and aid the development of comparative genetic/QTL maps for breeding efforts. It is expected that the benefits for improvement of lulo and tree tomato through comparative genomics is high since the Solanaceae family has not been subjected to large-scale duplication events (e.g. polyploidy) early in its radiation and as a result, macro and microsynteny conservation amongst

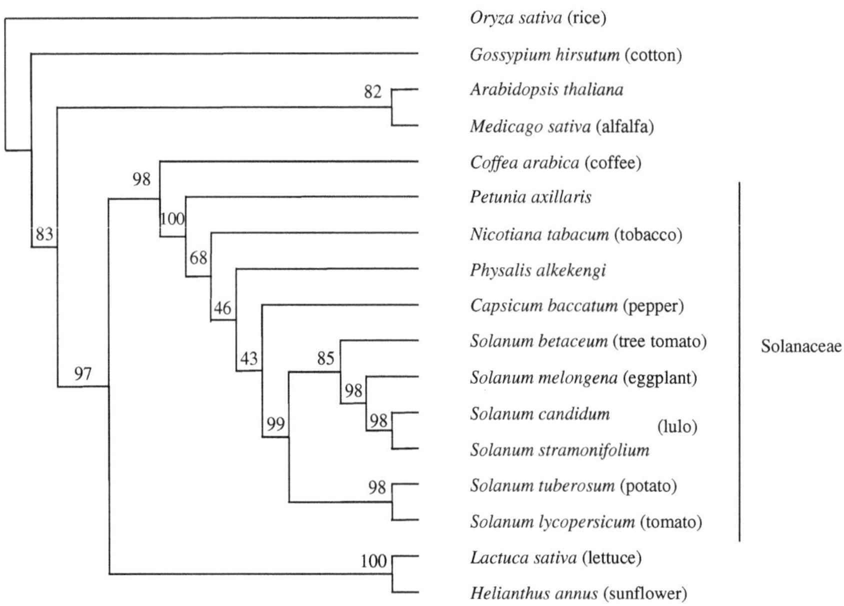

Fig. 19.2 Maximum likelihood tree based on ndhF sequences, including the tree tomato and 2 close relatives of lulo (*S. candidum* and *S. stramonifolium*) (Bohs, 2004). The numbers indicate bootstrap values. The rice sequence was used as out-group. Sequences were retrieved from Genebank. (Courtesy of F. Wu)

the genomes of Solanaceae species (tomato, potato, pepper, eggplant) is very high (Livingstone et al. 1999, Tanksley et al. 1992, Wang et al. 2006). Moreover, a significant proportion of the major QTLs for domestication traits (e.g. fruit weight, fruit shape, and anthocyanins) are conserved in tomato, potato, pepper, and eggplant (Doganlar et al. 2002). Lulo and tree tomato are very close relatives to these species (Fig. 19.2) and might also be conserved for some of these traits. The implications for their improvement might extend not only to yield and quality but also pest and disease resistance, which are major constraints in the Andean region. These traits might be efficiently improved in future studies since many genes conferring resistance to fungi, nematodes, aphids, bacteria and viruses have been cloned in the Solanaceae (tomato, potato, pepper, tobacco), and lulo and tree tomato are susceptible to many of the same pathogens (e.g. *Meloidogyne spp.*, *Phytophtora spp.*, *Fusarium spp.*).

19.7 Perspective

The expected wealth of genome sequence information that is forthcoming from two species in the genus *Solanum*, combined with the rich genetic resources of the related species, and the increasingly available genomic resources of tomato and potato, will increasingly facilitate research and crop improvement in the Solanaceae. COS is but one example of genomics tools (genome and EST sequence, microarrays, maps, markers, among others) that will make more efficient the future development of lesser known crops such as lulo and tree tomato. Tomato and potato, which have unique developmental attributes such as fleshy fruit and tubers that are not present in rice or *Arabidopsis*, will also provide increasingly useful model systems for research on fundamental biological systems.

Acknowledgments The authors would like to acknowledge www.ocati.com for permission to use color figures. We thank Mario Lobo (CORPOICA), and Lois Grant (OSU) for helpful comments on the manuscript. We gratefully acknowledge the National Science Foundation (NSF) and Colciencias for financial support of Colombian research.

References

Bosland PW (1996) Capsicums: Innovative uses of an ancient crop. In: J. Janick (ed.), Progress in new crops. ASHS Press, Arlington, VA p. 479–487

Centro Internacional de la Papa (CIP) (2007) CIP Potato Facts: Potato: Growth in Production Accelerates. http://www.cipotato.org/potato/facts/growth.asp (verified Aug. 11, 2007)

Doganlar S, Frary A, Daunay MC, Lester RN, Tanksley SD (2002) Conservation of gene function in the Solanaceae as revealed by comparative mapping of domestication traits in eggplant. Genetics 161: 1713–1726

Economic Research Service (ERS), USDA (2007) Vegetables and Melons Outlook /VGS-321/June 21. http://www.ers.usda.gov/publications/vgs/Tables/World.pdf (verified Aug. 11, 2007)

Facciola S (1990) Cornucopia - A Source Book of Edible Plants. Kampong Publications. ISBN 0–9628087-0–9

Flinn B, Rothwell C, Griffiths R, Lague M, DeKoeyer D, et al. (2005) Potato expressed sequence tag generation and analysis using standard and unique cDNA libraries. Plant Mol Biol 59:407–433

Fory P, Sánchez I, Bohórquez A, Medina, CI, Lobo M 2004. Caracterización molecular de la colección colombiana de lulo (*Solanum quitoense*) LAM. V Seminario Nacional e Internacional de Frutales p. 443

Fulton TM, Van der Hoeven R, Eannetta NT, Tanksley SD (2002) Identification, analysis, and utilization of conserved ortholog set markers for comparative genomics in higher plants. Plant Cell:1457–1467

Garcia P, Garcia R, Medina C, and Lobo M (2002) Variabilidad morfológica cualitativa en una colección de tomate de árbol Cyphomandra (Solanum) Betacea (Betaceum). In: Seminario Nacional de Frutales de Clima Frio Moderado. (4: 2002). Memorias del IV Seminario Nacional de frutales de clima frío moderado. p. 49–54

Ghislain MD, Andrade MD, Rodríguez F, Hijmans RJ and Spooner DM (2006) Genetic analysis of the cultivated potato *Solanum tuberosum* L. Phureja Group using RAPDs and nuclear SSRs. Theor Appl Genet 113:1515–1527

Gómez-Pompa A and Sosa, V (eds.) (1978) Flora de Veracruz

Grube RC, Radwanski ER, Jahn M (2000) Comparative genetics of disease resistance within the Solanaceae. Genetics 155:873–887

Hawkes JG (1990) The potato: evolution, biodiversity, and genetic resources. (Potato EGR) 182

Heiser C, Anderson G (1999) "New" Solanums. p. 379–384. In: J. Janick (ed.), Perspectives on new crops and new uses. ASHS Press, Alexandria, VA

Huamán Z, Spooner DM (2002) Reclassification of landrace populations of cultivated potatoes (*Solanum* sect. Petota). Amer J Bot 89:947–965

International Plant Genetic Resources Institute. New World Fruits Database [online]. (2006) <www.ipgri.cgiar.org/Regions/Americas/programmes/TropicalFruits/>

Isshiki S, Taura T (2003) Fertility restoration of hybrids between *Solanum melongena* L. and *S. aethiopicum* L. Gilo Group by chromosome doubling and cytoplasmic effect on pollen fertility. Euphytica 134:195–201

Kang BC, Yeam I, Frantz JD, Murphy JF, and Jahn MM (2005) The *pvr1* locus in Capsicum encodes a translation initiation factor eIF4E that interacts with Tobacco etch virus VPg. Plant J 42:392–405

Labate JA, Baldo AM (2005) Tomato SNP discovery by EST mining and resequencing. Mol Breeding:16:343–349

Lester RN (1986) Taxonomy of scarlet eggplants. *Solanum aethiopicum* L. ISHS Acta Horticulturae 182: I International Symposium on Taxonomy of Cultivated Plants (ed. L.J.G. van der Maesen). ISBN 978–90-66053–12-0 ISSN 0567–7572

Livingstone KD, Lackney VK, Blauth JR, van Wijk R, Jahn MK (1999) Genome mapping in *Capsicum* and the evolution of genome structure in the Solanaceae. Genetics 152:1183–1202

Lyons LA, Laughlin TF, Copeland NG, Jenkins NA, Womack JE, et al. (1997) Comparative anchor tagged sequences (CATS) for integrative mapping of mammalian genomes. Nature Genetics 15:47–56

Minna JD, Gazdar AF (1996) Translational research comes of age. Nature Med 2:974–975

National Research Council (1990) Lost Crops of the Incas: Little-Known Plants of the Andes with Promise for Worldwide Cultivation. (Ed. Popenoe H, et al) National Academy Press, Washington, DC ISBN 0–309-04264-X

Ori N, Eshed Y, Paran I, Presting G, Aviv D, et al. (1997) The I2C family from the wilt disease resistance locus I2 belongs to the nucleotide binding, leucine-rich repeat superfamily of plant resistance genes. Plant Cell 9:521–532

Pickersgill B (1997) Genetic resources and breeding of *Capsicum* spp. Euphytica 96:129–133

Ronning CM, Stegalkina SS, Ascenzi RA, Bougri O, Hart AL, et al. (2003) Comparative analyses of potato expressed sequence tag libraries. Plant Physiol 131:419–429

Shibata D (2005) Genome sequencing and functional genomics approaches in tomato. J Gen Plant Pathol 71:1–7

Spooner DM, Anderson DJ, Jansen RK (1993) Chloroplast DNA evidence for the interrelationships of tomatoes, potatoes, and pepinos (Solanaceae) Am J Bot 80:676–688

Stewart CB, Kang C, Liu K, Mazourek M, Moore SL, et al. (2005) The *Pun1* gene for pungency in pepper encodes a putative acyltransferase. Plant J 42: 675–688

Tanksley SD, Ganal MW, Prince JP, de-Vicente MC, Bonierbale MW, et al. (1992) High density molecular linkage maps of the tomato and potato genomes. Genetics 132:1141–1160

Tsugane T, Watanabe M, Yano K, Sakurai N, Suzuki H, et al. (2005) Expressed sequence tags of full-length cDNA clones from the miniature tomato (*Lycopersicon esculentum*) cultivar Micro-Tom. Plant Biotechnol 22:161–165

USDA, ARS, National Genetic Resources Program (2007) Germplasm Resources Information Network - (GRIN) [Online Database]. National Germplasm Resources Laboratory,Beltsville, Maryland. URL: http://www.ars-grin.gov/cgi-bin/npgs/html/taxon. pl?8904 (verified 11 August 2007)

Van Deynze A, Douches D, De Jong W, and Francis, D (2007) Summary of Solanaceae Coordinating Meetings. In: Spooner GM, Bohs l, Giovannoni J, Olmstead RG, Shibata D (eds), Solanaceae VI. Proc Sixth Intl Solan Conf. Madison, WI. Acta Hort.(ISHS) 745:533–536. http://www.actahort.org/books/745/745_40.htm

van Os H, Andrzejewski S, Bakker E, Barrena I, Bryan GJ, et al. (2006) Construction of a 10,000-marker ultradense genetic recombination map of potato: providing a framework for accelerated gene isolation and a genomewide physical map. Genetics 173:1075–1087

Wang Y, Tang C, Cheng Z, Mueller L, Giovannoni J, et al. (2006). Euchromatin and pericentromeric heterochromatin: comparative composition in the tomato genome. Genetics 172:2529–2540

Weese TL, Bohs L (2007) A three-gene phylogeny of the genus *Solanum* (Solonaceae). Systematic Bot 32:445–463

Wong KK, Tsang YT, Shen J, Cheng RS, Chang YM, et al. (2004) Allelic imbalance analysis by high-density single-nucleotide polymorphic allele (SNP) array with whole genome amplified DNA. Nucleic Acids Res 32(9):e69

Wu FN, Mueller LA, Crouzillat D, Petiard V, Tanksley SD (2006) Combining bioinformatics and phylogenetics to identify large sets of single-copy orthologous genes (COSII) for comparative, evolutionary and systematic studies: A test case in the euasterid plant clade. Genetics 174:1407–1420

Yang W, Bai X, Kabelka E, Eaton C, Kamoun S, et al. (2004) Discovery of single nucleotide polymorphisms in *Lycopersicon esculentum* by computer aided analysis of expressed sequence tags. Mol Breeding 14:21–34

Zhang LP, Lin GY, Niño-Liu D Foolad MR (2004) Mapping QTLs conferring early blight (*Alternaria solani*) resistance in a *Lycopersicon esculentum* × *L. hirsutum* cross by selective genotyping. Mol Breeding 12 (1):3–19

Zhang LP, Khan A, Nino-liu D, Foolad MR (2002) A molecular linkage map of tomato displaying chromosomal locations of resistance gene analogs. Genome 48: 391–400

Chapter 20
Genomics of Sorghum, a Semi-Arid Cereal and Emerging Model for Tropical Grass Genomics

Andrew H. Paterson, John E. Bowers, and F. Alex Feltus

Abstract Sorghum, an important failsafe crop in the global agroecosystem, is also emerging as a model for tropical grasses based on its small and well-characterized genome, low level of gene duplication, and close relationship to the larger and more complex genomes of maize and sugarcane. A whole-genome shotgun sequence of the sorghum genome is complete and being annotated. The sorghum sequence, together with the attributes of sorghum as a prospective functional genomics and association genetics system, has many implications for better understanding the structure, function, and evolution of cereal genomes. In addition, the sequence will raise to a new level the opportunities to engage genomics in the improvement of human livelihood in arid and semi-arid tropical regions in which sorghum is a staple. Already established as a seed-based ethanol crop, progress in understanding the genetic control of perenniality in sorghum makes it also promising as a cellulosic biofuels crop.

20.1 Introduction

Sorghum, *Sorghum bicolor* (L.) Moench, is one of the world's leading cereal crops, providing food, feed, fiber, fuel, and chemical/biofuel feedstocks across a range of environments and production systems. Its remarkable ability to produce a crop under adverse conditions, in particular with much less water than most other grain crops, makes sorghum an important "failsafe" source of food, feed, fiber, and fuel in the global agroecosystem.

Sorghum is a biofuel (ethanol) crop of growing importance. Already established as a seed-based ethanol crop, progress in understanding the genetic control of perenniality in sorghum (Paterson et al. 1995b; Hu et al. 2003) makes it promising as a cellulosic biofuels crop. Cellulosic biofuel production offers compelling advantages over seed-based production (Farrell et al. 2006), but will require greater utilization

A.H. Paterson
Plant Genome Mapping Laboratory, University of Georgia, 111 Riverbend Road Rm 228, Athens GA 30602
e-mail: paterson@uga.edu

© Springer 2008

of marginal lands to make the low per-unit value of biomass production economical and will be heavily dependent upon the use of perennials to be sustainable (Wagoner 1990; Scheinost 2001).

As a model for tropical grasses, sorghum is a logical complement to *Oryza* (rice). Sorghum is representative of tropical grasses in that it uses C4 photosynthesis, comprising complex biochemical and morphological specializations that improve carbon assimilation under limited water and at high temperatures. By contrast, rice is more representative of temperate grasses, using C3 photosynthesis. Sorghum's low level of gene duplication makes it, like rice, an attractive model for functional genomics. *Sorghum* is an excellent bridge from *Oryza* to leading tropical grass crops with much larger genomes and more gene duplication.

Its economic and scientific importance, together with progress in characterizing its genome (detailed below) have motivated the sequencing of an elite inbred line of *S. bicolor*, BTx623, to 8x genome coverage under the US Department of Energy Joint Genome Institute (JGI) "Community Sequencing Program" (CSP). The sequence is presently available in the NCBI trace archive and assembly is near completion.

20.1.1 Economic, Agronomic, and Societal Importance of Sorghum

Worldwide, sorghum is the fifth most important grain crop grown based on tonnage, after maize, wheat, rice, and barley (www.fao.org). Sorghum is unusually tolerant of low input levels, an essential trait for areas such as northeast Africa and the US Southern Plains that receive too little rainfall for most other grains. In the more arid countries of northeast Africa, such as Sudan, sorghum contributes 39% of the calories in the human diet (www.fao.org, 1999 statistics). Increased demand for limited fresh water supplies, increasing use of marginal farmland, and global climatic trends all suggest that dryland crops such as sorghum will be of growing importance to feed the world's expanding populations.

Despite the likely growing importance of sorghum, its improvement has lagged that of maize, wheat, and rice, each of which has more than doubled in average yield on a worldwide basis in the last 38 years, while sorghum yields have gained only 51% (average 1961–1963 compared to 1999–2001 – www.fao.org). In Sub-Saharan Africa, already home to many of the world's hungry and with a population projected to double over the next 40 years (US Census Bureau estimates 2002, http://www.census.gov), sorghum yields have gained only 6% over the last 38 years compared to 50% gains in wheat and maize (www.fao.org).

In the United States, sorghum was introduced over 200 years ago, possibly by Benjamin Franklin (Smith and Frederiksen 2000) and is now grown on 9–13 million acres. US sorghum is principally used as an animal feed, and therefore escapes direct notice by the general public, but is the 13th most valuable crop in the United States with a farm-gate value ranging from $0.8–2.0 billion/yr (USDA 1992–2001 statistics).

Sorghum is the number two crop (after maize) used in US ethanol production. Ethanol is the single largest value-added market for US grain sorghum producers (www.sorghumgrowers.com), presently consuming about 10% of the US sorghum crop. The generally lower water demands and market price for sorghum than maize, versus their equal per-bushel ethanol yields, suggests that sorghum will be of growing importance in meeting US biofuels needs. "Sweet sorghums" with high sugar content in stems, already grown for forage, silage, and sorghum molasses, may be especially promising. Further, good understanding of the genetics of perennation in sorghum may expedite development of perennial forms that improve sustainability of biomass production on marginal lands (Wagoner 1990; Scheinost 2001).

The *Sorghum* genus is also noteworthy in that it includes a major noxious weed. Vegetative dispersal by rhizomes (underground stems) and seed dispersal by disarticulation of the mature inflorescence ("shattering") cause "Johnsongrass" [*Sorghum halepense* (L.) Pers, $2n = 2x = 40$] to rank among the world's most noxious weeds (Holm et al. 1977). Johnsongrass is an interspecific hybrid of *Sorghum bicolor* and *S. propinquum*, the latter contributing rhizomatousness. Thus, *S. bicolor* and *S. propinquum* provide a system in which to dissect the genetic basis of rhizomatousness. The same features that make Johnsongrass such a troublesome weed are actually desirable in many forage, turf, and biomass crops that are genetically complex. Therefore, sorghum offers novel learning opportunities relevant to weed biology as well as to improvement of a wide range of other forage, turf, and biomass crops.

20.1.2 Sorghum as an Experimental Organism

The small genome of sorghum has long been an attractive model for advancing understanding of the structure, function, and evolution of cereal genomes. Sorghum is representative of tropical grasses in that it has C4 photosynthesis, using complex biochemical and morphological specializations to improve carbon assimilation at high temperatures. By contrast, rice is more representative of temperate grasses, using C3 photosynthesis.

Its lower level of gene duplication than many other tropical cereals makes sorghum, like rice, an attractive model for functional genomics. Evidence that a large fraction of "paleologs" may have undergone differential loss in *Oryza* and *Sorghum* since their divergence (Paterson et al. 2004) suggests that the preferred system in which to study a particular gene will vary on a case-by-case basis.

Sorghum is an excellent bridge from *Oryza* to leading tropical grass crops, with much larger genomes and much more gene duplication. *Sorghum* and *Zea* (maize, the leading US crop with a farm-gate value of \$15–20 billion/yr) diverged from a common ancestor ~12 mya (Gaut et al. 1997; Swigonova et al. 2004a) versus ~42 mya for rice and the maize/sorghum lineage (Paterson et al. 2004). *Saccharum* (sugarcane), arguably the most important biofuel crop worldwide, valued at ~\$30 billion including \$1 billion/yr in the US, may have shared ancestry with sorghum as little as 5 million years ago (Sobral et al. 1994), retains similar gene order (Ming et al. 1998), and even produces viable progeny in some intergeneric crosses (Dewet et al. 1976). *Zea* has undergone one whole-genome duplication since its divergence from *Sorghum* (Swigonova et al. 2004b), and *Saccharum* has undergone at least two (Ming et al. 1998).

Sorghum is extremely well-suited to association mapping methods (Hamblin et al. 2005). Its largely self-pollinating mating system tends to preserve linkage relationships for longer periods than in largely outcrossing crops such as maize. Self-pollination also obviates the need to develop inbred lines.

20.2 Genetic Mapping and Tagging in Sorghum

20.2.1 Linkage Mapping

Linkage mapping in sorghum takes advantage of the plant's straightforward diploid genetics, amenability to inbreeding, and high levels of polymorphism between *Sorghum* species and adequate levels within *S. bicolor*. High-density reference maps of one intraspecific *S. bicolor* (Xu et al. 1994; Bhattramakki et al. 2000; Klein et al. 2000; Menz et al. 2002) and one interspecific *S. bicolor* x *S. propinquum* (Chittenden et al. 1994; Bowers et al. 2003) cross provide about 2,600 sequence-tagged-sites (based on low-copy probes that have been sequenced), 2,454 AFLP, and ~1,375 sequence-scanned (based on sequences of genetically anchored BAC clones) loci. The two maps share one common parent (*S. bicolor* 'BTx623') and are essentially colinear (Feltus et al. 2006). Cytological characterization of the individual sorghum chromosomes has provided a generally adopted numbering system (Kim et al. 2005b).

More than 800 markers mapped in sorghum are derived from other taxa (hence serve as comparative anchors), and additional sorghum markers have been mapped directly in other taxa or can be plotted based on sequence similarity. Anchoring of the sorghum maps to those of rice (Paterson et al. 1995a; Paterson et al. 2004), maize (Whitkus et al. 1992; Bowers et al. 2003), sugarcane (Dufour et al. 1997; Ming et al. 1998), millet (Jessup et al. 2003), switchgrass (Missaoui et al. 2005), Bermuda grass (Bethel et al. 2006), and others provides for the cross-utilization of results to simultaneously advance knowledge of many important crops.

20.2.2 Gene Tagging

The linkage maps of sorghum have been employed in the "tagging" (mapping) of genes for a large number of diverse traits. The interspecific population has been especially useful for characterization of genes related to domestication, such as seed size, shattering (Paterson et al. 1995a), tillering, and rhizomatousness (Paterson et al. 1995b). Plant height and flowering time (Lin et al. 1995; Ulanch et al. 1996) have been a high priority. Similarly, the importance of hybrid sorghum motivated much research into the genetic control of fertility restoration (Klein et al. 2001; Klein et al. 2005; Wen et al. 2002). Resistance genes have been tagged for numerous diseases (Multani et al. 1998; Tao et al. 1998; McIntyre et al. 2004; McIntyre et al. 2005; Mutengwa et al. 2005; Totad et al. 2005; Singh et al. 2006; Wang et al. 2006), key insect pests (Agrama et al. 2002; Katsar et al. 2002; Tao et al. 2003; Nagaraj et al. 2005), and also the parasitic weed, striga (Haussmann et al. 2004; Mutengwa et al. 2005). Genes and QTLs have been identified that are related to abiotic stresses including post-reproductive stage drought tolerance (Stay-green) (Crasta et al. 1999; Subudhi et al. 2000; Xu et al. 2000; Haussmann et al. 2002); preharvest sprouting, (Lijavetzky et al. 2000; Carrari et al. 2003), and aluminum tolerance (Magalhaes et al. 2004). Numerous additional morphological characteristics have also been mapped in interspecific and/or intraspecific populations (Feltus et al. 2006).

20.3 Physical Characterization and Sequencing of the Genome

The small size of the sorghum genome facilitates its use as a tropical grass model, and sorghum was the first angiosperm for which a BAC library was made (Woo et al. 1994). Estimates of the physical size of the sorghum genome range from 700 Mbp based on Cot analysis (Peterson et al. 2002) to 772 Mbp based on flow cytometry (Arumuganathan and Earle 1991). This makes the sorghum genome about 60% larger than that of rice, but only about 1/4 the size of the genomes of maize or human. DNA renaturation kinetic analysis (Peterson et al. 2002) shows the sorghum genome to comprise about 16% foldback DNA, 15% highly repetitive DNA (with individual families occurring at an average of 5,200 copies per genome), 41% middle-repetitive DNA (avg. 72 copies), and 24% low-copy DNA. About 4% of the DNA remained single-stranded at very high Cot values and is assumed to have been damaged (thus the other percentages are slight under-estimates).

20.3.1 BAC Libraries and Physical Mapping

High-coverage BAC libraries are available for BTx623 (about 12X coverage from HindIII and 8X from BamHI), *S. propinquum* (13–14X coverage from EcoRI [~7X] and HindIII [~7X] and IS3620C (~9X coverage from HindIII). A total of 69,545 agarose-based fingerprints from BTx623 BACs are also anchored with 211,558 hybridization loci from 7,292 probes (about 2,000 of which are genetically mapped) In parallel, 40,957 agarose-based fingerprints from *S. propinquum* are anchored with 189,735 hybridization loci from 7,481 probes (2,000 genetically mapped). Each of these has been assembled into WebFPC-accessible physical maps (http://www.stardaddy.uga.edu/fpc/bicolor/WebAGCoL/WebFPC/ and http://www.stardaddy.uga.edu/fpc/propinquum/WebAGCoL/WebFPC/) for which earlier versions have been described in detail (Bowers et al. 2005). Additional resources include 20,000 HICF fingerprints (from genetically mapped contigs) and 6-D BAC pools (5X deep) from BTx623; and 10,000 HICF fingerprints and 6-D BAC pools (5X deep) from IS3620C. Targeted HICF of additional contig-terminal BACs has been used to fill gaps. About 456 *S. propinquum* and 303 *S. bicolor* BAC contigs (41% of BACs, 80% of single-copy loci) appear to be well-anchored to euchromatic regions, with the percentage of the genome attributable to euchromatin likely to rise with additional anchoring. The finding that 41% of BACs are already anchored to euchromatin, while only 24% of the sorghum genomic DNA is single- or low-copy (with an overall kinetic complexity of 1.64×10^8 (Peterson et al. 2002)), suggests that euchromatin includes a mixture of low-copy and repetitive DNA.

20.3.2 Genome Sequencing

The shotgun sequencing of a leading US sorghum inbred, BTx623, is now complete, and ~10.5 million reads (~8X coverage) have been deposited in the NCBI Trace Archive. Analysis of the preliminary assembly confirms that the sorghum genome sequence will be a suitable substrate for a complete and high quality annotation. In a preliminary assembly (that is expected to further improve with ongoing analysis), more than 97% of sorghum protein-coding genes (ESTs) were captured in the ~250 longest scaffolds. The vast majority of these can be linked, ordered, and oriented using the genetic and physical map to reconstruct complete chromosomes. Alignments of the preliminary assembly to sorghum methyl-filtered sequence; sorghum, maize, and sugarcane transcript assemblies; and the *Arabidopsis* and rice proteomes confirms the base-level accuracy of the assembly and correct local structure, allowing for the rough prediction of ~30,000–50,000 protein-coding loci (the lower figure stringently excludes retrotransposon-like sequences and requires significant support from homology and/or ESTs). Extensive conserved synteny with rice is clear, as expected from map comparisons.

20.4 Molecular Cytogenetics

Historically, sorghum has been problematic for cytogenetic analysis, due to its small chromosomes. Methodologies based on fluorescence- in situ hybridization (FISH) (Kim et al. 2002) have recently led to detailed molecular cytogenetic maps of most chromosomes (Islam-Faridi et al. 2002; Kim et al. 2005a; Kim et al. 2005c), and a community-accepted set of chromosome numbers (Kim et al. 2005b). Molecular cytogenetic data have also facilitated the isolation of centromere-specific repetitive DNA elements in sorghum (Miller et al. 1998); and the physical mapping of the rDNA (Sang and Liang 2000) and liguleless regions (Zwick et al. 1998). Finally, molecular cytogenetic evidence of ancient tetraploidy (Gomez et al. 1998) supported genetic evidence (Chittenden et al. 1994) and now has been shown to reflect an ancient duplication in a common ancestor of most if not all cereals (Paterson et al. 2004).

20.5 Functional Genomics Resources

20.5.1 Transcriptome Characterization

The sorghum gene space is presently represented by approximately 125,000 expressed sequence tags that have been clustered into $\sim$22,000 unigenes, representing more than 20 diverse libraries from several genotypes (Pratt et al. 2005).

About 500,000 methyl-filtered (MF) reads that provide an estimated 1x coverage of the MF-estimated gene space (Bedell et al. 2005) have been assembled into contigs (SAMIs, http://magi.plantgenomics.iastate.edu/). Another reduced-representation strategy, Cot-based cloning and sequencing (CBCS), was first demonstrated in sorghum in 2001 (Genbank accessions AZ921847-AZ923007), and further detailed subsequently (Peterson et al. 2002). This method offers the potential to further enhance gene space coverage beyond that offered by ESTs and MF, in a complementary manner as demonstrated for maize.

Progress in characterization of the transcriptome has been paralleled by identification of differential gene expression in response to biotic and abiotic factors, including greenbug feeding (Park et al. 2006), dehydration, high salinity and ABA (Buchanan et al. 2005), and methyl jasmonate, salicylic acid, and aminocyclopropane carboxylic acid treatments (Salzman et al. 2005).

20.5.2 Transformation

Sorghum transformation methods have been available since the early 1990s, initially for protoplasts (Battraw and Hall 1991) and cultured cells (Hagio et al. 1991), and subsequently in planta (Casas et al. 1993; Casas et al. 1997)), with both agrobacterium- and microprojectile-based protocols now available at substantially improved efficiencies (Zhao et al. 2000; Devi and Sticklen 2003; Tadesse et al. 2003; Carvalho et al. 2004; Gao et al. 2005a; Gao et al. 2005b; Howe et al. 2006).

20.5.3 Forward Genetics

Over some 30 years, the late K. Schertz collected ca. 400 *S. bicolor* mutants now under the curation of C. Franks (USDA-ARS, Lubbock TX). Most were spontaneous, although at least 27 are from irradiation experiments. The phenotypes of most have been verified, but they are otherwise largely unexplored (Webster 1964; Schertz and Stephens 1966)). Based on the affected growth stage, categories and brief descriptions of the mutants are: *Seedling (92 mutants)*: mainly chlorophyll deficiency, some are lethal (but live long enough to obtain DNA) but many are not; *Leaf (88)*: predominantly leaf-spotting and streaking, many similar to symptoms of foliar diseases such as zonate leaf spot and anthracnose; *Grain (30)*: includes sugary, waxy, high lysine, variegated pericarp ('Calico') and other pericarp and endosperm phenotypes;

Maturity (21): early or late flowering, also including lines used in seminal genetic analysis of this trait (Quinby 1974); *Linkage (12)*: dominant phenotypic markers used to determine linkage groups by early workers; *Reproductive (35)*: various forms of twin-seededness, 'Scaly' 'Deciduous' (shattering) and non-male forms of sterility ('Female' 'Random', 'Chimeral'); *Male sterility (20)*: most known forms of genetic male sterility (*Ms1, 2, 3, 7*, and *al*); *Miscellaneous (189)*: morphological/developmental, including plant and leaf architecture ('Lazy', 'ZigZag', 'Pineapple', 'Midget', 'Liguleless', 'Whorled Leaf', 'Rolled Leaf', others).

20.5.4 Reverse Genetics

Sorghum offers an opportunity to complement more extensive reverse genetics resources in *Oryza* and *Zea*, providing for the study of genes/gene families that are less tractable in maize or rice, and also for targeting functional analyses to specific sorghum genes implicated in key traits by association genetics or other approaches. To accelerate identification in a targeted manner of mutants useful to relate *Sorghum* genes to their functions, 1600 M3 annotated, individually pedigreed, mutagenized lines using ethyl methane sulfonate have been generated for sorghum genotype BTx623 and their preliminary characterization is in progress (Xin et al. 2007). To date, every M3 row inspected closely has been distinguishable from the original stock, and many have multiple mutant phenotypes (Z. Xin, personal communication). Additional M2 mutants are available for production of many thousands of additional M3 lines, if needed.

Cs1 is the first active transposable element isolated from sorghum. The *Cs1* element offers several advantages as an insertion mutagen in sorghum. *Cs1*-homologous sequences are present in low copy number in sorghum and other grasses, including sudangrass, maize, rice, teosinte, and sugarcane (Chopra et al. 1999). The low copy number and high transposition frequency of *Cs1* implies that this transposon could prove to be an efficient gene isolation tool in sorghum. Preliminary studies of *Cs1* as a mutagen (S. Chopra, personal communication) indicate the feasibility of using this transposon as a tagging tool.

20.6 Sequence and Marker Diversity

Thanks to efforts of the past few years, *Sorghum* has become a top echelon crop model for studying the genomic organization of allelic diversity. Sorghum is extremely well-suited to association mapping methods because of its medium-range patterns of linkage disequilibrium (Hamblin et al. 2005) and its self-pollinating mating system. Extensive *ex situ* sorghum germplasm collections exist within the U.S. National Plant Germplasm System and ICRISAT. Early characterization of complementary association genetics panels developed by a group of US scientists (Kresovich et al. 2005), and by Subprogram 1 of the Generation Challenge Program, is in progress. Much of the value of the sorghum sequence may be realized through better understanding of the levels and patterns of diversity in extant germplasm, which can contribute both to functional analysis of specific sorghum genes and to deterministic improvement of sorghum for specific needs and environments.

At present, more than 750 SSR alleles and 1402 SNP alleles discovered in 3.3 Mb of sequence (Schloss et al. 2002; Hamblin et al. 2004; Casa et al. 2005; Hamblin et al. 2005) are freely available from the *Comparative Grass Genomics Center* relational database (Gingle et al. 2007). Extensive studies of sequence variation in sorghum show that haplotype diversity is low, even when nucleotide diversity is high: for regions of average length 671 bp surveyed in 17 accessions, the median number of haplotypes was three and the mode was two (Hamblin et al. 2006). Common sequence variation can therefore be captured in a small sample of accessions.

20.7 Perspectives

The sorghum sequence, together with the attributes of sorghum as a prospective functional genomics and association genetics system, has many implications for better understanding the structure, function, and evolution of cereal genomes. Detailed physical maps that are also genetically anchored based on inclusion of genetically mapped, sequence-tagged sites provide a foundation upon which to overlay sequence assemblies, linking them to a rich history of genetics and genomics research. Much as the *Oryza* sequence provided an initial template for cereals, *Sorghum* will provide the first comparator, essential to unraveling structural and functional genomic divergence of the cereals from a common ancestral lineage that is thought to have existed ~42–47 million yrs ago (Paterson et al. 2004). Sorghum is an excellent bridge from rice to leading tropical grass crops such as maize and sugarcane with much larger genomes and much more gene duplication. Indeed, the fact that sorghum has not undergone a whole-genome duplication since the ancient one it shares with rice (Paterson et al. 2004) makes study of its genome critical to unraveling both global and gene-specific consequences of more recent duplications in leading crops such as maize (Swigonova et al. 2004c) and sugarcane (Ming et al. 1998).

Sequencing of sorghum will fill a key gap in plant biogeography in view of its African origin. It may also may prove to be an important vehicle for engagement of the African scientific community in genomics and its applications, in particular regarding documentation and analysis of *in situ* diversity that is presently inaccessible elsewhere.

To sustain coordination and communication in the genomic and postgenomic era for sorghum, a Sorghum Genomics Executive Committee was elected (Kresovich et al. 2005) to establish guidelines for future committee actions; collect, collate, and disseminate information; organize consortium meetings; and serve as an advocacy group for each country and area of research, with emphasis on cohesiveness of the community and importance of the crop.

Acknowledgments We thank the USDA-CSREES, NSF Plant Genome Research Program, and International Consortium for Sugarcane Biotechnology for funding relevant aspects of our research.

References

Agrama HA,, Wilde G., Reese J., Campbell L, Tuinstra M (2002) Genetic mapping of QTLs associated with greenbug resistance and tolerance in*Sorghum bicolor*. Theor Appl Genet 104:1373–1378

Arumuganathan K, Earle E (1991) Estimation of nuclear DNA content of plants by flow cytometry. Plant Mol Biol Reptr 9:208–218

Battraw M, Hall TC (1991) Stable transformation of *Sorghum-bicolor* protoplasts with chimeric neomycin phosphotransferase-Ii and beta-glucuronidase genes. Theor Appl Genet 82:161–168

Bedell JA, Budiman MA, Nunberg A, Citek RW, Robbins D, et al. (2005) Sorghum genome sequencing by methylation filtration. Plos Biol 3:103–115

Bethel CM, Sciara EB, Estill JC, Bowers JE, Hanna W, et al. (2006) A framework linkage map of bermudagrass (*Cynodon dactylon x transvaalensis*) based on single-dose restriction fragments. Theor Appl Genet 112:727–737

Bhattramakki D, Dong JM, Chhabra AK, Hart GE (2000) An integrated SSR and RFLP linkage map of *Sorghum bicolor* (L.) Moench. Genome 43:988–1002

Bowers J, Abbey C, Anderson S, Chang C, Draye X, et al. (2003) A high-density genetic recombination map of sequence-tagged sites for sorghum, as a framework for comparative structural and evolutionary genomics of tropical grains and grasses. Genetics 165:367–386

Bowers JE, Arias MA, Asher R, Avise JA, Ball RT, et al. (2005) Comparative physical mapping links conservation of microsynteny to chromosome structure and recombination in grasses. Proc Natl Acad Sci USA 102:13206–13211

Buchanan CD, Lim SY, Salzman RA, Kagiampakis L, Morishige DT, et al. (2005) Sorghum bicolor's transcriptome response to dehydration, high salinity and ABA. Plant Mol Biol 58:699–720

Carrari F, Benech-Arnold R, Osuna-Fernandez R, Hopp E, Sanchez R, et al. (2003) Genetic mapping of the *Sorghum bicolor vp1* gene and its relationship with preharvest sprouting resistance. Genome 46:253–258

Carvalho CH, Zehr UB, Gunaratna N, Anderson J, Kononowicz HH, et al. (2004) Agrobacterium-mediated transformation of sorghum: factors that affect transformation efficiency. Genet Mol Biol 27:259–269

Casa AM, Kononowicz AK, Zehr UB, Tomes DT, Axtell JD, et al. (1993) Transgenic sorghum plants via microprojectile bombardment. Proc Natl Acad Sci USA 90:11212–11216

Casa AM, Kononowicz AK, Haan TG, Zhang L, Tomes, DT, et al. (1997) Transgenic sorghum plants obtained after microprojectile bombardment of immature inflorescences. In Vitro Cell Dev Biol-Plant 33:92–100

Casa AM, Mitchell SE, Hamblin MT, Sun H, Bowers JE, et al. (2005) Diversity and selection in sorghum: simultaneous analyses using simple sequence repeats. Theor Appl Genet 111:23–30

Chittenden LM, Schertz KF, Lin YR, Wing RA, Paterson AH (1994) A detailed rflp map of *Sorghum-bicolor X S-Propinquum*, suitable for high-density mapping, suggests ancestral duplication of sorghum chromosomes or chromosomal segments. Theor App Genet 87:925–933

Chopra S, Brendel V, Zhang JB, Axtell JD, Peterson T (1999) Molecular characterization of a mutable pigmentation phenotype and isolation of the first active transposable element from *Sorghum bicolor*. Proc Natl Acad Sci USA 96:15330–15335

Crasta OR, Xu WW, Rosenow DT, Mullet J, Nguyen HT (1999) Mapping of post-flowering drought resistance traits in grain sorghum: association between QTLs influencing premature senescence and maturity. Mol Gen Genet 262:579–588

Devi PB, Sticklen MB (2003) In vitro culture and genetic transformation of sorghum by microprojectile bombardment. Plant Biosystems 137:249–254

Dewet JMJ, Gupta SC, Harlan JR, Grassl CO (1976) Cytogenetics of introgression from Saccharum into Sorghum. Crop Sci 16:568–572

Dufour P, Deu M, Grivet L, D'Hont A, Paulet F, et al. (1997) Construction of a composite sorghum genome map and comparison with sugarcane, a related complex polyploid. Theor Appl Genet 94:409–418

Farrell AE, Plevin RJ, Turner BT, Jones AD, O'Hare M, et al. (2006) Ethanol can contribute to energy and environmental goals. Science 311:506–508

Feltus FA, Hart GE, Schertz KF, Casa AM, Brown P, et al. (2006) Genetic map alignment and QTL correspondence between inter- and intra-specific sorghum populations. Theor Appl Genet 112:1295–1305

Gao ZS, Jayaraj J, Muthukrishnan S, Claflin L, Liang GH (2005a) Efficient genetic transformation of Sorghum using a visual screening marker. Genome 48:321–333

Gao ZS, Xie XJ, Ling Y, Muthukrishnan S, Liang GH (2005b) Agrobacterium tumefaciens-mediated sorghum transformation using a mannose selection system. Plant Biotech J 3:591–599

Gaut BS, Clark LG, Wendel JF, Muse SV (1997) Comparisons of the molecular evolutionary process at *rbcL* and *ndhF* in the grass family (Poaceae) Mol Biol Evol 14:769–777

Gingle AR, Huang YC, Yang H, Bowers JE, Kresovich S, Paterson AH (2007) CGGC: An integrated web resource for sorghum. Crop Sci in press

Gomez MI, Islam-Faridi MN, Zwick MS, Czeschin DG, Hart GE, et al. (1998) Tetraploid nature of *sorghum bicolor* (L.) Moench. J Heredity 89:188–190

Hagio T, Blowers AD, Earle ED (1991) Stable transformation of Sorghum cell-cultures after bombardment with DNA-coated microprojectiles. Plant Cell Rep 10:260–264

Hamblin MT, Mitchell SE, White GM, Gallego W, Kukatla R, et al. (2004) Comparative population genetics of the panicoid grasses: Sequence polymorphism, linkage disequilibrium and selection in a diverse sample of *Sorghum bicolor*. Genetics 167:471–483

Hamblin MT, Fernandez MGS, Casa AM, Mitchell SE, Paterson AH, et al. (2005) Equilibrium processes cannot explain high levels of short- and medium-range linkage disequilibrium in the domesticated grass *Sorghum bicolor*. Genetics 171:1247–1256

Hamblin MT, Casa AM, Sun H, Murray SC, Paterson AH, et al. (2006) Challenges of detecting directional selection after a domestication bottleneck. Genetics 173: 953–964

Haussmann BIG, Mahalakshmi V, Reddy BVS, Seetharama N, Hash CT, et al. (2002) QTL mapping of stay-green in two sorghum recombinant inbred populations. Theor Appl Genet 106:133–142

Haussmann BIG, Hess DE, Omanya GO, Folkertsma RT, Reddy BVS, et al. (2004) Genomic regions influencing resistance to the parasitic weed*Striga hermonthica* in two recombinant inbred populations of sorghum. Theor Appl Genet 109:1005–1016

Holm LG, Plucknett DL, Pancho JV, Herberger JP (1977) The world's worst weeds: distribution and biology. University Press of Hawaii, Honolulu, HI.

Howe A, Sato S, Dweikat I, Fromm M, Clemente T (2006) Rapid and reproducible Agrobacterium-mediated transformation of sorghum. Plant Cell Rep 25:784–791

Hu FY, Tao DY, Sacks E, Fu BY, Xu P, et al. (2003) Convergent evolution of perenniality in rice and sorghum. Proc Natl Acad Sci USA 100:4050–4054

Islam-Faridi MN, Childs KL, Klein PE, Hodnett G, Menz MA, et al. (2002) A molecular cytogenetic map of sorghum chromosome 1: Fluorescence in situ hybridization analysis with mapped bacterial artificial chromosomes. Genetics 161:345–353

Jessup RW, Burson BL, Burow G, Wang YW, Chang C, et al. (2003) Segmental allotetraploidy and allelic interactions in buffelgrass *(Pennisetum ciliare* (L.) Link syn. Cenchrus ciliaris L.) as revealed by genome mapping. Genome 46:304–313

Katsar CS, Paterson AH, Teetes GL, Peterson GC (2002) Molecular analysis of Sorghum resistance to the greenbug (Homoptera : Aphididae). J Econ Entomol 95:448–457

Kim JS, Childs KL, Islam-Faridi MN, Menz MA, Klein RR, et al. (2002) Integrated karyotyping of sorghum by in situ hybridization of landed BACs. Genome 45:402–412

Kim JS, Islam-Faridi MN, Klein PE, Stelly DM, Price HJ, et al. (2005a) Comprehensive molecular cytogenetic analysis of sorghum genome architecture: Distribution of euchromatin, heterochromatin, genes and recombination in comparison to rice. Genetics 171:1963–1976

Kim JS, Klein PE, Klein RR, Price HJ, Mullet JE, et al. (2005b) Chromosome identification and nomenclature of Sorghum bicolor. Genetics 169:1169–1173

Kim JS, Klein PE, Klein RR, Price HJ, Mullet JE, et al. (2005c) Molecular cytogenetic maps of sorghum linkage groups 2 and 8. Genetics 169:955–965

Klein PE, Klein RR, Cartinhour SW, Ulanch PE, Dong JM, et al. (2000) A high-throughput AFLP-based method for constructing integrated genetic and physical maps: Progress toward a sorghum genome map. Genome Res 10:789–807

Klein RR, Klein PE, Chhabra AK, Dong J, Pammi S, et al. (2001) Molecular mapping of the*rf1* gene for pollen fertility restoration in sorghum (*Sorghum bicolor* L.) Theor Appl Genet 102:1206–1212

Klein RR, Klein PE, Mullet JE, Minx P, Rooney WL, et al. (2005) Fertility restorer locus Rf1 of sorghum (*Sorghum bicolor* L.) encodes a pentatricopeptide repeat protein not present in the colinear region of rice chromosome 12. Theor Appl Genet 111:994–1012

Kresovich S, Barbazuk B, Bedell JA, Borrell A, Buell CR, et al. (2005) Toward sequencing the sorghum genome. A US National Science Foundation-Sponsored Workshop Report. Plant Physiol 138:1898–1902

Lijavetzky D, Martinez MC, Carrari F, Hopp HE (2000) QTL analysis and mapping of pre-harvest sprouting resistance in sorghum. Euphytica 112:125–135

Lin Y, Schertz K, Paterson A (1995) Comparative analysis of QTLs affecting plant height and maturity across the Poaceae, in reference to an interspecific sorghum population. Genetics 141:391–411.

Magalhaes JV, Garvin DF, Wang YH, Sorrells ME, Klein PE, et al. (2004) Comparative mapping of a major aluminum tolerance gene in sorghum and other species in the Poaceae. Genetics 167:1905–1914

McIntyre CL, Hermann SM, Casu RE, Knight D, Drenth J, et al. (2004) Homologues of the maize rust resistance gene *Rp1-D* are genetically associated with a major rust resistance QTL in sorghum. Theor Appl Genet 109:875–883

McIntyre CL, Casu RE, Drenth J, Knight D, Whan VA, et al. (2005) Resistance gene analogues in sugarcane and sorghum and their association with quantitative trait loci for rust resistance. Genome 48:391–400

Menz MA, Klein RR, Mullet JE, Obert JA, Unruh NC, et al. (2002) A high-density genetic map of*Sorghum bicolor* (L.) Moench based on 2926 AFLP (R), RFLP and SSR markers. Plant Mol Biol 48:483–499

Miller JT, Jackson SA, Nasuda S, Gill BS, Wing RA, et al. (1998) Cloning and characterization of a centromere-specific repetitive DNA element from *Sorghum bicolor*. Theor Appl Genet 96:832–839

Ming R, Liu SC, Lin YR, da Silva J, Wilson W, et al. (1998) Detailed alignment of Saccharum and Sorghum chromosomes: Comparative organization of closely related diploid and polyploid genomes. Genetics 150:1663–1682

Missaoui AM, Paterson AH, Bouton JH (2005) Investigation of genomic organization in switchgrass (*Panicum virgatum* L.) using DNA markers. Theor Appl Genet 110:1372–1383

Multani DS, Meeley RB, Paterson AH, Gray J, Briggs SP, et al. (1998) Plant-pathogen microevolution: Molecular basis for the origin of a fungal disease in maize. Proc Natl Acad Sci USA 95:1686–1691

Mutengwa CS, Tongoona PB, Sithole-Niang I (2005) Genetic studies and a search for molecular markers that are linked to *Striga asiatica* resistance in sorghum. African J Biotechnol **4**:1355–1361

Nagaraj N, Reese JC, Tuinstra M, Smith CM, St. Amand P, et al. (2005) Molecular mapping of sorghum genes expressing tolerance to damage by greenbug (Homoptera: Aphididae) J Econ Entomol 98:595–602

Park SJ, Huang YH, Ayoubi P (2006) Identification of expression profiles of sorghum genes in response to greenbug phloem-feeding using cDNA subtraction and microarray analysis. Planta 223:932–947

Paterson AH, Lin YR, Li ZK, Schertz KF, Doebley JF, et al. (1995a) Convergent domestication of cereal crops by independent mutations at corresponding genetic-loci. Science 269:1714–1718

Paterson AH, Schertz KF, Lin YR, Liu SC, Chang YL (1995b) The weediness of wild plants - molecular analysis of genes influencing dispersal and persistence of Johnsongrass, Sorghum halepense (L) Pers. Proc Natl Acad Sci USA 92:6127–6131

Paterson AH, Bowers JE, Chapman BA (2004) Ancient polyploidization predating divergence of the cereals, and its consequences for comparative genomics. Proc Natl Acad Sci USA 101:9903–9908

Peterson DG, Schulze SR, Sciara EB, Lee SA, Bowers JE, et al. (2002) Integration of Cot analysis, DNA cloning, and high-throughput sequencing facilitates genome characterization and gene discovery. Genome Res 12:795–807

Pratt LH, Liang C, Shah M, Sun F, Wang HM, et al. (2005) Sorghum expressed sequence tags identify signature genes for drought, pathogenesis, and skotomorphogenesis from a milestone set of 16,801 unique transcripts. Plant Physiol 139:869–884

Quinby JR (1974) Sorghum Improvement and the Genetics of Growth. Texas A&M University Press, College Station

Salzman RA, Brad, JA, Finlayson SA, Buchanan CD, Summer EJ, et al. (2005) Transcriptional profiling of sorghum induced by methyl jasmonate, salicylic acid, and aminocyclopropane carboxylic acid reveals cooperative regulation and novel gene responses. Plant Physiol 138:352–368

Sang YJ, Liang GH (2000) Comparative physical mapping of the 18S-5.8S-26S rDNA in three sorghum species. Genome 43:918–922

Scheinost P (2001) Perennial wheat: a sustainable cropping system for the Pacific Northwest. Amer J Altern Agric **16**, 147–151

Schertz KF, Stephens JC (1966) Compilation of gene symbols, recommended revisions, and summary of linkages for inherited characters of *Sorghum vulgare* Pers. Rep. No. 3. Texas A&M University, College Station

Schloss SJ, Mitchell SE, White GM, Kukatla R, Bowers JE, et al. (2002) Characterization of RFLP probe sequences for gene discovery and SSR development in *Sorghum bicolor* (L.) Moench. Theor Appl Genet 105, 912–920

Singh M, Chaudhary K, Singal HR, Magill CW, Boora KS (2006) Identification and characterization of RAPD and SCAR markers linked to anthracnose resistance gene in sorghum*Sorghum bicolor* (L.) Moench. Euphytica 149:179–187

Smith CW, Frederiksen RA (2000) Sorghum: origin, history, technology, and production. John Wiley and Sons, Hoboken

Sobral BWS, Braga DPV, Lahood ES, Keim P (1994) Phylogenetic analysis of chloroplast restriction enzyme site mutations in the Saccharinae Griseb subtribe of the Andropogoneae Dumort tribe. Theor Appl Genet **87**:843–853

Subudhi PK, Rosenow DT, Nguyen HT (2000) Quantitative trait loci for the stay green trait in sorghum (*Soughum bicolor* L. Moench): consistency across genetic backgrounds and environments. Theor Appl Genet 101:733–741

Swigonova Z, Lai J, Ma J, Ramakrishna W, Llaca V, et al. (2004a) Close split of sorghum and maize genome progenitors. Genome Res 14:1916–1923

Swigonova Z, Lai J, Ma JX, Ramakrishna W, Llaca M, et al. (2004b) On the tetraploid origin of the maize genome. Comp Func Genomics 5:281–284.

Swigonova Z, Lai JS, Ma JX, Ramakrishna W, Llaca V, et al. (2004c) Close split of sorghum and maize genome progenitors. Genome Res 14:1916–1923

Tadesse Y, Sagi L, Swennen R, Jacobs M (2003) Optimisation of transformation conditions and production of transgenic sorghum (*Sorghum bicolor*) via microparticle bombardment. Plant Cell Tissue Organ Culture 75:1–18

Tao YZ, Jordan DR, Henzell RG, McIntyre CL (1998) Identification of genomic regions for rust resistance in sorghum. Euphytica 103:287–292

Tao YZ, Hardy A, Drenth J, Henzell RG, Franzmann BA, et al. (2003) Identifications of two different mechanisms for sorghum midge resistance through QTL mapping. Theor Appl Genet 107:116–122

Totad AS, Fakrudin B, Kuruvinashetti MS (2005) Isolation and characterization of resistance gene analogs (RGAs) from sorghum (*Sorghum bicolor* L. Moench) Euphytica 143:179–188

Ulanch PE, Childs KL, Morgan PW, Mullet JE (1996) Molecular markers linked to *Ma*(1) in sorghum. Plant Physiol 111:709

Wagoner P (1990) Perennial grain development - past efforts and potential for the future. Crit Rev Plant Sci 9:381–408

Wang ML, Dean R, Erpelding J, Pederson G (2006) Molecular genetic evaluation of sorghum germplasm differing in response to fungal diseases: Rust (*Puccinia purpurea*) and anthracnose (*Collectotrichum graminicola*). Euphytica 148:319–330

Webster OJ (1964) Genetic studies in *Sorghum vulgare* (Pers.). Crop Sci 4:207–210.

Wen L, Tang HV, Chen W, Chang R, Pring DR, et al. (2002) Development and mapping of AFLP markers linked to the sorghum fertility restorer gene *rf*4. Theor Appl Genet 104:577–585

Whitkus R, Doebley J, Lee M (1992) Comparative genetic mapping of sorghum and maize. Genetics 132:1119

Woo S-S, Jiang J, Gill B, Paterson A, Wing R (1994) Construction and characterization of a bacterial artificial chromosome library of *Sorghum bicolor*. Nucleic Acids Res 22:4922–4931

Xin Z, Wang M, Barkley N, Franks C, Burow G, et al. (2007) Development of a Tilling population for sorghum functional genomics. In: International Plant and Animal Genome Conference, p W397, San Diego CA

Xu GW, Magill CW, Schertz KF, Hart GE (1994) A rflp linkage map of *Sorghum bicolor* (L) Moench. Theor Appl Genet 89:139–145

Xu WW, Subudhi PK, Crasta OR, Rosenow DT, Mullet JE, et al. (2000) Molecular mapping of QTLs conferring stay-green in grain sorghum (*Sorghum bicolor* L. Moench). Genome 43:461–469

Zhao ZY, Cai TS, Tagliani L, Miller M, Wang N, et al. (2000) Agrobacterium-mediated sorghum transformation. Plant Mol Biol 44:789–798

Zwick MS, Islam-Faridi MN, Czeschin DG, Wing RA, Hart GE, et al. (1998) Physical mapping of the liguleless linkage group in *Sorghum bicolor* using rice RFLP-selected sorghum BACs. Genetics 148:1983–1992

Chapter 21
Sugarcane: A Major Source of Sweetness, Alcohol, and Bio-energy

Angélique D'Hont, Glaucia Mendes Souza, Marcelo Menossi, Michel Vincentz, Marie-Anne Van-Sluys, Jean Christophe Glaszmann, and Eugênio Ulian

Abstract Sugarcane is an important tropical crop having C4 carbohydrate metabolism which, allied with its perennial nature, makes it one of the most productive cultivated plants. It is mostly used to produce sugar, accounting for almost two thirds of world production. Recently it has gained increased attention because of its important potential for bio-fuel production. However, sugarcane has one of the more complex crop genomes, which has long hampered the development of sugarcane genetics to support breeding for crop improvement programs. Sugarcane belongs to the genus *Saccharum* L, part of the Poaceae familly (Grasses) and the *Andropogonae* tribe, which encompasses only polyploid species. With the advent of molecular genomics, the sugarcane genome has become less mysterious, although its complexity has been confirmed in many aspects. Shortcuts to genomic analyses have been identified thanks to synteny conservation with other grasses, in particular sorghum and rice. Over time, new tools have become available for understanding the molecular bases behind sugarcane productivity and a renewed interest has surfaced in its genetics and physiology.

21.1 Introduction

21.1.1 Economic, Agronomic, and Societal Importance of Sugarcane

Sugarcane has been the main plant source of sweetener for humans for several millennia. It is able to partition carbon to sucrose in the stem, a vegetative organ, in contrast with other cultivated grasses that usually accumulate their reserve products

A. D'Hont
CIRAD, (Centre de Coopération Internationale en Recherche Agronomique pour le Développement), UMR1096, Avenue Agropolis, TA40/03, F-34398 Montpellier, France
e-mail: angelique.dhont@cirad.fr

© Springer 2008

in seeds. This almost unique feature was selected by man who first used its soft watery culm for chewing.

Sugarcane belongs to the genus *Saccharum* L., which is part of the Poaceae family (Grasses) and the *Andropogonae* tribe. The reference, domesticated species for sugarcane is *Saccharum officinarum* (also called noble cane). *S. officinarum* is a group of thick, juicy canes that were initially cultivated in Southeast Asia and the Pacific Islands before spreading over the inter-tropics between 1500 and 1000 BC (Daniels and Roach 1987). In China and India, *S. officinarum* crossed with wild relatives to form the natural hybrids *S. sinense* (Chinese canes) and *S. barberi* (North Indian canes) that were then selected and cultivated. Sugar extraction probably developed in India and China from such hybrids (Daniels and Daniels 1975). Sugar manufacturing appeared in Persia around 500 AD. A few clones of *S. barberi* or hybrids between *S. officinarum* and *S. barberi* were probably taken from India via Persia in the 6th century, arriving in the Mediterranean and Spain by the 8th century. From there the Portuguese took it to Madeira in the 15th century, from whence it spread to other islands and West Africa. Sugarcane reached the Americas in 1493 when Columbus took it to the Dominican Republic, and the Portuguese planted it in Brazil in the early 16th century. In the 16th century, sugar production for world trade progressively changed from cottage industries based on *S. sinense* and *S. barberi*, to plantation and factory industries based on selected clones of *S. officinarum* (Daniels and Roach 1987). Near the end of the 19th century, *S. spontaneum*, a wild species producing no sugar, with thin stalks, as well as a few "North Indian" sugarcanes were used in Java and India in breeding programs aimed at overcoming disease susceptibilities affecting *S. officinarum*. Interspecific hybridization was the major breakthrough in modern sugarcane breeding. Hybridization not only solved many of the disease problems but it also provided increased yields, improved ratooning ability, and adaptability for growth under various abiotic stresses (Roach 1972). All modern sugarcane cultivars have been derived essentially from a few rounds of intercrossing from those first interspecific hybrids (Arceneaux 1967; Price 1965).

Commercially, sugarcane is propagated vegetatively via stem cuttings. Germination of the lateral buds produces new plants that branch into stools consisting of a large number of tillers. Under good growth conditions, the plant will grow 4–5 meters in 12 months, with the extractable culms measuring 2–3 meters and containing 13–16% sucrose. Because it is a perennial crop, after harvest and under the right growing conditions, underground buds will sprout giving rise to a new crop. In most situations, four to six crops are harvested before the yields become economically unsustainable and the field is renewed with the planting of a new crop.

Sugarcane is currently cultivated on more than 20 million hectares in tropical and subtropical regions of the world, producing up to 1.3 billion metric tons of crushable stems. It is mostly used to produce sugar, accounting for almost two-thirds of world production. Recently, it has gained increasing attention since one of its products, ethanol, has been publicized as an important source of renewable bio-fuel, which could turn it into a global commodity and an important energy source. Ethanol is an alcohol that can be produced from a variety of agricultural products and by-products and is probably the best known biofuel. Brazil already diverts half of its sugarcane production to ethanol production and it will need to build more than 70 new mills and turn more than 2.5 million hectares of land over to sugarcane production to meet the demand for internal ethanol consumption (Pessoa et al. 2005). In addition, new technologies are emerging to convert cellulosic residues like bagasse and other agricultural byproducts, such as sugarcane trash (dry and green leaves and plant tops left in the field during harvest), into valuable commodities that would be degraded into small sugar molecules via either enzymatic or physical-chemical (or both) processes to be fermented into ethanol. These technologies are all in the scale-up phase and in the next few years will become commercial realities, changing the fate of cellulosic residues.

The economic importance of sugarcane and its main products to many countries in tropical and sub-tropical regions of the world has not always been met with significant investments for the research and development of new technologies to support the breeding programs and develop sugarcane genetics. One of the reasons for this is probably the complex nature of the sugarcane genome and the difficulties faced in selecting new, more productive cultivars in long selection programs that could take up to 15 years. With the advent of genomics, new tools have become available and a renewed interest in sugarcane genetics has surfaced (reviewed by D'Hont and Glaszmann 2001; Butterfield et al. 2001; Grivet and Arruda 2001; Ming et al. 2006).

21.1.2 Origin and Diversity of the Sugarcane Complex

The taxonomy of the sugarcane complex, based on morphology, chromosome numbers, and geographical distribution, has been controversial since the original classification of *Saccharum officinarum* by Linnaeus in 1753 (Daniels and Roach 1987; Daniels 1996; Irvine 1999). Recent molecular data are beginning to help trace the domestication and early evolution of sugarcane (review by Grivet et al. 2004, 2006) (Fig. 21.1).

A contribution by various genera other than *Saccharum*, particularly *Erianthus* (2n=20, 30, 40 and 60), *Miscanthus* (2n=38, 40, 76), *Sclerostachya* (2n=30), and *Narenga* (2n=30), to the emergence of sugarcane has been hypothesized by several sugarcane specialists (review in Daniels and Roach 1987). However, recent molecular data do not appear to confirm these hypotheses. Current extant species of the genera *Saccharum, Erianthus,* and *Miscanthus* are clearly distinct according to isozyme, nuclear, and cytoplasmic restriction fragment length polymorphism (RFLP) data (Glaszmann et al. 1989, 1990; Burnquist et al. 1992; Lu et al. 1994a; D'Hont et al. 1993, 1995; Besse et al. 1997), amplified fragment length polymorphism (AFLP) and simple sequence repeat (SSR) data (Selvi et al. 2004; Cai et al. 2005), and sequence data (Hodkinson et al. 2002). In addition, repeated species-specific sequences with multiple dispersed loci in the genome were cloned in *Miscanthus* and *Erianthus* and hybridized on the DNA of representatives of traditional cultivars and wild *Saccharum,* and no trace of these *Miscanthus* or *Erianthus* specific sequences was found in any of the individuals tested (Alix et al. 1998, 1999). Restriction fragment analysis of the chloroplast genome (Sobral et al. 1994)

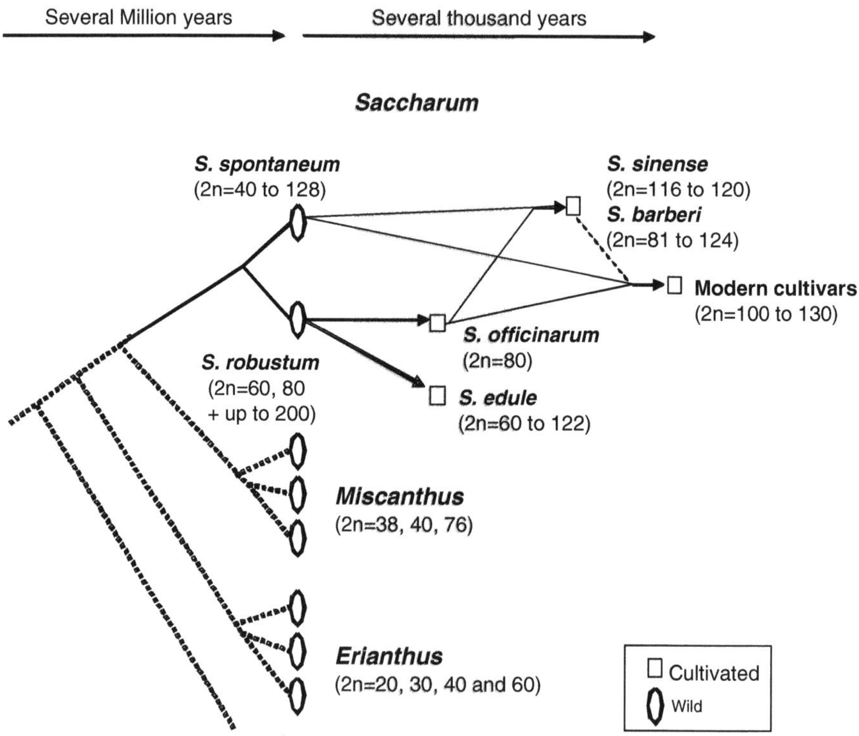

Fig. 21.1 Scenario compatible with available molecular data for sugarcane evolution and domestication (Adapted from Grivet et al. 2005)

and analysis of the nuclear repeated sequences (Alix et al. 1998, 1999) suggested that *Saccharum* is more closely related to *Miscanthus* than to *Erianthus*.

These data support the view that the genus *Saccharum* is a well-defined lineage that has diverged over a long period of evolution from the lineages leading to the *Erianthus* and *Miscanthus* genera (Grivet et al. 2006). Thus cultivated sugarcanes probably emerged from wild *Saccharum* species, and secondary introgressions with other genera are not likely pathways. However, this does not mean that natural intergeneric hybridizations are impossible and may not account for some local peculiarities. Artificial intergeneric hybrids with these genera have been produced (Chen et al. 1993; D'Hont et al. 1995; Piperidis et al. 2000).

The *Saccharum* genus includes six polyploid taxonomic groups often afforded species status: two wild species, *S. spontaneum* (2n=40 to 128) and *S. robustum* (2n= 60, 80 and up to 200); three groups of early cultivars, *S. officinarum* (2n= 80), *S. barberi* (2n=81–124), and *S. sinense* (2n=116–120); and the marginal sterile group, *S. edule* (2n= 2n = 60 to 122) (review in Daniels and Roach 1987; Sreenivasan et al. 1987).

S. spontaneum is characterized by thin stalks with no or very little sugar and has a huge geographic distribution from East Africa to Southeast Asia with a probable continental Asia origin. *S. robustum* is characterized by long, thick stalks with little or no sugar and has been reported as occurring in natural populations in the Indonesian islands of Kalimantan, Sulawesi, and Maluku, in New Guinea, and in the Bismarck, Solomon, and Vanuatu archipelagos. These two wild species display different structural organizations of their monoploid genome (basic set of chromosome). This was suggested by the presence of polyploid chromosome series based on multiples of eight and ten, respectively. This has been confirmed by cytogenetic mapping of the ribosomal RNAs 45S and 5S by fluorescent in situ hybridization (FISH), which established basic chromosome numbers of $x = 8$ for *S. spontaneum* (D'Hont et al. 1998; Ha et al. 1999) and $x = 10$ for *S. robustum* (D'Hont et al. 1998). Molecular diversity is much greater in *S. spontaneum* than in *S. robustum*. Allopatric populations of *S. spontaneum* and *S. robustum* are clearly differentiated at the DNA level. Indeed, *S. spontaneum* samples from Kalimantan and Sumatra and *S. robustum* from New Guinea and Halmahera are strongly differentiated by their nuclear RFLP (Glaszmann et al. 1990; Burnquist et al. 1992; Lu et al. 1994a), cytoplasmic RFLP (D'Hont et al. 1993), AFLP (Selvi et al. 2004) and randomly amplified polymorphic DNA (RAPD) (Nair et al. 1999). Data addressing relationships between sympatric populations of *S. spontaneum* and *S. robustum* are still sparse. In New Guinea, all *S. spontaneum* individuals observed have the same cytotype, $2n = 80$. D'Hont et al. (1998) showed that this cytotype is decaploid, with a typical *S. spontaneum* basic chromosome number of $x = 8$. However, field observations have shown a morphological continuum between extreme types, and some individuals presenting intermediate morphological characteristics between *S. spontaneum* and *S. robustum* are difficult to classify (Henty 1969). Moreover, a small sample of *S. spontaneum* individuals collected in New Guinea appears to be more closely related to *S. robustum* than to any other *S. spontaneum*, based on RFLP with nuclear low copy probes (Besse et al. 1997) and on the hybridization signal intensity of a repeated satellite sequence, SoCIR1 (Alix et al. 1998). This suggests that *S. spontaneum* populations from New Guinea are genetically closer to *S. robustum* than are the *S. spontaneum* populations west of Sulawesi.

Multiple lines of molecular evidence support a direct descent of *S. officinarum*, the domesticated sugarcane characterized by thick stalks, rich in sugar (also call Noble clones), from the wild species *S. robustum*. A single mitochondrial haplotype was detected among a series of *S. officinarum* clones (D'Hont et al. 1993). This haplotype is the most common haplotype detected in a collection of *S. robustum* clones from New Guinea and New Britain. It is also different from the six haplotypes revealed in a collection of *S. spontaneum* individuals sampled over a large geographic area. RFLP analysis of nuclear single copy DNA placed *S. officinarum* cultivars very close to *S. robustum*. The average similarity between a *S. officinarum* clone and a *S. robustum* clone is about the same as the average similarity between two *S. robustum* clones (Lu et al. 1994a). *S. officinarum* has a basic chromosome number of $x = 10$, as does *S. robustum* (D'Hont et al. 1998), and is octoploid like the most common cytotype ($2n = 80$) in the *S. robustum* wild species.

S. barberi and *S. sinense* have hybrid origins. RFLP with low copy nuclear DNA (Glaszmann et al. 1990; Burnquist et al. 1992; Lu et al. 1994a; Selvi et al. 2004) and genomic in situ hybridization (GISH) (D'Hont et al. 2002) clearly show that *S. barberi* and *S. sinense* cultivars are the result of interspecific hybridizations between representatives of the two genetic groups of the *Saccharum* genus, *S. spontaneun* on one side and *S. officinarum* or *S. robustum* on the other. Since the *S. barberi* and *S. sinense* clones have sweet stalks and the region where they were formerly cultivated is outside the natural distribution range of *S. robustum*, the scenario of Brandes (1956) provides the simplest explanation for their origins: *S officinarum* cultivars were probably transported by humans to mainland Asia, where they naturally crossed with local *S. spontaneum* giving rise to *S. barberi* and *S. sinense* in India and China, respectively. It is likely that these clones are early-generation hybrids because no, or very few, interspecific chromosome exchanges were detected using GISH (D'Hont et al. 2002). This contrasts with the observations of higher levels of interspecific chromosome exchange in modern cultivars. The *S. barberi* and *S. sinense* cultivars that were tested have the mitochondrial haplotype of *S. officinarum*, indicating that this species was the maternal parent and wild *S. spontaneum* the paternal parent in the founding crosses (D'Hont et al. 1993). Low copy nuclear RFLP suggests that each morpho-cytogenetic group represents a set of somatic mutants derived from a single founding interspecific hybrid event (D'Hont et al. 2002). The Pansahi group, alias *S. sinense*, is not particularly distinct from the other groups according to nuclear RFLPs. The *S. barberi* and *S. sinense* cultivars are thus all derived from similar processes involving an interspecific hybridization event, followed by morphological and genetic radiation through mutation, which may have occurred in different geographic regions of continental Asia.

Few molecular data are available for tracing the origin of marginal group of *S. edule*. This group is grown in subsistence gardens from New Guinea to Fiji for its edible, aborted inflorescence; its large, thick-stalked canes contain no sugar. The mitochondrial haplotype has been established for a single clone. It was the same as the *S. officinarum*, *S. barberi*, and *S. sinense* cultivars and most of the *S. robustum* (D'Hont et al. 1993). An independent investigation based on chloroplast RFLP markers from another clone led to a similar conclusion (Sobral et al. 1994). These sparse data support the hypothesis that *S. edule* corresponds to a series of mutant clones, which were identified in *S. robustum* populations and were preserved by humans.

21.2 Genome Structure and Molecular Diversity of Modern Cultivars

21.2.1 Chromosome Structure

The origin of modern sugarcane cultivars is well known. However, their precise genomic structure has only recently been elucidated, thanks mainly to molecular cytogenetics. Modern cultivars are derived from several artificial interspecific hybridizations between *S. officinarum*, used as the female, and *S. spontaneum* and, to a lesser extent, *S. barberi* as the pollen donor. F1 hybrids were then backcrossed to *S. officinarum* to recover a high-sugar-producing type species. This process was accelerated through the selection of hybrids derived from the 2n transmission of *S. officinarum* chromosomes (Bremer 1961). All present-day cultivars are derived from the interbreeding of these first interspecific hybrids. Altogether, it is estimated that 19 *S. officinarum* clones (four with high frequency), a few *S. spontaneum* (two with high frequency) clones, and one *S. barberi* clone were involved in these interspecific crosses (Arceneaux 1967).

Modern cultivars are thus highly polyploid and aneuploid, with about 120 chromosomes. GISH studies of chromosome preparations demonstrated that 15–25% of their chromosomes were inherited from *S. spontaneum*, and that the recombination between homoeologous chromosomes is possible (D'Hont et al. 1996; Piperidis and D'Hont 2001; Cuadrado et al. 2004). In cultivar 'R570', for example, 10% of the chromosomes are inherited in their entirety from *S. spontaneum*, 80% are inherited entirely from *S. officinarum,* and 10% are the result of recombination between chromosomes from the two ancestral species. In addition, as a consequence of the different basic chromosome numbers of *S. officinarum* and *S. spontaneum*, two distinct chromosome organizations coexist in current cultivars. The genome structure of a typical modern cultivar is represented in Fig. 21.2.

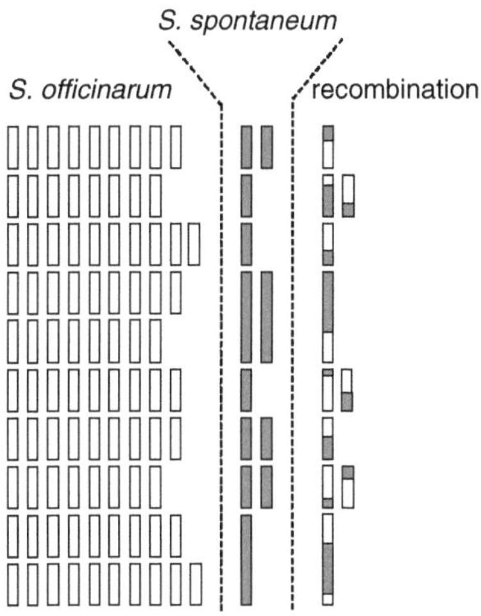

Fig. 21.2 Schematic representation of the genome of modern sugarcane cultivars as deduced from FISH and GISH experiments. Modern cultivars are highly polyploid and aneuploid with around 120 chromosomes. They are derived from interspecific hybridization between *S. officinarum* (2n=8x=80) and *S. spontaneum* (2n=5x to 16x=40 to 128). 10% to 20% of their chromosomes are inherited in their entirety from *S. spontaneum* (Grey bars); 70% to 80% are inherited entirely from *S. officinarum* (white bars) and around 10% are the result of recombination between chromosomes from the two ancestral species. In addition, as a consequence of the different basic chromosome numbers of *S. officinarum* (x=10) and *S. spontaneum* (x=8), two distinct chromosome organizations coexist in current cultivars

In diploids, genome sizes are generally given for the (non-replicated) gametic genome size (1C value), which in diploids corresponds to the size of the basic set of chromosomes (monoploid genome). However, in polyploids such as sugarcane, the gametic genome size (1C) value does not correspond to the size of the basic set of chromosomes. For this reason, in polyploids it seems more appropriate to refer to the genome size of (non-replicated) somatic cells (2C) or to the size of the monoploid genome.

The genome size in *S. officinarum* (2n=8x=80) has been estimated as 7.68 pg, which corresponds to 7440 mega base pairs (Mbp) for a somatic cell and to 926 Mbp for the monoploid genome (x=10) (D'Hont and Glaszmann 2001). For a *S. spontaneum* with 2n=8x=64, the genome size has been estimated as 6.30 pg, which corresponds to 6,080 Mbp for a somatic cell and to 760 Mbp for the monoploid genome (x=8) (D'Hont and Glaszmann 2001). With 760 to 926 Mbp, the size of the *Saccharum* monoploid (basic) genome is roughly double the monoploid genome size of rice (389 Mbp), similar to that of *Sorghum bicolor* Moench (760 Mbp), and significantly smaller than maize (2500 Mbp). The genome size of somatic cells (2C) of the typical modern cultivar R570 (2n= about 115) has been estimated as 10,000 Mb (D'Hont 2005).

21.2.2 Molecular Diversity

Low copy nuclear RFLP analysis showed that modern sugarcane cultivars are highly heterozygous, presenting multiple alleles at each locus (Lu et al. 1994b). Despite the low number of *S. offcinarum* involved in the interspecific hybridization at the origin of modern cultivars, more than 80% of the markers present in the whole *S. officinarum* samples studied by Jannoo et al. (1999a) and Selvi et al. (2005) were also found in modern cultivars. This is due probably to a high heterozygosity related to polyploidy (Jannoo et al. 1999a). Although the cultivars appear closely related to *S. officinarum* clones, the minor chromosome complement inherited from *S. spontaneum* seems to constitute the principal component of cultivar diversity (Lu et al. 1994b; Jannoo et al. 1999a). The sub-tropical cultivars seem to have retained a larger number of *S. spontaneum* markers than the tropical cultivars (Jannoo et al. 1999a; Selvi et al. 2005), reflecting the selection for different environments and resulting in numerous thinner stalks in subtropical regions and thicker stalks in tropical regions. SSR (microsatellite) markers are now routinely used in breeding programs for cultivar identification and progeny validation.

21.2.3 Linkage Disequilibrium

The small number of meiotic divisions since the first artificial crosses that gave rise to modern cultivars provided little opportunity to recombine founder chromosomes. Moreover, there were not many of these chromosomes, as there were only a few founder individuals involved in the origin of modern cultivars. Consequently, a high level of linkage disequilibrium is still expected among modern cultivars. This was suggested by Lu et al. (1994b) and confirmed in a sample of Mauritian cultivars in which some chromosome haplotypes are significantly conserved over regions as long as 10 cM (Jannoo et al. 1999b). This is an important finding because it may offer original and powerful perspectives to identify and locate genes that are involved in traits of interest. Polyploidy may greatly hamper this task, but the huge level of linkage disequilibrium, compared to that estimated for the human genome, for example, may offer some advantages. In particular, the density of markers needed for genetic mapping may be quite low. Recent work based on AFLP and diversity arrays technology (DArT) are paving the way to this type of application (Raboin, Pauquet, and Butterfield personal communication).

21.3 Genetic Mapping and Synteny

21.3.1 Genetic Maps

Polyploidy dictates particular constraints for mapping, which have been theoretically developed by Wu et al. (1992). When polyploidy is high and pairing is polysomic or irregular, such as in sugarcane, multiple bands are identified by a DNA probe or a pair of primers, and alleles with different dosage level segregate. In this context, alleles that are present as single copies are much more informative for the construction of genetic maps than any others. Using molecular marker technologies such as RAPD, AFLP, and RFLP markers, partial genetic maps have been produced for *S. spontaneum* (da Silva et al. 1993, 1995; Al-Janabi et al. 1993; Ming et al. 1998 *S. officinarum* (Guimarães et al. 1999; Ming et al. 1998 *S. robustum* (Guimarães et al. 1999; Ming et al. 1998 and modern cultivars (D'Hont et al. 1994; Grivet et al. 1996; Hoarau et al. 2001; Rossi et al. 2003; Aitken et al. 2005; Reffay et al. 2005; Raboin et al. 2006 (http://tropgenedb.cirad.fr/).

Co-dominant markers such as RFLP and SSR can reveal several alleles of the same locus and are thus very useful in the identification of co-segregating groups that correspond to hom(oe)ologous chromosomes. Developing a saturated, low-density genetic map for sugarcane requires much more work than for a diploid; for a given level of molecular diversity, the effort required to simultaneously distinguish ten or so haplotypes is much greater than that needed to distinguish only two. At the moment, none of the published genetic maps of sugarcane is saturated. About one half of the genome is estimated to be tagged onto the most refined maps (Rossi et al. 2003; Aitken et al. 2005). As for maps of current cultivars, marker coverage is uneven, with *S. spontaneum* chromosomes being covered more densely than those of *S. officinarum*.

At meiosis mainly bivalents are observed in *S. officinarum*, *S. robustum*, *S. spontaneum*, and interspecific cultivated clones (review by Sreenivasan et al. 1987). Mapping data suggest that pairing behavior probably does not fit any pre-established, clear-cut scheme, such as complete disomy or complete polysomy. In *S. robustum* (MOL5829, 2n=80), a high proportion of preferential pairing (50%) in the few co-segregation groups already defined was reported (Al-Janabi et al. 1994; Ming et al. 1998). In *S. officinarum,* some preferential pairing was also observed, whereas no preferential pairing was found in *S. spontaneum* (Al-Janabi et al. 1994; Ming et al. 1998). In the cultivar R570, Grivet et al. (1996) and Hoarau et al. (2001) observed a general polysomy with several cases of preferential pairing and suggested the possibility of complete local disomy. In R570, the preferential pairing detected concerned chromosomes of *S. officinarum, S. spontaneum,* as well as interspecific-recombinant origin. Jannoo et al. (2004) took advantage of a particular single copy probe (BNL 12.06) revealing 11 alleles by RFLP in cultivar R570. They determined the doses of the various BNL12.06 RFLP alleles among 282 progeny of R570 and estimated the mutual pairing frequencies among the corresponding homo- or homoeologous chromosomes using a maximum likelihood method. The result is an atypical picture, with pairing frequencies ranging from 0 to 40% and differential affinities leading to the identification of several chromosome subsets. It highlights a continuous range of pairing affinities between chromosomes and pinpoints a strong role of individual chromosome features, partly related to their ancestral origin, in the determination of these affinities (Jannoo et al. 2004).

21.3.2 Tagging Genes of Interest

An important application of genetic maps is the location on the genome of loci that contribute to the variation of phenotypic traits. Only three major genes have been mapped until now, two rust resistance genes (Daugrois et al. 1996; Raboin et al. 2006) and one gene responsible for stalk color (Raboin et al. 2006). Quantitative trait loci (QTL) detection is complicated by the potential for segregation of several (potentially up to 12 in a modern cultivar) alleles at a locus and by the lack of preferential pairing. As a consequence, different parental alleles are not mutually exclusive alternatives. For the subset of polymorphic alleles that show simplex segregation ratios, the effect of an allele can be estimated from the average phenotypic difference between the two possible genotypes (presence versus absence).

A QTL experiment was conducted with two interspecific *S. offinarum* x *S. spontaneum* crosses to investigate variation in the sucrose content of the stalk populations (Ming et al. 2001, 2002a). Many independent segregating alleles were identified and could be assigned to eight distinct loci. In several cases, the presence of several alleles at a locus that contributed to the variation of a trait, demonstrated that strong interaction effects between alleles could lead to an important buffering effect. Large-scale QTL mapping was also conducted for the modern cultivar R570 (Hoarau et al. 2002). The effects of individual QTLs were small for all of the traits investigated, always accounting for less than 7% of the phenotypic variance, and the size of the effects was not conserved across crop cycles. Additional QTL experiments using modern cultivars were conducted, revealing, in general, rather small effects of individual QTLs (Reffay et al. 2005; McIntyre et al. 2005). Polyploidy is going to challenge the application of markers in sugarcane more than in any other crop, and improved biometrical methods are needed to extract full information from the QTL-detection experiments.

Multiplex segregation at QTL loci may be partly responsible for phenotypic buffering, which is an important factor in the success of many autopolyploid crops. Non-additive gene action in multiple dose QTLs may also have contributed to evolutionary opportunities. For example, if a single copy of a gene/QTL is physiologically sufficient, the additional copies may be free to collect mutations, often becoming nonfunctional, perhaps occasionally resulting in a distinctive new function that improves fitness.

21.3.3 Synteny with Other Grasses

Due to its high polyploidy and the absence of diploid close relatives, the advantage of investigating synteny conservation between sugarcane and other grasses, in particular with other members of the Andropogoneae tribe, was realized early. The first comparisons were made with maize, which had, at that stage, a more advanced map. Gaut et al. (2000) confirmed synteny to be conserved (D'Hont et al. 1994; Grivet et al. 1994; Dufour et al. 1997), although quite perturbed by the duplicated structure of the maize genome and the presence of many rearrangements. Rice also showed relatively global simple synteny relationships with sugarcane with, however, many rearrangements explained by the large distance between the two species (Glaszmann et al. 1997). Rice remains an interesting model for sugarcane because the sequence of its genome is available (International Rice Genome Sequencing Project 2005) and because large numbers of rice mutants are being collected. To date, of the species studied, sorghum appears to have the simplest synteny relationship with sugarcane (Grivet et al. 1994; Dufour et al. 1997; Glaszmann et al. 1997; Guimarães et al. 1997; Ming et al. 1998), and thus more adapted to help with sugarcane studies (Asnaghi et al. 2000). Corresponding QTLs controlling plant height, stalk number, and flowering were found in sorghum and sugarcane (Ming et al. 2002b; Jordan et al. 2004). The availability of the entire sorghum genome sequence in the near future (Sorghum Genomics Planning Workshop Participants 2005) will be of great interest to sugarcane genomics, in particular for map-based gene isolation.

Co-linearity has been employed to help develop a fine map around a gene conferring resistance to brown rust (*Bru*1), which is the focus of a map based cloning approach using the cultivar R570. Sorghum and rice regions orthologous to the sugarcane target area were identified and markers derived from sorghum genetics (Boivin et al. 1999; Bowers et al. 2003), physical maps (http://genome.arizona.edu/genome/sorghum.html), and a comparison of sugarcane cDNAs with the rice orthologous sequence (http://www.genome.arizona.edu/fpc/rice/), were used to saturate the sugarcane map in the target region (Asnaghi et al. 2000, 2004; D'Hont unpublished data).

21.4 BAC Library Development and Utilization

A bacterial artificial chromosome (BAC) library of 103,296 clones, with a mean insert size of 130 kbp, has been constructed for the sugarcane cultivar R570 (Tomkins et al. 1999). On the basis of the monoploid genome size of sugarcane the coverage is estimated to be 14x. However, since sugarcane is highly polyploid and heterozygous, this represents only the coverage of 1.3x of the total genome of this sugarcane cultivar.

This BAC library is currently being used to develop a physical map of the region bearing the rust resistance gene *Bru*1 in cultivar R570 and to perform comparative genomic studies within sugarcane as well as with other grasses.

Two homoeologous BAC clones (97 kb and 126 kb), one derived from *S. spontaneum* and one from *S. officinarum,* corresponding to a region that has already been studied in several cereals (Ilic et al. 2003), were sequenced and compared (Jannoo et al. 2007). The results indicated that the two *Saccharum* species diverged by 1.5–2 mya from one another and 8–9 mya from sorghum. The two sugarcane homoeologous haplotypes showed perfect co-linearity and also high homology along the non-transcribed regions, apart from the insertion of a few retro-transposable elements. The gene distribution highlighted high synteny and co-linearity with sorghum and rice, and partial co-linearity with each homoeologous maize region, which became perfect when the sequences were combined. This first analysis of sugarcane haplotype organization at the sequence level suggested that the high ploidy in sugarcane did not induce generalized reshaping of its genome, thus challenging the idea that polyploidy quickly induces generalized rearrangement of genomes. These results also consolidated the fact that sorghum is a choice model for sugarcane.

21.5 Functional Genomics

21.5.1 EST Development

The large sugarcane genome, in which, on average, each single-copy gene is represented by ten alleles, represents a challenge for genetic analysis. Expressed sequence tag" (EST) collections may contribute significantly to identify candidate genes associated with important agronomical traits (i.e., tolerance to abiotic and biotic stress, mineral nutrition, and sugar content, amongst others) and more generally to provide relevant information for functional and evolutionary analyses.

Several sugarcane EST collections have been developed (Carson and Botha 2000, 2002; Casu et al. 2001, 2003; Ma et al. 2004; Bower et al. 2005), the largest one being the Brazilian sugarcane EST project, SUCEST (Vettore et al. 2003). All of the publicly available sugarcane sequence ESTs were assembled into tentative consensus sequences (virtual transcripts), singletons, and mature transcripts, referred to as the Sugarcane Gene Index (SGI; http://compbio.dfci.harvard.edu/tgi/cgi-bin/tgi/gimain.pl?gudb=s_officinarum). The SUCEST collection of ESTs was assembled into 43,141 putative unique sugarcane transcripts referred to as sugarcane assembled sequences (SAS) http://sucest.lbi.ic.unicamp.br/public/; (Vettore et al. 2003).

A comparative analysis between the SASs and the DNA and protein sequences from the model eudicotyledon plant *Arabidopsis* thaliana, model monocotyledon rice, and a set of other angiosperms was undertaken. Three main classes of sequences were identified, a core set of eudicot - monocot sequences that may represent angiosperm basic functions (75% of the SASs), monocot-specific sequences (14% of the SASs), and sequences restricted to sugarcane (13.5% of the SASs). A significant proportion of the monocot-specific sequences were found to represent fast-evolving sequences integrated in members of conserved angiosperm gene families. This observation was particularly relevant since a high rate of evolution may be related to a functional diversification that could be involved in the differentiation of specific evolutionary lineages. New protein architecture formed by monocot-specific motives or domains that were recruited into conserved eudicot-monocot proteins and long non-coding RNA (~500 nucleotides) were also identified among the monocot-specific sequences (Vincentz et al. 2004; Vincentz, unpublished results). The divergence between monocot and eudicot may therefore rely partly on functional diversification (generating new protein functions) from duplicated copies of conserved gene families. The extent to which the sugarcane-specific SASs represent novelties restricted to this organism remains unclear.

Several studies have described the SUCEST sequences and discussed their putative roles, in particular for genes that correspond to the general metabolism of sugarcane, signaling growth, development and stress responses (Grivet and Arruda 2001; Arruda 2001). Over 30% of the identified transcripts corresponded to genes with no sequence similarity to known genes. The identification of novel gene functions in such a complex organism is a formidable task. Ongoing efforts to associate putative functions with the sugarcane genes include gene expression profiling of tissues, identification of genes associated with sucrose content in cultivars showing contrasting Brix values, comparing different varieties submitted to biotic and abiotic stress, as well as mapping efforts to associate genes with phenotypes, as described above.

21.5.2 Tissue Profiling

The use of gene chips and cDNA micro-arrays allows both temporal and spatial gene expression data to be obtained. Custom designed cDNA micro-arrays constructed using the cDNA collection from the SUCEST database (http://sucest-fun.org), were used to determine the distribution of gene transcripts in sugarcane tissues and to define tissue-specific activities and ubiquitous genes. Using cDNA micro-arrays containing 1,280 distinct elements, the individual gene expression variation of plants grown in the field and transcript abundance in six plant organs (flowers, roots, leaves, lateral buds, 1st [immature] and 4th [mature] internodes), was analyzed (Papini-Terzi et al. 2005). The expression of 217 genes was found to be tissue-enriched, while 153 genes showed expression levels that were highly similar in all the tissues analyzed. A virtual profile matrix was constructed where the tissue expression levels can be compared amongst 24 tissue samples. Most of the genes characterized coded for signal transduction components, hormone biosynthesis, transcription factors, stress and pathogen response-related genes. A catalogue of sugarcane signal transduction and other regulatory genes can be found at the SUCAST Database (http://sucest-fun.org). The database integrates the gene catalog, the information generated by the phylogenetic categorization of the sugarcane kinome, tissue matrix, sequencing data, and analyses by the SUCEST Project. Tissue expression information can also aid in the identification of gene promoter sequences.

21.5.3 Exploitation of EST Resources for Functional Analysis

21.5.3.1 Sugar Synthesis, Transport, and Accumulation

Sucrose synthesized in sugarcane leaves is stored in the culms. Sugarcane has the striking ability of accumulating high levels of sucrose that can reach up to about 0.7 molar in mature internodes (Moore 1995), which corresponds to approximately 50% of the stalk dry weight. This physiological specialization makes sugarcane an interesting model for studies on sugar synthesis, transport, and accumulation.

Carbohydrates are synthesized in sugarcane leaves by CO_2 fixation during photosynthesis (Moore 1995; Lunn and Furbank 1999; Grof and Campbell 2001). In C4 plants, such as sugarcane, CO_2 is fixed by phosphenolpyruvate carboxylase in mesophyll cells, producing oxaloacetate, which is reduced to malate. In the bundle sheath cells, the malate is decarboxylated, the CO_2 being released and re-fixed by Rubisco in the photosynthetic carbon reduction (PCR) cycle. The resulting glyceraldehyde 3-phosphate molecules are the substrate for a multi-step pathway leading to the synthesis of sucrose, in which the activities of fructose-1,6-bisphosphatase (FBPase) and sucrose-phosphate synthase (SPS) play a major control role. Sucrose is thought to be synthesized exclusively in the mesophyll, and then transferred to phloem cells, where it is transported to the stem parenchyma cells (Grof and Campbell 2001). Besides these leaf reactions, Grof and Campbell (2001) highlighted three major rate limiting or co-limiting steps: the rate of transport to the stalks, including phloem loading; the rate of transport into the parenchyma cells and into their vacuoles; and the rate of sucrose mobilization used for the vegetative growth. It is supposed that once these limitations are solved, commercial yields could be doubled (Grof and Campbell 2001 and references cited therein).

It is known from several plant species that sugar transport relies on both symplastic and apoplastic steps (Patrick 1997; Lalonde et al. 2004). An evaluation of the SUCEST database revealed full-length genes encoding nine monosaccharide and four dissacharide transporters (Felix 2006). Casu et al. (2003) performed an EST survey comparing transcripts from immature and mature internodes. Several transcripts encoding proteins homologous to known sugar transporters were found, and all of them were more abundant in the mature internodes. Micro-array and northern blot analyses showed that a putative sugar transporter, type 2a, was highly expressed in mature internodes and absent in mature leaf and root. However, the sugar transported by this protein remains elusive. Rae et al. (2005) cloned the sugarcane *ShSUT1* gene, which encodes a transporter that may play a role in the transfer of sucrose from the vascular tissue to parenchyma cells of internodes. Since sink strength regulates photosynthesis in sugarcane (McCormick et al. 2006), the evaluation of sugar transporter genes in transgenic sugarcane plants will certainly help to assess their role in sugar accumulation in the internodes.

To analyze the genes expression during culm development, a collection of 7,409 ESTs from maturing sugarcane stems in combination with a smaller collection (1,089) of ESTs from immature stems (Casu et al. 2001, 2003, 2004) were analyzed by bio-informatics and using cDNA micro-arrays, allowing for the identification of genes that were differentially regulated with respect to stem maturity. These studies indicated that genes associated with sucrose metabolism were not abundantly expressed in stem tissues and that genes related to the synthesis and catalysis of sucrose were down-regulated as the stem matured and the sucrose concentration increased. In a similar approach, the use of sugarcane GeneChips from Affymetrix (Lockhart et al. 1996) containing approximately 6,024 distinct *S. officinarum* genes, led to the discovery that genes involved in cellulose synthesis, cell wall metabolism, and lignification were developmentally regulated during culm maturation (Casu et al. 2007).

Gene activity associated with internode development was also compared between a non-sucrose accumulating genotype from the species *S. robustum* and two high sucrose accumulating genotypes (an *S.officinarum* genotype and a hybrid cultivar) (Watt et al. 2005). Using nylon arrays containing 88 ESTs, gene expression variations associated with stalk development and those pertaining to sucrose accumulation were investigated. In mature internodes of all three genotypes, transcript-relative activities were found to decrease for the cell wall biosynthesis genes and increase for the sucrose metabolism-related genes. The most notable differences were represented by increased activity of sucrose synthase-1 and sucrose phosphatase. Mature and immature culm samples were also analyzed using cDNA micro-arrays containing 1,228 elements in common with the array used by Papini-Terzi et al. (2005) (mentioned above), plus an additional 317 elements including 229 kinases representatives of the sugarcane kinome. Sucrose accumulating internodes (sink tissues) were collected from field grown plants contrasting for Brix. In some cases, samples were collected throughout the year (Papini-Terzi et al. 2007). Genes identified as developmentally regulated during culm maturation included hormone signaling (auxin, ethylene, jasmonate, salicylic acid), stress responses, sugar transport, lignin biosynthesis and fiber content.

To identify genes associated with sucrose content, a strategy introduced by Jansen and Nap (2001), which involved the large-scale analysis of gene expression in a segregating population, was applied to sugarcane populations segregated for soluble solids content (Brix). Using cDNA micro-arrays containing 4,715 genes, 62 genes, mostly unrelated with sucrose metabolism, were identified as associated with sucrose content (Casu et al. 2005). In a similar work, cDNA micro-arrays containing 1,545 elements were used, and over 100 genes found to be differentially expressed for high sugar, as compared to low sugar genotypes (Papini-Terzi et al. 2007). The differentially expressed genes identified belonged to several functional categories including calcium signaling, stress responses, transcription, and ubiquitination. The categories with the highest number of hits included protein kinases from the SNF-related family of kinases, auxin hormone signaling, the Cyp family of cytochrome P450 monoxygenases, and other stress-related genes.

The expression profile of genes associated with signal transduction was also evaluated in leaves from high sugar and low sugar sugarcane plants from an F1 progeny selected from a cross between the sugarcane varieties SP 80–180 and SP 80–4966 (Felix 2006). Twenty-four genes were differentially expressed between high and low sugar plants. Five had higher transcript levels in high sugar plants, including an omega-3 fatty acid desaturase putatively involved in methyl jasmonate (MeJa) signaling, a putative receptor-like serine/threonine kinase, and an Myb domain transcription factor. Most of the genes had higher expressions in low sugar genotypes, such as those encoding three 14-3-3 like proteins and an SNF1-related protein. A homologue of this protein phosphorylates the enzyme SPS in vitro (Sugden et al. 1999), making it a putative target to interact with 14-3-3 proteins, which in turn reduces the SPS activity (Toroser et al. 1998; Huber et al. 1998).

The efficiency and control of carbon fixation and allocation, which is affected by sink strength (Watt et al. 2005), may be regulated at the source tissues. Ma et al. (2004) investigated gene expression in source tissues using EST analyses, and more recently, serial analysis of gene expression (SAGE) was used (Calsa and Figueira 2006). Sugarcane, as well as maize and sorghum, was considered to operate under the NADP-malic enzyme (NADP-ME) pathway (Bowyer and Leegood 1997), although a highly expressed photosynthesis-related phosphoenolpyruvate carboxykinase (PEPCK) had already been detected and validated in maize leaf bundle sheath cells (Furumoto et al. 1999, 2000). C4 grasses such as sugarcane, maize, and sorghum, contain anatomical and physiological adaptations to optimize CO_2 fixation for carbohydrate biosyntheses (Brown et al. 2005). Basically, three C4 photosynthetic primary carbon cycle pathways have been described that differ in the four-carbon organic acid intermediate transported from the mesophyll to the bundle sheath cells (malate and/or aspartate), in the three-carbon acid returned to the mesophyll cells (pyruvate or alanine), as well as the decarboxylation enzyme present in the bundle sheath cells, which can be either NADP-ME, NAD+malic enzyme (NAD-ME), or (PEPCK) (Taiz and Zeiger 1998). The combined SAGE and real time quantitative PCR (RT-qPCR) results (Calsa and Figueira 2006) suggested that PEPCK decarboxylation appeared to predominate over NADP-ME in mature field-grown sugarcane leaves, in contrast to the conventional NADP:ME model accepted for sugarcane, although both may occur.

21.5.3.2 Responses to Environmental Challenges

Plants react to changes in the environment through an array of cellular responses that are activated by stress stimuli, leading to plant defense and/or adjustment to adverse conditions. Physiological changes elicited by external signals can be modulated by transcriptional regulation resulting in the induction or repression of target genes. Studies have been conducted to unravel the responses of sugarcane to biotic and abiotic stress and the role of phytohormones in these processes.

Drought

Drought is a condition of special interest with respect to sugarcane, since water scarcity conditions prevent the expansion of this culture to vast areas in tropical regions. To identify differentially expressed genes in response to hydric stress, cDNA microarrays representing 1,545 genes were used (Rocha et al. 2007). Among the differentially expressed genes, regulators of drought-responsive genes such as the WRKY and MYC transcription factors (Abe et al. 1997) were identified. Cold and drought signaling overlap and many of the responses are mediated by the phytohormone ABA. Accordingly, low temperature-induced (LTI) proteins were seen to be up-regulated in response to lack of water in sugarcane, and genes induced by ABA have also been found to be induced by drought, including two delta-12 oleate desaturases, one S-adenosylmethionine decarboxylase, and a protein phosphatase ABI1/ABI2 (Tahtiharju and Palva 2001) that regulates stomatal closure. A sugarcane transcription factor homologous to rice DREB2 was induced and may represent an important transcription factor for the regulation of sugarcane drought responses, since the over-expression of DREB2A in *Arabidopsis* led to the development of plants tolerant to drought (Sakuma et al. 2006). Other genes identified include an S-adenosylmethionine decarboxylase, known to accumulate in response to salinity and drought (Li and Chen, 2000) and fatty acid desaturases (FAD2), directly related to drought tolerance (Zhang et al. 2005; Im et al. 2002).

Cold

Cold stress, which includes low temperatures above (chilling) and below (freezing) 0°C, causes severe losses of most crop plants, due to the formation of extracellular ice (Xin and Browse 2000). Sugarcane is considered to be a cold-sensitive crop (Tai and Lentini 1998), and although sugarcane fields are restricted to tropical and subtropical regions, cold stress is not unusual in these areas, decreasing sugar productivity. Nogueira et al. (2003) evaluated the gene expression profile in sugarcane plantlets exposed to 4°C. Thirty-four cold-inducible genes and 25 cold-repressed genes were found. Based on these data, a model of sugarcane response to cold stress was proposed. In their model, several transcription factors, such as an ABI3-interacting protein 2, an OsNAC6 and an OCSBF-1, regulate the transcription of proteins involved in the protection against oxidative stress, sugar transporters, protein degradation and cell wall synthesis. These genes could be good targets for study and to possibly improve sugarcane cold tolerance.

Phosphorus Deficiency

Phosphorus (P) is an essential nutrient because it is used in a large number of biological processes, from nucleic acids biosynthesis to the regulation of enzyme activities. Plants take up P as inorganic phosphate (Pi), and have developed several strategies to cope with the low availability of Pi in the soil, which is usually in the range from 2 to 10 mM (Raghothama 1999). Sugarcane is a crop that performs well in acid soils, indicating the use of strategies to overcome P deficiency. To access the expression profile of genes in response to P starvation, Rocha et al. (2007) evaluated sugarcane plantlets grown in the absence of P. Surprisingly, only genes repressed due to P starvation were found. The expression profile obtained pointed to changes in protein N-glycosylation and redox status due to an altered expression of an N-acetylglucosamine-1-phosphate transferase and two thioredoxins. Genes homologous to an MYB transcription factor and an ethylene insensitive-like (EIL) transcription factor putatively involved in the ethylene response were also repressed. These data indicated that under low levels of the nutrient, sugarcane roots might be under severe metabolic restraint, in line with the observations in *Arabidopsis*, where genes related to photosynthesis were repressed in response to Pi starvation (Wu et al. 2003).

Herbivory

Insect pests frequently challenge sugarcane productivity. Even though, over the last few decades, highly productive sugarcane cultivars with enhanced insect pest resistance have been developed in conventional breeding programs, modern cultivars appear to retain a lower degree of resistance when compared to wild-type genotypes. The availability of insect-control genes that could be genetically engineered to obtain pest resistant varieties is of significant interest. In a search for sugarcane orthologs of genes that are potential targets for the management of insect resistance, Falco et al. (2001) identified, among the SUCEST sequences ESTs coding for proteinase inhibitors, alpha-amylase inhibitors, lectins, chitinases, and polyphenol oxidases. In this study, putative systemic and constitutive wound response proteins were identified.

The sugarcane borer *Diatraea saccharalis* is the major sugarcane pest in Brazil, causing plant death due to apical bud death (dead heart) in up to four-month-old plants and damage to lateral bud development, aerial rooting, weight loss, and stalk breakage in older plants. The attack also allows for infection by opportunistic fungi, which results in production losses for both the sugar and alcohol industries (Braga et al. 2003). The expression profile of a variety highly susceptible to the borer was obtained in response to the insect attack using cDNA micro-arrays (Rocha et al. 2007). The expression data indicated a strong induction of a pathogenesis-related protein similar to thaumatin, 24 h after the onset of this stress. These proteins are important for plant defense mechanisms and may present anti-fungal action, endo-β1,3-glucanase activity, and trypsin or α-amylase inhibitory activity (Grenier et al. 1999; Franco et al. 2002). Further characterization of this sugarcane thaumatin-like protein should be carried out to define its activity and the defense mechanism that it may trigger against the sugarcane stalk borer.

Endophytic Bacteria

In Brazil, the long-term continuous cultivation of sugarcane with low N fertilizer inputs, without apparent depletion of the soil-N reserves, led to suggestions that N_2-fixing bacteria associated with the plants might be the source of agronomically significant N inputs for this crop. Years of study led to the conclusion that the diazotrophs that infected the interior of the plants, such as the 'endophytic diazotrophs' were responsible for the increased nitrogen contribution to Brazilian soils (Boddey et al. 2003). Diazotrophic acetobacters were also isolated from sugarcane roots or soil collected from four regions in Queensland, Australia (Li and Macrae 1991). However, biological nitrogen utilization seems to be restricted to some cultivars and regions. In South Africa for instance, it was shown that biological nitrogen fixation did not contribute to the nitrogen demand of a commercially grown cultivar (Hoefsloot et al. 2005).

In Brazil, sugarcane culture benefits considerably from its association with N_2-fixing endophytic bacteria (*Herbaspirillum seropedicae / Herbaspirillum rubrisubalbicans* and *Gluconacetobacter diazotrophicus*). Unlike rhizobium/leguminosae symbiosis, where the bacteria are restricted to nodules, *Herbaspirillum* spp. and *G. diazotrophicus* are endophytic, and colonize the intercellular spaces and vascular tissues of most plant organs, without causing damage to the host (James and Olivares 1998; Rheinhold-Hurek and Hurek 1998). These bacteria possibly promote plant growth by nitrogen fixation and also by the production of plant hormones (Sevilla et al. 2001). Despite the non-pathogenic aspects of this interaction, plants should limit bacterial growth inside their tissues to avoid disease development (Olivares et al. 1997). It is believed that sugarcane plants recognize these microorganisms and activate defense responses until the establishment of an efficient association (Vinagre et al. 2006). Using cDNA micro-arrays, four resistance gene analogs were found to be responsive to the endophytic association (Rocha et al. 2007). Plant disease resistance genes mediate specific recognition of pathogens via the perception of avirulence gene products (review by Ellis et al. 2000). Two resistance gene analogs were induced on account of the association with both *Herbaspirillum* and *Gluconacetobacter diazotrophicus*. Inoculation with *Gluconacetobacter* also led to the induction of a salicylic acid biosynthesis gene. Salicylic acid accumulates in plant tissues in response to pathogen attack, and is essential for the induction of systemic acquired resistance and for some responses mediated by resistance genes (Gaffney et al. 1993; Delaney et al. 1994; Mur et al. 1997). The expression of a PP2C and five transcription factors was altered when the plants were cultivated in association with endophytic bacteria. Amongst these, there were two zinc-finger transcription factors, one of which was up regulated by inoculation with either *Gluconacetobacter* or *Herbaspirillum*. A possible role for phosphatases and zinc-finger transcription factors in response to endophytic bacteria has also been pointed out by the in silico analysis of ESTs, that identified a SAS corresponding to these categories, exclusively or preferentially expressed in the cDNA libraries constructed from plants inoculated with *Gluconacetobacter* and *Herbaspirillum* (Vargas et al. 2003).

21.5.3.3 Phytohormone Signaling

Hormones such as methyl-jasmonate (MeJA) and abscisic acid (ABA) are key regulators of mechanisms that integrate plant responses to internal and external stimuli. Gene expression changes in response to these hormones have been evaluated in sugarcane (Bower et al. 2005; De Rosa et al. 2005; Rocha et al. 2007). MeJa induced several genes encoding protein homologues related to phytohormone signaling, including an MYB transcription factor and a receptor-like protein in sugarcane roots (Bower et al. 2005) and a zinc finger protein, a heat shock factor, and a protein kinase in young sugarcane leaves (De Rosa et al. 2005). In a survey of most of the sugarcane homologues for known genes related to hormone signaling, Rocha et al. (2007) found that MeJa induced the expression of protein kinases, an MYB transcription factor, and an NAC protein, and repressed another protein kinase. ABA induced the expression of genes encoding homologues to two receptor Ser/Thr kinases, a phosphatase and a small GTPase, while a protein kinase homologue was repressed. Schlögl et al. (2006) evaluated the expression profile of the whole set of known b-ZIP transcription factors in response to ABA and MeJa. Two bZIPs were induced by ABA and four were repressed, while two others were induced by MeJa.

21.5.3.4 Transposon Expression

Retrotransposons mobilize themselves through an RNA intermediate and are now considered one of the major forces driving genome expansion in plants (Piegu et al. 2006), while transposons usually move using either a cut/paste or a copy/paste mechanism. Recently, a hypothesis on the impact of transposable elements (TE) on genomic structure, gene regulation, and even on function has been proposed (Casacuberta and Santiago 2003; Kashkush et al. 2003; Bundock and Hooykaas 2005). Twenty-one different families of TEs were identified in the SUCEST collection, of which 54% correspond to classical transposons and 46% to retrotransposons (Rossi et al. 2001). Further studies to validate the expression profile of the TE families identified were developed, which confirmed that the callus is the tissue with more expressed TE families (Araujo et al. 2005). Although it has been proposed several times that tissue culture somaclonal variation could be a result of TE activity, this is the first report that demonstrates that callus is indeed a tissue where different TEs are expressed at the same time. Focus on particular sugarcane families highlighted the existence of lineages of elements with diverse levels of representation in the genome (Rossi et al. 2004).

21.6 Genetic Engineering

Genetic transformation has been extensively used to produce commercial varieties of a number of different crops such as soybeans, corn, and cotton, expressing traits such as herbicide and insect resistance, resulting in improvements in the farmers' incomes and a decrease in the use of pesticides (http://www.isaaa.org) over the last 10 years. Besides delivering such successful agricultural products, this technology also offers the possibility of studying the thousands of plant genes (with known and unknown functions) that have been produced by numerous genome programs conducted throughout the world (Dong et al. 2005). As a general rule these new genes are silenced or over-expressed, creating opportunities to study their function in the plant and to produce new phenotypes not possible through conventional breeding (Galun 2005; Mullerr 2006).

The first examples of the expression of exotic genes in sugarcane plants were obtained by the insertion of genes conferring antibiotic and herbicide resistance (Bower and Birch 1992; Gallo-Meagher and Irvine 1996). For many years now, the genetic transformation of sugarcane has been a reality in different laboratories around the world. The production of herbicide-resistant plants is now a common practice (Falco et al. 2000; Manickavasagam et al. 2004) and agronomical performance and inheritance studies of plants containing this trait have been performed in the field (Leibbrandt and Snyman 2003; Butterfield et al. 2002). Insect-resistant sugarcane plants were first produced by transformation with a truncated version of the *Bacillus thuringiensis cryIAb* gene and the plants produced very low amounts of the protein, presenting some larvicidal activity (Arencibia et al. 1997). Braga et al. (2001, 2003) reported the production of a number of transgenic events resistant to sugarcane borer in two commercial sugarcane cultivars. The gene used was a reconstructed version of *cryIAb* and the plants showed high resistance under greenhouse and field conditions. Recently, a truncated version of the *B. thuringiensis cryIAc* gene was expressed in sugarcane plants by Weng et al. (2006), and significant protein levels and insect resistance were obtained from at least two sugarcane clones. Other strategies have been used to obtain insect-resistant sugarcane plants using genes from various sources. Nutt et al. (1999) and Nutt (2005) obtained transgenic sugarcane plants expressing either the potato proteinase inhibitor II or the snowdrop lectin gene, which were able to reduce the weight of the cane grub larvae feeding on them. Transgenic plants containing the snowdrop lectin gene were also reported by Chen et al. (2004) with no information on insect resistance studies. Falco and Silva-Filho (2003) expressed the soybean Kunitz and Bowman-Birk trypsin inhibitors in sugarcane, obtaining a reduction in growth of the sugarcane borer larvae feeding on transgenic plants, with no mortality. Different groups have reported on resistance to viral diseases in sugarcane: for sugarcane mosaic virus (SCMV) (Joyce et al. 1998; Ingelbrecht et al. 1999); for Fiji disease virus (FDV) (McQualter et al. 2001); and for sugarcane yellow leaf virus (SCYLV) (Rangel et al. 2003). Even though sugarcane genetic engineering has demonstrated high potential, there has been no commercial release of transgenic sugarcane either due to intellectual property considerations or more probably to industrial concerns over public perception. These constraints on commercial release are likely to continue in the near future. The ease with which many sugarcane genotypes can now be transformed, together with the identification of the sequences of thousands of genes that this plant expresses, raise the possibility of altering the expression of specific genes in the plant and identifying the effects this modification has on the plant's phenotype. The development of RNAi and anti-sense technology allows for gene down-regulation even in a high polyploid situation. Non-flowering plants of a heavily flowering sugarcane variety were produced through the anti-sense expression of a single candidate flower development gene found in the Sucest database (Figueiredo 2003). Flowering is undesirable in sugarcane commercial fields. Wu and Birch (2007) showed that the expression of a heterologous sucrose isomerase gene directed towards the vacuole of transgenic sugarcane plants resulted in plants capable of doubling the total sugars stored in mature culms. In these plants, the amount of stored sucrose was the same as in control non-transgenic plants and the increase in total sugar was due to the accumulation of isomaltulose, a sucrose isomer. The transgenic plants with enhanced sugar accumulation also showed increased photosynthesis, sucrose transport and sink strength. Down-regulation and over-expression of the genes involved in carbohydrate metabolism is the topic of many studies, with the aim of increasing the content of sucrose and other metabolites.

In the last few years, sugarcane has also turned into a target for the production of novel products such as biopolymers (McQualter et al. 2005) and as a producer of pharmaceutical proteins with different properties (Wang et al. 2005). The plant is well-suited to these approaches due to some of its characteristics such as vegetative propagation, the absence of flowering in most commercial cultivars, the production of a large biomass, the large amount of carbon partitioned into sucrose (up to 42% of the stalk dry weight), and a mobile pool of hexose sugars throughout most of its life. The production of biopolymers was obtained by McQualter et al. (2005) in sugarcane plants, accumulating up to 7.3% and 1.5% dry weight of p-hydroxybenzoic acid in leaf and stem tissue, respectively. This product was quantitatively converted to glucose conjugates by endogenous uridine diphosphate-glucosyltransferases and presumably stored in the vacuole. Initial steps to produce pharmaceutical proteins were taken when Wang et al. (2005) successfully produced the human granulocyte macrophage colony-stimulating factor (GM-CSF) in sugarcane. Unfortunately, in the field, the plants were able to accumulate only 0.02% of the total soluble protein as GM-CSF.

In monocots, the most frequently used promoters are the maize ubiquitin Ubi-1 and the rice actin act1. Even though they maintain a relatively constant expression pattern, they show distinct expression patterns in different species, cell types, and cultivation conditions (Neuteboom et al. 2002). The identification of sugarcane tissue-specific promoters is an important step that will allow controlled gene expression so that novel products can be accumulated in the desired part of the plant, such as in the leaves or stem parenchyma cells. A few promoters from sugarcane genes have been tested in the past, with limited success (Birch et al. 1995; Wei et al. 1999, 2003; van der Merwe et al. 2003). Again, the large number of sugarcane genes available, coupled with studies to understand when and where they are expressed, become important tools to identify regulatory sequences that can be used to drive specific genes.

21.7 Perspectives

Sugarcane Mendelian genetics has literally started with the advent of molecular investigation techniques. The first monofactorial segregations and the first genetic linkage were oberved less than twenty years ago (Glaszmann et al. 1989). Many evolutionary questions have been addressed, confirming hypotheses or concluding earlier debates and occasionally throwing new light and improving overall understanding. The genome has become less mysterious, although its complexity has been confirmed for many aspects. Shortcuts have been identified thanks to synteny with other grasses, and the availability of the sorghum sequence will shortly initiate a new round of progress. The remarkable effort made with sugarcane ESTs is yielding a wealthy catalog of genes, whose documentation starts to open new perspectives for breeding better-adapted sugarcane. The worldwide realization of the central importance of bio-energy will undoubtedly foster attention on sugarcane physiology and genetics.

References

Abe H, Yamaguchi-Shinozaki K, Urao T, Iwasaki T, Hosokawa D, et al. (1997) Role of *Arabidopsis* MYC and MYB homologs in drought- and abscisic acid-regulated gene expression. Plant Cell 9:1859–1868

Aitken KS, Jackson PA, McIntyre CL (2005) A combination of AFLP and SSR markers provides extensive map coverage and identication of homo(eo)logous linkage groups in a sugarcane cultivar. Theor Appl Genet 110:789–801

Alix K, Baurens FC, Paulet F, Glaszmann JC, D'Hont A (1998) Isolation and characterization of a satellite DNA family in the *Saccharum* complex. Genome 41:854–864

Alix K, Paulet F, Glaszmann JC, D'Hont A (1999) Inter-Alu like species-specific sequences in the *Saccharum* complex. Theor Appl Genet 6:962–968

Al-Janabi SM, Honeycutt RJ, McClelland M, Sobral BWS (1993) A genetic linkage map of *Saccharum spontaneum* L. 'SES 208'. Genetics 134:1249–1260

Al-Janabi SM, Honeycutt RJ, Sobral BW (1994) Chromosome assortment in *Saccharum*. Theor Appl Genet 89:959–963

Araújo PG, Rossi M, Jesus EM, Saccaro-Junior N L, Kajihara D, et al. (2005) Transcriptionally active transposable elements in recent hybrid sugarcane. Plant J 44(5):707–17

Arceneaux G (1967) Cultivated sugarcanes of the world and their botanical derivation. Proc Int Soc Sugar Cane Technol 12:844–854

Arencibia A, Vazquez RI, Prieto D, et al. (1997) Transgenic sugarcane plants resistant to stem borer attack. Mol Breeding 3(4):247–255

Arruda P (2001) Sugarcane transcriptome. A landmark in plant genomics in the tropics. Genetics Mol Biol 24:1–296

Asnaghi C, Paulet F, Kaye C, Grivet L, Deu M, et al. (2000) Application of synteny across Poaceae to determine the map location of a sugarcane rust resistance gene. Theor Appl Genet 101:962–969

Asnaghi C, Roques D, Ruffel S, Kaye C, Hoarau J-Y, et al. (2004) Targeted mapping of a sugarcane rust resistance gene (*Bru*1) using bulked segregant analysis and AFLP markers. Theor Appl Genet 108:759–764

Besse P, McIntyre CL, Berding N (1997) Characterisation of Erianthus sect. Ripidium and Saccharum germplasm (Andropogoneae–Saccharinae) using RFLP markers. Euphytica 93:283–292

Birch RG, Bower R, Elliot A, Potier B, Franks T, et al. (1995) Expression of foreign genes in sugarcane. Proc Intl Soc Sugar Cane Technol 22:368–373

Boddey RM, Urquiaga S, Alves BJR, Reis V (2003) Endophytic nitrogen fixation in sugarcane: present knowledge and future applications. Plant and Soil 252:139–149

Boivin K, Deu M, Rami JF, Trouche G, Hamon P (1999) Towards a saturated sorghum map using RFLP and AFLP markers. Theor Appl Genet 98:320–328

Bower NI, Casu RE, Maclean DJ, Reverter A, Chapman SC, et al. (2005) Transcriptional response of sugarcane roots to methyl jasmonate. Plant Sci 168:761–772

Bower R, Birch RG (1992) Transgenic sugarcane plants via microprojectile bombardment. Plant J 2(3):409–416

Bowers JE, Abbey C, Anderson S, Chang C, Draye X, et al. (2003) A high-density genetic recombination map of sequence-tagged sites for sorghum, as a framework for comparative structural and evolutionary genomics of tropical grains and grasses. Genetics 165(1):367–86

Bowyer JR, Leegood RC (1997) Photosynthesis. In: Day PM, Harbone JB (eds) Plant biochemistry. Academic Press, San Diego, pp 49–110

Braga DPV, Arrigoni EDB Burnquist WL, Silva-Filho MC, Ulian EC (2001) A new approach for control of *Diatraea saccharalis* (Lepidoptera: Crambidae) through the expression of an insecticidal CryIa(b) protein in transgenic sugarcane. Proc Int Soc Sugar Cane Technol 24(2):331–336

Braga DPV, Arrigoni EDB, Silva-Filho MC, Ulian EC (2003) Expression of the Cry1Ab protein in genetically modified sugarcane for the control of *Diatraea saccharalis* (*Lepidoptera*: *Crambidae*). J New Seeds 5:209–222

Brandes EW (1956) Origin, dispersal and use in breeding of the Melanesian garden sugarcanes and their derivatives, *Saccharum officinarum* L. Proc Int Soc Sugar Cane Technol 9:709–750

Bremer G (1961) Problems in breeding and cytology of sugar cane. Euphytica 10:59–78

Brown NJ, Parsley K, Hibberd JM (2005) The future of C4 research-maize, Flaveria or Cleome? Trends Plant Sci 10:215–221

Bundock P, Hooykaas P (2005) An Arabidopsis hAT-like transposase is essential for plant development. Nature 436:282–284

Burnquist WL, Sorrells ME, Tanksley S (1992) Characterization of genetic variability in *Saccharum* germplasm by means of restriction fragment length polymorphism (RFLP) analysis. Proc Int Soc Sugar Cane Technol 21:355–365

Butterfield MK, D'Hont A, Berding N (2001) The sugarcane genome: a synthesis of current understanding, and lessons for breeding and biotechnology. Proc Soc Afr Sugarcane Technol Assn 75:1–5

Butterfield MK, Irvine JE, Garza MV, Mirkov E (2002) Inheritance and segregation of virus and herbicide resistance transgenes in sugarcane. Theor Appl Genet 104(5):797–803

Cai Q, Aitken KS, Piperidis G, Jackson PA, McIntyre CL (2005) A preliminary assessment of the genetic relationship between *Erianthus rockii* and the "Saccharum Complex" using microsatellite and AFLP markers. Plant Sci 169:976–984

Calsa Jr T, Figueira A (2007) Serial analysis of gene expression in sugarcane (*Saccharum* spp.) leaves revealed alternative C4 metabolism and putative antisense transcripts. Plant Mol Biol 63, 745–62

Carson DL, Botha FC (2000) Preliminary analysis of expressed sequence tags for sugarcane. Crop Sci 40:1769–1779

Carson DL, Botha FC (2002) Genes expressed in sugarcane maturing internodal tissue. Plant Cell Rep 20:1075–1081

Casacuberta JM, Santiago N (2003) Plant LTR-retrotransposons and MITEs: control of transposition and impact on the evolution of plant genes and genomes. Gene 5:311:1–11

Casu R, Dimmock C, Thomas M, Bower N, Knight D, et al. (2001) Genetic and expression profiling in sugarcane. Proc Int Soc Sugar Cane Technol 24:542–546

Casu RE, Grof CP, Rae AL, McIntyre CL, Dimmock CM, et al. (2003) Identification of a novel sugar transporter homologue strongly expressed in maturing stem vascular tissues of sugarcane by expressed sequence tag and microarray analysis. Plant Mol Biol 52:371–86

Casu RE, Dimmock CM, Chapman SC, Grof CP, McIntyre CL, et al. (2004) Identification of differentially expressed transcripts from maturing stem of sugarcane by in silico analysis of stem expressed sequence tags and gene expression profiling. Plant Mol Biol 54:503–17

Casu RE, Manners JM, Bonnett GD, Jackson PA, McIntyre CL, et al. (2005) Genomics approaches for the identification of genes determining important traits in sugarcane. Field Crops Res 92:137–147
Casu RE, Jarmey JM, Bonnett GD, Manners JM (2007) Identification of transcripts associated with cell wall metabolism and development in the stem of sugarcane by Affymetrix GeneChip sugarcane genome array expression profiling. Funct Integr Genomics 7:153–167
Chen PH, Lin MJ, Xue ZP, Chen RKA (2004) Study on genetic transformation of GNA gene in sugarcane. Acta Agriculturae Universitatis Jiangxiensis 26(5):740–743, 748
Chen Y, Chen C, Lo C (1993) Cytogenetic studies on *Saccharum-Miscanthus* nobilisation. Proc 7th Intl ongre SABRAO:223–233
Cuadrado A, Acevedo R, Moreno Dias de la Espina S, Jouve N, de la Torre C (2004) Genome remodelling in three modern *S. officinarum* x *S. spontaneum* sugarcane cultivars. J Exp Bot 55:847–854
da Silva J, Zorreéis ME, Burnquist W, Tanksley SD (1993) RFLP linkage map and genome analysis of *Saccharum spontaneum*. Genome 36:782–791
da Silva JA, Honeycutt RJ, Burnquist W, Al-Janabi, SM, Sorrells ME, et al. (1995) *Saccharum spontaneum* L. 'SE 208' genetic linkage map combining RFLP- and CR-based markers. Mol Breeding 1:165–179
Daniels C (1996) Vol 6 Biology and biological technology Part III Agro-industries and forestry. In: Needham J (Editor) Science and civilization in China. Cambridge University Press, Cambridge, United Kingdom
Daniels J, Daniels C (1975) Geographical, historical and cultural aspect of the origin of the Indian and Chinese sugarcanes *S. barberi* and *S. sinense*. Sugarcane Breeding newsletter 36:4–23
Daniels J, Roach BT (1987) Taxonomy and evolution in sugarcane, p 7–84. In: Heinz DJ (ed), Sugarcane improvement through breeding, Elsevier Press, Amsterdam
Daugrois J H, Grivet L, Roques D, Hoarau J Y, Lombard H, et al. (1996) A putative major gene for rust resistance linked with a RFLP markers in sugarcane cultivar 'R570'. Theor Appl Genet 92:1059–1064
De Rosa Jr VE, Nogueira FTS, Menossi M, Ulian EC, Arruda P (2005) Identification of methyl jasmonate responsive genes in sucarcane using cDNA arrays. Brazilian J Plant Physiol 17:131–136
Delaney TP, Uknes S, Vernooij B, Friedrich L, Weymann K, et al. (1994) A central role of salicylic acid in plant disease resistance. Science 266:1247–1250
D'Hont A (2005) Unravelling the genome structure of polyploids using FISH and GISH; examples of sugarcane and banana. Cytogenet Genome Res 109(1–3):27–33
D'Hont A, Glaszmann JC (2001) Sugarcane genome analysis with molecular markers, a first decade of research. Proc Int Soc Sugar Cane Technol 24:556–559
D'Hont A, Lu YH, Feldmann P, Glaszmann JC (1993) Cytoplasmic diversity in sugarcane revealed by heterologous probes. Sugar Cane 1:12–15
D'Hont A, Lu YH, Gonzàlez de Leòn D, Grivet L, Feldmann P, et al. (1994) A molecular approach to unraveling the genetics of sugarcane, a complex polyploid of the *Andropogoneae* tribe. Genome 37:222–230
D'Hont A, Grivet L, Feldmann P, Rao PS, Berding N, et al. (1995) Identification and characterization of sugarcane intergeneric hybrids, *Saccharum officinarum* x *Erianthus arundinaceus*, with molecular markers and DNA *in situ* hybridization. Theor Appl Genet 91:320–326
D'Hont A, Grivet L, Feldmann P, Rao PS, Berding, N, et al. (1996) Characterisation of the double genome structure of modern sugarcane cultivars (*Saccharun* spp) by molecular cytogenetics. Mol Gen Genet 250:405–413
D'Hont A, Ison D, Alix K, Roux C, Glaszmann JC (1998) Determination of basic chromosome numbers in the genus Saccharum by physical mapping of ribosomal RNA genes. Genome 41:221–225
D'Hont A, Paulet F, Glaszmann JC (2002) Oligoclonal interspecific origin of 'North Indian' and 'Chinese' sugarcanes. Chromosome Res 10:253–262

Dong Q, Lawrence CJ, Schluete, SD, Wilkerson MD, Kurtz S, et al. (2005) Comparative plant genomics resources at PlantGDB. Plant Physiol 139:610–618

Dufour P, Deu M, Grivet L, D'Hont A, Paulet F, et al. (1997) Construction of a composite sorghum genome map and comparison with sugarcane, a related complex polyploid. Theor Appl Genet 94:409–418

Ellis J, Dodds P, Pryor T (2000) Structure, function and evolution of plant disease resistance genes. Curr Opin Plant Biol 3:278–284

Falco MC, Silva-Filho MC (2003) Expression of soybean proteinase inhibitors intransgenic sugarcane plants: effects on natural defense against *Diatrae saccharalis* Plant Physiol Biochem 41:761–766

Falco MC, Neto AT, Ulian EC (2000) Transformation and expression of a gene for herbicide resistance in a Brazilian sugarcane. Plant Cell Rep 19:1188–1194

Falco MC, Marbach PAS, Pompermayer P, Lopes FCC, Silva-Filho MC (2001) Mechanisms of sugarcane response to herbivory. Genetics Mol Biol 24:113–122

Felix J M (2006) Análise da expressão gênica envolvida no metabolismo de sacarose em cana-de-açúcar (*Saccharum* spp) PhD thesis defended March 10th, 2006 Universidade Estadual de Campinas, Campinas, São Paulo, Brasil http://libdigi.unicamp.br/document/?view=vtls000380284

Figueiredo LHM (2003) Caracterização do gene LEAFY de Saccharum ssp e análise filogenética entre diferentes espécies vegetais utilizando as seqüências da família LEAFY/Floricaula. Master's thesis 137 p University of São Paulo, Ribeirão Preto, Brazil

Franco OL, Rigden DJ, Melo FR, Grossi-De-Sa MF (2002) Plant alpha-amylase inhibitors and their interaction with insect alpha-amylases. Eur J Biochem 269:397–412

Furumoto T, Hata S, Izui K (1999) cDNA cloning and characterization of maize phosphoenolpyruvate carboxykinase, a bundle sheath cell-specific enzyme. Plant Mol Biol 41:301–311

Furumoto T, Hata S, Izui K (2000) Isolation and characterization of cDNAs for differentially accumulated transcripts between mesophyll cells and bundle sheath strands of maize leaves. Plant Cell Physiol 41:1200–1209

Gaffney T, Friedrich L, Vernooij B, Negrotto D, Nye G, et al. (1993) Requirement of salicylic acid for the induction of systemic acquired resistance. Science 261:754–756

Gallo-Meagher M, Irvine JE (1996) Herbicide resistant transgenic sugarcane plants containing the bar gene. Crop Sci 36 (5):1367–1374

Galun E (2005) RNA silencing in plants. In Vitro Cell Dev-Pl 41 (2):113–123

Gaut BS, Le Thierry d'Ennequin M, Pekk AS, Sawkins MC (2000) Maize as a model for the evolution of plant nuclear genomes. Proc Nat Acad Sci USA 97:7008–7015

Glaszmann JC, Fautret A, Noyer JL, Feldmann P, Lanaud C (1989) Biochemical genetic markets in sugarcane. Theor Appl Genet 78:537–543

Glaszmann JC, Lu YH, Lanaud C (1990) Variation of nuclear ribosomal DNA in sugarcane. J Genet Breed 44:191–198

Glaszmann JC, Dufour P, Grivet L, D'Hont A, Deu M, et al. (1997) Comparative genome analysis between several tropical grasses. Euphytica 96:13–21

Grenier J, Potvin C, Trudel J, Asselin A (1999) Some thaumatin-like proteins hydrolyse polymeric beta-1,3-glucans. Plant J 19:473–480

Grivet L, Arruda P (2001) Sugarcane genomics: depicting the complex genome of an important tropical crop. Curr Opin Plant Biol 5:122–127

Grivet L, D'Hont A, Dufour P, Hamon P, Roques D, Glaszmann JC (1994) Comparative genome mapping of sugar cane with other species within the *Andropogoneae* tribe. Heredity 73:500–508

Grivet L, D'Hont A, Roques D, Feldmann P, Lanaud C, et al. (1996) RFLP mapping in cultivated sugarcane (*Saccharum* spp): genome organization in a highly polyploid and aneuploid interspecific hybrid. Genetics 142:987–1000

Grivet L, Daniels C, Glaszmann JC, D'Hont A (2004) A review of recent molecular genetics evidence for sugarcane evolution and domestication. Ethnobot Res Applic 2:9–17

Grivet L, Glaszmann JC, D'Hont A (2006) Molecular evidences for sugarcane evolution and domestication. In Motley T, Zerega N, Cross H (eds) Darwin's Harvest. New approaches to the origins, evolution, and conservation of crops. Columbia University Press, USA

Grof CPL, Campbell JA (2001) Sugarcane sucrose metabolism: scope for molecular manipulation. Aust J Plant Physiol 28:1–12;

Guimarães CT, Sills GR, Sobral BWS (1997) Comparative mapping of Andropogoneae: *Saccharum* L. (sugarcane) and its relation to sorghum and maize. Proc Natl Acad Sci, USA 94:14261–14266

Guimarães CT, Honeycutt RJ, Sills GR, Sobral BWS (1999) Genetic maps of *Saccharum officinarum* L. and *Sacharum robustum* Brandes & Jew. Ex Grassl Genetics Mol Biol 22:125–132

Ha S, Moore PH, Heinz D, Kato S, Ohmido N, et al. (1999) Quantitative chromosome map of the polyploid *Saccharum spontaneum* by multicolor fluorescence *in situ* hybridization and imaging methods. Plant Mol Biol 39:1165–1173

Henty EE (1969) A manual of the grasses of New Guinea. Bot Bull 1, Lae, New Guinea

Hoarau JY, Offmann B, D'Hont A, Risterucci AM, Roques D, et al. (2001) Genetic dissection of a modern cultivar (*Saccharum* spp). I. Genome mapping with AFLP. Theor Appl Genet 103:84–97

Hoarau JY, Grivet L, Offmann B, Raboin LM, Diorflar JP, et al. (2002) Genetic dissection of a modern sugarcane cultivar (*Saccharum* spp). II. Detection of QTLs for yield components. Theor Appl Genet 105:1027–1037

Hodkinson TR, Chase MW, Lledo MD, Salamin N, Renvoize SA (2002) Phylogenetics of *Miscanthus*, *Saccharum* and related genera (Saccharinae, Andropogoneae, Poaceae) based on DNA sequences from ITS nuclear ribosomal DNA and plastid trnL intron and trnL-F intergenic spacers. J Plant Res115(5):381–92

Hoefsloot G, Termorshuizen AJ, Watt DA, Cramer MD (2005) Biological nitrogen fixation is not a major contributor to the nitrogen demand of a commercially grown South African sugarcane cultivar. Plant and Soil 277:85–96

Huber SC, Toroser D, Winter H, Athwal GS, Huber JL (1998) Regulation of plant metabolism by protein phosphorylation. Possible regulation of sucrose-phosphate synthase by 14-3-3 proteins. XIth Intl Photosynthesis Congress, Budapest, Hungary, Kluwer Academic Publishers

Ilic K, SanMiguel PJ, Bennetzen JL (2003) A complex history of rearrangement in an orthologous region of the maize, sorghum, and rice genomes. Proc Natl Acad Sci USA 100:12265–12270

Im YJ, Han O, Chung GC, Cho BH (2002) Antisense expression of an Arabidopsis omega-3 fatty acid desaturase gene reduces salt/drought tolerance in transgenic tobacco plants. Mol Cells 13:264–271

Ingelbrecht IL, Irvine JE, Mirkov TE (1999) Posttranscriptional gene silencing in transgenic sugarcane. Dissection of homology-dependent virus resistance in a monocot that has a complex polyploid genome Plant Physiol 119(4):1187–98

International Rice Genome Sequencing Project (2005) The map-based sequence of the rice genome. Nature 436: 793–800

Irvine J (1999) Saccharum species as horticultural classes. Theor Appl Genet 98: 186–194

James EK, Olivares FL (1998) Infection and colonization of sugar cane and other graminaceous plants by endophytic diazotrophs. Crit Rev Plant Sci 17:77–119

Jannoo N, Grivet L, Seguin M, Paulet F, Domaingue R, et al. (1999a) Molecular investigation of the genetic base of sugarcane cultivars. Theor Appl Genet 99:171–184

Jannoo N, Grivet L, Dookun A, D'Hont A, Glaszmann JC (1999b) Linkage disequilibrium among modern sugarcane cultivars. Theor Appl Genet 99:1053–1060

Jannoo N, Grivet L, David J, D'Hont A, Glaszmann JC (2004) Differential chromosome pairing affinities at meiosis in polyploid sugarcane revealed by molecular markers. Heredity 93(5):460–7

Jannoo N, Grivet L, Chantret N, Garsmeur O, Glaszmann J C, et al. (2007) Orthologous comparison in a gene-rich region among grasses reveals stability in the sugarcane polyploid genome. Plant J 50(4):574–585

Jansen RC, Nap JP (2001) Genetical genomics: the added value from segregation. Trends Genetics 17:388–391
Joyce PA, McQualter RB, Bernad MJ, Smith GR (1998) Engineering for resistance to SCMV in sugarcane. Acta Hort 461:385–391
Jordan DR, Casu RE, Besse P, Carroll BC, Berding N, et al. (2004) Markers associated with stalk number and suckering in sugarcane colocate with tillering and rhizomatousness QTLs in sorghum. Genome 47:988–993
Kashkush K, Feldman M, Levy AA (2003) Transcriptional activation of retrotransposons alters the expression of adjacent genes in wheat. Nat Genet 33(1):102–6
Lalonde S, Wipf D, Frommer W (2004) Transport mechanisms for organic forms of carbon and nitrogen between source and sink. Annu Rev Plant Biol 55:341–372
Leibbrandt NB, Snyman SJ (2003) Stability of gene expression and agronomic performance of a transgenic herbicide-resistant sugarcane line in South Africa. Crop Sci 43(2):671–677
Li RP, Macrae IC (1991) Specific association of diazotrophic acetobacters with sugarcane. Soil Biol Biochem 23:999–1002
Li ZY, Chen SY (2000) Differential accumulation of the S-adenosylmethionine decarboxylase transcript in rice seedlings in response to salt and drought stresses. Theor Appl Genet 100:782–788
Lockhart DJ, Dong H, Byrne MC, Follettie MT, Gallo MV, et al. (1996) Expression monitoring by hybridization to high-density oligonucleotide arrays. Nat Biotechnol 14:1675–1680
Lu YH, D'Hont A, Walker DIT, Rao PS, Feldmann P, et al. (1994a) Relationships among ancestral species of sugarcane revealed with RFLP using single copy maize nuclear probes. Euphytica 78:7–8
Lu YH, D'Hont A, Paulet F, Grivet L, Arnaud M, et al. (1994b) Molecular diversity and genome structure in modern sugarcane varieties. Euphytica 78:217–226
Lunn JE, Furbank RT (1999) Sucrose biosynthesis in C-4 plants. New Phytol 143:221–237
Ma H-M, Schulze S, Lee S, Yang M, Mirkov E, et al. (2004) An EST survey of the sugarcane transcriptome. Theor Appl Genet 108:851–863
Manickavasagam M, Ganapathi A, Anbazhagan VR, Sudhakar B, Selvaraj N, et al. (2004) Agrobacterium-mediated genetic transformation and development of herbicide-resistant sugarcane (*Saccharum* species hybrids) using axillary buds. Plant Cell Rep 23(3):134–143
McCormick AJ, Cramer MD, Watt DA (2006) Sink strength regulates photosynthesis in sugarcane. New Phytol 171:759–770
McIntyre CL, Whan VA, Croft B, Magarey R, Smith GR (2005) Identification and validation of molecular markers associated with Pachymetra root rot and brown rust resistance in sugarcane using map- and association- based approaches. Mol Breeding 16:151–161
McQualter RB, Harding RM, Dale JL, Smith GR (2001) Virus derived transgenes confer resistance to Fiji disease in transgenic sugarcane plants. Proc Int Soc Sugar Cane Technol 24(2):584–585
McQualter RB, Chong FB, Meyer K, Dyk DE, van O'Shea MG, et al. (2005) Initial evaluation of sugarcane as a production platform for p-hydroxybenzoic acid. Plant Biotechnol J 3(1): 29–41
Ming R, Liu SC, Lin YR, Braga D, da Silva J, et al. (1998) Alignment of *Sorghum* and *Saccharum* chromosomes: comparative organization of closely-related diploid and polyploid genomes. Genetics 150:1663–1882
Ming R, Liu SC, Moore PH, Irvine JE, Paterson AH (2001) QTL analysis in a complex autopolyploid: genetic control of sugar content in sugarcane. Genome Res 11:2075–2084
Ming R, Wang YW, Dryer X, Moore PH, Irvine JE, et al. (2002a) Molecular dissection of complex traits in autopolyploid: mapping QTLs influencing sugar yield and related traits in sugarcane. Theor Appl Genet 105:332–345
Ming R, DelMonte T, Moore PH, Irvine JE, Paterson AH (2002b) Comparative analysis of QTLs affecting plant height and flowering time among closely-related diploid and polyploid genomes. Genome 45:794–803
Ming R, Moore PH, Wu KK, D'Hont A, Glassman JC, et al. (2006) Sugarcane improvement through breeding and biotechnology. In: Janick J (ed) Plant Breeding Reviews 27:15–118 J Wiley & Sons, Inc, NY

Moore PH (1995) Temporal and spatial regulation of sucrose accumulation in the sugarcane stem. Austr J Plant Physiol 22:661–679

Mullerr AE (2006) Applications of RNA interference in transgenic plants. CAB Reviews: Perspectives in Agriculture, Veterinary Science, Nutrition and Natural Resources 1 (024): 13 pp

Mur LA, Bi YM, Darby RM, Firek S, Draper J (1997) Compromising early salicylic acid accumulation delays the hypersensitive response and increases viral dispersion during lesion establishment in TMV-infected tobacco. Plant J 12:1113–1126

Nair NV, Nair S, Sreenivasan TV, Mohan M (1999) Analysis of genetic diversity and phylogeny in *Saccharum* and related genera using RAPD markers. Genet Res Crop Evol 46:73–79

Neuteboom LW, Kunimitsu WY, Webb D, Christopher DA (2002) Characterization and tissue-regulated expression of genes involved in pineapple (*Ananas comosus* L.) root development. Plant Sci 163:1021–1035

Nogueira FTS, Rosa Jr VE, Menossi M, Ulian EC, Arruda P (2003) RNA expression profiles and data mining of sugarcane response to low temperature. Plant Physiol 132:1811–1824

Nutt KA (2005) Characterisation of proteinase inhibitors from canegrubs for possible application to genetically engineer pest-derived resistance into sugarcane. PhD: Queensland University of Technology, Brisbane, Australia

Nutt KA, Allsopp PG, McGhie TK, Shepherd KM, Joyce PA, et al. (1999) Transgenic sugarcane with increased resistance to canegrubs Proc Austr Soc Sugar Cane Technol 171–176

Olivares FL, James EK, Baldani JI, Döbereiner J (1997) Infection of mottled stripe disease-susceptible and resistant sugarcane varieties by the endophytic diazotrophs *Herbaspirillum.* New Phytol 135:723–727

Papini-Terzi, F.S., Felix, J. M., Rocha, F. R., Waclawovsky, A. J., Ulian, E. C., Chabregas, S., Falco, M. C., Nishiyama-Jr, M. Y., Vêncio, R. Z. N., Vicentini, R., Menossi, M. e Souza, G. M. (2007). The SUCEST-FUN Project: identifying genes that regulate sucrose content in sugarcane plants. Proc. Int. Soc. Sugar Cane Technol. 26.

Papini-Terzi FS, Rocha FR, Vêncio RZN, Oliveira KC, Felix JM, et al. (2005) Transcription profiling of signal transduction-related genes in sugarcane tissues. DNA Res 12:27–38

Patrick JW (1997) Phloem unloading: sieve element unloading and post sieve element transport. Annu Rev Plant Physiol Plant Mol Biol 48:191–222

Pessoa Junior A, Roberto IC, Menossi M, Santos RR, Ortega Filho S, et al. (2005) Perspectives on bioenergy and biotechnology in Brazil. Appl Bioch Biotech 121:59–70

Piegu B, Guyot R, Picault N, Roulin A, Saniyal A, et al. (2006) Doubling genome size without polyploidization: Dynamics of retrotransposition-driven genomic expansions in *Oryza australiensis*, a wild relative of rice. Genome Res 10:1262–1269

Piperidis G, D'Hont A (2001) Chromosome composition analysis of various *Saccharum* interspecific hybrids by genomic *in situ* hybridisation (GISH). Proc Int Soc Sugar Cane Technol 11:565–566

Piperidis G, Christopher MJ, Carroll BJ, Berding N, D'Hont A, (2000) Molecular contribution to selection of intergeneric hybrids between sugarcane and the wild species *Erianthus arundinaceus*. Genome 43:1033–1037

Price S (1965) Interspecific hybridization in sugarcane breeding. Proc Int Soc Sugar Cane Technol 12:1021–1026

Raboin LM, Oliveira KM, Lecunff L, Telismart H, Roques D, et al. (2006) Genetic mapping in sugarcane, a high polyploid, using bi-parental progeny: identification of a gene controlling stalk colour and a new rust resistance gene. Theor Appl Genet 112(7):1382–1391

Rae AL, Perroux JM, Grof CP (2005) Sucrose partitioning between vascular bundles and storage parenchyma in the sugarcane stem: a potential role for the ShSUT1 sucrose transporter. Planta 220:817–825

Raghothama KG (1999) Phosphate acquisition. Annu Rev Plant Physiol Plant Mol Biol 50:665–693

Reffay N, Jackson PA, Aitken KS, Hoarau J-Y, D'Hont A, et al. (2005) Characterisation of genome regions incorporated from an important wild relative into Australian sugarcane. Mol Breeding 15:367–381

Reinhold-Hurek B, Hurek T (1998) Life in grasses: diazotrophic endophytes. Trends Microbiol 6:139–144

Roach BT (1972) Nobilisation of sugarcane. Proc Int Soc Sugar Cane Technol 14:206–216

Rocha FR, Papini-Terzi FS, Nishiyama-Jr MY, Vêncio RZN, Vicentini R, et al. (2007) Signal transduction-related responses to phytohormones and environmental challenges in sugarcane. BMC Genomics 8:71

Rossi M, Araújo PG, Van Sluys MA (2001) Survey of transposable elements in sugarcane expressed tags (ESTs). Genetics Mol Biol 24:147–154

Rossi M, Araujo PG, Paulet F, Garsmeur O, Dias VM, et al. (2003) Genomic distribution and characterization of EST-derived resistance gene analogs (RGAs) in sugarcane. Mol Gen Genomics 269:406–419

Rossi M, Araújo PG, Jesús EM Varani AM Van Sluys MA (2004) Comparative analysis of sugarcane Mutator-like transposases. Mol Gen Genomics 272:194–203

Sakuma Y, Maruyama K, Osakabe Y, Qin F, Seki M, et al. (2006) Functional analysis of an Arabidopsis transcription factor, DREB2A, involved in drought-responsive gene expression. Plant Cell 18:1292–1309

Selvi A, Nair N, Noyer JL, Singh NK, Balasundaram N, et al. (2004) AFLP analysis of the phenetic organization and genetic diversity in the sugarcane complex, *Saccharum* and *Erianthus*. Genetic Res Crop Evol 53(4):831–842

Selvi A, Nair NV, Noyer JL, Singh NK, Balasundaram N, et al. (2005) Genomic constitutionand genetic relationship among the tropical and subtropical Indian sugarcane cultivars revealed by AFLP. Crop Sci 45: 1750–1757

Sevilla M, Burris RH, Gunapala N, Kennedy C (2001) Comparison of benefit to sugar cane plant growth and $^{15}N_2$ incorporation following inoculation of sterile plants with *Acetobacter diazotrophicus* wild-type and Nif-mutant strains. Mol Plant-Microbe Interact 14:358–366

Sobral BWS, Braga DPV, LaHood ES, Keim P (1994) Phylogenetic analysis of chloroplast restriction enzyme site mutations in the *Saccharinae* Griseb. subtribe of the *Andropogoneae* Dumort tribe. Theor Appl Genet 87:843–853

Sorghum Genomics Planning Workshop Participants (2005) Toward sequencing the sorghum genome. A US National Science Foundation-Sponsored Workshop Report. Plant Physiol 138:1898–1902

Sreenivasan TV, Ahloowalia BS, Heinz DJ (1987) Cytogenetics. p 211–253 In: Heinz DJ (ed), Sugarcane improvement through breeding. Elsevier, Amsterdam

Sugden C, Donaghy PG, Halford NG, Hardie DG (1999) Two SNF1-related protein kinases from spinach leaf phosphorylate and inactivate 3-hydroxy-3-methylglutarylcoenzyme A reductase, nitrate reductase, and sucrose phosphate synthase in vitro. Plant Physiol 20:257–74

Tahtiharju S, Palva T (2001) Antisense inhibition of protein phosphatase 2C accelerates cold acclimation in *Arabidopsis thaliana*. Plant J 26:461–470

Tai PYP, Lentini RS (1998) Freeze damage of Florida sugarcane. In Anderson DL (ed) Sugarcane Handbook, Ed 1. Florida Cooperative Extension

Taiz L, Zeiger E (1998) Plant physiology. Sinauer Associates Inc Publishers, Sunderland MS, pp 214–215

Tomkins JP, Yu Y, Miller-Smith H, Frisch DA, Woo SS, et al. (1999) A bacterial artificial chromosome library for sugarcane. Theor Appl Genet 99:419–424

Toroser D, Athwal GS, Huber SC (1998) Site-specific regulatory interaction between spinach leaf sucrose-phosphate synthase and 14-3-3 proteins. FEBS Letters 435:110–114

van der Merwe MJ, Groenewald JH, Botha FC (2003) Isolation and evaluation of a developmentally regulated sugarcane promoter Proc South African Sugar Cane Technol 77:146–169

Vargas C, de Pádua V, Nogueira EM, Vinagre F, Masuda HP, et al. (2003) Signaling pathways mediating the association between sugarcane and endophytic diazotrophic bacteria: a genomic approach. Symbiosis 35:159–180

Vettore AL, da Silva FR, Kemper EL, Souza GM, da Silva AM, et al. (2003) Analysis and functional annotation of an expressed sequence tag collection for tropical crop sugarcane. Genome Res 13:2725–35

Vinagre F, Vargas C, Schwarcz K, Cavalcante J, Nogueira EM, et al. (2006) SHR5: a novel plant receptor kinase involved in plant-N2-fixing endophytic bacteria association. J Exp Bot 57:559–569

Vincentz V, Cara FAA, Okura V K, da Silva FR, Pedrosa GL, et al. (2004) Evaluation of monocot and eudicot divergence using the sugarcane transcriptome. Plant Physiol 134:951–959

Wang ML, Goldstein C, Su W, Moore PH, Albert HH (2005) Production of biologically active GM-CSF in sugarcane: a secure biofactory Transgenic Res 14 167–178

Watt D, McCormics A, Govender C, Crame, M, Huckett B, et al. (2005) Increasing the utility of genomics in unraveling sucrose accumulation Field Crop Res 92:149–158

Wei H, Albert HH, Moore PH (1999) Differential expression of sugarcane polyubiquitin genes and isolation of promoters from two highly-expressed members of the gene family. J Plant Physiol 155:513–519

Wei H, Moore PH, Albert HH (2003) Comparative expression analysis of two sugarcane polyubiquitin promoters and flanking sequences in transgenic plants J Plant Physiol 160:1241–1251

Weng L, Deng H, Xu J, Li Q, Wang L, et al. (2006) Regeneration of sugarcane elite breeding lines and engineering of stem borer resistance. Pest Management Sci 62 (2):178–187

Wu KK, Burnquist W, Sorrells ME, Tew TL, Moore PH, et al. (1992) The detection and estimation of linkage in polyploids using single-dose restriction fragments. Theor Appl Genet 83:294–300

Wu L, Birch R G (2007) Doubled sugar content in sugarcane plants modified to produce a sucrose isomer Plant Biotechnol J 5:109–117

Wu P, Ma L, Hou X, Wang M, Wu Y, et al. (2003) Phosphate starvation triggers distinct alterations of genome expression in arabidopsis roots and leaves. Plant Physiol 132:1260–1271

Xin Z, Browse J (2000) Cold comfort farm: the acclimation of plants to freezing temperatures. Plant Cell Environ 23:893–902

Zhang M, Barg R, Yin M, Gueta-Dahan Y, Leikin-Frenkel A, et al. (2005) Modulated fatty acid desaturation via overexpression of two distinct omega-3 desaturases differentially alters tolerance to various abiotic stresses in transgenic tobacco cells and plants. Plant J 44:361–371

Chapter 22
Genomics of Wheat, the Basis of Our Daily Bread

Manilal William, Peter Langridge, Richard Trethowan, Susanne Dreisigacker, and Jonathan Crouch

Abstract Wheat, being an important source of calories across the Americas, Europe, North Africa and Asia, is the most widely grown food crop in the world. Wheat yields have undergone a spectacular rise over the last half century, contributing to the Green Revolution in Asia. However, productivity increases appear to have reached a plateau in recent years and many consider that new advances in genomics will be essential to delivery the rates of productivity increases necessary to prevent hunger. New molecular tools will enhance on-going wheat breeding, offering the plant breeder considerable advantages in time, cost, and response to selection. Perhaps most importantly, it is believed that genomics tools will also facilitate much more efficient utilization of new sources of genetic variation for important agronomic traits from wild species. This chapter provides an overview of the botany and conventional breeding of wheat including a summary of past successes, the current primary breeding targets, and the major constraints to achieving those goals. We then focus on genomic advances in bread wheat and durum wheat during the past decade and the implications of these advances for increasing resilience, stability and productivity in tropical, sub-tropical and semi-arid production systems across the world. This includes the use of genomics to improve the search for, and the characterization of, new beneficial genetic variation and the identification of molecular markers to facilitate the efficient manipulation of that variation in breeding programs. Finally, we provide a list of the currently available trait markers and a perspective on likely future trends and challenges in wheat molecular breeding.

22.1 Introduction

Wheat is the most widely grown food crop in the world, occupying 216 million hectares (mha), producing 600 million tonnes (mt) of grain, compared to 153 mha of rice and 147 mha of maize (FAO 2006). It is one of the first domesticated food

M. William
Genetic Resources and Enhancement Unit, International Maize and Wheat Improvement Center (CIMMYT), Carretera Méx-Veracruz, El Batán, Texcoco, México CP56130
j.crouch@cgiar.org

© Springer 2008

Fig. 22.1 The production in Afghanistan of wheat, the country's staple crop, has risen significantly in recent years, but average yields remain relatively low—on the order of 2.0–2.5 tons per hectare—with landholdings in many areas being small and not amenable to mechanization. (photo by CIMMYT, used with permission) (See color insert)

species and has been the major source of calories in Europe, West Asia, and North Africa since the inception of organized farming. It is widely grown across the temperate regions of Central Asia, Europe, and North America and is a major crop in many developing countries across the sub-tropical regions of the world including India (26 mha sown per annum-p.a.), Pakistan (8 mha p.a.), Iran (6 mha p.a.), Brazil (2 mha p.a.), Syria (2 mha p.a.), Egypt and Ethiopia (both over 1 mha p.a.) (see Fig. 22. 1). Wheat is an important source of calories across the world due to its wide agronomic adaptability, ease of grain storage, and the wide range of diverse food products that can be made from its flour. Bread wheat flour can be used for leavened bread due to the specific viscoelastic properties conferred by gluten, an elastic form of protein that traps the CO_2 emitted during fermentation, causing the dough to rise. Wheat flour can also be used to make flat bread which is more popular in South Asia, North Africa, and the Middle East, while it is used for making noodles in China and Japan, and for making biscuits across the world. In contrast, flour from durum wheat is used for pasta in the western world and for local products like couscous in North Africa. Dramatic increases in global wheat production have taken place during the last 50 years primarily due to increased productivity, rather than expansion of the cultivated area (Curtis 2002). Average global yields have risen from 1 t/ha in the 1950s to about 2.5 t/ha at the turn of the century (Curtis 2002). With world population projected to reach 7.9 billion by 2025 (US Census Bureau 1998), and assuming no changes in consumption patterns, significant increases in wheat production must be made to meet the expected demand for food. However, productivity increases appear to have reached a plateau in recent years and many consider that new advances in genomics will be essential to delivery the necessary rates of productivity increases. New molecular tools will enhance ongoing wheat breeding programs, offering breeders considerable advantages in time, cost, response to selection, and opportunities to address new goals.

Wheat and its relatives comprise diploid (2n = 2x = 14), tetraploid (2n = 4x = 28), and hexaploid (2n = 6x = 42) forms. Of the diploid wheats, einkorn wheat (*Triticum monococcum*), a possible donor of the A-genome, can still be found in limited cultivation with its wild form ssp. *aegilopoides* widely distributed across the Middle East. The tetraploid wheats, also known as emmer wheats, have two distinct forms; widely grown *T. turgidum* with genome AABB (2n = 4x = 28) and *T. timopheevii* with AAGG (2n = 4x = 28); both are found across the Fertile Crescent (Gill and Friebe 2002). The widely cultivated free-threshing, non-fragile, tetraploid form of *Triticum* spp., popularly known as durum or macaroni wheat, has the genome AABB. *T. dicoccum* (AABB) was most likely the first cultivated form of wheat. Cytological, archeological, and molecular genetic studies suggest that *T. dicoccoides* (AABB) arose by hybridization between the *T. urartu* (AA) and an unknown diploid with genome composition similar to the Sitopsis section of the genera *Aegilops* about 10,000 years ago (Zohary and Hopf 1993).

The hexaloid wheats are composed of two types, *Triticum aestivum,* also known as bread wheat or common wheat (AABBDD: 2n = 6x = 42), and *T. zhukovskyi* (AAAAGG: 2n = 6x = 42). The *Triticum aestivum* wheats arose from hybridization between tetraploid AABB species and *Aegiolops squarosa* (2n = 2x = 14; DD) (McFadden and Sears 1946). *T. zhukovskyi* possibly resulted from hybridization between *T. timopheevii* (AAGG) and *T. monococcum* (AA) (Upadhya and Swaminathan 1963). It is likely that these hexaploid forms arose during cultivation of the tetraploid progenitors in close proximity with diploid relatives, since there is no evidence of wild forms of hexaploid wheat. Despite the allopolyploid nature of bread wheat and durum wheat, both show disomic inheritance. Pairing between homoeologous chromosomes is mainly regulated by pairing homolog genes *Ph1* (Riley and Chapman 1958) and *Ph2* (Mello-Sampayo 1971). The allopolyploid nature of wheat allows it to withstand a variety of chromosome substitutions and additions, which has been widely exploited by researchers to develop cytogenetic stocks of most wheat chromosomes. These genetic stocks have been fundamental to many advances in wheat genetics and genomics during the last half century.

This chapter focuses on genomic advances in bread wheat and durum wheat during the past decade and their implications for increasing resilience, stability and productivity in tropical, sub-tropical, and semi-arid production systems across the world.

22.2 Progress in Conventional Breeding

Wheat productivity has undergone a spectacular rise over the last half century. Initial increases were due to the introduction and dissemination of high yielding, fertilizer responsive varieties with short stature, generally known as the Green Revolution varieties. The introduction of dwarfing genes and subsequent improvements in harvest index increased grain yield, reduced crop lodging, and allowed farmers to apply higher rates of nitrogen fertilizer. Shuttle breeding, originally initiated by the International Maize and Wheat Improvement Center (CIMMYT) and based on growing alternate generations in two diverse environments in Mexico followed by international germplasm exchange and global testing networks, has made significant contributions to global advances in yield (Ortiz et al. 2006). The use of two growing seasons per year has facilitated rapid genetic gains and selection in these contrasting environments has led to the development of improved germplasm with wide adaptation, since the two locations differ in rainfall, temperature, photoperiod, and soil type (Rajaram et al. 2002). CIMMYT's shuttle breeding efforts resulted in the production of wheat lines with relative insensitivity to photoperiod, broad adaptability, and broad spectrum resistance to a number of important biotic stresses, primarily the rust diseases (Rajaram et al. 2002; Ortiz et al. 2007). These changes led to substantial improvements in wheat productivity; first in Asia and then in much of the rest of the developing world (Borlaug 1968; Trethowan et al. 2007). Consequently, many wheat breeding programs across the world have adopted multilocation testing in contrasting environments as an integral part of their breeding philosophy (Rajaram et al. 2002).

Since the Green Revolution, wheat breeders have maintained an average yield advance of 1% per annum (Byerlee and Moya 1993; Sayre et al. 1997), although improved crop management practices have also made significant contributions (Bell et al. 1995). Average national wheat yields have grown faster in developing countries than in the high income countries throughput the past 40 years. However, wheat yield increases have shown some leveling off in recent years (Reynolds et al. 1996).

22.2.1 Utilization of Genetic Variability from Wild Relatives

The primary gene pool of wheat includes the cultivated and landrace forms of hexaploid bread wheat and tetraploid durum wheat as well as the diploid A genome donor (*T. monococcum var. urartu*) and the diploid D genome donor (*Ae. squarossa*). The secondary gene pool includes polyploid relatives belonging to the genus *Triticum* and *Aegilops* that have at least one genome in common with wheat. The tertiary gene pool is composed of species with varying ploidy levels with genomes that are not homologous to those of cultivated wheat. Generally, crosses involving primary and secondary gene pools do not require special cytogenetic manipulation other than embryo rescue and culture to produce F_1 hybrids. Genomes of species in the tertiary gene pool often show homeologous relationships with the A, B, and D genomes of cultivated wheat. The primary gene pool of wheat represents a valuable source of genetic diversity that has been used in a number of wheat improvement programs. The landraces and wild species of the diploid A and D genomes also possess many novel genes and can be readily crossed with durum and bread wheat breeding lines. Accessions of *T. tauschii* ($2n = 2x = 14$; DD) have been widely used in crosses with *T. durum* (2n = 4x = 28; AABB) for the "artificial" resynthesis of the hexaploid genomes of cultivated bread wheat (2n = 6x = 42; AABBDD) (Mujeeb Kazi and Rajaram 2002). The resultant F_1 hybrid embryos are then rescued and grown on culture media, followed by chromosome doubling using colchicine. CIMMYT has produced nearly 1,500 resynthesized hexaploid wheat lines. These have been extensively used, particularly in CIMMYT's rainfed wheat breeding program, to incorporate superior levels of drought tolerance in wheat lines targeted for marginal environments. However, these resynthesized synthetic hexaploid wheats have a number of undesirable agronomic traits such as shattering, tall stature, and late maturity, and not all durum wheat germplasm can be crossed with *T. tauchii* accessions due to hybrid necrosis. Fortunately, following backcrossing to agronomically elite breeding lines, the resultant "synthetic derivatives" have shown recovery of these deleterious traits and improved disease resistance and abiotic stress tolerance (Mujeeb-Kazi et al. 1998; Lage and Trethowan 2007). Direct introgression of D genome variation has also been accomplished by crossing hexaploid wheat directly with *Triticum tauschii* (Gill and Raupp 1987).

Wide ranges of whole chromosomal substitutions and partial chromosomal translocations have been made using species from the secondary and tertiary gene pools (reviewed by Sharma and Gill 1983; Jiang et al. 1994). Not all species in the tertiary gene pool can be successfully crossed with bread wheat, primarily due to the chromosomal differentiation. The wheat cultivar Chinese Spring has been used in many intergeneric crosses because it has recessive alleles at the "crossability" loci *kr1*, *kr2* and *kr3* (Falk and Kasha 1983). A number of disease resistance genes have been transferred to bread wheat through interspecific and intergeneric crosses (McIntosh et al. 2003). One particularly successful example involves the rye chromosome 1R. Wheat lines carrying the 1BL/1RS translocation are present in many high yielding wheat cultivars with wide adaptation (Rajaram et al. 1983).

22.2.2 Advances in Genetics and Breeding of Agronomic Traits

Wheat is cultivated from the tropics to the fringes of the Arctic and from sea level to over 3,000 m elevations on the Andean plateau. This wide range of adaptation has been made possible by the presence of genes controlling vernalization, photoperiod response, and early maturity, thus enabling wheat breeders to tailor cultivars to different agro-ecological regions. Vernalization response is conditioned by a homoelogous set of genes designated as *Vrn-A1* (*Vrn1*), *Vrn-B1* (*Vrn2*), and *Vrn-D1* (*Vrn3*) located on the short arms of homoelogous chromosomes 5A, 5B, and 5D, respectively (Worland, 1996). Response to day length is primarily conditioned by another homoelogous set of genes, *Ppd-D1* (*Ppd1*), *Ppd-B1* (*Ppd2*), and *Ppd-A1* (*Ppd3*) located on the short arms of chromosomes 2D, 2B, and 2A, respectively. However, other genes located on other chromosomes may also play a role in vernalization and photoperiod response (Law et al. 1998). The genetic control of early maturity, considered to be conditioned by genes conferring earliness per se, is less well documented.

Recent increases in wheat yields have been associated with dramatic reductions in plant height resulting in significant increases in harvest index. Although a tall plant can compete with weeds more effectively, a plant with shorter stature is more efficient in partitioning assimilates to the grain and tends to be more lodging tolerant. Twenty-one genes controlling plant height have been described in wheat (McIntosh et al. 2003). The two most important height-controlling genes are *Rht-B1* and *Rht-D1*. located on chromosomes 4BL and 4DL, respectively (Gale et al. 1975; McVittie et al. 1978), with most semi-dwarf wheat germplasm possessing alleles *Rht-B1b* or *Rht-D1b* which are mutants insensitive to gibberllic acid (Peng et al. 1999). These two genes acting alone can reduce plant height by an average of 18 cm while at the same time significantly increasing spikelet fertility in high input environments (Flintham et al. 1997).

More recently, doubled haploids have been used to improve the speed and precision of wheat breeding (Aung et al. 1995; Tuvesson et al. 2003). Doubled haploid systems allow rapid generation of homozygous lines which improves breeding efficiency by decreasing the amount of time required to develop fixed lines. They also allow the breeder to select among fixed lines at the maximum level of genetic variability, viz. at the first generation after crossing. Wheat doubled haploids can be generated through anther or microspore culture (Konzak and Zhou 1991) or by using a maize pollen induction system (Laurie and Bennett 1986). In European breeding programs, thousands or even tens of thousands of doubled haloids are produced as part of the annual breeding process (Dayteg et al. 2007). Doubled haploid systems also enable easy integration of molecular markers in breeding programs as well as facilitating mapping and genetic studies within breeding populations (Dayteg et al. 2007; Howes et al. 1998).

22.3 Structural Genomics

22.3.1 Molecular Cytogenetics and Physical Mapping

In situ hybridization techniques, developed in the late 1960s, allow the detection of DNA sequences directly in cytological preparations on glass slides. Although originally developed using radioactive probes, these techniques were later optimized for utilization with non-radioactive labels such as biotin-dUTP and digoxiginin (reviewed in Jiang and Gill 1994). More recently, fluorochromes have been used for fluorescent in situ hybridization (FISH) with increased sensitivity and precision while facilitating the detection of multiple targets on the same chromosome preparation (Mukai et al. 1993; Oliver et al. 2006). Modified forms of traditional chromosome banding techniques (C-banding and N-banding) coupled with in situ hybridization procedures have also been used to detect and characterize alien translocations and multi-copy DNA sequences (Jiang and Gill 1993). These techniques can also be used in phylogenetic and evolutionary studies (Badaeva et al. 2002). Recent advances in molecular cytogenetics procedures have been reviewed by Jiang and Gill (2006).

Endo and Gill (1996) have developed a set of deletion stocks of specific wheat chromosomes. These deletion stocks can be used to physically locate genes and expressed sequence tags (EST) on specific chromosomal regions, such as the control of homologous pairing gene *Ph1* on chromosome 5BL (Gill et al. 1993), vernalization response gene *Vrn A-1* on chromosome 5AL (Sarma et al. 1998), and the grain hardness locus, *Ha,* on chromosome 5DS (Sarma et al. 2000). Establishment of the relationship between genetic and physical maps by mapping a series of microsatellite markers on to deletion bins has also been accomplished (Sourdille et al. 2004).

22.3.2 Molecular Markers as Tools

The allohexaploid nature of bread wheat, with three distinct genomes makes it the largest of the cultivated cereals. The haploid complement of bread wheat has approximately 40 times more DNA (16×10^9 bp) than rice (4×10^8 bp). Genetic characterization studies have established that about 95%–99% of the hexaploid wheat genome is not transcribed (Sandhu and Gill 2002a). Most of the transcribed genes in wheat seem to exist in clusters spanning physically small chromosomal regions that are designated as gene rich regions (Sandhu and Gill 2002b).

There are a number of marker technologies available for genetic characterization in wheat and each system has its advantages and disadvantages (Langridge et al. 2001). Restriction fragment length polymorphism (RFLP) markers are valuable in comparative genetic analysis and synteny mapping, but are not suitable for routine marker-assisted selection (MAS). Random amplified polymorphic DNA (RAPD) markers are also no longer commonly used in wheat due to lack of reliability and robustness, although some RAPD markers linked to important genes of interest have been converted to more robust sequence tagged site (STS) markers (http://maswheat.ucdavis.edu). More recently, microsatellite markers (also known as simple sequence repeats; SSR) have become popular due to their robustness as an assay system, plus their highly polymorphic and co-dominant nature of inheritance (Somers et al. 2004). Diversity array technology (DArT) is a microarray-based hybridization technique that allows simultaneous genotyping of several hundred polymorphic loci distributed across the genome (Jaccoud et al. 2001). The large number of loci that can be genotyped simultaneously makes DArT technology an efficient method of low cost, high-throughput genotype fingerprinting and map construction (Akbari et al. 2006; Semagn et al. 2006). However, its potential as a tool in marker-assisted selection is still not clear. More recently, single nucleotide polymorphism (SNP)-based markers are beginning to be developed in wheat (Ravel et al. 2006; Somers et al. 2003). SNPs are highly abundant in all genomes, and SNP markers are highly amenable to automation offering dramatic increases in throughput potential and unit cost efficiency. However, the frequency of SNP polymorphisms in wheat breeding populations is surprisingly low (Ravel et al. 2006). Therefore, at the present time, SSR markers remain the assay of choice for marker-assisted selection in wheat. ESTs have also become valuable in SNP discovery and for developing SSR markers. Since ESTs are derived from expressed gene sequences, they provide an efficient route for the development of candidate gene-based markers (see section 22.4.1).

The development of linkage maps in bread wheat and durum wheat has been generally slow compared to other important crops such as rice, maize, barley, and soybean. This is partly due to the large genome size of wheat and the resulting large number of linkage groups that require molecular characterization (21 linkage groups in bread wheat as opposed to 10 in rice, 12 in maize, and 7 in barley). In addition, wheat has a low level of detectable polymorphism with most marker systems. A number of linkage maps are available in hexaploid bread wheat (e.g. Roder et al. 1998; Somers et al. 2004; Semagn et al. 2006; Akbari et al. 2006) and on a lesser scale for durum wheat (Blanco et al. 1998; Elouafi and Nachit 2004). The International Triticeae Mapping Initiative (ITMI) generated the most comprehensive publicly available linkage map in wheat based on a single seed descent-derived population originating from a cross between the cultivar Opata85 and a resynthesized hexaploid wheat (W7984) developed at CIMMYT (http://wheat.pw.usda.gov/). Attempts have been made to develop consensus linkage maps in wheat, the latest having over 4,000 loci (Appels 2003). Somers et al. (2004) developed a high density consensus linkage map by using common SSR markers on each chromosome in four different mapping populations. However, even the most comprehensive consensus wheat linkage map lacks uniform marker coverage across all chromosomes, particularly the D genome.

22.3.3 Genome Diversity Analysis

The assessment of genetic diversity among cultivars is indispensable for plant breeding purposes since it provides a means for analyzing variation available in germplasm collections. Measures of genetic diversity were initially based on co-ancestry and pedigree records (Van Beuningen 1997; Kim and Ward 1997). Pedigree records are relatively abundant in wheat; however, they often lack detail, especially when large numbers of breeding lines or cultivars are being assessed. Furthermore, the underlying assumptions of co-ancestry are rarely met as selection ensures that gene frequency is not random, thus coefficients of parentage remain a theoretical estimate of the identity by descent (Cox et al. 1985; Graner et al. 1994). Molecular markers have enabled the estimation of genetic variation at the molecular level. Molecular marker profiles can be used to follow the effects of selection and genetic drift (which take place over breeding cycles), leading to more accurate estimates of the relationships among genotypes. Among the different marker systems currently available, SSRs are most commonly used for genetic diversity analysis. However, new platforms based on DArT and SNP markers have the potential for simultaneous screening of whole genome haplotypes and will make detailed analysis of genetic diversity relatively straightforward and cost effective (Jaccoud et al. 2001; Rafalski 2002).

A popular opinion is that the intensive selection practiced by modern plant breeders over the last decades has dramatically reduced the genetic diversity among cultivars, narrowing the germplasm base and limiting future advances from breeding (Tanksley and McCouch 1997). Extensive cultivation of germplasm with a narrow genetic base creates a significant genetic vulnerability risk because mutations in disease or insect populations or changes in environmental conditions may result in drastic crop losses. This risk has been highlighted by the outbreak of a new virulent strain of stem rust resistance (*Puccinia graminis,* Ug99) in southwest Uganda (http://www.globalrust.org/).

Characterization of CIMMYT bread wheat breeding lines from 1950–2003 showed a significant decrease of genetic diversity in the improved CIMMYT lines of the 1980s. However, this was followed by an increase in genetic diversity in lines from the 1990s through to 2003, largely due to substantial increases in the use of landraces and synthetic derivatives in breeding nurseries during this period (Reif et al. 2005; Warburton et al. 2006). CIMMYT breeders have been using landraces and synthetic derivatives as new sources of resistance to diseases and tolerance of abiotic stresses. This trait-driven approach has clearly also had positive effects on the overall levels of genetic diversity in breeding material without causing detrimental effects on progress in yield improvement. However, other molecular marker studies analyzing individual regional breeding programs over time have provided conflicting conclusions on the effect of selection on overall genetic diversity (Donini et al. 2000; Christiansen et al. 2002; Roussel et al. 2004, 2005; Khan et al. 2005; Fu et al. 2005, 2006). This is likely to be due to differences in size and structure of the populations studied, differences in the type of marker and statistical analysis applied, and differences in breeding strategies and goals. Nevertheless, maintaining a high level of genetic diversity in CIMMYT's global breeding programs is considered important to ensure good progress in the adaptive breeding by end-user national and regional programs while minimizing the chance of homogeneity effects across large wheat breeding areas creating unacceptable levels of risk of large scale disease epidemics. Thus, for CIMMYT wheat breeding programs, the emphasis on introduction of novel sources of variation for important agronomic traits has an important spillover on overall genetic diversity which should be of benefit to most other breeding programs and target cropping systems.

Significant screening of old and unimproved germplasm as well as materials from the primary and secondary gene pools maintained at gene banks has also been conducted at the molecular level. Examining wheat landraces has revealed high levels of genetic diversity and major genetic differences between landraces and improved materials demonstrating that selective pressure from evolution and modern plant breeding has formed two independent gene pools (Hao et al. 2006, Reif et al. 2005; Dreisigacker et al. 2005; Zhang et al. 2005b, 2006). The characterization of species from the primary and secondary gene pools allows the discovery of additional genetic variability. The level of variation available within the species of the bread wheat progenitors such as *T. dicoccum* and *T. tauschii* etc. has been shown to be extensive and considerably higher than in the AB and D genome of wheat, respectively (Lage et al. 2003; Li et al. 2003). Results are generally closely related to the eco-geographical origin of the examined accessions, indicating that genetic diversity is highly correlated to geographic distribution. This may mean that geographical information systems (GIS) data could be sufficient for coarse level stratification of wheat genetic resources within some species. However, for some species such as *T. dicoccoides,* where there is a substantial amount of variation within populations, this approach is less likely to be effective. Novel alleles observed in germplasm collections can be introduced into cultivated wheat via marker-assisted intergeneric hybridization followed by introgression or by genetic transformation (Rajaram and van Ginkel et al. 2001).

22.3.4 Association Genetics

Association analyses in plants detect quantitative trait loci (QTL) based on the strength of the correlation between variation in a trait phenotype and a marker genotype (Zondervan and Cardon 2004). Association mapping offers greater precision in determining QTL location than family-based linkage analysis and should lead to more efficient marker-assisted selection tools and gene discovery programs. Association analysis also promises to help connect sequence diversity with heritable phenotypic differences. Unlike family-based linkage analysis, association analyses do not require family or pedigree information and can be applied to a range of experimental and non-experimental populations (Kraakman et al. 2004). Collections of homozygous wheat cultivars are particularly suitable for association analyses as multiple tests over years and environments can be used to generate high quality phenotype data for a wide range of traits (Morgante and Salamini 2003).

Various methods of association analyses have been developed (reviewed by Mackay and Powell, 2007). For association analyses to be possible, LD must be present in the population under study. LD can simply be defined as the "non-random association of alleles at different loci". It is the correlation between genetic polymorphisms (detected by SSRs or SNPs, etc.) that are the consequence of a shared history of mutation and recombination. In addition, population structure including several factors such as genetic drift, selection, and admixture can also cause LD between markers and traits (Flint-Garcia et al. 2003). Thus, association analyses must take care to remove these circumstantial correlations that cause false positive results.

Knowledge of the extent of LD in plants is limited (Flint-Garcia et al. 2003). LD in the out-crossing species maize decays within a few hundred base pairs in diverse samples (Tenallion et al. 2001), though the extent of LD increases when narrower selections of germplasm or products of artificial selection are analyzed (Jung et al. 2004; Remington et al. 2001). In self-pollinating species, such as wheat, levels of long-range LD are expected because the rate of effective recombination is reduced by the breeding system. A recent genome-wide study in Arabidopsis has shown that LD at most loci decays within 250 kb (Nordborg et al. 2002). High LD at distances up to 10 cM was found among AFLP loci in barley cultivars (Kraakman et al. 2004). In wheat, LD should be equally extensive, as it is predominantly self-pollinating and has undergone severe bottlenecks in its evolution and strong selection pressures throughout its breeding history. Within subgroups of 134 durum wheat accessions characterized with 70 SSRs, high levels of LD were reported for tightly to moderately linked locus pairs ($<$20 cM), but LD levels were greatly reduced for loosely linked (more than 50 cM) and independent locus pairs (Maccaferri et al. 2005). In a population of 149 soft winter wheat cultivars, Breseghello and Sorrells (2005) determined LD in chromosome 2D and part of 5A with 62 SSRs. Consistent LD on chromosome 2D was $<$ 1 cM, whereas in the centromeric region of 5A, LD extended for $\sim$5 cM. In the same study significant associations between kernel traits and SSR markers were found in agreement with previous QTL studies and alleles potentially useful for selection were identified.

Large-scale EST sequencing projects (section 22.4.1) allow direct analyses of DNA sequence polymorphisms and the identification of haplotypes representing several linked SNPs (Caldwell et al. 2004; Gu et al. 2004; Giles et al. 2006). Furthermore, detection of SNP polymorphisms resulting in a dramatic change of phenotype can be crucial if new alleles are to be rapidly and easily identified (Ravel et al. 2006). The development of high-throughput SNP and DArT genotyping platforms will allow cost effective genome-wide association analysis, thereby enabling more efficient allele mining.

22.3.5 Genetic Characterization of Traits

Dense linkage maps with markers well distributed across the genome and associated information on sequence variation are invaluable resources for determining the expression of large numbers of genes in synteny mapping and gene characterization. Characterization of a range of simply inherited qualitative traits as well as dissection of complex traits in to Mendelian components have been reported for a range of traits such as yield, vernalization, photoperiod response, tolerance to abiotic stresses, maturity, and agronomic parameters associated with quality (see Hoisington et al. 2002 for a review). However, the precision of field phenotype data and the size and appropriateness of mapping populations, continue to be the most rate limiting factors for successful marker identification and subsequent applications in wheat breeding. Bulked segregant analysis (BSA: Michelmore et al. 1991), using pools of the extreme genotypes from the phenotypic distribution of the target trait, has also been used in wheat to characterize simply inherited traits (Eastwood et al. 1994) and to identity quantitatively inherited genes of large effect (William et al. 2003; Shen et al. 2003). The success of this approach, although considerably less expensive compared to linkage map construction, is highly dependent upon the quality of the phenotype data. One disadvantages of this approach is a reduced probability of identifying markers for QTLs of small effect. However, these are increasingly seen as of minimal importance for subsequent practical application in molecular breeding, as it is currently difficult to devise efficient breeding systems for pyramiding large numbers of small effect QTLs for an individual trait. Markers identified through BSA must still be mapped to establish their genomic location. Nevertheless, BSA offers a rapid and cost effective process for identifying a small number of the most important markers which can then be screened across the entire population for precision mapping. Public databases such as Graingenes (http://wheat.pw.usda.gov/cgi-bin/graingenes) provide frequently updated information on mapped traits in wheat (see Table 22.1 for a current overview).

In addition to use of traditional marker-based approaches in genetic characterization of traits of importance, comparative genomics tools enable researchers to make cross-genome comparisons of structure and function at the molecular level among different species. The information derived from these studies makes it possible to transfer genetic information from model species, where a wealth of genomic information is available, to other species which are more complex at the molecular level and have less genomic characterization (Gale and Devos 1998; Feuillet and Keller 1999; Freeling 2001). Successful application of comparative genomics can facilitate the identification and characterization of genes conditioning target traits in the species of interest. For example, rice with an extensively studied small genome is the model species for cereal crops. Although extensive macrosynteny has been observed between rice and wheat, there are numerous discontinuities in microsynteny due to evolutionary events. This often complicates the transfer of information between species (Sorrels et al. 2003). Thus, for complex agronomic traits, comparative genomics may not identify all the important loci in the target species. Nevertheless, synteny mapping involving species such as rice, barley, and *Triticum monococcum,* and map-based cloning, have been used successfully to clone wheat *Vrn-A1* gene (Kato et al. 1999; Yan et al. 2003). Similarly, synteny mapping involving Arabidopsis, rice, maize, and wheat has enabled the successful isolation of important alleles of major dwarfing genes *Rht-B1b* and *Rht-D1b* (Peng et al. 1999); perfect markers were subsequently developed for the *Rht* genes by Ellis et al. (2002). Another successful application of synteny mapping was the identification of the wheat grain protein locus *Gpc-6B1* on chromosome 6B, which was found to be highly co-linear with a 350 kb region on rice chromosome 2; candidate genes identified in rice were used to saturate the wheat linkage group *Gpc-6B1*. These efforts led to the development of a codominant PCR marker for this trait (Distelfeld et al. 2006). Other recent examples of positional cloning based on comparative

Table 22.1 Markers reported to be associated with genes in wheat (updated from Hoisigton et al. 1998)

Trait	Locus	Source	Marker type	Chr.	Reference
Fungal Disease Resistance					
Leaf rust	*Lr1*	*T. aestivum*	RFLP/STS	5DL	Feuillet et al. 1995
	Lr9	*Ae. umbellulata*	RAPD/STS RFLP	6BL	Schachermayr et al. 1994; Autrique et al. 1995
	Lr10	*T. aestivum*	RFLP/STS	1AS	Schachermayr et al. 1997
	Lr13	*T. aestivum*	RFLP	2BS	Seyfarth et al. 1998
	Lr19	*Ag. Elongatum*	STS	7DL	Prins et al. 2001
	Lr20	*T. aestivum*	RFLP	7AL	Neu et al. 2002.
	Lr21	*T. tauschii*	RFLP	1DS	Huang and Gill 2001
	Lr23	*T. turgidum*	RFLP	2BS	Nelson et al. 1997
	Lr24	*Ag. elongatum*	RFLP RAPD/STS RAPD/SCAR	3DL	Autrique et al. 1995; Schachermayr et al. 1995; Dedryver et al. 1996
	Lr25	*S. cereale*	RAPD	4BL	Procunier et al. 1995
	Lr27	*T. aestivum*	RFLP	3BS	Nelson et al. 1997
	Lr29	*Ag. elongatum*	RAPD	7DS	Procunier et al. 1995
	Lr31	*T. aestivum*	RFLP	4BL	Nelson et al. 1997
	Lr32	*T. tauschii*	RFLP	3DS	Autrique et al. 1995
	Lr35	*Ae. Speltoides*	SCAR	2B	Gold et al. 1999
	Lr37	*Ae. Ventricosa*	STS/CAPS	2A	Helguera et al. 2003
	Lr39	*T. Tauschii*	SSR	2DS	Raupp et al. 2001
	Lr47	*T.speltoides*	CAPS	7A	Helguera et al. 2000
	Lr50	*T. timopheevii*	SSR		Brown-Guedira et al. 2003
	Lr51	*T. speltoides*	STS		Helguera et al. 2005
Stem rust	*Sr2*	*T. turgidum*	STS	3BS	Hayden et al. 2004
	Sr22	*T. monococcum*	RFLP	7AL	Paull et al. 1995
	Sr24	*Ag. elongatum*	STS	3DL	Mago et al. 2005
	Sr26	*Ag. elongatum*	STS	6A	Mago et al. 2005
	Sr38	*Ae. Ventricosa*	STS/CAPS	2A	Helguera et al. 2003
	Sr39	*Ae. speltoides*	STS	2B	http://maswheat.ucdavis.edu
	Sr R	*Secale cereale*	STS	1B/1D	Mago et al. 2002
Stripe rust	*Yr5*	*T. spelta*	STS	2BL	Yan et al. 2003; Chen et al. 2003
	Yr10	*T. aestivum*	SSR	1BS	Wang et al. 2002
	Yr15	*T. dicoccoides*	SSR	1B	Peng et al. 2000
	Yr17	*Ae. Ventricosa*	STS/CAPS	2A	Helguera et al. 2003
	Yr26	*H. Villosa*	SSR	6A	Ma et al. 2001
	Yr28	*T. aestivum*	RFLP	4DS	Sing et al. 2000
	YrH52	*T. dicoccoides*	SSR	1B	Peng et al. 2000
Powdery mildew	*Pm1*		RFLP	7AS	Ma et al. 1994
	Pm2		RFLP	5D	Ma et al. 1994
	Pm3		RFLP	1A	Ma et al. 1994,
	Pm4a		RAPD		Li et al. 1995
	Pm4b		AFLP		Hartl et al. 1998

Table 22.1 (continued)

Trait	Locus	Source	Marker type	Chr.	Reference
	Pm12	*Ae. speltoides*	RFLP	6B/6S	Jia et al. 1994
	Pm13	*Ae. longissima*	STS	3S	Cenci et al. 1999
	Pm18		RFLP	7AL	Hartl et al. 1995
	Pm21	*Haynaldia villosa*	SCAR	6VS, 6AL	Liu et al. 1999
	Pm25	*T. monococcum*	RAPD	1A	Shi et al. 1998
	Pm26	*T. turgidum*	RFLP	2BS	Rong et al. 2000
	H9		RAPD		Dweikat et al. 1994
	H21	*Secale cereale*	RAPD	2RL	Seo et al. 1997
	H23, H24	*T. tauschii*	RFLP	6D, 3DL	Ma et al. 1993
	H25	*Secale cereale*	SSR	4A	http://maswheat.ucdavis.edu/protocols/
	H31	*T. turgidum*	STS	5B	http://maswheat.ucdavis.edu/protocols/
Pest Resistance					
Russian	*Dn2*		SSR		Miller et al. 2001
Wheat	*Dn4*		SSR		Liu et al. 2002
Aphid	*Dn6*		SSR		Liu et al. 2002
Quality traits					
Kernel hardness	*Ha*	*T.aestivum*	STS	5B/5D	Giroux and Morris 1997
High protein	Gpc-B1	*T. dicoccoides*	ASA	6B	Distelfeld et al. 2006
LMW glutenins		*T. turgidum*		1B	D'Ovidio and Porceddu 1996
HMW glutenins	*Glu -D1 -1*	*T. aestivum*	ASA	1DL	D'Ovidio and Anderson 1994
Other Traits					
Heterodera avenae reistance	*Cre1*	*T. aestivum*	STS	2BL	Ogbonnaya et al. 2001
	Cre3	*T.tauschii*	STS	2DL	Ogbonnaya et al. 2001
Stature	*Rht-B1b*	*T. aestivum*	STS	4B	Ellis et al. 2002
	Rht-D1b	*T. aestivum*	STS	4D	Ellis et al. 2002
	Rht8	*T. aestivum*	SSR	2B	Korzun et al. 1998
Virus	*Bdv2*	*Ag. intermedium*	STS	7DL	Stoutjesdijk et al., 2001
Cadmium uptake		*T. turgidum*	RAPD		Penner et al. 1995
Meiotic pairing	*ph1b* deletion		STS	5BL	Qu et al. 1998
Vernalization	*Vrn-A1*	*T. aestivum*	STS	5A	Sherman et al. 2004

genomics include the identification of candidate genes associated with a QTL for Fusarium Head Blight resistance (Shen et al. 2006) and with a locus conferring sensitivity to Tan Spot toxin (Lu et al. 2006). Several web-based genomic resources that can be used in comparative genetics and synteny mapping are also available (e.g. http://www.gramene.org).

22.4 Functional Genomics

22.4.1 EST Development

EST development in wheat and other members of the Triticeae was lagging well behind many other plant species during the 1990s. Consequently, the global wheat research community, through the International Triticeae Mapping Initiative (ITMI), launched a collaborative effort to improve genomic resources for wheat, barley, rye, and wild relatives. As a first stage the International Triticeae EST Cooperative was established (http://wheat.pw.usda.gov/genome/). This group encourages laboratories each to contribute 1,000 or more ESTs; over 25,000 ESTs were accumulated within the first six months, and now wheat has the largest public EST database (over 850,000) of any plant species. Table 22.2 provides an overview of the number of ESTs publicly available for various members of the Triticeae at the time of writing. Key to the utilization of these EST resources is the availability of suitable database structures that facilitate the retrieval of relevant EST and related information. GrainGenes (http://wheat.pw.usda.gov/GG2/index.shtml) has been the most widely used database for wheat and barley genetic and genomic information for many years (Matthews et al. 2003) and it continues to provide access to mapped and annotated ESTs. A comprehensive wheat EST database with annotations can be downloaded from http://harvest.ucr.edu/. More specific databases were assembled to support the development of the Affymetrix wheat gene chip and to provide information on EST assemblies. BarleyBase (http://www.barleybase.org/) has been one of the most important of these (Shen et al. 2005). There are also databases that link the requirements of crop scientists with EST resources and provide some valuable tools for wheat researchers such as CR-EST (Kunne et al. 2005). These extensive wheat EST resources have proven highly valuable in analyzing the expression of wheat genes and provide a tool for rapid gene expression profiling. A clear description of this application was recently provided by Mochida et al. (2006) based on ESTs derived from a set of 21 cDNA libraries. A more extensive, but less well structured set of libraries, was used by Chao et al. (2006) to provide an expression profiling resource. More specifically, Ciaffi et al. (2005) used EST resources to study spikelet development to identify possible candidates for more detailed analysis, while Ogihara et al.

Table 22.2 Number of Triticeae EST available in the public databases (1st Sept 2006)

Species	Number of ESTs
Triticum aestivum (wheat)	854,015
Hordeum vulgare subsp. vulgare (barley)	437,321
Hordeum vulgare subsp. spontaneum	24,150
Triticum monococcum	11,190
Secale cereale	9,195
Triticum turgidum subsp. durum	8,924
Aegilops speltoides	4,315
Triticum turgidum	1,938

(2003) used the expression profiles to group functional genes. Expression analysis in polyploid wheat has led to some surprising results. The analysis of expression patterns of homoeologous genes in wheat, based around the use of ESTs generated from diverse tissues, has confirmed that homoeologous genes can be expressed in just one genome and silent in one or both of the remaining genomes (Mochida et al. 2004). Further, the tissue specificity of homoeologous genes was also found to vary. It was particularly surprising to find that 72% of the homoeoloci studied showed genome-specific expression.

The EST collections are also proving valuable resources in supporting positional cloning projects in wheat. The large scale mapping of wheat ESTs carried out through a large NSF-funded project in the USA has provided a resource that is being used by wheat researchers around the world (http://wheat.pw.usda.gov/NSF/). The USA study used 7,104 EST in Southern hybridizations against wheat aneuploid stocks and a deletion line series to assign ESTs to specific chromosome bins. Each EST detected an average of 4.8 restriction fragments and 2.8 loci. The resultant map placed over 16,000 loci into their respective chromosome bins (Qi et al. 2004). The bin maps not only place a large number of genes onto the wheat physical and genetic maps but also provide a means for comparative genomics across the cereals. Resources such as these are valuable tools in comparative studies (for example, Hattori et al. 2005).

The large size of the wheat EST collections has provided opportunities for the development of several important resources. A clear application has been the development of microarray platforms. One of the earliest was a cDNA-based array (Wilson et al. 2004). However, oligo arrays have also been produced. The most widely used is the Affymetrix wheat Genechip (http://www.affymetrix.com/products/arrays/specific/wheat.affx) which represents over 55,000 transcripts. It is anticipated that transcript profiling datasets based on this array and other systems will be publicly available for wheat researchers in the near future, similar to those already available for barley (http://www.barleybase.org/). A reference dataset for wheat based on the Affymetrix GeneChip is currently under development and is likely to be released soon. This dataset will match a tissue series already developed for barley (Druka et al. 2006).

The wheat EST databases have also been used to develop SSR and SNP markers (reviewed by Varshney et al. 2005). There are several reports describing the development and mapping of such markers and comparing them to SSRs derived from other techniques (for example, Gadaleta et al. 2006; Yu et al. 2004). The collection of EST-derived SSRs is now extensive and they have proven useful in linking wheat genetic maps to maps from other cereals based on orthology to the genes from which the SSRs were derived (Tang et al. 2006; Zhang et al. 2005a). The EST-derived SSRs appear to be more readily transferable between species than previously developed SSRs, although the number of alleles detected and the level of variation tends to be lower. Nevertheless, EST-derived SSRs have proved useful for diversity studies (Zhang et al. 2006) and are suitable for determining variation and mapping in the wild relatives of wheat (Mullen et al. 2005). The development of SNPs from EST resources has been slower than EST-SSR discovery.

However, a large-scale effort is underway through an NSF-funded project in the USA (http://wheat.pw.usda.gov/NSF/). A database of primers, SNPs, and the status of the program can be found at http://rye.pw.usda.gov/snpworld/Search).

22.4.2 TILLING

Mutagenesis has been widely used in crop improvement since the 1950s and many modern cultivars carry mutations induced by chemical mutagenesis or ionizing radiation. The recent discovery of enzymes capable of cleaving single base mismatches provides a tool for high throughput screening of single base differences in mutant populations and allows mutant alleles to be found in a target gene. The technique, referred to as targeting induced local lesions in genomes (TILLING), has revitalized mutation research as it provides a method to knock out genes and allows the generation of variation without the need for transformation, greatly simplifying the regulatory process. The method and background has been recently reviewed by Slade and Knauf (2005) and by Comai and Henikoff (2006).

Concerns have been expressed regarding the utility of this technique in wheat since it was felt that polyploidy would hide mutations and complicate both the screening and the phenotypic assessment of mutant lines. However, the technique has proved highly successful in wheat (Slade et al. 2005; Weil 2005). Polyploidy appears to allow wheat to tolerate a far higher mutation load than diploid crops and this reduces the number of mutant families that must be screened. Therefore, Slade et al. (2005) were able to recover 246 alleles in the waxy genes from a screen of only 1,920 mutagenised lines. Given that wheat has only two functional waxy genes (granule-bound starch synthase I) this represents a surprisingly high success rate. Several groups around the world are now developing mutant or TILLING populations for bread and durum wheat, and this is likely to become a widely used technique in functional analysis of candidate genes.

22.4.3 Transformation as a Tool in Genomics

The success of genetic transformation depends on the proper introduction and insertion of the target gene into the nuclear genome and ensuring its expression in a heritable manner (Shewry and Jones 2005; Jones 2005). Usually, soft explant tissue derived from immature embryos is used as the source material for wheat transformation. Micro-projectile bombardment (Sparks and Jones 2004) has been extensively used in the past as the means of delivery of gene constructs. However, Agrobacterium-mediated transformation systems are preferred as they enable the delivery of single copy insertions (Wan and Layton 2006; Wu et al. 2006) and are subject to a lower frequency of transgene silencing (Hu et al. 2003). Alternative transformation methods are being investigated in an attempt to circumvent the tight intellectual property controls associated with biolistic and Agrobac-

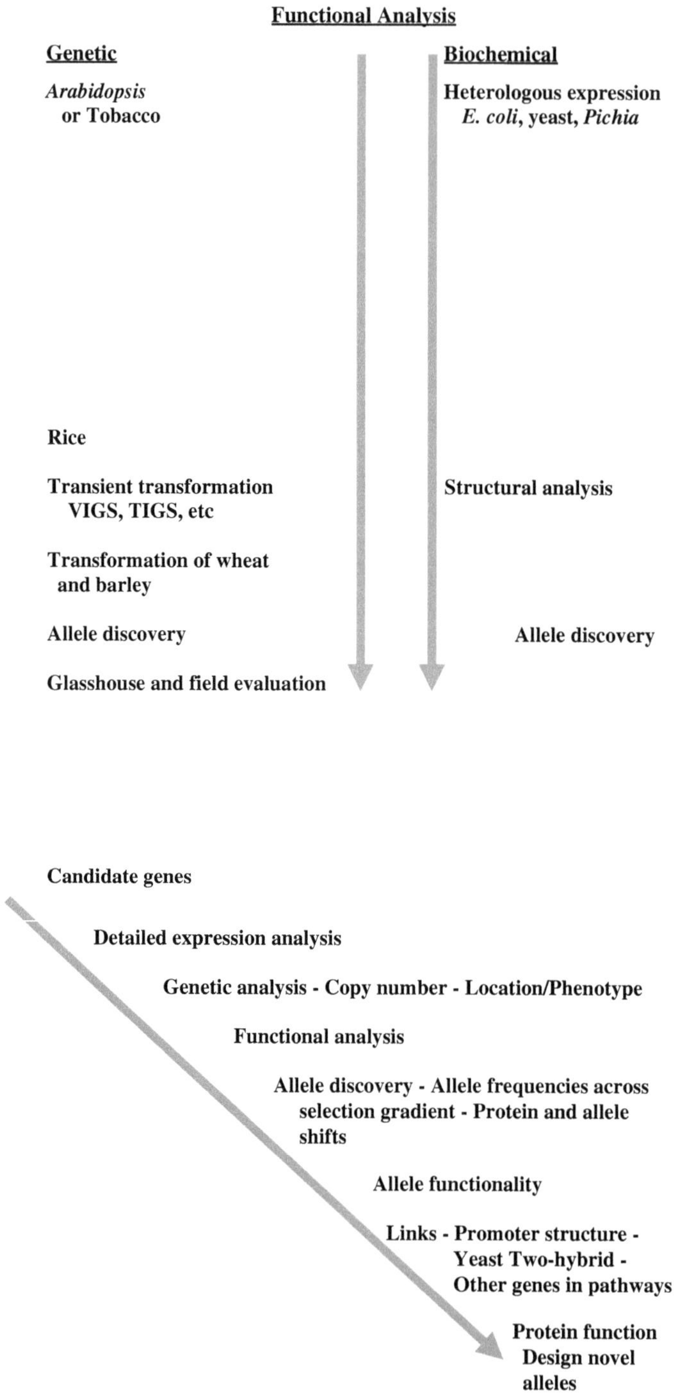

Fig. 22.2 An outline of various processes involved in functional analysis of genes and alleles

terium transformation protocols (for example, Badr et al. 2005). A summary of the selectable markers that are suitable for use in wheat transformation can be found in Goodwin et al. (2005).

Transformation has become an important tool in functional analysis either through ectopic expression of the transgenes or through gene silencing or reduced expression. There are now many examples in the literature where transformation has been used for functional analysis. One recent example was the characterization of the "Q" gene from wheat (Simons et al. 2006). This gene is responsible for the free-threshing character that was crucial for wheat domestication. Ectopic expression allowed both silenced and over-expressed phenotypes to be observed and was important in confirming the identity of the cloned gene.

Transient transformation has also been useful in functional analysis; an illustration is given by Srichumpa et al. (2005). In this case transient single cell transformation was used to study several alleles of the powdery mildew resistance locus, *Pm3*. A related technique has been the use of virus-induced gene silencing (VIGS) to specifically knock down the expression of candidate genes. In wheat this method uses barley stripe mosaic virus (BSMV) (Holzberg et al. 2002), but it has not yet been widely used for functional analysis. Nevertheless, Scofield et al. (2005) were able to use VIGS for functional analysis of the wheat leaf rust resistance gene, *Lr21*. An outline of various processes involved in functional analysis of genes and alleles is given in Fig. 22.2.

22.5 Applications of Genomics in Wheat Breeding

22.5.1 Developing an Effective Integrated Marker-Assisted Selection System

Once a marker is identified through linkage or association mapping analysis, its utility as an indirect selection tool must be validated in appropriate breeding populations. Validation failures can be due to an absence of polymorphisms at the locus in the target germplasm, different recombination patterns in the target germplasm causing loss of linkage between the marker and the target locus, or confounding effects of new epistatic interactions between the marked locus and the genetic background of the target germplasm. The practical value of a marker depends on how successfully it can be integrated into a breeding program and how easily it can be applied on a large scale in modern breeding programs. Marker systems such as RFLPs or AFLPs do not meet these criteria due to the laborious nature of their application. Thus, molecular breeding programs should focus on PCR-based assay systems such as STS, SSR, and SNP markers.

Simply inherited disease resistance is a common target for marker-assisted selection (MAS), particularly where breeding programs do not have ready access to disease hot spots and where there is a need to pyramid resistance genes. In wheat, there are a number of inter-chromosomal translocations from related species that carry useful genes for which markers are available; these markers allow the translocated segment containing the target gene to be easily introduced into elite lines (http://maswheat.ucdavis.edu/; McIntosh et al. 2003). However, molecular dissection of loci that contribute to complex traits such as yield and abiotic stess tolerance remains a considerable challenge, even with the newly available marker technologies. MAS applications for complex traits are limited because of the rarity of QTLs of large effect with good stability across cropping environments and diverse genetic backgrounds.

Public wheat breeding programs generally use pedigree breeding methods or modifications thereof. Individual plants are selected in early generations to increase the frequency of simply inherited alleles for the target traits such as resistance to diseases, plant height and stature, agronomic type, etc. At more advanced generations, when a sufficient level of homozygosity is present, selections are then made among families for quantitatively inherited traits such as yield, drought, heat and salinity. The exact profile of target traits will be specific to the target region. For example, many wheat breeding programs in Australia use markers for race specific leaf and stripe rust resistance genes, where these genes are still effective. In contrast, CIMMYT's wheat improvement strategy is based on durable or race non-specific genes. Race-specific genes are avoided because the rust pathogen overcomes these resistances with time. However, race-specific genes are more effective if deployed in combination and breeding programs in many regions have attempted to do this. An example is the effort to pyramid race-specific resistance genes to counter a recent outbreak of a new strain of stem rust in eastern Africa (http://globalrust.org).

A number of wheat breeding programs have begun to use marker-assisted selection on a modest scale. Breeding programs have to develop pragmatic approaches to integrating MAS. The breeding strategies used will be dependent on breeding objectives, resource availability, and information from genetic characterization of different traits. For example, MAS may not be justified for simply inherited traits that can be reliably screened under field conditions such as disease resistance, unless there are extenuating circumstances. Thus, it may not be possible to reliably pyramid disease resistance genes without the use of markers. Similarly, it may be important to carry out MAS to retain disease resistance loci during single seed descent programs. Markers are also valuable for characterizing potential parental genotypes in order to assist in designing crosses.

Marker-assisted breeding strategies can also be designed to rapidly and efficiently generate fixed lines for a target gene or combination of genes. Considering the relatively high cost of DNA extraction and subsequent marker assays, it is important to identify the optimum points for MAS interventions in the breeding process to increase the efficiency and effectiveness of the breeding program. Coordinated programs for MAS have been established in various countries:, the National Wheat Molecular Marker Program (NWMMP) in Australia, established in 1996 (Eagles et al., 2001); the national wheat MAS consortium in the USA, established in 2001 (Dubcovsky 2004); similar initiatives in Canada (R. DePauw and C. Pozniak, pers. comm.); and cooperatives established among breeding companies in Europe (Koebner and Summers 2003). Target traits for MAS include a range of disease and pest resistances and quality traits.

Although genetically modified (GM) forms of wheat are not currently in commercial use, ongoing gene discovery projects will likely find candidate genes with potential for future transformation programs. Some cultivars are clearly more receptive to transformation than others. When the cultivar with the best agronomic type is not the most receptive to transformation, it is possible to transform a more receptive cultivar (Pellegrineschi et al. 2002) and then introgress the gene into the target background using diagnostic markers for the transgene. This type of MAS aided line conversion can be accomplished for any crop species including wheat. Marker-assisted introgression of transgenes into a range of desired backgrounds is commonly practiced in the private sector for crops such as maize.

22.5.2 Marker-Assisted Selection in the CIMMYT Wheat Breeding Program

An important feature of CIMMYTs marker implementation program is the systemic integration of molecular genotyping with field-based screening. Currently, reliable markers are used for a limited number of traits. These traits are relevant to the breeding program's goals and therefore justify the investment in MAS. To keep the number of assays manageable within the breeding program, markers are used once in the early generations to favorably skew allele frequency and again on the advanced progeny to confirm the presence of the target alleles in the genetically fixed material. When two or more genes are targeted using markers, the segregating progeny are usually screened using MAS at the F_1 top-cross or F_2 generations. Tissue sampling is delayed as long as possible in the field to allow the breeder to first select for disease reaction and agronomic type; materials are then screened for presence/absence of the target alleles using markers. This strategy reduces the time available to run large numbers of marker assays as tissue sampling occurs later in the growth cycle and the breeder requires the gene profiles before harvest; an alternative strategy is to sample plants in the seedling stage, this extends the time available to provide the marker data but results in the screening of many plants with unsuitable background genes and agronomic type. Only fixed lines positive for the target markers are advanced to expensive replicated multilocational yield and quality evaluation (William et al. 2007). The extent of MAS investment at CIMMYT is determined by the importance of the target trait to the breeding program, the reliability of alternative phenotypic screens, and the additional selective power provided by the assays.

22.6 Key Challenges for Molecular Breeding of Wheat

The improvements in wheat yields attributable to the Green Revolution were achieved by radically changing the crop architecture to maximize yield under high-input conditions. It is unlikely that a similar modification for any other single trait will lead to such dramatic increases in yield again. Thus, it is expected that continued progress in productivity will come through incremental improvements. Yield potential and crop adaptation are constrained by a number of factors including: available genetic variability for yield enhancing traits; the complexity of inheritance of economically important traits such as yield potential and drought tolerance; climate change and erosion of productivity in many farming systems. The large-scale use of resynthesized wheat lines in the CIMMYT breeding programs has led to dramatic improvements in both yield potential and adaptation to multiple stresses (Trethowan et al. 2005a) and adaptation around the globe (Dreccer et al. 2007; Lage and Trethowan 2007). At CIMMYT, improvements in drought stress adaptation attributable to resynthesized wheat were achieved by improving the heritability of drought screening procedures (Trethowan and Reynolds 2006) and the understanding of the physiological basis of adaptation to drought (Reynolds et al. 2007).

It is likely that the negative effects of climate change on wheat production in countries at lower latitudes such as India and Pakistan will be much greater than in developed countries where production may even increase as lands at high latitude are brought into production (Rosenzweig and Hillel 1995). According to Trethowan et al. (2005b), the area in India currently regarded as close to optimal for wheat production will halve over the next 40 to 50 years as temperatures increase. Wheat breeding can help mitigate some of the effects of climate change, largely by improving adaptation to higher temperatures and increasing drought tolerance and/or water use efficiency.

Many farmers have introduced conservation agriculture (reduced or zero-tillage and crop residue retention) to reduce erosion, improve crop water use, and reduce costs, thereby improving overall profitability and sustainability of farming. These changes have significant implications for wheat breeders. For example, the spectrum of wheat diseases changes with stubble retention, such as diseases like tan spot (*Pyrenophora tritici-repentis*) and crown rot (*Fusarium pseudograminearum*), become more prevalent (Duveiller and Dubin 2002; Mezzalama et al. 2001). In addition, evidence also exists of a cultivar x tillage practice interaction for both yield and quality (Gutierrez 2006). Although characters such as coleoptile length do explain some of the variation in crop emergence and establishment in these systems (Trethowan et al. 2005a), most of the variation remains unexplained. Clearly, the key traits required for good performance in resource conservation systems must be identified if cultivars are to be bred that are better adapted to such farming systems.

22.6.1 Future Prospects for Wheat Molecular Breeding

If the rates of advance in wheat yields are to be maintained or even increased, our understanding and ability to manipulate the underlying genetic control of complex characters such as yield and abiotic stresses must be improved. The search for QTLs influencing yield and stress tolerance has been confounded by poor quality phenotypic data, the inappropriate nature and size of mapping populations, or the inadequate density of molecular markers. Genotype x year interactions are frequently the single largest source of variation in the analysis of multi-environment trials. Therefore, it is not surprising that many QTLs are not consistent across seasons, locations, or populations.

Traditional QTL mapping using genetic populations generated by crossing two genotypes contrasting for a trait of interest has been useful in establishing the putative genomic location of the genetic factors contributing to the trait and for partitioning the variation in to single Mendelian genetic factors. However, this mapping approach is slow and expensive and can elucidate only the relative effects of the two alleles contributed by the two parental genotypes. Moreover, the resultant markers are often population dependent, thus suffering a substantial level of redundancy when validated in breeding populations. Association analysis has the potential to overcome these problems and improve the cost efficiency and speed of marker identification for certain important agronomic traits. Although linkage mapping in biparental populations is likely to remain important for some traits and where fine mapping is required.

The existence of phenotypically well characterized breeding populations combined with new cost effective genome-wide scan technologies (such as DArT) and association analysis approaches offers powerful new opportunities. For example, advanced CIMMYT wheat breeding lines have been distributed annually to around 100 global locations for the past half century. Yield and agronomic data have been collected from these trials and returned to CIMMYT for analysis and collation in public access databases. Seed of all these materials was kept in the CIMMYT gene bank and is now being used for genotyping and pilot testing of association analysis using breeding material (Crossa et al. 2007). It is hoped that this approach will identify genomic regions with a putative influence on yield potential and other complex agronomic traits.

The large-scale use of markers in wheat breeding is still limited due to a lack of markers for high value traits and the absence of low cost high throughput analytical platforms appropriate to the needs of wheat molecular breeding. Marker detection through capillary electrophoresis offers significant incremental advances in throughput and unit costs, but dramatic progress will have to await appropriate SNP-based systems. Large-scale EST sequencing projects will undoubtedly lead to the generation a large number of SNP gene-based markers. SNP markers developed in this way will then provide an important source of candidate gene-based markers for molecular breeding and allele mining. There are a number of potential high throughput platforms for large-scale low cost simultaneous genotyping of less than one hundred SNP markers, which may be appropriate for the next generation of wheat molecular breeding applications scenarios: (i) Luminex (http://www.appliedcytometry.com/starsupport/docs/STarBase.pdf) which currently offers simultaneous detection of up to around 50 SNP polymorphisms per DNA sample based on bead hybridization and detection coupled with flow cytometry; (ii) SNPWave (http://www.keygene.com/techs-apps/technologies_snpwave.htm) which is based on highly multiplexed allele discrimination using capillary electrophoresis and may allow selective simultaneous detection of nearly 100 SNPs; (iii) TaqMan (http://www. appliedbiosystems.com) which is based on allele discrimination using RT-PCR technology based on 5' nuclease activity that has been adapted for high throughput applications (Ranade et al. 2001). Recent advances of the technology have enabled deployment of 384-well based platforms; (iv) MassARRAY (www.sequenom.com) technology combines primer extension reaction chemistry with mass spectrometry based on MALDI-TOF for rapid and cost effective characterization of SNP polymorphisms. In the human diagnostics arena, researchers have been able to reduce the average cost of SNP genotyping from US$1 to 10 cents per data point (Roses 2002). Although this is based on intensive investment in optimization of a range of candidate SNP markers, similar advances will ultimately be possible for wheat molecular breeders. The added advantage of SNP-based marker systems is the avoidance of gel-based allele separation for visualization and their potential for automation in high throughput assay platforms. This ongoing research will inevitably lead to the development of more robust, simple and cost effective high throughput assays (Jenkins and Gibson 2002). The challenge is establishing an intimate and iterative collaboration between molecular biologists and wheat breeders such that the results of whole genome scanning and association genetics can be rationalized and deployed in wheat breeding programs. These techniques have the potential to substantially improve parent selection for crossing, the rate of genetic gain, and the time taken to develop new cultivars.

References

Akbari M, Wenzel P, Caig V, Carlig J, Xia L, et al. (2006) Diversity arrays technology (DArT) for high-throughput profiling of the hexaploid wheat genome. Theor Appl Genet 113:1409–1420

Appels R (2003). A consensus molecular genetic map for wheat – a cooperative international effort. In: Pogna NE, Romano M, Pogna EA, Galterio G (eds) Proc 10th Intl Wheat Genetics Symp Pasteum, Italy. Rome: Instituto Sperimentale per la Cerealicoltura. 1:211–214

Autrique E, Singh RP, Tanksley SD, Sorrells ME (1995) Molecular markers for four leaf rust resistance genes introgressed into wheat from wild relatives. Genome 38:75–83

Aung T, Howes NK, McKenzie RIH, Towney-Smith TF (1995) Application of the maize pollen method for wheat doubled haploid (DH) generation in western Canadian spring wheat breeding programs. Ann Wheat Newsl 41:70

Badaeva ED, Amosova AV, Muravenko OV, Samatadze TE, Chikida NN, et al. (2002) Genome differentiation in *Aegilops*, 3. Evolution of the D genome cluster. Plant Syst Evol 231:163–190

Badr YA, Kereim MA, Yehia MA, Fouad OO, Bahieldin A (2005) Production of fertile transgenic wheat plants by laser micropuncture. Photochem Photobiol Sci 4:803–807

Bell MA, Fischer RA, Byerlee D, Sayre K (1995) Genetic and agronomic contributions to yield gains: a case study for wheat. Field Crops Res 44:55–56

Borlaug NE (1968) Wheat breeding and its impact on world food supply. In Proc. 3rd Intl. Wheat Genetics Symp. Australian Academy of Science, Canberra, Australia. pp. 1–36

Blanco A, Bellomo MP, Cenci A, De Giovanni C, D'Ovidio R, et al. (1998) A genetic linkage map of wheat. Theor Appl Genet 97:721–728

Breseghello F, Sorrells ME (2005) Association mapping of kernel size and milling quality in wheat (*Triticum aestivum* L.) cultivars. Genetics 172:1165–1177

Brown-Guedira GL, Singh S, Fritz AK (2003) Performance and mapping of leaf rust resistance transferred to wheat from *Triticum timopheevii* ssp. *armeniacum*. Phytopathology, 93:784–789

Byerlee D, Moya P (1993) Impacts of international wheat breeding research in the developing world, 1966–1990. CIMMYT, Mexico, D.F.

Caldwell KS, Dvorak,J, Lagudah ES, Akhunov E, Luo M-C, et al. (2004) Sequence polymorphism in polyploid wheat and their D-genome ancestor. Genetics 167:941–947
Cenci A, D'Ovidio R, Tanzarella OA, Ceoloni C, Porceddu E (1999) Identification of molecular markers linked to *Pm13*, an *Aegilops longissima* gene conferring resistance to powdery mildew in wheat. Theor Appl Genet 98:448–454
Chao S, Lazo GR, You F, Crossman CC, Hummel DD, et al. (2006) Use of a large-scale *Triticeae* expressed sequence tag resource to reveal gene expression profiles in hexaploid wheat (*Triticum aestivum* L.). Genome 49:531–544
Chen X, Marcelo A, Soria A, Guiping YS, Dubcovsky J (2003) Development of sequence tagged site and cleaved amplified polymorphic sequence markers for wheat stripe rust resistance gene *Yr5*. Crop Sci 43:2058–2064
Christiansen MJ, Andersen SB, Ortiz R (2002) Diversity changes in an intensively bred wheat germplasm during the 20th century. Mol Breed 9:1–11
Ciaffi M, Paolacci AR, D'Aloisio E, Tanzarella OA, Porceddu E (2005) Identification and characterization of gene sequences expressed in wheat spikelets at the heading stage. Gene 346:221–230
Comai L, Henikoff S (2006) TILLING: practical single-nucleotide mutation discovery. Plant J 45:684–94
Cox TS, Hkiang YT, Gorman MB, Rogers DM (1985) Relationships between coefficient of parentage and genetic indices in soybean. Crop Sci 25:529–532
Crossa J, Burgueno J, Dreisigacker S, Vargas M, Herrera-Foessel SA, Lillemo M, Singh RP, Trethowan R, Warburton M, Franco J, Renolds M, Crouch J, Ortiz R (2007) Association analysis of historical bread wheat germplasm using additive genetic covariance of relatives and population structure. Genetics doi:10.1534/genetics.107.078659
Curtis B (2002) Wheat in the world. In: Curtis BC, Rajaram S, Gomez Macpherson H (eds). Bread wheat – Improvement and production. Plant production and protection series No. 30 pp. 1–17
D'Ovidio R, Anderson OD (1994) PCR analysis to distinguish between alleles of a member of a multigene family correlated with wheat bread-making quality. Theor Appl Genet 88:759–763
D'Ovidio R, ad Porceddu E (1996) PCR-based assay for detecting 1B-genes for low molecular weight glutenin subunits related to gluten quality properties in durum wheat. Plant Breeding 115:413–415
Dayteg C, Tuvesson S, Merker A, Jahoor A, Kolodinska-Brantestam A (2007) Automation of DNA marker analysis for molecular breeding in crops: practical experience of a plant breeding company. Plant Breeding 126:410–415
Dedryver F, Jubier M-F, Thouvenin J, Goyeau H (1996) Molecular markers linked to the leaf rust resistance gene *Lr24* in different wheat cultivars. Genome 39:830–835
Distelfeld A, Uauy C, Fahima T, Dubcovsky J (2006) Physical map of the wheat high-grain protein content gene Gpc-B1 and development of a high-throughput molecular marker. New Phytol 169:753–763
Donini P, Law JR, Koebner RM, Reeves JC, Cooke RJ (2000) Temporal trends in the diversity of UK wheat. Theor Appl Genet 100:912–917
Dreccer MF, Borgognone MG, Ogbonnaya FC, Trethowan RM, Winter B (2007) CIMMYT-selected synthetic bread wheats for rainfed environments: yield evaluation in Mexico and Australia. Field Crops Res 100:218–228
Dreisigacker S, Zhang P, Warburton ML, Skovmand B, Hoisington D, et al. (2005) Genetic diversity among and within CIMMYT wheat landrace accessions investigated with SSRs and implications for plant genetic resources management. Crop Sci 45:653–661
Druka A, Muehlbauer V, Druka I, Caldo R, Baumann U, et al. (2006) An atlas of gene expression from seed to seed through barley development; Funct Int Genom 6:202–211
Dubcovsky J (2004) Marker assisted selection in public breeding programs: the wheat experience. Crop Sci 44:1895–1898
Duveiller E, Dubin HJ (2002) Helminthosporium leaf blights: spot blotch and tan spot. In: Curtis BC, Rajaram S, Gomez Macpherson H (eds) 'Bread wheat improvement and production' Plant Production and Protection Series No. 30. FAO. pp. 285–299

Dweikat I, Ohm H, Mackenzie S, Patterson F, Cambron S, et al. (1994) Association of a DNA marker with Hessian fly resistance gene *H9* in wheat. Theor Appl Genet 89:964–968
Eagles HA, Bariana HS, Ogbonnaya FC, Rebetzke GJ, Hollamby GH, et al. (2001) Implementation of markers in Australian wheat breeding. Aust J Agric Re. 52:1349–1356
Eastwood RF, Lagudah ES, Appels R (1994) A directed search for DNA sequences tightly linked to cereal cyst nematode resistance genes in *Triticum tauschii*. Genome 37:311–319
Ellis MH, Spielmeyer W, Gale KR, Rebetzke GJ, Richards RA (2002) "Perfect" markers for the Rht-B1b and Rht-D1b dwarfing genes. Theor Appl Genet 105:1038–1042
Elouafi I, Nachit MM (2004) A genetic linkage map of the Durum x *Triticum dicoccoides* backcross population based on SSRs and AFLP markers, and QTL analysis for milling traits. Theor Appl Genet 108:401–413
Endo TR, Gill BS (1996) The deletion stocks of common wheat. J Hered 87:295–307
Falk DE, Kasha KJ (1983) Genetic studies of the crossability of hexaploid wheat with rye and *Hordeum bulbosum*. Theor Appl Genet 64:303–307
FAO (2006) Food and Agriculture Organization of the United Nations, Global Crop Production Statistics. http://faostat.fao.org/site/336/default.aspx
Feuillet C, Messmer M, Schachermayr G, Keller B (1995) Genetic and physical characterisation of the *Lrl* leaf rust resistance locus in wheat (*Triticum aestivum* L.). Mol Gen Genet 248: 553–562
Feuillet C, Keller B (1999) High gene density is conserved at syntenic loci of small and large grass genomes. Proc Natl Acad Sci USA 96:8265–8270
Flint-Garcia SA, Thornsberry JM, Buckler ES IV (2003) Structure of linkage disequilibrium in plants. Annu Rev Plant Biol 54:357–374
Flintham JE, Borner A, Worland AJ, Gale MW (1997) Optimizing wheat grain yield: effects of Rht (gibberellin-insensitive) dwarfing genes. J Agric Sci Cambridge 128:11–25
Freeling M (2001) Grasses as a single genetic system. Reassesment 2001. Plant Physiol 125:1191–1197
Fu Y-B, Peterson GW, Richards KW, Somers D, DePauw RM, et al. (2005) Allelic reduction and genetic shift in the Canadian hard red spring wheat germplasm released from 1845 to 2004. Theor Appl Genet 110:1505–1516
Fu Y-B, Peterson GW, Yu JK, Gao L, Jia J, et al. (2006) Impact of plant breeding on genetic diversity of the Canadian hard red spring wheat germplasm as revealed by EST-derived SSR markers. Theor Appl Genet 112:1239–1247
Gadaleta A, Mangini G, Mule G, Blanco A (2007) Characterization of dinucleotide and trinucleotide EST-derived microsatellites in the wheat genome. Euphytica 153:73–85
Gale MD, Devos K (1998) Comparative genetics in the grasses. Proc Natl Acad Sci USA 95:1971–1974
Gale MD, Law CN, Worland AJ (1975) The chromosomal location of a major dwarfing gene from Norin 10 in new British semi dwarf wheats. Heredity 35:417–421.
Giles RJ, Brown TA (2006) Glu allele variations in *Aegilops tauschii* and *Triticum aestivum*: implications for the origins of hexaploid wheats. Theor Appl Genet112:1563–1572
Gill BS, Friebe B (2002) Cytogenetics, phylogeny and evolution of cultivated wheats. In: Curtis BC, Rajaram S, Gomez Macpherson H (eds) Bread wheat – Improvement and production. Plant production and protection series No. 30. pp. 71–88
Gill BS, Raupp WJ (1987) Direct gene transfers from *Aegilops squarrosa* L. to hexaploid wheat. Crop Sci 27:445–450
Gill KS, Gill BS, Endo TR, Mukai Y (1993) Fine physical mapping of *Ph1*, a chromosome pairing regulator gene in polyploidy wheat. Genetics 134:1231–1236
Giroux MJ, Morris CF (1997) A glycine to serine change in puroindoline b is associated with wheat grain hardness and low levels of starch surface friabilin. Theor Appl Genet 95:857–864
Gold J, Harder D, Townley-Smith F, Aung T, Procunier J (1999) Development of a molecular marker for rust resistance genes *Sr39* and *Lr35* in wheat breeding lines. Electronic J Biotechnol 2:(1)

Goodwin JL, Pastori GM, Davey MR, Jones HD (2005) Selectable markers - Antibiotic and herbicide resistance. In: Pena L (ed) "Transgenic Plants: Methods and Protocols. Methods in Molecular Biology". pp. 191–201

Graner A, Ludwig WF, Melchinger AE (1994) Relationships among European barley germplasm: II. Comparison of RFLP and pedigree data. Crop Sci 34:1199–1205

Gu YQ, Coleman-Derr D, Kong X, Anderson OD (2004) Rapid genome evolution revealed by comparative sequence analysis of orthologous regions from four *Triticeae* genomes. Plant Physiol 135:459–470

Gutierrez A (2006) Adaptation of bread wheat to different tillage practices and environments in Mexico. PhD Thesis, Chapingo University, Texcoco, Estado de Mexico, Mexico

Hao CY, Zhang XY, Wang LF, Dong YS, Shang XW, et al. (2006) Genetic diversity and core collection evaluations in common wheat germplasm from the northwestern spring wheat region in China. Mol Breed 17:69–77

Hartl L, Weiss S, Stephan U, Zeller FJ, Jahoor A (1995) Molecular identification of powdery mildew resistance genes in common wheat (*Triticum aestivum* L). Theor Appl Genet 90:601–606

Hartl L, Mori S, Schweizer G (1998) Identification of a diagnostic molecular marker for the powdery mildew resistance gene *Pm4b* based on fluorescently labelled AFLPs. Proc 9th Intl Wheat Genet Symp pp 111–113

Hattori J, Ouelle, T, Tinker NA (2005) Wheat EST sequence assembly facilitates comparison of gene contents among plant species and discovery of novel genes. Genome 48:197–206

Hayden MJ, Kuchel H, Chalmers KJ (2004) Sequence tagged microsatellites for the *Xgwm533* locus provide new diagnostic markers to select for the presence of stem rust resistance gene *Sr2* in bread wheat (*Triticum aestivum* L). Theor Appl Gene. 109:1641–1647

Helguera M, Khan IA, Dubcovsky J (2000) Development of PCR markers for wheat leaf rust resistance gene *Lr47*. Theor Appl Genet 101:625–631

Helguera M, Khan IA, Kolmer J, Lijavetzky D, Zhong-qi L, et al. (2003) PCR assays for the*Lr37-Yr17-Sr38* cluster of rust resistance genes and their use to develop isogenic hard red spring wheat lines. Crop Sci 43:1839–1847

Helguera M, Vanzetti L, Soria M, Khan IA, Kolmer J, et al. (2005) PCR Markers for *Triticum speltoides* leaf rust resistance gene *Lr51* and their use to develop isogenic hard red sring wheat lines. Crop Sci 45:728–734

Hoisington D, Bohorova N, Fennell S, Khairallah M, Pellegrineschi A, et al. (2002) The application of biotechnology to wheat improvement: New tools to improve wheat productivity. In: Curtis BC, Rajaram S, Gomez Macpherson H (eds) Bread wheat improvement and production. Plant production and protection series No. 30. pp. 175–198

Holzberg S, Brosio P, Gross C, Pogue GP (2002) Barley stripe mosaic virus-induced gene silencing in a monocot plant. Plant J 30:315–27

Howes NK, Woods SM, Townley-Smith TF (1998) Simulation and practical problems of applying multiple marker-assisted selection and doubled haploids to wheat breeding programs. Euphytica 100:225–230

Hu T, Metz S, Chay C, Zhou HP, Biest N, et al. 2003. Agrobacterium mediated large-scale transformation of wheat (*Triticum aestivum* L.) using glyphosate selection. Plant Cell Rep 21:1010–1019

Huang L, Gill BS (2001) An RGA like marker detects all known *Lr21* leaf rust resistance gene family members in *Aegilops tauschii* and wheat. Theor Appl Genet 103:1007–1013

Jaccoud D, Peng K, Feinstein D, Kilian A (2001) Diversity Arrays: a solid state technology for sequence information independent genotyping. Nucleic Acids Res 29:1–7

Jenkins S, Gibson N (2002) High-throughput SNP genotyping. Comp Funct Genom 3:57–66

Jia J, Devos KM, Chao S, Miller TE, Reader SM, et al. (1994) RFLP-based maps of the homoeologous group-6 chromosomes of wheat and their application in the tagging of *Pm12*, a powdery mildew resistance gene transferred from *Aegilops speltoides* to wheat. Theor Appl Genet 92:559–565

Jiang J, Gill BS (1993) Sequential chromosome banding and in situ hybridization analysis. Genome 36:792–795
Jiang J, Gill BS (1994) Nonisotopic in situ hybridization and plant genome mapping: the first 10 years. Genome 37:717–725
Jiang J, Gill BS (2006) Current status and the future of fluorescence in situ hybridization (FISH) in plant genome research. Genome 49:1057–1068
Jiang J, Friebe B, Gill BS (1994) Recent advances in alien gene transfer in wheat. Euphytica 72:199–212
Jones HD (2005) Wheat transformation: current technology and applications to grain development and composition. J. Cereal Sci 41:137–147
Jung M, Ching A, Bhattramakki D, Dolan M, Tingey S, et al. (2004) Linkage disequilibrium and sequence diversity in a 500-kbp region around the adh1 locus in elite maize germplasm. Theor Appl Genet 109:681–689
Kato K, Miura H, Sawada S (1999) Comparative mapping of the wheat *Vrn-A1* region with the rice Hd-6 region. Genome 42:204–209
Khan IA, Awan FS, Ahmad A, Fu YB, Iqbal A (2005) Genetic diversity of Pakistan wheat germplasm as revealed by RAPD markers. Genet Resour Crop Evol 52:239–244
Kim HS, Ward RW (1997) Genetic diversity in eastern U.S. soft winter wheat (*Triticum aestivum* L. em. Thell.) based on RFLPs and coefficient of parentage. Theor Appl Genet 94:472–479
Koebner RMD, Summers W (2003) 21st century wheat breeding: plot selection or plate detection? Trends Biotechnol 21:59–63
Konzak CF, Zhou H (1991) Anther culture methods for double haploid production in wheat. Cereal Res Comm 19:147–164
Korzun V, Roder MS, Ganal MW, Worland AJ, Law CN (1998) Genetic analysis of the dwarfing gene (*Rht8*) in wheat. Part 1. Molecular mapping of *Rht8* on the short arm of chromosome 2D of bread wheat (*Triticum aestivum* L.). Theor Appl Genet 96:1104–1109
Kraakman ATW, Niks RE, Van den Berg P, Stam P, Van Eeuwijk FA (2004) Linkage disequilibrium mapping of yield and yield stability in modern spring barley cultivars. Genetics 168: 435–446
Kunne C, Lange M, Funke T, Miehe H, Thiel T, et al. (2005) CR-EST: a resource for crop ESTs. Nucl Acids Res 33:D619–D621
Lage J, Warburton ML, Crossa J, Skovmand B, Andersen SB (2003) Assessment of genetic diversity in synthetic hexaploid wheats and their *Triticum dicoccum* and *Aegilops tauschii* parents using AFLPs and agronomic traits. Euphytica 134:305–317
Lage J, Trethowan RM (2007) Synthetic hexaploid wheat improves bread wheat adaptation to rainfed environments globally. Aust J Ag Res (In press)
Langridge P, Lagudah ES, Holton TA, Appels R, Sharp PJ, et al. (2001) Trends in genetic and genome analysis in wheat: a review. Aust J Agric Sci 52:1043–1077
Laurie DA, Bennett MD (1986) Wheat x maize hybridization. Can J Genet Cytol 28:313–316
Law CN, Suarez E, Miller TE, Worland AJ (1998) The influence of the group 1 chromosomes of wheat on ear-emergence times and their involvement with vernalization and day length. Heredity 80:83–91
Li S, Zhang Z, Wang B, Zhong Z, Yao J (1995) Tagging the Pm4a gene in NILs by RAPD analysis. Acta Genet Sin 22:103–108
Li Y-C, Fahima T, Röder MS, Kirzhner VM, Beiles A, et al. (2003) Genetic effects on microsatellite diversity in wild emmer wheat (*Triticum dicoccoides*) at the Yehudiyya microsite, Israel. Heredity 90:150–156
Liu XM, Smith CM, Gill BS (2002) Identification of microsatellite markers linked to Russian wheat aphid resistance genes *Dn4* and *Dn6*. Theor Appl Genet 104:1042–1048
Liu Z, Sun Q, Ni Z, Yang T (1999) Development of SCAR markers linked to *Pm21* gene conferring resistance to powdery mildew in common wheat. Plant Breeding 118:215–219
Lu HJ, Fellers JP, Friesen TL, Meinhardt SW, Faris JD (2006) Genomic analysis and marker development for the Tsn1 locus in wheat using bin-mapped ESTs and flanking BAC contigs. Theor Appl Genet 112:1132–1142

Ma JX, Zhou,R, Dong Y, Wang L, Wang X, et al. (2001) Molecular mapping and detection of the yellow rust resistance gene *Yr26* in wheat transferred from *Triticum turgidum* L. using microsatellite markers Euphytica 120:219–226
Ma Z-Q, Gill BS, Sorrells ME, Tanksley SD (1993) RFLP markers linked to two Hessian fly-resistance genes in wheat (*Triticum aestivum* L.) from *Triticum tauschii* (coss.) Schmal. Theor Appl Genet 85:750–754
Ma ZQ, Sorrells ME, Tanksley SD (1994) RFLP markers linked to powdery mildew resistance genes *Pm1, Pm2, Pm3* and*Pm4* in wheat. Genome 37:871–875
Maccaferri M, Sanguineti MC, Noli E, Tuberosa R (2005) Population structure and long-range linkage disequilibrium in a durum wheat elite collection. Mol Breed 15:271–289
Mackay I, Powell W (2007) Methods for linkage disequilibrium mapping in crops. Trends Plant Science 12:57–63
Mago R, Spielmeyer W, Lawrence GL, Lagudah ES, Ellis GJ, et al. (2002) Identification and mapping of molecular markers linked to rust resistance genes located on chromosome 1RS of rye using wheat-rye translocation lines. Theor Appl Genet 104:1317–1324
Mago R, Bariana HS, Dundas IS, Spielmeyer W, Lawrence GL, et al. (2005) Development of PCR markers for the selection of wheat stem rust resistance genes *Sr24* and *Sr26* in diverse wheat germplasm. Theor Appl Agenet 111:496–504
Matthews DE, Carollo VL, Lazo GR, Anderson OD (2003) GrainGenes, the genome database for small-grain crops. Nucl Acids Res 31:183–186
McFadden ES, Sears ER (1946) The origin of Triticum spelta and its free-threshing hexaploid relatives. J Hered 37:81–89
McIntosh RA Yamazaki Y, Devo, KM, Dubkowsky J, Rogers WJ, et al. (2003) Catalogue of gene symbols for wheat. In: Pogna NE, Romano M, Pogna EA, Galterio G (eds) Proc 10th Intl Wheat Genetics Symp Pasteum, Italy. Rome: Instituto Sperimentale per la Cerealicoltura. 4: 1–34
McVittie JA, Gale MD, Marshall GA, Westcott B (1978). The interchromosomal mapping of the Norin 10 and Tom Thumb dwarfing genes. Heredity 40:67–70
Mello-Sampayo T (1971) Genetic regulation of meiotic chromosome pairing by chromosome 3D of *Triticum aestivum*. Nature New Biol 230:22–23
Mezzalama M, Sayre KD, Nicol J (2001) Monitoring root rot diseases on irrigated, bed-planted wheat. In: Reeves J, McNab A, Rajaram S (eds). Proc Warren E. Kronstad Symp CIMMYT pp. 148–151
Michelmore RW, Paran I, Kesseli RV (1991) Identification of markers linked to disease-resistance genes by bulked segregant analysis: a rapid method to detect markers in specific genomic regions by using segregating populations. Proc Natl Acad Sc. USA 88:9828–9832
Miller CA, Altinkut A, Lapitan NLV (2001) A microsatellite marker for tagging Dn2, a wheat gene conferring resistance to the Russian wheat aphid. J Phytopath 149:641–648
Mochida K, Yamazaki Y, Ogihara Y (2004) Discrimination of homoeologous gene expression in hexaploid wheat by SNP analysis of contigs grouped from a large number of expressed sequence tags. Mol Gen Genom 270:371–377
Mochida K, Kawaura K, Shimosaka E, Kawakami N, Shin-I T, et al. (2006) Tissue expression map of a large number of expressed sequence tags and its application to in silico screening of stress response genes in common wheat. Mol Gen Genet 276:304–312
Morgante M, Salamini F (2003) From plant genomics to breeding practice. Curr Opin Biotechnol 14:214–219
Mujeeb-Kazi A, Gilchrist LI, Fuentes-Davila G, Delgado R (1998) Production and utilization of D genome synthetic hexaploids in wheat improvement. In: Jaradat AA (ed) Proc 3rd Intl Triticeae Symp ICARDA, Science Publishers, pp. 369–374
Mujeeb-Kazi A, Rajaram S (2002) Transferring alien genes from related species and genera for wheat improvement. In: Curtis BC, Rajaram S, Gomez Macpherson H (eds) Bread wheat – Improvement and production Plant production and protection series No. 30. pp. 71–88
Mukai Y, Nakahara Y, Yamamoto M (1993) Simultaneous discrimination of the three genomes in hexaploid wheat by multicolor fluorescence in situ hybridization using total genomic and highly repeated DNA probes. Genome 36:489–494

Mullen DJ, Platteter A, Teakle NL, Appels R, Colmer TD, et al. (2005) EST-derived SSR markers from defined regions of the wheat genome to identify *Lophopyrum elongatum* specific loci. Genome 48:811–822

Nelson JC, Singh RP, Autrique JE, Sorrells ME (1997) Mapping genes conferring and suppressing leaf rust resistance in wheat. Crop Sci 37:1928–1935

Neu C, Stein N, Keller B (2002) Genetic mapping of the *Lr20-Pm1* resistance locus reveals suppressed recombination on chromosome arm 7AL in hexaploid wheat. Genome 45:737–744

Nordborg M, Borevitz JO, Bergelson J, Berry,CC, Chory J, et al. (2002) The extent of linkage disequilibrium in Arabidopsis thaliana. Nat Genet 30:190–193

Ogbonnaya FC, Subrahmanyam NC, Moullet O, Majnik J de, Eagles HA, et al. (2001) Diagnostic DNA markers for cereal cyst nematode resistance in bread wheat. Aust J Agric Res 52:1367–1374

Ogihara Y, Mochida K, Nemoto Y, Murai K, Yamazaki Y, et al. (2003) Correlated clustering and virtual display of gene expression patterns in the wheat life cycle by large-scale statistical analyses of expressed sequence tags. Plant J 33:1001–1011

Oliver RE, Xu SS, Snack RW, Friesen TL, Jin Y, et al. (2006) Molecular cytogenetic characterization of four partial wheat-*Thinopyrum ponticum* amphiploids and their reactions to Fusarium head blight, tan spot and Stagonospora nodorum blotch. Theor Appl Genet 112: 1473–1479

Ortiz R, Mowbray D, Dowswell C, Rajaram S (2007) Dedication ~ Norman E. Borlaug: The humanitarian plant scientist who changed the world. Plant Breeding Rev 28:1–37

Ortiz R, Trethowan R, Ortiz Ferrara G, Iwanaga M, Dodds JH, et al. (2007) High yield potential, shuttle breeding and new international wheat improvement strategy. Euphytica (in press)

Paull JG, Pallotta MA, Langridge P, The TT (1995) RFLP markers associated with *Sr22* and recombination between chromosome 7A of bread wheat and the diploid species *Triticum boeoticum*. Theor Appl Genet, 89:1039–1045

Pellegrineschi A, Noguera LM, McLean S, Skovmand B, Brito RM, et al. (2002). Identification of highly transformable wheat genotypes for mass production of fertile transgenic plants. Genome 45:421–430

Peng J, Richards DE, Hartley NM, Murphy GP, Devos KM, et al. (1999) 'Green revolution' genes encode mutant gibberellin response modulators. Nature 400:256–261

Peng JH, Fahima T, Roeder MS, Huang QY Dahan A, et al. (2000) A High-density molecular map of chromosome region harboring stripe-rust resistance genes *YrH52* and *Yr15* derived from wild emmer wheat, *Triticum dicoccoides*.. Genetica 109:199–210

Penner GA, Clarke J, Bezte LJ, Leisle D (1995) Identification of RAPD markers linked to a gene governing cadmium uptake in durum wheat. Genome 38:543–547

Prins R, Groenewald JZ, Marais GF, Snape JW, Koebner RMD (2001) AFLP and STS tagging of *Lr19,* a gene conferring resistance to leaf rust in wheat. Theor Appl Genet 103:618–624

Procunier JD, Townley-Smith TF, Fox S, Prashar S, Gray M, et al. (1995) PCR-based RAPD/DGGE markers linked to leaf rust resistance genes *Lr29* and *Lr25* in wheat (*Triticum aestivum* L.). J Genet Breed 49:87–92

Qi LL, Echalier B, Chao S, Lazo GR, Butler GE, et al. (2004) A chromosome bin map of 16,000 expressed sequence tag loci and distribution of genes among the three genomes of polyploid wheat. Genetics 168:701–712

Qu LJ, Foote TN, Roberts MA, Money TA, Aragon-Alcaide L, et al. (1998) A simple PCR-based method for scoring the *ph1b* deletion in wheat. Theor Appl Genet 96:371–375

Rafalski A (2002) Applications of single nucleotide polymorphisms in crop genetics. Curr Opin Plant Biol 5:94–100

Rajaram S, Mann CHE, Ortiz-Ferrara G, Mujeeb-Kazi A (1983) Adaptation, stability and high yield potential of certain 1B/1R CIMMYT wheats. In: Sakamoto S (ed) Proc. 6th Int. Wheat Genetics Symp pp. 613–621

Rajaram S, van Ginkel M (2001) Mexico, 50 years of international wheat breeding. (Chapter 22). In: Bonjean AP, Angus WJ (eds) The World Wheat Book, A History of Wheat Breeding, pp. 579–608. Lavoisier Publishing, Paris

Rajaram S, Borlaug NE, van Ginkel M (2002) CIMMYT International wheat breeding. In: Curtis BC, Rajaram S, Gomez Macpherson H (eds) Bread wheat – Improvement and production Plant production and protection series No. 30. pp. 103–117

Ranade K, Chang M-S, Ting C-T, Pei D, Hsiao C-F, et al. (2001) High throughput genotyping with single nucleotide polymorphisms. Genome Res 11:1262–1268

Raupp WJ, Sukhwinder-Singh, Brown-Guedira GL, Gill BS (2001) Cytogenetic and molecular mapping of the leaf rust resistance gene *Lr39* in wheat. Theor Appl Genet 102: 347–352

Ravel C, Praud S, Murigneux A, Linossier L, Dardevet M, et al. (2006) Identification of Glu-B1-1 as a candidate gene for the quantity of high-molecular-weight glutenin in bread wheat (*Triticum aestivum* L.) by means of an association study. Theor Appl Genet 112: 738–743

Reif JC, Zhang P, Dreisigacker S, Warburton ML, van Ginkel M, et al. (2005) Trends in genetic diversity during the history of wheat domestication and breeding. Theor Appl Genet 110: 859–864

Remington DL, Thornsberry JM, Matsuoka Y, Wilson LM, Whitt SR, et al. (2001) Structure of linkage disequilibrium and phenotypic associations in the maize genome. Proc Natl Acad Sci USA 98:11479–11484

Reynolds MP, van Beem J, van Ginkel M, Hoisington D (1996) Breaking the yield barriers in wheat: a brief summary of the outcomes of an international consultation. In: Reynolds MP, Rajaram S, McNab A (eds) Increasing yield Potential in Wheat: Breaking the Yield Barriers' CIMMYT. pp. 1–11

Reynolds M, Dreccer F, Trethowan R (2007) Drought adaptive mechanisms from wheat landraces and wild relatives. J Exp Bot 58:177–186

Riley R, Chapman V (1958) Genetic control of the cytologically diploid behavior of hexaploid wheat. Nature 182:713–715

Röder MS, Korzun V, Wendehake K, Plaschke J, Tixier M-H, et al. (1998) A microsatellite map of wheat. Genetics 149:2007–2023

Rong JK, Millet E, Manisterski J, Feldman M (2000) A new powdery mildew resistance gene: introgression from wild emmer into common wheat and RFLP based mapping. Euphytica 115:121–126

Rosenzweig C, Hillel D (1995) Potential impacts of climate change on agriculture and food supply. Consequences, Vol. 1, No. 2

Roses AD (2002) Pharmacogeentics place in modern medical science and practice. Life Sci 70:1471–1480

Roussel V, Koenig J, Beckert M, Balfourier F (2004) Molecular diversity in French bread wheat accessions related to temporal trends and breeding programmes. Theor Appl Genet 108: 920–930

Roussel V, Leisova L, Exbrayat F, Stehno Z, Balfourier F (2005) SSR allelic diversity changes in 480 European wheat varieties released from 1840 to 2000. Theor Appl Genet 111: 162–170

Sandhu D, Gill BS (2002a) Gene-containing regions of wheat and other grass genomes. Plant Physiol 128:803–811

Sandhu D, Gill BS (2002b) Structural and functional organization of the '1so.8 gene-rich region' in the Triticeae. Plant Mol Biol 48:791–804

Sarma RN, Fish L, Gill BS, Snape JW (2000) Physical characterization of the homoeologous group 5 chromosomes of wheat in terms of rice linkage blocks and physical mapping of some important genes. Genome 43:191–198

Sarma RN, Gill BS, Sasaki T, Galiba G, Sutjk J, et al. (1998) Comparative mapping of the wheat chromosome 5A. Vrn A-1 region with rice and its relationship to QTL for flowring time. Theor Appl Genet 97:103–109

Sayre KD, Rajaram S, Fischer RA (1997) Yield potential progress in short bread wheats in northwest Mexico. Crop Sci 37:36–42

Schachermayr G , Messmer MM, Feuillet C, Winzeler H, Winzeler M, et al. (1995) Identification of molecular markers linked to the *Agropyron elongatum*-derived leaf rust resistance gene*Lr24* in wheat. Theor Appl Genet 90:982–990

Schachermayr G, Siedler H, Gale MD, Winzeler H, Winzeler M, et al. (1994) Identification and localization of molecular markers linked to the *Lr9* leaf rust resistance gene of wheat. Theor Appl Genet 88:110–115

Schachermayr G, Feuillet C, Keller B (1997) Molecular markers for the detection of the wheat leaf rust resistance gene Lr10 in diverse genetic backgrounds. Mol Breeding 3:65–74

Scofield SR, Huang L, Brandt AS, Gill BS (2005) Development of a virus-induced gene-silencing system for hexaploid wheat and its use in functional analysis of the Lr21-mediated leaf rust resistance pathway. Plant Physiol 138:2165–2173

Semagn K, Bjonstad A, Skinnes H, Maroy AG, Tarkegne Y, et al. (2006) Distribution of DArT, AFLP, and SSR markers in a genetic map of a doubled-haploid hexaploid wheat population. Genome 49:545–555

Seo YW, Johnson JW, Jarret RL (1997) A molecular marker associated with the *H21* Hessian fly resistance gene in wheat. Mol Breeding 3:177–181

Seyfarth R, Feuillet C, Keller B (1998) Development and characterization of molecular markers for the adult leaf rust resistance genes *Lr13* and *Lr35* in wheat. Proc 9th Intl Wheat Genet Symp 3:154–155

Sharma HC, Gill BS (1983) Current status of wide hybridisation in wheat. Euphytica 32:17–31

Shen LH, Gong J, Caldo RA, Nettleton D, Cook D, et al. (2005) BarleyBase - an expression profiling database for plant genornics. Nucl Acids Res 33:D614–D618

Shen X, Zhou M, Lu W, Ohm H (2003) Detection of fusarium head blight resistance QTL in a wheat population using bulked segregant analysis. Theor Appl Genet 106:1041–1047

Shen XR, Francki MG, Ohm HW (2006) A resistance-like gene identified by EST mapping and its association with a QTL controlling Fusarium head blight infection on wheat chromosome 3BS. Genome 49:631–635

Sherman JD, Yan L, Talbert L, Dubcovsky J (2004) A PCR marker for growth habit in common wheat based on allelic variation at the *VRN-A1* gene. Crop Sci 44:1832–1838

Shewry PR, Jones HD (2005) Transgenic wheat: where do we stand after the first 12 years? Ann Appl Biol 147:1–14

Shi AN, Leath S, Murphy JP (1998) A major gene for powdery mildew resistance transferred to common wheat from wild einkorn wheat. Phytopath 88:144–147

Simons KJ, Fellers JP, Trick HN, Zhang ZC, Tai YS, et al. (2006) Molecular characterization of the major wheat domestication gene Q. Genetics 172:547–555

Singh RP, Nelson JC, Sorrels ME (2000) Mapping *Yr28* and other genes for resistance to stripe rust in wheat. Crop Sci 40:1148–1155

Slade AJ, Knauf VC (2005) TILLING moves beyond functional genomics into crop improvement. Transgenic Res 14:109–115

Slade AJ, Fuerstenberg SI, Loeffler D, Steine MN, Facciotti D (2005) A reverse genetic, nontransgenic approach to wheat crop improvement by TILLING. Nature Biotech 23:75–81

Somers DJ, Kirkpatrick R, Moniwa M, Walsh A (2003) Mining single nucleotide polymorphisms from hexaploid wheat ESTs. Genome 46:431–437

Somers DJ, Isaac P, Edwards K (2004) A high-density microsatellite consensus map for bread wheat (*Triticum aestivum* L.). Theor Appl Genet 109:1105–1114

Sorrells ME, La Rota M, Bermudez-Kandianis CE, Greene RA, Kantety R, et al. (2003) Comparative DNA sequence analysis of wheat and rice genomes. Genome Res 13: 1818–1827

Sourdille P, Singh S, Cadalen T, Brown-Guedira GL, Gay G, et al. (2004) Microsatellite-based deletion bin system for the establishment of genetic-physical map relationships in wheat (*Triticum aestivum* L.). Funct Integr Genomics 4:12–25

Sparks CA, Jones HD (2004) Transformation of wheat by biolistics. In: Curtis IS (ed) "Transgenic Crops of the World: Essential Protocols". pp. 19–34

Srichumpa P, Brunner S, Keller B, Yahiaoui N (2005) Allelic series of four powdery mildew resistance genes at the Pm3 locus in hexaploid bread wheat. Plant Physiol 139: 885–895

Stoutjesdijk P, Kammholz SJ, Kleven S, Matssy S, Banks PM, et al. (2001) PCR-based molecular marker for the *Bdv2 Thinopyrum intermedium* source of barley yellow dwarf virus resistance in wheat. Aust J Agric Res 52:1383–1388

Tang J, Gao L, Cao Y, Jia J (2006) Homologous analysis of SSR-ESTs and transferability of wheat SSR-EST markers across barley, rice and maize. Euphytica 151:87–93

Tanksley SD, McCouch SR (1997) Seed banks and molecular maps: Unlocking genetic potential from the wild. Science 277:1063–1066

Tenaillon MI, Sawkins MC, Long AD, Gaut RL, Doebley JF (2001) Patterns of DNA sequence polymorphism along chromosome 1 of maize (*Zea mays* ssp.*mays* L.). Proc Natl Acad Sci USA 98:9161–9166

Trethowan RM, Reynolds MP, Sayre KD, Ortiz-Monasterio I (2005a) Adapting wheat cultivars to resource conserving farming practices and human nutritional needs. Ann Appl Biol 146:404–413

Trethowan RM, Hodson D, Braun HJ, Pfeiffer WH (2005b) Wheat breeding environments. In: Dubin J, Lantican MA, Morris ML (eds) 'Impacts of International Wheat Breeding Research in the Developing World, 1988–2002. CIMMYT pp. 4–11

Trethowan RM, Reynolds MP, Ortiz-Monasterio JI, Ortiz R (2007) The Genetic Basis of the on-going Green Revolution in wheat production. Plant Breed Rev 28:39–58

Trethowan RM, Reynolds MP (2007) Drought resistance: genetic approaches for improving productivity under stress. In: Buck HT, Nisi JE, Salom6n N (eds), Wheat Production in Stressed Environments. Series: Developments in Plant Breeding, Vol 12:289–299

Tuvesson S, v Post R, Ljungberg A (2003) Wheat anther culture. In: Maluszynski M, Kasha KJ, Forster BP, Szarejko I (eds) Doubled haploid production in crop plants – a manual. pp 71–76. Kluwer Academic Publishers, Dordrecht/Boston/ London

Upadhya MD, Swaminathan MS (1963) Genome analysis in *Triticum zhukovskyi*, a new hexaploid wheat. Chromosoma 14:589–600

Van Beuningen LT, Bush RH (1997) Genetic diversity among North American spring wheat cultivars: I. Analysis of the coefficient of parentage matrix. Crop Sci 37:570–579

Varshney RK, Graner A, Sorrells ME (2005) Genic microsatellite markers in plants: features and applications. Trends Biotech 23:48–55

Wan YC, Layton J (2006) Wheat (*Triticum aestivum* L.), In: Wang K (ed) "Agrobacterium Protocols, 2nd Edition, Vol 1 Methods in Molecular Biology" pp. 245–253

Wang L, Ma J, Zhou R, Wang X, Jia J (2002) Molecular tagging of the yellow rust resistance gene *Yr10* in common wheat, PI 178383 (*Triticum aestivum* L). Euphytica 124:71–73

Warburton ML, Crossa J, Franco J, Kazi M, Trethowan R, et al. (2006) Bringing wild relatives back into the family: recovering genetic diversity in CIMMYT bread wheat germplasm. Euphytica 149:289–301

Weil C (2005) Single base hits score a home run in wheat. Trends Biotechnol 23:220–222

William HM, Singh RP, Huerta-Espino J, Ortiz-Islas S, Hoisington D (2003) Molecular marker mapping of leaf rust resistance gene Lr46 and its association with stripe rust resistance gene Yr29 in wheat. Phytopathology 93:153–159

William HM, Trethowan R, Crosby-Galvan EM (2007) Wheat breeding assisted by markers: CIMMYT's experience. Euphytica (In press)

Wilson ID, Barker GLA, Beswick RW, Shepherd SK, Lu CG, et al. (2004) A transcriptomics resource for wheat functional genomics. Plant Biotech J 2:495–506

Worland AJ (1996) The influence of flowering time genes on environmental adaptability in European wheats. Euphytica 89:49–57

Wu H, Sparks C, Jones H (2006) Characterisation of T-DNA loci and vector backbone sequences in transgenic wheat produced by Agrobacterium-mediated transformation. Mol Breeding 18:195–208

Yan L, Loukoianov A, Tranquilli G, Helguera M, Fahima T, et al. (2003) Positional cloning of the wheat vernalization gene Vrn-1. Proc Natl Acad Sci 100:6263–6268

Yu JK, Dake TM, Singh S, Benscher D, Li WL, et al. (2004) Development and mapping of EST-derived simple sequence repeat markers for hexaploid wheat. Genome 47:805–818

Zhang LY, Bernard M, Leroy P, Feuillet C, Sourdille P (2005a) High transferability of bread wheat EST-derived SSRs to other cereals. Theor Appl Genet 111:677–687

Zhang P, Dreisigacker S, Melchinger AE, van Ginkel M, Hoisington D, et al. (2005b) Quantifying novel sequence variation in CIMMYT synthetic hexaploid wheats and their backcross-derived lines using SSR markers. Mol Breeding 12:1–10

Zhang LY, Ravel C, Bernard M, Balfourier F, Leroy P, et al. (2006) Transferable bread wheat EST-SSRs can be useful for phylogenetic studies among the Triticeae species. Theor Appl Genet 113:407–418

Zohary D, Hopf M (1993) Domestication of plants in the old world, 2nd ed. Oxford, UK, Calrendon Press

Zondervan KT, Cardon LR (2004) The complex interplay among factors that influence allelic association. Nat Rev Genet 5:86–100

Chapter 23
Genomics of Yams, a Common Source of Food and Medicine in the Tropics

Hodeba D. Mignouna, Mathew M. Abang, and Robert Asiedu

Abstract Yams (*Dioscorea* spp., Dioscoreaceae), grown either for their starchy tubers or medicinal properties, are important crops in the tropics and subtropics. Yams broaden the food base and provide food security and income to over 300 million people. They are vegetatively propagated and comprise both diploid and polyploid species. Despite their economic and socio-cultural importance, very little is known about the genetics and genomics of yams due to research neglect and several biological constraints. Consequently, conventional breeding efforts have been severely hampered. Research to unravel the apparent complexity of the yam genome will have far-reaching implications for genetic improvement of this important tuber crop. Nevertheless, progress has been made recently towards understanding *Dioscorea* phylogeny and phylogenetic relationships within the genus. Also, improved molecular technologies have been developed for genome analysis, including germplasm characterization, cytogenetics, genetic mapping and tagging, and functional genomics. Genetic linkage maps have been constructed for *D. rotundata* and *D. alata*, and quantitative trait loci associated with resistance to *Yam mosaic virus* in *D. rotundata* and anthracnose (*Colletotrichum gloeosporioides*) in *D. alata* have been identified. In addition, candidate random amplified polymorphic DNA markers associated with major genes controlling resistance to *Yam mosaic virus* and anthracnose have been identified. These markers could be converted to sequence-characterized amplified regions and used for marker-assisted selection for resistance to diseases. An initial cDNA library has been constructed to develop expressed sequence tags for gene discovery and as a source of additional molecular markers. Genetic engineering offers a powerful tool, complementing conventional breeding approaches, for yam improvement. Methods for yam transformation, including in vitro plant regeneration, gene delivery, selection of transformed tissues, and recovery of transgenic plants have been developed but still need improvements. This chapter reviews advances made in yam molecular marker development for genome

H.D. Mignouna
African Agricultural Technology Foundation (AATF), Nairobi, Kenya
e-mail: h.mignouna@aatf-africa.org

P.H. Moore, R. Ming (eds.), *Genomics of Tropical Crop Plants*

© Springer 2008

analysis, phylogeny, molecular cytogenetics, characterization of genetic diversity, genetic mapping and tagging, and progress in functional genomics.

23.1 Introduction

Yams are classified in the genus *Dioscorea*, a genus widely reported as comprising around 600 species (Burkill 1960). More recent estimates indicate that approximately 200 species are distributed throughout the tropics and subtropics (Ayensu 1972). Plants of the genus *Dioscorea* are angiosperms that belong to the monocotyledon order Dioscoreales. Interestingly, the order Dioscoreales is characterized by several dicotyledonous features, such as reticulate-veining, stalked net-nerving leaves, circularly arranged vascular bundles in the stem, and the lateral position of the pistil. Yams show a second vestigial cotyledon, which renders them intermediate with respect to the phylogenetic relationships between mono- and dicotyledonous plants, even though the traditional division of the angiosperms in mono- and dicotyledonous plants was formally discontinued with the introduction of the Magnoliopsida as a distal class of the angiosperms (Frohne and Jensen 1998). Yam plants are herbaceous or woody climbing plants with tuberous, starch-rich storage organs. The aerial storage organ of Dioscoreaceae is the bulbil. They are perennial plants with a strongly marked annual cycle of growth (Coursey 1983). In the southern United States the name yam is used for sweet potato (*Ipomoea batatas*, L. Poir.) and elsewhere for the edible tubers of aroids (Frohne and Jensen 1998; Purseglove 1988). More generally, and in the present chapter, the term yam is confined to plants of the genus *Dioscorea*. Guinea yams (*D. rotundata* and *D. cayenensis*) were domesticated in West Africa, while the water or greater yam (*D. alata*) probably originated from the southeast Asian-Oceanian region (Malapa et al. 2005). *D. alata* was previously considered to be a possible cultigen (Barrau 1965), but it is now known to be a true species with normal sexuality (Lebot et al. 1998; Malapa et al. 2005).

In West and Central Africa, where Guinea yams were domesticated about 7000 years ago, farmers selected genotypes that best suited their needs and thus have generated a large number of traditional cultivars. Yam production has increased steadily in the last decade, from 18 million metric tonnes in 1990 to recent estimates of over 39 million (FAO 2006). This increase has been achieved mainly through the planting of traditional landraces and can be explained by the rapid increase in acreage of yam fields into marginal lands and into non-traditional yam growing areas. This expansion highlights the need to provide farmers with improved varieties that combine high yields with pest and disease resistance and acceptable tuber quality.

Collaborative evaluations of International Institute of Tropical Agriculture (IITA)-derived breeding lines with national yam programs in Africa have led to the official release of a number of white yam varieties having multiple pest and disease resistance, wide adaptability, and good organoleptic attributes. However, this progress has been difficult, time-consuming, and laborious due to biological constraints that impede the elucidation of the genetics of important traits in yam. Genetic improvement of yam has been hampered by a long growth cycle (lasting about eight months or more), dioecy, poor to no flowering, asynchronous flowering of male and female parents, polyploidy, vegetative propagation, high heterozygosity, and poor knowledge of the crop's genetic diversity (Asiedu et al. 1998). Yam is cultivated in widely varying agroecological zones and the performance of genotypes is disparate across regions, thereby multiplying breeding goals.

Molecular markers that are linked to genes controlling economic traits would be useful in selection at an early stage of the plant's growth, thereby enhancing the speed and efficiency of selection. Biotechnology not only provides an alternative approach, but also complements the efforts in conventional breeding (Mignouna et al. 2003a). This chapter will review yam molecular marker development for genome analysis, phylogeny, cytogenetics, characterization of genetic diversity, genetic mapping and tagging, and progress in functional genomics.

23.1.1 Economic, Agronomic, and Societal Importance of Yams

Yam is produced throughout the tropical and sub-tropical regions of the world. Guinea yams are the most popular and economically important yams in West and Central Africa where they are indigenous, while water or greater yam is the most widely distributed species globally. The majority of global yam production is in Africa. West Africa accounts for about 95% of world production and 96% of the area (FAO 2006). Yam production globally reached 39.85 million Mt harvested from 4.44 million ha in 2005 (FAO 2006). The largest producer was Nigeria with 26.59 million Mt, followed by Ghana (3.89), Côte d'Ivoire (3.00), and Benin (2.56). The profitability of yam production, the value of yams in local trade (Hahn et al. 1987; Nweke et al. 1991), as well as the current and potential revenue from their export to ethnic markets in Europe and Northern America are often underestimated. In many parts of West Africa, for instance southeastern Nigeria, yams rank first among the major food crops in terms of cash income per hectare (IITA 1988; Nweke et al. 1991).

Food yams are grown principally for the carbohydrate they provide. The tubers, which are the only edible part, have a tremendous capacity to store food reserves. They broaden the food base and bring food security to 300 million people in the low-income, food-deficit countries of the tropics, providing them with about 200 kilocalories daily. The net dietary protein calorie content in yams is about 4.6%, which compares well with 4.7% in maize (Hahn et al. 1987; FAO 1999). Socioeconomic surveys conducted in Nigeria indicated that there was a positive elasticity of demand for yams at all expenditure levels, and that production research towards increasing yam supply will consequently increase quantities consumed at low-income levels in sub-Saharan Africa (Nweke et al. 1992).

In West Africa, yam tubers are typically boiled and pounded into dough for easy swallowing. In Madagascar, tubers of some species can be eaten raw (e.g., *D. soso*, *D. nako*, and *D. fandra*). Others are simply boiled or baked (e.g., *D. alata*), while others need extensive preparation such as immersion in running water for 1–3 days or drying in the sun (e.g., *D. antaly*). *Dioscorea* species are not only known for their food value but also for their secondary metabolites. They contain steroidal saponins, diterpenoids, and alkaloids, which have been exploited for making poisons (Neuwinger 1996) and pharmaceutical products (Chu and Figueiredo-Ribeiro 1991).

23.1.2 Yam as an Experimental Organism

The genus *Dioscorea* has been considered to be an attractive model for investigating ploidy events and chromosome evolution in wild and cultivated species in relation to vegetative propagation and the process of domestication (Bousalem et al. 2006). Yam, though an "orphan" crop, can provide a good model for traits not possessed by other model crops. For instance, the tuber is an important ecological (and economic) trait possessed by only a few models: potato may serve for eudicots, but we have little basis to judge how suitable it might be as a model for monocots. In other words, we do not know how general the tuberization process is in angiosperms. Knowledge of gene expression at the appropriate stages in a tuberous monocot (e.g., *Dioscorea*, yams), matched with a candidate gene approach, would allow us to address this question. Phylogenetic morphology studies reveal that the "monocot" mode of leaf development typifies a nested group. However, not all monocots have this mode of leaf development; some have either dicot or intermediate modes of development. The grass models may serve taxa with monocot modes; but other taxa (e.g., *Dioscorea*) may be needed to understand other developmental modes (Bharathan 1996).

Given its dioecious nature with different morphologies of staminate and pistillate plants in some species; its dicot-like leaf structure (net-veined and petiolate) with early development intermediate between dicot and monocot modes (Bharathan 1996); distinct changes in shoot apical meristem (SAM) structure and phyllotaxy during phase transition from juvenile to adult (Burkill 1960); tuber formation and dormancy; small C-value and widespread polyploidy (Dansi et al. 2001; Egesi et al. 2002; Bousalem et al. 2006), *Dioscorea* offers a system in which to raise general biological questions that cannot be addressed in many other species. It thus holds great promise of yielding important clues to explain differences between eudicot and grass models (e.g., non-orthology of KNOX genes controlling SAM indeterminacy [Bharathan et al. 1999]) and offering examples of biological phenomena such as dioecy, tuberization, and modes of vine twining.

Tuber dormancy is an important field adaptive mechanism that also helps to maintain organoleptic quality during storage, but it creates a major problem for plant breeders. This is because harvested tubers remain dormant (i.e. incapable of developing an internal shoot bud or external shoot bud/sprout) for 30 to 150 d (Orkwor and Ekanayake 1998), only one crop cycle is possible per year, which slows progress in yam improvement. Knowledge gained from yams may lead to the elucidation and successful manipulation of tuber dormancy in other plant species. Elucidation of the molecular changes taking place in yams during post-harvest storage will help in understanding the process of tuber dormancy (Kone-Coulibaly et al. 2003).

23.2 Development of Molecular Markers for Genome Analysis

Yams are monocots, but very distantly related to the grasses. Thus there is no convenient model system for yam genomics. Initial efforts in yam genomics sought to exploit heterologous DNA sequences as a source of RFLP markers (Terauchi et al. 1992). Later, the approach of using uncharacterized DNA sequences was adopted as a source of genetic markers. AFLP was the molecular marker of choice (Mignouna et al. 1998). RAPD and AFLP polymorphism was high among diverse yam species, with AFLP revealing the highest polymorphism. Sixty-four AFLP primer combinations were tested for their potential use in assessment of genetic diversity in white Guinea yam (Mignouna et al. 1998). Although RAPD markers were adequate for genetic diversity studies (Dansi et al. 2000a), the level of polymorphism detected in mapping populations was low; therefore, RAPD was not considered a good marker-system for mapping purposes. Contrary to RAPDs, the high level of polymorphism revealed by AFLP markers, coupled with their robustness, made AFLP a more reliable and reproducible marker-system for yam genome analysis (Mignouna et al. 1998; Mignouna et al. 2003b; Malapa et al. 2005).

As progress was being made in yam genomics, co-dominant molecular markers such as microsatellites or simple sequence repeats (SSRs) were required because of their expected high polymorphism, co-dominant inheritance, high abundance and even distribution across the genome. In a study of a natural population of *D. tokoro*, a wild diploid East Asian yam species ($2n = 20$), Terauchi and Konuma (1994) detected microsatellite polymorphisms. A high number of polymorphic alleles was detected per microsatellite locus, suggesting that these microsatellite primers could be transferable to other *Dioscorea* species. Unfortunately, when the *D. tokoro* microsatellite primers were applied to other yam species, they failed to amplify any DNA, indicating that these primer sequences are not conserved among *Dioscorea* species. However, the study demonstrated the potential usefulness of these markers for yams. Microsatellite markers were later developed for food yams in a collaborative project between IITA and the University of Saskatchewan, Canada, and used to assess genetic diversity in *D. rotundata* (Mignouna et al. 2003b). A few microsatellite markers were characterized by several authors, but because of the relatively small number of markers developed (six in *D. tokoro* [Terauchi and Konuma 1994] and nine in *D. rotundata* [Mignouna et al. 2003b]) and the low level of polymorphism detected in mapping populations, microsatellites were not considered a good marker system for mapping purposes.

Increased interest in yam genomics and the need for robust molecular and genetic tools for genome analysis led to the development of 10 microsatellite markers in *D. japonica* (Mizuki et al. 2005). Tostain et al. (2006) developed and characterized

16 new SSR markers in different species of yam, several of which were transferable to species of other *Dioscorea* sections. Transferability was higher among species belonging to the same botanical section (*Enantiophyllum*. Within the Enantiophyllum section, the patterns differed for the African species on one hand and the Asian-Oceanian species *D. alata* and *D. nummularia* on the other. Similarly, Hochu et al. (2006) developed 20 microsatellite markers in American yam (*D. trifida*) and found high cross-species amplification involving four additional *Dioscorea* species: the cultivated *D. alata, D. cayenensis–D. rotundata,* and the two African wild yams, *D. praehensilis* and *D. abyssinica.* The four species tested are classified into the botanical section *Enantiophyllum* that is phylogenetically distant from the section *Macrogynodium* to which *D. trifida* belongs. This large cross-species applicability indicated that the primers will be useful for additional studies within the *Dioscorea* genus.

23.3 Phylogeny, Molecular Cytogenetics, and Genetic Diversity

23.3.1 Yam Phylogeny

Phylogenetic relationships of yams have not been well established because of difficulties in species identification due to a high level of polymorphism in morphological characters. Although all species in the genus are dioecious, some species have different species names for its male and female plants. Recent analyses of morphological and molecular data sets have indicated relationships within Dioscoreaceae R. Br. (Caddick et al. 2002a), and a formal reclassification of the family has been presented (Caddick et al. 2002b). Dioscoreaceae now contains four distinct genera, *Dioscorea, Stenomeris, Tacca* (previously in Taccaceae), and *Trichopus*. The dioecious Dioscoreaceae genera, *Borderea, Epipetrum, Nanarepenta, Rajania, Tamus,* and *Testudinaria*, are nested within *Dioscorea* in phylogenetic analyses (Caddick et al. 2002a), and are therefore sunk into it.

Wilkin et al. (2005) conducted phylogenetic analysis of yams based on sequence data from the plastid genes *rbcL* and *matK*, using 67 species of *Dioscorea* and covering all the main Old World and selected New World lineages. They found that the main Old World groups (such as the right-twining *Dioscorea* section *Enantiophyllum* to which most edible yams belong) are monophyletic and that there are two distinct lineages among the endemic Malagasy taxa. These findings have important consequences for character evolution, intrageneric classification, and the origins of diversity in *Dioscorea*. Earlier, Kawabe et al. (1997) had examined the phylogenetic relationship of six species (*D. gracillima*, *D. nipponica*, *D. quinqueloba*, *D. septemloba*, *D. tenuipes*, and *D. tokoro*), in the section *Stenophora* of the genus *Dioscorea*, based on nucleotide sequence variation in 1073 bp of the coding region of the phosphoglucose isomerase locus. They found that *D. tenuipes* and *D. tokoro* belonged to a monophyletic clade, while the other species formed a separate monophyletic group. These studies point to the possibility of greatly simplifying the classification of yams proposed by Knuth and Burkill (Chaïr et al. 2005).

Based on RFLP analysis of the chloroplast and nuclear ribosomal DNA, Terauchi et al. (1992) found four different taxonomic groups with*D. rotundata* and *D. cayenensis* being classified in the same chloroplast DNA-defined group as the wild species *D. praehensilis, D. abyssinica,* and*D. liebrechtsiana*. The other three classes identified among the wild species comprised *D.minutiflora, D.burkilliana, D.smilacifolia,* and *D.togoensis*. Cluster analysis based on the enzyme system 6-PGD revealed a tendency towards separation of the annual species (*D. abyssinica, D. praehensilis, D. rotundata*) from the perennial species (*D. burkilliana, D. smilacifolia, D. minutiflora*) and their derivative (*D. cayenensis*) (Mignouna et al. 2003c). This indicated that 6-PGD may be useful in phylogenetic studies in yam.

23.3.2 Molecular Dissection of the D. cayenensis-rotundataComplex

Ayensu and Coursey (1972), Martin and Rhodes (1978), and Miège (1982a, b) proposed merging of Guinea yams, *D. cayenensis* and *D. rotundata*, into a species complex based on a comparison of their morphological characteristics. However, the taxonomy and evolution of the *D. cayenensis-rotundata* complex remains controversial (Dansi et al. 1999), with different authors considering Guinea yam to be represented either by one species, two species, or a species complex (Martin and Rhodes 1978; Miège 1982a, b; Onyilagha and Lowe 1985; Hamon and Touré 1990a, b; Hamon et al. 1992; Terauchi et al. 1992; Asemota et al. 1996). Cluster analysis of 467 Guinea yam accessions based on seven polymorphic enzyme systems clearly separated the *D. rotundata* (white yam) and the *D. cayenensis* (yellow yam) accessions (Dansi et al. 2000b). This clear partition into two groups was consistent with the concept that the two forms of Guinea yam represent different genetic entities which may be treated as two separate taxa, supporting the view of Onyilagha and Lowe (1985).

Molecular markers have been used to delineate species boundaries surrounding *D. rotundata* and *D. cayenensis* (Terauchi et al. 1992; Mignouna et al. 1998; Mignouna et al. 2005a, b; Chaïr et al. 2005). On the basis of RFLP analysis of chloroplast and nuclear ribosomal DNA, Terauchi et al. (1992) proposed that *D. rotundata* was domesticated from one of the wild species that shared the same chloroplast genotype, and that *D. cayenensis* is of hybrid origin and should be considered as a variety of *D. rotundata*. Similar results were obtained by Chaïr et al. (2005), who reported that *D. cayenensis* and *D. rotundata* share the same cpSSR haplotype. However, Ramser et al. (1997) used four molecular marker systems (RAPD, microsatellite-primed PCR random amplified microsatellite polymorphism, and a comparative sequence analysis of three noncoding chloroplast DNA sequences) to confirm the separation of Guinea yams into two distinct species, *D. rotundata* and *D. cayenensis*. Mignouna et al. (1998) used two AFLP primer combinations to generate a total of 87 polymorphic loci across 20 Guinea yam cultivar groups. Phylogenetic analysis of the data revealed five major cultivar groups among which the group that corresponded to *D. cayenensis* was genetically distant from the varietal groups of *D. rotundata*, as found in other molecular studies. In another study with RAPD and double stringency PCR markers (Mignouna et al. 2005a), accessions of Guinea yam, which were classified into seven morphotypes/cultivar groups, could be clearly separated into two major groups corresponding to *D. rotundata* and *D. cayenensis*. It was proposed, based on these results, that cultivars classified into *D. cayenensis* should be considered as a taxon separate from *D. rotundata*. Mignouna et al. (2005a) considered that the discrepancy between their results and those of Terauchi et al. (1992) probably arose from the fact that they scanned the entire genome using PCR-based markers while the RFLP analysis of Terauchi et al. (1992) was based on the rDNA gene. Although useful for inferring phylogenetic relationships, the rDNA gene represents only a small fraction of the total genome and there are risks of recreating gene trees rather than species trees.

23.3.3 Molecular Cytogenetics

Identification of the most common gametic ploidy level of each accession in a polyploid species, such as yams, is necessary for efficient hybridization. It is of practical importance for yam breeders to determine the ploidy status of clones, especially of new introductions, before they can be utilized in a breeding program, to enable matching of ploidy levels as well as facilitate ploidy manipulations in intraspecific crosses. The existence of various ploidy levels and the lack of a diploid relative to the cultivated polyploid yams have greatly complicated genetic studies of the crop. Unlike most plants, differences in ploidy levels in yam plants are not reflected by any characteristic morphological feature. Phenotypic differences are expectedly greater within than between ploidy levels as also observed in other species (Dessauw 1988). Thus, cytological irregularities leading to erratic flowering and reproductive behavior are expected. Observations have been restricted in most cases to the determination of chromosome numbers and chromosome pairing from mitotic (Sharma and De 1956; Raghavan 1958, 1959; Ramachandran 1968; Essad 1984) and meiotic (Abraham and Nair 1990; Abraham 1998) cells. However, because yam chromosomes are small, generally dot-like, and most often clumped, determining ploidy levels by counting chromosomes is tedious and difficult (Baquar 1980; Zoundjihekpon et al. 1990).

Our current knowledge of yam ploidy is based on the basic chromosome number of 10 or nine, with a high frequency of polyploid species (Essad 1984; Zoundjihekpon et al. 1990; Gamiette et al. 1999; Dansi et al. 2000c, 2001; Egesi et al. 2002). Tetraploid species are the most frequent, followed by 6x and 8x forms in similar proportions. The base chromosome number $x = 10$ is reported in all the Asian species, but is found in only 52% of the African species and 13% of the American species examined so far. The remaining African and American species are considered to have a basic number of $x = 9$ (Essad 1984). In segregating populations of water yam (*D. alata*) and white Guinea yam (*D. rotundata*) ($2n = 4x = 40$), the observed segregation of AFLP markers reflected a disomic inheritance

(Mignouna et al. 2002a, b). These results indicated an allotetraploid structure for *D. rotundata* and *D. alata*. However, segregation analysis using isozyme and microsatellites markers led to the conclusion that *D. rotundata*, belonging to the botanical section *Enantiophyllum*, is a diploid species (Scarcelli et al. 2005). *D. trifida* was considered to be an octoploid species with 80 chromosomes ($x = 10$) (Esad 1984). In microsatellite segregation analysis, individual patterns showed a maximum of four alleles, strongly suggesting that *D. trifida* is a tetraploid species with $2n = 4x = 80$ chromosomes (Hochu et al. 2006). Bousalem et al. (2006) used cytogenetic evidence to show that the species is autotetraploid with a basic chromosome number of $x = 20$. Interestingly, Segarra-Moragues et al. (2004) concluded that the two species of the Bordera section, *D. pyrenaica* and *D. chouardii* (Caddick et al. 2002b) endemic to the Pyrenees (Spain and France), are allotetraploid with the chromosome base number of $x = 6$, which was not previously reported within the Dioscoreaceae. The finding of two new basic chromosome numbers, $x = 6$ (Segarra-Moragues and Catalán 2003; Segarra-Moragues et al. 2004) and $x = 20$ (Scarcelli et al. 2005), raises questions on the validity of the current ploidy data in the genus *Dioscorea*. If these new basic chromosome numbers are confirmed in a larger number of yam species, that should lead us to reconsider the basic chromosome number of yams on a more general level and, as a consequence, to decrease the level of ploidy of at least some species.

23.3.4 Genetic Diversity

Molecular markers are increasingly being used to examine the genetic diversity of cultivated and wild yam species (Mignouna et al. 2005b). Dansi et al. (1999) used a comparative morphological study to establish linkages between Guinea yam morphotypes/cultivar groups and their wild relatives. RAPD markers showed considerable variability when used for cultivar identification of Jamaican yam cultivars belonging to five food yam species: *D. alata*, *D. cayenensis*, *D. esculenta*, *D. rotundata*, and *D. trifida* (Asemota et al. 1996). The usefulness of RAPD as a discriminative and informative marker system in yam was also demonstrated by Ramser et al. (1996) using 23 *D. bulbifera* accessions collected from different geographic locations in Africa, Asia, and Oceania. That study also provided evidence in support of an earlier proposal of the independent domestication of this species in Africa and Asia.

Mignouna et al. (1998) found one varietal group among germplasm originating from Cameroon clustered separately from all other West African genotypes, indicating that this group constitutes a separate gene pool, which could be useful for genetic improvement of West African *Dioscorea* germplasm. A study to investigate the genetic relationships among West and Central African *D. rotundata* germplasm revealed a low level of genetic similarity between the yam accessions, with each genotype being identified as a unique individual using the three marker assays (Mignouna et al. 2003b). This study confirmed the high intraspecific variation within *D. rotundata* reported by Asemota et al. (1996), Mignouna et al. (1998), and Dansi et al. (2000a, b). Tostain et al. (2006) surveyed the diversity at 10 microsatellite loci for 146 *D. rotundata* accessions from Benin and the diversity of six microsatellite loci on 56 others. A significant excess of heterozygotes was observed at nine of the 15 polymorphic loci, which is expected in this vegetatively propagated crop. The significant excess of homozygotes, estimated at two loci, could be explained by the presence of null alleles.

Malapa et al. (2005) showed that *D. alata* is a heterogeneous species that shares a common genetic background with *D. nummularia*. Cluster analysis, using UPGMA (unweighted pair group method with arithmetic mean) based on AFLP profiles, revealed the existence of three major groups of genotypes within *D. alata*, each assembling accessions from distant geographical origins and different ploidy levels. Lebot et al. (1998) found no correlations between morphotypes, chemotypes, and zymotypes of 269 cultivars of *D. alata* (originating from the South Pacific, Asia, Africa, and the Caribbean), which were analyzed with four enzyme systems, including 6-PGD. The existing genetic variation is believed to be due to sexual recombination imposed by outcrossing (Lebot et al. 1998; Malapa et al. 2005).

Mignouna et al. (2005a) investigated genetic relationships among wild and cultivated yams in Nigeria and found that *D. rotundata* cultivars appeared most closely related to *D. praehensilis* and *D. liebrechtsiana* De Wild. *D. abyssinica* was widespread in the northern savannahs of the country. Similar to the situation with *D. praehensilis*, cultivars classified in 10 cultivar groups were morphologically very similar to *D. abyssinica* and might have been domesticated from this species (Chaïr et al. 2005). Isozyme analysis of wild yam species from Côte d'Ivoire revealed three groups: annual, semi-perennial, and perennial. Some cultivated accessions clustered with annual wild species, whereas others clustered with semi-perennial or perennial species (Hamon 1987). For Miège (1968), *D. burkilliana* and *D. minutiflora* are two morphologically very close species that differ only by the characteristics of their below-ground parts. However, Mignouna et al. (2003c) used 6-PGD isozyme analysis to show that the two species are genetically distinct. The principal species associations revealed by cluster analysis were *D. abyssinica/D. praehensilis*, *D. liebrechtsiana/D. praehensilis*, *D. mangenotiana/D. praehensilis*, *D. rotundata/D. praehensilis*, *D. cayenensis/D. burkilliana*.

There is unanimity among farmers and considerable agreement in research findings (Hamon 1987; Terauchi et al. 1992) that all the cultivated forms of the *D. cayenensis/D. rotundata* complex are the products of an ancient, or more or less recent, domestication of the four major wild species *D. abyssinica* Hochst, *D. praehensilis* Benth., *D. burkilliana* Miège, and *D. mangenotiana* Miège) a process that is still in progress in certain parts of West and Central Africa (Dumont and Vernier 2000; Mignouna and Dansi 2003; Scarcelli et al. 2006a, b). Mignouna and Dansi (2003) called for a revision of the taxonomy of *Dioscorea* species because they found it difficult to understand how individuals identified in the wild as *D. praehensilis* or *D. abyssinica* can directly become *D. rotundata* or *D. cayenensis* following "domestication" without any genetic change. In fact, Mignouna and Dansi (2003) showed that predomesticated yam plants could not always be clearly identified as belonging to either wild or cultivated species.

To assess the effect of farmers' practices on the diversity of *D. cayenensis–D. rotundata* cultivars, Scarcelli et al. (2006a) used AFLP analysis of a total of 213 yam accessions consisting of predomesticated yams, *D. cayenensis–D. rotundata*, *D. abyssinica,* and *D. praehensilis*. Of the 32 predomesticated accessions, 16% clustered with *D. praehensilis*, 37% with *D. abyssinica,* and the remaining 47% with *D. cayenensis–D. rotundata*. They thus demonstrated the use of wild plants by farmers in their domestication process and showed that through domestication farmers influence and increase the genetic diversity in yam by using sexual reproduction of wild and possibly cultivated yams. In a related study on the impact of ennoblement of spontaneous yams on the genetic diversity of yam in Benin, Scarcelli et al. (2006b) used 11 microsatellite markers to analyze yam tubers from a small village in northern Benin and demonstrated that wild × cultivated hybrids are spontaneously formed. Many of the spontaneous yams collected by farmers from surrounding savannah areas for ennoblement were shown to be wild and hybrid genotypes. They demonstrated that some yam varieties have a wild or hybrid signature and performed a broader-ranging genetic analysis on yam material from throughout Benin, which revealed that ennoblement is practiced in different ecological and ethno-linguistic regions. By maintaining a mixed yam propagation system (sexual cycle and asexual propagation), farmers ensure widespread cultivation of the best genotypes while preserving the potential for future adaptation. The mechanism underlying phenotypic modifications during "domestication" is unknown. They could result from phenotypic plasticity, epigenetic modifications, or somatic mutations. The latter two explanations are compatible with the fact that morphological changes are maintained through vegetative multiplication.

23.4 Genetic Mapping and Tagging in Yam

Molecular genetic maps and marker-aided analysis of complex traits can be used to elucidate the genetic control of yield potential and tuber quality and to locate genes of pest and disease resistance, nutrient use efficiency, tuberization, and flowering. For these reasons, a concerted effort to map the yam genome and dissect the inheritance of complex traits was initiated at IITA. It was anticipated that cultivated yams would have their origin from a cross between genetically distinct individuals, so the alleles derived from each parent may be different. One general approach to mapping plants of this type is to examine the genotypes of selfed progeny; however, this is not feasible for dioecious yams, so the approach taken was to generate multiple F_1 individuals derived from crosses between the same parents, male or female. F_1 mapping populations of *D. alata* and *D. rotundata* were subjected to in vitro micropropagation based on techniques developed by Ng (1992). *D. rotundata* populations segregated components of resistance to *Yam mosaic virus* (YMV), genus *Potyvirus* (Mignouna et al. 2001b), while the *D. alata* populations segregated for yam anthracnose disease resistance (Mignouna et al. 2001a).

YMV is a major limiting factor for stable production of yams and *D. rotundata* is particularly susceptible to the virus (Thouvenel and Dumont 1990). A study of the genetic control of YMV resistance in three *D. rotundata* cultivars to a Nigerian isolate of YMV showed that resistance is manifested differentially as the action of a single dominant gene in simplex condition or a major recessive gene in duplex condition (Mignouna et al. 2001b). The dominant locus that contributes to YMV resistance was tentatively named *Ymv-1* until tests of allelism are conducted. Anthracnose disease, caused by *C. gloeosporioides* (Abang et al. 2003), is a major constraint to the production of yam worldwide (Winch et al. 1984; McDonald et al. 1998), with *D. alata*, the most widely distributed species, being particularly susceptible to the disease. Initial genetic inheritance studies showed that resistance to yam anthracnose in *D. alata* is dominantly but quantitatively inherited (Mignouna et al. 2001a). A single major dominant locus controlling resistance in the breeding line TDa 95/00328 was tentatively designated *Dcg-1* until allelism is investigated. The efficiency and effectiveness of breeding for YMV and anthracnose resistance will be greatly improved by marker-assisted selection based on genetic mapping of major genes controlling the resistance.

23.4.1 Linkage Mapping

Chromosome pairing in tetraploids can occur such that only homologues pair or such that any two homeologues may pair. These two types of pairing have very different consequences for segregation patterns so that these plants may, in the extreme, exhibit either diploid or tetraploid genetics. Intermediate types of behavior may also occur. Thus it was important to establish which type of segregation was being observed in the cultivated yams. Genes controlling important traits such as yield, tuber quality, and pest and disease resistance are usually distributed among several quantitative trait loci (QTLs), which may not be linked, thus making these traits difficult to manipulate using conventional breeding methods. The recessive nature of YMV resistance in some *D. rotundata* genotypes means that such resistance cannot be easily tracked at the phenotypic level, demanding refined diagnostic procedures such as molecular mapping for detailed genetic localization of specific genes (Mignouna et al. 2001b). Screening by molecular markers linked to QTLs has the advantage of selecting pairs of parents with genes at different loci for the same trait (Solomon-Blackburn and Barker 2001).

Genetic mapping using AFLP led to construction of the first, separate, comprehensive, molecular linkage maps of *D. rotundata* and *D. alata* (Mignouna et al. 2002c, d). The *D. rotundata* map was based on 341 co-dominantly scored AFLP markers segregating in an intraspecific F_1 cross (Mignouna et al. 2002d). Separate maternal and paternal linkage maps were constructed, comprising 12 and 13 linkage groups, respectively. The mapping population was produced by crossing a landrace, TDr 93-1, as female parent and a breeder's line, TDr 87/00211, as the male parent. The markers segregated like a diploid cross-pollinator population, suggesting that the

D. rotundata genome is an allotetraploid (2n = 4x = 40). More recent findings have confirmed that *D. rotundata* is a diploid species (Scarcelli et al. 2005). Three QTLs with effect on resistance to YMV were identified on the maternal linkage map, while one QTL for YMV was detected on the paternal linkage map (Mignouna et al. 2002d). These results showed that both parents contributed to resistance in the progeny.

Similarly, a genetic linkage map of the water yam (*D. alata*) genome was constructed based on 469 co-dominantly scored AFLP markers segregating in an intraspecific F_1 cross (Mignouna et al. 2002c). The F_1 was obtained by crossing two improved breeding lines, TDa 95/00328 as female parent and TDa 87/01091 as the male parent. The 469 markers were mapped on 20 linkage groups with a total map length of 1,233 cM. Again, the markers segregated as in a diploid cross-pollinator population, suggesting that the water yam genome is allotetraploid (2n=4x=40). One QTL located on linkage group 2 was found to be associated with anthracnose resistance, explaining 10% of the total phenotypic variance (Mignouna et al. 2002c).

Conservative estimates put the genome coverage of the *D. rotundata* and *D. alata* maps at 56% and 65%, respectively. There are several reasons why the maps may not give complete coverage. The most obvious is that the two parents may have some common ancestry so that segments of the linkage maps may be devoid of polymorphism and thus cannot be identified in genetic analysis.

One approach towards gaining insights on this issue would be to align the *D. alata* and *D. rotundata* maps. This would give us additional confidence in the general map structures and enable the development of suitable markers for genomic surveys of other populations. An attempt was made to derive gene sequence-based markers, but unfortunately the cDNA library used for this analysis contained an unexpectedly high proportion of rRNA sequences. Nevertheless, this remains a viable objective, and would also permit the alignment of these maps with that recently presented for diploid *D. tokoro,* 2n=2x=20 (Terauchi and Kahl 1999). Both maps provide useful tools for further genetic analysis of agronomically important traits in yam. While AFLPs continue to be identified and used for mapping the yam genome, efforts are geared towards saturating the map with simple sequence repeats (SSRs) and expressed sequence tags (ESTs), for greater ease of application in yam breeding.

23.4.2 Gene Tagging

Bulked segregant analysis has been shown to be efficient for initial identification of disease resistance-linked markers. The approach has been successfully applied in yams for identification of YMV and anthracnose resistance genes (Mignouna et al. 2002a, b). Two RAPD markers. $OPW18_{850}$ and $OPX15_{850}$, closely linked in coupling phase with the dominant YMV-resistance locus *Ymv*-1 were identified. These markers successfully identified the resistance gene in resistant genotypes among a sample of 12 *D. rotundata* varieties (Mignouna et al. 2002b). Similarly, a single locus, *Dcg-1*, that contributes to anthracnose resistance was identified in

D. alata. Two RAPD markers, $OPI17_{1700}$ and $OPE6_{950}$, closely linked in coupling phase with *Dcg-1* were identified (Mignouna et al. 2002a). Both markers successfully identified *Dcg-1* in resistant *D. alata* genotypes among 34 breeding lines, indicating their potential use in marker-assisted selection (MAS). The RAPD markers identified in these studies will be made more reliable and specific and easier to apply for indirect selection by converting them into co-dominant PCR-based sequence-characterized amplified regions. Further AFLP mapping is planned to identify additional QTLs and strengthen existing marker-QTL linkages. Candidate gene analyses are yet to be employed to investigate a variety of traits. To date, significant associations have been demonstrated for disease resistance in numerous crops. The yam breeding program at IITA plans to use MAS for selecting parental lines for breeding purposes. It is likely that as QTL experiments are expanded, additional genes will be identified for use in breeding.

23.5 Functional Genomics

The development of genomic resources and technology is a major focus in the yam genetics and breeding community. A cDNA library, produced from male flowers, was constructed in Bluescript vector and used for EST analysis (H. Mignouna, unpublished data). This approach has proven to be efficient for gene identification, gene expression profiling, and cataloging. It also provides markers and resources for the development of cDNA microarrays. Microarrays are not yet available for yams, mainly because the number of available gene sequences is still very small. Two cDNA libraries, one each for *D. alata* genotypes resistant or susceptible to yam anthracnose disease, have also been constructed recently (based on total RNA isolated from young leaves) towards identification of clones that are differentially expressed in the two genotypes (Narina et al. 2007). The libraries from the resistant and susceptible genotypes now have 10,000 and 6,000 cDNA clones, respectively, which are being sequenced.

Another reliable and potentially powerful way to identify candidate loci controlling agronomic traits in yam is application of the cDNA/AFLP technique, which generates polymorphic transcript-derived fragments (TDFs) between the parents of a mapping cross. cDNA generated from total RNA was subjected to cDNA-AFLP techniques to gain molecular insights and identify differentially expressed genes up-regulated and down-regulated during the dormancy in yam tubers (Kone-Coulibaly et al. 2003). Two primer pairs were identified that had equal potential for producing the same number of TDFs in dormant yam samples. The resulting TDFs from postharvest-treated tubers will aid in the selection of putative up- and down-regulated fragments during yam dormancy. Once candidate genes have been identified, they can be employed in gene tagging and QTL mapping studies to look for associations between the candidate gene and the trait in question. The availability of a BAC library and the development of an effective system for transforming yam with large DNA fragments will provide conclusive evidence of the contribution of the candidate gene through complementation studies.

23.5.1 EST Development

The genome size of *D. rotundata* was estimated by Feulgen-stained root tip nuclei to be 0.8 pg per haploid nucleus, and thus is equivalent to the genome size of species such as rice, soybean, and spinach (Conlan et al. 1995). The current *D. rotundata* map covers a minimum of 56% of the yam genome. Based on the haploid nuclear DNA content of *D. rotundata* of 800 Mbp/1C, the physical distance per map unit could be estimated at 400 kb per cM, making map-based gene cloning feasible (Mignouna et al. 2002d). We have generated 1100 ESTs from cDNA clones randomly picked from libraries constructed from male flowers. However, most of the sequenced ESTs were either ribosomal or housekeeping genes. To understand the physiological complexity of the yam genome, expression and/or functional gene analyses need to be undertaken. Northern analysis and differential display PCR techniques could be used, but these techniques have limitations in the number of genes that can be analyzed simultaneously. There is a need to develop approaches such as the use of cDNA microarrays. Other plant microarrays could be evaluated for use. As pointed out earlier, the development of a large number of ESTs will allow larger scale expression analysis.

23.5.2 Transformation

Attempts have been made to develop *in vitro* breeding strategies (such as somatic hybridization and gene insertion techniques) to overcome breeding barriers and to hasten the genetic improvement of food yams. For instance, Mantell (1994) fused protoplast mixtures between disease-sensitive and disease-resistant clones of *D. alata* in attempts to develop somatic hybrids with increased tolerance to anthracnose. There is considerable scope for introducing specific genes encoding resistance to fungal diseases (i.e., glucanase, chitinase, and antimicrobial protein gene constructs) and to nonpersistently transmitted potyviruses (i.e., sense and antisense genes of the coat protein of yam mosaic viruses). Three prerequisites for applying genetic transformation for plant improvement are: (1) a reliable regeneration system that is compatible with transformation methods allowing regeneration of transgenic plants; (2) an efficient way to introduce DNA into the regenerable cells; and (3) a procedure to select and regenerate transformed plants at a satisfactory frequency (Birch 1997).

Early plant transformation experiments on yam were hampered by false positive transformants that were found to be due to endophytic bacteria which exist within aseptically micropropagated shoot cultures and which express β-glucuronidase (Tor et al. 1992). Eventually, Tor et al. (1993) successfully demonstrated stable genetic transformation of *D. alata* embryogenic cell suspensions using biolistic insertion methods. However, biolistic approaches have a number of disadvantages such as the production of chimeric colonies containing mixtures of transformed and non-transformed cells and the instability of such colonies to retain inserted genes once antibiotic and/or herbicide selection conditions are withdrawn following plant regeneration. Later efforts gave rise to successful yam protoplast culture leading to cell regeneration and direct gene transfer into yam protoplasts (Tor et al. 1998). Embryogenic cell suspension protoplasts of *D. alata* cv. Oriental Lisbon were successfully transformed using a standard polyethylene glycol-mediated uptake method. The availability of a protoplast system for transient gene expression studies in yams is expected to speed efforts towards the transformation of these tuber crops. The functional expression of valuable disease resistance genes, such as viral coat protein genes of yam mosaic viruses in either sense or anti-sense configurations, and combinatorial chitinase, glucanase, and anti-microbial protein genes driven by a range of either dicot promoters (NOS and CaMV35S) or monocot promoters such as ubiquitin, actin, ricin, and *emu,* needs to be investigated.

A number of host defense genes that could be good candidates for use in yam transformation have been characterized. Five chitinase isoforms, designated A, E, F, H1, and G, from yam tuber have been purified and characterized (Arakane et al. 2000). Chitinases E, F, and H1 had the highest lytic activity against the pathogen *Fusarium oxysporum*, while chitinase E was shown to be a possible biocontrol agent against strawberry powdery mildew (*Spherotheca humuli*) (Karasuda et al. 2003). Yam chitinase E has a similar amino acid sequence to a reported family 19 chitinase from *D. japonica* (Araki et al. 1992). Mitsunaga et al. (2004) cloned and sequenced a class IV chitinase from yam (*D. opposita*). The deduced amino acid sequence showed 50 to 59% identity to class IV chitinases from other plants. The yam chitinase, however, had an additional sequence of eight amino acids (a C-terminal extension) following the cysteine that was reported as the last amino acid for other class IV chitinases; this extension is perhaps involved in subcellular localization. A homology model based on the structure of a class II chitinase from barley suggested that the class IV enzyme recognizes an even shorter segment of the substrate than class I or II enzymes. This might explain why class IV enzymes are better suited to attack against pathogen cell walls.

23.6 Perspectives

The development and application of biotechnology tools are necessary to complement field breeding of yams. Molecular approaches have the potential to make yam breeding more efficient to reduce the cost and time required to produce new varieties. However, understanding and exploiting the complexity of the yam genome for improved yield and quality of yams remains a huge challenge. Large-scale gene identification and mapping have taken place in a number of model plants (e.g., *Arabidopsis* and *Medicago*) as well as some important food crops (e.g., rice, soybean, tomato, and maize). Whole genome sequencing and expression analyses have been conducted in *Arabidopsis* and rice and offer opportunities to understand the biological complexity of other plant genomes. However, these advances are yet to benefit under-researched tropical food crops such as yams (Nelson et al. 2004).

Completed genome sequences provide templates for the design of genome analysis tools in "orphan" crops lacking sequence information. Feltus et al. (2006) have shown that conserved-intron scanning primers are an effective means to explore poorly characterized genomes. Genes involved in many biochemical pathways and processes are similar across the plant kingdom (Thorup et al. 2000). Functions such as gene regulation, general metabolism, nutrient acquisition, disease resistance, general defense, flowering time, and flower development are largely conserved across taxa. Comparative mapping studies reveal that gene order is conserved for chromosomal segments among grass species (Devos and Gale 2000), with weaker chromosomal colinearity between monocots and dicots (Bennetzen 2000). Given the unique position of yams between monocots and dicots, it is doubtful how the work on models such as *Arabidopsis* and *Medicago* will benefit the species (e.g., Conlan et al. 1995). Although *Dioscorea* is a complex and highly variable genus, with several aspects of its biology still unresolved, we consider that there is a case for the adoption of yam as a "model" for plant genomics.

Efforts in yam genetics and genomics should be pursued and we believe the following specific areas need to be addressed in the near future. There is still a paucity of information, and some of the reports are conflicting, on yam phylogeny and the evolution of *Dioscorea* based on morphological, cytological, and molecular data. In this regard, the importance of non heritable or heritable epimutations in the development of yams should be investigated. Also, there is need for comparative analysis of the genomes of potato (dicotyledon) and yam (monocotyledon). The relationship between monoecious plants of *D. rotundata* (Scarcelli et al. 2005) and their normally dioecious relatives deserves further examination, as well as the nature of spontaneous hybrids in sympatric populations of wild and cultivated yams in Africa (Scarcelli et al. 2006). Selection and domestication of other annual yam species, including several indigenous West African and Malagasy species, should be undertaken before the natural populations disappear. Intraspecific hybridization between genetically distant landraces should be continued; for instance, between early and late maturing varieties of *D. rotundata* or between *D. alata* with and without bulbils. Hybrids obtained from these crosses do not require embryo culture.

Genetic linkage mapping of the two most important yam species (*D. rotundata* and *D. alata*) should be pursued. Denser genetic maps of each species and a consensus map for both must be constructed for practical breeding and germplasm enhancement purposes. QTL mapping should be reactivated with the initial identification of markers linked to disease resistance genes. Candidate gene identification using microarray and other approaches should be conducted to pin down the genes or QTLs involved in important agronomic traits. BAC library construction should be initiated, and efforts towards establishing a system for yam transformation should now be given more impetus (Tör et al. 1998). Embryo rescue will enable yam breeders to successfully make wide crosses with a greater number of related species of wild yams and have access to a much wider range of genes that can be used for the genetic improvement of yams. Wide crosses and embryo culture hold great promise for the transfer of tolerance to biotic and abiotic stresses from wild relatives to cultivated yams. Research to better understand the biology and agronomy of wild relatives will greatly facilitate efforts aimed at unlocking the genetic potential hidden in the wild yam germplasm.

Acknowledgments The authors would like to acknowledge the financial support from Gatsby Charitable Foundation, UK, through funds to support yam genome analysis at IITA. We thank Prof. Stephen Kresovich and the staff of the Institute for Genomic Diversity of Cornell University for their technical assistance in developing genomics tools for yam genome analysis.

References

Abang MM, Winter S, Mignouna HD, Green KR, Asiedu R (2003) Molecular taxonomic, epidemiological and population genetic approaches to understanding yam anthracnose disease. African J Biotechnol 2:486–496

Abraham KA (1998) Occurrence of hexaploid males in *Dioscorea alata* L. Euphytica 99:5–7

Abraham KA, Nair PG (1990) Vegetative and pseudogamous parthenocarpy in *Dioscorea alata*. J Root Crops 16:58–60

Arakane Y, Hoshika H, Kawashima N, Fujiya-Tsujimoto C, Sasaki Y, et al. (2000) Comparison of chitinase isozymes from yam tuber enzymatic factor controlling the lytic activity of chitinases. Biosci Biotechnol Biochem 64:723–730

Araki T, Funatsu J, Kuramoto M, Konno H, Torikata T (1992) The complete amino acid sequence of yam (*Dioscorea japonica*) chitinase. A newly identified acidic class I chitinase. J Biol Chem 267:19944–19947

Asemota HN, Ramser J, Lopez-Peralta C, Weising K, Kahl G (1996) Genetic variation and cultivar identification of Jamaican yam germplasm by random amplified polymorphic DNA analysis. Euphytica 92:341–351

Asiedu R, Ng SYC, Bai KV, Ekanayake IJ, Wanyera NMW (1998) Genetic Improvement. In: Orkwor GC, Asiedu R, Ekanayake IJ (eds) Food yams: Advances in research. Ibadan, Nigeria: IITA and NRCRI pp 63–104

Ayensu ES (1972) Dioscoreales. In: Metcalfe CR (ed) Anatomy of the monocotyledons. Clarendon Press, Oxford, UK, pp 182

Ayensu ES, Coursey DG (1972) Guinea yams. The botany, ethnobotany, use and possible future of yams in West Africa. Econ Bot 26:301–318

Baquar SR (1980) Chromosome behaviour in Nigerian yams (*Dioscorea*). Genetica 54:1–9

Barrau J (1965) Histoire et prehistoire horticole de l'Oceanie tropicale. J Soc Oceaniste 21:55–78

Bennetzen JL (2000) Comparative sequence analysis of plant nuclear genomes: Microcolinearity and its many exceptions. Plant Cell 12:1021–1029

Bharathan G (1996) Does the monocot mode of leaf development characterize all monocots? Aliso 14:271–27

Bharathan G, Janssen B-J, Kellogg EA, Sinha N (1999) Phylogenetic relationships and evolution of the KNOTTED class of plant homeodomain proteins. Mol Biol Evol 16:553–563

Birch RG (1997) Plant transformation: problems and strategies for practical application. Annu Rev Plant Physiol Mol Biol 48:297–326

Bousalem M, Arnau G, Hochu I, Arnolin R, Viader V, et al. (2006) Microsatellite segregation analysis and cytogenetic evidence for tetrasomic inheritance in the American yam *Dioscorea trifida* and a new basic chromosome number in the *Dioscorea*e. Theor Appl Genet 113:439–451

Burkill IH (1960) The organography and the evolution of the *Dioscoreaceae*, the family of the yams. J Linn Soc (Bot) London 56:319–412

Caddick LR, Rudall PJ, Wilkin P, Hedderson TAJ, Chase MW (2002a) Phylogenetics of Dioscoreales based on combined analyses of morphological and molecular data. Bot J Linn Soc 138:123–144

Caddick LR, Wilkin P, Rudall PJ, Hedderson TAJ, Chase MW (2002b)Yams reclassified: a recircumscription of Dioscoreaceae and Dioscoreales. Taxon 51:103–114

Chaïr H, Perrier X, Agbangla C, Marchand JL, Dainou O, et al. (2005) Use of cpSSRs for the characterisation of yam phylogeny in Benin. Genome 48:674–684

Chu EP, Figueiredo-Ribeiro RCL (1991) Native and exotic species of *Dioscorea* used as food in Brazil. Econ Bot 45:467–479

Conlan SR, Griffiths LA, Napier JA, Shewry PR, Mantell S, et al. (1995) Isolation and characterization of cDNA clones representing the genes encoding the major tuber storage protein (dioscorin) of yam (*Dioscorea cayenensis* Lam). Plant Mol Biol 28:369–380

Coursey DG (1983) Yams. In: Chan HC (ed) Handbook of tropical foods. Marcel Dekker Inc, New York, USA

Dansi A, Mignouna HD, Zoundjihekpon J, Sangare A, Asiedu R, et al. (1999) Morphological diversity, cultivar groups and possible descent in the cultivated yams (*Dioscorea cayenensis–D. rotundata* complex) of Benin Republic. Genet Resourc Crop Evol 46:371–388

Dansi A, Mignouna HD, Zoudjihekpon J, Sangare A, Asiedu R, et al. (2000a) Identification of some Benin Republic's Guinea yam (*Dioscorea cayenensis/Dioscorea rotundata*) cultivars using randomly amplified polymorphic DNA. Genet Resourc Crop Evol 47:619–625

Dansi A, Mignouna HD, Zoundjihékpon J, Sangaré A, Asiedu R, et al. (2000b) Using isozyme polymorphism to assess genetic variation within cultivated yams (*Dioscorea cayenensis/Dioscorea rotundata* complex) of the Benin Republic. Genet Resourc Crop Evol 47:371–383

Dansi A, Pillay M, Mignouna HD, Mondeil F, Dainou O (2000c) Ploidy level of the cultivated yams (*Dioscorea cayenensis/D. rotundata* complex) from Benin Republic as determined by chromosome counting and flow cytometry. African Crop Sci J 8:355–364

Dansi A, Mignouna HD, Pillay M, Zok S (2001) Ploidy variation in the cultivated yams (*Dioscorea cayenensis-D. rotundata* complex) from Cameroon as determined by flow cytometry. Euphytica 119:301–307

Dessauw D (1988) Etude des facteurs de la sterilité du bananier (*Musa* spp) et des relations cytotaxonomiques entre *M. acuminata* et *M. balbisiana* Colla. Fruits 43:539–700

Devos KM, MD Gale (2000) Genome relationships: The grass model in current research. Plant Cell 12:637–646

Dumont R, Vernier P (2000) Domestication of yams (*Dioscorea cayenensis-rotundata* complex) within the Bariba ethnic group in Benin. Outlook Agric 29:137–142

Egesi CN, Pillay M, Asiedu R, Egunjobi JK (2002) Ploidy analysis in water yam, *Dioscorea alata* L. germplasm. Euphytica 128:225–230

Essad S (1984) Variation géographique des nombres chromosomiques de base et polyploïdie dans le genre *Dioscorea* à propos du dénombrement des espèces transversa Brown, pilosiuscula Bert et trifida. Agronomie 4:611–617

FAO (1999) FAO's Position Paper. Food and Agriculture Organization of the United Nations, Rome, Italy http://www.fao.org/

FAO (2006) FAOSTAT Agricultural database: agricultural production, crops primary, yams. Food and Agriculture Organization, Rome, Italy (http://www.fao.org)

Feltus FA, Singh HP, Lohithaswa HC, Schulze SR, Silva TD, et al. (2006) A comparative genomics strategy for targeted discovery of single-nucleotide polymorphisms and conserved-noncoding sequences in orphan crops. Plant Physiol 140:1183–1191

Frohne D, Jensen U (1998) Systematik des Pflanzenreichs: unter besonderer Berücksichtigung chemischer Merkmale und pflanzlicher Drogen. Wissenschaftliche Verlagsgesellschaft, Stuttgart, Germany

Gamiette F, Bakry F, Ano G (1999) Ploidy determination of some yam species (*Dioscorea* spp) by flow cytometry and conventional chromosomes counting. Genet Resourc Crop Evol 46:19–27

Hahn SK, Osiru DSO, Akoroda MO, Otoo JA (1987) Yam production and its future prospects. Outlook Agric 16:105–110

Hamon P (1987) Structure, origine génétique des ignames cultivées du complexe D. cayenensis-rotundata et domestication des ignames en Afrique de l'Ouest. Thèse de Doctorat es-Sciences, Université Paris XI, Centre d'Orsay: 223 pp

Hamon P, Touré B (1990a) Characterisation of traditional yam varieties belonging to the *Dioscorea cayenensis–rotundata* complex by their isozymic patterns. Euphytica 46:101–107

Hamon P, Touré B (1990b) The classification of the cultivated yams (*Dioscorea cayenensis–rotundata* complex) of West Africa. Euphytica 47:179–187

Hamon P, Brizard JP, Zoundjihekpon J, Duperray C, Borgel A (1992) Etude des index d'ADN de huit espèces d'ignames (*Dioscorea* species) par cytométrie en flux. Can J Bot 70:996–1000

Hochu I, Santoni S, Bousalem M (2006) Isolation, characterization and cross-species amplification of microsatellite DNA loci in the tropical American yam *Dioscorea trifida*. Mol Ecol Notes 6:137–140

IITA (1988) IITA Strategic Plan 1989–2000. IITA, Ibadan. 108 pp

Karasuda S, Tanaka S, Kajihara H, Yamamoto Y, Koga D (2003) Plant chitinase as a possible biocontrol agent for use instead of chemical fungicides. Biosci Biotechnol Biochem 67:221–224

Kawabe A, Miyashita NT, Terauchi R (1997) Phylogenetic relationship among the section *Stenophora* in the genus *Dioscorea* based on the analysis of the nucleotide sequence variation in the phosphoglucose isomerase (pgi) gene. Genes Genet Syst 72:253–262

Kone-Coulibaly S, Egnin M, He G, Prakash CS (2003) Profiling differentially expressed gene in yam (*Dioscorea rotundata* Poir) during dormancy. In Vitro Cell Dev Biol 39(4):27A

Lebot V, Trilles B Noyer JL, Modesto J (1998) Genetic relationships between *Dioscorea alata* L cultivars. Genet Resourc Crop Evol 45:499–509

Malapa R, Arnau G, Noyer JL, Lebot V (2005) Genetic diversity of the greater yam (*Dioscorea alata* L) and relatedness to *D. nummularia* Lam. and *D. transversa* Br. as revealed with AFLP markers. Genet Resourc Crop Evol 52:919–929

Mantell SH (1994) Summary of the Final report of EU Contract TS2-A-117: Development of anthracnose disease resistant *Dioscorea* yams using somatic fusion techniques. In: Risopoulos S (ed) Projets de recherche 1987 – 1991 Vol 1. CTA/DGXII Joint Publication pp 69–75

Martin FW, Rhodes AM (1978) The relationship of *Dioscorea cayenensis* and *D. rotundata*. Trop Agric (Trinidad) 55:193–206

McDonald FD, Alleyne AT, Ogarro LW, Delauney AJ (1998) Yam anthracnose in the English-speaking islands of the Eastern Caribbean–successes and research advances in disease management. Trop Agric 75:53–57

Miège J (1968) *Dioscorea*ceae. In: Hepper FN (ed) Flora of West Tropical Africa. J Hutchinson & J M Dalziel Vol 3, Millbank, London, UK, pp 144–154

Miège J (1982a) Etude chimiotaxonomique de dix cultivars de Côte d'Ivoire relevant du complexe *D. cayenensis-D. rotundata*. In: Miège J, Lyonga SN (eds) Yams-Ignames. Claredon Press, Oxford pp 197–231

Miège J (1982b) Notes sur les espèces *Dioscorea cayenensis* Lamk. et *D. rotundata* Poir. In: Miège J, Lyonga SN (eds) Yams-Ignames. Oxford University Press, Oxford, UK, pp 367–375

Mignouna HD, Dansi A (2003) Yam (*Dioscorea* spp) domestication by the Nago and Fon ethnic groups in Benin. Genet Resourc Crop Evol 50:519–528

Mignouna HD, Ellis NTH, Asiedu R, Ng QN (1998) Analysis of genetic diversity in Guinea yams (*Dioscorea* spp) using AFLP fingerprinting. Trop Agric (Trinidad) 75:224–229

Mignouna HD, Abang MM, Green KR, Asiedu R (2001a) Inheritance of resistance in water yam (*Dioscorea alata*) to anthracnose (*Colletotrichum gloeosporioides*). Theor Appl Genet 103:52–55

Mignouna HD, Njukeng P, Abang MM, Asiedu R (2001b) Inheritance of resistance to *Yam mosaic virus*, genus *Potyvirus*, in white yam (*Dioscorea rotundata*). Theor Appl Genet 103:1196–1200

Mignouna HD, Abang MM, Onasanya A, Asiedu R (2002a) Identification and application of RAPD markers for anthracnose resistance in water yam (*Dioscorea alata*). Ann Appl Biol 141:61–66

Mignouna HD, Abang MM, Onasanya A, Agindotan B, Asiedu R (2002b) Identification and potential use of RAPD markers linked to *Yam mosaic virus* resistance in white yam (*Dioscorea rotundata* Poir). Ann Appl Biol 140:163–169

Mignouna HD, Mank RA, Ellis THN, van den Bosch N, Asiedu R, et al. (2002c) A genetic linkage map of water yam (*Dioscorea alata* L) based on AFLP markers and QTL analysis for anthracnose resistance. Theor Appl Genet 105:726–735

Mignouna HD, Mank RA, Ellis THN, van den Bosch N, Asiedu R, et al. (2002d) A genetic linkage map of Guinea yam (*Dioscorea rotundata* L) based on AFLP markers. Theor Appl Genet 105:716–725

Mignouna HD, Abang MM, Asiedu R (2003a) Harnessing modern biotechnology for tropical tuber crop improvement: Yam (*Dioscorea* spp) molecular breeding. African J Biotechnol 2:478–485

Mignouna HD, Abang MM, Fagbemi SA (2003b) A comparative assessment of molecular marker assays (AFLP, RAPD and SSR) for white yam (*Dioscorea rotundata* Poir) germplasm characterisation. Ann Appl Biol 142:269–276

Mignouna HD, Dansi A, Asiedu R (2003c) 6-phosphoglucose dehydrogenase (6-PGD) in yam (*Dioscorea* spp): variation and potential in germplasm characterization and classification. Plant Genet Resourc Newsl 133:27–30

Mignouna HD, Abang MM, Wanyera NW, Chikaleke VA, Asiedu R, et al. (2005a) PCR marker-based analysis of wild and cultivated yams (*Dioscorea* spp) in Nigeria: genetic relationships and implications for *ex situ* conservation. Genet Resourc Crop Evol 52:755–763

Mignouna HD, Abang MM, Dansi A, Asiedu R (2005b) Morphological, biochemical and molecular approaches to yam (*Dioscorea* spp) genetic resource characterization. In: Thangadurai D, Pullaiah T, Pinheiro de Carvalho MA (eds) Genetic Resources and Biotechnology Vol 1. Regency Publications, New Delhi, pp 162–185

Mitsunaga T, Iwase M, Ubhayasekera W, Mowbray SL, Koga D (2004) Molecular cloning of a genomic DNA encoding yam class IV chitinase. Biosci Biotechnol Biochem 68:1508–1517

Mizuki I, Tani N, Ishida K, Tsumura Y (2005) Development and characterization of microsatellite markers in a clonal plant, *Dioscorea japonica* Thunb. Mol Ecol Notes 5:721–723

Narina SSS, Andebrhan T, Mohamed A, Asiedu R, Mignouna HD (2007) Development of genomic tools for improvement of yam (*Dioscorea alata* L). Plant Animal Genome Conf, P21, p 106

Nelson RJ, Naylor RL, Jahn MM (2004) The role of genomics research in improvement of "orphan" crops. Crop Sci 44:1901–1904

Neuwinger HD (1996) African ethnobotany: poisons and drugs: chemistry, pharmacology, toxicology. Chapman and Hall, London UK

Ng SYC (1992) Micropropagation of white yam (*Dioscorea rotundata* Poir) In: Bajai VPS (Ed) Biotechnology in Agriculture and Forestry, High-tech and Micropropagation III, Vol 19. Springer-Verlag Berlin, Heidelberg. pp 135–159

Nweke FI, Ugwu BO, Asadu CLA, Ay P (1991) Production costs in the yam-based cropping systems of S.E. Nigeria. RCMD Research Monograph No 6. RCMD, IITA, Ibadan, 29 pp

Nweke FI, Okorji EC, Njoku JE, King DJ (1992) Elasticities of demand for major food items in a root and tuber-based food system: emphasis on yam and cassava in southeastern Nigeria. RCMP Research Monograph No 11, International Institute of Tropical Agriculture, Ibadan. pp 11–19

Onyilagha JC, Lowe J (1985) Studies on the relationship of *Dioscorea cayenensis* and *D. rotundata* cultivars. Euphytica 35:733–739

Orkwor GC, Ekanayake IJ (1998) Growth and development. In: Orkwor GC, Asiedu R, Ekanayake IJ (eds) Food yams: Advances in research. Ibadan, Nigeria: IITA and NRCRI pp 39–62

Purseglove JW (1988) Tropical crops: monocotyledons. Longman Scientific and Technical, Harlow, UK

Raghavan SR (1958) A chromosome survey of Indian *Dioscorea*. Proc Indian Acad Sci Sec B 48:59–63

Raghavan SR (1959) A note on some South Indian species of the genus *Dioscorea*. Curr Sci 28:337–338

Ramachandran K (1968) Cytological studies in *Dioscoreacea*. Cytologia 33:401–410
Ramser J, Lopez-Peralta C, Wetzel R, Weising K, Kahl G (1996) Genomic variation and relationships in aerial yam (*Dioscorea bulbifera* L) detected by random amplified polymorphic DNA. Genome 39:17–25
Ramser J, Weising K, Lopez-Peralta C, Terhalle W, Terauchi R, et al. (1997) Molecular marker-based taxonomy and phylogeny of Guinea yam (*Dioscorea rotundata-D. cayenensis*). Genome 40:903–915
Scarcelli N, Daïnou O, Agbangla C, Tostain S, Pham JL (2005) Segregation patterns of isozyme loci and microsatellite markers show the diploidy of African yam *Dioscorea rotundata* (2n = 40). Theor Appl Genet 111:226–232
Scarcelli N, Tostain S, Vigouroux Y, Agbanla C, Daïnou O, et al. (2006a) Farmers' use of wild relative and sexual reproduction in a vegetatively propagated crop. The case of yam in Benin. Mol Ecol 15:2421–2431
Scarcelli N, Tostain S, Mariac C, Agbangla C, Daïnou O, et al. (2006b) Genetic nature of yams (*Dioscorea* sp) domesticated by farmers in Benin (West Africa). Genet Resourc Crop Evol 53:121–130
Segarra-Moragues JG, Catalán P (2003) Life history variation between species of the relictual genus Borderea (*Dioscorea*ceae): phylogeography, genetic diversity, and population genetic structure assessed by RAPD markers. Biol J Linn Soc 80:483–498
Segarra-Moragues JG, Palop-Esteban M, Gonza'Lez-Candelas F, Catalán P (2004) Characterization of seven $(CTT)_n$ microsatellite loci in the Pyrenean Endemic Borderea pyrenaica (*Dioscorea*ceae): remarks on ploidy level and hybrid origin assessed through allozymes and microsatellite analyses. J Hered 95:177–183
Sharma AK, De DN (1956) Polyploidy in *Dioscorea*. Genetica 28:112–120
Solomon-Blackburn RM, Barker H (2001) Breeding virus resistant potatoes (*Solanum tuberosum*): a review of traditional and molecular approaches. Heredity 86:17–35
Terauchi R, Kahl G (1999) Mapping of the *Dioscorea tokoro* genome: AFLP markers linked to sex. Genome 42:752–762
Terauchi R, Konuma A (1994) Microsatellite polymorphism in *Dioscorea tokoro*, a wild yam species. Genome 37:794–801
Terauchi R, Chikaleke V, Thottappilly G, Hahn SK (1992) Origin and phylogeny of Guinea yams as revealed by RFLP analysis of chloroplast DNA and nuclear ribosomal DNA. Theor Appl Genet 83:743–751
Thorup TA, Tanyolac B, Livingstone KD, Popovsky S, Paran I, et al. (2000) Candidate gene analysis of organ pigmentation loci in the Solanaceae. Proc Natl Acad Sci USA 97:11192–11197
Thouvenel JC, Dumont R (1990) Perte de rendement de l'igname infectée par le virus de la mosaique en Côte d'Ivoire. L'Agron Trop 45:125–129
Tör M, Mantell S H, Ainsworth C C (1992) Endophytic bacteria expressing β-glucuronidase cause false positives in transformation of *Dioscorea* species. Plant Cell Rep 11:452–456
Tör M, Ainsworth C, Mantell S H (1993) Stable transformation of the food yam *Dioscorea alata* L by particle bombardment. Plant Cell Rep 12:468–473
Tör M, Twyford CT, Funes I, Boccon-Gibod J, Ainsworth CC, et al. (1998) Isolation and culture of protoplasts from immature leaves and embryogenic cell suspensions of *Dioscorea* yams: tools for transient gene expression studies. Plant Cell Tiss Organ Cult 53:113–125
Tostain S, Scarcelli N, Brottier P, Marchand JL, Pham J-L, et al. (2006) Development of DNA microsatellite markers in tropical yam (*Dioscorea* sp). Mol Ecol Notes 6:173–175
Wilkin P, Schols P, Chase MW, Chayamarit CA, Huysmans S, et al. (2005) A plastid gene phylogeny of the Yam genus, *Dioscorea*: roots, fruits, and Madagascar. Syst Bot 30:736–749
Winch JE, Newhook FJ, Jackson GVH, Cole JS (1984) Studies of *Colletotrichum gloeosporioides* disease on yam, *Dioscorea alata*, in Solomon Islands. Plant Pathol 33:467–477
Zoundjihékpon J, Essad S, Touré B (1990) Dénombrement chromosomique dans dix groupes variétaux du complexe *Dioscorea cayenensis-rotundata*. Cytologia 55:115–120

Subject Index

ACC oxidase gene, 216
Acidothermus cellulolyticus, 352
Acyl-acyl carrier protein (ACP) desaturase enzyme, 326
Aegilops, 517, 518
AFLP markers, *see* Amplified fragment length polymorphism markers
Agrobacterium-mediated transformation
  for banana, 101, 104
  for cacao, 162–163
  chickpea, 176, 182–183
  for citrus spp., 196–197
  of *Coffea* sp., 214–217
  for cowpea, 249–250
  for maize, 352–353
  for oilpalm, 375
  for peanut, 426–427
  Phaseolus species, 133–134
  for wheat, 528
Agrobacterium rhizogenes, 133, 216
Alectra vogelii, 231, 238
α-*Al1*gene, 218
Allele mining, of maize, 338–339
Allozyme diversity, 118
Alpha-amylase inhibitor, 126–127
Alpinia galanga, 302
Alt_{SB} gene, 47
Amaranthus hypochondriacus, 74–75
1-Aminocyclopropane-1-carboxylic acid (ACC) synthase, 68
Amplified fragment length polymorphism markers
  Ananas species, 441
  chickpea, 175
  citrus, 191
  cocoa, 149–150
  coffee, 205
  cowpea, 231–232, 245
  Eucalyptus, 266, 269, 272, 275
  macadamia, 317
  oil palm, 381
  P. vulgaris, 118, 124–126
  papaya, 405–406, 408–409
  peanut, 422
  sugarcane, 483–484, 487
  wheat, 529
  yams, 549, 552
AMT, *see Agrobacterium*-mediated transformation
Ananas comosus, *see* Pineapple
Arabidopsis, 12, 75, 189, 209, 347, 429
Arabidopsis thaliana, 130, 209, 218, 388, 411
Arachis hypogaea, *see* Peanut
Ascochyta rabiei, 175
Aspergillus spp., 423
Association mapping, *see* Linkage disequilibrium (LD) mapping

BAC end sequencing
  chickpea, 177
  Musa spp., 95–96
  maize, 345
  papaya, 410
Bacillus thuringiensis, 67, 352
  cryIAb gene, 503
Bacterial artificial chromosome (BAC) cloning, 13, 270
  in chickpea
    BAC libraries, development, 176–177
    physical mapping, 177–178
    whole-BAC sequencing, 178–179
  citrus, 191–192
  in coffee species, 207
  in genus *Phaseolus*
    BAC libraries development, 124–126
    whole-BAC mapping and sequencing, 126–127

Bacterial (*cont.*)
in maize, 345–349
in *Musa* spp., 89–92
for oil palm, 389–390
physical mapping, 160
in sugarcane, 494
Banana and plantain (*Musa* spp.), 83
BAC cloning and utilization, 88–91
economic, agronomic, and societal importance, 84–85
functional genomics and gene validation
cDNA libraries and ESTs, 98–100
genetic transformation, 103–104
gene trapping, 100–101
mutagenesis, TILLING, 101–103
genetic diversity, 86–87
genetic mapping, 88
genome sequencing
BAC end sequencing, 95–96
reduced representation sequencing, 97
whole BAC sequencing, 96–97
molecular cytogenetics, 91–92
nuclear genome, 92–94
retro-elements and BSV, 94–95
Banana streak virus, 94–95
Bar genes, 215, 218
Barley stripe mosaic virus, 533
BAT93 x Jalo EEP558 RI population, for beans, 117, 120
be2s1 gene, 132
Bean common mosaic necrosis virus (BCMNV), 123
Bean golden yellow-mosaic virus, 122, 133
Beauveria bassiana, 210
Bertholletia excelsa, 74
BES, *see* BAC end sequencing
Best linear unbiased predictor, 377
BGYMV, *see* Bean golden yellow-mosaic virus
BIBAC (binary BAC) libraries, 176
Bill & Melinda Gates Foundation, 28, 56–57, 241
BioCASE, 50
Biodiversity, geographic distribution, 8–10
Biofortification, 72
BioMoby, 50
Blackeye cowpea mosaic virus (B1CMV), 240
BLUP, *see* Best linear unbiased predictor
Bng122-D0140-Bng 171-Bng173 markers, 115–116
Brabeiumstellatifolium, 316
Brep1 repetitive DNA family, 93
Brown midrib *(bm)* mutants, in maize, 350
BSMV, *see* Barley stripe mosaic virus
BSV, *see* Banana streak virus

C4BAM library, 89
Cacao germplasm, molecular markers for, 148–153
CAD, *see* Cinnamyl alcohol dehydrogenase
Callosobruchus maculatus, 231, 250
Capsicum spp., *see* Pepper
Carica papaya, *see* Papaya
Cassava, 230
Cassava mosaic disease, 54
Catharanthus roseus, 101
Cauliflower mosaic virus, 219
cDNA
libraries, banana, 98–100
microarrays, 562
CentiMcClintocks (cMC), of maize, 337
chiA gene, 162
Chickpea, 171
BAC libraries, development, 176–177
BAC physical mapping, 177–178
economic, agronomic and societal importance of, 172–173
EST development, 180–181
genetic mapping and tagging in, 174–175
molecular cytogenetics, 179
sequence and marker diversity, 175–176
TILLING, 181–182
transformation, 182–183
whole-BAC sequencing, 178–179
Chocolate tree, *see Theobroma cacao*
Cicer arietinum, *see* Chickpea
Cinnamyl alcohol dehydrogenase, 288
Citrus, genomics, 187
BAC cloning, 192
economic, agronomic, and societal importance, 188
EST development, 194–195
gene tagging, 191
genome sequencing, 193–194
linkage mapping in, 190–191
microarrays, 196
molecular cytogenetics, 193
molecular markers, 197
transformation systems, 196–197
Clavatal1 (clv1), maize, 350–351
Cleaved amplified polymorphisms (CAPs), 207, 427, 459
Clonal forestry, of *Eucalyptus*, 265–268
See also Eucalyptus

Clones identification method, in *Eucalyptus*, 268–270
CMD, *see* Cassava mosaic disease
Coffee, 203
  Coffea spp., 204, 206–210, 215–216, 218
  economic and societal importance of, 205–206
  EST exploitation, 213–214
  EST resources, 210–212
  genetic transformation
    coffee promoters, identification, 219–220
    direct gene transfer, 214–215
    indirect gene transfer, 215–217
    transformed tissue selection, 217–218
    transgenic coffee plants, testing, 218
  linkage mapping, 207–208
  marker diversity, 207
  molecular cytogenetics and BAC cloning, 208–209
Coffee berry borer, 205, 218
COK-4 gene, 126
Collaborative Crop Research Program (CCRP), 28
Common bean, *see Phaseolus vulgaris*
Common-parent-specific (CPS) markers, 343
Community Sequencing Program (CSP), 470
Composite interval mapping (CIM), 243
Conarachin, in peanut, 431–432
Conserved ortholog set (COS) markers, 459, 461
Consortium for Maize Genomics (CMG), 346
Consultative Group on International Agricultural Research (CGIAR) system, 25–27, 29, 35, 44
Coordinated Agricultural Project (CAP) program, 459
Cot-based cloning and sequencing (CBCS), 97, 475
Cotton, 230
Cowpea, 227
  breeding, 236
    common strategies for, 237–239
    germplasm collections, 237
    for improved nutritional quality, 241
    molecular markers and MAS in, 245–248
    for regional preference in seed type, 241–242
    for resistance to biotic stresses, 239–240
    for tolerance to abiotic stresses, 240–241
  economic, agronomic, and social importance, 229–231
  genetic maps, 242–245
  genomic resources for, 248–249
  taxonomic relationships, 231–232
    molecular phylogeny and genome organization, 234–236
    origin and diversity, of cultivated forms, 232–234
  transgenic, transformation systems for, 249–250
Cowpea aphids *(Aphis craccivora)*, 231, 240
Cowpea mottle carmovirus, 243
CP4 *epsps*, 66
CPMoV, *see* Cowpea mottle carmovirus
CpTI trypsin inhibitor gene, 67
Crop domestication, agricultural origins centers and, 9–11
Crop improvement, transgenics for, 75
  tropical crops and, 76–77
crtI gene, 73
Cry protein, 67
Cucumber mosaic cucumovirus, 247
Curcuma longa, *see* Turmeric

Derived CAPs (dCAPs) anchor markers, 427
Diatraea saccharalis, *see* Sugarcane borer
Differential display reverse transcription (DDRT), 180
Dihydroflavonol, 4–reductase (DFR), 119–120
Dioscorea cayenensis-rotundata complex, molecular dissection, 555–556
Dioscorea spp, *see* Yams
Diversity array technology (DArT), 42, 521
Drought, sugarcane and, 499

Eco-TILLING, 42, 53, 389
Ecozones, 3, 5
Eggplant, 453, 457–458
  domesticated relatives of, 458
Elaeis guineensis Jacq, *see* Oil palm
EMBRAPA, Brazilian Agricultural Research Corporation, 32, 47, 49, 54, 99
EMS, *see* Ethyl methyl sulfonate
Endophytic bacteria, sugarcane and, 501
3–Enolpyruvateshikimate-, 5–phosphate synthase, 66
Ensete, 86
EPSPS, *see* 3–Enolpyruvateshikimate-, 5–phosphate synthase
Erwinia uredovora, 73
Escherichia coli, 88, 389, 411
ESTs, *see* Expressed sequence tags

Ethylene insensitive-like (EIL) transcription factor, 500
Ethyl methyl sulfonate, 181–182, 388
Eucalyptus, 259
association mapping in, 281–283
biology and domestication, 262–265
breeding, 263–264
by transgenic technology, 288–289
clonal forestry, 265–268
EST sequencing, gene discovery and, 283–285
gene expression and detection of expression-QTLs, analysis, 285–286
genetic resources for, 273–274
marker-assisted management of genetic variation in
breeding populations, 270–273
clones identification, 268–270
mating and deployment designs, 271–273
varietal protection, 270
MAS in, 278–280
molecular markers and maps for, 274–276
physical mapping and genome structure, 287–288
QTL mapping in, 276–278
Eucalyptus camaldulensis, 263–264
Eucalyptus globulus, 263–264, 267–271, 274
Eucalyptus grandis, 261, 263–264, 267–271, 274, 284
Eucalyptus urophylla, 264, 267, 271–274
European corn borer (ECB), 358, 359
Expressed sequence tags, 13, 75, 119, 152, 249, 429, 449, 460
banana, 98–100
cacao plants and, 160–161
citrus species and, 194–195
coffee, resources and exploitation, 213–217
database, for turmeric and ginger, 305–309
development
chickpea, 180–181
sugarcane, 494–495
in wheat, 529–531
yams, 563
maize and, 348–349
oil palm, 392
papaya, 411–412
resources exploitation, for functional analysis
enviromental changes and, 498–501
phytohormone signaling, 502
sugar synthesis, transport, and accumulation, 496–498
transposon expression, 502
sequence, *P. vulgaris*, 128–130
sequencing, gene discovery in *Eucalyptus* and, 283–285
Expression (eQTLs), 285–286, 351–352

Family and individual selection (FIS) systems, 377
FAO, *see* Food and Agricultural Organization
Fehi cultivars, of banana, 86
Fiji disease virus (FDV), 503
Finger-print contigs (FPC), 95
Flanking sequence tags, 348
Flow cytometry, 91
Fluorescence *in situ* hybridization (FISH) technique, 89, 91–92, 128, 193, 337, 391, 413, 487, 520
Food and Agricultural Organization, 15–16, 22, 229
Food security, 55–56
Ford Foundation (FF), 24
Fructose-1, 6–bisphosphatase (FBPase), 496
FSTs *see* Flanking sequence tags
Full Length Insert Sequencing (FLIS), 337
Fusarium oxysporum, 102, 175

Ganoderma boninensis, 377
"Garbanzo, *see* Chickpea
Generation Challenge Program (GCP), 34, 96
capacity building and enabling delivery, 53–55
crop improvement and, 47–49
gene discovery, genomics for, 43–47
genetic diversity and, 39–43
genetic resources, genomic, and crop information systems, 50–53
mission and research strategy, 35–37
scientific approach of, 37–39
Gene-space sequence (GSS) clones, 249
Gene tagging
in chickpea, 174–175
in citrus, 191–192
in maize, 341–342
P. vulgaris, 122–124
in sorghum, 472–473
in sugarcane, 492–493
in yams, 561–562
Genetic diversity analysis
of maize, 338–340
of wheat, 522–523
Genetic fingerprinting, 382–383
Genetic mapping

in chickpea, 174–175
of cowpea, 242–245
and genomics, in oil palm, 379–381
BAC libraries development, 389–390
diversity analysis, 383–385
genetic fingerprinting, 382–383
genome structure, 391–392
linkage mapping, 385
mutation stocks, 388–389
physical mapping and genome sequencing, 390–391
QTL analysis, 385–387
SNP markers for, 387–388
in *Musa*, 88
P. vulgaris, 120–121
in papaya, 409–410
in sugarcane, 491–492
and tagging, in peanut
markers and genetic linkage mapping, 425–426
markers and phenotypic analysis, 427
resistance genes, identification, 427–428
Gene trapping, banana, 100–101
GENOLYPTUS project, 274–275, 284–286
Genomic *in situ* hybridization, 92, 208–209, 391, 488–489
Genotyping Support Service (GSS), 54
Geographical information systems (GIS) data, 524
Germplasm, molecular markers for, 148–153
GFP marker gene, 162
GFSelector program, 97
Ginger and turmeric
EST database, 306–309
genomics, 304–305
importance and uses, 299–301
phylogenetic analysis of, 300
ploidy levels and genetic diversity in, 303
taxonomy, 301–302
[6]-Gingerol, 300
GISH, *see* Genomic *in situ* hybridization
Global *Musa* Genomics Consortium, 99
Glufosinate/glyphosate-tolerant crops, 66
Golden Rice, 1 (*GR1*), 72–73
Gossypium sp., *see* Cotton
*Gpc-6B, 1*wheat grain protein locus, 528
Grand Challenges in Global Health (GCGH) project, 73
Granulocyte macrophage colony-stimulating factor (GM-CSF), in sugarcane, 504
Green fluorescent protein, 218
Green Revolution, 21, 24–25, 517–518
Groundnut, *see* Peanut
GUS reporter gene, 219

HarvEST:Citrus database, 196
Helicoverpa armigera, 182
Hemileia vastatrix, 205
Herbicide tolerant crops, 65–66
High information content fingerprinting (HICF), 346
High molecular weight (HMW) DNA, 88–89
Hordeum vulgare, 379
Hypothenemus hampei, *see* Coffee berry borer

Insect resistance, 66–67
See also Transgenic plant products
Interfering RNA (RNAi), 133
International Food Policy Research Institute, 33
International Group for Genetic Improvement of Cocoa (INGENIC), 147–148
International Maize and Wheat Improvement Center (CIMMYT), 24, 28
International Program on Rice Biotechnology (IPRB), 30–31
INTERPROSCAN program, 307
Interretroelement amplified polymorphism (IRAP) method, 95
Inter simple sequence repeat (ISSR), 175, 235
Inverse sequence-tagged repeat (ISR), 207
Isozyme, 272, 380
for *Ananas* species, 444
for macadamia, 318–319

Kalpatharu, *see* Banana
Kanamycin, 217
Kanamycin selectable marker gene, 162
Klebsiella pneumoniae, 66
KNOX genes, 552

Laser capture microdissection (LCM), 349
Leaf area index, 5
Leptosphaeria maculans, 325
Lima bean, *see Phaseolus lunatus*
LINK2PALM program, 386
Linkage disequilibrium (LD) mapping, 41–42, 524
in *Eucalyptus*, 281–283
in maize, 343–344
in sugarcane, 490–491
Linkage mapping
Arachis, 425–426
cacao, 153–156
citrus, 189–190

Linkage mapping (*cont.*)
Coffea spp., 207–208
macadamia, 321–323
maize, 340–341
oil palm, 385
sorghum, 472
yams, 560–561
Linkage mapping, 153–156
Long terminal repeats (LTRs), 94
Lulo fruit, 456, 463

Macadamia, 314, 327–329
cytogenetics, 316
domestication, 315–316
gene sequencing, 323–327
genetic markers, 316
AFLP, 319
isozyme, 317–318
RAF and RAMiFi, 319–320
RAPD and STMS, 318
SSRs, 320–321
linkage mapping, 321–323
Macadamia integrifolia, 315–316, 321, 325, 327, 329–330
Macadamia tetraphylla, 315, 321, 327, 329–330
Maize, 230, 335
allele mining, 339–340
BAC DNA sequencing, 346–347
BAC libraries, development, 345–346
BAC physical mapping, 345
databases and tools, 353–354
functional
EST development, 348–349
gene cloning, 349–350
insertional mutation, 347–348
TILLING, 353
transcription profiling, 351–352
transformation, 352–353
gene tagging/QTL mapping, 341–342
genetic diversity analysis, 338–340
genomics-assisted breeding of tropical, 354
abiotic stresses, 358–359
biotic stresses, 358–359
quality, 357
yield and heterosis, 356–357
linkage disequilibrium (LD) mapping, 339–340
linkage mapping, 340–341
MAS applications for, 343–344
molecular cytogenetics, 333–334
Maize Mapping Project (MMP), 341, 345
Maize TILLING Project (MTP), 353

MALDI-TOF analysis, *see* Matrix-assisted laser desorption/ionization time-of-flight analysis
Male specific region of Y chromosome (MSY), in papaya, 413
physical mapping of, 414–415
Manihot esculenta, see Cassava
Marker-assisted selection, 14–15
chickpea, 175
CIMMYT wheat breeding program, 534–535
coffee, 213
common bean, 122
cowpea breeding, 245–248
Eucalyptus, 274, 276, 278–280
macadamia, 321, 323
maize, 344–345, 356
peanut, 427
wheat, 521, 533–535
Maruca pod borer *(Maruca vitrata)*, 231, 250
MAS, *see* Marker-assisted selection
Matrix-assisted laser desorption/ionization time-of-flight analysis, 432, 533
McKnight Foundation, 28
Medicago truncatula, 12, 15, 115, 178, 180, 429
Methionine-rich 2S albumin protein, 74
Methyl jasmonate (MeJa) signaling, 498, 502
MiAMP1 protein, 326
Microarrays
cacao plants and, 161–162
for citrus spp., 196
for maize, 351–352
Molecular breeding
in Eucalyptus
association mapping in, 281–283
EST sequencing, gene discovery and, 283–285
gene expression and detection of expression-QTLs, analysis, 285–286
genetic resources for, 273–274
marker-assisted selection in, 278–281
molecular markers and maps for, 274–276
physical mapping and genome structure, 287–288
QTL mapping in, 276–278
by transgenic technology, 288–289
in wheat, 535–538
Molecular cytogenetics
chickpea, 179
in citrus, 193–194

fluorescence *in situ* hybridization (FISH),
in coffee, 208–209
maize, 336–337
Musa species, 91–92
Phaseolus sp., 127–128
of sex chromosomes, in papaya, 414
sorghum, 474
wheat, 520
yams, 556–557
Moniliophthora perniciosa, 146, 153, 157
Moniliophthora roreri, 146, 156, 158
Musa acuminata, 86–89, 91–93, 99, 304
Musa balbisiana, 86, 92–93, 99
Musa beccarii, 91
Musa schizocarpa, 86
Musa spp., *see* Banana and plantain (*Musa* spp)
Musa textilis, 85
Musella, 86
Mycosphaerella fijiensis, 102

NADP-malic enzyme (NADP-ME) pathway, 498
Narcissus pseudonarcissus, 73
National agricultural research systems (NARS), 31–32, 39, 54, 57
National Wheat Molecular Marker Program (NWMMP), 534
Near-infrared spectroscopy (NIR), 277–278, 281
Neolithic Revolution, 3
Net primary productivity (NPP), 8
N-methyl transferase (NMT) gene, 219
Noble cane, *see* Sugarcane
Non-governmental organizations (NGOs), 25
nptII, kanamycin resistance gene, 214–215
Nucleolus organizing region (NOR), 92
Nucleotide binding site (NBS) domain genes, 428

Oil palm, 367
bioinformatics and informationm, from other plant systems, 393–394
biotechnology of, 378–379
economic, agronomic, and societal importance, 373–374
EST development, 392
gene mapping and genomics, 379–382
BAC libraries development for, 389–390
diversity analysis, 383–385
genetic fingerprinting, 382–383
genome structure, 391–392
linkage mapping, 385
mutation stocks, 388–389
physical mapping and genome sequencing, 390–391
QTL analysis, 385–387
SNP markers for, 387–388
genetics and breeding of, 375–377
metabolomics, 393
proteomics, 392–393
transcriptomics, 392
Opaque 2 (o2) mutant, 342, 357
Oryza sativa, *see* Rice
Ostrinia nubilalis, *see* European corn borer (ECB)

Palm oil mill effluent, 374
Papaya, 405
cytogenetics, 408–409
economic, agronomic, and societal importance, 406
EST resources, 411–412
genetic diversity, 407–408
genetic mapping, 409–410
genome organization, 411
pest and pathogen resistance, genetic engineering for, 416–417
physical mapping, 410, 414–415
sex chromosomes, *see* Sex chromosomes, in papaya
Papaya ringspot virus, 68, 408, 409, 416–417
Pathogen-derived resistance (PDR), 68
Peanut, 421
biodiversity and markers, 432–434
bioinformatics, 434
cytogenetics, markers, and species identification, 424–425
economic, agronomic, and societal importance, 422
gene expression analysis, 429–430
gene sequencing, 429
genetic mapping and tagging
markers and genetic linkage mapping, 425–426
markers and phenotypic analysis, 427
resistance genes, identification, 427–428
geographic origin of, 422–423
large-insert libraries and physical mapping, 428
proteomics and allergen genes, 431–432
TILLING, for gene function analysis and mutant phenotypes generation, 430
transformation, 430–431

PeanutMap, 434
Pearl millet *(Pennisetum glaucum)*, 230
Peas *(Pisum sativum)*, 171–172
PEPC genes, *see* Phosphoenolpyruvate carboxylase gene
Pepper, 454, 461
  domesticated relatives of, 458
Petunia hybrida, 101
PFGE, *see* Pulse Field Gel Electrophoresis
6-PGD isozyme analysis, for yams, 558
Phaseolus acutifolius, 117
Phaseolus beans, 113
  BAC cloning and utilization, 124–127
  economic, agronomic, and societal importance of, 114–115
  EST development, 128–130
  genetic mapping and tagging, in *P. vulgaris*, *see Phaseolus vulgaris*
  genetic resources, 116–117
  marker and sequence diversity, 118–120
  molecular cytogenetics, 127–128
  TILLING, 130–132
  transformation, 132–134
Phaseolus coccineus, 117, 130, 133
Phaseolus dumosus, 117
Phaseolus lunatus, 117, 133
Phaseolus vulgaris, 114, 116–118, 126, 133, 172, 230
  EST sequences, 129–130
  gene tagging in, 122–124
  genetic mapping in, 120–121
Phosphinothricin acetyl transferase (PAT), 66, 132–133
Phosphoenolpyruvate carboxylase gene, 74
Photosynthetic carbon reduction (PCR) cycle, 496
Physalis, 454
  minor crops in, 458–459
Physical mapping
  BAC libraries and, sorghum, 473–474
  in chickpea, 177–178
  in *Eucalyptus*, 287–288
  and genome sequencing, in maize
    BAC DNA sequencing, 346–347
    BAC libraries development, 345–346
    BAC physical mapping, 345
  and genome sequencing, in oil palm, 390–391
  papaya, 410, 414–415
  peanut, 428
  in wheat, 520
Phytophthora megakarya, 146
Phytophthora palmivora, 156–157
Pineapple, 441
  bioinformatics resource, 448
  genetic diversity, 443
  genetic map of, 446–447
  molecular markers for, 443–445
  species of, 442
Plants, as bioreactors, 73–74
Polymerase chain reaction (PCR)-based markers, 120
POME, *see* Palm oil mill effluent
Poncirus trifoliata, 191
Post-transcriptional gene silencing, 417
Post-World War II period, 22–23
Potato, 453, 460
  cultivated tuber-bearing, diversity, 457
PPP model, *see* Public-private partnerships model
PRSV, *see* Papaya ringspot virus
Pseudananas sagenarius, 442
  See also Pineapple
Psilanthus genus, 204
PTGS, *see* Post-transcriptional gene silencing
Public-private partnerships model, 32–34
Pulse Field Gel Electrophoresis, 287

Quantitative trait loci (QTLs), 14
  analysis, in oil palm, 383–385
  chickpea, 175
  citrus, 190
  cowpea, 245
  genome regulation and, 46
  mapping
    in *Eucalyptus*, 276–278, 285–286
    in maize, 341–342
    in *T. cacao*, 149, 156–160
  P. vulgaris, 117, 123
  peanut, 427
  sugarcane, 492–493
  wheat, 524, 526, 537
  yams, 561
Quantitative trait nucleotide (QTN), 281

RACE, *see* Rapid amplification of cDNA ends
Radiation-based mutation induction method, 102
Random amplified polymorphic DNA markers
  for *Ananas* species, 444
  cacao, 149–150
  chickpea, 174
  citrus, 191, 197
  coffee, 207
  cowpea, 234
  in *Eucalyptus*, 268–269, 271–272, 274, 276
  macadamia, 319

oil palm, 381, 385
P. vulgaris, 118, 120–122
papaya, 409
peanut, 425–426, 427
sugarcane, 487–488, 491
wheat, 521
yams, 553, 557
Randomly amplified DNA fingerprinting (RAF) marker system, for macadamia, 320–321, 323
Randomly amplified microsatellite fingerprinting (RAMiFi) marker system, for macadamia, 322
RAPD markers, *see* Random amplified polymorphic DNA markers
Rapid amplification of cDNA ends, 351
Reciprocal recurrent selection (RRS) programs, 357
Recombinant inbred (RI) populations, 117
Recombinant inbred line (RIL) populations, 174
RescueMu, transposon tag, 350
Resistance-gene analogs (RGA)-based markers, 121, 175, 350, 428
Restriction fragment length polymorphisms markers
Ananas species, 443
cacao, 149
citrus, 197
coffee, 207
cowpea, 234
eggplant, 461
maize, 341
oil palm, 385
P. vulgaris, 118, 120–121, 128
peanut, 425–426
sugarcane, 487–490, 492
tomato, 459
wheat, 521, 533
yams, 553
Retro-elements, in banana, 94–95
Retrotransposons, 380, 502
Reverse transcription polymerase chain reaction (RT-PCR), 193
RFLP markers, *see* Restriction fragment length polymorphisms markers
Rhizobium leguminsarum, 173
Rhizobium tropici, 130
Rice, 96, 130
Rice Genome Automated Annotation System (RiceGAAS), 96–97
RNA interference (RNAi) technology, 216
Rockefeller Foundation (RF), 23–24, 56
for rice improvement, 29–31
Root-knot nematode *(Meloidogyne arenaria)*, 205, 239, 246–247, 427, 433
Rsg2–1 gene, 246
Runner bean, *see Phaseolus coccineus*

Saccharum officinarum, *see* Sugarcane
S-adenosylmethionine (SAM), 68
SAGE, *see* Serial analysis of gene expression
SAS, *see* Sugarcane assembled sequences
SCMV, *see* Sugarcane mosaic virus
SCYLV, *see* Sugarcane yellow leaf virus
Sequence-characterized amplified region (SCAR) markers, 246, 409, 412, 414, 427
Sequence tagged microsatellite site markers, 174–175, 179
for macadamia, 318
Sequence tagged site (STS) markers, in wheat, 521
Sequence-Tagged Sites/Sequence-Characterized Amplified Regions (STS/SCAR), 120–121
Sequencing the Maize Genome Project (STMG), 347
Serial analysis of gene expression, 498
Sex chromosomes in papaya
molecular cytogenetics of, 414
MSY region, physical mapping of, 414–415
primitive, identification of, 412–413
X- and Y-BACs sequencing, 415–416
Shell-thickness *(Sh)* genes, 376–377
ShSUT1 gene, sugarcane, 497
Simple sequence repeats markers
cacao, 149–152, 159
chickpea, 175
citrus, 191
cowpea, 234
macadamia, 321
maize, 339
Musa spp., 95
oil palm, 383
P. vulgaris, 118–121
papaya, 409–410
peanut, 426, 434
tomato, 459
wheat, 521
yams, 553–554
Single feature polymorphism (SFP) analysis, 339
Single-nucleotide polymorphisms (SNPs), 14, 42, 119, 152, 175–176, 278, 282, 339, 460, 461

Single-nucleotide (*cont.*)
  markers
    in maize, 341–343, 344
    in oil palm, 387–389
    papaya, 409
    in wheat, 521–522, 537
Solanaceae Coordinated Agricultural Project (SolCap), 459
Solanaceae Genomics Network (SGN), 213
Solanum aethiopicum, *see* Eggplant
Solanum betaceum, *see* Tree tomato fruit
Solanum lycopersicum, *see* Tomato
Solanum muricatum (pepino dulce), 456
Solanum quitoense, *see* Lulo fruit
Solanum tuberosum, *see* Potato
Solonaceous food crops, 453
  breeding and genetics of, markers for, 461–462
  cultivated tuber-bearing potatoes diversity, 457
  economic importance of, 454
  eggplant, domesticated relatives of, 457–458
  functional genomics resources
    eggplant, 461
    pepper, 461
    potato, 460
    tomato, 459–460
  genetics, 454–455
  "minor", 455–457
    in genus *Physalis*, 458–459
  pepper *(C. annuum)*, domesticated relatives of, 458
  Solanum genus, taxonomy of, 455
  translational genomics, 459
    for emerging Andean solanum species, 462–465
Sorghum, 230, 469
  BAC libraries and physical mapping, 473–474
  economic, agronomic, and societal importance, 470–471
  functional genomics resources, 475–476
  gene tagging, 472–473
  genome sequencing, 474
  linkage mapping, 472
  marker diversity, 476–477
  molecular cytogenetics, 474
Sorghum bicolor, *see* Sorghum
SSRs markers, *see* Simple sequence repeats markers
Stable transformation frequency (STF), 104
STMS markers, *see* Sequence tagged microsatellite site markers
Streptomyces viridichromogenes, 66
Striga gesnerioides, 231, 238, 244
Striga Race, 1/3 (SG1/3), 246
Subtropical ecozones
  dry, tropics and, 6
  with winter rains, 5–6
  with year-round rain, 6
Subtropical Horticulture Research Station (SHRS), 160
Sucrose-phosphate synthase (SPS), 496
Sugarcane
  assembled sequences, 495
  BAC library development and utilization, 494
  borer, 500
  chromosome structure, 489–490
  economic, agronomic, and societal importance, 483–484
  EST development, 494–495
  EST resources exploitation, for functional analysis, 496–502
  genetic maps, 491–492
  genetic transformation, 502–504
  linkage disequilibrium (LD), 490–491
  molecular diversity, 490
  origin and diversity of, 485–489
  synteny with other grasses, 493
  tagging genes, 492–493
  tissue profiling, 495–496
Sugarcane mosaic virus, 503
Sugarcane yellow leaf virus, 503
Sweet potato, 550
Syngenta Foundation, 28

Targeting induced local lesions in genomes, 15, 45
  banana, 101–103
  for chickpea, 181–182
  common bean, 130–132
  for gene function analysis and mutant phenotypes generation, in peanut, 430
  maize, 353
  for wheat, 531–532
Tepary bean, *see Phaseolus acutifolius*
Theobroma cacao
  economic, agronomic, and societal importance of, 145–146
  genetics and breeding, 147
  genomic resources, 147–148
    BAC libraries development, 160

ESTs, 160–161
genetic transformation system, 162–163
informatics databases, 163
microarrays, 161–162
germplasm, molecular markers for, 148–153
linkage mapping, 153–156
QTLs mapping, 156–160
Thrips *(Megalurothrips sjostedti)*, 231
TILLING, *see* Targeting induced local lesions in genomes
Tissue profiling, in sugarcane, 495–496
Tomato, 453, 459–460
Trade-Related Intellectual Property Rights (TRIPS), 14
Transcript-derived fragments (TDFs), 285
Transgenic plant products
commercialized
disease resistance, 67–68
herbicide tolerance, 65–66
insect resistance, 66–67
post-harvest quality, improved, 68
developing, 68–70
nutritional quality, 72–73
plants as bioreactors, 73–74
post-harvest quality, 71
tropical germplasm as source and, 74–75
Transgenic technology , for breeding in *Eucalyptus*, 288–289
Tree nut crop, *see* Macadamia
Tree tomato fruit, 456, 463
Tripsacum, 34
Triticum aestivum, *see* Wheat
Tropical crops
and food security, genomics of, 12
genetic improvement of, 13–15
poverty and, 15–18
transgenic approaches and, 71–74, 76
Tropical ecozones, 4
dry subtropics and, 6
with summer rains, 6–7
with year-round rain, 7
Turmeric and ginger, *see* Ginger and turmeric

UidA gene, 215, 219
UN Development program (UNDP), 25

Vasconcellea, 404
Vernalisation, genes for (*Vrn1* and *Vrn2*), 388, 519
Vigna unguiculata, *see* Cowpea
Vrn-A1 gene, wheat, 388, 519, 528

W.K. Kellogg Foundation, 28
Watermelon mosaic virus, 68
Water use efficiency, 17
Wheat, 388, 515
association genetics, 524–525
breeding
conventional, 517–520
genomics in, 531–533
molecular, challenges for, 535–537
EST development in, 529–531
genetic characterization, of traits, 525–528
genome diversity analysis, 522–523
molecular cytogenetics and physical mapping, 520
molecular markers as tools, 521–523
TILLING, 531
transformation, 532
Whole BAC sequencing
chickpea, 178–179
Musa genome, 96–97
Phaseolus sp., 126–127
Whole genome shotgun (WGS) sequencing project, 95
WMV, *see* Watermelon mosaic virus
World Cocoa Foundation, 146
WUE, *see* Water use efficiency

X- and Y-BACs sequencing, in papaya, 415–416
Xylose isomerase *(XylA)* gene, 218
Xylotrechus quadripes, 205

YAC, *see* Yeast artificial chromosome
Yam mosaic virus, 559–560
Yams, 549
D. cayenensis-rotundata complex, 555–556
economic, agronomic, and societal importance, 551–552
EST development, 563
gene tagging, 561–562
genetic diversity, 557–559
linkage mapping, 560–561
molecular cytogenetics, 556–557
molecular markers development, for genome analysis, 553–554
phylogeny, 554–555
transformation, 563–564
Year bean, *see Phaseolus dumosus*
Yeast artificial chromosome, 88, 176
YMV, *see Yam mosaic virus*

Zea mays, *see* Maize
Zingiber officinale, *see* Ginger
ZmPox3, peroxidase, 350
Zucchini yellow mosaic virus (ZYMV), 68

MIX
Papier aus verantwortungsvollen Quellen
Paper from responsible sources
FSC® C105338

If you have any concerns about our products,
you can contact us on
ProductSafety@springernature.com

In case Publisher is established outside the EU,
the EU authorized representative is:
Springer Nature Customer Service Center GmbH
Europaplatz 3, 69115 Heidelberg, Germany

Printed by Libri Plureos GmbH
in Hamburg, Germany